AF789562

inverse problems
and theoretical imaging

P.C. Sabatier (Ed.)

Inverse Methods in Action

Proceedings of the Multicentennials Meeting on Inverse Problems, Montpellier, November 27th – December 1st, 1989

With 125 Figures

Springer-Verlag Berlin Heidelberg New York
London Paris Tokyo Hong Kong

Professor Dr. Pierre C. Sabatier
Laboratoire de Physique Mathématique,
Université des Sciences et Techniques du Languedoc, Place Eugène Bataillon,
F-34060 Montpellier Cedex, France

ISBN-13:978-3-642-75300-8 e-ISBN-13:978-3-642-75298-8
DOI: 10.1007/978-3-642-75298-8

This work is subject to copyright. All rights are reserved, whether the whole or part of the material is concerned, specifically the rights of translation, reprinting, reuse of illustrations, recitation, broadcasting, reproduction on microfilms or in other ways, and storage in data banks. Duplication of this publication or parts thereof is only permitted under the provisions of the German Copyright Law of September 9, 1965, in its current version, and a copyright fee must always be paid. Violations fall under the prosecution act of the German Copyright Law.

© Springer-Verlag Berlin Heidelberg 1990
Softcover reprint of the hardcover 1st edition 1990

The use of registered names, trademarks, etc. in this publication does not imply, even in the absence of a specific statement, that such names are exempt from the relevant protective laws and regulations and therefore free for general use.

2157/3150(3011)-543210 – Printed on acid-free paper

Preface

This volume contains the Proceedings of a meeting held at Montpellier from November 27th to December 1st 1989 and entitled "Inverse Problems Multicentennials Meeting". It was held in honor of two major centennials: the foundation of Montpellier University in 1289 and the French Revolution of 1789. The meeting was one of a series of annual meetings on interdisciplinary aspects of inverse problems organized in Montpellier since 1972 and known as "RCP 264". The meeting was sponsored by the Centre National de la Recherche Scientifique (contract GR 264) and by the Direction des Recherches et Etudes Techniques (contract 88 CO 283).

The Proceedings are presented by chapters on different topics, the choice of topic often being arbitrary. The chapter titles are "Tomographic Inverse Problems", "Distributed Parameters Inverse Problems", "Spectral Inverse Problems (Exact Methods)", "Theoretical Imaging", "Wave Propagation and Scattering Problems (Imaging and Numerical Methods)", "Miscellaneous Problems", "Inverse Methods and Applications to Nonlinear Problems". In each chapter but the first, the papers have been sorted alphabetically according to author*. In the first chapter, a set of theoretical papers is presented first, then more applied ones. There are so many well-known and excellent lectures that I will not try to refer to them all here (the reader will be easily convinced by reading the Table of Contents). My comments at the conference are summarized by the short scientific introduction at the beginning of the volume.

A comparison of the workshop program and the table of contents shows that only a few contributions are missing. For three, only the abstracts are available. The program of the 1988 meeting is also listed at the back. That of previous meetings can be found in references 1, 2, and 3 of my introduction.

I wish to acknowledge with thanks Mrs. O. Albernhe and Mrs. F. Duceau who organized the meeting; Mrs. J. Cellier who was responsible for obtaining the manuscripts and retyping and resetting them where necessary; and my colleagues and friends of "Laboratoire de Physique Mathématique de l'Université de Montpellier" who did their utmost to assist me. Finally, I must say how much I appreciated the presence of so many distinguished scientists in Montpellier, in

---

*Received by Springer-Verlag January 31, 1990.

particular those who travelled a great distance to be at the meeting. I and my colleagues thank them, and also all our friends from the European Universities, for the skilful presentation of the results of their work. We hope the reader will share this pleasure.

Montpellier, December 1989 *P.C. Sabatier*

In Memoriam

We just learned the news of the untimely death of *André Roger* in a mountain accident. *André Roger* was one of the very few who knew both mathematics and numerical problems connected with scattering phenomena. Furthermore, he was interested not only in theoretical problems of electromagnetic scattering but also in practical problems of geophysical prospecting, and in all his domains of interest, he published very good papers. He already attended our annual meeting in 1977, when he was only 25 years old, and we still remember his beautiful lecture on the determination of the index profile of a dielectric plate from scattering data. We became accustomed to his bright remarks and questions, always put forth with a smile. His absence is strongly felt by us, and by all those who had the privilege of knowing him and working with him. As a testimony of our sympathy for his family and all who loved him, we dedicate this book, which contains his last paper, to his memory.

Contents

Part V Theoretical Imaging

Part VI Wave Propagation and Scattering Problems (Imaging and Numerical Methods)

For technical reasons the contribution “Operator Algebra and Inverse Methods” by Prof. Dr. V.A. Marchenko could not be included into this volume. It will be published in the book “What Is Integrability?” edited by V.E. Zakharov to appear soon with Springer-Verlag.

Modelling or Solving Inverse Problems?

P.C. Sabatier

Laboratoire de Physique Mathématique, U.S.T.L.,
Place Eugène Bataillon, F-34095 Montpellier Cédex 5, France

The first meeting sponsored by the "Centre National de la Recherche Scientifique" through the contract "RCP 264" was held in Montpellier in 1972, and was renewed every year since that time. Collections of unpublished Proceedings[1,2] and published ones[3], were regularly issued. Meanwile, a large set of monographs and a journal[4] completed the definition of our field of interest, from inverse problems to the nonlinear partial differential equations which can be solved by inverse methods. However, the relation between physical modelling and solving inverse problems is not clear for everybody, and sometimes felt as an opposition ! We think that the best way to introduce the present work goes through discussing this point a little. According to "Adam's Diary"[5], Eve used to solve the Inverse Problem of Eden by shape recognition. "The new creature names everything that comes along, before I can get in a protest. And always that same pretext is offered - it *looks* like the thing. There is the dodo, for instance. Says the moment one looks at it one sees at a glance that it "looks like a dodo". It will have to keep that name, no doubt". But after the Fall, Eve was not long to discover experimental physics together with modelling and interpolating[5] : "At first it used to vex me because, with all my watching, I was never smart enough to be around when the water was running up-hill ; but now I do not mind it. I have experimented and experimented until now I know it never does run uphill, except in the dark. I know it does in the dark, because the pool never goes dry ; which it would, of course, if the water didn't come back in the night. It is best to prove things by actual experiment ; then you *know* ; whereas if you depend on guessing and supposing and conjecturing, you will never get educated".

Six thousand years after our fall from Eden, I am not quite sure that every physicist really knows that a trivial use of logarithmic scales turns the shape of a pig into that of a baboon, that modelling is an art which cannot be separated from a deep study of the associated direct and inverse problems, and that choosing a way of interpolating data is in most problems equivalent to choosing a model. Let us now proceed to harder "science".

I - MODELLING IN PHYSICS

I.1 Definitions

"A model is a way of relating parameters that describe a physical system to all possible measurements results".

This definition is commonly accepted but is ambiguous, because there are at least three extreme ways of describing a "physical system" :

(a) *"Extensive" description.* In this usual way, all the values are given for all parameters, as functions of the spatial coordinate x.
(b) *"Decisive" description.* A few theoretical measurements of a physical system - for instance values of the total mass, or other moments of a distribution -, may be a sufficient information for further decisions. Their values give a "decisive" description of the system.
(c) *"Comprehensive" description.* It is a way of describing the system which emphasizes the properties through which the system can be compared with others and which can be really measured ("essential physical properties"). Thus the parameters first will be shifted into two sets - those which are irrelevant for measurements, ie do not modify their values within the noise level, and those which are relevant. Then the set of relevant parameters is divided into equivalence classes (parameters being equivalent if they yield the same data), and a representative of each class is selected so as to correspond to an "essential physical property". Hence the system is described only by its "essential parameters".

I.2 Typical research of a model

The following sketch of such a research shows how modelling and solving inverse problems are related. We go through the successive steps :

(a) *Data.* Experimental measurements collect data.
(b) *First Model.* It is suggested by "Physics", ie by previous studies of the system, and it is represented by a simple mathematical proposal-typically a mapping $\mathcal{M}$ from a wide set of parameters $\mathcal{C}$ to a wide set of results $\mathcal{E}$. We call $\mathcal{C}$, resp. $\mathcal{E}$, a "wide set" because it is chosen in such a way that it contains all the possible values of parameters, resp. data, *and* is a simply defined mathematical set (in most cases a normed space) e.g. L^2 or $\mathbb{R}^N$. We assume in the following $\mathcal{C}$ and $\mathcal{E}$ are normed spaces. The norms must be chosen in such a way that they can represent physical deviations or errors on parameters, resp. data. If it is not so, "stability theorems" do no mean anything. Giving values of $\mathcal{M}x$ in $\mathcal{E}$ as x runs through $\mathcal{C}$ is called "solving the direct problem". These values are called "computed" or "synthetic" results, and their set is the image $\mathcal{M}(\mathcal{C})$.
(c) *Signal Processing.* This word stands for various ways of classing the collected data. It is not known by every physicist that such an art depends on the model and, in particular, on how the measurements at various points and times can be compared to each other.
(d) *The way of modelling.* This fourth step is often overlooked : one must choose the way of modelling - extensive, decisive, or comprehensive.
(e) *Ill-posedness.* It is then time to set the mathematical questions of illposedness for sets of real data - The answer must be so detailed that it paves the way for the following steps.

(f) *A priori information.* Restricting assumptions and structures are setted on $\mathcal{C}$ or $\mathcal{M}$. Sometimes called "giving a priori information" they are guided by the following rules :

(f 1) they must be justified either by physics or by mathematical consistency. In other words they are omitted pieces of information, not only convenient constraints.

(f 2) they set an order in $\mathcal{C}$. One can call this order a quality criterion for the parameters (or for the solutions). It can be either given "a priori" or justified by a statistical description of $\mathcal{C}$. Thus the parameters may be ordered as "admissible" or not, or with respect to their "likelihood" or "reliability" etc. This is generally obtained by "cost functionals", such that c_1 is "better" than c_2 if $F(c_1) < F(c_2)$.

(f 3) they set an order in $\mathcal{E}$. The most usual "a priori" assumption is that a computed result $\mathcal{M}(c)$ is "better" if it yields a smaller misfit $\|\mathcal{M}(c) - e\|$ to the result e of measurements.

Setting the points f1,f2,f3 is nothing but redefining the model, which is no longer the "first model" but a "regularized model".

(g) *Regularized model.* If the "regularized model" is good (this is related with compactness properties of the sets, etc), "solutions" of the inverse problem can be defined, ie stable ways to go back from $\mathcal{E}$ to $\mathcal{C}$ by making a compromise between the quality criteria ordering these sets. A "solution" is stable with respect to experimental errors if it is strongly continuous for the norm in $\mathcal{E}$ and robust with respect to casual perturbations at collecting data.

(h) *Algorithms.* One can then define *algorithms*, ie constructive methods to derive solutions, designed in such a way that they are also numerically stable.

(i) *Decision tests.* What we call the first "cycle" for modelling is then achieved by processing the data, checking solutions, and *deciding* according to the following rules : either

(i 1) *The I.P. has no solution* - go back to (b), the model must be redefined and a "new cycle" begins

(i 2) *the I.P. has unstable solutions* - go back to (f) and reset the "a priori" information

(i 3) the I.P. has a *quite large set of admissible solution* ("equivalent" parameters) ("strongly ill posed model") - go back to (d) and define "theoretical measurements" which are decisive and well-posed *(well-posed questions)* and go back to (a).

(i 4) the I.P. has only *narrow sets of equivalent solutions*. This the only case where the research is closed : the I.P. is solved, since any algorithm giving a point interior to the set of equivalent solution gives a good solution.

(i 5) the I.P. has sets of equivalent solutions which are of medium size or disconnected. This is a very common case in physical problems with experimental data, and each author reacts by defining his own algorithm ! of course, one should rather go back to (a) and (e), and (d) (in particular one should try to describe the sets by means of physical properties and extract the "essential" parameters), and if nothing can be improved therein, there is no escape but describing the sets of equivalent solutions by means of the usual techniques of data analysis.

I.3 Conclusion

Modelling and solving the I.P. are together a joint analysis of information, which should be conducted as a musical duo. The various steps of this analysis will all have developments or consequences in the lectures to be presented. A particular emphasis is always put on "classical inverse problems", whose solution reduces the problem to exactly known diffeomorphisms between classes of parameters and classes of data, so that "essential parameters" are well known, together with exact conditions to be satisfied on data collected in the step (a) in order to end the discussion "cycle" in the case (i 4) ("essential data").

II - SCANNING THE LECTURES

The various schemes which are presented are related more or less closely to our previous analysis.

II.1 Classical Inverse Problems

Studies of these problems show the essential parameters and essential data between which there are bijections and $\mathcal{M}^{-1}$ is continuous (but not always for the most "reasonable" norm). Classical problems exactly solved by the Gelfand-Levitan and the Marchenko method belong to the class. One find also the problems whose solution is unique, stable, and *proved* to be obtained by minimizing a convenient cost functional. Studies of the solution may be directed either

(a) to show their intrinsic interest

(b) to modifying and adapting the method of solution to sets of real data (incomplete, redundant, with error, etc).

(c) or as the basis of an "inverse method" to solve nonlinear partial differential equations.

Related lectures are those on

<u>Vibration problems</u> (points a and b) : Andersson, Chadan, Gladwell, Hald, McLaughlin.

<u>Boundary measurement problems</u>, where distributed parameters of a partial differential equation should be reconstructed (points a and b) : Chavent, Crosta, Davies, Engl, Kunisch, Nachman, Petkov, Pidcock, Rundell, Stefanov.

<u>Scattering problems</u> (points a and b) : Bachelot, Blazek, Brander, Corones, de Facio, de Hoop, Fiedeldey & Lipperheide, Fiddy, Kirst, Newton, Tabbara, Tchamitchian, Tijhuis, Somersalo, Vandermee, Weston.

Spectral problems of inverse methods (point c) :
These studies are more formal, and so are the connected studies of Backlund transformations, and of the various ways to make a joint exploration of $\mathcal{C}$ and $\mathcal{E}$ (Marchenko, Martinez-Alonso, Ragnisco, Santini).

General studies of classical inverse problems (point a) :
Anger, Attouch, Fischman, Maas, Schnitzer, Spearman, Stark.

II.2 Imaging

Imaging is representing extensively the model by one value at each point, e.g. the "most likely one" over the set of possible approximate solutions. Hence it is an approximate extensive description of the model, with the following consequences :
(a) when the set of equivalent models is "narrow", imaging is relevant, and can be implemented by minimizing simple cost functions, or solving the direct problem for samples of models. Automatic processing may be designed in certain cases.
(b) if an extensive description of the model cannot be reached, imaging is irrelevant. This is so if very large classes of equally acceptable models do exist (e.g. problems of gravity measurements) - or if they form unconnected sets.
(c) between these two extreme cases, imaging is an art where more or less scruples are involved, and whose relevance depends on with how much care the "duo" described in § (I.2) has been performed.

In most "applied" lectures, minimizing a cost functional is done without a complete study of the mathematical problem. Thus they are more relevant to imaging than to inverse problems, but authors knowing inverse problems can be told among others by their care. The corresponding lectures are given by Bolomey, Jordan, Ladouceur, Pierri, Ruelle, Tarantola.

The case of tomography

In this imaging technique one tries to construct a model from measurements *that are correlated to each other because of their locations*. Linear inverse problems appear when the "sounding" rays are straight ones (X-ray, γ-ray, etc). Mathematics are well understood by means of the singular value decomposition, which exhibits the "essential parameters" ("degrees of freedom" of the model). Imaging algorithms are available, their relevance to the inverse problem is known and most of them are sound. Extension to slightly curved rays is possible, as well as to diffuse scattering. The most general case of arbitrary paths, giving an idea of the future problems, is treated in Grünbaum's lecture. More classical tomographies, and the difficulties associated to sparse data, shadow ranges, etc, appear in the lectures of Dolveck-Guilpart, Franchois, Langenberg, Louis, Natterer, Seatzu, Solmon.

II.3 Signal Processing

Work during the eighties has shown that signal processing should depend on the model of propagation : weak nonlinearities show after long times of propagation and it follows that the invariant structures which you like to identity at any point of observation are no longer obtained by a simple Fourier spectrum. It is quite remarkable that *universal* models of propagation which deal with weak nonlinearities *exist*, can be managed by the *inverse transform*, and can be related to *invariant spectra* as in the linear case ! Let us give some more details.

Self modulations due to weak nonlinearities

One-dimensional wave propagation shows "typical signals", e.g. "plane waves" or "pulses". When weak nonlinearity modulates such a signal, the resulting signal is propagated in almost all cases by a special evolution equation - Korteveg de Vries for a pulse, Non-linear Schrödinger for a plane wave :

Plane wave. The signal $\exp[i\ (kx - \omega t)]$ is a solution of the linear equation $Du = 0$:

$$\begin{cases} D = \dfrac{\partial^2}{\partial t^2} + \displaystyle\sum_{j=0}^{J} (-1)^j\ b_j\ \dfrac{\partial^j}{\partial x^{2j}} \\ \omega^2(k) = \displaystyle\sum_{j=0}^{J} b_j\ k^{2j} \in \mathbb{R}^+ \quad \text{for} \quad k \in \Omega \end{cases} \tag{1}$$

We know that a smooth wave packet whose spectrum is concentrated around k travels with the group velocity $v = d\omega/dk$ and slowly disperses. Now consider the equation

$$\begin{cases} Du = \displaystyle\sum_{m=2}^{M} \epsilon^{m-1}\ F^{(m)}\ [u,\ u_x, u_t, \ldots] \\ F^{(m)}[u, \ldots] = \displaystyle\sum_{\mu=0}^{m} (u_t)^{\mu} \sum_{j_1, j_2, \ldots, j_{m-\mu}} C^{m,\mu}_{j_1, j_2, \ldots, j_{m-\mu}}\ u^{(j_1)} \ldots u^{(j_{m-\mu})} \end{cases} \tag{2}$$

where the symbols u_x or $u^{(1)}$ stand for the partial x-derivative, u_t the time derivative of u. The right hand side of (2) is a *weak* nonlinear perturbation of (1) if $\epsilon \to 0$. One seeks a solution u that reduces to $\exp[i\delta]$ $(\delta = kx - t\omega(k))$ as $\epsilon \to 0$, by using the following ansatz :

$$u(x,t) = \sum_{n=-\infty}^{n=+\infty} \epsilon^{\gamma_n}\ \exp[in\delta]\ \psi_n(\xi, \tau) \qquad \psi_n = \psi^*_{-n} \tag{3}$$

where

$$\xi = \epsilon(x - vt)\ , \quad \tau = \epsilon^2 t\ , \quad \gamma_n = ||n| - 1| \quad . \tag{4}$$

If ξ, τ, are non-infinitesimal values, we observe the wave motion on a long-time scale around the moving wave packet, whose envelope is described by ψ_1. The outstanding result is that one can match all the coefficients of all terms $\epsilon^{n'} \exp[in''\delta]$ provided the ψ_n's are constructed from $\psi_1 = \psi$ which satisfies the nonlinear Schrödinger equation (unless the special coefficient B_1 turns out to vanish) :

$$2i\omega \frac{\partial\psi}{\partial\tau} - B_1 \frac{\partial^2\psi}{\partial\xi^2} + \eta|\psi|^2 \psi = 0 \quad . \tag{5}$$

<u>Pulse</u>. A pulse $u = F(x-t)$ is solution of a plane wave equation which can be perturbed as above into the following one

$$\left(\frac{\partial}{\partial t} + \frac{\partial}{\partial x}\right) u_\epsilon(x,t) = \epsilon\, P\left(\frac{\partial}{\partial x}, \frac{\partial}{\partial t}, \epsilon\right) u_\epsilon(x,t) \quad . \tag{6}$$

Setting then $\xi = x-t$, $\tau = \epsilon t$, for analytic forms of P, and $\epsilon \to 0$, we obtain for u_ϵ the limit equation as a Burgers-Korteweg de Vries equation :

$$\frac{\partial u}{\partial\tau} = \alpha u \frac{\partial u}{\partial\xi} + \beta \frac{\partial^2 u}{\partial\xi^2} + \gamma \frac{\partial^3 u}{\partial\xi^3} \tag{7}$$

which again, is an almost universal form of the nonlinearity perturbed pulse propagation. There may be solutions of (5) and (7) (e.g. solitons) propagating in a typical nonlinear way.

<u>Signal processing.</u> The (almost) universal limit forms of propagation enable one to define invariant spectra as in the linear case. We explain it only on the "KdV" case, where the "linearized" equation would be

$$\eta_t + c_o\, \eta_x + \beta\, \eta_{xxx} = 0 \tag{8}$$

whose solutions can be analyzed by a standard Fourier signal processing, centered on the Fourier spectrum $\tilde{\eta}(k,t) = \int_{-\infty}^{+\infty} \eta(x,t)\, e^{-ikx}\, dx$, whose support $(-\infty, +\infty)$ is invariant. The evolution of η according to (8) is equivalent for $\tilde{\eta}$ to a very simple evolution :

$$\tilde{\eta}(k,t) = \tilde{\eta}(k,0) \exp[-i\,\omega(k)\, t] \tag{9a}$$

$$\omega(k) = k\left(c_o - \beta\, k^2\right) \tag{9b}$$

provided $|\eta(x,0)| \in L^1(\mathbb{R})$ (consistency condition). The analysis can be read on the following scheme :

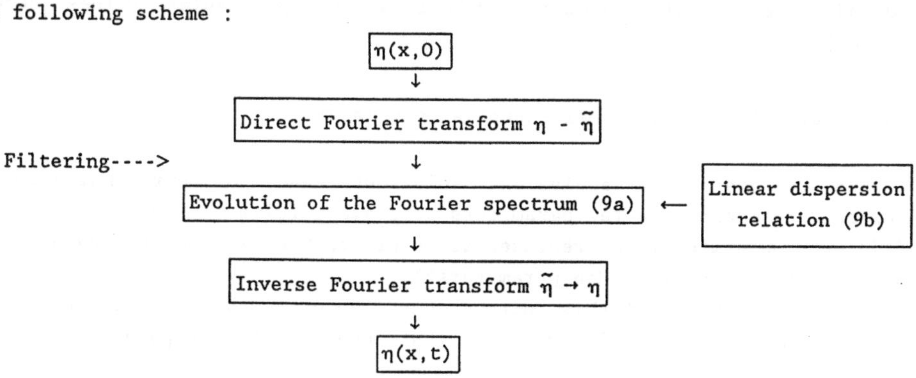

The inverse (spectral) transform yields exactly the same scheme to process the exact KdV equation :

$$\eta_t + c_o \eta_x + \alpha \eta \eta_x + \beta \eta_{xxx} = 0 \quad . \tag{10}$$

The spectral transform $\tilde{\eta}(k,t)$ is now made for $k \in \mathbb{R}$ of the reflection coefficient $r(k,t)$ which comes in the scattering problem of the Schrödinger equation on the line :

$$\psi_{xx} + \lambda[\eta(x,t) + k^2]\psi = 0 \qquad (\lambda = \alpha/6\beta) \tag{11}$$

and for Re $k = 0$, the normalizing coefficients c_n of the discrete spectrum $k \in \{i\, \kappa_n\}$ of (11). Again the support of the spectrum or (11) is the motion invariant. Again the evolution of $\tilde{\eta}(k,t)$ that corresponds to (10) is trivial.

$$r(k,t) = r(k,0)\ \exp[-it\ \omega(k)] \qquad \omega(k) = k\left(c_o - \beta\ k^2\right) \tag{12a}$$

$$c_n(t) = c_n(0)\ \exp[t\ \Omega_n] \qquad \Omega_n = \kappa_n\left(c_o + 4\beta\ \kappa_n^2\right) \quad . \tag{12b}$$

Again, the scheme we drawed above will hold. Calculating the direct Fourier transform becomes solving the direct problem $\eta(x,0) \to \tilde{\eta}(k,0)$. The evolution (12) yields $\tilde{\eta}(k,t)$ which gives back $\eta(x,t)$ by solving the inverse problem by means of the Faddeyev-Marchenko equation. The validity conditions are a little harder than in the linear case. As $\eta \to 0$, the "spectral transform" goes over to the "Fourier transform", but for $\eta \neq 0$, the "structures" that correspond to the discrete spectral values κ_n and their weights $c_n(t)$ are specifically non-linear ones. These "solitons" can thus identified by applying such a nonlinear processing to long internal waves in open sea. Filtering and other special signal modifications can be done as in the linear case. (Unfortunately Prof. Osborne was missing. See the lecture by Calogero).

II.4 Applications of Inverse Problems to improve measurements

Inversion Theory gives a way to analyze information and thus to extract more information of measure instruments. This has led the design of several improvements to confocal microscopes (Bertero, de Mol, Pike). Improvements are possible because the corresponding (linear) inverse problem is fully understood and its essential parameters can be shown up (by singular values decomposition).

II.5 Direct selection of model essential parameters

When the classes of equivalent solutions (exact *or* approximate) can be described by means of a few parameters, the ambiguities are *reduced*, ie it is sufficient determining a representative of each class and the "parameters" describing the class ("hidden parameters").

A common example of this direct selection of equivalence classes, to be expected in many physical cases, is that where parts of parameters can be "homogeneized" ie small variations on either side of an average value do not show up in the measurements (more difficult cases are well known e.g. gravity measurements or scattering at fixed energy). The only way to ascertain that homogeneizing or generalized homogeneizing is possible goes through

(a) a good definition of the equivalence classes

(b) an alternate analysis of direct and inverse problem constructions in order to refine the result

(c) a precise definition of $\mathcal{M}$: the reduction of a model to its "essential parameters" always depends on $\mathcal{M}$, ie, in a physical problem, to the ways used for sounding, filtering data, etc.

Lectures by Corones and Attouch deal with homogenization problem. Although I will not reproduce it in the present proceedings, I may cite also at this point my own work during the last three years on discontinuous impedances[6]. After reducing the direct problem to a chain of Schrödinger equations, both in the one-dimensional and in the three-dimensional case, and deriving the Born and quadratic approximations of the scattering amplitude in closed form, we have been able to study two questions :

(1) Is it possible to replace a distribution of "small reflectors" by "continuous parameters" ? The answer in the one-dimensional case is NO if results and analysis are exact, YES if data were made through a window-filter and if errors are allowed. In the n-dimensional case, there is no general answer.

(2) Dealing with restricted problems in the n-dimensional case - e.g. scattering at fixed k, fixed $|k|$, or fixed transfer $|k - k'|$, is it possible to balance a deformation of discontinuities surface by a perturbation of continuous part ? Again no general answer as yet.

II.6 Inverse methods and nonlinear structures

We already recalled the simplest solitons. The lectures of the present volume show the last years evolution towards symmetrical inverse methods and multidimensional structures. Symmetrical inverse methods are related to analyses of inverse problems where $\mathcal{C}$ and $\mathcal{E}$ are treated on equal foot, described by essential elements, and the analysis reduces to a joint study of routes among them [7]. So as to explain our point, first recall the Lax method, where the Schrödinger operator L is imposed a t-dependence such that it is isospectral. In this scheme the KdV equation (the simplest realization of the isospectral constraint), the direct Schrödinger problem (whose asymptotics suggest the "spectral transform") and the inverse problem, play quite different roles. Now compare to the latest avatar of AKNS-D Bar method, as it has been used by Léon et al. ("dispersion relation methods").

One starts with the "essential" matrix-valued function $R(k,x,t)$, $k \in \mathcal{D}$ whose evolution in $\mathcal{E}$ is defined by the dispersion functions Ω and Λ :

$$\frac{\partial R}{\partial t} = [R, \Omega] \quad ; \quad \frac{\partial R}{\partial x} = [R, \Lambda] \tag{13}$$

R *yields* a "wave function" ψ by solving the $\bar{\partial}$-problem

$$\frac{\partial}{\partial \bar{k}} \psi(k,x,t) = \psi(k,x,t)\ R(k,x,t) \quad \text{(+ boundary conditions)} \tag{14}$$

and the "potentials" (in $\mathcal{C}$) :

$$U = \frac{\partial \psi}{\partial x}\psi^{-1} - \psi \Lambda \psi^{-1} \qquad V = \frac{\partial \psi}{\partial t}\psi^{-1} - \psi \Omega \psi^{-1} \quad . \tag{15}$$

The "evolution equation" is then given simultaneously in $\mathcal{C}$ and $\mathcal{E}$, which *play exactly the same role :*

$$\frac{\partial \Lambda}{\partial t} - \frac{\partial \Omega}{\partial x} + [\Omega, \Lambda] = 0 \iff \frac{\partial U}{\partial t} - \frac{\partial V}{\partial x} + [U, V] = 0 \quad . \tag{16}$$

Furthermore, U and V can be determined by solving the $\bar{\partial}$-problems :

$$\frac{\partial U}{\partial \bar{k}} = - \psi \frac{\partial \Lambda}{\partial \bar{k}} \psi^{-1} \quad ; \quad \frac{\partial V}{\partial \bar{k}} = -\psi \frac{\partial \Omega}{\partial \bar{k}} \psi^{-1} \quad . \tag{17}$$

Various choices of R, Ω, Λ, yield the well-known integrable evolution equations together with solutions corresponding to the values of R : hence what was the inverse problem in the Lax method is solved in a direct way !

The generalization of inverse methods to multidimensional problems was made possible by using also $\bar{\partial}$-methods. The most spectacular result as yet obtained is the "two-dimensional soliton" discovered in our laboratory by Boiti, Léon, Martina, Pempinelli[8].

Lectures on various problems connected with inverse methods and nonlinear evolution equations are given by Cornille, Degasperis, Lambert, Latifi, Marchenko, Martinez-Alonso, Matveev, Musette, Pempinelli, Ragnisco, Santini, Sung, Zacharov.

III - FOR THOSE WHO LIKE EXERCICES

Among the original requirements of the RCP264 contracts, I had to give in our workshop a few pedagogical comments, or anything like that. Hence, in the present proceedings, I show now how quite elementary calculations can produce the famous two dimensional "coherent structure" of Boiti, Léon, Martina, Pempinelli.This remarkable solution of Davey-Stewardson equations (in particular), travels at constant velocity, with its phase depending linearly on all variables, is exponentially confined in all directions, and would be a genuine soliton if not for nonvanishing asymptotic conditions that appear in the term $A^{(o)}$ introduced below. But all its other properties emphasize it among any other solutions, including the compactly sized one introduced for instance by Cornille. Fokas & Santini have proposed to call the BLMP solution a dromion. Everybody acknowledges this result as a beautiful success of inverse methods. But after our derivation of a method which applies as well to N-dimensional equations, the BLMP solution shows up as an almost trivial example ! We shall explain this point and say a few words about generalizations.

We look for a special solution of the equation

$$i \frac{\partial q}{\partial t} + \Delta q + Aq = 0 \tag{18}$$

where

$$A(x_1,\dots,x_N,t) = \alpha_o \int \dots \int_Q \Delta(|q|^p)\, dx'_1 \dots dx'_N - A^{(o)}(x_1,\dots,x_N,t) \quad (19a)$$

$$Q = \{x'_1,\dots,x'_N \,|\, x'_1 \leqslant x_1,\dots,x'_N \leqslant x_N\} \quad (19b)$$

and $A^{(o)}$ is a sum of real functions depending only on $p \leqslant N-1$ coordinates, so that

$$\frac{\partial^N A^{(o)}}{\partial x_1 \dots \partial x_N} = 0 \quad . \quad (20)$$

Let $q = f \exp[-i\,\alpha]$, where f and α are real functions of $x_1,\dots,x_N,t$. Assume the first and second derivatives of f^{-p} are Lebesgue-integrable in $\mathbb{R}^N$, and define $g = \exp[u]$ such that

$$\frac{\partial_N \log g}{\partial x_1 \dots \partial x_N} \equiv \frac{\partial^N u}{\partial x_1 \dots \partial x_N} = C_o\, f^{-p} \quad . \quad (21)$$

It follows from (19) and (21) that

$$A = s_o\, \Delta u - B^{(o)} \quad ; \quad s_o = \alpha_o / C_o \quad (22)$$

where $B^{(o)}$ satisfies (20). We write $B^{(o)}$ in the case $N = 2$ and $N = 3$:

$$s_o^{-1}\, B^{(o)} = \frac{\partial^2}{\partial x_1^2} u(x_1, -\infty) + \frac{\partial^2}{\partial x_2^2} u(-\infty, x_2) + s_o^{-1} A^{(o)} \quad (23)$$

$$s_o^{-1}\, B^{(o)} = \Delta_{1,3}\, u(x_1, -\infty, x_3) + \Delta_{1,2}\, u(x_1, x_2, -\infty) + \Delta_{2,3}\, u(-\infty, x_2, x_3)$$

$$- \frac{\partial^2}{\partial x_1^2} u(x_1, -\infty, -\infty) - \frac{\partial^2}{\partial x_2^2} u(-\infty, x_2, -\infty) - \frac{\partial^2}{\partial x_3^2} u(-\infty, -\infty, x_3) + s_o^{-1} A^{(o)} \quad (24)$$

The function q is a solution of (18) if

$$- f^{-1} \frac{\partial f}{\partial t} - \Delta\alpha + 2\, f^{-1}\, \mathrm{grad}\, f \,.\, \mathrm{grad}\, \alpha = 0 \quad (25)$$

$$s_o \Delta u - f^{-1} \Delta f + 2 f^{-2}\, \mathrm{grad}\, f.\mathrm{grad}\, f + \frac{\partial \alpha}{\partial t} - \mathrm{grad}\, \alpha.\mathrm{grad}\, \alpha - B^{(o)} = 0 \quad . \quad (26)$$

The solution we seek is "special" first because we assume its phase depends linearly on all variables :

$$\alpha(x_1,\dots,x_N,t) = \beta + \gamma t + \sum_1^N \eta_i\, x_i \quad . \quad (27)$$

The equation (25) is solved provided f depends on time only through the generalized coordinates

$$X_i = x_i + 2\eta_i\, t \quad (28)$$

and (because of (21)), so should do u. Hence f is a solution of (26) if it satisfies

$$s_o \, \Delta(\log g) - f^{-1} \Delta f + 2 f^{-2} \text{ grad } f . \text{grad } f = a_o \tag{29}$$

where

$$a_o = B^{(o)} - \gamma + \sum_{i=1}^{N} \eta_i^2 \tag{30}$$

and the differential operators are now meant with respect to the X_i's. We obtain a still more special solution of (18) in the case $p = 2$, $s_o = 2$ and appropriate $A^{(o)}$, if we are able to find a function f such that (21) holds with $f = g$ and that $\frac{\Delta f}{f} = a_o$ satisfies (20), this last condition following from (29). Now, for $N = 2$, $p = 2$, $f = g$, the condition grad $C_o = 0$ yields from (21) the two equations (where we use the notation X,Y instead of X_1, X_2) :

$$\frac{\partial}{\partial Y}\left[f^{-1} \frac{\partial^2 f}{\partial X^2}\right] - \frac{\partial}{\partial X}\left[f^{-1} \frac{\partial^2 f}{\partial Y^2}\right] = 0 \quad . \tag{31}$$

The general solution of (31) depends on two arbitrary functions $\lambda(X)$, $\mu(Y)$. Once they are chosen, we define the linearly independent solutions $F_1(X)$ $F_2(X)$ resp. $G_1(Y)$, $G_2(Y)$, of

$$F''(X) - \lambda^2(X) \, F(X) = 0 = G''(Y) - \mu^2(Y) \, G(Y) \tag{32}$$

and the solution of (31) is

$$\begin{aligned} f(X,Y) = {} & a_{11} F_1(X) G_1(Y) + a_{12} F_1(X) G_2(Y) + a_{21} F_2(X) G_1(Y) \\ & + a_{22} F_2(X) G_2(Y) \end{aligned} \tag{33}$$

where the coefficients a_{ij} are almost arbitrary. Indeed,

$$f^2 \frac{\partial^2}{\partial X \, \partial Y} \log f = (a_{11} a_{22} - a_{12} a_{21}) \, W(F_1, F_2) \, W(G_1, G_2) \tag{34}$$

must not be cancelled by our choice. Furthermore, q is regular only if f has no zero. Notice that $f^{-1} \Delta f$ reduces to $\lambda^2(X) + \mu^2(Y)$, which satisfies (20). Now, the solution introduced by Boiti et al. corresponds to the case of constant λ and μ, and

$$a_{11} = 1 + r \quad , \quad a_{12} = a_{21} = a_{22} = 1 \quad , \tag{35}$$

where r is strictly positive, so that

$$f(X,Y) = 4 \text{ ch } \lambda X \text{ ch } \mu Y + r \exp[\lambda X + \mu Y] \tag{36}$$

and it is clear that $f^{-2} = |q|^{-2}$ is "exponentially" confined.

Three-dimensional generalizations

In the case $N = 3$, the study is not easy and a simpler case is that of $p = 1$, where we reset (21) as

$$g^3 \frac{\partial^3 \log g}{\partial X \, \partial Y \, \partial Z} = C_o \, h = C_o \, g^3 \, f^{-1} \quad . \tag{37}$$

For $s_o = 12$, the equation (26) reduces then to

$$g\, g^{-1}\, \Delta g + h^{-1}\, \Delta h - 6\, g^{-1}\, h^{-1} \text{ grad } g.\text{grad } h = a_o \quad . \tag{38}$$

We have been able to find several classes of solutions of (37)-(38) but not exponentially confined. An amusing example is

$$f = h\, g^{-3} = \frac{2\, abc\, sh(X - X_o)\, sh\,(Y - Y_o)\, sh(Z - Z_o)}{[a\, ch(X - X_o) + b\, ch(Y - Y_o) + c\, ch(Z - Z_o)]^3} \tag{39}$$

which is not confined in the direction (1,1,1) but is exponentially confined in any other direction. Needless to say, $(X - X_o)$ could be replaced by $\mu(X - X_o)$ and the parameters $A^{(o)}$, α_o, have to be fitted. Another amusing example can be built out of a linear convex combination of e^{X+Y+Z}, e^{-X-Y+Z}, e^{X+Y-Z}, but unfortunately the fourth term of the permutation, e^{X-Y+Z}, which would ensure the confinement, cannot be included. Similar remarks hold for the "hamiltonian" two-dimensional case, where (30) must hold with $B^{(o)}$ replaced by

$$B^{(H)} = \frac{1}{2}\, s_o \left[\frac{\partial^2\, u(-\infty,y)}{\partial y^2} + \frac{\partial^2\, u(\infty,y)}{\partial y^2} + \frac{\partial^2\, u(x,-\infty)}{\partial x^2} + \frac{\partial^2\, u(x,\infty)}{\partial x^2}\right] \tag{40}$$

and for further generalizations of our simple device.

REFERENCES

(1) RCP meeting "Etude Interdisciplinaire des Problèmes Inverses" 1972, unpublished report of "Laboratoire de Physique Mathématique", Montpellier.

(2) Cahiers Mathématiques de Montpellier n°s 24 (1981), 28, (1982), 31, (1983), 32, (1984), 35, (1985).

(3) "Applied Inverse Problems", P.C. Sabatier Ed. Lecture Notes in Physics n° 85, Springer. Berlin, Heidelberg, New York 1978.
"Problèmes Inverses, Evolution non linéaire", P.C. Sabatier Ed. Editions du C.N.R.S. Paris 1980.
"Inverse Problems : an Interdisciplinary Study", P.C. Sabatier Ed. - Suppl. 19 to Advances in Electronics and Electron Physics - Academic Press - London, Orlando, New York, Sydney, Tokyo, Toronto (1987)
"Topics on Inverse Problems"
"Some Topics on Inverse Problems", P.C. Sabatier Ed. World Scientific. Singapore, New Jersey, Hong Kong (1988).

(4) See for example current issues of the journal "Inverse Problems" published by the Institute of Physics (Bristol, London) since 1985.

(5) Mark Twain "Adam's Diary", "Eve's Diary" Harper & Brothers New York. N.Y.

(6) P.C. Sabatier. On modelling discontinuous data. Three-dimensional scattering. J. Math. Phys. 30, 2585-2598 (1989).

(7) P.C. Sabatier. A few geometrical features of inverse and ill-posed problems, in "Inverse and Ill-Posed Problems" H.W. Engl & C.W. Groetsch ed. Ac. Press 1987, p.1-18.

(8) M. Boiti, J. Leon, L. Martina, F. Pempinelli, "Scattering of localized solitons in the plane" Phys. Lett. A 132, 432-439.

Part I

Tomographic Inverse Problems

Tomography with Diffusion

F.A. Grünbaum

Department of Mathematics, University of California, Berkeley, CA 94720, USA

Foreword.

On the occasion of the celebration of seven centuries in the life of the University of Montpellier it is appropriate for the participants in the traditional Reucontres on Inverse Problems to thank its "soul and spirit" Professor Pierre Sabatier for many years of organizational and scientific leadership, past and future.

His efforts in bringing together a spectrum of theoretical and practical workers dealing with inverse problems in the physical sciences and the importance he assigns to getting these groups to talk and learn from each other are essential for the further progress of science both applied and pure.

Pierre: Congratulations on Montpellier's traditions and many, many thanks!!!

Introduction and statement of the problem.

The purpose of this paper is to present a method for obtaining images of the interior of an object by means of radiation that is attenuated and scattered on its way through the object.

If one ignores the possibility of a photon changing direction as it travels through the object and restricts attention to the phenomenon of absorption one recovers "standard X-ray tomography". Each pixel or voxel is then characterized by an unknown attenuation coefficient and the mathematical problem is to obtain these coefficients from a collection of "*straight* line integrals". Here we take the position that a photon no longer travels necessarily on a straight line. Rather, on entering a voxel a photon either gets absorbed or else emerges by preserving its incoming direction, or reversing it or finally by making a ninety-degree turn into one of four possible adjacent voxels. This picture requires voxels which are little parallelograms with edges aligned with the coordinate axis x, y, z, an assumption that we make from now on. Four of the six possibilities are illustrated below. It is clear that many other discretizations are possible without any deep change in the model.

Partially supported by NSF under grant DMS-87-20007 and by AFOSR under contract AFOSR-88-0250.

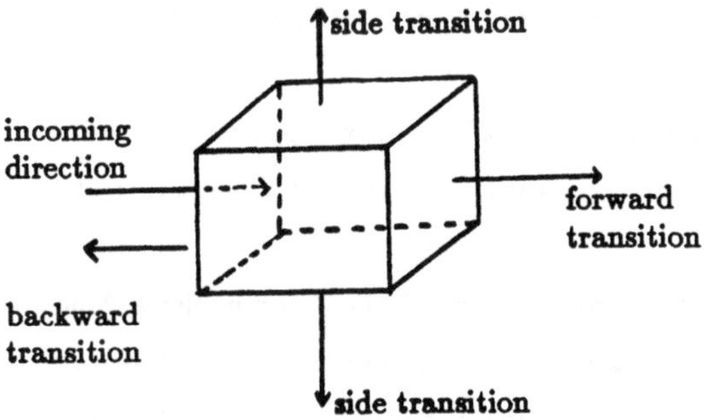

We go further and assume that the object is a cube broken up into smaller cubical voxels, very much like the "Rubik cube". We put N small cubes on each dimension, so that we get a total of N^3 voxels.

If a voxel is parametrized by the labels i, j, k then v_{ijk} will denote its absorption probability, while f_{ijk}, b_{ijk} and s_{ijk} denote the probabilities of a forward, backward and sidewise transition respectively. Notice that we have the "conservation law"

$$f_{ijk} + b_{ijk} + 4s_{ijk} + v_{ijk} = 1.$$

This is an assumption of our model, eliminating phenomena like fluorescence.

We can now formulate our "imaging problem". Pick any one of the $6N^2$ external faces of the voxels on the surface of our object and place a source of photons orthogonal to it. Each individual photon is assumed to move unperturbed by the presence of other photons. After a complicated sequence of transitions a photon will either have died somewhere along the way or will emerge on one of the exposed faces of the surface voxels, where a detector is ready to pick it up. We are assuming that we cannot measure "time of flight" information and that in a reasonable short time the system has reached equilibrium and all emerging photons have made it to their respective detectors.

Assume for simplicity that we place sources and detectors at each one of the $6N^2$ "exposed faces" of the little cubes, thus getting a total of $(6N^2)^2 = 36N^4$ measurements. A moment's thought shows that exchanging source and detectors has no effect on these quantities, so that one has

$$(36N^4 - 6N^2)/2 + 6N^2 \cong 18N^4$$

measurements.

PROBLEM: Find the $3N^3$ independent quantities v_{ijk}, f_{ijk}, b_{ijk}, s_{ijk} from these measurements.

Although the problem is overdetermined—especially so in dimension three—I think that one's natural reaction is that this problem cannot be solved either in theory or at least certainly in practice. The purpose of this note is to shake up a bit people holding such a sensibly pessimistic position.

We will concentrate on some few analytical results leaving a battery of supporting numerical evidence for a separate paper. All these results point in a very encouraging, albeit surprising, direction.

Before we describe our approach to this problem we give a word about terminology. Since the problem is three-dimensional from scratch, it is probably inappropriate to use the word tomography: we are not necessarily reconstructing a "section or thome" at a time. The word diffusion is probably a misnomer too. It is clear that we are doing some sort of inverse transport theory. We are really dealing with a two-step Markov process since the transition mechanism depends not only on the present state but also on the preceding one. Thus we could get a Markov process in an enlarged state space. In spite of these caveats we stick to the name "diffusion tomography"; if this field produces useful results someone will certainly come up with a better name for it. If this does not happen, nobody will care about the name anyway.

The direct problem.

For each detector position let

$$P^{\alpha}_{ijk}$$

denote the probability that a photon that is alive at voxel ijk having arrived at it from the direction α, $(\alpha = 1, \dots, 6)$ will make it to the detector in question.

These quantities will be related by a set of linear equations involving the unknowns f_{ijk}, b_{ijk}, s_{ijk}, v_{ijk}. These equations are the *core* of the method.

Notice that most of these quantities are not measurable: for each one of the detector positions there is a set of $6N^2$ of these quantities that can be measured. We call these variables the "source" variables.

The remaining $6N^3 - 6N^2$ are called the "internal" variables. To get a closed set of linear equations we need to have labels for the detectors. This is most conveniently done by adding an extra layer of variables on each face of the cube. These variables P^{α}_{ijk} have one of the lower indices either equal to zero or to $N + 1$, and they give a convenient way of keeping track of what detector is "turned on". We refer to these $6N^2$ variables as "detector" variables.

For notational convenience denote by X a vector made up of these "detector" variables, by Z a vector with the "internal" variables and finally Y a vector with the "source" variables.

With appropriate matrices A, B, C, D the linear equations pertaining to external or internal voxels are given by

$$Y + CZ + DX = 0$$

$$AZ + BX = 0$$

respectively.

This gives

$$Y = (CA^{-1}B - D)X.$$

This expression embodies the solution of the direct problem.

The probability of going from a source to a detector is given by one of the entries of the square matrix given above. We notice in passing that this combination is exactly the "Schur complement" of the matrix

$$\begin{bmatrix} A & B \\ C & D \end{bmatrix}.$$

Our inverse problem is one of reconstructing this matrix from this "Schur complement". I am sure that in general this is not possible or even meaningful. However, our matrix has a lot of extra structure.

We close this section by giving one simple example of the entries of this "transfer matrix" in the case $N = 2$. We give the expression for P_{14}, the

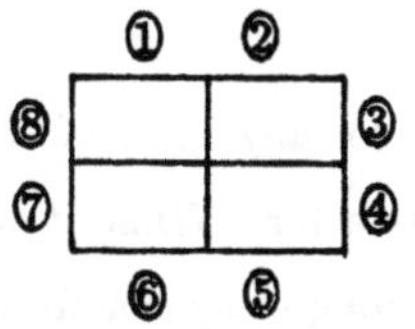

probability of being caught at detector ④ if one is implanted at source location ①.

One gets the rather compact expression

$$\begin{aligned}
P_{14} = (&s_{11}^2 s_{12}^2 s_{21} s_{22}^2 - b_{11} f_{11} s_{12}^2 s_{21} s_{22}^2 - b_{12}^2 s_{11}^2 s_{21} s_{22}^2 \\
&+ b_{11} b_{12}^2 f_{11} s_{21} s_{22}^2 - b_{12} f_{11} s_{21} s_{22}^2 + f_{11} f_{22} s_{11} s_{12} s_{21}^2 s_{22} \\
&- b_{11} f_{22} s_{11} s_{12} s_{21}^2 s_{22} - b_{22} f_{11} s_{11} s_{12} s_{21}^2 s_{22} \\
&+ b_{11} b_{22} s_{11} s_{12} s_{21}^2 s_{22} - b_{21}^2 f_{11} f_{22} s_{11} s_{12} s_{22} \\
&+ b_{11} b_{21}^2 f_{22} s_{11} s_{12} s_{22} - b_{21} f_{22} s_{11} s_{12} s_{22} + b_{21}^2 b_{22} f_{11} s_{11} s_{12} s_{22} \\
&- b_{21} f_{11} s_{11} s_{12} s_{22} - b_{11} b_{21}^2 b_{22} s_{11} s_{12} s_{22} + b_{21} b_{22} s_{11} s_{12} s_{22} \\
&+ b_{11} b_{21} s_{11} s_{12} s_{22} - s_{11} s_{12} s_{22} - b_{22} f_{22} s_{11}^2 s_{12}^2 s_{21} \\
&+ b_{11} b_{22} f_{11} f_{22} s_{12}^2 s_{21} + b_{12}^2 b_{22} f_{22} s_{11}^2 s_{21} - b_{12} f_{22} s_{11}^2 s_{21} \\
&- b_{11} b_{12}^2 b_{22} f_{11} f_{22} s_{21} + b_{12} b_{22} f_{11} f_{22} s_{21} + b_{11} b_{12} f_{11} f_{22} s_{21} \\
&- f_{11} f_{22} s_{21}) \,/ \\
(&s_{11}^2 s_{12}^2 s_{21}^2 s_{22}^2 - b_{11}^2 s_{12}^2 s_{21}^2 s_{22}^2
\end{aligned}$$

$$
\begin{aligned}
&- b_{12}^2 s_{11}^2 s_{21}^2 s_{22}^2 + b_{11}^2 b_{12}^2 s_{21}^2 s_{22}^2 - b_{11} b_{12} s_{21}^2 s_{22}^2 \\
&- b_{21}^2 s_{11}^2 s_{12}^2 s_{22}^2 + b_{11}^2 b_{21}^2 s_{12}^2 s_{22}^2 - b_{11} b_{21} s_{12}^2 s_{22}^2 \\
&+ b_{12}^2 b_{21}^2 s_{11}^2 s_{22}^2 - b_{11}^2 b_{12}^2 b_{21}^2 s_{22}^2 + b_{11} b_{12} b_{21}^2 s_{22}^2 \\
&+ b_{11} b_{12}^2 b_{21} s_{22}^2 - b_{12} b_{21} s_{22}^2 - 2 s_{11} s_{12} s_{21} s_{22} - b_{22}^2 s_{11}^2 s_{12}^2 s_{21}^2 \\
&+ b_{11}^2 b_{22}^2 s_{12}^2 s_{21}^2 + b_{12}^2 b_{22}^2 s_{11}^2 s_{21}^2 - b_{12} b_{22} s_{11}^2 s_{21}^2 \\
&- b_{11}^2 b_{12}^2 b_{22}^2 s_{21}^2 + b_{11} b_{12} b_{22}^2 s_{21}^2 + b_{11}^2 b_{12} b_{22} s_{21}^2 - b_{11} b_{22} s_{21}^2 \\
&+ b_{21}^2 b_{22}^2 s_{11}^2 s_{12}^2 - b_{21} b_{22} s_{11}^2 s_{12}^2 - b_{11}^2 b_{21}^2 b_{22}^2 s_{12}^2 \\
&+ b_{11} b_{21} b_{22}^2 s_{12}^2 + b_{11}^2 b_{21} b_{22} s_{12}^2 - b_{11} b_{22} s_{12}^2 - b_{12}^2 b_{21}^2 b_{22}^2 s_{11}^2 \\
&+ b_{12} b_{21}^2 b_{22} s_{11}^2 + b_{12}^2 b_{21} b_{22} s_{11}^2 - b_{12} b_{21} s_{11}^2 + b_{11}^2 b_{12}^2 b_{21}^2 b_{22}^2 \\
&- b_{11} b_{12} b_{21}^2 b_{22}^2 - b_{11} b_{12}^2 b_{21} b_{22}^2 + b_{12} b_{21} b_{22}^2 - b_{11}^2 b_{12} b_{21}^2 b_{22} \\
&+ b_{11} b_{21}^2 b_{22} - b_{11}^2 b_{12}^2 b_{21} b_{22} + 2 b_{11} b_{12} b_{21} b_{22} - b_{21} b_{22} \\
&+ b_{11} b_{12}^2 b_{22} - b_{12} b_{22} + b_{11}^2 b_{12} b_{21} - b_{11} b_{21} - b_{11} b_{12} + 1)
\end{aligned}
$$

Similar expressions can be derived for the remaining terms. I guarantee the reader hours of unmatched enjoyment.

The method given above produces for any value of N the values of the probabilities of a transition from any source to any detector. While it would be unwise to give the general formulas in question, the problem of computing exactly these probabilities once the values of the parameters f_{ijk}, etc. have been fixed is straightforward and numerically feasible.

The inverse problem.

The only remaining problem is to solve these

$$\frac{1}{2}(36N^4 - 6N^2) + 6N^2 = 18N^4 + 3N^2$$

equations in terms of the $3N^3$ quantities of interest. Let the reader beware.

It is not hard to think of cases where this cannot be done, i.e. the equations have a nonunique solution.

a) if $b_{ijk} \equiv s_{ijk} \equiv 0$ then as soon as $N \geq$ dim space one does not have enough measurements to determine v_{ijk}. This is "ordinary tomography" with too few directions, i.e. before the days of G. Hounsfield and A. Cormack.

b) a more interesting case comes up when one has a very symmetric object, as below

α	β	α
β	γ	β
α	β	α

Here α denotes the common (unknown) value of f_{ij}, b_{ij}, s_{ij} at each corner. Likewise β denotes the common (unknown) value of these quantities at the indicated pixels and finally γ the common (unknown) value at the center. One can see that this example leads to a continuum of solutions.

In a separate place we have given a 2×2 example showing that when $s_{ij} \equiv 0$ we can give an explicit solution for

$$f_{ij},\, b_{ij} \qquad 1 \leq i, j \leq 2$$

at least in the case of "general position". We do not repeat that example here. Rather we give a new example for an even simpler geometry, namely

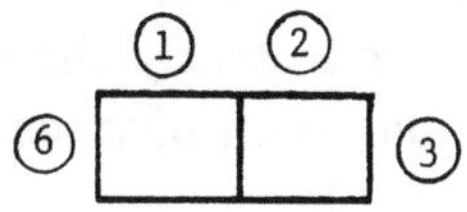

Here we assume that we have sources and detectors only at the indicated positions, but we make *no restrictions* on the values of f_{ij}, b_{ij}, s_{ij}.

With the usual meaning for the quantities P_{ij}, $i, j \in \{1, 2, 3, 6\}$ we get

$$s_{11} = \frac{P_{13}P_{26} - P_{16}P_{23}}{P_{23}P_{36} - P_{26}P_{33} - P_{23}}$$

$$f_{11} = \frac{P_{36}(P_{13}P_{26} - P_{16}P_{23})}{P_{13}(P_{23}P_{36} - P_{26}P_{33} - P_{23})}$$

$$b_{11} = \frac{P_{23}P_{36}s_{11} - P_{13}P_{26}}{P_{13}(P_{23}P_{36} - P_{26}P_{33})}.$$

By an obvious symmetry we can get s_{12}, f_{12}, b_{12}.

Once again these formulas are only useful in "general position".

References

F. A. Grünbaum — Backscattering comes to the rescue, AMS Conference on the Radon Transform, Arcata, 1989, to appear in Contemporary Mathematics, T. Quinto, editor.

Medical Imaging: State-of-the-Art and Future Development

A.K. Louis

Fachbereich Mathematik, Technische Universität Berlin, D-1000 Berlin 12, Germany

1 Introduction

The aim of medical imaging is to provide in an non – invasive way morphological information about a human patient. The information is obtained by performing an " experiment " where the interaction of a source of radiation anf the tissue under consideration is measured. From the measured data the desired information has to be computed, hence we face an inverse problem. It is always ill – posed in the sense that small errors in the data can be amplified to large changes in the reconstruction. For developing efficient and stable software we have to study the mathematical model; i.e., the description of the experiment based on physical and engeneering knowledge. In optimal situations it is possible to derive " inversion formulas " which relate in a constructive way the data to the searched – for information. Reconstruction algorithms can be found by discretisizing these formulas. But of course we have to perform a stability analysis in order to design the software such that the influence of the data noise is reduced as much as possible. If such inversion formulas are unknown or cannot be discretisized in an accurate way direct discretization and iterative methods are used for the computation.

The techniques in medical imaging can be grouped into two categories, namely in transmission computerized tomography (TCT) and emission computerized tomography (ECT). The basic differences between TCT and ECT are wether the source of radiation is outside or within the patient and whether the directions of the emanating rays are a – priori known or not, respectively. Another possibility of subdividing these techniques in different groups is the used wavelength, namely short or large wavelength.

In the next chapters we describe different technologies and the underlying mathematical models. First we discuss methods which are presently used and then we mention possibilities for the future.

2 Present Day Imaging Technologies

In this assembly of modern technologies in medical imaging the oldest is x – ray computerized tomography. The so – called CAT (computer assisted tomography) scanners are installed in all major hospitals. The basic idea is the following. If one considers the result of the standard x – ray examinations, where shadow pictures are produced, one realizes that soft tissue is hidden by dense tissues as bones for example. When searching for a tumor in the brain this method is obviously not applicable because of the skull. Instead one considers the examined part of the body slice by slice. Through a fixed slice x rays are sent in many different directions. When travelleing through the body the denser the tissue the more photons are absorbed. The photons leaving the body are counted in a detector sitting on the opposite side of the x – ray tube. These measurements are then repeated for different slices, for example in a typical head scan ten slices are examined.

The present day scanners consist essentially of an x – ray tube and a set of detectors, usually 700 or more, which are mounted on a gantry moving around the patient. The number of positions where the x – ray tube is fired is between 400 and 1400, thus giving a total of around half a million data per slice measured in less than ten seconds. The desired result, the density distribution in the measured slice through the body, is then visualized on a TV screen with again some 512×512 picture elements, so – called " pixels ", showing gray levels which represent different densities.

With the physical assumption that the x rays travel on straight lines, the attenuation being proprotional to the intensity and the path length the mathematical model is, after introducing the proportionality factor f, the x – ray attenuation coefficient,

$$\int f \; dt = -\ln \frac{I_l}{I_0}$$

where I_0 is the intensity of the ray when leaving the x – ray tube, and I_L is the measured intensity at the detector for the path L. The mathematical model is the Radon transform, which is, after introducing the parametrization with $s \in I\!R$, the signed distance of the ray from the origin, and $\omega \in S^1$, the direction perpendicular to the ray,

$$Rf(\omega, s) = \int_{I\!R} f(s\omega + t\omega^\perp)dt.$$

An inversion formula, relating the data Rf with the searched – for density f is given in two steps allowing for parallel processing,

$$f(x) = \frac{1}{4\pi} \int_0^\pi q(\omega, \omega \cdot x) d\omega$$

where

$$q(\omega, s) = \int_{I\!R} (\widehat{Rf})(\omega, \sigma) |\sigma| e^{\imath \sigma s} ds$$

with $(\widehat{Rf})$ denoting the Fourier transform with respect to the second variable.

When the data are sampled in a regular manner, discretizations of this formula are used. For more details see Natterer [20]. In incomplete data problems data completion methods, see [3,10,11,15], or projection methods are used [1,4,13,20].

A newer medical imaging system is magnetic resonance imaging (MRI) or, as it was named in its early days, nuclear magnetic resonance (NMR). The basic physical idea is the following. If a constant magnetic field B_0 is applied on a tissue the spin of atomic nuclei processes around the direction of B_0 and resonance occurs at the Lamor frequency depending on the strenght of the field. In order to achieve a spatial resolution instead of a constant field a spatially varying field is applied where planes or lines are picked out. A first crucial point is the design of these fields. But then, in a rather complicated way, the Fourier transform of the searched – for distribution is measured on a rectangular grid, hence the inversion is nothing but a fast Fourier transform. For more details see for example Hinshaw – Lent [5].

In emission computerized tomography (ECT) the source of radiation is within the patient. This is achieved by injecting radiopharmaceuticals. The object is to find the distribution of the radioactive tracers in the body. In principle we distinguish between two methods. In single photon emission computerized tomography (SPECT) one can use as radioactive tracer any radioisotop that emits γ rays, emitted as single photons. A severe problem is here to define the direction of a ray, which is needed for the reconstruction algorithm. To this end collimators are used which have the drawback that most of the rays are discarded. This leads to a low signal – to – noise ratio. The physical assumption that the attenuation of the intensity is proportional to the intensity and to the density of the tissue in which the ray is travelling leads for each ray to the ordinary differential equation

$$\dot{I}(t) = -\mu(t)I(t) \ , \ I(t_0) = I_0$$

where $\mu(t)$ denotes the attenuation coefficient of the tissue and I_0 is the original intensity at the point t_0. Note that in contrast to x – ray CT, where we have the same differential equation, the point t_0 is not outside the body. The correlation between the measured data $g(\omega, s)$ and the activity distribution f is then given by the line integral

$$\begin{aligned} g(\omega, s) &= R_\mu f(\omega, s) \\ &= \int_{I\!R} f(s\omega + t\omega^\perp) \exp(-D\mu(\omega, s\omega + t\omega^\perp))dt \end{aligned}$$

where

$$D\mu(\omega, x) = \int_0^\infty \mu(x + \tau\omega^\perp)d\tau.$$

The transform R_μ is denoted as the attenuated Radon transform. Inversion formulas are known only for constant μ, hence in any practical application approximation formulas have to be used. A crude method consists in neglecting the attenuation, note that $\mu = 0$ reduces to the imaging problem in x – ray CT, and then corrections for known cupping effects and so on are added. A sophisticated inversion based on the characterization of the range is reported in Natterer [19].

In positron emission tomography (PET) radioisotopes that decay by emitting a positron are used. The annihilation radiation consists of two electromagnetic photons (511 keV photons) that are emitted in time coincidence and are oppositely directed. Then one performs coincidence measurements, hence no collomators are necessary as in SPECT which results in a much better singal – to – noise ratio. The drawback of this method is that, because of the short half lives of the isotopes involved, a very expensive cyclotron production facility has to be located near the hospital. As mathematical model we get

$$R_{\mu,c} f(\omega, s) = \int_{L(\omega,s)} f(x) \exp[-(D\mu(\omega, x) + D\mu(-\omega, x))]dx$$

where $L(\omega, s)$ denotes the line $x \cdot \omega = s$. Because of

$$D\mu(\omega, x) + D\mu(-\omega, x) = R\mu(\omega, x \cdot \omega)$$

we finally get

$$R_{\mu,c}f(\omega,s) = e^{-R\mu(\omega,s)}\, Rf(\omega,s)$$

and the reconstruction problem is after dividing by the exponential term reduced to the one from x – ray CT. For additional information we refer for example to Knoll [6].

3 Future Developments

Since a long time ultrsound techniques are studied in medical imaging. Before the introduction of MRI they offered a possibility for a system without hazards for the patient. The critical point is, even when all the technical problems are solved, the reconstruction of the searched – for information from the data. In contrast to the methods discussed so far, ultrasound is not travelling on straight lines, it is diffracted, hence we face a nonlinear problem.

When performing time – of – flight measurements, the mathematical model is described by the Helmholtz equation

$$(\Delta + K^2)\psi = 0$$

where

$$K^2 = \frac{\omega^2}{c^2} + \frac{1}{2\rho}\Delta\rho - \frac{3}{4}\left(\frac{1}{\rho}\nabla\rho\right)^2.$$

Here P denotes the acoustic pressure, ρ the density and $\psi = P\rho^{-1/2}$. When rewriting as a Schrödinger equation we have to find the potential V in

$$(\Delta + k^2)\psi = k^2 V\psi$$

when measurements of ψ are taken. Theoretical inversion formulas are given by Nachman [18] and Ramm [21]. Both methods are based on a non – physical solution ψ_θ of

$$(\Delta + k^2)\psi_\theta = 0$$

which behaves as

$$\psi_\theta(x) = e^{ik\theta x}\left(1 + v_\theta(x)\right)$$

where $\theta \in \mathbb{C}^3$ with $\theta \cdot \theta = 1$ and small v_θ. In both methods this function ψ_θ is used for computing the Fourier transform of the searched – for potential V. Nachman applies Green's formula and defines a Dirichlet – to – Neumann map for relating the data to the potential. Ramm determines an expansion of ψ_θ in terms of scattering solutions for plane wave insonification. But so far no numerical realization of either of the methods is known and it seems still a lot to be done. Computations with linearizations, Born or Rytov, do not lead to the desired and expected accuracy.

In impedance tomography the internal conductivity or impedance in the body is searched. To this end boundary measurements of voltages and corresponding current fluxes are performed. Besides morphological informations one hopes for a way of measuring non – invasively the temperature, which is very important for example in hyperthermia treatment where the tumor has to be overheated to some $42.5°C$ and the healthy tissue must be kept below a certain level, see [7]. Monitoring the heating process is very important.

The mathematical problem is modelled by

$$\nabla(\gamma \nabla u) = 0$$

in the body. Here u denotes the voltage potential and γ the medium conductivity which has to be determined.

Even further in the future are techniques where for example the biomagnetism is measured by so – called squids near the body. The strength of these fileds is 10^{-8} of the one from the earth, so at the moment it is not suitable for imaging techniques but it gives some indications for example in the case of epilepsy, see [16].

References

[1] Dolveck - Guilpart, B.: Deux problèmes de représentation et d'identification d'un milieu a partir de sondages, Thesis, Montpellier, 1989

[2] Greenleef, J.F.: Computerized tomography with ultrasound. Proc. IEEE **71**, 356-372, 1983

[3] Grünbaum, F.A.: Reconstruction with arbitrary directions : dimensions two and three. In : Herman, G.T., Natterer, F.(eds.): Mathematical aspects of computerized tomography, Springer LNMI 8, 112-126, 1981

[4] Herman, G.T.: Image reconstruction from projections. The fundamentals of computerized tomography. Academic Press, New York, 1980

[5] Hinshaw, W.S., Lent, A.H.: An introduction to NMR imaging : from Bloch equation to imaging equation, Proc. IEEE **71**, 338-350, 1983

[6] Knoll, G.F.: Single-photon emission computed tomography. Proc. IEEE **71**, 320-329, 1983

[7] Kremer, J., Louis, A.K.: About the mathematical foundation of hyperthermia treatment. Math. Meth. Appl. Sci. in press, 1990

[8] Langenberg, K.J.: Introduction to the special issue of inverse problems. Wave Motion **11**, 99-112, 1989

[9] Lewitt, R.M.: Reconstruction algorithms : transform methods. Proc. IEEE **71**, 390-408, 1983

[10] Louis, A.K.: Picture reconstruction from projections in restricted range. Math. Meth. Appl. Sci. **2**, 209-220, 1980

[11] Louis, A.K.: Incomplete data problems in x-ray computerized tomography, I: Singular value decomposition of the limited angle transform. Numer. Math. **48**, 251-262, 1986

[12] Louis, A.K.: Inverse und Schlecht Gestellte Probleme, Teubner, Stuttgart, 1989

[13] Louis, A.K., Maaß,P.: A mollifier method for linear operator equations of the first kind. Inverse Problems, to appear

[14] Louis, A.K., Natterer, F. : Mathematical problems of computerized tomography. Proc. IEEE, **71**, 379-389, 1983

[15] Louis, A.K., Rieder, A.: Incomplete data problems in x-ray computerized tomography, II: Truncated projections and region-of-interest tomography. Numer. Math. **56**, 371-383, 1989

[16] Louis, A.K., Schwierz, G.: Rekonstruktionsverfahren in der medizinischen Bildgebung. ZAMM **70** 1990

[17] Macovski, A.: Physical problems of computerized tomography. Proc. IEEE **71**, 373-378, 1983

[18] Nachman, A.I.: Reconstructions from boundary measurements. Annals of Mathematics **128**, 531-576, 1988

[19] Natterer, F.: Computerized tomography with unknown sources. SIAM J. Appl. Math. **43**, 1201-1212, 1983

[20] Natterer, F.: The mathematics of computerized tomography. Teubner - Wiley, Stuttgart - New York, 1986

[21] Ramm, A.G.: Recovery of the potential from fixed-energy scattering data. Inverse Problems **4**, 877-886, 1988

[22] Sabatier, P.C.: Tomography and inverse problems. Adam Hilger, Bristol, 1987

[23] Santosa, F., Vogelius, M.: A backprojection algorithm for electrical impedance imaging. SIAM J. Appl. Math.

Inverse Problems for Discrete Evolution Models

F. Natterer

Institut für Numerische und Instrumentelle Mathematik,
Westfälische Wilhelms-Universität Münster, Einsteinstr. 62,
D-4400 Münster, Fed. Rep. of Germany

ABSTRACT. *We consider first and second order recursions for a sequence u^ℓ of vectors involving a tridiagonal matrix T. The problem is to recover T from the known initial state and from the observation of one component of u^ℓ.*

0. INTRODUCTION

There has recently been much interest in discrete versions of exact inversion methods. See Case and Kak [3], Landau [5], Gladwell [4] for the spectral inverse problem, Yagle [6], Burridge and Bube [2], Bruckstein and Kailath [1] for the time-domain approach. In the present paper we make an attempt to understand the problem from a purely linear algebra point of view, concentrating on computational issues rather than on the physical aspects. The algorithms should have immediate applications to discrete versions of 1D and multi-dimensional inverse problems for parabolic and hyperbolic partial differential equations.

1. FIRST ORDER EVOLUTION PROBLEMS

Let T be the (n+1, n+1) tridiagonal matrix

$$T = \begin{pmatrix} a_0 & b_0 & & & \\ c_0 & a_1 & \cdot & & \\ & & \cdot & & \\ & & & \cdot & b_{n-1} \\ & & & c_{n-1} & a_n \end{pmatrix} .$$

Consider the evolution model

$$u^{\ell+1} = Tu^\ell \; , \quad \ell = 0,1,\ldots \; , \quad u^\ell = \begin{pmatrix} u_0^\ell \\ \vdots \\ u_n^\ell \end{pmatrix} . \tag{1.1}$$

The problem is to recover T from the values

$$g^{\ell} = u_0^{\ell} \quad , \quad \ell = 0,1,\ldots \quad , \qquad (1.2)$$

the initial state u^0 being known. We shall identify various cases in which this is possible to a certain extent.

1.1. Focused initial state

Here we assume that $u^0 = e^0 = (1,0,\ldots,0)^t$. Introducing the vectors

$$v^{\ell+1} = T^t v^{\ell}, \quad \ell = 0,1,\ldots \quad , \quad v^0 = e^0 \qquad (1.3)$$

we have

$$g^{\ell+k} = e^{0t} u^{\ell+k} = e^{0t} T^{\ell+k} e^0 = (T^{t\ell} e^0)^t (T^k e^0) = v^{\ell t} u^k \ .$$

With the (n+1,n+1)-matrix

$$G = \begin{pmatrix} g^0 & g^1 & \cdots & g^n \\ \vdots & \vdots & & \vdots \\ g^n & g^{n+1} & & g^{2n} \end{pmatrix}, \quad U = (u^0,\ldots,u^n), \quad V = (v^0,\ldots,v^n)$$

this can be rewritten as

$$G = V^t U \qquad (1.4)$$

Note that V and U are upper triangular. This is a consequence of T being tridiagonal.

We distinguish between several cases, depending on which information on T is available.

1.1.1. T is symmetric, the elements outside the diagonal being positive, i.e. $c_i = b_i > 0, \quad i = 0,\ldots,n-1$.

In that case, U = V, and U is upper triangular with positive diagonal elements. Thus, U is uniquely determined by (1.4) and can be otained numerically by a Cholesky decomposition of G. Once U is known we can compute T recursively from (1.1): We have

$$\left.\begin{aligned} u_\ell^{\ell+1} &= c_{\ell-1}\, u_{\ell-1}^\ell + a_\ell\, u_\ell^\ell \\ u_{\ell+1}^{\ell+1} &= c_\ell\, u_\ell^\ell \end{aligned}\right\} , \qquad \ell = 0,1,\dots,n-1 . \tag{1.5}$$

where $c_{-1} = u_{-1}^\ell = 0$. Since $u_\ell^\ell \neq 0$ this determines $c_\ell = b_\ell$ and a_ℓ for $\ell = 0,\dots,n-1$. Hence T is found except for the (n,n)-element a_n.

1.1.2. T is arbitrary except that the off-diagonal elements b_i, c_i are non-zero, and the elements c_i below the diagonal are known .

In that case we know the diagonal of the non-singular upper triangular matrix U. Thus U and V are uniquely determined by (1.4) and can be computed numerically by an LU-decomposition of G. Once V is known we can compute T recursively from (1.3): We have

$$\left.\begin{aligned} v_\ell^{\ell+1} &= b_{\ell-1}\, v_{\ell-1}^\ell + a_\ell\, v_\ell^\ell \\ v_{\ell+1}^{\ell+1} &= b_\ell\, v_\ell^\ell \end{aligned}\right\} , \qquad \ell = 0,1,\dots,n-1 \tag{1.6}$$

where $b_{-1} = v_{-1}^\ell = 0$. Since $v_\ell^\ell \neq 0$, this determines a_ℓ , b_ℓ for $\ell = 0,\dots,n-1$. Hence T is found except for the (n,n)-element a_n.

1.1.3. An analoguous case occurs if, instead of the elements c_i below the diagonal the elements b_i above the diagonal are known.

Then the diagonal of V is known, and U, V can be computed from (1.4). Once U is found (1.5) is used to compute T except for a_n.

1.1.4. T is arbitrary except that the off-diagonal elements b_i, c_i are non-zero.

In that case the diagonal of either of the matrices V, U is unknown.

Let D be the diagonal of V. D is invertible, and

$$G = (DV^t)^{-1}(D\,U) \quad . \tag{1.7}$$

Now DV^t is a lower triangular matrix all of whose diagonal elements are 1. Hence the upper triangular non-singular matrix W = DU is uniquely determined by (1.7) and can be computed numerically by an LU-decomposition of G.

With w^ℓ the columns of W, (1.1) reads

$$w^{\ell+1} = D^{-1} \, TDw^\ell \quad , \qquad \ell = 0,\dots,n-1 \quad .$$

From this we can compute $D^{-1}TD$ except for the last column exactly in the same fashion as we computed T from the vectors u^ℓ in (1.5). Thus the diagonal entries of T, except for the last one, can be computed, whereas nothing can be said about the off-diagonal entries.

1.2. Non focused initial state

Now we assume $u^0 = e$ where e is a known vector with non-zero components. Then (1.4) still holds, but U is no longer upper triangular. Therefore we do an LU-decomposition of U,

$$U = LR \quad , \quad \ell_{ii} = 1, \qquad i = 0,\dots,n \quad .$$

Then (1.4) assumes the form

$$G = (V^tL)R \quad . \tag{1.8}$$

Now let us consider the case in which T is arbitrary, except that the off-diagonal entries b_i, c_i are non-zero, and the entries c_i below the diagonal are known.

Then the diagonal of U, hence the diagonal of R is known, and the lower triangular matrix V^tL as well as the upper triangular matrix R are non-singular. Hence V^tL, R are uniquely determined by (1.8) and can be obtained by an LU-decomposition of G. With r^ℓ the columns of R, (1.1) reads

$$\begin{aligned} r^{\ell+1} &= S\, r^\ell \quad , \qquad \ell = 0,1,\dots,n-1 \quad , \\ S &= L^{-1}\, T\, L \quad . \end{aligned} \tag{1.9}$$

Since R is upper triangular, it follows from (1.9) and $r^\ell_\ell \neq 0$ that S must be upper Hessenberg, i.e. $s_{ij} = 0$ for $i > j + 1$. S can be computed from (1.9) recursively except for the last column. On the other hand, $LS = TL$. With ℓ^j the rows of L this reads

$$\ell^j S = c_{j-1}\, \ell^{j-1} + a_j\, \ell^j + b_j\, \ell^{j+1} \quad , \qquad j = 0,\dots,n \tag{1.10}$$

where $c_{-1} = b_n = 0$ and $\ell^{-1} = \ell^{n+1} = 0$. Since L is lower triangular and $\ell_j^j = 1$ it follows that S is also lower Hessenberg, i.e. S is tridiagonal as is T. In particular, $\ell^j S$ can be computed for $j = 0,\dots,n-2$ without using the unknown non-zero elements in the last column of S, if we only know ℓ^j. Note that $e = Lr^0$ is the zero column of L, i.e. the elements ℓ_0^j, are known and non-zero. Now we can compute a_j, b_j for $j = 0,\dots,n-2$ from (1.10): For $j = 0$ we have

$$\ell^0 S = a_0 \ell^0 + b_0 \ell^1 .$$

Since $\ell^0 = e_0^t$ is known and $\ell_1^1 = 1$, this determines b_0, a_0 and hence ℓ^1. Suppose we have already determined $a_0,\dots,a_{j-1}$, $b_0,\dots,b_{j-1}$ and $\ell^0,\dots,\ell^j$ for some $j \leq n-2$. Then we can obtain b_j from (1.10) since $\ell_{j+1}^{j+1} = 1$. Knowing b_j we can use the zero component of (1.10) to determine a_j since $\ell_0^j \neq 0$ and ℓ_0^{j+1} are known. This in turn determines ℓ^{j+1}, i.e. we now know $a_0,\dots,a_j$, $b_0,\dots,b_j$, $\ell^0,\dots,\ell^{j+1}$. Thus all these quantities are determined by induction until $j = n-2$. We conclude that it is possible to determine T except for a_{n-1}, a_n, b_{n-1}.

1.3. Block-tridiagonal matrices

The entries of T need not be scalars. Assume that a_i, b_i, c_i, u_i^ℓ are matrices of dimension (p_i, p_i), (p_i, p_{i+1}), (p_{i+1}, p_i) and (p_i, p_i), respectively. Then (1.1) makes sense. We mention very briefly the necessary modifications for the matrix case.

A focused initial state is now characterized by $u_0^0 = I$, the (p_0, p_0) unity matrix, and $u_i^0 = 0$ for $i > 0$. Then, (1.4) holds also in the matrix case.

For T symmetric, the discussion in 1.1.1 can be extended immediately to the matrix case if, in addition, $c_i = b_i$ are positive definite commuting matrices. In particular, all the p_i coincide in this case. Then the matrices u_ℓ^ℓ are positive definite, making the block Cholesky decomposition and the solution of the recursion (1.5) unique.

For T arbitrary but with either the c_i's or the b_i's known, the discussion in 1.1.2 and 1.1.3 required that c_i and b_i do not vanish. In the matrix case we have to require that the matrices

$$u_\ell^\ell = c_{\ell-1} \cdot \ldots \cdot c_0 \quad , \qquad v_\ell^\ell = b_{\ell-1}^t \cdot \ldots \cdot b_0^t$$

have full rank. In particular, the rank of c_i, b_i, $i < \ell$, has to be at least p_ℓ. Hence $p_0 \geq p_1 \geq \ldots \geq p_\ell$. Once U, V are determined we compute the elements of T, except for a_n, from (1.5) if the c_i are given. If the b_i are given we have to use (1.6) with a_i, b_i replaced by a_i^t, b_i^t.

2. SECOND ORDER EVOLUTION PROBLEMS

In view of the applications we consider the following special recursion

$$u^{\ell+1} + u^{\ell-1} = Tu^\ell \quad , \qquad \ell = 0,1,\ldots \tag{2.1}$$

rather than a general 3-term recursion. Again the problem is to recover T from

$$g^\ell = u_0^\ell \quad , \qquad \ell = 0,1,\ldots \tag{2.2}$$

and the known initial states u^0, u^{-1}.

The solution of (2.1) satisfies

$$u^{\ell+k} = T_k\, u^\ell - T_{k-1}\, u^{\ell-1}, \quad k,\ell = 0,1,\ldots \tag{2.3}$$

where

$$T_{\ell+1} + T_{\ell-1} = TT_\ell \qquad \ell = 0,1,\ldots \ , \quad T_0 = I,\ T_{-1} = 0. \tag{2.4}$$

This can easily be verified either by induction on k or by the relation of T_k to the Chebyshev polynomials.

We consider only the special focused initial state

$$u^0 = e^0 = (1,0,\ldots,0)^t \ , \quad u^{-1} = 0 \ .$$

In that case,

$$g^{k+\ell} = e^{0t}u^{\ell+k} = e^{0t}(T_k u^\ell - T_{k-1}u^{\ell-1}) = (T_k^t e^0)^t u^\ell - (T_{k-1}e^0)^t u^{\ell-1}$$

$$= v^{kt}u^\ell - v^{(k-1)t}u^{\ell-1} \tag{2.5}$$

where $v^k = T_k^t\, e^0$ is a solution of

$$v^{k+1} + v^{k-1} = T^t v^k \ , \qquad k = 0,1,\dots \ ,$$
$$v^0 = e^0, \quad v^{-1} = 0 \ . \tag{2.6}$$

Note that the matrices $U = (u^0,\dots,u^n)$, $V = (v^0,\dots,v^n)$ are upper triangular. Depending on the information available on T we consider various cases.

2.1. T is symmetric, the off-diagonal elements b_i, c_i being positive.

In this case, (2.5) reads

$$g^{k+\ell} = u^{kt}\, u^\ell - u^{(k-1)t}\, u^{\ell-1} \ .$$

Since U is upper triangular we get for $\ell \geq k$

$$g^{k+\ell} = u_0^k u_0^\ell + \dots + u_k^k u_k^\ell - u_0^{k-1} u_0^{\ell-1} - \dots - u_{k-1}^{k-1} u_{k-1}^{\ell-1} \ . \tag{2.7}$$

From these relations U can be computed in a Cholesky-type manner: for $k = 0$ we get (as we already know) $u_0^\ell = g^\ell$, hence the zero row of U is known. Suppose we have already computed rows $0,\dots,k-1$ of U. Then, putting $\ell = k$ in (2.7) and solving for $(u_k^k)^2$, we get

$$(u_k^k)^2 = g^{2k} - (u_0^k)^2 - \dots - (u_{k-1}^k)^2 + (u_0^{k-1})^2 + \dots + (u_{k-1}^{k-1})^2 \ .$$

Since

$$u_k^k = b_0 \cdot \dots \cdot b_{k-1} > 0 \ ,$$

u_k^k is determined uniquely. Once u_k^k is known, the remaining elements in row k of U are computed by solving (2.7) for u_k^ℓ.

The computation of T from U can now be done exactly in the same way as for first order evolutions: (1.5) still holds and can be used to determine T except for the (n,n)-element.

2.2. T is arbitrary except that the off-diagonal elements b_i, c_i are non-zero, and the elements c_i below the diagonal are known.

In this case U, V are computed from (2.5) in a Crout-type fashion. This is possible since the diagonal elements

$$v_k^k = c_0 \cdot \ldots \cdot c_{k-1}$$

are known and non-zero. The zero row of U is already known (or can be obtained from (2.5) with k = 0). Likewise, the zero row of V can be found from (2.5) by putting $\ell = 0$. (Note that the zero rows of U and V coinside). Assume that rows 0,...,k-1 of U, V are already computed. For $\ell \geq k$ (2.5) reads

$$g^{k+\ell} = v_0^k u_0^\ell + \ldots + v_{k-1}^k u_{k-1}^\ell + v_k^k u_k^\ell - v_0^{k-1} u_0^{\ell-1} - \ldots - v_{k-1}^{k-1} u_{k-1}^{\ell-1} .$$

Since $v_k^k \neq 0$ this can be solved for u_k^ℓ. Interchanging the role of U, V in (2.5) we get

$$g^{k+\ell} = u_0^k v_0^\ell + \ldots + u_{k-1}^k v_{k-1}^\ell + u_k^k v_k^\ell - u_0^{k-1} v_0^{\ell-1} - \ldots - u_{k-1}^{k-1} v_{k-1}^{\ell-1} .$$

Since $u_k^k \neq 0$ this can be solved for v_k^ℓ. Hence row k of U and V can be computed. Now we can compute T except for the (n,n)-element, by using (1.5).

2.3. The case in which the elements b_i below the diagonal are known is treated in the same way.

2.4. T is arbitrary except that the off-diagonal elements are non-zero.

Let D be the (unknown) diagonal of V. Then (2.5) reads

$$g^{k+\ell} = (D^{-1}v^k)^t D u^\ell - (D^{-1}v^{k-1})^t D u^{\ell-1} .$$

The diagonal elements of $D^{-1}V$ are all 1 and hence known. Thus we have precisely the situation of 2.2, with the matrices U and V replaced by W = DU and $D^{-1}V$. Hence W is determined uniquely. With w^ℓ the columns of W, (2.1) reads

$$w^{\ell+1} + w^{\ell-1} = D^{-1}TDw^\ell , \qquad \ell = 0,\ldots,n-1 .$$

From this we can compute $D^{-1}TD$, except for columns n, exactly in the same way as in 2.2. Thus we can obtain the diagonal elements of T except for a_n.

REFERENCES

[1] BRUCKSTEIN, A. - KAILATH, T.: Some matrix factorization identities for discrete inverse scattering. *Linear Algebra and its Applications* 74 , 157-172 (1986).

[2] BUBE, K.P. - BURRIDGE, R.: The one-dimensional inverse problem of reflection seismology. *SIAM Review* 25 , 497-559 (1983).

[3] CASE, K.M. - KAK, M.: A discrete version of the inverse scattering problem. *J. Math. Phys.* 4 , 594-603 (1983).

[4] GLADWELL, G.M.L. - WILLMS, N.B.: A discrete Gel'fand-Levitan method for band-matrix inverse eigenvalue problems. *Inverse Problems* 5 , 165-179 (1989).

[5] LANDAU, H.J.: The inverse problem for the vocal tract and the moment problem. *SIAM J. Math. Anal.* 14 , 1019-1035 (1983).

[6] YAGLE, A.E.: One-dimensional inverse scattering problems: An asymmetric two-component wave system framework. *Inverse Problems* 5, 641-665 (1989).

Recent Developments in Electrical Impedance Tomography

M.K. Pidcock

Applied Analysis Research Group, Department of Computing
and Mathematical Sciences, Oxford Polytechnic, Headington, Oxford OX3 0BP, UK

The reconstruction problem for Electrical Impedance Tomography is an ill-posed non-linear inverse problem. It is highly sensitive to modelling and measurement errors and iterative reconstruction algorithms are very computationally demanding. In this paper we describe some of our work to overcome these difficulties by investigating the effect on reconstructions of geometric errors and of our attempts to speed up the reconstruction algorithms using parallel processing techniques.

1. Introduction

Electrical Impedance Tomography (EIT) is a technique of medical imaging which exploits the contrast in the electrical conductivity of different body tissues to produce an image of the material distribution within a part of the body. The data used to obtain these images are measurements taken on the surface of an object of the electrical potential which is induced in the object by the application of known electrical currents to that surface. Several potential medical applications of this technique have been identified.

The problem of EIT can be posed in the following way. Suppose that an object Ω with boundary $\partial\Omega$ consists of an isotropic Ohmic material with conductivity distribution σ. If ϕ is the electrical potential in Ω then

$$\nabla.(\sigma\nabla\phi) = 0 \qquad \text{in } \Omega$$

where ϕ is known (measured) on $\partial\Omega$

$$-\sigma\frac{\partial\phi}{\partial n} = j \qquad \text{is known (applied) on } \partial\Omega$$

The aim of EIT is to determine σ from the boundary data. It has been shown [1-5] that this problem has a solution for many reasonable conductivity distributions and reconstruction algorithms have been described [6-11]. However, the reconstruction problem of EIT is a highly non-linear inverse problem and it has been demonstrated numerically [7] that the linearised problem is extremely ill-posed. The inversion procedure is therefore very sensitive to the possible errors associated with this technique.

There is a wide range of such errors and a detailed study of them is essential if EIT is to become a useful diagnostic tool. In this paper we will consider the difficulties caused by imperfect knowledge of boundary shape. In EIT the electric current is applied to the surface via a number of electrodes positioned on the boundary but in this work we will assume that the current is adjustable at any point on the boundary. In Section 2 we will describe the basic problem and the ideas behind our calculations and in Section 3 we will discuss a scheme for overcoming the difficulties described in Section 2.

The reconstruction algorithm which we have developed at Oxford Polytechnic is highly iterative and consequently extremely demanding in computer power. It is unlikely that EIT will be able to produce high resolution images but if it is to exploit its various advantages [6] over other medical imaging modalities we must be able to produce images in a reasonable time. In Section 4 we will describe some work which we have been doing to speed our reconstructions. We use polar coordinates (r,θ) throughout.

2. The model problem

A simple two dimensional example often considered in EIT is that of distinguishing between two conductivity distributions σ_1, σ_2 on the unit disc, $0 \le r \le 1$ defined by

$$\sigma_1(r,\theta) = 1 \qquad 0 \le r \le 1$$

$$\sigma_2(r,\theta) = \begin{cases} \sigma & 0 \le r \le R < 1 \\ 1 & R < r \le 1 \end{cases}$$

It is straightforward to show that if the applied current $j(\theta)$ is given by

$$j(\theta) = \sum_{n=1}^{\infty} a_n\cos(n\theta) + b_n\sin(n\theta)$$

then

$$\phi(r,\theta;\sigma_i,j) = \sum_{n=1}^{\infty} \lambda_n^i\, r^n\Big[a_n\cos(n\theta) + b_n\sin(n\theta)\Big]$$

where $\lambda_n^1 = \dfrac{1}{n}$, $\lambda_n^2 = \dfrac{1}{n}\left[\dfrac{1 - \mu R^{2n}}{1 + \mu R^{2n}}\right]$ and $\mu = \dfrac{\sigma - 1}{\sigma + 1}.$

As we are able to choose the coefficients $\{a_n,b_n\}$ freely, we can ask if there are combinations which are better than others in some sense. Various criteria for the choice of optimal currents have been proposed [12],[13] but we confine our attention to those in [12].

For the case of the problem described above, these optimal currents are $j_m(\theta) = \cos(m\theta), \sin(m\theta)$ for integer m, and the corresponding value of $\delta_m \equiv \delta_{j_m}$ is $\frac{2}{m}\frac{\mu R^{2m}}{(1+\mu R^{2m})}$. This is the basis of our calculations. We will calculate δ_m arising when these optimal currents are applied to a body whose shape is thought to be the unit disc and which consists of material with a uniform conductivity distribution, σ_1. If the boundary were correctly known and all the measurements were accurate then the value of δ_m would be zero. The error in the boundary description will mean that a non-zero value of δ_m will be obtained and we could infer that there is a circular anomaly of the form σ_2.

3. Boundary Distortion

Consider the situation where the polar description of the boundary curve is $r = 1 + \varepsilon F(\theta)$ rather than $r = 1$. If we write

$$\phi^{\varepsilon}(r,\theta) = \sum_{n=1}^{\infty} \varepsilon^k \phi^k(r,\theta)$$

then we have shown previously [14] that perturbation techniques can be used to find $\{\phi^k(r,\theta),\ k=0,1,\ldots\}$ and hence an approximation to δ_m. For example, suppose that we have the case $r(\theta) = 1 + \varepsilon\cos(k\theta)$, i.e. $F(\theta) = \cos(k\theta)$, then if we apply the current $j_m(\theta) = \sin(m\theta)$ we find after considerable computation that

$$k\ {}_k\Delta_m^{\varepsilon} = \frac{\varepsilon}{4} + O(\varepsilon^3) \qquad m = k$$

$$= \frac{k}{|m^2 - k^2|}\left(\frac{m^2 + k^2}{2}\right)^{\frac{1}{2}}\varepsilon + O(\varepsilon^3) \qquad m > k$$

$$= \frac{k}{2}\left[\frac{1}{(m+k)^2} + \left(2 + \frac{1}{(k-m)}\right)^2\right]^{\frac{1}{2}}\varepsilon + O(\varepsilon^3) \qquad m < k$$

It is clear that there is a direct relationship between the detectable error in boundary shape and the measurement accuracy of the system. It is interesting to note that if we identify δ_m and ${}_k\Delta_m^{\varepsilon}$ then we can interpret our measurements as predicting a circular anomaly of radius R for some fixed conductivity contrast σ [14]. We give some typical results in Table 1.

These results are rather disturbing at first sight as they seem to suggest the complete impossibility of EIT. This is not the correct interpretation however but it is likely that it weakens the case of those who suggest that EIT should not attempt to produce images but merely analyse

Table 1 Radius of the circular anomaly inferred from the data found in the problem of Section 2. $\varepsilon = 0.01$, $\sigma = 4$.

k \ m	1	2	3	4
1	0.05	0.31	0.44	0.53
2	0.10	0.21	0.48	0.55
3	0.10	0.40	0.36	0.60
4	0.10	0.38	0.58	0.46

signals and blow a whistle when some calculated quantity exceeds a critical value. In these calculations it appears that the major problem is that we have concentrated all of the signal information into the solution of a highly ill-posed equation for R. Nevertheless, it seems that the boundary shape problem is serious and that its resolution could be crucial to the success of EIT.

3.General Fourier Perturbations

It is interesting to note that in the special case described in the previous section and for small ε, the relationship between ${}_k\Delta_m^{\varepsilon}$ and ε is essentially linear. It is also known that the conductivity contrast between various body tissues decreases as the temporal frequency of the applied currents diminishes. This encourages us to think that more general perturbations which can be described in terms of a Fourier expansion might be identified. Such an expansion should be very appropriate in the case of the human body where the relative smoothness of the body surface should lead to an economical description in terms of trigonometric series.

Consider, therefore, the more general boundary perturbation given by

$$F(\theta) = \frac{1}{2}A_0 + \sum_{n=1}^{\infty} A_n \cos(n\theta)$$

If we apply one of the optimal boundary currents $j_m(\theta) = \sin(m\theta)$ the analysis suggested in the previous section gives (to first order in ε)

$$\frac{\partial\phi^0}{\partial r}(1,\theta) = \sin(m\theta)$$

$$\frac{\partial \phi^1}{\partial r}(1,\theta) + F(\theta)\frac{\partial^2 \phi^0}{\partial r^2}(1,\theta) - F'(\theta)\frac{\partial \phi^0}{\partial \theta}(1,\theta) = 0$$

and it follows that since $\phi^0(r,\theta)$ and $\phi^1(r,\theta)$ are harmonic in Ω

$$\phi^0(r,\theta) = \frac{r^m}{m}\sin(m\theta)$$

$$\phi^1(r,\theta) = -\sum_{k=1}^{\infty} \frac{r^k}{k} B_k^m \sin(k\theta).$$

where $B_k^m = \frac{1}{2}\left[(k-1)\, A_{|m-k|} + (k+1)\, A_{m+k} \right].$

Hence, the potential actually measured differs from that expected if there were no boundary distortion by an amount

$$\delta\phi_m(\theta) = \phi^0(1+\varepsilon F(\theta),\theta) + \varepsilon\, \phi^1(1+\varepsilon F(\theta),\theta) - \phi^0(1,\theta) + O(\varepsilon^2)$$

$$= \varepsilon \sum_{k=1}^{\infty} C_k^m \sin(k\theta) + O(\varepsilon^2)$$

where $C_k^m = \frac{1}{2k}\left[A_{|m-k|} - (2k+1)\, A_{m+k} \right].$

We now have a scheme to determine the Fourier coefficients. Apply a series of currents $\sin(m\theta)$ to the object and measure resulting voltages at a number of points. Use a Discrete Fourier Transform to express the measured voltage $\delta\phi_m(\theta)$ in terms of its Fourier components and use the above expression (3) as a system of linear equations to determine $\{ A_k \}$.

The system of linear equations which determine the boundary Fourier coefficients is quite ill-conditioned and the weighting term (2k+1) which appears to emphasise the higher order terms would suggest that we are unlikely to be able to determine the shape with any great precision. Nevertheless,we have implemented this scheme and in Figures 1 and 2 we present examples of results obtained using computer simulations. The data used in these calculations is accurate except for the errors introduced by the use of the Finite Element Method to solve the perturbed problem. They do seem to indicate that provided the boundary remains reasonably smooth it should be possible to obtain useful information about its shape by applying currents of sufficiently high temporal frequency. The performance of this approach with data containing realistic errors and of its interaction with a more sophisticated modelling of a finite number of electrodes have not yet been studied. Further details will be given elsewhere.

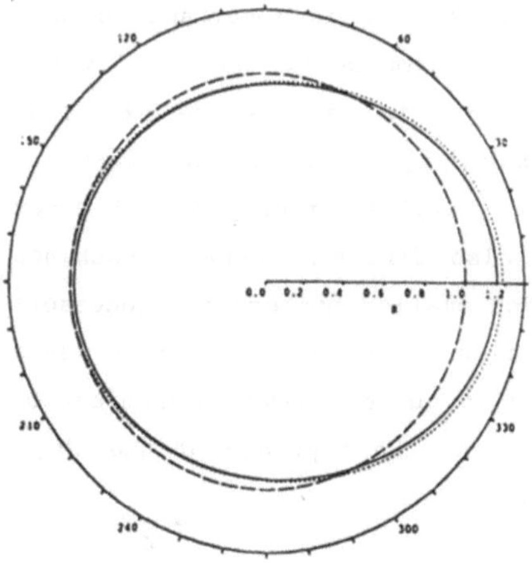

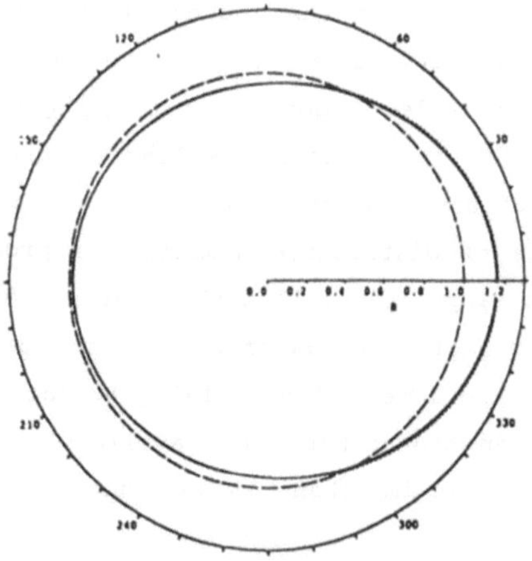

Figures 1 and 2. Results obtained using the method of Section 3. Dashed curve - assumed boundary, solid curve - actual boundary, the dotted curve - predicted curve using $j(\theta) = \sin(m\theta)$. Figure 2 m=1, Figure 3 m=4. $\varepsilon = 0.05$.

4. Parallelism

The reconstruction algorithm developed by the group at Oxford Polytechnic has been described elsewhere [6-8] but an essential feature is that the non-linear problem is linearised and an iterative procedure is used to calculate σ. At each stage of this iteration optimal currents are calculated and these too are evaluated using an iterative process [6]. At each iteration step we must solve a regularised system of linear equations of the form $Ax = b$ where normally x is a vector of nodal conductivities in a Finite Element discretisation of Ω and b is a vector of voltages. The solution of these systems constitutes a significant proportion of the reconstruction calculation and any methods which can speed this process make the success of EIT much more likely. One way is to exploit the parallelism in EIT using transputer arrays.

The Inmos Transputer is a single chip computer which includes 4K bytes of fast RAM and an interface to 4 Gbytes of external memory. The T800 transputer has a floating point unit capable of 2.5 megaflops. It also has 4 bidirectional links which mean that groups of transputers can be joined together in various topologies. Boards containing groups of transputers are available for many microcomputers at a relatively modest price.

The main advantage of these chips is that computational processes can be split into separate subprocesses which can then be run in parallel on different transputers. The precise way in which this splitting is done has a great effect on the increased computation speed and it is important to note that a parallel version of the best serial algorithm is not necessarily the best parallel algorithm for any given task. This is because the execution speed of an algorithm is less strongly determined by the number of algebraic operations required than by the communication overhead moving data between units of distributed memory. The problem is also different between machines with large and small numbers of processors. For smaller numbers of processors the relative speeds of a floating point operation and data transmission means that the level of parallel granularity should be quite coarse. Communication between transputers, the availability of data and the topology of the chips all become important issues which the programmer must address.

We have chosen to connect our four transputers in a unidirectional ring thereby ignoring two of the links on each chip. This is mainly because this configuration is quite efficient in several ways and safe from races and deadlocks while at the same time it enables us to write programs containing a parameter which is the number of transputers. Increasing the number of transputers is not a major problem as it might be with other topologies. The ring topology is also very suitable for linear algebra algorithms as they can often be written in such a fashion that only nearest neighbour transputers need communicate. This means that the relatively large diameter of this topology is not a major problem.

Our work so far has been concerned with the solution of the linear systems mentioned above and to test our algorithms a matrix typical of those met in EIT was constructed. More specifically a 300x208 matrix with condition number 112 and exponentially decreasing singular values was generated and the linear system solved using Choleski decomposition, QR factorisation and the conjugate gradient method. We found that each of these methods can be implemented efficiently in parallel Fortran and that Choleski Decomposition was easily the fastest in our tests. The conjugate gradient method was hopelessly uncompetitive unless a suitable preconditioner was applied. A preconditioner with linearly increasing singular values and condition number 1/112 was used but even then the method was well beaten by Choleski decomposition. Some data relating to these tests is given in Table 2 and it is interesting to note that almost linear speedup with respect to the number of transputers was obtained. It is very striking that using the transputer together withthese parallel processing techniques, calculations which were taking about 15 minutes on a SUN3 workstation now take about 20 seconds on a 4 transputer board in a PC host. A full report will appear elsewhere.

Table 2. Times(seconds) for the solution of the linear system using different algorithms using n transputers.

n	CH	QR	Conj (440 it)	PreConj (60 it)	Speedup
1	11	81	308	42	1.0
2	6	43	164	19	1.9
3	4	29	113	15	2.7
4	3	22	85	12	3.6

5. Acknowledgements

I would like to thank the SERC/DTI transputer initiative for the loan of transputer equipment and Kevin Paulson for supplying the data in Table 2.

References.

1. Kohn R V and Vogelius M 1984 *Commun.Pure Appl.Math.* **37** 289-298
2. Kohn R V and Vogelius M 1984 *SIAM-AMS Proc.* **14** 113-123
3. Kohn R V and Vogelius M 1985 *Commun.Pure Appl.Math.* **39** 644-667
4. Sylvester J and Uhlmann G 1986 *Commun.Pure Appl.Math.* **39** 91-112
5. Sylvester J and Uhlmann G 1987 *Annals of Math.* **125** 153-169
6. Breckon W R and Pidcock M K 1988 *Some Topics on Inverse Problems* ed P C Sabatier (World Scientific) 255-265
7. Breckon W R and Pidcock M K 1988 *Information Processing in Medical Imaging* ed C N de Graaf and M A Viergever (Plenum) 235-244
8. Breckon W R and Pidcock M K 1986 *Proc.NATO-ASI 'Mathematics and Computer Science in Medical Imaging'* **39** 204-215
9. Yorkey T and Webster J 1987 *Clin.Phys.Physiol.Meas.* **8A** 55-62
10. Barber D, Brown B and Seagar A 1985 *Clin.Phys.Physiol.Meas.* **6** 109-121
11. Barber D and Seagar A 1986 *Clin.Phys.Physiol.Meas* **8A** 47-54
12. Gisser D, Isaacson D and Newell J 1987 *Clin. Phys. Physiol.Meas.* **8A** 39-46
13. Breckon W R to be published.
14. Pidcock M K and Breckon W R, Oxford Polytechnic Report.

Two Inverse Problems for the Exponential Radon Transform

D.C. Solmon

Department of Mathematics, Oregon State University, Corvallis, OR 97331, USA

Introduction

Let φ, $0 \le \varphi < 2\pi$, be an angle measured counterclockwise from the positive x-axis in the plane. Let $\omega = (\cos\varphi, \sin\varphi)$ be the direction corresponding to φ and $\omega^\perp = (-\sin\varphi, \cos\varphi)$. The exponential Radon transform of a function f in the plane is defined by

$$\mathcal{R}_\mu f(\varphi,s) = \int_{-\infty}^{\infty} f(s\omega + t\omega^\perp)e^{\mu t}dt. \tag{1.1}$$

This operator arises in single photon emission computed tomography. Physically f represents the location and intensity of radioactive sources inside a convex medium of constant attenuation μ. (When $\mu = 0$ $\mathcal{R}_\mu$ is the Radon transform, which we denote by $\mathcal{R}$.) See Budinger et.al. [2] for background. The inverse problem at hand is to determine f from the data $\mathcal{R}_\mu f$. If μ is known a priori, then the goal is to invert the integral transform $\mathcal{R}_\mu$; the inversion problem. However, if in addition μ is unknown, then one first must find μ and then find f. This is the identification problem. For simplicity we assume throughout that f is C^1 with compact support.

Section 2 contains a discussion of the inversion problem. The main result, which is joint work with A. Hertle, is the solution to the identification problem given by Theorem 3.2. An outline of the proof of Theorem 3.2 is contained in section 3. A complete proof will be given in [11].

2. The Inversion Problem

The inversion problem was first solved by Tretiak and Metz [15]. Other papers that have addressed the inversion of $\mathcal{R}_\mu$ (or equivalently the exponential x-ray transform in the plane) are those of Markoe [13], Clough and Barrett [4] and Hazou and Solmon [6,7]. Hertle [9] has shown that $\mathcal{R}_\mu$ is injective on the space of distributions with compact support. It is assumed throughout this section that μ is known.

Suppose that f is radial (that is f(x) depends only on |x|, the distance of the point x in the plane to the origin) and has support in B(0,R), the closed disc centered at the origin with radius $R > 0$. Then $f(x) = F(|x|)$ where F is a function of one variable and has support in [0,R]. An obvious change of variable in (1.1) gives

$$\mathcal{R}_\mu f(\varphi,s) = \mathcal{T}_\mu F(s) = 2\int_{|s|}^{R} F(r)\cosh[\mu(r^2-s^2)^{1/2}](r^2-s^2)^{-1/2} r\,dr. \tag{2.1}$$

The integral transform $\mathcal{T}_\mu$ has been studied by Clough and Barrett [4]. It is essentially their attenuated Abel transform. Clough and Barrett show that for all real μ the transform (2.1) and the transform

$$\mathcal{T}_\mu^{-1}G(r) = -\pi^{-1}\int_r^\infty G'(s)\cos[\mu(s^2-r^2)^{1/2}](s^2-r^2)^{-1/2}ds \tag{2.2}$$

are inverses of each other. Thus (2.2) provides an inversion formula for $\mathcal{R}_\mu$ on the space of radial functions.

Now let us obtain an inversion formula for $\mathcal{R}_\mu$ in the general case. The formal adjoint, $\mathcal{R}_\mu^t$, of $\mathcal{R}_\mu$ is defined by

$$\mathcal{R}_\mu^t g(x) = \int_0^{2\pi} g(\varphi,\langle x,\omega\rangle)\exp(\mu\langle x,\omega^\perp\rangle)d\varphi, \tag{2.3}$$

where $\langle , \rangle$ denotes the standard inner product in the plane. In analogy with the radial case discussed above, if g is independent of φ and even in t, so $g(\varphi,t) = G(|t|)$, then (2.3) becomes

$$\mathcal{R}_\mu^t g(x) = \mathcal{B}_\mu G(r) = 2\int_0^r G(s)\cosh[\mu(r^2-s^2)^{1/2}](r^2-s^2)^{-1/2}ds,\ r = |x|. \tag{2.4}$$

The integral equation above arises in radiation therapy and has been inverted by Brahme, Roos and Lax [1]. See also [6]. Indeed, one has

$$K(s) = \mathcal{B}_\mu^{-1}E(s) = (2\pi)^{-1}(d/ds)\int_0^s E(t)\cos[\mu(s^2-t^2)^{1/2}](s^2-t^2)^{-1/2}\,t\,dt. \tag{2.5}$$

Approximate inversion formulas and inversion formulas for $\mathcal{R}_\mu$ can be derived via classical analysis and the above formula. Indeed, one readily computes that for any locally integrable function $K(\varphi,t)$

$$f * \mathcal{R}^t_{-\mu}K = \mathcal{R}^t_{-\mu}(\mathcal{R}_\mu f * K), \tag{2.6}$$

where $*$ denotes convolution. Note that the convolution on the left above is two dimensional, while that on the right is one dimensional, i.e. $(\mathcal{R}_\mu f * K)(\varphi,s) = \int_{-\infty}^{\infty} \mathcal{R}_\mu f(\varphi,s-t)K(\varphi,t)dt$. The right hand side of (2.6) is computable from $\mathcal{R}_\mu f$, while the left hand side will be an approximation of f provided $\mathcal{R}^t_{-\mu}K$ is an approximate delta function; i.e. an approximate convolution identity. Choose a radial integrable function E such that $\int E(x)dx = 1$ and for $\rho > 0$ set $E_\rho(x) = \rho^{-2}E(x/\rho)$. Let K_ρ be the solution to the equation $\mathcal{R}^t_{-\mu}K_\rho = E_\rho$ provided by (2.5) when E is replaced by E_ρ. The function K_ρ is the "filter" corresponding to the "point spread function" E_ρ. Now (2.6) becomes

$$f * E_\rho = \mathcal{R}^t_{-\mu}(\mathcal{R}_\mu f * K_\rho). \tag{2.7}$$

The right hand side of (2.7) provides a filtered backprojection formula for computing the approximation $f * E_\rho$ to f from the data $\mathcal{R}_\mu f$. The point spread function E and dilation parameter ρ can be chosen a priori to suit the problem at hand and the corresponding filter K_ρ computed via (2.5).

Standard results in classical analysis show that the left hand side of (2.7) converges to f as $\rho \to 0$, and so the right hand side must also. Computation of this limit on the right gives the Tretiak and Metz inversion formula [15],

$$f = (4\pi)^{-1}\mathcal{R}^t_{-\mu}\Lambda_\mu \mathcal{R}_\mu f, \tag{2.8}$$

where $(\Lambda_\mu g)(s) = [(d/ds)\mathcal{H}_\mu g](s)$, and the singular integral operator $\mathcal{H}_\mu$ is defined by $\mathcal{H}_\mu g(s) = \pi^{-1}\int_{-\infty}^{\infty} g(t)\cos[\mu(s-t)]/(s-t)dt$. Details of this analysis are given in [6]. Numerical examples, other formulas for, and explicit computations of filters are given in [7].

Tretiak and Metz [15] derive (2.8) via (2.6) also, but they begin with an explicit family of filters K_ρ and then prove that $f * \mathcal{R}^t_{-\mu} K_\rho \to f$ as $\rho \to 0$. In the case of the Radon transform, the relation $E = \mathcal{R}^t K$ between the filter K and point spread function E goes back to Cho [3]. This relation was first exploited heavily by Davison and Grünbaum [5] in their analysis of the limited angle problem for the Radon transform.

3. The Identification Problem

Now let us assume that the attenuation μ and the function f are unknown and address the question whether μ can be determined from the data $\mathcal{R}_\mu f$. This problem has been studied by Hertle [10]. In a related setting Natterer [14] has shown that a variable attenuation μ can be determined as far as can be expected when f consists of a finite number of point sources.

Suppose that f is radial. From (2.1) it is clear that $\mathcal{R}_\mu f = \mathcal{R}_{-\mu} f$, and that $\mathcal{R}_\mu f$ is an even function of s which is independent of φ. The latter shows that $\mathcal{R}_\mu f$ trivially satisfies Helgason's moment conditions [8]. Hence there exists a radial function g vanishing outside of $B(0,R)$ such that $\mathcal{R}_\mu f = \mathcal{R} g$. Thus μ cannot be distinguished from 0 or $-\mu$. The formula of Clough and Barrett, (2.2), shows that the situation is much worse.

Theorem 3.1. *Let* $f \in C^1(\mathbb{R}^2)$ *be a fixed but arbitrary radial function with support in* $B(0,R)$ *and let* μ *be a fixed but arbitrary real number. Then for each real number* ν*, there exists a radial function* $g_\nu \in C^1(\mathbb{R}^2)$ *with support in* $B(0,R)$ *such that* $\mathcal{R}_\nu g_\nu(\varphi,t) = \mathcal{R}_\mu f(\varphi,t)$ *for each* φ *and* t.

Proof. Simply define $g_\nu(x) = \mathcal{T}_\nu^{-1} G(|x|)$ where $G(s) = \mathcal{R}_\mu f(\varphi,s)$, $s \geq 0$, and μ is replaced by ν in (2.2).

Thus, when f is an unknown radial function, no conclusion can be drawn about the attenuation μ from the data $\mathcal{R}_\mu f$. In contrast, the next theorem gives a formula for computing μ from the data $\mathcal{R}_\mu f$ whenever f is not radial.

Theorem 3.2. *The attenuation* μ *and the function* f *are uniquely determined by the data* $\mathcal{R}_\mu f$ *if and only if* f *is not a radial function. More precisely, for each nonnegative integer* n *define*

$$\alpha_n(\varphi) = d/d\varphi \int s^{2n}\mathcal{R}_\mu f(\varphi,s)ds \quad \textit{and} \quad \beta_n(\varphi) = \int s^{2n+1}\mathcal{R}_\mu f(\varphi,s)ds.$$

a) *Suppose that there exists an integer* $n \geq 0$ *and an angle* φ_o *such that either* $\alpha_n(\varphi_o) \neq 0$ *or* $\beta_n(\varphi_o) \neq 0$. *Let* k *be the smallest such integer.*

i) *If* $\alpha_k(\varphi_o)\beta_k(\varphi_o) = 0$, *or* $\alpha_k(\varphi_o+\pi)/\beta_k(\varphi_o+\pi) = -\alpha_k(\varphi_o)/\beta_k(\varphi_o)$, *then* $\mu = 0$.

ii) *Otherwise* $\mu = -(2k+1)\alpha_k(\varphi_o)/\beta_k(\varphi_o)$.

b) $\alpha_n(\varphi) = \beta_n(\varphi) = 0$ *for all* φ *and all* $n \geq 0$, *if and only if* f *is radial, and in this case no information can be obtained about* μ *from the data* $\mathcal{R}_\mu f$.

Theorem 3.2 is a consequence of the following result.

Theorem 3.3. *The formula* $\mu\beta_k(\varphi) = -(2k+1)\alpha_k(\varphi)$, *or equivalently*

$$\mu \int s^{2k+1}\mathcal{R}_\mu f(\varphi,s)ds = -(2k+1)d/d\varphi \int s^{2k}\mathcal{R}_\mu f(\varphi,s)ds, \tag{3.4}$$

holds for all μ *when* $k = 0$ *and for* $\mu \neq 0$ *whenever* $k > 0$ *and*

$$\beta_j = \int s^{2j+1}\mathcal{R}_\mu f(\varphi,s)ds = 0 \textit{ for all } \varphi \textit{ and } j = 0,1,...k-1.$$

Theorem (3.2) follows from a recent result of Kuchment and Ya.L'vin characterizing the range of the exponential Radon transform. A proof of Theorem 3.3 that is independent of the Kuchment Ya.L'vin result will be given in [11]. We will give the idea behind the proof in [11] and verify (3.4) in the case $k = 1$. First let us use (3.4) to prove Theorem 3.2.

Proof of Theorem 3.2. If $\beta_n(\varphi) = \alpha_n(\varphi) = 0$ for all n and φ, then $\mathcal{R}_\mu f$ is both even and independent of φ. Clough and Barrett's inversion formula, (2.2), can be used to define a radial function h such that $\mathcal{R}_\mu h = \mathcal{R}_\mu f$ and Hertle's injectivity theorem [9] gives that $f = h$. Hence f is radial. This, together with Theorem 3.1, proves Theorem 3.2 b.

When $\mu = 0$, Helgason's range characterization theorem, [8], for the Radon transform gives that for each $n \geq 0$, $\alpha_n(\varphi)$ is an even trigonometric polynomial of degree $\leq 2n$ and $\beta_n(\varphi)$ is an odd trigonometric polynomial of degree $\leq 2n+1$. Thus $\alpha_k(\varphi_o+\pi)/\beta_k(\varphi_o+\pi) = -\alpha_k(\varphi_o)/\beta_k(\varphi_o)$ when both $\alpha_k(\varphi_o)$ and $\beta_k(\varphi_o)$ are nonzero and $\mu = 0$. On the other hand, (3.4) shows that this cannot hold when $\mu \neq 0$. Also (3.4) shows that when $\mu \neq 0$, the product $\alpha_k(\varphi_o)\beta_k(\varphi_o)$ cannot be zero unless $\alpha_k(\varphi_o) = \beta_k(\varphi_o) = 0$. This proves a).

Remark. Assume that it is known a priori that $\mu \neq 0$. Then μ is uniquely determined by (3.4) provided that $\beta_n(\varphi) \neq 0$ for some n and φ. If $\beta_n(\varphi) = 0$ for all n and φ, then f is radial and no conclusion can be drawn about μ.

Remark. Using Fourier transforms the formula for μ becomes

$$\mu = -(2k+1)i[\partial^{2k+1}(\mathcal{R}_\mu f)^\wedge/\partial\sigma^{2k}\partial\varphi)](\varphi,0)/[\partial^{2k+1}(\mathcal{R}_\mu f)^\wedge/\partial\sigma^{2k+1}](\varphi,0).$$

Recast this way, Theorem 3.2 holds for distributions with compact support.

We now sketch the ideas behind Theorem 3.3. Introduce the operators

$$\mathcal{R}_\mu^j f(\varphi,s) = \int_{-\infty}^{\infty} f(s\omega + t\omega^\perp)e^{\mu t}t^j dt,\ j = 0,1,2,\dots . \tag{3.5}$$

Differentiate (3.5) with respect to φ and integrate by parts to obtain

$$(\partial/\partial\varphi)\mathcal{R}_\mu^j f(\varphi,s) = -(d/ds)\mathcal{R}_\mu^{j+1}f(\varphi,s) - js\mathcal{R}_\mu^{j-1}f(\varphi,s) - \mu s\mathcal{R}_\mu^j f(\varphi,s). \tag{3.6}$$

Multiply both sides of (3.6) by s^n and integrate with respect to s to get

$$d/d\varphi\int\mathcal{R}_\mu^j f(\varphi,s)s^n ds + \mu\int\mathcal{R}_\mu^j f(\varphi,s)s^{n+1}ds$$

$$= n\int\mathcal{R}_\mu^{j+1}f(\varphi,s)s^{n-1}ds - j\int\mathcal{R}_\mu^{j-1}f(\varphi,s)s^{n+1}ds, \tag{3.7}$$

with both (3.6) and (3.7) valid for $j = 0,1,2,\dots$, $n = 0,1,2,\dots$ and all μ. The case j

$= n = 0$ gives (3.4) when $k = 0$, a result due to Hertle [10]. Theorem 3.3 can be proved using (3.7) and several inductive lemmas. We conclude by showing how (3.7) can be used to obtain (3.4) when $k = 1$. Assume that $\mu \neq 0$ and that

$$\int \mathcal{R}_\mu f(\varphi,s)s ds = 0 \quad \textit{for all } \varphi. \tag{3.8}$$

Set $j = 0$ and $n = 2$ in (3.7) to obtain

$$d/d\varphi \int \mathcal{R}_\mu f(\varphi,s)s^2 ds + \mu \int \mathcal{R}_\mu f(\varphi,s)s^3 ds = 2\int \mathcal{R}^1_\mu f(\varphi,s)s ds. \tag{3.9}$$

Set $j = 0$ and $n = 1$ in (3.7), apply (3.8) and differentiate to obtain

$$\mu d/d\varphi \int \mathcal{R}_\mu f(\varphi,s)s^2 ds = d/d\varphi \int \mathcal{R}^1_\mu f(\varphi,s) ds. \tag{3.10}$$

Next, in light of (3.8), setting $j = 1$ and $n = 0$ in (3.7) gives

$$d/d\varphi \int \mathcal{R}^1_\mu f(\varphi,s) ds = -\mu \int \mathcal{R}^1_\mu f(\varphi,s)s ds. \tag{3.11}$$

Since $\mu \neq 0$, (3.11) and (3.10) together imply that

$$\int \mathcal{R}^1_\mu f(\varphi,s)s ds = -d/d\varphi \int \mathcal{R}_\mu f(\varphi,s)s^2 ds,$$

and substituting this in (3.9) gives (3.4) when $k = 1$.

Acknowledgement. This work was supported by USA NSF Grant DMS-8805921-01. The author is indebted to Adel Faridani for several helpful conversations.

References

1. Brahme, A., Roos, J-E., Lax, I.: Solution of an integral equation encountered in rotation therapy, Phys. Med. Biol. 27-10, 1221-1229, (1982).
2. Budinger, T.F., Gullberg, G.T., Huseman, R.H.: Emission computed tomography. In: G. T. Herman (ed.): Image reconstruction from projections. Topics in Applied Physics. Vol. 32, 147-246, New York: Springer 1979.
3. Cho, Z. H.: General views on 3-D image reconstruction and computerized transverse axial tomography, IEEE Trans. Nucl. Sci. NS-21(3), 44-71, (1974).
4. Clough, A.V., Barrett, H.H.: Attenuated Radon and Abel transforms. J. Opt. Soc. Am. 73 (11), 1590-1595, (1983).

5. Davison, M.E., Grünbaum, F. A.: Tomographic reconstruction with arbitrary directions, Comm. Pure Appl. Math 34 No. 1, 77-119, (1981).
6. Hazou, I.A., Solmon, D.C.: Inversion of the exponential x-ray transform I: Analysis. Math. Meth. Appl. Sci. 10, 561-574, (1988).
7. Hazou, I.A., Solmon, D.C.: Inversion of the exponential X-ray transform II: Numerics, to appear in Math. Meth. Appl. Sci.
8. Helgason, S.: The Radon transform of Euclidian spaces, compact two-point homogeneous spaces and Grassmann manifolds. Acta Math. 113, 153-180, (1965).
9. Hertle, A.: On the injectivity of the attenuated Radon transform. Proc. AMS 92, 201-205, (1984).
10. Hertle, A.: The identification problem for the constantly attenuated radon transform. Math. Z. 197, 13-19, (1988).
11. Hertle, A., Solmon, D.C.: The identification problem for the exponential Radon transform, in preparation.
12. Kuchment, P.A., Ya. L'vin, S.: Paley-Wiener theorem for the exponential Radon transform. preprint.
13. Markoe, A.: Fourier inversion of the attenuated x-ray transform. SIAM J. Math. Anal. 15(4), 718-722, (1984).
14. Natterer, F.: Computed tomography with unknown sources. SIAM J. Appl. Math. 43 (5), 1201-1210, (1983).
15. Tretiak, O., Metz, C: The exponential Radon transform, SIAM J. Appl. Math. 39 (2), 341-354, (1980).

Tomographic Image Reconstruction from a Limited Set of Projections Using a Natural Pixel Decomposition

B. Dolveck-Guilpart

Laboratoire Spectroscopie Nucléaire, Université de Nantes,
F-44072 Nantes Cédex 03, France

In this paper we present a practical implementation of a method proposed by Buoconore [1], for two-dimensional tomographic image reconstruction from a limited set of real projections. The convergence of the iterative reconstruction method is proved and tested on real reconstructions.

I- Introduction

In tomographic imaging systems, it is often wished to obtain images when the projections are limited in number or restricted in angle. These situations arise when the angular range of the projections is restricted by obstacles or when it is not possible to make many measurements. In these case, the reconstruction problem is an inverse ill posed problem. The ill-posedness comes altogether from the finite number of measurements and from experimental errors. In many practical situations, one is able to obtain some "a priori" information about the object (shape, location). But if one has no "a priori" knowledge, is it possible to construct an image of the object, and what is its accuracy?
To deal with this question, we present here a method for 2D tomographic image reconstruction from a limited set of real projections. Our aim is to determine what information one can obtain about the object when the set of data is limited, and we study only real data. The method used has been proposed by Buoconore and al. [1] in 1981 under the title " a natural pixel decomposition for two dimensional image reconstruction", but it has been used only in few cases ([2],[3],[4]) and our work [5] is, up to our knowledge, the first one using real data.

II- Modeling the problem

The unknown object is represented by a distribution f(x,y) that we try to reconstruct. The geometry used for collecting measurements data is a fan-beam geometry (fig.1).

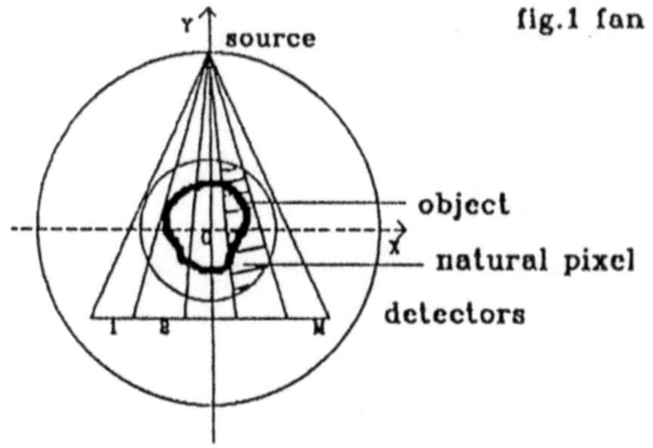

fig.1 fan beam geometry

For each source position, a set of caracteristic functions $\{\phi_i(x,y)\}$ is defined for representing integration domains of f(x,y) between the source and the (ith) detector.

$$\phi_i(x,y) = \begin{cases} 1 & \text{if } (x,y) \text{ is on the path between the source and the ith detector in the region of interest} \\ 0 & \text{otherwise} \end{cases} \qquad (2.1)$$

The functions $\phi_i(x,y)$ are called "the natural pixels", and they are uniquely defined by the paths of measurement beams. There are as many natural pixels as measurements. Thus the region of interest is decomposed on the basis of functions $\{\phi_i(x,y)\}$. These functions are orthogonal only if they correspond to the same source position, otherwise the natural pixels overlap (fig.2).

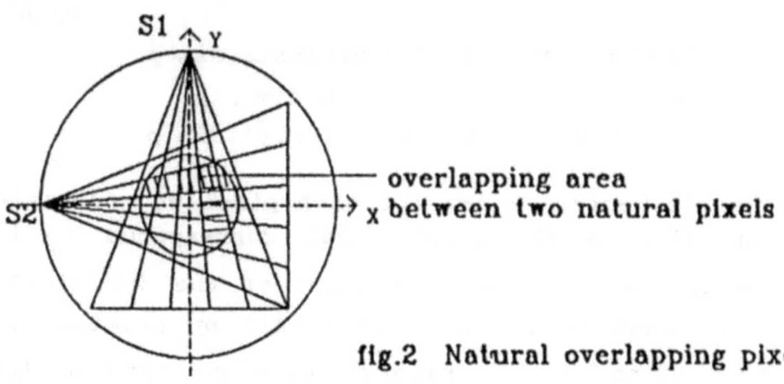

fig.2 Natural overlapping pixels representation

Let H be the matrix defined by the formula

$$H_{ij} = \iint \phi_i(x,y)\ \phi_j(x,y)\,dxdy \tag{2.2}$$

The elements of H can be interpreted as so:

$$H_{ij} = \begin{cases} \text{area of beam path} & i=j \\ \text{area of intersection} & i \neq j \end{cases}$$

The measurements (Y_i) are the moments of f:

$$Y_i = \iint f(x,y)\,\phi_i(x,y)\,dxdy \qquad i=1,..,p \tag{2.3}$$

where p is the number of measurements.

In order to construct an acceptable image of f(x,y) from the data Y_i, we assume it can be written as a linear combination of the caracteristic functions $\phi_i(x,y)$:

$$f(x,y) = \sum_{j=1}^{p} X_j . \phi_j(x,y) \tag{2.4}$$

Then inserting (2.4) in (2.3) we obtain a linear system of equations which gives a relation between the measurements Y_i and the coefficients X_j of the equation (2.4).

$$Y_i = \sum_{j=1}^{p} H_{ij} . X_j \qquad i=1,..,p \tag{2.5}$$

Let X be the vector of the X_j coefficients, the linear system (2.5) can be rewritten:

$$Y = HX \tag{2.6}$$

where Y is the vector of measurements.dim p
H is the projection matrix.dim p p
X is the state vector of f(x,y).dim p

The meaning of the "state vector" X is: for each pixel we associate a value X_j, constant through the pixel, that corresponds to an "average" value of f(x,y) within the boundaries of the jth pixel. The value of f(x,y) at each point (x,y) is obtain by summing all the X_j values corresponding to the pixels which go through the point (x,y)(see equation (2.4)).

III- Algorithm

The iterative reconstruction algorithm of the X vector is:

Input

Let D the diagonal matrix: $D = \mathrm{diag}(H_{ii})$

$\hat{X}^0 = 0$ first estimate of X

Current step for $k \geq 0$ (given H,D,Y,N)

$$\hat{X}^{k+1} = (I - \frac{1}{N} D^{-1} H)\hat{X}^k + \frac{1}{N} D^{-1} Y$$
$$= \sum_{n=0}^{k} (I - \frac{1}{N} D^{-1} H)^n \Delta^0 \qquad \Delta^0 = \frac{1}{N} D^{-1} Y$$

I is the identity matrix

N is the number of different source positions.

If the iterative method converges, the limit X satisfies:

$$X = X - \frac{1}{N} D^{-1}HX + \frac{1}{N} D^{-1}Y \qquad \text{i.e.} \quad HX=Y$$

So as to prove the convergence we need two theorems

Theorem 1

If Y is such that the equation in X : $X = X - \frac{1}{N} D^{-1}HX + \frac{1}{N} D^{-1}Y$ has a solution, then the iterative reconstruction method

$$\hat{X}^{k+1} = (I - \frac{1}{N} D^{-1} H)\hat{X}^k + \frac{1}{N} D^{-1} Y, \qquad k \geq 0 \text{ and } \hat{X}^0 \text{ given}$$

converges to a limit, which may depend on $\hat{X}^0$, if and only if the matrix $B = (I - \frac{1}{N} D^{-1} H)$ can be written as

$$B = P \begin{bmatrix} I_q & 0 \\ 0 & K \end{bmatrix} P^{-1}$$

I_q is the identity in $\mathbb{R}^q$ ($q \leq p$)

K is a square matrix of order p-q with a spectral radius $\rho(K) < 1$

This theorem is proved in reference [6].

Theorem 2 (an extension of the Ostrowski-Reich's theorem, due to H.B. Keller 1965)

If H is a p order hermitian matrix, decomposed as $H = M - N$, where M is an inversible matrix and $M^* + N$ is positive definite, then $B = M^{-1} N$ can be written as

$$B = P \begin{bmatrix} Iq & \emptyset \\ \emptyset & K \end{bmatrix} P^{-1}$$

Iq is the identity in $\mathbb{R}^q$ $(q \leqslant p)$

K is a square matrix of order p-q with a spectral radius $\rho(K) < 1$

P is an inversible matrix of order p

if and only if H is semi-positive definite.
This theorem is proved in reference [7].

Now it is easy to show that the projection matrix H is semi-positive definite. For the decomposition of H we choose:

$M = N.D$ where N is the number of different source positions and D the diagonal matrix previously defined.

$N = ND-H$

M is inversible and $M^* + N$ is positive definite. Thus

$$B = M^{-1}N = (I-\frac{1}{N}D^{-1}H) = P \begin{bmatrix} Iq & \emptyset \\ \emptyset & K \end{bmatrix} P^{-1}$$

This completes the proof.

IV- Actual reconstructions from an incomplete set of experimental data

We try to reconstruct a slice of an hollow aluminium cube (fig.3-a). The data have been recorded with a laboratory tomograph, working in fan-beam geometry, and using gamma ray radiation. Thus we deal with noisy real data and the distribution f(x,y) is that of the linear absorption coefficients in the object. We are interested by two situations of incomplete set of data: i) the projections are only a few but equally spaced around the object, ii) the projections may be more numerous but are recorded in a limited angular range.

For all reconstructions, we assume the object is located in the region of interest but we have no other a priori knowledge about it (its shape, its localisation or its densities are quite precise unknown).

Figure 3-a is the image of the test object. Figure 3-b is a reconstruction obtained from 15 projections and 39 samples in each

3-a Test object

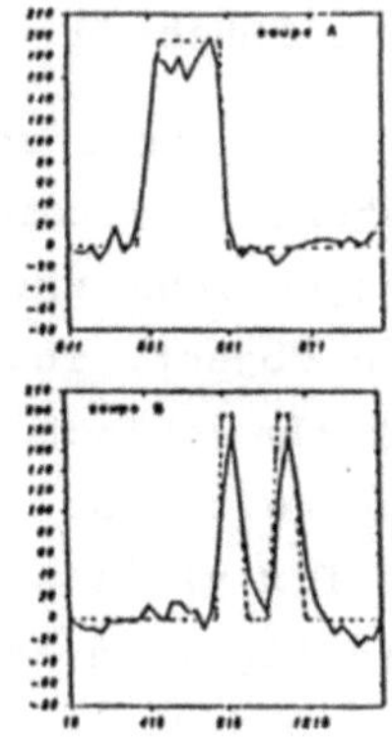

3-b Reconstruction from
15 equally spacedprojections

3-c
Convergence criterium evolution
respect to the number of iterations

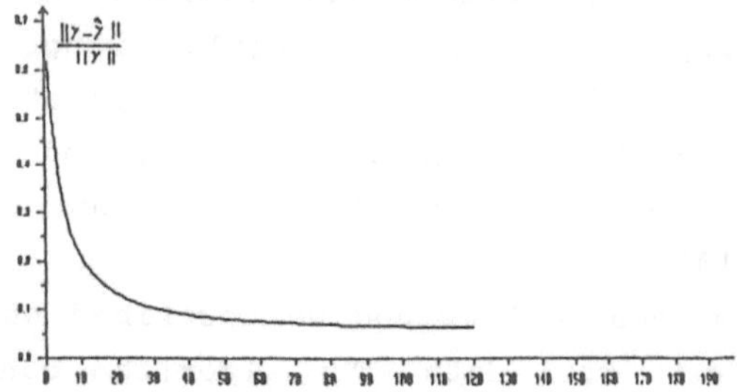

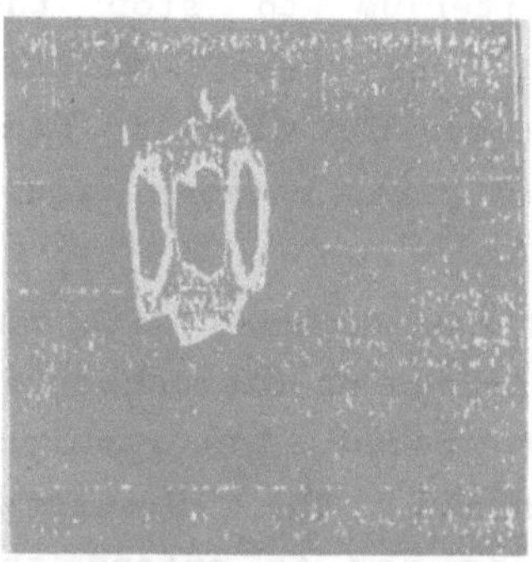

3-d Reconstruction from 15
projections equally spaced
in the limited angular
range [48°,132°]

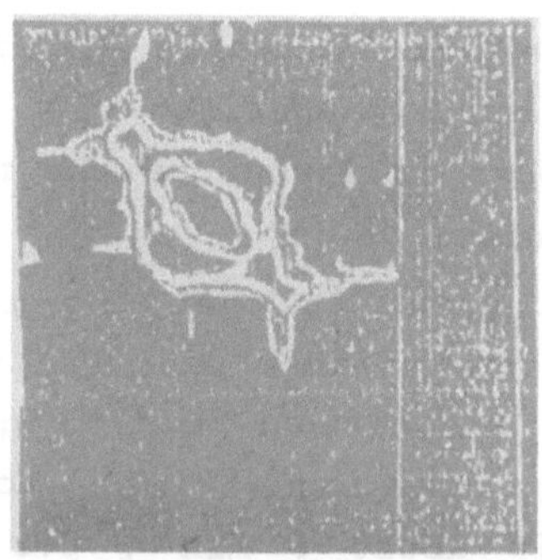

3-e Reconstruction from 15
projections equally spaced
in the limited angular
range [0°,84°]

Figure 3

projection, equally spaced around the object. We can see also numerical comparisons of the true and the reconstructed values of f(x,y). Figure 3-c gives the convergence criterium evolution with respect to the number of iterations. Figure 3-d is a reconstruction obtained from 15 projections and 39 samples in each projection, recorded in the limited angular range [48°,132°]. Figure 3-e is a reconstruction obtained from 15 projections and 39 samples in each projections, recorded in the limited angular range [0°,84°].

V- Discussion

These results lead us to say that it is possible to obtain interesting tomographic image reconstructions from a limited set of projections when the projections are almost equally spaced around the object. The reconstruction on figure 3-b gives a good estimate of the object density, shape and localisation in the region of interest. The error betweem the true and the reconstructed values of f(x,y) is about 10% . This result is interesting since the data are highly noisy (noise about 8%) and no "a priori" information on the object is given. But as it can be seen on figures 3-d and 3-e, when the projections are limited in angle ,the reconstructions strongly depend on the angular range where the projections are recorded. If the object is not symmetric we cannot obtain a good estimate of the object without introducing further information. Of course these limitations are related to the problem ill-posedness [9],[10].

The mathematical method we described shows some advantages when we work with real data. The algorithm converges and we can define a simple convergence criterium (i.e. a criterium to stop the algorithm). The method is stable and robust, and these qualities can be improved [5]. A comparison with ART algorithm with the same data has been always in favour of the natural pixel method, except for the time calculations [5].

Another important point is that the method can be used with any experimental geometry, either parallel or fan-beam or disconnected, and even if the projections are inequally spaced. The required IBM/3090 CPU processing time to obtain the presented reconstructions is about 5 min . Our aim is now to extend the method to 3 dimensional cases.

References

[1] M.H. BUOCONORE, W.R. BRODY and A. MACOVSKI, " A Natural Decomposition for two Dimensional Reconstruction", IEEE Trans. on Biomedical Engineering, vol. BME 28, NO.2, Feb.1981.

[2] L. GARNERO "Reconstruction d'Images Tomographiques à partir d'un Ensemble Limité de Projections", Thèse de Doctorat d'état, université d'orsay 1987.

[3] N. REJEEVAN and K. RAMAKRISHNA, " A Fast Algorithm for Limited Range CT Image Reconstruction", ISCAS'84, vol.3.

[4] L. GARNERO, J.P. HUGONIN, N. De BEAUCOUDRAY, " Limited-Angle Tomographic Imaging using a Constrained Natural-Pixel Decomposition", proceeding du deuxième colloque image, CESTA, NICE, APRIL 1986.
[5] B. DOLVECK-GUILPART "Deux Problèmes de Représentation et d'Identification d'un Milieu à partir de Sondages", Thèse de Doctorat, Université de montpellier, 1989.
[6] P.G. CIARLET "Introduction à l'Analyse Numérique Matricielle et à l'Optimisation",pp 95-109, MASSON 1985.
[7] H.B. KELLER "On the Solution of Singular and Semi-Definite Linear Systems by Iteration", J.Siam, Numer. Anal., ser.B, vol.2, NO.2, 1965.
[8] F. NATTERER, "The Mathematics of Computerized Tomography", WILEY 1986.
[9] A.K. LOUIS, "Incomplete Data Problems in X-Ray Computerized Tomography", Numer. Math. 48, 251-262 (1986).
[10] P.C. SABATIER, Ed., "Basic Methods of Tomography and Inverse Problems", Adam Hilger, Bristol and philadelphia, 1987.

Application of the Simulated Annealing Technique to Microwave Tomography: Preliminary Results

A. Franchois[1], *L. Garnero*[2], *Ch. Pichot*[1], *and J.P. Hugonin*[2]

[1]Equipe Electromagnétisme, Laboratoire des Signaux et Systèmes (CNRS-ESE), F-91192 Gif-sur-Yvette Cédex, France

[2]Institut d'Optique, URA 14, CNRS, BP 147, F-91403 Orsay, France

An iterative algorithm based on simulated annealing, for the reconstruction of the permittivity and conductivity distributions of a section of a strongly inhomogeneous cylindrical object, is presented. It has been tested, using simulated noise-free data, for some small objects. The effect of simulated annealing is compared to a conventional iterative method.

Introduction

Active microwave imaging is a fairly recent imaging technique. Because of its sensitivity to dielectric properties, it can be ranged as a complementary technique among ultrasound, infra-red, optical, magnetic resonance and x-ray imaging. It consists, first, of exposing an object to a low power microwave radiation field and measuring the scattered field somewhere outside the object, second, of reconstructing from these measurements parameters characterizing the object. This is an inverse problem.

Our work is situated in the field of microwave tomography. It concerns the reconstruction of the complex permittivity distribution of a section S of a strongly inhomogeneous object which we supposed cylindrical (Fig. 1). A possible application in the biomedical field is non invasive temperature control in hyperthermia treatment of deep lying tumors, since temperature and complex permittivity are related [1-5].

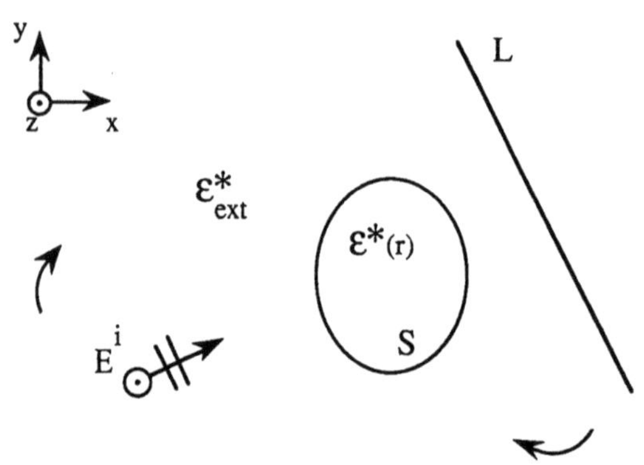

Fig. 1: Microwave tomography

The incident electrical field E^i is a plane wave polarized along Oz and the scattered field is measured on a line L, which is perpendicular to the propagation direction of the incident field. These measurements are repeated for different rotation angles of the line and the incident field. Before, we used a spectral reconstruction algorithm based on so-called "diffraction tomography" [6-9]. It retrieves the mean equivalent current distribution of the section from its Fourier transform which is obtained, by filling the spectral plane, from data corresponding to different projections. This method proved to be efficient for quasi homogeneous objects, but for objects of high contrasts, artefacts are introduced due to diffraction effects. In this paper we present an iterative reconstruction algorithm based on simulated annealing [10-12]. It solves the inverse problem by means of a sequence of direct problems. In each iteration, the present estimate of the complex permittivity distribution is slightly modified and this new estimate is then accepted or rejected according to the rules of simulated annealing. First we discuss the direct problem. Then we give an outline of simulated annealing and apply it to our problem. Finally we present some preliminary results and compare simulated annealing to a local optimization method.

The Direct Problem

The direct problem solves for the electrical field, in particular on the measurement line, given the incident field E^i, i.e. the field in the absence of an object, and given the complex permittivity of the section $\varepsilon^*(r)$ and of the surrounding medium ε^*_{ext}. For a time dependence $\exp(-j\omega t)$ the complex permittivity is defined as

$$\varepsilon^*(r) = \varepsilon_0\varepsilon_r(r) + j\sigma(r)/\omega, \tag{1}$$

where $\varepsilon_r(r)$ is the relative permittivity and $\sigma(r)$ the conductivity at a point r (x,y). The field is a solution of the Helmholtz equation for a homogeneous medium with induced sources in the right member

$$\Delta E + k^2_{ext}E = -\left(k^2(r) - k^2_{ext} \right)E, \tag{2}$$

with $k^2(r) = \omega^2\mu_0\varepsilon^*(r)$. The exact solution of (2) is represented by the integral equation

$$E(r) = E^i(r) + \iint_S \left(k^2(r') - k^2_{ext} \right) E(r')\, G(r;r')\, dr', \tag{3}$$

where G is the free-space Green's function

$$G(r;r') = (j/4)\, H_0^{(1)}(k_{ext}\, |r-r'|). \tag{4}$$

Equation (3) shows that the field in a point is given by the incident field in that point plus an integral which contains the field in the section. Once the field is known in the section, it can be calculated anywhere else, in particular on the measurement line.

The field in the section is calculated with the moment method [13,14]. The complex permittivity distribution is discretized into N pixels and the integral equation is expressed in the centre of each pixel $r_1,\ldots, r_n$. A system of linear equations of the Nth order is then obtained

$$\left[-(k^2(r_j)-k^2_{ext}) \iint_{S_j} G(r_i;r')\, dr' + \delta_{ij} \right] \begin{bmatrix} E(r_1) \\ \vdots \\ E(r_n) \end{bmatrix} = \begin{bmatrix} E^i(r_1) \\ \vdots \\ E^i(r_n) \end{bmatrix}. \tag{5}$$

The unknowns are the fields in the centre of each pixel. We solve this system with a direct Gauss-Jordan algorithm.

Simulated Annealing

Simulated annealing (SA) is an optimization algorithm. It is a form of the Metropolis algorithm [11] and it has been applied to the optimization of problems where many parameters are involved (circuit design [15], image reconstruction [16-19],...). SA searches for the optimum state of a system, which is defined as the global minimum of a cost function $\mathcal{E}$ associated to the system. Starting from an arbitrary initial estimate, it introduces in each iteration a slightly modified estimate, for which the cost is calculated and this new estimate is then accepted or rejected. Moving from estimate to estimate, it tries to find the global minimum of the cost function in a way similar to the physical annealing process of an amorph hot solid of high energy towards a monocristal of minimum energy, where the temperature is lowered slowly enough as to reach thermodynamic equilibrium at each temperature. If the temperature were lowered abruptly, the solid would freeze into a suboptimum state.

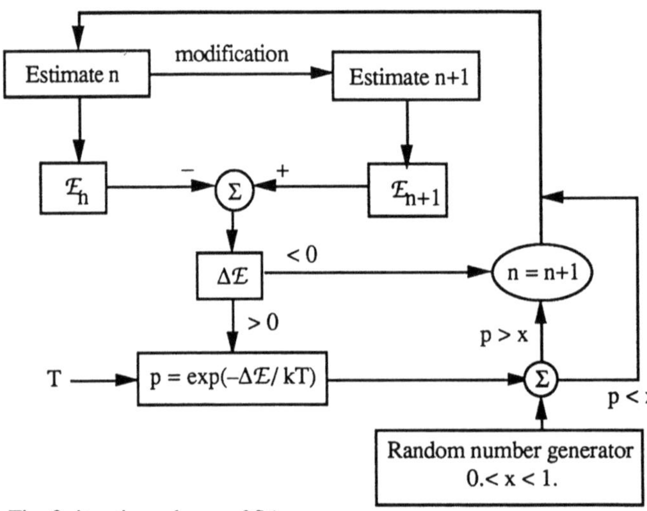

Fig. 2: iteration scheme of SA

Figure 2 illustrates the principle of SA. A modification of the estimate introduces a variation $\Delta\mathcal{E} = \mathcal{E}_{n+1} - \mathcal{E}_n$ of the cost function. For $\Delta\mathcal{E}$ negative, the modified cost is lower than the previous cost and the new estimate approaches the optimum, so it is accepted. If $\Delta\mathcal{E}$ is positive, the modified estimate won't be rejected necessarily, as is usually done in conventional methods of local optimization. Such a worse estimate will be accepted with a Boltzman probability distribution

$$p = \exp(-\Delta\mathcal{E} / kT) \tag{6}$$

where k is a constant, usually set to one and where T is a parameter called temperature. A generator of random numbers between zero and one is used for this purpose. For T large, p tends to one, so each worse estimate will be accepted, while for T small, p tends to zero and only better estimates will do. The temperature is chosen high in the beginning and it is lowered at equilibrium. By equilibrium is meant that the number of accepted worse estimates equals on the average the number of better estimates [17]. T equal to zero corresponds to a conventional method of local optimization, which often guides a system to

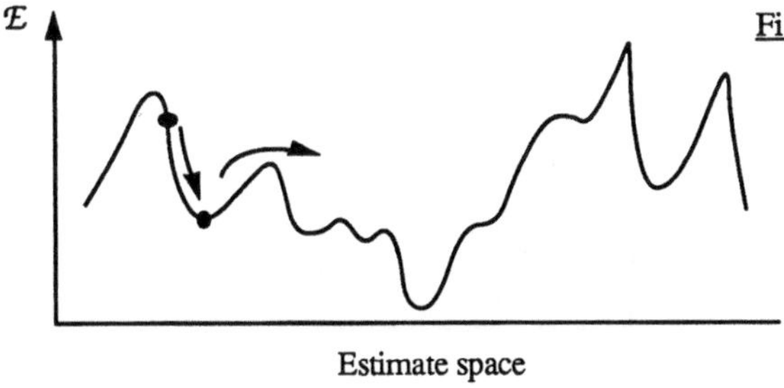

Fig. 3: SA searches the global minimum

a local minimum of the cost function.The virtue of SA and thus of introducing the probability (6) is that it gives the system a chance to escape from local minima (Fig. 3).

Summarized, the use of SA requires the definition of a cost function, the choice of an initial estimate and of a way of modifying the estimate, the definition of a temperature profile and finally the definition of a criterion to stop the algorithm.

We applied SA to the problem of reconstructing the complex permittivity of a section. The initial estimate is chosen arbitrarily. For estimate n we calculate the field on the measurement line (i.e. the direct problem) and we compare it to the measured field. The cost function, which we want to minimize, is defined as the Normalized Root Mean Square Error (NRMSE) of the calculated field E and the measured field E^m on the line, for the different views

$$\mathcal{E} = \sqrt{\sum (E - E^m)^2} \; / \; \sqrt{\sum (E^m)^2}. \tag{7}$$

The estimate is modified as follows: one pixel is chosen at random and its values of ε and σ are changed arbitrarily by adding to them or subtracting from them a grain which we chose proportional to $\mathcal{E}$. We change the temperature T every fixed number of iterations instead of lowering it at equilibrium. When at the end of an interval of constant temperature the system is close to equilibrium, we lower the temperature. If it is still far from equilibrium, we raise T. This is decided from the ratio of the number of accepted worse estimates to the number of better estimates counted during the interval.

Examples

The algorithm has been tested for some small biological objects, consisting of bone and muscle and immersed in water. Table 1 represents values of ε_r and σ at 3 GHz and 23°C. We used simulated noise-free data of the scattered field on the measurement line for different views. One example concerns an object of four pixels, half bone, half muscle and where we used four positions of the incident field and line (Fig. 4). Another example concerns an object of nine pixels, one pixel of bone in the middle and muscle around and where we used eight views (Fig. 9). For the reconstruction of the permittivity and conductivity distributions of the object, we started from an initial estimate with values for ε_r and σ of 10 per cent below the real values given in Table 1 (41.2 instead of 46 etc...). A non-negativity constraint has been imposed on ε_r and σ. We compared SA ($T \neq 0$) to a conventional method ($T = 0$). In the case of SA, the initial temperature was chosen 10^{-3} and it was adapted every 400 iterations. Curves of the

3 GHz, 23°C	ε_r	$\sigma\ (\Omega m)^{-1}$
water	74.	2.39
muscle	46.	2.
bone	8.	0.22

Table 1: Values of the relative permittivity and of the conductivity

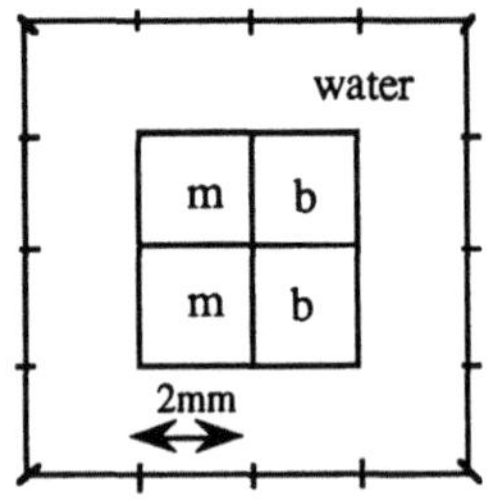

Fig. 4: four pixels of muscle and bone immersed in water and four positions of the measurement line. The locations of the measurements are marked

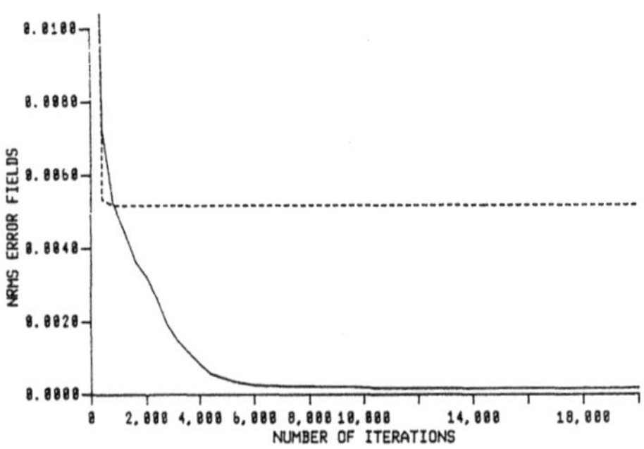

Fig. 5: The NRMS error of the fields for 4 pixels. ——— : with SA, $T_{initial} = 10^{-3}$
---- : without SA, T = 0

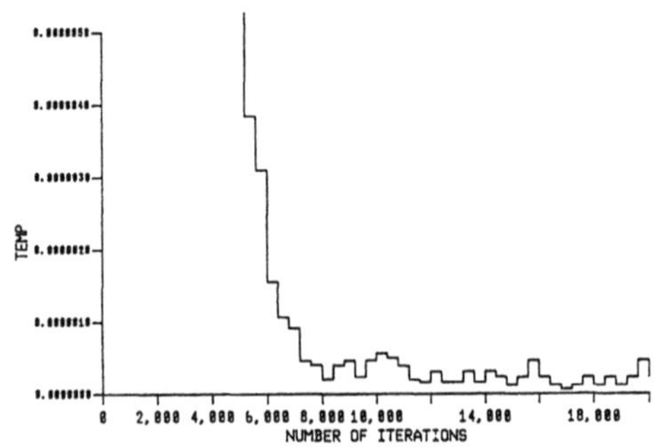

Fig. 6: The temperature T for 4 pixels. $T_{initial} = 10^{-3}$ and is adapted every 400 iterations

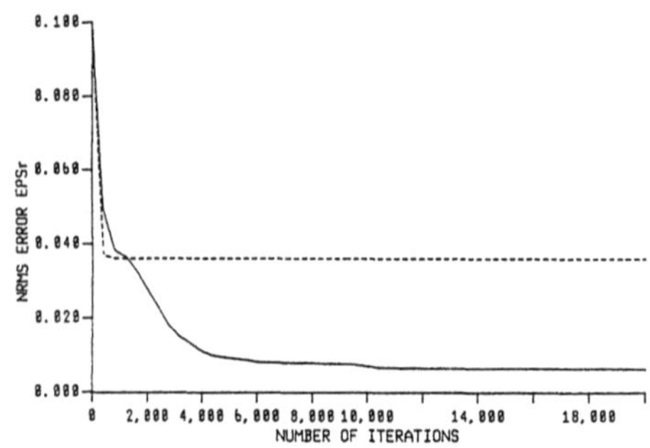

Fig. 7: The NRMS error of ε_r for 4 pixels.
——— : with SA, $T_{initial} = 10^{-3}$
---- : without SA, T = 0

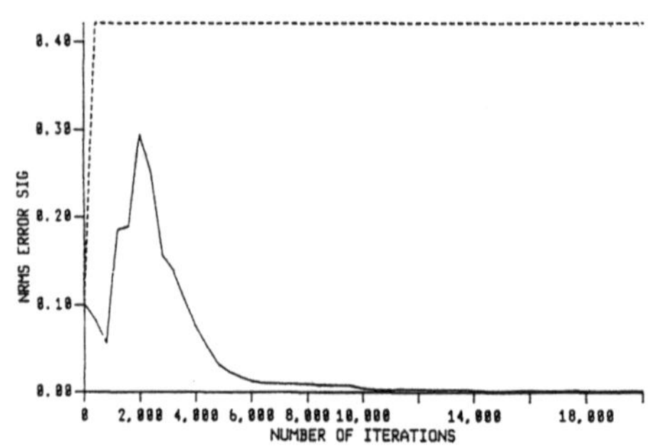

Fig. 8: The NRMS error of σ for 4
——— : with SA, $T_{initial} = 10^{-3}$
---- : without SA, T = 0

NRMSE of the fields, of the NRMSE of ε_r and σ with respect to their real values and of the temperature T as a function of the number of iterations are presented for the four-pixel case (Figs. 5-8) and for the nine-pixel case (Figs. 10-13). They represent results every 400 iterations.

For the four-pixel object, the relative errors of initially 10% have dropped to 2.5%, 0.9%, 0.4% and 0.1% for $\varepsilon_{r,\,bone}$, σ_{bone}, $\varepsilon_{r,muscle}$ and σ_{muscle} resp. after 20,000 iterations of SA. Figures 5, 7 and 8

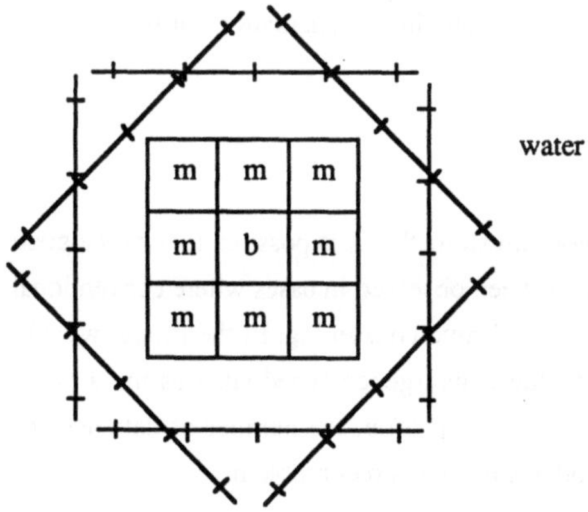

Fig. 9: nine pixels of muscle and bone immersed in water and eight positions of the measurement line

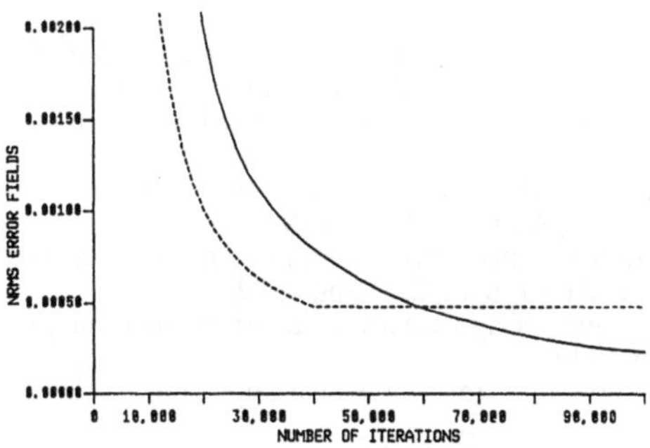

Fig. 10: The NRMS error of the fields for 9 pixels. ——— : with SA, $T_{initial} = 10^{-3}$
- - - - : without SA, T = 0

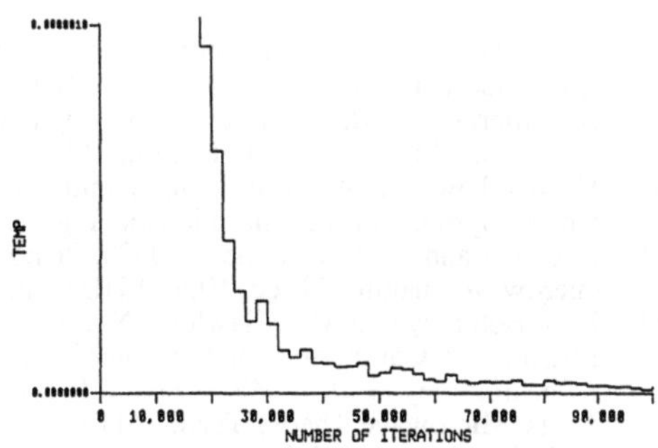

Fig. 11: The temperature T for 9 pixels. $T_{initial} = 10^{-3}$ and is adapted every 400 iterations

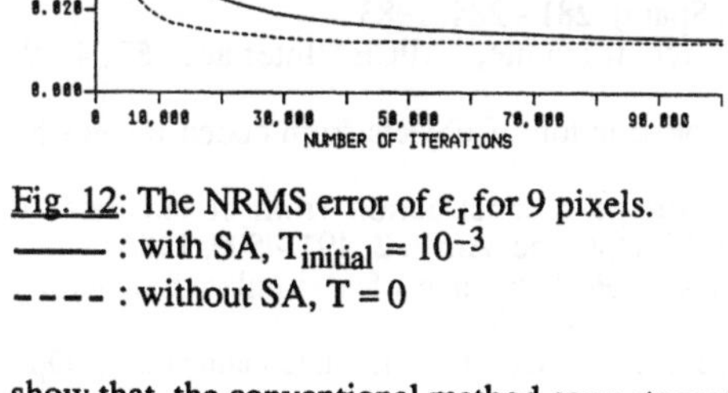

Fig. 12: The NRMS error of ε_r for 9 pixels.
——— : with SA, $T_{initial} = 10^{-3}$
- - - - : without SA, T = 0

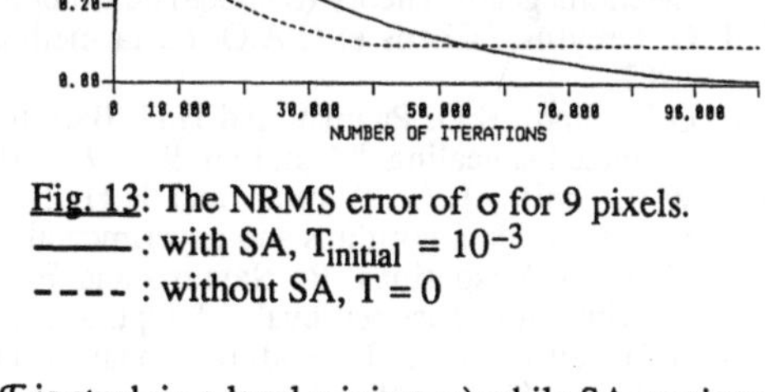

Fig. 13: The NRMS error of σ for 9 pixels.
——— : with SA, $T_{initial} = 10^{-3}$
- - - - : without SA, T = 0

show that the conventional method soon stagnates ($\mathcal{E}$ is stuck in a local minimum) while SA continues to converge slowly.

For the nine-pixel object, the relative errors have dropped below 1% for $\varepsilon_{r,muscle}$ and σ_{muscle}, after 100,000 iterations of SA, while $\varepsilon_{r,\,bone}$ does not yet give satisfaction. We will need to examine more carefully our definition of the grain. Figures 10, 12 and 13 again show a stagnation of the NRMSE when no SA was used.

Conclusion

Preliminary results obtained with our reconstruction algorithm show that it is possible to obtain a good convergence. Positive influence of simulated annealing has been observed in cases where conventional methods fail. The main disadvantage though, is that it is CPU time consuming. In the future, we will concentrate on the influence of different parameters on the convergence speed such as the a priori knowledge, the way of modifying the estimate, the temperature profile, the number of data and the influence of noise and also on a faster approximate method to solve the direct problem.

Bibliography

[1] J.Ch. Bolomey, G. Peronnet and L. Joffre, "On the possible use of active microwave imaging for remote thermal sensing," IEEE Trans. Microwave Theory Tech. , MTT **31**, 777 - 781, 1983.

[2] G. Gaboriaud *et al.* , "Microwave tomography in non-invasive control of hyperthermia," Proc. 9-th IEEE EMB Int. Symp., Boston, 1618 - 1619, Nov. 1987.

[3] M. Miyakawa, "A study of the temperature dependency of the permittivity of biological tissues," 5th Int. Symp. on Hyperthermic Oncology, 287, Kyoto, Aug. 29 - Sept. 3 1988.

[4] G. Gaboriaud, J.Ch. Bolomey and Ch. Pichot, "Control of deep hyperthermia by means of active microwave imaging," Proc. 10th ESHO Conference, Amsterdam, 22, Sept. 1989.

[5] J.Ch. Bolomey and M.S. Hawley, "Non-invasive control of hyperthermia, in clinical thermology, edited by M. Gautherie, Vol **3**, chapter 2, Springer 1990.

[6] J.Ch. Bolomey *et al.* , "Microwave diffraction tomography for biomedical applications," IEEE Trans. Microwave Theory Tech. , MTT **30**, 1998 - 2000, 1982.

[7] W. Tabbara *et al.* , "Diffraction tomography: contribution to the analysis of some applications in microwaves and ultrasonics," Inverse Problems **4**, 1129 - 1142, 1988.

[8] M. Azimi and A.C. Kak, "Distortion in diffraction tomography caused by multiple scattering," IEEE Trans. Med. Imaging, MI **2**, 176 - 195, 1983.

[9] M. Slaney, A.C. Kak and E. Larsen, "Limitations of imaging with first order diffraction tomography," IEEE Trans. Microwave Theory Tech. , MTT **32**, 860 - 874, 1984.

[10] S. Kirkpatrick, C.D. Gelatt. Jr and M.P. Vecchi, "Optimization by simulated annealing," Science **220**, 671 - 680, 1983.

[11] N. Metropolis *et al.* , "Equation of state calculations by fast computing machines,"J. Chem. Phys. **21**, 1087 - 1092, 1953.

[12] E. Bonomi and J.L. Lutton, "Le recuit simulé," Pour la Science **129**, 68 - 77, 1988.

[13] M.J. Hagmann, O.P. Gandhi, D.K. Ghodgaonkar, "Application of moment-methods to electromagnetic imaging, " Int. 1981 IEEE MTT-Symp. Digest, 482.

[14] M. Ney *et al.* , "Electromagnetic imaging using moment methods," URSI Int. Symp. Dig. on Electromagnetic Theory (Santiago de Compostela, Spain), 281 - 284, 1983.

[15] G. Dreyfus, "Chaos et C.A.O. ou la méthode du recuit simulé," AFCET/Interfaces **53**, 4 - 9, 1987.

[16] G.R. Gindi, R.G. Paxman and H.H. Barrett, "Reconstruction of objects from coded images by simulated annealing, " Opt. Lett. **9**, 199 - 201, 1983.

[17] W.E. Smith, R.G. Paxman and H.H. Barrett, "Image reconstruction from coded data: I. reconstruction algorithms and experimental results," J. Opt. Soc. Am A **2**, 491 - 500, 1985.

[18] M. Nieto-Vesperinas, R. Navarro and F.J. Fuentes, "Performance of a simulated-annealing algorithm for phase retrieval, " J. Opt. Soc. Am A **5**, 30 - 38, 1988.

[19] H. Haneishi *et al.* , "Three-dimensional blood vessel reconstruction by simulated annealing," Opt. Lett. **14**, 1095 - 1097, 1989.

Elastic Wave Diffraction Tomography

K.J. Langenberg, T. Kreutter, K. Mayer, P. Fellinger, and M. Berger

Dept. Electrical Engineering, FB 16, University of Kassel,
D-3500 Kassel, Fed. Rep. of Germany

Following the guidelines of a recently developed unified theory of scalar inverse scattering within the linearizing Born or Kirchhoff approximations — culminating in a 3D imaging system for ultrasonic NDT applications — we extend its basic ideas to elastic wave inverse scattering, especially for the case of a scatterer with stress-free boundaries being illuminated by a plane pressure or shear wave. Two inversion schemes, both yielding the same result, i.e. the "visible" part of the singular function of the scattering surface, are proposed: far-field inversion, provided the data are collected in the remote region of the defect, or, far-field inversion combined with a near-field far-field transformation of the measured data for the case of defects closed to a measurement surface. The latter is computationally very effective in terms of Fourier transforms provided the mesurement surface is planar or circular cylindrical. In addition, we point out why the powerful idea of backpropagation resulting in the scalar Porter-Bojarski integral equation is not very appropriate for the elastodynamic case. Our algorithms are checked against simulations and discussed as elastodynamic extensions of the scalar SAFT-algorithm in terms of a mode-matched SAFT; their use in advanced imaging systems is presently only limited due to the requirement of tangential as well as normal components of the displacement vector to be measured on the material surface.

1. INTRODUCTION

If the differential equation for the displcement vector $\mathbf{u}(\mathbf{R},t)$ as function of position $\mathbf{R}$ and time t

$$(\lambda+2\mu)\nabla\nabla\cdot\mathbf{u}(\mathbf{R},t)-\mu\nabla\times\nabla\times\mathbf{u}(\mathbf{R},t)-\varrho\frac{\partial^2}{\partial t^2}\mathbf{u}(\mathbf{R},t)=0 \tag{1}$$

— λ and μ denote Lamé's constants of the homogeneous and isotropic bulk material of density ϱ — is "scalarized" in terms of the potentials

$$\mathbf{u}(\mathbf{R},t)=\nabla\Phi(\mathbf{R},t)+\nabla\times\Psi(\mathbf{R},t)\text{ with }\nabla\cdot\Psi(\mathbf{R},t)=0 \tag{2}$$

by concentrating only on the scalar potential $\Phi(\mathbf{R},t)$, which is approximately correct if we just consider an incident pressure wave and observe only in the scattered pressure wave mode, we can formulate a unified and concise theory of scalar inverse scattering for typical canonical scattering defects, i.e. penetrable defects like material inclusions, and perfectly scattering defects like voids, which are characterized by soft or rigid boundary conditions [1,2]. This theory could be called a generalized diffraction tomography, because, first, it applies only if the Born or Kirchhoff approximations hold, which linearize the inverse scattering problem, and second, it reduces to conventional diffraction tomography in terms of the Fourier Diffraction Slice Theorem if planar measurement surfaces are under concern for monofrequent multiple angle plane wave excitation. But, as a matter of fact, it also holds for the case of frequency diversity, either for pitch-catch or pulse-echo experiments, and, interesting enough, it allows the evaluation of explicit time domain backpropagation schemes, thus providing a rigorous diffraction theory of the otherwise heuristically proposed SAFT-algorithm [1], which has found wide application in ultrasonic nondestructive testing [3] (compare Fig. 1 for a flowchart, which combines all the different algorithms). That way, Fourier transform alternatives to SAFT-processing of time domain synthetic aperture data have been developed for planar apertures [4,1,5] and circular cylindrical apertures [6], which even allow 3D ultrasonic imaging [7]. An example is given in Fig. 2, where the 3D image of a steel specimen is displayed, which contains a T-shaped drilled hole, where chemical liquids under high pressure flow through; therefore, crack-fields appear close to the drilling, and their location and shape is intuitively imaged in 3D. (details of the

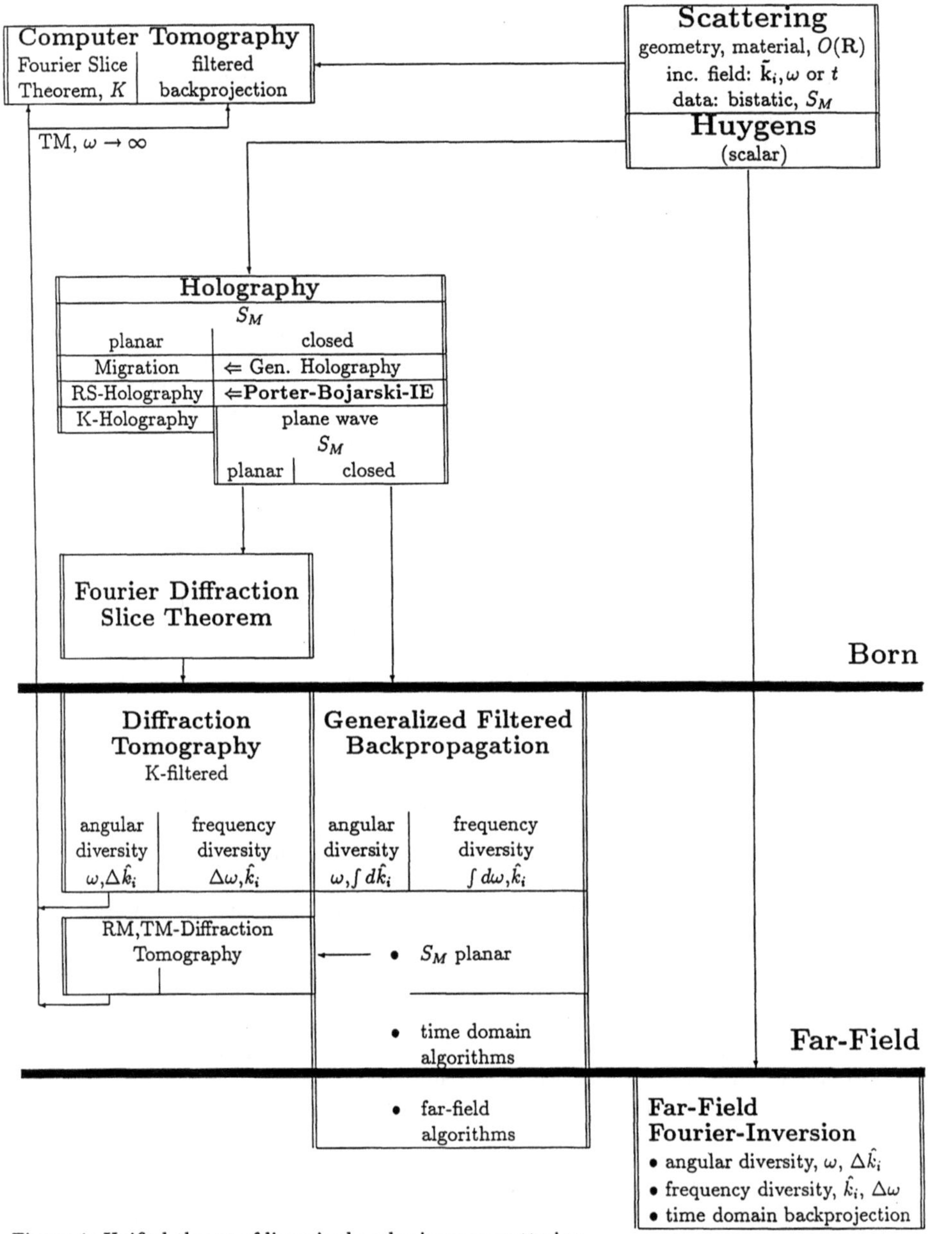

Figure 1: Unified theory of linearized scalar inverse scattering

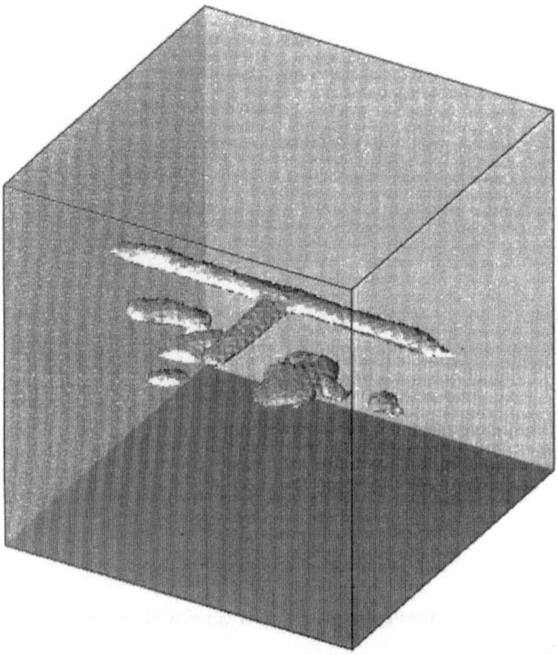

Figure 2: 3D iso-contour plot image of a T-shaped drilled hole and crack-fields as obtained with the GhK-TET 3D ultrasonic imaging system based on FT-SAFT

system can be found in [7]). Another example for a 2D image of the circular cylindrical cross-section of a turbine shaft is given in Fig. 3; it was obtained via an appropriate nearfield-farfield transformation in combination with far-field Fourier inversion (see below), and it exhibits mm-sized defects around the center of the shaft.

The above-mentioned theory allows three approaches:

- Writing down the solution for $\Phi(\mathbf{R},\omega)$ — ω denoting circular frequency — in the far-field of the scatterer, i.e. assuming far-field measurements on a closed measurement surface S_M, the far-field scattering amplitude can be made the basis of a spatial Fourier space inversion scheme (compare Fig. 1), if the spatial Fourier vector $\mathbf{K}$ is defined in terms of experimental parameters according to

$$\mathbf{K} = k_P(\hat{\mathbf{R}} - \hat{\mathbf{k}}_i) \tag{3}$$

  where k_P is the pressure wave number, and $\hat{\mathbf{k}}_i$ is the unit vector of propagation of the incident plane wave, whereas $\hat{\mathbf{R}}$ denotes the unit vector to the point of observation.

- For S_M in the near-field of the scatterer, the scattering amplitude can be defined alternatively in terms of a Huygens-type integral extending over S_M [8,9,2], which provides a near-field far-field transformation; then it is used as above. This approach is particularly effective, if S_M is planar or circular cylindrical, allowing the computation of the Huygens integral via Fourier transforms.

- Instead of defining the scattering amplitude in terms of a Huygens integral *outside* the measurement surface, a generalized holographic field can be defined *inside* that surface applying the principle of backpropagagtion through insertion of the complex conjugate of Green's function. Manipulation of this representation yields the so-called Porter-Bojarski integral equation (compare Fig. 1) relating the equivalent sources defining the "pressure wave" defect [10,11] and the holographic field, to which

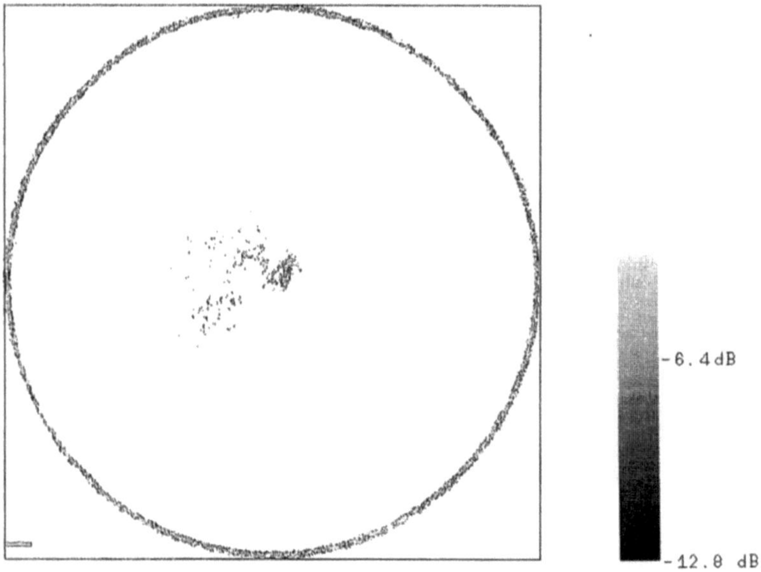

Figure 3: 2D grey scale image of the circular cylindrical cross-section of a steel turbine shaft as obtained with an FT-SAFT based algorithm

one has access via measurements. This integral equation gives rise to the above-mentioned unified theory of linearized inverse scattering if it is integrated with regard to ω or with regard to $\hat{\mathbf{k}}_i$.

In fact, all three approaches are equivalent with regard to the results, they only differ in algorithmic effectivity, and in computational usefulness to verify relationships between different algorithms. Hence, in order to obtain the full elastodynamic counterpart of this theory, we are free to choose that version, which is most appropriate. It turns out, that the near-field far-field transformation together with the far-field inversion gives rise to the easiest access, whereas elastodynamic backpropagation seems somewhat cumbersome.

2. ELASTODYNAMIC FAR-FIELD INVERSION FOR SCATTERERS WITH STRESS-FREE BOUNDARIES

A Huygens-type solution of equ. 1 for the displacement $\mathbf{u}_s$ scattered by a defect with a stress-free surface S_c is given by

$$\mathbf{u}_s(\mathbf{R},\omega) = \int\!\!\int_{S_c} \mathbf{n}'\mathbf{u}(\mathbf{R}',\omega) : \mathbf{\Sigma}(\mathbf{R}-\mathbf{R}',\omega)\, dS' \tag{4}$$

where $\mathbf{n}'$ denotes the outer normal on S_c [12]; we do not give the third rank Green's tensor $\mathbf{\Sigma}$ explicitly, because we only need its far-field approximation

$$\begin{aligned}\mathbf{\Sigma}^{far}(\mathbf{R}-\mathbf{R}',\omega) &= -\frac{j\mu k_S^3}{\varrho\omega^2}\left[\hat{\mathbf{R}}\mathbf{I} + (\hat{\mathbf{R}}\mathbf{I})^{213} - 2\hat{\mathbf{R}}\hat{\mathbf{R}}\hat{\mathbf{R}}\right] G_S^{far}(\mathbf{R}-\mathbf{R}',\omega) \\ &\quad -\frac{jk_P^3}{\varrho\omega^2}(\lambda\mathbf{I}\hat{\mathbf{R}} + 2\mu\hat{\mathbf{R}}\hat{\mathbf{R}}\hat{\mathbf{R}})G_P^{far}(\mathbf{R}-\mathbf{R}',\omega) \\ &= \frac{e^{jk_SR}}{4\pi R}e^{-jk_S\hat{\mathbf{R}}\cdot\mathbf{R}'}\mathbf{\Sigma}_S(\hat{\mathbf{R}},\omega) + \frac{e^{jk_PR}}{4\pi R}e^{-jk_P\hat{\mathbf{R}}\cdot\mathbf{R}'}\mathbf{\Sigma}_P(\hat{\mathbf{R}},\omega)\end{aligned} \tag{5}$$

with

$$G_{P,S}^{far}(\mathbf{R}-\mathbf{R}',\omega) = \frac{e^{jk_{P,S}R}}{4\pi R}e^{-jk_{P,S}\hat{\mathbf{R}}\cdot\mathbf{R}'} \quad . \tag{6}$$

Here, S is the "shear wave index", and $\mathbf{I}$ is the dyadic identity operator. We obtain the far-field approximation of equ. 4 through

$$\begin{aligned}
\mathbf{u}_s^{far}(\mathbf{R},\omega) &= \frac{e^{jk_SR}}{4\pi R}\iiint_{V_c}\mathbf{U}_c(\mathbf{R}',\omega):\mathbf{\Sigma}_S(\hat{\mathbf{R}},\omega)e^{-jk_S\hat{\mathbf{R}}\cdot\mathbf{R}'}\,d^3\mathbf{R}' + \\
&\quad \frac{e^{jk_PR}}{4\pi R}\iiint_{V_c}\mathbf{U}_c(\mathbf{R}',\omega):\mathbf{\Sigma}_P(\hat{\mathbf{R}},\omega)e^{-jk_P\hat{\mathbf{R}}\cdot\mathbf{R}'}\,d^3\mathbf{R}' \\
&= \frac{e^{jk_SR}}{R}\mathbf{C}_S(\hat{\mathbf{R}},\omega) + \frac{e^{jk_PR}}{R}\mathbf{C}_P(\hat{\mathbf{R}},\omega) \\
&= \mathbf{u}_{sS}^{far}(\mathbf{R},\omega) + \mathbf{u}_{sP}^{far}(\mathbf{R},\omega) \qquad (7)
\end{aligned}$$

where we have introduced dyadic equivalent surface sources $\mathbf{U}_c(\mathbf{R}',\omega)$ through

$$\mathbf{U}_c(\mathbf{R}',\omega) = \gamma(\mathbf{R}')\mathbf{n}'\mathbf{u}(\mathbf{R}',\omega) \qquad (8)$$

with the singular function $\gamma(\mathbf{R})$ of S_c [13]; the introduction of the latter allows the extension of the surface integral in equ. 4 to a volume integral over the volume V_c of the scatterer. The quantities $\mathbf{C}_\beta(\hat{\mathbf{R}},\omega)$ with $\beta = P, S$ play the role of vectorial scattering amplitudes in the β-mode, and it is easily verified that the following properties hold

$$\begin{aligned}
\mathbf{C}_P(\hat{\mathbf{R}},\omega)\times\hat{\mathbf{R}} &= 0 \qquad (9)\\
\mathbf{C}_S(\hat{\mathbf{R}},\omega)\cdot\hat{\mathbf{R}} &= 0 \qquad (10)
\end{aligned}$$

This permits computational selection of a special mode in the far-field

$$\mathbf{u}_{s\beta}^{far}(\mathbf{R},\omega) = \frac{e^{jk_\beta R}}{R}\mathbf{C}_\beta(\hat{\mathbf{R}},\omega) \qquad (11)$$

to be considered separately for inversion. We linearize in terms of physical elastodynamics (PE) [14]

$$\begin{aligned}
\mathbf{U}_c(\mathbf{R}',\omega) &\Rightarrow \mathbf{U}_{c\alpha}^{PE}(\mathbf{R}',\omega)\gamma_u(\mathbf{R}') \\
&= \mathbf{U}_{0\alpha}(\mathbf{n}',\hat{\mathbf{k}}_i,\omega)\gamma_u(\mathbf{R}')e^{jk_\alpha\hat{\mathbf{k}}_i\cdot\mathbf{R}'} \qquad (12)
\end{aligned}$$

with

$$\mathbf{U}_{0\alpha}(\mathbf{n}',\hat{\mathbf{k}}_i,\omega) = \mathbf{u}_{0\alpha}(\mathbf{n}',\hat{\mathbf{k}}_i,\omega)\mathbf{n}' \qquad (13)$$

where k_α is the wave number of the incident plane wave in the mode $\alpha = P, S$; $\mathbf{u}_{0\alpha}(\mathbf{n}',\hat{\mathbf{k}}_i,\omega)$ denotes the surface displacement "induced" by the incident wave as computed by PE, whence the occurrence of only *that* part of the singular function, which is illuminated by that wave, i.e. $\gamma_u(\mathbf{R}')$, the index u indicating multiplication of $\gamma(\mathbf{R}')$ with a step-function. We obtain

$$\mathbf{u}_{s\alpha\beta}^{far}(\mathbf{R},\omega) = \frac{e^{jk_\beta R}}{4\pi R}\iiint_{V_c}\gamma_u(\mathbf{R}')\mathbf{U}_{0\alpha}(\mathbf{n}',\hat{\mathbf{k}}_i,\omega):\mathbf{\Sigma}_\beta(\hat{\mathbf{R}},\omega)e^{-j(k_\beta\hat{\mathbf{R}}-k_\alpha\hat{\mathbf{k}}_i)\cdot\mathbf{R}'}\,d^3\mathbf{R}' \qquad (14)$$

Computing a "stationary phase normal" $\mathbf{n}' = \mathbf{n}_{\alpha\beta}$ [2] according to

$$\mathbf{n}_{\alpha\beta} = \frac{k_\beta\hat{\mathbf{R}} - k_\alpha\hat{\mathbf{k}}_i}{|k_\beta\hat{\mathbf{R}} - k_\alpha\hat{\mathbf{k}}_i|} \qquad (15)$$

we get rid of the $\hat{\mathbf{R}}$-dependence of the source term in the integral 14 replacing $\mathbf{n}'$ by $\mathbf{n}_{\alpha\beta} = \mathbf{n}_{\alpha\beta}(\hat{\mathbf{R}},\hat{\mathbf{k}}_i)$ to end up with

$$\mathbf{u}_{s\alpha\beta}^{far}(\mathbf{R},\omega) = \frac{e^{jk_\beta R}}{4\pi R}\mathbf{U}_{0\alpha\beta}(\hat{\mathbf{R}},\hat{\mathbf{k}}_i,\omega):\mathbf{\Sigma}_\beta(\hat{\mathbf{R}},\omega)\iiint_{V_c}\gamma_u(\mathbf{R}')e^{-j\mathbf{K}\cdot\mathbf{R}'}\,d^3\mathbf{R}' \qquad (16)$$

with the Fourier vector

$$\mathbf{K} = k_\beta\hat{\mathbf{R}} - k_\alpha\hat{\mathbf{k}}_i \qquad (17)$$

and

$$\mathbf{U}_{0\alpha\beta}(\hat{\mathbf{R}},\hat{\mathbf{k}}_i,\omega) = \mathbf{u}_{0\alpha}(\mathbf{n}_{\alpha\beta},\hat{\mathbf{k}}_i,\omega)\mathbf{n}_{\alpha\beta} \qquad (18)$$

The double contraction of the dyad $\mathbf{U}_0$ with the triad $\mathbf{\Sigma}_\beta$ yields a vector

$$\mathbf{U}_{0\alpha\beta} : \mathbf{\Sigma}_\beta = \mathbf{C}_{0\alpha\beta}(\hat{\mathbf{R}}, \hat{\mathbf{k}}_i, \omega) \tag{19}$$

Hence, in shorthand notation equ. 16 reads

$$\mathbf{u}_{s\alpha\beta}^{far}(\mathbf{R}, \omega) = \frac{e^{jk_\beta R}}{4\pi R} \mathbf{C}_{0\alpha\beta}(\hat{\mathbf{R}}, \hat{\mathbf{k}}_i, \omega)\, \tilde{\gamma}_u(\mathbf{K}) \tag{20}$$

where $\tilde{\gamma}_u(\mathbf{K})$ denotes the threedimensional Fourier transform of $\gamma_u(\mathbf{R})$. Equ. 20 is our desired far-field inversion scheme, because, due to the knowlegde of $\mathbf{C}_{0\alpha\beta}(\hat{\mathbf{R}}, \hat{\mathbf{k}}_i, \omega)$ from experiments, we can easily compute the Fourier transform of the "visible" singular function from

$$4\pi R e^{-jk_\beta R} \frac{\mathbf{C}_{0\alpha\beta} \cdot \mathbf{u}_{s\alpha\beta}^{far}}{|\mathbf{C}_{0\alpha\beta}|^2} = \tilde{\gamma}_u(\mathbf{K}) \tag{21}$$

and, therefore, a threedimensional Fourier inversion will get us back to the spatial domain. Obviously, the result of the far-field inversion scheme 21 is neither dependent on the incoming mode nor dependent on the observed wave mode, hence, it might be called a mode-matched inversion scheme. From the scalar theory, we know that the time domain counterpart of 21 is a backprojection algorithm [1], as it is the case for far-field SAFT, so the name mode-matched FT-SAFT for equ. 21 seems appropriate, where "FT" indicates that spatial Fourier transforms are involved.

In case we do not account for the compensation of the elastodynmaic nature of the experiment by taking explicitly care of $\mathbf{C}_{0\alpha\beta}(\hat{\mathbf{R}}, \hat{\mathbf{k}}_i, \omega)$ we could instead compute

$$\hat{\mathbf{R}} \cdot \mathbf{u}_{s\alpha P}^{far} = \frac{e^{jk_P R}}{4\pi R} \hat{\mathbf{R}} \cdot \mathbf{C}_{0\alpha P}(\hat{\mathbf{R}}, \hat{\mathbf{k}}_i, \omega)\, \tilde{\gamma}_u(\mathbf{K}) \tag{22}$$

$$\hat{\mathbf{R}} \times \mathbf{u}_{s\alpha S}^{far} = \frac{e^{jk_S R}}{4\pi R} \hat{\mathbf{R}} \times \mathbf{C}_{0\alpha S}(\hat{\mathbf{R}}, \hat{\mathbf{k}}_i, \omega)\, \tilde{\gamma}_u(\mathbf{K}) \tag{23}$$

which would, via Fourier inversion, result in a weighted singular function depending on the selection of the outgoing mode, which, in a real experiment, could be realized for instance by time gating. At least the inversion according to equ. 22 is strictly scalar, and, in fact, it is nothing but a "mode-dependent" P-wave FT-SAFT, i.e. a "scalarized" FT-SAFT, if, for example, α is chosen as P. Fig. 4 shows computer

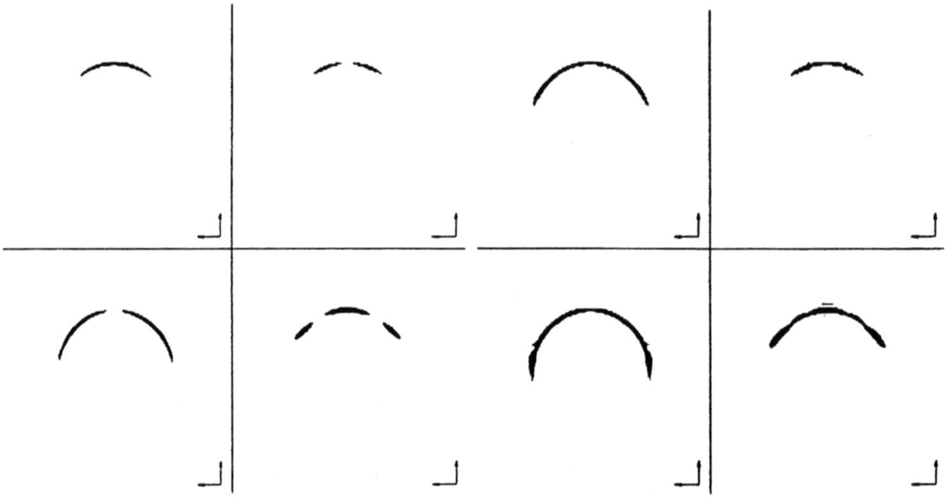

Figure 4: Left: Result of the mode-dependent far-field FT-SAFT inversion as applied to a circular cylinder with stress-free boundary for an incident plane pressure or shear wave coming from the top; top: result of $\hat{\mathbf{R}} \cdot \mathbf{u}_{sPP}^{far}$ and $\hat{\mathbf{R}} \cdot \mathbf{u}_{sSP}^{far}$; bottom: result of $\hat{\mathbf{R}} \times \mathbf{u}_{sPS}^{far}$ and $\hat{\mathbf{R}} \times \mathbf{u}_{sSS}$. Right: Result of the mode-matched far-field FT-SAFT inversion as applied to a circular cylinder with stress-free boundary for an incudent plane pressure or shear wave coming from the top; clockwise: $\alpha\beta = PP, SP, PS, SS$

simulations of the output of either inversion scheme 21 or inversion schemes 22, and 23, respectively; the twodimensional example of a circular cylinder with stress-free boundary condition was chosen to compute the scattered field for every $\alpha\beta$-combination, and then, both inversion algorithms have been applied. Obviously, in both cases, the visible part of the singular function is imaged, but, as predicted, the mode-matched algorithm compensates for the angular weighting of the induced surface displacement.

3. ELASTODYNAMIC NEAR-FIELD FAR-FIELD TRANSFORMATION

For measurement surfaces S_M in the near-field of the scatterer we can apply the complete version of the elastodynamic Huygens principle, which, in contrast to equ. 4, also involves the stress tensor $\mathbf{T}$ and a dyadic Green's function $\mathbf{G}$, because no particular boundary conditions are imposed:

$$\mathbf{u}_s(\mathbf{R},\omega) = \int\!\!\int_{S_c} \{\mathbf{u}(\mathbf{R}',\omega)\cdot[\mathbf{n}'\cdot\mathbf{\Sigma}(\mathbf{R}-\mathbf{R}',\omega)] - [\mathbf{T}(\mathbf{R}',\omega)\cdot\mathbf{n}']\cdot\mathbf{G}(\mathbf{R}-\mathbf{R}',\omega)\}\, dS' \tag{24}$$

For observation points $\mathbf{R}$ in the far-feld of S_M we compute the β-mode scattering amplitude from equ. 24 as

$$\mathbf{C}_\beta(\hat{\mathbf{R}},\omega) = \frac{1}{4\pi}\int\!\!\int_{S_M} [\mathbf{u}\mathbf{n}' : \mathbf{\Sigma}_\beta - (\mathbf{T}\cdot\mathbf{n}')\cdot\mathbf{G}_\beta]\, e^{-jk_\beta \hat{\mathbf{R}}\cdot\mathbf{R}'}\, dS' \tag{25}$$

with

$$\mathbf{\Sigma}_P(\hat{\mathbf{R}},\omega) = -\frac{jk_P^3}{\varrho\omega^2}(\lambda\mathbf{I}\hat{\mathbf{R}} + 2\mu\hat{\mathbf{R}}\hat{\mathbf{R}}\hat{\mathbf{R}}) \tag{26}$$

$$\mathbf{\Sigma}_S(\hat{\mathbf{R}},\omega) = -\frac{j\mu k_S^3}{\varrho\omega^2}\left[\hat{\mathbf{R}}\mathbf{I} + (\hat{\mathbf{R}}\mathbf{I})^{213} - 2\hat{\mathbf{R}}\hat{\mathbf{R}}\hat{\mathbf{R}}\right] \tag{27}$$

$$\mathbf{G}_P(\hat{\mathbf{R}},\omega) = \frac{k_P^2}{\varrho\omega^2}\hat{\mathbf{R}}\hat{\mathbf{R}} \tag{28}$$

$$\mathbf{G}_S(\hat{\mathbf{R}},\omega) = \frac{k_S^2}{\varrho\omega^2}(\mathbf{I} - \hat{\mathbf{R}}\hat{\mathbf{R}}) \tag{29}$$

If the data have been produced with an α-mode, we can explicitly account for that adding that index, which yields

$$\mathbf{u}_{s\alpha\beta}^{far} = \frac{e^{jk_\beta R}}{R}\mathbf{C}_{\alpha\beta}(\hat{\mathbf{R}},\omega) \tag{30}$$

where

$$\mathbf{C}_{\alpha\beta}(\hat{\mathbf{R}},\omega) = \frac{1}{4\pi}\int\!\!\int_{S_M} [\mathbf{u}_\alpha\mathbf{n}' : \mathbf{\Sigma}_\beta - (\mathbf{T}_\alpha\cdot\mathbf{n}')\cdot\mathbf{G}_\beta]\, e^{-jk_\beta \hat{\mathbf{R}}\cdot\mathbf{R}'}\, dS' \tag{31}$$

Equ. 30 can then be used as input into the inversion scheme 21.

4. ELASTODYNAMIC BACKPROPAGATION

Similar to the scalar case, we can define a displacement vector holographic field in terms of the elastodynamic Huygens principle introducing complex conjugate Green's tensors, thus accounting for elastodynamic backpropagation. The resulting Porter-Bojarski equation can be integrated with regard to frequency, provided the approximation of physical elastodynamics is introduced, but the subsequent dyadic inversion turns out to be very cumbersome; hence, we prefer the above inversion schemes.

ACKNOWLEDGEMENTS

The test specimen of Fig. 2 has been provided to us by the BASF company in Ludwigshafen, FRG, and the measurements for the turbine shaft image (Fig. 3) were performed in the Fraunhofer Institute for Nondestructive Testing in Saarbrücken, FRG.

References

[1] G.T. Herman, H.K. Tuy, K.J. Langenberg, P. Sabatier. *Basic Methods of Tomography and Inverse Problems*. Adam Hilger, Bristol 1987

[2] K.J. Langenberg. Wave Motion 11 (1989) 99

[3] V. Schmitz, W. Müller, G. Schäfer. In: *Review of Progress in QNDE*, Plenum Press, New York 1986

[4] K.J. Langenberg, M. Berger, T. Kreutter, K. Mayer, V. Schmitz. NDT International 19 (1986) 177

[5] K. Nagai. Proc. IEEE 72 (1984) 748

[6] T. Kreutter, K.J. Langenberg. In: *Review of Progress in QNDE*, Plenum Press, New York 1990

[7] K. Mayer, R. Marklein, K.J. Langenberg, T. Kreutter. Ultrasonics (1989) (to appear)

[8] A.J. Devaney, G. Beylkin. Ultrasonic Imaging 6 (1984) 181

[9] M. Cheney, J.H. Rose. Inverse Problems 4 (1988) 435

[10] K.J. Langenberg. Proc. 12th World Conf. NDT, Elsevier, Amsterdam 1989

[11] K. Mayer, R. Marklein, K.J. Langenberg, T. Kreutter. In: *Review of Progress in QNDE*, Plenum Press, New York 1990

[12] Y.H. Pao, V, Varatharajulu. J. Acoust. Soc. Am. 59 (1976) 1361

[13] N. Bleistein. Wave Motion 11 (1989) 113

[14] J.D. Achenbach. *Wave Propagation in Elastic Solids*, North-Holland, Amsterdam 1973

Part II

Distributed Parameters Inverse Problems

A Non-linear Least-Square Theory for Inverse Problems

G. Chavent

CEREMADE, Université Paris-Dauphine, F-75775 Paris Cédex, France, and
INRIA, Domaine de Voluceau-Rocquencourt, BP 105,
F-78153 Le Chesnay Cédex, France

1 - Q-well posedness of Non Linear Least Square Problems

Because of modeling and measurement errors, most inverse problems can only be solved in the least square sense :

find $\hat{x} \in C$ such that $J(x) = \|\varphi(x)-z\|^2_F$ = Min over C [1.1]

where :

F is an Hilbert space (observation space)

E is a Banach space (parameter space)

$C \subset E$ is a closed convex set (admissible parameters) [1.2]

$\varphi : C \rightarrow F$ is a $\mathcal{C}^2$ mapping, with φ, φ' and φ'' bounded

$z \in F$ is a given point (data to be inverted)

As the only way of solving actually [1.1] is by running an optimization code in order to attempt to minimize J over C, it would be highly desirable that J has no local minima. Hence we introduce the :

Definition 1.1 : The NLLS problem [1.1] is said to be quadratically well posed (Q-wellposed in short) on some open neighborhood $\mathcal{V}$ of $\varphi(C)$ for the $\| \|_E$ norm on C iff :

i) for any $z \in \mathcal{V}$, there exists a unique solution $\hat{x}$ of [1.1]

ii) the $z \rightarrow \hat{x}$ mapping is Lipschitz continuous from $(\mathcal{V}, \| \|_F)$ to $(C, \| \|_E)$

iii) for any $z \in \mathcal{V}$, J has no local minima distinct from $\hat{x}$, and any minimizing sequence converges to $\hat{x}$ for the $\| \|_E$ norm.

We sketch in this paper a sufficient condition for the Q-wellposedness of NLLS problems, based on a geometrical study of the set $\varphi(C)$: we introduce the notion of strictly quasiconvex sets and two sufficient "size × curvature" conditions. We try only to give a very brief overview of the theory in this paper, referring the interested reader to the reference (Chavent 1989a, 1989b, 1989c) for a detailed presentation of the material. We conclude by studying the Q-wellposedness of the regularized problem for the estimation of a distributed diffusion parameter in an elliptic equation.

2 - Strictly Quasiconvex sets

Let $D \subset F$ be given we suppose that D has been equipped with a family $\mathcal{P}$ of pathes which shall play for the set D the role played by the segments for a convex set (in the applications, D will be $\varphi(C)$ and $\mathcal{P}$ the family of the images by φ of the segments of C !) :

Definition 2.1 : A mapping $P : [0,L] \to D$ is a path iff :

$v \to P(v)$ is of class $\mathcal{C}^2$ [2.1]

$\|P'(v)\|_F = 1 \qquad \forall v \in F$ [2.2]

Definition 2.2 : A set $\mathcal{P}$ of pather is a collection of pathes iff :

$\mathcal{P}$ is made of pathes [2.3]

$\mathcal{P}$ is complete, i.e. $\forall X,Y \in D, \exists P \in \mathcal{P}$ such $P(0) = X, P(L) = Y$ [2.4]

$\mathcal{P}$ is stable with respect to restriction [2.5]

Notice that condition [2.2] implies that v is the arc-length along the path P. The notion of quasiconvex sets is relative to a set D equipped with pathes $\mathcal{P}$:

Definition 2.3 : A set $(D,\mathcal{P})$ is quasiconvex iff :

i) $\mathcal{P}$ is a collection of pathes.

ii) There exists a neighborhood $\mathcal{V}$ of D in F, and a l.s.c. function $\varepsilon : \mathcal{V} \to]0,+\infty]$ such that :

$$\begin{matrix} z \in \mathcal{V} \\ 0 < \eta < \varepsilon(z) \end{matrix} \Rightarrow k(z,\eta) < 1 \qquad [2.6]$$

where :

$$k(z,\eta) = \underset{P\in\mathcal{P}(z,\eta)}{\text{Sup}} \ \underset{v\in[0,L]}{\text{Max}} < z-P(v),P''(v) >_F \qquad [2.7]$$

$$\mathcal{P}(z,\eta) = \left\{ P\in \mathcal{P} \mid \|P(j)-z\| \le d(z,D)+\eta \quad j=0,L \right\} \qquad [2.8]$$

Définition 2.4 : A set $(D,\mathcal{P})$ is strictly quasiconvex iff :

i) $(D,\mathcal{P})$ is quasiconvex for some neighborhood $\mathcal{V}$ and function $z \to \varepsilon(z)$.

ii) $\mathcal{V}$ and ε satisfy :

$$\begin{matrix} z \in \mathcal{V} \\ P \in \mathcal{P} \\ d(z,P) < d(z,D)+\varepsilon(z) \end{matrix} \Rightarrow \begin{matrix} \text{the "distantce to z"} \\ \text{function is strictly} \\ \text{quasiconvex along the path P} \end{matrix} \qquad [2.9]$$

The interest of strictly quasiconvex sets comes from the fact that they behave quite like convex sets as far as the projection is concerned :

Theorem 2.1 : (Projection on a closed and strictly quasiconvex set). If :

D is closed in F [2.10]

$(D,\mathcal{P})$ is strictly quasiconvex on a neighborhood $\mathcal{V}$ [2.11]

Then :

The "projection on D" problem is Q-wellposed on $\mathcal{V}$ for the $\mathcal{P}$-arc-length distance $\delta(X,Y)$ on D [2.12]

where, for any two distinct points X and Y of D :

$$\delta(X,Y) = \underset{\substack{P\in\mathcal{P} \\ P(0)=X,\ P(L)=Y}}{\text{Sup}} L \qquad [2.13]$$

This result shows that strict quasiconvexity of set is a useful tool for studying NLLS problems where $D=\varphi(C)$, provided one can "come back" from D to C.
However, the definition of strict-quasiconvexity of sets is quite technical, and give now simpler sufficient conditions.

3 - Size × curvature conditions

We first introduce the notion of global radius of curvature $\rho_G(\nu,\nu')$ of P at ν seen from ν', which is the distance of $P(\nu)$ to the intersection of outer half-spaces normal to P at ν and ν', and which can be easily computed by the :

Lemma 3.1 : Given a path $P\in\mathcal{P}$ and $\nu\neq\nu'\in[0,L]$, one has :

$$\rho_G(\nu,\nu') = \begin{cases} 0 & \text{if } NUM \leq 0 \\ NUM & \text{if } NUM \geq 0 \text{ and } \langle v,v'\rangle \leq 0 \\ NUM/\sqrt{1-\langle v,v'\rangle^2} & \text{if } NUM \geq 0 \text{ and } \langle v,v'\rangle \geq 0 \end{cases} \quad [3.1]$$

where :

$$v = P'(\nu),\ v' = P'(\nu') \quad [3.2]$$

$$NUM = \mathrm{Sgn}(\nu'-\nu) \langle P(\nu')-P(\nu), v' \rangle \quad [3.3]$$

Remember that the usual (local) radius of curvature $\rho(\nu)$ of P at ν is given by :

Lemma 3.2 : Given a path $P\in\mathcal{P}$ and $\nu\in[0,L]$, one has :

$$\rho(\nu) = \|P''(\nu)\|^{-1}_F \quad [3.4]$$

We introduce now the smallest radii of curvature along all pathes of $\mathcal{P}$:

Definition 3.1 : Given a set $(D,\mathcal{P})$ we define :

$$R_G(D) = \inf_{P\in\mathcal{P}} \inf_{\nu,\nu'\in[0,L]} \rho_G(\nu,\nu') \quad \text{(smallest global radius of curvature)} \quad [3.5]$$

$$R(D) = \inf_{P\in\mathcal{P}} \inf_{\nu\in[0,L]} \rho(\nu) \quad \text{(smallest radius of curvature)} \quad [3.6]$$

which satisfy necessarily :

$$0 \leq R_G(D) \leq R(D) \quad [3.7]$$

The interest of the notion of global radius of curvature comes from the :

Proposition 3.3 : Let $(D,\mathcal{P})$ be given :

If :

$$R_G(D) > 0 \qquad (R_G\text{ - size × curvature condition}) \quad [3.8]$$

Then :

$(D,\mathcal{P})$ is strictly quasiconvex, with $\mathcal{V}$ and ε given by :

$$\mathcal{V} = \{z\in F \mid d(z,D) < R_G(D)\} \quad [3.9]$$

$$\forall z\in\mathcal{V},\ \varepsilon(z) = R_G(D) - d(z,D) > 0 \quad [3.10]$$

The above sufficient condition is sharp in the sense that it recognizes exactly all strictly quasiconvex sets made of a curve.

Formula [3.1] for the global radius of curvature are easy to compute numerically, but difficult to use for paper-and-pencil calculations. Hence we introduce now a weaker sufficient condition by looking for a lower bound to $R_G(D)$.

Let us introduce the deflection $\theta(\nu,\nu')$ of the path P between ν and ν', which is simply the angle (in $[0,\pi]$) between the tangents to P at ν and ν', and the arc-length $\delta(\nu,\nu')$ along P between ν and ν'.

Definition 3.2 : Given a set $(D,\mathcal{P})$ we define :

$$\Theta(D) = \sup_{P\in\mathcal{P}} \sup_{\nu,\nu'\in[0,L]} \theta(\nu,\nu') \qquad \text{(largest deflection)} \qquad [3.11]$$

$$\Delta(D) = \sup_{P\in\mathcal{P}} \sup_{\nu,\nu'\in[0,L]} \delta(\nu,\nu') \qquad \text{(largest path-length)} \qquad [3.12]$$

Then the following proposition gives bounds to $\Theta(D)$ and $R_G(D)$:

Proposition 3.4 : Let $(D,\mathcal{P})$ be given :

Then an upper bound to $\Theta(D)$ is given by :

$$\Theta(D) \le \sup_{P\in\mathcal{P}} \int_0^L \frac{d\nu}{\rho(\nu)} \qquad [3.13]$$

and a lower bound to $R_G(D)$ is given by :

$$R_G(D) \ge R_G \qquad [3.14]$$

where :

$$R_G = \begin{cases} R & \text{if } 0 \le \Theta \le \pi/2 \\ R\sin\Theta+(\Delta-R\Theta)\cos\Theta & \text{if } \pi/2 \le \Theta \le \pi \end{cases} \qquad [3.15]$$

where R is any (positive) lower bound to R(D), and Θ and Δ are any upper bound to $\Theta(D)$ and $\Delta(D)$.

As a corollary of propositions 3.3 and 3.4 we have the :

Proposition 3.5 : Let $(D,\mathcal{P})$ be given, and let R be a lower bound to R(D), and Θ,Δ be upper bounds to $\Theta(D)$ and $\Delta(D)$. If :

$$0 \le \Theta \le \pi/2 \qquad [3.16]$$

or :

$$\pi/2 \le \Theta \le \pi \qquad \text{and } \Delta/R \le \Theta-\tan\Theta \qquad [3.17]$$

Then :

$(D,\mathcal{P})$ is strictly quasiconvex, with $\mathcal{V}$ and ε given by :

$$\mathcal{V} = \{z\in F / d(z,D) < R_G\} \qquad [3.18]$$

$$\forall z\in \mathcal{V},\ \varepsilon(z) = R_G\text{-}d(z,D) > 0 \qquad [3.19]$$

where R_G is given by [3.15].

Notice that, using [3.13], all quantities Θ,Δ,R can be computed only in terms of the radius of curvature $\rho(\nu)$ and the arc-length ν along a path P.

4 - Application to Non-Linear Least Squares

Coming back to the non-linear least square problem [1.1], we notice first, as φ is of class $\mathcal{C}^2$, that there exists β such that :

$$\|\varphi''(x)(y,y)\|_F \leq \beta\|y\|^2 \qquad \forall x\in C,\ \forall y\in E, \tag{4.1}$$

and we make the additional hypothesis that the derivative $\varphi'(x)$ is uniformly invertible, i.e. that there exists α_m and α_M such that :

$$\alpha_m\|y\|_E \leq \|\varphi'(x).y\|_F \leq \alpha_M\|y\|_E \qquad \forall x\in C,\ \forall y\in E, \tag{4.2}$$

We may then apply the results of paragraph 3 to the set $D=\varphi(C)$ equipped with the pathes :

$$\mathcal{P} = \{\varphi([x,y]),\ x,y \in C\} \tag{4.3}$$

One checks easily (cf.[Chavent 1989c]) that, given a path $P=\varphi([x,y])$, $x,y\in C$, one has :

$$v(t) = \int_0^t \|\varphi'(x_t)(y-x)\|_F\, dt \geq t\,\alpha_m\|x-y\|_E \tag{4.5}$$

$$\rho(v) \geq \frac{\|\varphi'(x_t)(y-x)\|_F^2}{\|\varphi''(x_t)(y-x,y-x)\|_F} \geq \frac{\alpha_m^2}{\beta} \tag{4.6}$$

where :

$$x_t = (1-t)x + ty \qquad \forall t\in [0,1] \tag{4.7}$$

From [4.5],[4.6] we find that :

$$\int_0^L \frac{dv}{\rho(v)} \leq \int_0^1 \frac{\|\varphi''(x_t)(y-x,y-x)\|_F}{\|\varphi'(x_t)(y-x)\|_F}\, dt \leq \frac{\beta}{\alpha_m}\|y-x\|_E \tag{4.8}$$

These formula give us the following lower and upper bounds for the attributes of the pathes of $D=\varphi(C)$:

$$\begin{aligned} &R = \alpha_m^2/\beta && \text{(lower bound to radii of curvature)}\\ &\Delta = \alpha_m \operatorname{diam} C && \text{(upper bound to path length)}\\ &\Theta = (\beta/\alpha_m) \operatorname{diam} C && \text{(upper bound to path deflection)} \end{aligned} \tag{4.9}$$

We deduce then from Proposition 3.5 and Theorem 2.1 the :

Proposition 4.1 : Let hypothesis [1.2],[4.1] and [4.2] hold, and R,Δ,Θ and R_G be defined by [4.9] and [3.15]. If :

$$\operatorname{diam} C \leq \frac{\pi}{2}\,(\alpha_m/\beta) \tag{4.10}$$

or :

$$\operatorname{diam} C \leq \pi\ (\alpha_m/\beta) \quad \text{and} \quad (\alpha_M/\alpha_m-1)\Theta < |\tan \Theta| \tag{4.11}$$

Then :

The Non-Linear Least square problem [1.1] is Q-well-posed on the neighborhood $\mathcal{V}$ of $\varphi(C)$ defined in [3.18] for the $\|\ \|_E$ norm on C, and the following stability estimate holds :

$$\left.\begin{array}{l}\alpha_m\|\hat{x}_1-\hat{x}_2\| \le \left(1-\frac{d}{R}\right)^{-1}\|z_1-z_2\| \\ \text{as soon as} \\ \|z_1-z_2\| + \max_{j=1,2} d(z_j,\varphi(C)) \le d < R_G\end{array}\right\} \qquad [4.12]$$

This result can be used (see reference [Chavent 1989c]) to investigate the wellposedness of regularized problems, in particular in order to check if the regularizing parameter has been taken large enough to compensate for the non-linearity of φ. We illustrate the kind of result which can be obtained along this line by a simple exemple :

Consider the one dimensional elliptic PDE :

$$\left.\begin{array}{l}-\frac{d}{dx}\left(a(u)\frac{du}{dx}\right) = f \quad \text{in } \Omega =]0,1[\\ u(0)=0,\ a\frac{du}{dx}(1)=0\end{array}\right\} \qquad [4.13]$$

in which one tries to estimate the diffusion coefficient a from a measurement z of u. Hence we define, given $0 < \underline{a} \le \bar{a}$, $a_0 = (\underline{a} + \bar{a})/2$, and $M \ge 0$:

$$\begin{array}{l}E = H^1(0,1) \\ C = \{a \in H^1(01) | \underline{a} \le a(x) \le \bar{a} \text{ a.e., } \|a-a_0\|_{H^1(0.1)} \le M\} \\ F = L^2(01) \\ \varphi : a \in C \to u = \text{solution of [4.13]} \\ z \in L^2(01) = \text{given observation of u}\end{array} \qquad [4.14]$$

The inversion of φ being known to be unstable, we consider only the regularized problem :

find $\hat{a}_\varepsilon \in C$ such that $J_\varepsilon(a)$ = min over C [4.15]

where :

$$J_\varepsilon(a) = \|\varphi(a)-z\|^2_{L^2(01)} + \varepsilon^2 \|a-a_0\|^2_{H^1(01)} \qquad [4.16]$$

Then using proposition 4.1 adapted to regularized problems (namely proposition 5.1 of [Chavent 1989c]) yields immediately the following results :

<u>Proposition 4.2</u> : Let hypothesis [4.14] hold. If :

$$\varepsilon > \frac{4M\ \|f\|_{L^2(01)}}{\pi\ \underline{a}^3} \qquad [4.17]$$

Then :

The regularized parameter estimation problem [4.15], [4.16] is Q-well posed on the neighborhood :

$$\mathcal{V} = \left\{z \in L^2(01) \mid d(z,\varphi(C)) < \varepsilon\sqrt{\varepsilon^2\underline{a}^6 - M^2}\right\} \qquad [4.18]$$

for the $H^1(0,1)$ norm on the parameter set C.

References

Guy Chavent 1989 a. "Quasiconvex sets and Size × Curvature Condition ; Application to Non-Linear Least Square Problems", INRIA Report 1017, April 89, Submitted to Jamo.

Guy Chavent 1989b. "New Size × Curvature Conditions for Strict Quasiconvexity of Sets", INRIA Report 1020, November 1989.

Guy Chavent 1989c. "A New Sufficient Condition for the Wellposedness of Non Linear Least Square Problems arising in Identification and Control", INRIA Report 1121, November 1989, 9th Int. Conf. on Analysis and Optimization of Systems, Antibes, June 12-15, 1990.

Circulant Preconditioners for Non-negatively Constrained Deconvolution Problems

A.R. Davies

Department of Mathematics, University College of Wales,
Aberystwyth SY23 3BZ, UK

1. Introduction

Algorithms for computing optimally regularized solutions of first kind integral equations under non-negativity constraints necessarily require far more computer time than algorithms which do not implement such constraints, particularly in higher dimensions. Whenever the structure of the kernel of the integral equation allows, therefore, it is important to take full advantage of special properties to maximize efficiency. In this paper we shall present some ideas for improving efficiency in the case of non-negatively constrained deconvolution problems. The algorithm we propose to modify is originally due to Wahba [1], and uses quadratic programming to achieve constrained Tikhonov regularization while using cross-validation to optimize the regularization parameter.

Although the use of cross-validation to determine the regularization parameter in unconstrained problems is widely discussed in the literature, it will be useful to review the pertinent details in the context of Fourier methods for deconvolution so that we may understand more readily the nature of the modifications proposed for the non-negatively constrained case. For simplicity we confine our discussion to one space dimension, although our modifications become far more cost-effective in higher dimensions.

2.Tikhonov regularization and cross-validation for unconstrained deconvolution

In many physical deconvolution problems the supports of the kernel (point spread function), the data function (image) and solution (object), are compact or essentially compact. This means that even if the interval over which the convolution takes place is infinite, the problem may be truncated to a finite interval without significant error. Moreover, a finite interval can always be chosen which is consistent with an artifical periodic continuation of the functions involved, thereby allowing the convenient properties of discrete Fourier series to be exploited fully without aliasing.

Let us assume a scaling of the space variable so that the appropriate finite interval is [0,1]. Thus we may consider the Fredholm convolution integral equation

$$(Kf)(x) \equiv \int_0^1 k(x-y)f(y)dy = g(x), \qquad 0 \leqslant x \leqslant 1, \tag{2.1}$$

where the kernel k, data function g and solution f are periodic with period 1, and g is measured discretely with white noise. Let the sampled data be

$$y_n = g(x_n) + e_n, \quad n = 0, \ldots, N-1, \tag{2.2}$$

with uniform sampling points $x_n = n/N$.

For convenience let N be even and $M = N/2$. The discrete Fourier transform (DFT) of a real N-vector $\mathbf{u}_N = (u_0, \ldots, u_{N-1})^T$ is given by

$$\hat{\mathbf{u}}_N = N^{-1}\Psi^H \mathbf{u}_N \tag{2.3}$$

where Ψ is the NxN DFT matrix with elements $\psi_{rs} = \exp(2\pi i \, rs/N)$, $r,s = 0, \ldots, N-1$. The inverse DFT is then

$$\mathbf{u}_N = \Psi \hat{\mathbf{u}}_N, \tag{2.4}$$

and has a real-valued trigonometric interpolant

$$u_N(x) = \sum_{q=-M}^{M}{}'' \hat{u}_{N,q} \exp(i\,\omega_q x), \quad 0 \leqslant x \leqslant 1, \qquad \omega_q = 2\pi q, \tag{2.5}$$

where $\hat{u}_{N,q}$ denote the elements of $\hat{\mathbf{u}}_N$ and have the cyclic property

$$\hat{u}_{N,-q} = \hat{u}_{N,N-q} = \overline{\hat{u}_{N,q}}, \quad 0 \leqslant q \leqslant M. \tag{2.6}$$

It is clear that the most convenient space for calculating the regularized solutions of (2.1) is T_M, the space of real-valued trigonometric polynomials of degree at most M and period 1. Under pth order Tikhonov regularization we therefore seek

$$\min_{f \in T_M} \left[N^{-1} \sum_{n=0}^{N-1} [(Kf)(x_n) - y_n]^2 + \lambda \| f^{(p)} \|_{L^2} \right] \tag{2.7}$$

where $f^{(p)}$ denotes the pth derivative of f and λ denotes the regularization parameter, which is to be chosen in advance. (For a discussion of methods for choosing λ optimally see, for example, [2], [3].) Equivalently, (2.7) may be written

$$\min_{f \in \mathbb{C}^N} \left[| \hat{K}_N \hat{f}_N - \hat{y}_N |^2 + \lambda \hat{f}_N^H J \hat{f}_N \right] \tag{2.8}$$

where

$$\hat{K}_N = \operatorname{diag}(\hat{k}_0, \hat{k}_1, \ldots, \hat{k}_M, \hat{k}_{-M+1}, \ldots, \hat{k}_{-1}), \tag{2.9}$$

$$J = \operatorname{diag}(0, \omega_1^{2p}, \ldots, \omega_M^{2p}, \omega_{M-1}^{2p}, \ldots, \omega_1^{2p}), \tag{2.10}$$

and $\hat{k}_q$ in (2.9) denote the *exact* Fourier coefficients of k , i.e.

$$\hat{k}_q = \int_0^1 k(x)\exp(-i\,\omega_q x)dx.$$

It is easily shown that the minimizer of (2.7) is

$$f_{N;\alpha}(x) = \sum_{q=-M}^{M}{}'' \hat{f}_{N,q;\alpha}\exp(i\,\omega_q x)\,, \qquad 0 \leqslant x \leqslant 1, \tag{2.11}$$

which depends on the parameter set $\alpha = (p,\lambda)$.The coefficients $\hat{f}_{N,q;\alpha}$ are the elements of

$$\hat{\mathbf{f}}_{N;\alpha} = \hat{A}(\alpha)\hat{K}_N\hat{\mathbf{y}}_N \tag{2.12}$$

where

$$\hat{A}(\alpha) = \mathrm{diag}(1, Z_{1,\alpha}, \ldots, Z_{M,\alpha}, Z_{M-1,\alpha} \ldots, Z_{1,\alpha})$$

and

$$Z_{q,\alpha} = (1 + \lambda\omega_q/|k_q|^2)^{-1}, \qquad -M \leqslant q \leqslant M. \tag{2.13}$$

The numbers $Z_{q;\alpha}$ constitute a discrete *regularization filter* depending on α. It is this filter which effects the stabilization of the inversion process.

Finally, from $f_{N;\alpha}$ we may define the predictive data function $g_{N;\alpha} = Kf_{N;\alpha}$. This is interpolated by the vector

$$\mathbf{g}_{N;\alpha} = A(\alpha)\mathbf{y}_N \tag{2.14}$$

where

$$A(\alpha) = N^{-1}\Psi A(\alpha)\Psi^H \tag{2.15}$$

is known as the *influence matrix* of the regularized problem.

We have stated earlier that λ is to be chosen by the method of cross-validation. In fact, p may also be chosen this way. It is sufficient to minimize with respect to α the cross-validation function

$$V_{CV}(\alpha) = \frac{N^{-1}\|(I - A(\alpha))\mathbf{y}_N\|^2}{[N^{-1}\mathrm{Trace}(I - A(\alpha))]^2}\,. \tag{2.16}$$

Usually it is more convenient to minimize $V_{CV}(\alpha)$ with respect to λ for several choices of p and then choose the pair α corresponding to the smallest minimizing value of $V_{CV}(\alpha)$.

We note from (2.16) that function evaluations of $V_{CV}(\alpha)$ are extremely easy in this unconstrained case of deconvolution. This is due to the simple fact that $A(\alpha)$ is circulant, and so (2.16) may be evaluated efficiently using FFTs of length N. In the next section we see that this structure is partly or totally lost when non-negativity constraints are introduced.

3. Non-negatively constrained deconvolution

Wahba [1] has shown how to modify the above approach when (2.7) is minimized subject to the non-negativity constraint $\mathbf{f}_N \geqslant \mathbf{0}$, where $\mathbf{f}_N = (f(x_0), \ldots, f(x_{N-1}))^T$. For given α there is a unique minimizer $\mathbf{f}_{N;\alpha}$. Let I denote the set of indices $\{n_0, \ldots, n_{L-1}\}$ corresponding to inactive constraints, i.e. $f_{N;\alpha}(x_n) > 0$, j = 0, ..., L-1. Furthermore, let E be the NxL indicator matrix associated with I, having a unit element in the kth row and jth column if $k = n_j$, j = 0, ..., L-1, and consisting of zeros elsewhere. Thus, given any NxN matrix A, the matrix E^HAE is the LxL matrix obtained when rows $k \neq n_j$ are removed from A together with the corresponding columns. Similarly, given any LxL matrix A, the matrix EAE^H is the NxN matrix obtained when rows and columns of zeros are introduced in positions $k \neq n_j$.

Following Wahba it is possible to show that the constrained minimizer of (2.7) interpolates the vector

$$\mathbf{f}_{N;\alpha} = E(E^H\Psi K^H K \Psi^H E + \lambda E^H \Psi J \Psi^H E)^{-1} E^H \Psi K^H \Psi^H \mathbf{y}_N \tag{3.1}$$

and that there is an NxN influence matrix satisfying (2.14) but with (2.15) replaced by

$$A(\alpha) = \Psi K_N \Psi^H E(\Sigma_K + \lambda \Sigma_J)^{-1} E^H \Psi K^H \Psi^H, \tag{3.2}$$

where

and

$$\left.\begin{aligned} \Sigma_K &= E^H \Psi K^H \Psi^H E, \\ \Sigma_J &= E^H \Psi J \Psi^H E. \end{aligned}\right] \tag{3.3}$$

Although the influence matrix is no longer circulant it nevertheless has the simplifying property

$$\mathrm{Trace}(I - A(\alpha)) = N - L + \lambda \mathrm{Trace} B \tag{3.4}$$

where B is the LxL matrix satisfying

$$(\Sigma_J + \lambda \Sigma_K) B^T = \Sigma_J. \tag{3.5}$$

Wahba argues that the cross-validation function is correspondingly modified to

$$V_{CV}(\alpha) = \frac{\| \hat{\hat{K}}\mathbf{f}_{N;\alpha} - \hat{\mathbf{y}}_N \|^2}{[N^{-1}(N - L + \lambda \mathrm{Trace} B)]^2}. \tag{3.6}$$

Unfortunately, (3.6) is no longer a simple function of α to minimize. B must first be found by solving the LxL systems (3.5). Thus TraceB depends in a complicated way on the indicator matrix E which is itself implicitly dependent on α. In fact, for fixed p, (3.6) is not even a continuous function of λ, in general.

For fixed p, Wahba suggests the following algorithm for determining $f_{N;\alpha}$ iteratively.

Step 0. Guess a λ value. (Possibly the unconstrained λ).

Step 1. Compute $f_{N;\alpha}$ and its inactive set I using a quadratic programming (QP) algorithm. (See, for example, Chapter 10 of [4] for a good description of QP algorithms).

Step 2. Set up the matrices Σ_K and Σ_J and solve (3.5) for B. Exploiting the symmetry of the matrices means that at $O(L^3)$ operations are required.

Step 3. Compute $V_{CV}(\alpha)$.

Step 4. Adjust the value of λ, if necessary, and return to Step 1. Otherwise accept $f_{N;\alpha}$ as optimal.

This is, of course, a time-consuming procedure. One way of improving its efficiency is to calculate or estimate TraceB without solving the systems (3.5). This would be extremely difficult without the aid of structure in the Σ matrices. We examine the nature of this structure and exploit it in the next section.

4. Circulant Matrices and Preconditioners

First we note that the matrices $\Psi\hat{K}_N^H\hat{K}_N\Psi^H$ and $\Psi J\Psi^H$ are real, symmetric, and circulant. When rows and columns are removed, however, the circulant structure is, in general, destroyed. From (3.3) we see that the Σ matrices are real and symmetric. However, if the index set I consists of strings of integers in arithmetic order, i.e. $n_{j+1} = n_j + 1$, which start with 0 or end in N-1, then the Σ matrices remain circulant (of length L). This will be the case when the active constraint $f(x) = 0$ occurs in end intervals of the form [0,a] and [b,1], while $f(x) > 0$ in (a,b). This occurs frequently in practice and so is an important special case.

In this case, TraceB may be computed directly by FFTs of length L. The diagonalization of Σ may be achieved in as little as O(LlogL) operations if the calculations are done in parallel. Writing

$$\Sigma_K = L^{-1}\Psi_L D_K \Psi_L^H, \quad \Sigma_J = L^{-1}\Psi_L D_L \Psi_L^H, \tag{4.1}$$

where the diagonal matrices D contain the eigenvalues of Σ, it requires only a further O(L) operations to calculate

$$\text{TraceB} = \text{Trace}(D_K + \lambda D_J)^{-1} D_J. \tag{4.2}$$

More generally, provided I consists of no more than a few disjoint strings in arithmetical order, the structure of the Σ matrices remains *almost* circulant. The diagonal in each case consists of a single repeated element, while off-diagonals consist of strings of repeated elements, the number of different elements in each off-diagonal depending on the number of disjoint strings in I. This suggests the splitting

$$\Sigma = C - R,$$

where C is a circulant matrix of length L generated from the L most common elements in the central bands of Σ. (Other sensible generating rules for a circulant C matrix are, of course, possible). R then consists mainly of zeros and a few off-diagonal strings of repeated elements representing differences of elements of Σ. The crucial point is that replacing Σ_K and Σ_J in (4.1) by C_K and C_J, respectively, an estimate of TraceB is still given by (4.2). But how good is the estimate?

Clearly, the matrix $(C_K + \lambda C_J)^{-1}$ is an obvious (circulant) preconditioner for the systems (3.5). A first correction to formula (3.8) for TraceB is then given by

$$\begin{aligned} &\mathrm{Trace}(C_K + \lambda C_J)^{-1}[(R_K + \lambda R_J)(C_K + \lambda C_J)^{-1}C_J - R_J] \\ &= L^{-1}\mathrm{Trace}\{(D_K + \lambda D_J)^{-1}[(P_K + \lambda P_J)(D_K + \lambda D_J)^{-1}D_J - P_J]\} \end{aligned} \qquad (4.3)$$

where the diagonal matrices P represent the diagonal parts of $\Psi_L R \Psi_L^H$, and again the diagonal matrices D contain the eigenvalues of C. With care, (4.3) may also be calculated in no more than a further O(L) operations once given the D matrices. Higher corrections, of course, would be more expensive.

In each of a variety of numerical deconvolution experiments, we have found that provided the index set I does not contain more than about four disjoint strings of successive integers, the use of the estimate (4.2) in the cross-validation function (3.6) proves most effective in determining an optimal regularization parameter. Inclusion of the first correction (4.3) in (3.6) made a negligible contribution in determining the minimizer. Of course, there may well be undiscovered pathological cases where this approach breaks down, but we are hopeful that modifications of similar type to those we have described will prove robust in cases where the active constraints hold mainly (although not entirely) in end regions.

In principle, further efficiencies could be effected by inclusion of similar modifications in the quadratic programming procedure of Step 1, but we have not yet explored such a possibility.

Finally, in cases where the active constraints occupy several disjoint subintervals within the region, so that I consists of several disjoint strings, it is possibly worth replacing direct elimination methods for solving the systems (3.5) by a few cycles of a preconditioned conjugate gradient algorithm, again using a circulant preconditioner. Ideas contained in [5] may prove useful here, but there is considerable scope for further research in this area.

Acknowledgement

I am grateful to Dr. M. Iqbal for his numerical experiments on earlier aspects of this work [6].

References

1. G. Wahba. Constrained regularization for ill-posed operator equations, with applications in meteorology and medicine. In *Ststistical Decision Theory and Related Topics III,* Vol. 2, eds. S.S. Gupta and J. O. Bergen (Academic Press, New York) 1982.

2. A. R. Davies and R. S. Anderssen. J. Aust. Math. Soc. Ser. B, **28** (1986) 114-133.

3. A. R. Davies and M. F. Hassan. Optimality in the regularization of ill-posed inverse problems. In *Inverse Problems: An Interdisciplinary Study,* ed. P. C. Sabatier (Academic Press, London) 1987, pp 553-562.

4. R. Fletcher. Practical Methods of Optimization, Vol. 2. John Wiley, Chichester. 1981.

5. R. H. Chan and G. Strang. The asymptotic Toeplitz-Circulant eigenvalue problem. Research Report, CMA-R21-87. Centre for Mathematical Analysis, Australian National University.

6. M. Iqbal. The Numerical Solution of Fredholm Integral Equations of the First Kind. Ph.D. Thesis, University of Wales. 1989.

Tikhonov Regularization of Nonlinear Differential-Algebraic Equations

H.W. Engl[1], *M. Hanke*[2], *and A. Neubauer*[1]

[1]Institut für Mathematik, Johannes-Kepler-Universität, A-4040 Linz, Austria
[2]Sektion Mathematik, Humboldt-Universität,
DDR-1086 Berlin, German Democratic Republic

Abstract: Using recent results about Tikhonov regularization of nonlinear ill–posed problems, we show that nonlinear index–2 differential–algebraic equations can be solved in a stable way via Tikhonov regularization in various ways. Convergence rates for the regularized solutions are derived.

1. Introduction, Abstract Results

In the last years there has been much activity in the numerical solution of differential–algebraic equations. This term has been coined to comprise differential equations with algebraic constraints and singular implicit differential equations. In the present paper we deal with semi–explicit systems

$$\begin{aligned} u' - h(u,v,\cdot) &= q_1 \\ g(u,\cdot) &= q_2 \end{aligned} \tag{1.1}$$

subject to boundary conditions

$$r\big(u(0),u(1)\big) = q_3. \tag{1.2}$$

Here, $u(t) \in \mathbb{R}^m$, $v(t) \in \mathbb{R}^n$, and $h : \mathbb{R}^m \times \mathbb{R}^n \times [0,1] \to \mathbb{R}^m$, $g : \mathbb{R}^m \times [0,1] \to \mathbb{R}^n$, $r : \mathbb{R}^m \times \mathbb{R}^m \to \mathbb{R}^l$ are sufficiently smooth functions of their arguments. In (1.1) and below, we mean by $h(u,v,\cdot) := h\big(u(\cdot),v(\cdot),\cdot\big)$ etc. We assume that (1.1)–(1.2) has a solution $(u_0,v_0) \in H^1(0,1)^m \times L^2(0,1)^n$, where $H^k(0,1)$ denote the usual Sobolev spaces (cf. [1]). Adopting the convention to denote partial derivatives by subscripts we introduce the notation

$$h_{u_0}(t) := h_u\big(u_0(t),v_0(t),t\big), h_{v_0}(t) := h_v\big(u_0(t),v_0(t),t\big), g_{u_0}(t) := g_u\big(u_0(t),t\big),$$
$$r_{00} = r_{u(0)}\big(u_0(0),u_0(1)\big), r_{10} = r_{u(1)}\big(u_0(0),u_0(1)\big).$$

Assumption (I): For all $t \in [0,1]$, $g_{u_0}(t)h_{v_0}(t)$ is nonsingular.

(I) assures that (1.1) is a so–called index–2 system at least locally around (u_0,v_0) ([7], [14]). Equations of this type arise naturally in many applications, e.g., dynamical systems subject to balance invariants and constrained mechanical systems of rigid bodies (cf. [5], [14], [6]).

Since the second equation of (1.1) already includes boundary conditions, namely $g\big(u(0),0\big) = q_2(0)$, the appropriate number l of additional boundary conditions (1.2) is less than m. Note that Assumption (I) implies that $m \geq n$. Therefore, we require

Assumption (BC): Let $l = m - n$ and assume that the homogeneous equation

$$\begin{aligned} u' - h_{u_0}u - h_{v_0}v &= 0, \\ g_{u_0}u &= 0, \\ r_{00}u(0) + r_{10}u(1) &= 0 \end{aligned}$$

has only the trivial solution.

Note that the equation in (BC) is the (homogeneous) linearization of (1.1)–(1.2) in (u_0, v_0); differentiation of its second equation and of the second equation in (1.1) with respect to t lead to a linear and nonlinear index–1 differential–algebraic equation, where the first one is the linearization of the latter. The Implicit Function Theorem, applied to these index–1 problems, yields that if (BC) holds, then (1.1)–(1.2) has no further solution in a neighbourhood of (u_0, v_0). Throughout this paper we assume that (I) and (BC) hold.

In the linear case it is known that under Assumption (I), (1.1) represents an ill–posed problem, i.e., the solution (u, v) does not depend continuously on the data (q_1, q_2, q_3) ([11], [12], [10]). This is due to the fact that (1.1) contains differentiation problems as a special case. This ill–posedness is also preserved in the nonlinear case: It can easily be seen from (1.1) that the solution depends not only on q_1, q_2, q_3, but also on q_2' if Assumption (I) is fulfilled. Thus, the need for some kind of regularization arises. See [21], [16], [17], [8], [3] for various aspects of regularization methods for solving ill–posed problems.

In the linear case of (1.1)–(1.2), a regularization is constructed in [11], [12] via a singularly perturbed boundary value problem. Moreover, this regularization has properties which are very similar to the well–known Tikhonov regularization methods. This motivates the use of Tikhonov regularization for the nonlinear problem (1.1)–(1.2): Using recent results on Tikhonov regularization of nonlinear problems ([4], [19], [20]) we will derive conditions under which convergence of regularized solutions of (1.1)–(1.2) is assured and convergence rates with respect to the noise level in the data q_1, q_2, q_3 are obtained.

An alternative way to treat the boundary value problem numerically is the direct application of finite difference methods, e.g., backward differentiation formulae ([14], [2], [15]) or Runge–Kutta methods ([9]). The convergence of these methods can be shown but they may become unstable. In view of the ill-posedness of the problem this fact is not surprising. Fortunately, this instability is not very severe, and a very careful design of error control and stepsize selection can overcome this difficulty to some extent; this stepsize selection can also be interpreted as a regularization method.

Now, we describe Tikhonov regularization for a nonlinear problem in the abstract setting of [4]: Consider the equation

$$F(x) = y_0 \tag{1.3}$$

where $F : D(F) \subseteq X \to Y$ is a nonlinear operator between Hilbert spaces X and Y. As notion of a solution of problem (1.3) we choose the concept of an "x^*–minimum–norm solution x_0", i.e.,

$$F(x_0) = y_0 \quad \text{and} \quad \|x_0 - x^*\| = \min\{\|x - x^*\| \mid F(x) = y_o\}.$$

We will assume that an x^*–minimum–norm solution exists for the data y_0. The choice of x^* is of course crucial: Available a–priori information about the location of the solutions of (1.3) has to enter into the selection of x^*. In the case of multiple solutions, x^* plays the role of a selection criterion: By the choice of x^*, we can influence which solution we want to approximate.

In Tikhonov regularization, a solution of problem (1.3) is approximated by solutions of the minimization problem

$$\min \|F(x) - y_0\|^2 + \alpha\|x - x^*\|^2 \quad \text{over} \quad D(F), \tag{1.4}$$

where $\alpha > 0$ is a small parameter. Instead of the exact right–hand side y_0 of (1.3), one usually has only approximate data y_δ and an estimate $\|y_\delta - y_0\| \leq \delta$ at hand. Moreover, since (1.4) is a nonlinear optimization problem, we cannot expect to be able to solve it exactly (with y_0 replaced by y_δ). Thus, we assume $x_\alpha^{\delta,\eta}$ to be a solution of the problem ($\alpha > 0$, $\eta \geq 0$)

$$\|F(x_\alpha^{\delta,\eta}) - y_\delta\|^2 + \alpha\|x_\alpha^{\delta,\eta} - x^*\|^2 \leq \|F(x) - y_\delta\|^2 + \alpha\|x - x^*\|^2 + \eta \quad \text{for all} \quad x \in D(F). \tag{1.5}$$

In the following we assume that F is continuous and weakly (sequentially) closed. In this case, (1.5) has a solution even for $\eta = 0$.

In [4], it is shown that for $\alpha > 0$, the set of minimizers $\{x_\alpha^{\delta,\eta}\}$ depends continuously on y_δ in an appropriate set–valued sense, so that the solution of (1.5) really stabilizes the ill–posed problem (1.3). Also, convergence of $\{x_\alpha^{\delta,\eta}\}$ to x^*–minimum–norm solutions with $\alpha \to 0$ is shown there under the conditions that $\frac{\delta^2}{\alpha} \to 0$ and $\frac{\eta}{\alpha} \to 0$. Moreover, in [4], [19] the following convergence rates results are proved:

Theorem 1:
Let F be weakly closed and $D(F)$ be convex, let $y_\delta \in Y$ with $\|y_\delta - y_0\| \leq \delta$, let x_0 be an x^*–minimum–norm solution of (1.3), and $x_\alpha^{\delta,\eta}$ be defined by (1.5).

a) *Let the following conditions hold:*
   (i) *F is Fréchet–differentiable,*
   (ii) *there exists an $L \geq 0$ such that $\|F'(x_0) - F'(z)\| \leq L\|x_o - z\|$ for all $z \in D(F)$,*
   (iii) *there exists an $\omega \in Y$ satisfying $x_0 - x^* = F'(x_0)^*\omega$,*
   (iv) *$L\|\omega\| < 1$.*

   Then for the choices $\alpha \sim \delta$ and $\eta = O(\delta^2)$, we obtain

$$\|x_\alpha^{\delta,\eta} - x_0\| = O(\delta^{1/2}).$$

b) Let the hypotheses of part a) hold and assume, moreover that

(v) $x_0 \in int\ D(F)$,

(vi) there exists $\epsilon > 0$ *such that* $\|F'(x) - F'(z)\| \leq L\|x - z\|$ *for all* $x, z \in D(F) \cap U_\epsilon(x_0)$,

(vii) $x_o - x^* \in R(F'(x_0)^* F'(x_0))$,

(viii) $L\|\omega\| < \frac{27}{10+7\sqrt{7}}$, *where* $\omega \in N(F'(x_0)^*)^\perp$ *is such that* $x_0 - x^* = F'(x_0)^*\omega$.

Then for the choices $\alpha \sim \delta^{2/3}$ *and* $\eta = O(\delta^2)$, *we obtain*

$$\|x_\alpha^{\delta,\eta} - x_0\| = O(\delta^{2/3}).$$

Note that some conditions of Theorem 1 can be weakened ([4], [19]).

In order to apply Theorem 1 to (1.1)–(1.2), there are various possibilities to define the operator F and the underlying Hilbert spaces X and Y. We consider three different approaches. The use of different formulations of (1.1)–(1.2) as an operator equation leads to different regularization algorithms. By changing X, we obtain different smoothing terms $\alpha\|x - x^*\|^2$ in (1.4) and convergence of $x_\alpha^{\delta,\eta}$ towards x_0 with respect to different norms. On the other hand, changing Y leads to different error norms $\|y_\delta - y_0\|$ indicating in which norm the data noise is considered small. In the first approach we use (1.1) "as is". In the second approach, we replace the first equation of (1.1) by its integrated form. The third approach is based on the fact that the ill-posedness of (1.1)–(1.2) is essentially due to the second equation of (1.1). Therefore, F will be given by this second equation while the first one is used to determine u as a function of v; this approach allows only errors in q_2.

For the sake of simplicity we assume F to be defined on the whole space X in the following. Obviously, it is easy to give local versions of our results. Moreover, assuming all functions involved to be sufficiently smooth we discuss mainly the assumptions (iii), (iv) and (vii), (viii), respectively, of Theorem 1.

Note that [19] contains also results about the realization of (1.5) in finite–dimensional spaces. These results could also be applied to our setting.

We denote by $H^k(0,1)$ the usual Sobolev spaces ([1]) with the norms $\|\cdot\|_k$. $\|\cdot\|$ and $|\cdot|$ will denote the usual Hilbert space norms in $L^2(0,1)$ and $\mathbb{R}^{m-n}$, respectively. We denote convergence in norm and weak convergence by "$\rightarrow$" and "$\rightharpoonup$", respectively.

2. Application to Differential–Algebraic Equations: Approach 1

Let $X = H^1(0,1)^m \times H^1(0,1)^n$, $Y = L^2(0,1)^m \times L^2(0,1)^n \times \mathbb{R}^{m-n}$,

$$F : X \longrightarrow Y \quad ,$$

$$F(u,v) := \begin{pmatrix} u' - h(u,v,\cdot) \\ g(u,\cdot) \\ r(u(0),u(1)) \end{pmatrix} \tag{2.1}$$

and $y_0 = (q_1, q_2, q_3)$.

The Tikhonov functional (1.4) to be minimized reads

$$\|u' - h(u, v, \cdot) - q_1^\delta\|^2 + \|g(u, \cdot) - q_2^\delta\|^2 + |r(u(0), u(1)) - q_3^\delta|^2 + \alpha(\|u - u^*\|_1^2 + \|v - v^*\|_1^2). \tag{2.2}$$

With this functional, we use the notation $x_\alpha^{\delta,\eta} =: (u_\alpha^{\delta,\eta}, v_\alpha^{\delta,\eta})$ for solutions of the problem (1.5). Note that δ is now a bound for $(\|q_1 - q_1^\delta\|^2 + \|q_2 - q_2^\delta\|^2 + |q_3 - q_3^\delta|^2)^{1/2}$. Let $x_0 = (u_0, v_0)$ denote a solution of $F(x) = y_0$ with $y_0 = (q_1, q_2, q_3)$. It is easy to see that F is Fréchet–differentiable with Lipschitz continuous derivative provided that h, g, and r are sufficiently smooth. Moreover,

$$F'(u_0, v_0)(u, v) = \begin{pmatrix} u' - h_{u_0}u - h_{v_0}v \\ g_{u_0}u \\ r_{00}u(0) + r_{10}u(1) \end{pmatrix} \tag{2.3}$$

It is straightforward to show that F is weakly (sequentially) continuous using the fact that $H^1(0,1)$ is compactly imbedded into $C[0,1]$. Unfortunately, we could not use $X = H^1(0,1)^m \times L^2(0,1)^n$, since on this space, F need not be weakly closed (as a Nemytskij operator with respect to v). For $(u, v) \in X$ and $\omega = (\zeta, \xi, \gamma) \in Y$,

$$F'(u_0, v_0)^*\omega = \begin{pmatrix} D^*\rho + B^{-1}(g_{u_0}^T\xi - h_{u_0}^T\zeta) + \rho_0 \cdot r_{00}^T\gamma + \rho_1 \cdot r_{10}^T\gamma \\ -B^{-1}h_{v_0}^T\zeta \end{pmatrix}. \tag{2.4}$$

Here, $B : D(B) := \{\varphi \in H^2(0,1) \mid \varphi'(0) = \varphi'(1) = 0\} \to L^2(0,1)$ is defined by $B\varphi = -\varphi'' + \varphi$. Note that B^{-1} is the adjoint of the imbedding operator from $H^1(0,1)$ into $L^2(0,1)$ (cf., e.g., [18,p.560]). B^{-1} is to be understood conponentwise if applied to a vector function. Moreover,

$$\rho_0(t) = \frac{1}{e^2 - 1}(e^t + e^{2-t}), \quad \rho_1(t) = \frac{e}{e^2 - 1}(e^t + e^{-t}), \quad t \in [0,1].$$

ρ_0 and ρ_1 are computed via the reproducing kernel of $H^1(0,1)$. Finally, D^* denotes the adjoint of $D : H^1(0,1)^m \to L^2(0,1)^m$, $Du = u'$. D^* is characterized by

$$D^*\zeta = ((D^*\zeta)' - \zeta)', \quad (D^*\zeta)' - \zeta \in H^2(0,1)^m \cap H_0^1(0,1)^m.$$

Now, let $x^* := (u^*, v^*)$ be the element used in the definition of the regularized solutions via (1.5). In view of Theorem 1, it must be such that $x_0 = (u_0, v_0)$ is the x^*–minimum–norm solution (1.3) (with F, X, Y, y_0 as above). Thus, x^* must be some a–priori estimate for x_0. Now, we reformulate the conditions of Theorem 1 within the present setting:

We have $x_0 - x^* = F'(x_0)\omega$ iff

$$D^*\zeta + B^{-1}(g_{u_0}^T\xi - h_{u_0}^T\zeta) + \rho_0 \cdot r_{00}^T\gamma + \rho_1 \cdot r_{10}^T\gamma = u_0 - u^* \tag{2.5}$$

$$-B^{-1}h_{v_0}^T\zeta = v_0 - v^*. \tag{2.6}$$

From this system, we eliminate $D^*\zeta$ using $D^*\zeta = ((D^*\zeta)' - \zeta)'$. Hence, (2.5) is eqivalent to

$$T' = u_0 - u^* - B^{-1}(g_{u_0}^T\xi - h_{u_0}^T\zeta) - \rho_0 \cdot r_{00}^T\gamma - \rho_1 \cdot r_{10}^T\gamma$$

$$T \in H_0^1(0,1)^m \cap H^2(0,1)^m,$$

where

$$T = (u_0 - u^*)' - \left[B^{-1}(g_{u_0}^T \xi - h_{u_0}^T \zeta)\right]' - \rho_o' \cdot r_{00}^T \gamma - \rho_1' \cdot r_{10}^T \gamma - \zeta.$$

Inserting the definition of B, we obtain

$$\left[(u_0 - u^*)' - \zeta\right]' - h_{u_0}^T \zeta + g_{u_0}^T \xi = u_0 - u^*.$$

Assuming now that $u_0 - u^* \in H^2(0,1)^m$, we finally obtain

$$\zeta' + h_{u_0}^T \zeta - g_{u_0}^T \xi = -\widetilde{B}(u_0 - u^*), \tag{2.7}$$

where $\widetilde{B} : H^2(0,1)^m \to L^2(0,1)^m$ and $\widetilde{B}\varphi = -\varphi'' + \varphi$. On the other hand, (2.6) is equivalent to $v_0 - v^* \in D(B)$ and

$$h_{v_0}^T \zeta = -B(v_0 - v^*). \tag{2.8}$$

Since by Assumption (I), $g_{u_0} h_{v_0}$ is nonsingular, so is $h_{v_0}^T g_{u_0}^T$. Hence, (2.7)–(2.8) is a linear index–2 system of the form (1.1).

Lemma 2: *The linear system*

$$\begin{array}{rl} u' - \quad H_1 u - \quad H_2 v & = q_1, \\ K u \qquad\qquad & = q_2 \end{array} \tag{2.9}$$

with $K(t)H_2(t)$ being nonsingular for all $t \in [0,1]$ is equivalent to the system

$$\begin{aligned} Ru &= H_2(KH_2)^{-1} q_2, \\ \left[(I-R)u\right]' + \left(R' - (I-R)H_1\right)\left(Ru + (I-R)u\right) &= (I-R)q_1, \\ v &= (KH_2)^{-1} K\{q_1 - u' + H_1 u\} \end{aligned}$$

with $R = H_2(KH_2)^{-1}K$.

Proof: See, e.g., [12].

If $q_1 \in H^i(0,1)^m$ and $q_2 \in H^j(0,1)^n$, then, for the solution (u,v) of (2.9), $u \in H^{k+1}(0,1)^m$, and $v \in H^k(0,1)^n$ with $k = \min(i, j-1)$ provided that H_1, H_2, and K are sufficiently smooth; more precisely, one needs that H_2 and K lie in $W^{k,\infty}$ (cf. [1]) with $k = \max(j, \min(i, j+1))$ and that $H_1 \in W^{k,\infty}$ with $k = \min(i,j)$. If we apply Lemma 2 to (2.7)–(2.8), the necessary smoothness requirements on $H_1 := -h_{u_0}^T$, $H_2 := g_{u_0}^T$, $K := h_{v_0}^T$ stem from smoothness requirements on the coefficient function h and g and on u_0, v_0. In the results stated below, we will explicitely formulate only the smoothness requirements on u_0 and v_0, since these are requirements for the unknown solution. From the coefficient functions h and g we will demand "sufficient smoothness"; the exact degrees of smoothness needed could be extracted from the considerations above.

By applying Lemma 2 to (2.7)–(2.8), we now obtain from Theorem 1:

Theorem 3:

Let (I) and (BC) hold for (1.1)–(1.2) and g, h be sufficiently smooth with g, h having Lipschitz continuous first order derivatives. Moreover, let $u_0, v_0 \in W^{1,\infty}(0,1)$ and let the following conditions hold:

i) $v_0 - v^* \in D(B) \cap H^3(0,1)^n$,

ii) $u_0 - u^* \in H^2(0,1)^m$

iii) $[(u_0 - u^*)' - \zeta]|_{t=0} = r_{00}^T\gamma, \quad [(u_0 - u^*)' - \zeta]|_{t=1} = -r_{10}^T\gamma,$

iv) $(u_0 - u^*)' - [B^{-1}(g_{u_0}^T\xi - h_{u_0}^T\zeta)]' - \zeta \in H^2(0,1)^m$,

v) $\|u_0 - u^*\|_2$, $\|v_0 - v^*\|_3$ *are sufficiently small.*

Here, $\omega = (\zeta, \xi, \gamma)$ is a solution of (2.7)–(2.8). Then for the choices $\alpha \sim \delta$ and $\eta = O(\delta^2)$, we obtain

$$\|u_\alpha^{\delta,\eta} - u_0\|_1 = O(\delta^{1/2}), \qquad \|v_\alpha^{\delta,\eta} - v_0\|_1 = O(\delta^{1/2}).$$

The hypotheses of Theorem 3 contain smoothness conditions on the solution (u_0, v_0) and the initial guess (u^*, v^*) together with some boundary conditions. Note that ζ and ξ depend on $v_0 - v^*$ and $u_0 - u^*$ as well. So (iii) and (iv) are boundary and smoothness conditions, respectively, again, but this time for $u_0 - u^*$ and $v_0 - v^*$. Thus, these conditions put restrictions on the quality of the a–priori estimates (u^*, v^*) of (u_0, v_0) in the sense that the errors must be sufficiently smooth (and small: (v)). Note, however, that due to the need to apply Lemma 2, we need sufficient smoothness not only for $u_0 - u^*$ and $v_0 - v^*$, but also for (u_0, v_0) (and of course for the coefficient functions g and h), as has been noted above. By the definition of B, (iv) is fulfilled if $u_0 - u^* \in H^3(0,1)^m$ and $\xi \in H^1(0,1)^n$, $\zeta \in H^2(0,1)^m$. Hence, (iv) holds if $u_0 - u^* \in H^3(0,1)^m$ and $v_0 - v^* \in H^4(0,1)^n$. But the latter conditions are even sufficient for the convergence rate $O(\delta^{2/3})$ (with a different parameter choice strategy $\alpha = \alpha(\delta)$):

Theorem 4:

Let the hypotheses of Theorem 3 hold. Let, moreover, $u_0, v_0 \in W^{2,\infty}(0,1)$ and let (i) and (ii) be replaced by

i)' $v_0 - v^* \in D(B) \cap H^4(0,1)^n$,

ii)' $u_0 - u^* \in H^3(0,1)^m$,

(iv) can be omitted.

Then, for the choices $\alpha \sim \delta^{2/3}$ and $\eta = O(\delta^2)$, we obtain

$$\|u_\alpha^{\delta,\eta} - u_0\|_1 = O(\delta^{2/3}), \qquad \|v_\alpha^{\delta,\eta} - v_0\|_1 = O(\delta^{2/3}).$$

Proof:

$$\begin{pmatrix} u_0 - u^* \\ v_0 - v^* \end{pmatrix} \in R(F'(u_0, v_0)^* F'(u_0, v_0))$$

iff there exist

$$\omega = (\zeta, \xi, \gamma) \quad \text{with} \quad \begin{pmatrix} u_0 - u^* \\ v_0 - v^* \end{pmatrix} = F'(u_0, v_0)^*\omega$$

and

$$(u,v) \quad \textit{with} \quad \omega = F'(u_0,v_0)(u,v).$$

The latter equation holds iff

$$\begin{aligned} u' - h_{u_0}u - h_{v_0}v &= \zeta, \\ g_{u_0}u \qquad &= \xi, \\ r_{00}u(0) + r_{10}u(1) &= \gamma. \end{aligned}$$

Because of (BC) and Lemma 2, this system is solvable for every $\gamma \in \mathbb{R}^{m-n}$, $\xi \in H^1(0,1)^n$, *and* $\zeta \in L^2(0,1)^m$ *with* $u \in H^1(0,1)^m$, $v \in L^2(0,1)^n$. *Since* $v = (g_{u_0}h_{v_0})^{-1}g_{u_0} \cdot (\zeta - u' + h_{u_0}u)$, $v \in H^1(0,1)^n$ *if* $\zeta \in H^1(0,1)^m$ *and* $\xi \in H^2(0,1)^n$. *Inserting* $\omega = (\zeta,\xi,\gamma) \in H^1(0,1)^m \times H^2(0,1)^n \times \mathbb{R}^{m-n}$ *into (2.7)–(2.8) yields* $\begin{pmatrix} u_0 - u^* \\ v_0 - v^* \end{pmatrix} \in R(F'(u_0,v_0)^* F'(u_0,v_0))$ *if (i)', (ii)', (iii) hold. The assertion follows now from Theorem 1, part b.*

◇

Remark:

(i) The condition (v) of Theorem 2 represents the condition (iv) of Theorem 1 on ω to have sufficiently small norm. The conditions (iv) and (viii) of Theorem 1 can be replaced by weaker conditions leading to estimates of $x_0 - x^*$ in weaker norms ([4], [19]). We conjecture that this extension of Theorem 1 can also be applied here and allows to use weaker norms in condition (v) of Theorem 2.

(ii) The results given above are not restricted to semi–explicit systems (1.1). In the proofs, we used essentially the fact that the linearization and its adjoint are linear semi–explicit differential- algebraic equations with index 2. The linearization of general differential–algebraic equations being "tractable with index 2" ([7]) is a linear differential–algebraic equation being tractable with index 2 whose adjoint is so in some cases. The techniques used above are then applicable in these cases, too.

3. Application to Differential–Algebraic Equations: Approach 2

Approach 1 has the disadvantage that v has to be differentiable in order to obtain a weakly closed operator. Since (1.1) does not contain v', this is a restriction. In an attempt to weaken this condition we integrate the first equation of (1.1) once and define

$$\begin{aligned} X &= H^1(0,1)^m \times L^2(0,1)^n, Y = L^2(0,1)^m \times L^2(0,1)^n \times \mathbb{R}^{m-n}, \\ F : X &\longrightarrow \qquad Y \qquad , \\ F(u,v) &:= \begin{pmatrix} u - u(0) - \int_0^{\bullet} h(u(s),v(s),s)\,ds \\ g(u,\cdot) \\ r(u(0),u(1)) \end{pmatrix}. \end{aligned} \tag{3.1}$$

The equation to be solved is now $F(x) = y_0$ with $y_0 = (\overline{q}_1, q_2, q_3)$ and $\overline{q}_1(t) = \int_0^t q_1(s)\,ds$. The Tikhonov functional (1.4) is now given by

$$\|u - u(0) - \int_0^{\bullet} h(u(s), v(s), s)\, ds - \overline{q}_1^{\delta}\|^2 + \|g(u,\cdot) - q_2^{\delta}\|^2 + |r(u(0), u(1)) - q_3^{\delta}|^2$$
$$+ \alpha(\|u - u^*\|_1^2 + \|v - v^*\|^2).$$

With this functional we use the notation $x_\alpha^{\delta,\eta} := (u_\alpha^{\delta,\eta}, v_\alpha^{\delta,\eta})$ for solutions of the problem (1.5); δ is now a bound for the error $\left(\|\overline{q}_1 - \overline{q}_1^{\delta}\|^2 + \|q_2 - q_2^{\delta}\|^2 + |q_3 - q_3^{\delta}|^2\right)^{1/2}$, which is a weaker notion of noise than in Approach 1.

Obviously, F is weakly (sequentially) continuous (and hence weakly closed). Again, let $x_0 = (u_0, v_0)$ be a solution of (1.1)–(1.2). Then

$$F'(u_0, v_0)(u, v) = \begin{pmatrix} u - u(0) - \int_0^{\bullet} (h_{u_0} u(s) + h_{v_0}(s))\, ds \\ g_{u_0} u \\ r_{00} u(0) + r_{10} u(1) \end{pmatrix}, \tag{3.2}$$

$$F'(u_0, v_0)^*(\zeta, \xi, \gamma) = \begin{pmatrix} B^{-1}(\zeta - h_{u_0}^T \int_{\bullet}^{1} \zeta(s)\, ds + g_{u_0}^T \xi) + \rho_0 \int_0^1 \zeta(s)\, ds + \rho_0 \cdot r_{00}^T \gamma + \rho_1 \cdot r_{10}^T \gamma \\ -h_{v_0}^T \int_{\bullet}^{1} \zeta(s)\, ds \end{pmatrix}.$$

Here, B is as in Approach 1. A simple calculation shows now that $\begin{pmatrix} u_0 - u^* \\ v_0 - v^* \end{pmatrix} = F'(u_0, v_0)^* \omega$, $\omega = (\zeta, \xi, \gamma) \in L^2(0,1)^m \times L^2(0,1)^n \times I\!R^{m-n}$ iff

$$T := \rho_0 \int_0^1 \zeta(s) + \rho_0 \cdot r_{00}^T \gamma + \rho_1 \cdot r_{10}^T \gamma - (u_0 - u^*) \in D(B) \tag{3.3}$$
$$v_0 - v^* \in H^1(0,1)^n, \qquad (v_0 - v^*)(1) = 0$$

and

$$\begin{aligned} -\varphi' - h_{u_0}^T \varphi + g_{u_0}^T \xi &= \widetilde{B}(u_0 - u^*) \\ -h_{v_0}^T \varphi &= v_0 - v^*, \end{aligned} \tag{3.4}$$

where $\varphi(t) = \int_t^1 \zeta(s)\, ds$. Because of Assumption (I), (3.4) is again a linear index–2 differential–algebraic equation, and Lemma 2 applies.

Before stating the final result, we derive conditions for $x_0 - x^* \in R(F'(u_0, v_0)^* F'(u_o, v_0))$. Analogously to the proof of Theorem 4, $\omega = (\zeta, \xi, \gamma) = F'(u_0, v_0)(u, v)$ for some $(u, v) \in H^1(0,1)^m \times L^2(0,1)^n$ iff (cf. (3.2))

$$\begin{aligned} u' - h_{u_0} u - h_{v_0} v &= \zeta' \\ g_{u_0} u &= \xi, \\ r_{00} u(0) + r_{10} u(1) &= \gamma. \end{aligned}$$

Because of (I) and (BC), this system is solvable for (u, v) iff $\zeta \in H^1(0,1)^m$ and $\xi \in H^1(0,1)^n$. Summarizing, Theorem 1 yields:

Theorem 5:

Let (I) and (BC) hold for (1.1)–(1.2) and g, h be sufficiently smooth with g, h having Lipschitz continuous first order derivatives. Moreover, let $u_0, v_0 \in W^{1,\infty}(0,1)$ and let the following conditions hold:

(i) $v_0 - v^* \in H^1(0,1)^n$,

(ii) $u_0 - u^* \in H^2(0,1)^m$,

(iii) $(v_0 - v^*)(1) = 0$,

(iv) $T \in D(B)$ *with* T *given by (3.3),*

(v) $\|v_0 - v^*\|_1$, $\|u_0 - u^*\|_2$ *sufficiently small.*

Then for the choices $\alpha \sim \delta$ *and* $\eta = O(\delta^2)$, *we obtain*

$$\|u_\alpha^{\delta,\eta} - u_0\|_1 = O(\delta^{1/2}), \quad \|v_\alpha^{\delta,\eta} - v_0\| = O(\delta^{1/2}).$$

If, instead of (i), (ii), even

(i)' $v_0 - v^* \in H^2(0,1)^n$,

(ii)' $u_0 - u^* \in H^3(0,1)^m$

hold and $u_0, v_0 \in W^{2,\infty}(0,1)$, *then, for the choice* $\alpha \sim \delta^{2/3}$ *and* $\eta = O(\delta^2)$, *we obtain*

$$\|u_\alpha^{\delta,\eta} - u_0\|_1 = O(\delta^{2/3}), \quad \|v_\alpha^{\delta,\eta} - v_0\| = O(\delta^{2/3}).$$

Again, the conditions (i), (ii) and (i)', (ii)', respectively, represent smoothness conditions. Since $\rho_1, \rho_2 \in C^\infty[0,1]$, (iii) and (iv) are both boundary conditions.

Note that compared to Approach 1, we now obtain convergence rates for the v–components of $x_\alpha^{\delta,\eta}$ in a weaker norm, namely in L^2 instead of H^1.

4. Application to Differential–Algebraic Equations: Approach 3

The third approach is based on the observation that the ill–posedness of (1.1)–(1.2) is rooted in the second equation of (1.1) (cf. also Lemma 2): Whereas the solution depends continuously on q_1 and q_3, it depends on the derivative of q_2 yielding the ill-posedness. This gives rise to the following setting:

$$\begin{aligned} F : D(F) \subseteq X &\to Y, \\ F(v) &:= g\big(u(v), \cdot\big), \end{aligned} \tag{4.1}$$

where $u = u(v)$ is defined as the solution of the boundary value problem

$$\begin{aligned} u' - h(u, v, \cdot) &= q_1 \\ r\big(u(0), u(1)\big) &= q_3 \\ g\big(u(0), 0\big) &= q_2(0) \end{aligned} \tag{4.2}$$

for fixed v.

Since we assumed (1.1) to be solvable, $q_2 \in H^1(0,1)^n$ and $q_2(0)$ is well–defined.

The Tikhonov functional (1.4) is now given by

$$\|g(u(v),\cdot) - q_2\|_Y^2 + \alpha\|v - v^*\|_X^2.$$

With this functional, we use the notation $v_\alpha^{\delta,\eta}$ for solutions of the problem (1.5). Note that δ is now merely a bound for $\|q_2 - q_2^\delta\|_Y$; errors in $q_2(0)$ and in q_1 and q_3 cannot be treated directly in this setting, since this would lead to changes in $u = u(v)$ and hence in F. This would require versions of the results in [4] and [19] with possible perturbations also in F, which are under investigation. But note that since the ill–posedness stems mainly from perturbations in q_2, it is not too severe to restrict attention to the effect of perturbations in q_2, since this approach has the advantage that we can choose the underlying Hilbert spaces nearly arbitrarily. We consider two possiblities:

(1) $X = L^2(0,1)^n$, $Y = L^2(0,1)^n$, h linear with respect to v,

(2) $X = H^1(0,1)^n$, $Y = H^1(0,1)^n$.

The linearity of h with respect to v in (1) is needed for showing that F is weakly closed. In many applications, however, v represents some kind of Lagrange multiplier which appears linearly in h (cf. [5], [6]).

As $D(F)$ in (4.1), we may take any convex weakly closed subset of X on which F is defined and continuous (cf. Lemma 7 below). The next Lemma shows that such a set exists under the following assumption:

Assumption (BC3): The homogeneous equation $u' - h_{u_0}u = 0$, $r_{00}u(0) + r_{10}u(1) = 0$, $g_{u_0}u(0) = 0$ has only the trivial solution.

Lemma 6: *Let (BC3) hold. Then F is well–defined by (4.1)–(4.2) and continuous in a neighbourhood of v_0 in both settings (1) and (2).*

Proof: follows from the Implicit Function Theorem, applied to the mapping $H : H^1(0,1)^m \times X \to L^2(0,1)^m \times \mathbb{R}^m$ with

$$H(u,v) = \begin{pmatrix} u' - h(u,v,\cdot) - q_1 \\ r(u(0),u(1)) - q_3 \\ g(u(0),0) - q_2(0) \end{pmatrix} : \tag{4.3}$$

The continuous Fréchet differentiability of H follows if h, r, and g are (as always assumed) sufficiently smooth. The fact that the partial Fréchet derivative with respect to u is continuously invertible in (u_0, v_0) follows from (BC3) with standard arguments from the theory of linear boundary value problems.

◇

As a consequence of (4.3) we obtain

$$\frac{du}{dv}(v_0)v = w,$$

where w is the solution of

$$\begin{aligned} w' - h_{u_0} w &= h_{v_0} v, \\ r_{00} w(0) + r_{10} w(1) &= 0, \\ g_{u_0} w(0) &= 0. \end{aligned} \tag{4.4}$$

Lemma 7: *In both settings (1) and (2), F is weakly closed if $D(F)$ is weakly closed.*

Proof:

(1) Let $\{v_n\} \subseteq L^2(0,1)^n$ such that $v_n \rightharpoonup v$ and $F(v_n) \rightharpoonup y$. Since $\{v_n\}$ is bounded in $L^2(0,1)^n$, $u_n = u_n(v_n)$ given by (4.2) is bounded in $H^1(0,1)^m$. Hence, there exist a subsequence $\{u_{n_i}\}$ and a $w \in H^1(0,1)^m$ such that $u_{n_i} \rightharpoonup w$ in $H^1(0,1)^m$. Hence, $u_{n_i} \to w$ in $C[0,1]^m$. Since h is linear in v, $h(u_{n_i}, v_{n_i}, \cdot) \rightharpoonup h(w, v, \cdot)$ in $L^2(0,1)^m$. Moreover, $r(u_{n_i}(0), u_{n_i}(1)) \to r(w(0), w(1))$ and $g(u_{n_i}(0), 0) \to g(w(0), 0)$. Hence, w is the solution of

$$u' = h(u, v, \cdot), \quad r(u(0), u(1)) = q_3, \quad g(u(0), 0) = q_2(0).$$

Since $g(u_{n_i}, \cdot) \to g(w.\cdot)$ in $L^2(0,1)^m$, $y = F(v)$.

(2) Let $\{v_n\} \subseteq H^1(0,1)^n$ such that $v_n \rightharpoonup v$ and $F_n \rightharpoonup y$. Hence, $v_n \to v$ in $L^2(0,1)^n$ and, furthermore, $u_n \to w$ in $H^1(0,1)^m$ by Lemma 6, where w is the solution of $u' = h(u, v, \cdot)$, $r(u(0), u(1)) = q_3$, $g(u(0), 0) = q_2(0)$. This implies $u_n \to w$ in $C[0,1]^m$, hence $F(v_n) = g(u_n, \cdot) \to g(w, \cdot) = F(v)$ in $L^2(0,1)^n$. Furthermore,

$$\begin{aligned} \frac{d}{dt} F(v_n) &= \frac{d}{dt} g(u_n, \cdot) = g_u(u_n, \cdot) u_n' + g_t(u_n, \cdot) \\ &\to g_u(w, \cdot) w' + g_t(w, \cdot) = \frac{d}{dt} F(v) \end{aligned}$$

in $L^2(0,1)^n$, and the assertion follows. ◇

Note, that in case (2), F is even weakly continuous and compact (as can be seen from the proof of Lemma 6).

We consider now the case (1) only. Case (2) can be treated similarly. Let G denote the Green's function of (4.4). Then

$$F'(v_0) v = g_{u_0} \int_0^1 G(\cdot, s) h_{v_0}(s) v(s)\, ds$$

and

$$F'(v_0)^* \xi = h_{v_0}^T \int_0^1 G(t, \cdot)^T g_{u_0}^T(t) \xi(t)\, dt.$$

Now, let v^* be the initial estimate for v_0 used in the Tikhonov functional. In order to verify condition (iii) of Theorem 1, we have to consider solutions of

$$F'(v_0)^* \xi = v_0 - v^*. \tag{4.5}$$

Since $G(t,s)^T$ is the Green's function of the boundary value problem adjoint to (4.4), (4.5) is equivalent to

$$\begin{aligned} -\zeta' - h_{u_0}^T\zeta - g_{u_0}^T\xi &= 0 \\ h_{v_0}^T\zeta \qquad &= v_0 - v^* \end{aligned} \tag{4.6}$$

subject to the adjoint boundary conditions

$$R_0^*\zeta(0) + R_1^*\zeta(1) = 0 \tag{4.7}$$

(cf., e.g., [13, §B.6.2/3]). (4.6) is again a linear index–2 problem covered by Lemma 2. It is now straightforward to derive conditions implying that $v_0 - v^* \in R\big(F'(v_0)^*F(v_0)\big)$. These considerations and Theorem 1 imply

Theorem 8:

Let (I), (BC), and (BC3) hold for (1.1)–(1.2) and g, h be sufficiently smooth with g, h having Lipschitz continuous first order derivatives. Consider the case (1). Moreover, let $u_0, v_0 \in W^{1,\infty}(0,1)$ and let the following conditions hold:

(i) $v_0 - v^* \in H^1(0,1)^n$,

(ii) the adjoint boundary conditions (4.7) are consistent with (4.6).

(iii) $\|v_0 - v^*\|_1$ is sufficiently small.

Then, for the choices $\alpha \sim \delta$ and $\eta = O(\delta^2)$, we obtain

$$\|v_\alpha^{\delta,\eta} - v_0\| = O(\delta^{1/2}), \quad \|u(v_\alpha^{\delta,\eta}) - u_0\|_1 = O(\delta^{1/2}).$$

If, instead of (i), even

(i)' $v_0 - v^* \in H^2(0,1)^n$

holds, and $u_0, v_0 \in W^{2,\infty}(0,1)$, then, for the choices $\alpha \sim \delta^{2/3}$ and $\eta = O(\delta^2)$, we obtain

$$\|v_\alpha^{\delta,\eta} - v_0\| = O(\delta^{2/3}), \qquad \|u(v_\alpha^{\delta,\eta}) - u_0\|_1 = O(\delta^{2/3}).$$

We remark in closing that for all our approaches, convergence as $\alpha, \delta \to 0$ can immediately be obtained from [4, Theorem 2.3] as soon as the respective F is continuous and weakly closed without further assumptions. Thus, in all our setups, convergence holds if $\frac{\delta^2}{\alpha} \to 0$ and $\frac{\eta}{\alpha} \to 0$. All the additional assumptions in Theorems 3,4,5, and 8 were used only for obtaining convergence rates and are therefore needed only locally around the appropriate solution. See [4, Remark 2.5] for further discussion.

References:

[1] R.Adams, Sobolev Spaces; Academic Press, New York 1975

[2] K.E. Brenan, L.R.Engquist, Backward differentiation approximations of nonlinear differential algebraic systems; Math.Comp. 51 (1988), pp. 659–676

[3] H.W.Engl, C.W.Groetsch (eds.), Inverse and Ill–Posed Problems, Academic Press, Orlando 1987

[4] H.W.Engl, K.Kunisch, A.Neubauer, Convergence rates for Tikhonov regularization of nonlinear ill-posed problems; Inverse Problems 5 (1989), pp. 523–540

[5] C.W.Gear, Maintaining solution invariants in the numerical solution of ODE's, SIAM J. Sci. Comput. 7 (1986), pp. 734–743

[6] C.W.Gear, G.K.Gupta, B.Leimkuhler, Automatic integration of Euler–Lagrange equations with constraints; J. Comp. Appl. Math. 12 & 13, (1985), pp. 77–90

[7] E.Griepentrog, R.März, Differential–Algebraic Equations and their Numerical Treatment, Teubner–Texte zur Math. 88, Teubner, Leipzig 1986

[8] C.W.Groetsch, The Theory of Tikhonov Regularization for Fredholm Equations of the First Kind, Pitman, Boston 1984

[9] E.Hairer, Ch.Lubich, M.Roche, The Numerical Solution of Differential–Algebraic Systems by Runge–Kutta Methods, Lecture Notes in Mathematics, Springer, Berlin, Heidelberg, New York 1989

[10] M.Hanke, Linear differential–algebraic equations in spaces of integrable functions; J. Differ.Equations 79 (1989), pp. 14–30

[11] M.Hanke, R.März, A.Neubauer, On the regularization of linear differential–algebraic equations; in [3], pp. 523–540

[12] M.Hanke, R.März, A.Neubauer, On the regularization of a class of nontransferable differential–algebraic equations; J. Differ. Equations 73 (1988), pp. 119–132

[13] E.Kamke, Differentialgleichungen: Lösungsmethode und Lösungen I, 8.Auflage; Akad. Verlagsgesellschaft Geest & Portig, Leipzig 1967

[14] P.Lötstedt, L.R.Petzold, Numerical solution of nonlinear differential equations with algebraic constraints I, SIAM J.Sci.Stat. Comput. 7(1986), pp.720–733

[15] R.März, Higher Index Differential–Algebraic Equations: Analysis and Numerical Treatment, Banach Center Publ. XIV, Warszawa, Polish Scientific Publishers 1989

[16] V.A.Morozov, Methods for Solving Incorrectly Posed Problems, Springer, New York 1976

[17] M.Z.Nashed (ed.), Generalized Inverses and Applications, Academic Press, New York 1976

[18] A.Neubauer, When do Sobolev spaces form a Hilbert scale? Proc.Amer.Math.Soc. 103 (1988), pp. 557–562

[19] A.Neubauer, Tikhonov regularization for nonlinear ill–posed problems: optimal convergence rates and finite dimensional approximation, Inverse Problems 5 (1989), pp. 541–557

[20] T.I.Seidman, C.R.Vogel, Well–posedness and convergence of some regularization methods for nonlinear ill–posed problems, Inverse Problems 5 (1989), pp. 227–238

[21] A.N.Tikhonov, V.Y.Arsenin, Solution of Ill–Posed Problems, Wiley, New York 1977

Acknowledgements:
H.W. Engl and A. Neubauer acknowledge support from the Austrian Fonds zur Förderung der wissenschaftlichen Forschung (project S32/03), M. Hanke from the Ministerium für Hoch– und Fachschulwesen (GDR), which made possible his visit to Linz where this paper was written.

On the Choice of the Regularization Parameter Using Model Functions*

M. Kroller[1] *and K. Kunisch*[2]

[1]Institut für Mathematik, Karl-Franzens-Universität Graz, Elisabethstr. 16, A-8010 Graz, Austria
[2]Institut für Mathematik, Technische Universität Graz, Kopernikusgasse 24, A-8010 Graz, Austria

1. INTRODUCTION

The purpose of this contribution is to discuss the use of model functions to solve nonlinear illposed inverse problems and to illustrate their success in numerical computations. Specifically, the problems of estimating the amount of noise in the data and the choice of an "optimal" regularization parameter will be addressed. The main ideas will be explained by means of the specific example of determining the diffusion coefficient in elliptic partial differential equations from knowledge of the state. We refer to [**IK2**] where the techniques of this paper are analyzed for an abstract class of problems. The introduction of model functions in the context of inverse problems was inspired by the role that model functions play in modern numerical techiques for nonlinear programming problems, see e.g. [**DS**]. The differential equation under consideration is given by

$$\begin{aligned} -\text{div}(a \text{ grad } u) + cu = f \quad \text{on} \quad \Omega, \\ \left.\frac{\partial u}{\partial n}\right|_{\partial\Omega} = 0, \end{aligned} \tag{1}$$

where $f \in L^2(\Omega)$, $c \in L^\infty(\Omega)$, $c(x) \geq \kappa > 0$, and Ω is a bounded domain in $\mathbf{R}^n$, $n = 1,2$ or 3, with Lipschitz continuous boundary $\partial\Omega$ if $n = 2$ or 3, or $\Omega = (0,1)$, if $n = 1$. The problem consists in determining the diffusion coefficient a from knowledge of a possibly error corrupted observation z^δ corresponding to the state $u(a^*)$ of (1) for some "true" coefficient a^*. This will require, in some sense, to invert the parameter-to-solution mapping $a \to u(a)$. This mapping is nonlinear, nonconvex, not radially unbounded, and not injective. It is compact when considered, for example, as a mapping from

$$Q_{ad} = \{a \in H^* : a(x) \geq \nu > 0\} \subset H^*$$

to $H^1(\Omega)$, where $H^* = H^1(0,1)$ if $n = 1$, and $H^* = H^2(\Omega)$, if $n = 2$ or $n = 3$. Several specific examples for lack of continuous invertibility of $a \to u(a)$ are given in the literature, see [**BK**, Chapter 4] and the references given there. Thus numerical procedures to solve the nonlinear illposed inverse problem of determining a^* from z^δ require some type of regularization. One of the possibilities is given by considering a regularized least squares problem of the type

$$\min \frac{1}{2} \left|u(a) - z^\delta\right|_{L^2} + \frac{\beta}{2} N(a) \quad \text{over} \quad a \in Q_{\text{ad}}, \tag{P^β}$$

where N is the square of the seminorm given by $N(a) = |a|^2_{H^*} - |a|^2_{L^2}$, and β is a "small" positive parameter. In (P^β), z^δ denotes the data which may contain measurement error. We shall also

*Work supported in part by the Fonds zur Förderung der wissenschaftlichen Forschung, Austria, S3206.

refer to the error free data z^*. In case of no modelling error, it is reasonable to assume that z^* is attainable, i.e. that there exists $a^* \in Q_{\text{ad}}$ with $u(a^*) = z^*$; in the case of modelling error there may exist no such a^*. It is simple to argue the existence of a solution a^β of (P^β) for any $\beta > 0$. We refer to **[A,CK1,CK2]** for results on the stability of the solutions of (P^β) with respect to perturbations in the observation z^δ. In **[IK1]** conditions are given which guarantee that a^β is even directionally differentiable with respect to β, z^δ and f. Moreover in **[EKN,N]** it is shown that under appropriate conditions $a^\beta \to a^*$ like $O(\beta^{1/2})$ (respectivley $O(\beta^{2/3})$) as $z^\delta \to z^* = u(a^*)$ in $L^2(\Omega)$.

From a practical point of view the choice of the regularization parameter β for a given observation z^δ is of paramount importance. Numerical experiments show that for the problem under consideration there is a unique best regularization parameter (Plot 1). However, it is by no means simple to determine this papameter in an efficient manner. In accordance with the wellknown Morozov principle, which is especially wellunderstood for linear inverse problems [M], one might want to choose β such that

$$|u(a^\beta) - z^\delta|_{L^2} = \delta, \tag{2}$$

where z^δ satisfies $|z^\delta - z^*|_{L^2} \leq \delta$. We observe that (2) is a nonlinear equation for β and any iterative solution would require to solve (P^β) and subsequently (1) for several values of a at each iteration. Moreover (2) depends on knowledge of δ, which may be unrealistic in practical problems. In numerical computations we also observed that even with exact $\delta = |z^\delta - z^*|_{L^2}$ the Morozov principle frequently overestimated the best (in the sense of the L^2-error $\|a^* - a^\beta\|_{L^2}$) β value. -We shall address these aspects in the next section.

2. A CLASS OF MODEL FUNCTIONS

Let us introduce the minimal value function of (P^β) with respect to $\beta > 0$:

$$F(\beta) = \min\{\frac{1}{2}|u(a) - z^\delta|^2_{L^2} + \frac{\beta}{2}N(a) : a \in Q_{\text{ad}}\}.$$

We shall derive a four parameter family of functions $m(\beta)$ that approximate $F(\beta)$. It can be shown that $\beta \to F(\beta)$ is monotonically nondecreasing. To assert further properties of $F(\beta)$ we require a hypothesis in terms of the Lagrangian functional $\mathcal{L}$ associated with (P^β):

$$\mathcal{L}(a) = \frac{1}{2}|u(a) - z^\delta|^2_{L^2} + \frac{\beta}{2}N(a) + \langle \eta^*, \nu - a\rangle_H.$$

where η^* is the Lagrange multipier associated with the affine constraint $a \geq \nu$. Henceforth we restrict our attention to an intervall V of β values for which the Hessian of $\mathcal{L}$ with respect to a is positive definite, i.e. there exists $\kappa > 0$ such that

$$\mathcal{L}''(a^\beta)(h,h) = |u'(a^\beta)h|^2_{L^2} + \langle u''(a^\beta)(h,h), u(a^\beta) - z^\delta\rangle_{L^2} + \beta N(h) \geq \kappa |h|^2_H. \tag{3}$$

for all $h \in H^*$ and $\beta \in V$. Here primes denote differentiation with respect to a. For a discussion of (3) we refer to **[CK1,CK2]**. With (3) holding the solutions a^β are locally unique and depend continuously on z^δ. It can also be shown [IK] that

$$F'(\beta) = \frac{1}{2}N(a^\beta) \quad \text{for all} \quad \beta \in V. \tag{4}$$

Moreover, $\beta \to a^\beta$ is directionally differentiable with right-derivative denoted by $\dot{a}^\beta$. Consequently, the right-derivative of F' denoted by F'', is given by

$$F''(\beta) = \langle a^\beta, \dot{a}^\beta \rangle_N, \quad \text{for} \quad \beta \in V, \tag{5}$$

where $\langle v, w \rangle_N = \langle v, w \rangle_{H^*} - \langle v, w \rangle_{L^2}$. Henceforth it will be assumed that $a^\beta \in \text{int } Q_{\text{ad}}$ for $\beta \in V$. In this case $\beta \to a^\beta$ and $\beta \to F'(\beta)$ are continuously differentiable and the necessary optimality condition is given by

$$\langle u(a^\beta) - z^\delta, u'(a^\beta)h \rangle_{L^2} + \beta \langle a^\beta, h \rangle_N = 0 \quad \text{for all} \quad h \in H^*. \tag{6}$$

For $\beta \in V$ one can differentiate (6) with respect to β to obtain

$$\begin{aligned} &\langle u(a^\beta) - z^\delta, u''(a^\beta)(\dot{a}^\beta, h) \rangle_{L^2} + \langle u'(a^\beta)(\dot{a}^\beta), u'(a^\beta)h \rangle_{L^2} \\ &\quad + \langle a^\beta, h \rangle_N + \beta \langle \dot{a}^\beta, h \rangle_N = 0 \quad \text{for all} \quad h \in H^* \quad \text{and} \quad \beta \in V. \end{aligned} \tag{7}$$

Evaluating (7) for $h = \dot{a}^\beta$ leads to

$$F''(\beta) = -\mathcal{L}''(a^\beta)(\dot{a}^\beta, \dot{a}^\beta) \leq -\kappa |\dot{a}^\beta|_{H^*}^2 \quad \text{for all} \quad \beta \in V. \tag{8}$$

Combining (4) and (8) one concludes that F is monotonically increasing and convex downwards for $\beta \in V$.

We reconsider (7) and evaluate this equality for $h = a^\beta$. Using (4) one obtains

$$\langle u(a^\beta) - z^\delta, u''(a^\beta)(\dot{a}^\beta, a^\beta) \rangle_{L^2} + \langle u'(a^\beta)(\dot{a}^\beta), u'(a^\beta)(a^\beta) \rangle_{L^2} + 2F'(\beta) + \beta F''(\beta) = 0 \tag{9}$$

for $\beta \in V$. The heuristic step in the derivation of the proposed class of model functions follows next. We observe that the first two terms in (9) contain bilinear forms that are evaluated at $(\dot{a}^\beta, a^\beta)$. In view of (5) we shall approximate these two terms by functions of the class $(c_1\beta + c_2)F''$, with $c_i \in \mathbf{R}$. We shall distinguish between the function F and the model functions m that arise due to this approximative step. From (9) it follows that m satisfies the following ordinary differential equation

$$(c_1\beta + c_2)m'' + 2m' + \beta m'' = 0. \tag{10}$$

Solving (10) one arrives at the following four parameter family of functions

$$m(\beta) = e - \frac{b}{(\beta + c)^d}, \quad (e, b, c, d) \in \mathbf{R}^4,$$

where d is related to c_1 by $d = \frac{2}{1+c_1} - 1$. While m was derived as a local model function, we hope that for appropriately chosen parameters (e, b, c, d) it is a good approximation to F for a large range of β-values.

Let us proceed with some comments concerning the model functions. Solving (P^β) once for some $\beta > 0$ one obtains values for $F(\beta)$ as well as for $F'(\beta) = \frac{1}{2}N(a^\beta)$, which are used as values for $m(\beta)$ and $m'(\beta)$. Consequently, solving (P^β) for two values of β gives four conditions which can be used to determine (e, b, c, d) in the model function. The model functions will be used in an iterative process starting with "large" values for β and decreasing β until a stopping criterion is reached. Thus one always computes in a range of stable β-values and stops decreasing β before the problems (P^β) become "too illposed" .- Conditions on the range of (e, b, c, d) can be given that guarantee that m has the same monotonicity and convexity properties as F. For example, let us assume that $c_1 < 1$. Then $d > 0$ and m will not have a pole for positive β if $c \geq 0$. Further $m' > 0$ and $m'' < 0$ for $\beta > 0$ if $b > 0$.

Next we turn to the questions raised at the end of section 1. Let us assume that z^* is attainable, i.e. $z^* \in \{u(a) : a \in Q_{\text{ad}}\}$. Then $F(0)$ can be used to estimate δ from below: $F(0) \leq \frac{1}{2}|z^* - z^\delta|_{L^2}^2 \leq \frac{1}{2}\delta^2$, and hence

$$\sqrt{2F(0)} \leq \delta \quad \text{if} \quad |z^* - z^\delta|_{L^2} \leq \delta \quad \text{and if} \quad z^* \quad \text{is attainable.} \tag{11}$$

In Plot 2 we demonstrate that $\sqrt{2m(0)}$ can be used efficiently to provide an estimate for the amount of noise in the data for our model problem associated with the estimation of a in (1). Turning to the question of determining a best regularization parameter, we observe that for $\beta \in V$, the Morozov principle can be expressed as $2(F(\beta) - \beta F'(\beta)) = \delta^2$. This will be approximated by means of the model function as

$$2(m(\beta) - \beta m'(\beta)) = \delta^2. \tag{12}$$

Once the model function is determined a solution of (12) can be sought without recourse to the partial differential equation. As already mentioned, the Morozov principle generally overestimated the best regularization parameter in our examples and hence we also used a different class of criteria to determine the regularization parameter which is given by

$$|u(a^\beta) - z^\delta|_{L^2}^2 + \beta^\gamma N(a) = \delta^2, \tag{13}$$

where γ is chosen in $[1,2]$. These criteria can again be expressed easily by means of the model function:

$$2[m(\beta) + (\beta^\gamma - \beta)m'(\beta)] = \delta^2. \tag{14}$$

The model function m, determined from evaluation of (P^β) for two values of β together with approximations (12) or (14) of principles such as (2) or (13) can be used in an iterative process. In an initialization phase one determines β_0 and β_1 and calculates a first model function m_1. Using m_1 the nonlinear equation (12) or (14) is solved for β_2. Then β_0 is dropped and m_2 is calculated from β_1 and β_2. This process is repeated until consecutive values for β_i satisfy a stopping criterion. Once consecutive values of β_i become close, determining the parameters in the model functions become numerically increasingly more difficult. In our implementation of the iterative process we therefore use an appropriate combination of the last three values for β_i, once these values become close to each other. In most of our calculations the termination criterion was reached after 4-6 iterations.

3. A NUMERICAL EXAMPLE

As a specific numerical example we consider the estimation of a in

$$\begin{gathered} -(au_x)_x + u = f \quad \text{in} \quad (0,1), \\ au_x(0) = 1, au_x(1) = 2. \end{gathered} \tag{15}$$

The true parameter is chosen to be $a^* = 1+x^2$ and $f = 1-x$. The solution of (15) is $u(a^*) = 1+x$ in this case, and its singular set $S = \{x : u(a^*)(x) = 0\}$ is empty. The discretization of (15) was carried out by a Galerkin scheme using linear spline functions on the uniform grid $\{\frac{i}{17}\}_{i=0}^{17}$. The L^2-output criterion was replaced by the discrete criterion

$$J_d(a) = \sum_{i=0}^{17} |u(a)(\frac{i}{17}) - z^\delta(\frac{i}{17})|^2,$$

where z was obtained from $u(a^*)$ by fitting a cubic spline through

$$u(a^*)(\frac{i}{8}) + \hat{\delta} * \mathrm{rand}(i), \quad i = 0, \ldots, 8,$$

Plot 1

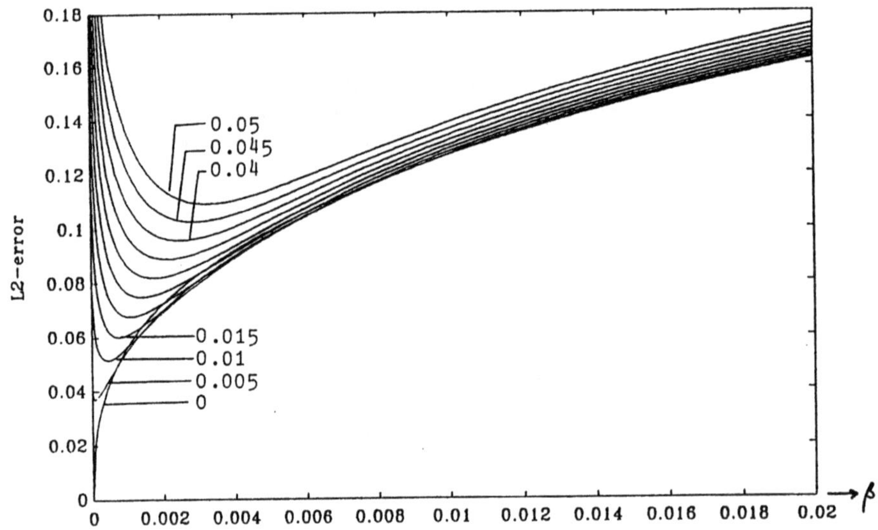

Plot 2

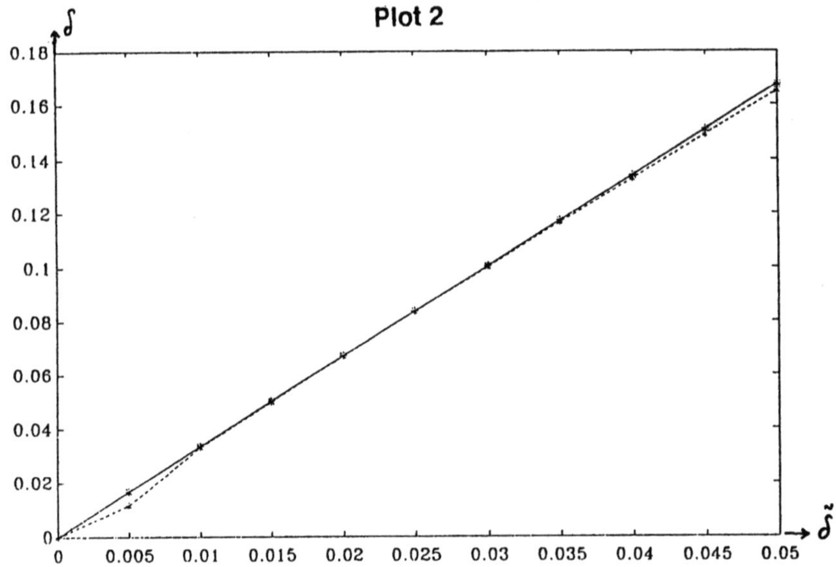

where rand(i) are uniformly distributed random numbers in $[-1,1]$ and $\hat{\delta}$ is a chosen error level. For further details of the algorithm and many additional examples we refer to [**KK**]. Plot 1 shows $|a^* - a^\beta|_{L^2}$ as a function of β for various choices of $\hat{\delta}$, where a^β is a solution of (P^β) with the L^2-criterion replaced by J_d above. One observes the existence of a best β-value (in the L^2-sense) which increases with $\hat{\delta}$. In Plot 2 we show the ability of our algorithm to estimate the noise in the data. The abscissa gives the $\hat{\delta}$-values. For each $\hat{\delta}$ there is an associated value for $\delta = \sqrt{\sum_{i=0}^{17} |u(a^*)(\frac{i}{17}) - z^\delta(\frac{i}{17})|^2}$ which gives the solid line in Plot 2. The dotted line shows the values for $\sqrt{2m(0)}$ which is the noise level estimate of our algorithm.

Plot 3

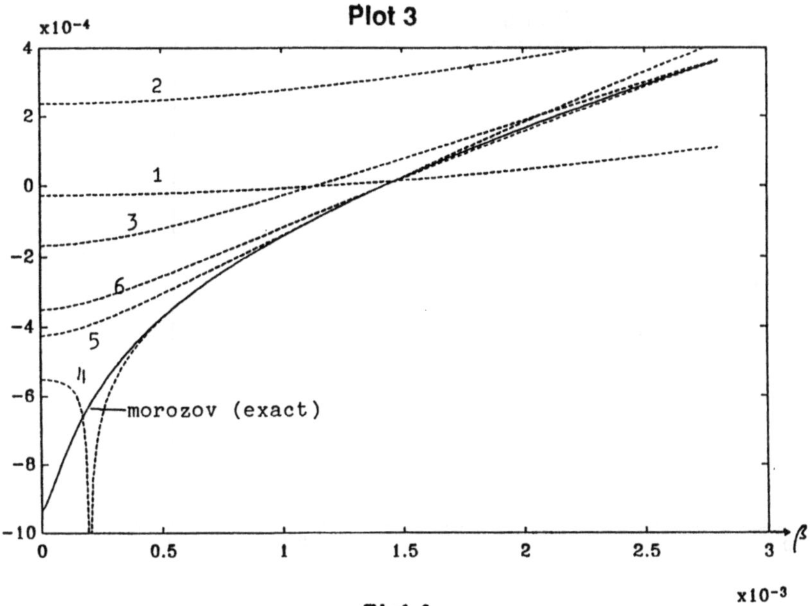

Plot 4

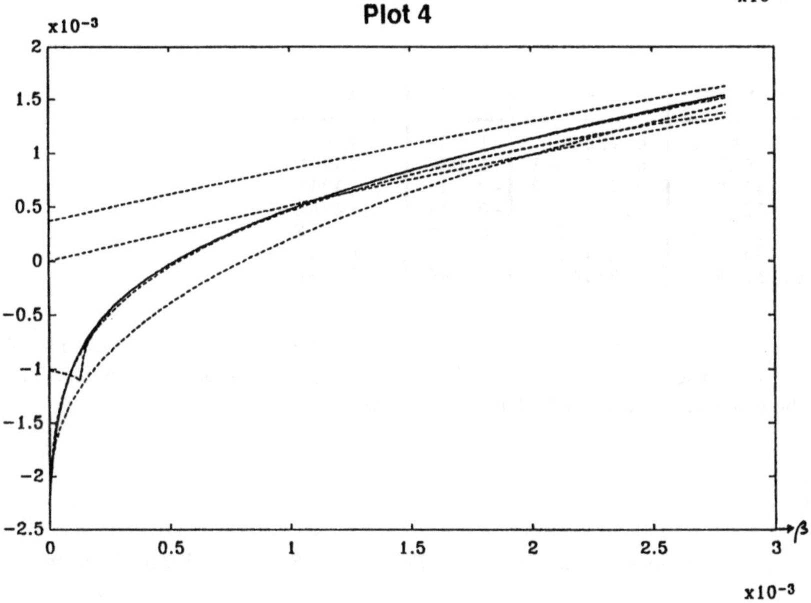

Plot 3 shows a zoom-view of the iterates (in the sense explained at the end of section 2) of the Morozov curves $2(m - \beta m') - \delta^2$ in the neighbourhood of the root (dotted curves). The solid line gives the exact Morozov function $2(F - \beta F') - \delta^2$. Here we used the exact value for δ. Replacing δ by our estimate $\sqrt{2m(0)}$ only changes these curves insignificantly. The dotted lines of Plot 4 give the first 5 iterates of the curves $2m(\beta) - \delta^2$ based on (13) and (14) with $\gamma = 1$. We observe that these curves approximate the solid line $2F(\beta) - \delta^2$ very well especially in the neighbourhood of the root. Plot 5 shows the best β value (*) for different values of δ, and the β-values based on the Morozov principle (o) as well as those based on (13) (+). It can be seen that the Morozov principle overestimates the best β-values while the β-values based on (13) underestimates them. Finally in Table 1 we give the values for the coefficients (e_k, b_k, c_k, d_k) of

Plot 5

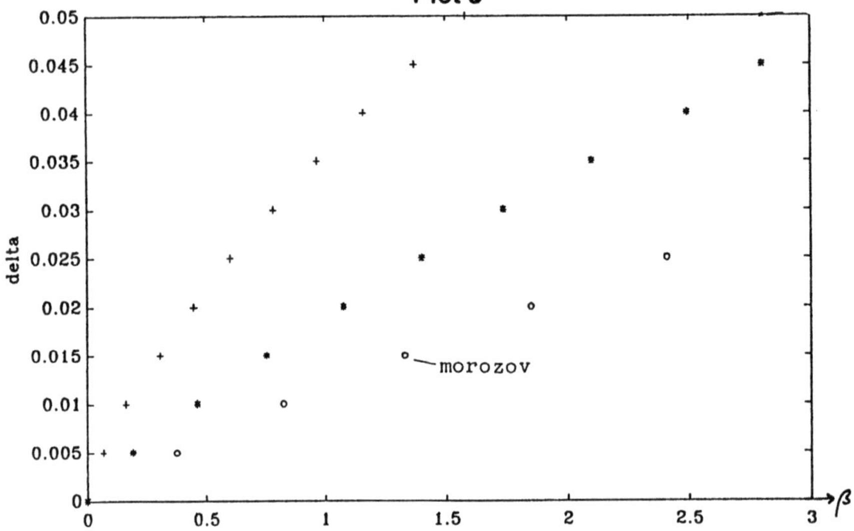

Table 1

e	b	c	d	β
0.1048E-1	0.2394E-3	0.2362E-1	1.006	0.7283E-2
0.1046E-1	0.2265E-3	0.2626E-1	1.038	0.3641E-2
-0.9017E-2	-0.3442E-1	0.4747E-2	-0.242	0.1821E-2
-0.3268E-3	-0.6205E-1	0.5503E-3	-0.616	0.6476E-3
0.2286E-3	-0.8871E-1	-0.9307E-4	-0.715	0.8235E-3
-0.2442E-3	-0.4245E-1	0.2675E-3	-0.552	0.8231E-3
-0.2553E-3	-0.4116E-1	0.2674E-3	-0.546	0.8235E-3

the model function as well as of the β- values, which resulted from iterative use of the model functions and Morozov's priciple with $\delta = 0.01$. We observe convergence of the β-values. The best β-value (in the sense of L^2-error) is 0.000462 in this case.

REFERENCES

[A] W. Alt, *Stability for Parameter Estimation in Two Point Boundary Value Problems*, preprint.

[BK] H.T.Banks and K.Kunisch, "Estimation Techniques of Distributed Parameter Systems," Birkhäuser, Boston, 1989.

[CK1] F. Colonius and K. Kunisch, *Output least squares stability in elliptic systems*, Appl. Math. and Optimization **19** (1989), 33-63.

[CK2] F. Colonius and K. Kunisch, *Stability of Perturbed Optimization Problems with Application to Parameter Estimation*, submitted to SIAM J. Control and Optimization.

[DS] J.E.Dennis and R.B.Schnabel, "Numerical Methods for Unconstrained Optimization and Nonlinear Equations," Prentice Hall, Englewood Cliff, 1983.

[EKN] H.W.Engl, K.Kunisch and A.Neubauer, *Tikhonov Regularization for the Solution of Nonlinear Illposed Problemns*, Inverse Problems **5** (1989), 523-540.

[IK1] K.Ito and K.Kunisch, *Sensitivity Analysis of the Solutions to Optimization Problems in Hilbert Spaces with Applications to Optimal Control and Estimation*, submitted.

[IK2] K.Ito and K.Kunisch, *On the Choice of the Regularization Parameter in Nonlinear Inverse Problems*, manuscript.

[KK] M. Kroller and K. Kunisch, *On the Choice of the Regularization Parameter Using Model Functions: the Algorithm and Numerical Examples*, Technical University and University of Graz, Technical Report (1990).

[M] V.A.Morozov, "Methods for Solving Incorrectly Posed Problems," Springer, New-York, 1984.

[N] A.Neubauer, *Tikhonov Regularization for Nonlinear Illposed Problems: Optimal Convergence Rates and Finite Dimensional Approximation*, Inverse Problems **5** (1989), 541-558.

Exact Reconstruction Procedures for Some Multidimensional Inverse Scattering and Inverse Boundary Value Problems

A.I. Nachman

Department of Mathematics, University of Rochester,
Rochester, NY 14627, USA

This talk reviewed recent progress in understanding various multidimensional inverse problems from an analytic point of view.

To describe the kind of results now available, we discuss a problem arising, for instance, in medical imaging by ultrasound. Briefly, the experimental apparatus consists of a water tank (large enough so that its boundary may be considered, for practical purposes, to be at infinity) and a number of transducers (which can act both as emitters and receivers) surrounding the tissue to be imaged. Mathematically, we thus take the water tank to be all of $\mathbf{R}^3$ and let Ω denote the bounded region enclosed by the transducers, $\partial\Omega$ its surface. We allow essentially arbitrary geometry for the surface $\partial\Omega$ of emitters and receivers. The pressure field at x generated by a point source at y oscillating harmonically with frequency ω satisfies

$$\Delta_x p_\omega(x,y) + \frac{\omega^2}{c^2(x)} p_\omega(x,y) = -\delta(x-y) \tag{1}$$

throughout $\mathbf{R}^3$ and is outgoing at infinity. Here $c(x)$ is the speed of sound, which outside Ω is equal to a known constant value c_0. The function $c(x)$ is assumed bounded away from zero (this condition can be relaxed); it is allowed to be discontinuous. The problem we consider is to reconstruct $c(x)$ inside Ω from measurements of $p_\omega(x,y)$ at all receivers x on $\partial\Omega$, for all sources y on $\partial\Omega$, at one fixed frequency ω.

Until recently, very little had been known about exact methods for the solution of genuinely multidimensional problems such as the one described above. In practice, one had to resort to approximate methods. Even theoretically, the question of whether the surface data measured at one frequency was sufficient to determine $c(x)$ uniquely, had not been answered. Conventional intuition would in fact suggest that, since one is working at a fixed frequency, one cannot expect to recover details much smaller than the corresponding wavelength. The following theorem therefore came as somewhat of a surprise, at least to the author.

THEOREM. *Let Ω be a bounded domain in $\mathbf{R}^n$, $n \geq 3$. Assume that the boundary $\partial\Omega$ is of class $C^{1,1}$ and that $\mathbf{R}^n \setminus \overline{\Omega}$ is connected. Suppose $c(x) \in L^\infty(\mathbf{R}^n)$ is real valued with a positive lower bound and $c(x) \equiv c_0$ outside $\overline{\Omega}$. Let ω be a fixed frequency chosen so that ω^2 is not a Dirichlet eigenvalue of $-c^2(x)\Delta$ or of $-c_0^2\Delta$ in Ω. Let $p_\omega(x,y)$ be the solution of (1) satisfying the Sommerfeld radiation condition. Then $c(x)$ is uniquely determined by the surface data $p_\omega(x,y)$ given at all $x \in \partial\Omega$, $y \in \partial\Omega$.*

The proof of this theorem, given in [1], is in fact constructive: a method is introduced, whereby, by solving certain integral equations on the surface $\partial\Omega$ one can, in theory, recover $c(x)$ exactly. The procedure does not rely on simplifying approximations or analytic continuation. We do not claim however that the method, as it stands, is ready for practical applications.

The theorem above and its proof suggest that, when multiple scattering is fully taken into account, one may be able to recover details much finer than the wavelength. The limitations on the resolution appear to have less to do with the frequency ω than with the margin of error in the boundary measurements. This opens the door to applications where only lower frequency probing may be available or desirable.

Experimental evidence suggests that the variations in the density as well as those in the compressibility are significant for tissue characterization. We thus also considered

$$\nabla_x \cdot \left(\frac{1}{\rho(x)} \nabla_x p_\omega(x,y) \right) + \omega^2 \kappa(x) p_\omega(x,y) = -\delta(x-y) \qquad (2)$$

and showed that one can determine both $\rho(x)$ and $\kappa(x)$ inside Ω from knowledge of $p_\omega(x,y)$ for all x, y on the surface $\partial\Omega$ at <u>two</u> frequencies. The precise statement and method of reconstruction are given in [1].

To justify the title of our talk, let us mention that other topics presented were

- the inverse boundary value problem of recovering an isotropic conductivity inside a body from electric measurements at its surface
- the inverse scattering problem at fixed energy for compactly supported potentials in $L^{n/2}(\mathbf{R}^n)$.

The constructive solutions given for these problems are related to the procedure leading to the theorem described above, and rely on the $\overline{\partial}$-method for multidimensional inverse scattering.

The interested reader is referred to [1], [2] for full details and further references.

This research was supported in part by the NSF under grants DMS-8602033 and DMS-8705610.

[1] A. Nachman, Reconstructions from boundary measurements, *Annals of Mathematics,* **128** (1988), 531–576.

[2] R. Lavine and A. Nachman, Global uniqueness in inverse problems with singular potentials (in preparation).

On the Discretization and Regularization of Selfadjoint Ill-Posed Problems

R. Plato

MA 6-2, Technische Universität Berlin, Straße des 17. Juni 135,
D-1000 Berlin 12, Germany

In this paper we consider a class of regularization methods for a discretized version of an selfadjoint operator equation (which includes the case that the problem is ill-posed) with approximately given right-hand side. We propose an a posteriori-parameter choice which is similar to the discrepancy principle of Ivanov-Morozov. Convergence results and (under additional conditions) convergence rates are given in the main Theorem 2.1 which is a result of joint work with G. Vainikko.

1 Introduction

Let X be a Hilbert space and $A \in L_b(X)$, i.e., $A : X \to X$ is a bounded linear operator. We suppose $A = A^* \geq 0$, i.e., A is a nonnegative selfadjoint operator, and consider the equation

$$Ax = y. \tag{1.1}$$

Here is $y \in R(A) = \{Ax : x \in X\}$. We assume that only an approximation $y_\epsilon \in Y$ to y is available with $\|y - y_\epsilon\| \leq \epsilon$, where $\epsilon > 0$ is a known error bound. To get an approximation to the minimum norm solution x_* of (1.1) we have to discretize the problem. For $h > 0$, let X_h be a finite-dimensional linear subspace of X and $P_h : X \to X$ the orthogonal projection onto X_h. We define the finite-dimensional version of A by

$$A_h : X_h \to X_h, \; x_h \mapsto P_h A x_h.$$

In the following we will assume that

$$\|A(I - P_h)\| \to 0 \;\; (h \to 0). \tag{1.2}$$

It is sufficient and necessary for condition (1.2) to hold that A is a compact operator and that $P_h \to I$ $(h \to 0)$ pointwise on $N(A)^\perp$, i.e., $P_h x \to x$ $(h \to 0)$ for all x which are elements of the orthogonal complement of the nullspace $N(A)$ of A.

If equation (1) is ill-posed we have to use a regularization method. For example, we may reguralize by discretization, i.e., we may choose h in dependence of ϵ (in particular not too small) and take the solution $x_h^* \in X_h$ of the equation $A_h x_h = P_h y_\epsilon$ (we assume for the moment that it exists and is unique) as approximation to x_*. A more favorable way is the use of regularization methods which are generated by continuous functions

$$g_r : [0, a] \to I\!R,$$

$r \geq 0$, $\|A\| \leq a$. We assume that the functions g_r satisfy the following conditions:

$$\sup_{0\leq t\leq a} t^p|1 - tg_r(t)| \leq \gamma_p r^{-p}, \quad r > 0, \quad 0 \leq p \leq p_0, \tag{1.3}$$

$$\sup_{0\leq t\leq a} |g_r(t)| \leq \gamma r, \quad r \geq 0, \tag{1.4}$$

where $p_0 > 0$, γ_p and γ are constants.

Now we suppose h and ϵ to be fixed and generate a family $(x^r)_{r\geq 0}$ of elements in X_h by

$$x^r = g_r(A_h)P_h y_\epsilon. \tag{1.5}$$

Then, with an appropriate parameter choice r (see Section 2), an approximation to the solution x_* of (1.1) is given by x^r. First let us give some examples for methods which are generated by such functions g_r.

1. *The method of Lavrentiev*: We have to determine the solution $x^r \in X_h$ of the equation

$$(A_h + r^{-1}I)x = P_h y_\epsilon.$$

The method is of the form (1.5) with $g_r(t) = (t + r^{-1})^{-1}$, $r > 0$. Conditions (1.3) and (1.4) hold with $p_0 = 1$.

2. *The generalized method of Lavrentiev*: Let $q \geq 0$. We have to determine the solution $x^r \in X_h$ of the equation

$$(A_h^{q+1} + r^{-q-1}I)x = A_h^q P_h y_\epsilon.$$

The method is of the form (1.5) with $g_r(t) = t^q/(t^{q+1} + r^{-q-1})$, $r > 0$. Conditions (1.3) and (1.4) hold with $p_0 = q + 1$.

The next two examples are iteration procedures. Here the parameter r corresponds to the number of iteration steps. Looking for a good parameter choice now means to give an appropriate stopping rule.

3. The method of successive approximation (explicit scheme): Let $0 < \mu < \frac{2}{a}$. The algorithm is given by $x^0 = 0$,

$$x^r = (I - \mu A_h)x^{r-1} + \mu P_h y_\epsilon, \quad r = 1, 2, ...$$

The method is of the form (1.5) with $g_r(t) = \frac{1}{t}[1-(1-\mu t)^r]$, $t \neq 0$. Conditions (1.3) and (1.4) hold for any $p_0 > 0$.

4. Implicit scheme: Let $0 < \mu$ be constant. The algorithm is given by $x^0 = 0$,

$$(A_h + \mu I)x^r = \mu x^{r-1} + P_h y_\epsilon$$

The method is of the form (1.5) with $g_r(t) = \frac{1}{t}[1 - (\frac{\mu}{\mu+t})^r]$, $t \neq 0$. Conditions (1.3) and (1.4) hold for any $p_0 > 0$.

2 Parameter choice

Again let h and ϵ be fixed. Let the conditions of Section 1 hold and suppose that one knows a $\xi_h \geq 0$ such that $\|A(I - P_h)\| \leq \xi_h$. Then we propose the following discrepancy principles.

Rule 1. Let $1 < d_1 \leq d_2$.
1. If $\|P_h y_\epsilon\| \leq d_1\epsilon$, then take $x_{h,\epsilon} = 0$ as approximation to x_*.
2. If $\|P_h y_\epsilon\| > d_1\epsilon$, then:
   a) Choose $0 < r \leq \xi_h^{-1} =: r_{max}$, such that

$$d_1\epsilon \leq \|A_h x_r - P_h y_\epsilon\|, \tag{2.1}$$
$$d_2\epsilon \geq \|A_h x_r - P_h y_\epsilon\|. \tag{2.2}$$

   b) If there is no $r \leq r_{max}$, such that (2.2) holds, choose $r = r_{max}$.
Then we take $x_{h,\epsilon} = x^r$ as approximation.

At first we want to show that this parameter selection rule is practicable, if some further conditions are fulfilled. For that we have to investigate the behaviour of the defect function $r \mapsto \|A_h x^r - P_h y_\epsilon\|$. Here some remarks.

1. Suppose that $|1 - tg_r(t)|$ is decreasing in r for any $t \geq 0$. Then the defect $\|A_h x^r - P_h y_\epsilon\|$ is decreasing in r. This can be shown by spectral methods.
2. From (1.4) we know that $x^0 = 0$, hence $\|A_h x^0 - P_h y_\epsilon\| = \|P_h y_\epsilon\|$.
3. Further, if $|1 - tg_r(t)|$ is continuous in r, then $\|A_h x_r - P_h y_\epsilon\|$ is continuous in r.
4. One can show that

$$\lim_{r\to\infty} \|A_h x^r - P_h y_\epsilon\| \leq \epsilon + \xi_h\|(I - P_h)x_*\|. \tag{2.3}$$

It follows that if $\xi_h = 0$ (i.e., no discretization is taken into account; in this case Rule 1 turns out to be similar to the discrepancy principle of Ivanov-Morozov), then there is really a parameter r such that (2.2) hold. Therefore Rule 1 works. Finally, it should be mentioned that if the intersection of X_h with the nullspace of A is trivial, i.e., $N(A) \cap X_h = \{0\}$, then (2.3) can be sharpened to

$$\lim_{r \to \infty} \|A_h x^r - P_h y_\epsilon\| = 0.$$

However, if $\|A(I - P_h)\| \neq 0$, then choosing the parameter r just by conditions (2.1) and (2.2) in Rule 1 do not provide convergence in general, which is due to the ill-posedness of the original problem. Therefore we have to give one more condition on r, namely condition 2 b) in Rule 1.

The next parameter selection rule may be applied to iteration procedures. The parameter r may be restricted to the set of nonnegative integers.

Rule 2. Let $1 < d,\ 0 < \theta < 1$.
1. If $\|P_h y_\epsilon\| \leq d\epsilon$, then choose $x_{h,\epsilon} = 0$ as approximation to x_*.
2. If $\|P_h y_\epsilon\| > d\epsilon$, then:
   a) Choose $0 < r \leq \xi_h^{-1} =: r_{max}$, such that there is a $s \in [\theta r, r]$ with

$$d\epsilon \leq \|A_h x^s - P_h y_\epsilon\|, \tag{2.4}$$
$$d\epsilon \geq \|A_h x^r - P_h y_\epsilon\|. \tag{2.5}$$

   b) If there is no $r \leq r_{max}$, such that (2.5) holds, choose $r = r_{max}$ or $r = [r_{max}] + 1$. $[x]$ denotes the largest integer not greater than x.

Then take $x_{h,\epsilon} = x^r$ as approximation.

Now we will state the main result. In part 1 of the following theorem a convergence result is given. There the notation "$x_{h,\epsilon} \to x_*$ $(h \to 0,\ \epsilon \to 0)$" means that for sequences of nonnegative numbers $(h_n)_{n \in \mathbb{N}}$ and $(\epsilon_n)_{n \in \mathbb{N}}$ with $h_n + \epsilon_n > 0$ we have "$x_{h_n,\epsilon_n} \to x_*$ $(n \to \infty)$". The second part provides convergence rates for solutions of (1.1) which are smooth in some sense.

Theorem 2.1 *Let $A \in L_b(X)$, $A = A^* \geq 0$, $\|A\| \leq a$, $y \in R(A)$, x_* the minimum norm solution of $Ax = y$ and $\|y - y_\epsilon\| \leq \epsilon$. Let $P_h \in L_b(X)$ be an orthogonal projection and $A_h : X_h \to X_h$, $x_h \mapsto P_h A x_h$. Suppose that $\|A(I - P_h)\| \leq \xi_h$. Suppose that conditions (1.3) - (1.5) hold for functions g_r with $p_0 > 1$ in (1.3) and that $r \mapsto |1 - tg_r(t)|$ is decreasing for any $t \geq 0$. Let the approximation $x_{h,\epsilon}$ be chosen according to Rule 1 or Rule 2.*

1. *If* $\xi_h \to 0$ $(h \to 0)$, *then* $x_{h,\epsilon} \to x_*$ $(h \to 0,\ \epsilon \to 0)$.

2. *If* $0 < p \leq p_0 - 1,\ x_* = A^p z,\ \|z\| \leq \rho$, *then*

$$\|x_* - x_{h,\epsilon}\| \leq e_p(\ (\rho\epsilon^p)^{\frac{1}{p+1}} + \rho\xi_h^{min\{p,1\}}\ ).$$

e_p *is independent of* ϵ, h *and* ρ.

Remark 1. More details (including the proof of Theorem 2.1) are given in [1].

2. Equation (1.1) with a non-selfadjoint operator A is considered in [2].

References

1. R. Plato and G.M. Vainikko, On the regularization of the Ritz-Galerkin method for solving ill-posed problems, Uch. Zap. Tartu Gos. Univ. 863, 3-18 (1989)

2. R. Plato and G.M. Vainikko, On the regularization of projection methods for solving ill-posed problems, to appear in "Numerische Mathematik"

Age Structured Population Dynamics*

M. Pilant and W. Rundell

Department of Mathematics, Texas A&M University, College Station, TX77843, USA

1. Introduction

This talk considers the problem of determining unknown coefficients in a first order hyperbolic operator subject to both initial and boundary data We wish to recover either this data or coefficients that occur in the equation itself or in the boundary conditions. To be more specific, we shall consider the single equation

$$\rho_t(a,t) + \rho_a(a,t) + \lambda\rho(a,t) = 0. \tag{1.1}$$

Equations such as (1.1) are common throughout the physical sciences, but our choice of independent variables indicates a particular application we have in mind; a stands for *age*, t for *time* and the density $\rho(a,t)$ denotes the number of individuals of age a at a particular time t. As will be explained in the next section the function λ will correspond to a "death function", and the different types of boundary condition will arise from different assumptions on the birth process.

Although we are using the language of *age structured population dynamics,* our " species" could be also be mechanical parts. In this case, $\rho(a,t)$ would denote the number of parts of age a still in service at time t, and λ would be age dependent rate of failures. Our results are by no means be limited to these applications – there are obvious extensions to more complex equations or systems of such equations, but constraints on both time and space prohibits our discussion of any of these topics.

2. The Direct Problem

Let $\rho(a,t)$ represent the number of individuals of age a alive at time t. We let L denote the maximum life span of the species. The quantity $\rho(a,t)$ is far more important, from a demographic standpoint, than the total population $P(t)$ (which is equal to $\int_0^L \rho(a,t)da$). The latter quantity would tell us the total number of individuals alive in the year 2000, the former would provide information such as the number of number of children entering school for the first time or the number of individuals of a pensionable age.

WHAT INFORMATION DO WE NEED TO RECOVER $\rho(a,t)$?

- An initial population structure — $\phi(a) \equiv \rho(a,0)$.
- A knowledge of the birth and death processes.

---

* This research was supported in part by AFOSR and NSF through grant #DMS8701338 and by ONR through grant #ONR N00014-89-J1008.

DEATHS: Let $\lambda(a)da$ denote the probability that an individual of age a will die in the interval a to $a+da$. The function λ is known as the *death function.* It is convenient to introduce the *life table*

$$\pi(a) = e^{-\int_0^a \lambda(s)ds}.$$

The quantity $\pi(a)$ represents the probability that an individual will still be alive at age a. It is a monotone nonincreasing function of a with $\pi(0)=1$ and $\pi(L)=0$, where L denotes the *maximum life span* of the species.

BIRTHS: We could prescribe the value of $B(t) = \rho(0,t)$ – the *birth rate* at time t. It is, however, more usual to assume that this function satisfies a *birth law* of the form

$$B(t) = \rho(0,t) = \int_0^L \beta(a)\rho(a,t)\,da.$$

The function $\beta(a)$ is called the *birth function* and $\beta(a)da$ denote the probability that an individual of age a will give birth in the interval a to $a+da$.

Typical examples of these functions are given in the figure below

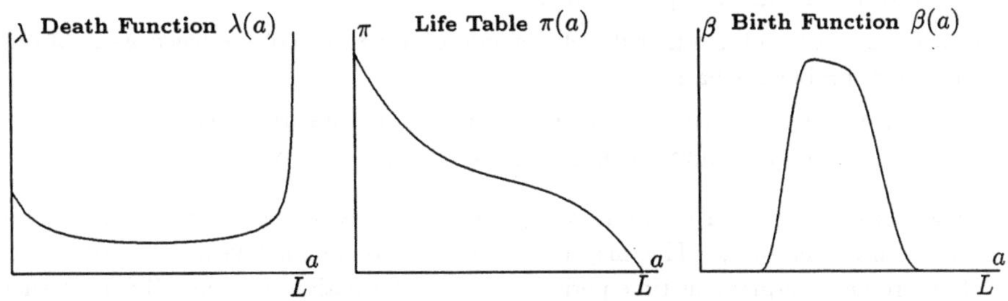

We have shown a birth function β with compact support in the interval $[a_1, a_2] \subset [0, L]$. This is realistic for human populations, but many species are able to give birth throughout old age. This can be for physiological reasons or simply because premature death to predation curtails the maximum life span to be less than obtainable in a more hospitable environment. This leads to the following *direct problem.*

GIVEN β, λ AND ϕ, RECOVER $\rho(a,t)$.

There are two essentially equivalent formulations of this problem. The first introduction of age structure in a mathematical formulation was given some seventy years ago by Lotka, [3]. He showed there exists a unique solution given (mild) assumptions on the functions β, λ, ϕ. The main idea is to show that $\rho(a,t)$ can be recovered once $B(t)$ is known. $B(t)$ is obtained as the solution of a second kind Volterra integral equation whose kernel and free term depend on β, π and ϕ. The second, due to McKendrick [4], identifies $\rho(a,t)$ as the solution of a hyperbolic initial-boundary value problem with non-local boundary condition.

It is this latter formulation that we will be of interest here. The relevant equations are

$$\rho_t(a,t) + \rho_a(a,t) + \lambda\rho(a,t) = 0 \tag{2.1}$$

$$\rho(0,t) = \int_0^L \beta(a)\rho(a,t)\,da \tag{2.2}$$

$$\rho(a,0) = \phi(a). \tag{2.3}$$

One must be careful in the interpretation of equation (1.2). From (2.3) we see that $\rho(0^+,0) = \phi(0)$, while (2.2) gives the value $\rho(0,0^+) = B(0) = \int_0^L \beta(a)\phi(a)da$. The function $\rho(a,t)$ will thus have a discontinuity at the origin unless the initial data is constrained to be consistent with the birth process, $\phi(0) = B(0)$, a condition we certainly do not wish to require. From (2.1) it then follows that this initial discontinuity will propagate along the characteristic line $t = a$ for $t < L$.

These models really have to be considered as first approximations. Their principal defect is that they neglect any effects due to overcrowding. We should expect (at least) a dependence on the total population P, $\lambda = \lambda(a,P)$ $\quad \beta = \beta(a,P)$. Existence and uniqueness results can be obtained for this case, and were first done so relatively recently. (Independently by Gurtin and McCamy [1], and Hoppensteadt [2].)
A better model might perhaps assume $\beta = \beta(a, \int_0^L w(a,s)\rho(s,t)ds)$ where the function $w(a,s)$ is relatively large for s near a and small otherwise, so that an individual of age a_0 will have a higher probability of giving birth if there are a relatively large number individuals of similar age, but a much reduced chance of giving birth otherwise regardless of the total population. Clearly, there are many possible extensions to this model, even before account is taken of such issues as the interaction of other species and the influence of spatial features. The book by Webb [8], provides further examples of the direct problem.

We will, however, look in other directions and examine some of the basic assumptions on which the direct problem is based.

ARE WE REALLY ABLE TO MEASURE THE DATA $\rho(a,0) = \phi(a)$
AS WELL AS THE BIRTH AND DEATH FUNCTIONS β AND λ?

In many cases the answer is definitely negative. The measurement of initial data assumes that a census has been taken. This may not be possible, or we may wish the reconstruct the population profile for a previous time period but lack information regarding the age structure at the start of this period. Accurate knowledge of the birth function β, even for the case where this function depends only on age, requires a monitoring of each birth to determine the age of the mother and this will only be possible under exceptional circumstances. Provided the death function λ depends only on age the experimenter may have more success in the reconstruction of this function from direct measurement, but what if it has a strong dependence on the total population? Thus a fundamental question is

CAN WE RECOVER THE BIRTH, DEATH, AND INITIAL FUNCTIONS?
IF SO, HOW?

Clearly, we must provide additional information and the exact nature of this must be dependent on the specific problem. We reformulate the previous question as

IF SOME OF THESE FUNCTIONS MUST BE CONSIDERED AS UNKNOWN,
WHAT CONSTITUTES REASONABLE OVERSPECIFIED DATA?

We shall consider two possibilities. First, assuming we can perform a census, we do so at certain time periods; that is, we measure $\rho(a,T_1)$, $\rho(a,T_2)$, etc, for certain discrete values of time T_1, T_2, ... Second, we measure the total population $P(t)$ over some suitable time interval. This last requirement might be met more easily. For example, one could monitor the total population of a herd of antelope from aerial observations, whereas the determination of the age of individual births and deaths might be impossible.

3. The Inverse Problem

From the discussion in the previous section we have some indication of the vast range of inverse problems that we might propose. To date, most of the work on recovering the birth and death processes have been based on parameter identification. Here one assumes a given form for the unknown function and adjusts a small number (2 or 3) of parameters in order to find a "best fit" with some measured data. Our approach is to determine the unknown as a function. Although in any numerical scheme such a function will depend only on a finite number of values, we do not insist that that any restriction be placed on this number. We would also like to answer the modeling issues of what constitutes sufficient data, and this requires an analysis of the map from the space of unknown coefficients to the space of overposed data. We therefore attempt to answer the following questions

UNIQUENESS: What constitutes sufficient data?

STABILITY: To what extent do the unknowns depend continuously on the data?

EXISTENCE: Can we characterize necessary conditions for the overposed data? This will be a hopeless thing to attempt in all but the simplest problems.

RECONSTRUCTION: Assuming that we are presented with reasonably accurate data that came from an actual situation and that our model is correct, can we find a numerically feasible algorithm for the recovery of the unknown coefficient or term? We must therefore seek methods that do not rely on proof by contradiction or on analytic continuation.

To date we have made no attempt to recover more than one unknown, and we also require that this unknown depend on a single argument. In this talk we shall restrict ourselves to three problems for which we have obtained results.

[1] The initial age structure $\phi(a)$ is unknown. Thus, given the birth and death rates determine $\phi(a)$ from overposed data consisting of either:

  a. The results of a census at a later time T, that is, the function $\psi(a) = \rho(a,T)$ is measured for $0 \leq a \leq L$.

  b. The total population $P(t)$ over a life span L, that is, measure $P(t) = \int_0^L \rho(a,t)da$.

[2] The birth function $\beta(a)$ is unknown. It is to be recovered from knowledge of $\phi(a)$, $\lambda(a,P)$ and overposed data consisting $P(t)$ for $0 \leq t \leq L$.

[3] The death function λ is known to depend on both age and total population, and that the dependence on age has already been determined. The birth function and initial age structure are known. The overposed data is $P(t)$ for $0 \leq t \leq T$.

  a. The death rate is a simple sum of the natural and environmental rates, that is, λ has the form $\lambda_N(a) + \lambda_E(P)$, where λ_N is the death rate due to natural causes and λ_E is the death rate due to the environment.

  b. Environmental pressure is greater on the age group with a high natural death rate, $\lambda(a,P) = \lambda_N(a)[1 + \lambda_E(P)]$.

In both cases we expect that λ_E is small for relatively small values of P, so that $\lambda \approx \lambda_N(a)$ and that the process is well understood in this case. The problem is to extrapolate this knowledge to attempt to recover the dependence of λ on P when the latter may strongly affect the death process.

4. Summary of Results

We shall summarize our results for these problems. Our intention is not to give precise statements but to show the flavor of what can be feasibly recovered. In each case we have done some numerical experiments to test the validity of our approaches. Details can be found in [5, 6, 7].

PROBLEM [1.A]: Given the birth and death functions λ and β, recover the initial age structure $\phi(a) = \rho(a,0)$ from

$$\rho_t(a,t) + \rho_a(a,t) + \lambda\rho(a,t) = 0 \quad 0 < a < L,\ 0 < t < T$$

$$\rho(0,t) = \int_0^L \beta(a)\rho(a,t)\,da \quad 0 \le t \le T$$

$$\int_0^L \rho(a,t)\,da = P(t). \quad 0 \le t \le T \qquad \text{(overposed data)}$$

ASSUMPTIONS: The crucial condition concerns the rate of growth of the death function as $a \to L$. It will be useful to cast this in terms of the behavior of the life table $\pi(a)$ at the point $a = L$. By definition this function must vanish at L and be strictly positive for $a > L$, but at what rate does $\pi(a)$ tend to zero as a approaches L? Specifically, we assume that there is an integer m such that the mth derivative $\pi^{(m)}(L) \neq 0$, but $\pi^{(r)}(L)$ is zero for $0 \le r < m$. We require some smoothness on the data, basically $\pi(a)$, $\beta(a)$ should have $2m+1$ continuous derivatives and $P(t)$ should have $m+2$ continuous derivatives.

Under these assumptions it can be shown that the problem can be reduced to a second kind Fredholm integral equation for the unknown $\phi(a)$. The kernel of this equation depends only on the birth and death functions, and does so in continuous manner. Thus in this linear case we obtain both existence and uniqueness for almost all β and λ. One can give a relationship between the birth and death functions that will guarantee uniqueness but this is sufficiently complex that one would not use it in practice.

REMARK 1: The above assumptions are not sufficient to guarantee that the initial function $\phi(a)$ will be nonnegative, which is the only admissible class of such functions. From a practical standpoint this is not a serious objection since any physically observed $P(t)$ must come from a nonnegative initial distribution.

REMARK 2: One must provide the values of the overposed data on an interval of length L. If this interval is $[\tau, L+\tau]$ then one can recover $\rho(a,\tau)$. When $\tau = 0$ this recovers the initial population structure $\rho(a,0) \equiv \phi(a)$.

REMARK 3: The degree of ill-conditioning of the inverse problem is determined completely by the value of the integer m. We can expect that the inverse problem will have m "loss of derivatives", that is, the we will be able to control the supremum norm of ϕ provided we control the supremum norm of $P^{(m)}(t)$. For small values of m we can expect the recovery problem to be reasonably straightforward. From numerical experiments we have found that good results are obtained for $m = 1$, with rapid degradation for higher values.

REMARK 4: We get a Fredholm equation formulation here because the unknown ϕ has its variables in the *age* direction, while the overposed data P has variables in the orthogonal *time* direction.

PROBLEM [1.B] Recover the initial age structure $\phi(a) = \rho(a,0)$ from

$$\rho_t(a,t) + \rho_a(a,t) + \lambda\rho(a,t) = 0 \quad 0 < a < L,\ 0 < t < T$$

$$\rho(0,t) = \int_0^L \beta(a)\rho(a,t)\,da \quad 0 \le t \le T$$

$$\rho(a,T) = \psi(a). \quad 0 \le a \le L \qquad \text{(overposed data)}$$

We introduce the *net maternity function* $b(a) \equiv \beta(a)\pi(a)$. A typical value of this function is shown below.

Net Maternity Function $\beta(a)$

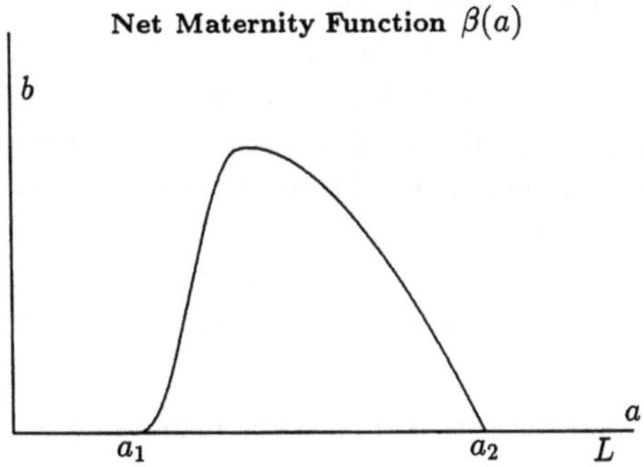

ASSUMPTIONS: Now the crucial condition is the behavior of the net maternity at the point a_2. Let m be the least integer such that the mth derivative $b^{(m)}(a_2)$ is nonzero. Of course we need some smoothness on the data; $\pi(a)$, $\beta(a)$ and $\psi(a)$ should be $m+1$ times continuously differentiable on $[0, L]$, except in the case that $T < L$ when we must allow ψ to have a simple discontinuity at $a = T$.

With this we can show that the problem can be reduced to a second kind Volterra integral equation for the function $\phi(a)$ if $T < L$, and thus existence and uniqueness holds. We should note the following.

REMARK 5: If $T > L - a_2$ then we can never recover those values of $\phi(a)$ on the range (a_2, L). (Those individuals in the initial population whose age is greater then a_2 do not participate in the birth process and have all died within a time period of length $L - a_2$.)

REMARK 6: The degree of ill-conditioning of the inverse problem is determined by the behavior of the birth function $b(a)$ in a neighborhood of its maximum support. If the species is able to give birth until age L then the index m is determined by the life table function as in the previous case.

REMARK 7: If $T > L$ then we can reduce the problem from the time interval $[T - L, T]$ to one on the interval $[T - 2L, T - L]$ etc. However, if $T > nL$ then we must solve n first kind Volterra integral equations in order to do this. Each one of these will entail a "derivative loss" of degree m_L where now m_L is the least integer such that the m_Lth derivative of b does not vanish at the maximum life span L. For the case when $a_2 < L$ $m_L = \infty$ and this implies that the problem will be severely ill conditioned. Thus, the message here is clear; although uniqueness of the solution is guaranteed for all values of T, if you wish to recover initial data from a later census, do so before a life span has elapsed.

PROBLEM [2] Recover the birth function $\beta(a)$ from

$$\rho_t(a,t) + \rho_a(a,t) + \lambda(a,P)\rho(a,t) = 0 \quad 0 < a < L,\ 0 < t < L$$

$$\rho(0,t) = \int_0^L \beta(a)\rho(a,t)\,da \quad 0 \le t \le L$$

$$\int_0^L \rho(a,t)\,da = P(t) \quad 0 \le t \le L \qquad \text{(overposed data)}$$

ASSUMPTIONS: Once again there is a not–so–obvious condition that should be imposed. We *require* that the initial data *not* be compatible with the birth process, $\phi(0) \neq B(0)$. This condition can be expressed in terms of the given data. The overposed data should have two continuous derivatives, the non-negative functions ϕ and λ should have one.

With this we can obtain a second kind Fredholm equation for the unknown β. Despite the fact that the direct problem may be nonlinear (since λ may depend on P) the inverse problem is linear in β. The fredholm alternative applies, and the solvability condition involves not only the functions λ and ϕ but also the overposed data $P(t)$. The continuous solution β (when unique) depends continuously in the supremum norm on the second derivative of the overposed or measured data $P(t)$ so that the problem is mildly ill-conditioned. Examples show that one cannot require less, and that the condition $B(0) \neq \phi(0)$ is, in general, essential for existence and uniqueness.

REMARK 8: We obtain a Fredholm equation here since the data $P(t)$ and the unknown $\beta(a)$ have variables that lie in orthogonal directions.

PROBLEM [3.A] Recover the "environmental death rate" $\lambda_E(P(t))$ from

$$\rho_t(a,t) + \rho_a(a,t) + \lambda\rho(a,t) = 0 \quad 0 < a < L,\ 0 < t < T$$

$$\rho(0,t) = \int_0^L \beta(a)\rho(a,t)\,da \quad 0 \le t \le T$$

$$\rho(a,0) = \phi(a). \quad 0 \le a \le L$$

$$\int_0^L \rho(a,t)\,da = P(t) \quad 0 \le t \le T \qquad \text{(overposed data)}$$

where $\lambda(a,P) = \lambda_N(a)[1 + \lambda_E(P)]$ and $\lambda_N(a)$ is known. The case $\lambda(a,P) = \lambda_N(a) + \lambda_E(P)$ is similar and in fact simpler.

ASSUMPTIONS: We of course require some smoothness and positivity requirements on the data. The overposed data must satisfy certain inequality constraints.

We can show that the inverse problem for recovery of λ_E can be reduced to showing that a certain mapping $\mathbf{T}$ has a fixed point. The contraction mapping theorem can be invoked to obtain uniqueness, existence and continuous dependence on the data. It is feasible to numerically reconstruct the solution from successive iterates of the process $\lambda_E^{(n+1)} = \mathbf{T}\lambda_E^{(n)}$ with $\lambda_E^{(0)} = 0$.

PROBLEM [3.B] As in the previous case except the birth law is of the form

$$\rho(0,t) = B(t)$$

where $B(t)$ is given data.

This is an easier case and here one can show that the mapping $\mathbf{T}$ is isotone so that the iterates converge monotonically to the solution.

References

[1] Gurtin, M. E and MacCamy, R. C. Nonlinear age dependent population dynamics, *Archives for Rational Mechanics and Analysis.* **54**, (1974), 281–300.

[2] Hoppensteadt, F. An age dependent epidemic model, *Journal Franklin Institute*, **297**, (1974), 325–333.

[3] Lotka, A. J. (1956) *Elements of Mathematical Biology*, New York: Dover.

[4] McKendrick, A. G. Applications of mathematics to medical problems, *Proceedings Edinburgh Mathematics Society*, **44**, (1926), 98–130.

[5] Pilant, M and Rundell, W. Determining the initial age distribution for an age structured population, *Submitted.*

[6] Pilant, M and Rundell, W. Determining a coefficient in a first order hyperbolic equation, *Submitted.*

[7] Rundell, W. Determining the birth function for an age structured population, *Mathematical Population Studies*, **4**, (1989)

[8] Webb, G. F. *Theory of Nonlinear Age-dependent Population Dynamics*, New York:Marcel Dekker (1985).

Ball Stabilization in Inverse Problems

M. Ciulli[1], *S. Ciulli*[1], *and T.D. Spearman*[2]

[1]Laboratoire de Physique Mathématique, University of Montpellier 2, F-34095 Montpellier Cedex 05, France

[2]School of Mathematics, Trinity College, Dublin 2, Ireland

1. Introduction

We take as starting point a continuous, bijective map $\mathcal{M}$ from a Banach space C to a Banach space S.[1] The inverse problem starts from an initial data function $D \equiv D(s)$, which is a point in S, and seeks, by using a properly defined generalized inverse map $\mathcal{M}_Q^{-1}$ to obtain as solution a well-defined function $\phi \equiv \phi(t)$ belonging to C. Continuity of $\mathcal{M}$ does not ensure that the inverse map $\mathcal{M}^{-1}$ will be continuous (see Fig. I below); more generally the stabilization of the generalized (quasisolution) inverse map $\mathcal{M}_Q^{-1}$ is a central element in solving the inverse problem. The need to introduce quasisolutions arises because D, although it belongs to S, may not be in $\mathcal{M}(C)$.

A standard method of approach to this problem, due to Tykhonov /1963/, is to restrict the space C to some compact subset C_K. This is illustrated in Figure 1. The continuity of

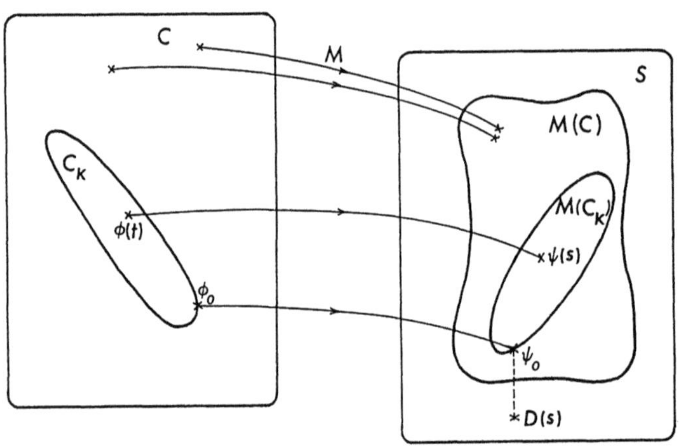

Figure 1: *Continuous bijective map $\mathcal{M}$ from C to S. Note that $\mathcal{M}^{-1}$ will not, in general, be continuous. $D \equiv D(s)$ is an input data function which is in S but may not be in $\mathcal{M}(C)$. C_K is a compact set in C. ψ_0 is the nearest point (see discussion in text) in $\mathcal{M}(C_K)$ to D.*

[1]P.C. Sabatier /Sabatier, 1978/uses the notation C and E for these two spaces, and this convention hs been widely adopted. We have translated the French *Espace* into an English *Space*. The space C is usually referred to as a *Control* space.

$\mathcal{M}$ means that for any compact set $C_K \subset C$ the image of $C_K, \mathcal{M}(C_K) \subset S$, will also be compact. Assuming that D is outside $\mathcal{M}(C_K)$, the compactness of $\mathcal{M}(C_K)$ allows one to prove the existence of a nearest point $\psi_0 \in \mathcal{M}(C_K)$ to D. Provided that this nearest point is unique, which depends on the geometry of $\mathcal{M}(C_K)$ (for example if $\mathcal{M}(C_K)$ is convex ψ_0 will be unique), then it also follows from compactness that ψ_0 will vary continuously with D. This leads to the following procedure. Starting from D one obtains a nearest point ψ_0 in $\mathcal{M}(C_K)$. Since C_K is compact, the inverse map $\mathcal{M}^{-1}$ from $\mathcal{M}(C_K)$ to C_K will be continuous. Thus provided that the geometry of $\mathcal{M}(C_K)$ is such that ψ_0 is unique,the map $\mathcal{M}_Q^{-1}$ defined by the chain of operations

$$D \in S \rightarrow \psi_0 \in \mathcal{M}(C_K) \rightarrow \phi_0 \in C_K \tag{1}$$

will be continuous.

The solution thus obtained is strongly dependent on the choice of the compact set C_K. If the solution ϕ_0 is to be relevant to the physical problem under consideration then the restriction of the control space C to C_K should be based on some physical condition which it is appropriate to incorporate into the problem, not just the arbitrary imposition of some compact set, achieved for example by some cut-off procedure[2]. The difficulty in this is that typical physical conditions cannot readily be represented by the restriction to a compact set. It is much more likely that physical constraints take the form of a bound of some sort, represented by a ball condition in some suitably defined norm on C.

We define the ball B_δ, for some finite radius δ, by the condition $\|\phi\| \leq \delta$. The type of bound which this represents will depend on the choice of norm - for example, in the L^∞ norm this corresponds to a direct bound on $|\phi|$ whereas in L^2 it takes the form $\int |\phi(t)|^2 \sigma(t) dt \leq \delta^2$, for some positive weight function $\sigma(t)$. There is no reason to restrict B_δ to be centred at the origin; it is simply for convenience that we write it so. Unfortunately, a ball B_δ is not compact (unless the space C is finite-dimensional) and so the Tykhonov procedure cannot be applied directly. The main purpose of this paper is to analyse the implications of imposing a ball constraint in C and to examine to what extent, if any, this can contribute towards stabilizing the inverse problem.

In order to proceed we shall assume that $\mathcal{M}$ is a linear map defined through an integral kernel $K(s,t)$ as follows:

$$\psi_n(s) = \int_\Gamma K(s,t)\phi_n(t)dt \quad t \in \Gamma, s \in \gamma. \tag{2}$$

[2]One method which is frequently used to restrict C to a compact subset is to make some sort of expansion and then to truncate this so as to retain only a finite number of terms. This has the effect of reducing the space to a finite dimensional one in which any bounded closed set will be compact. The resulting solution, although stable for a particular truncation procedure, will generally be rather sensitive to the choice of truncation which is adopted and one cannot readily justify the selection of a particular mode of truncation on directly physical grounds.

The functions $\phi_n(t)$ are assumed to be L^p on the closed interval Γ, for some $p(1 \leq p \leq \infty)$. $K(s,t)$ will be required to be continuous in s and bounded ($|K(s,t)| < M$), though at times we shall relax this requirement and only insist that $K(s,t)$ be Hilbert-Schmidt, i.e. $\int_\gamma ds \int_\Gamma dt |K(s,t)|^2 < M$.

To go further we need some mathematical results which will be introduced in the next section. These will enable us to show that subject to certain conditions M(B_δ) is compact.

2. Compactness of $\mathcal{M}(B_\delta)$ from the Arzelà-Ascoli and Alaoglu Banach theorems

The Arzelà-Ascoli theorem states that *any infinite family of uniformly bounded equicontinuous functions $\psi_n(s)$, defined on a closed interval γ, will have a sub-sequence $\{\psi_{n'}(s)\}$ which converges uniformly on γ. The limit $\psi_0(s)$ need not however belong to the family $\psi(s)$*[3]

The functions $\psi_n(s)$ defined by eq. (2) above, for $\phi_n(t) \in L^p(1 \leq p \leq \infty)$ and $K(s,t)$ continuous in s and bounded, satisfy the conditions for the theorem and hence converge uniformly on the closed interval γ on which they are defined. It follows immediately that $\mathcal{M}(B_\delta)$ is relatively compact i.e. that $\overline{\mathcal{M}}(B_\delta)$ is compact. Uniform convergence means convergence in the L^∞ norm, so the relative compactness of $\mathcal{M}(B_\delta)$ is with respect to the L^∞ norm or, a fortiori, with respect to any other L^p norm.

It is important to emphasise at this stage that the usefulness of the relative compactness of $\mathcal{M}(B_\delta)$ is, on its own, distinctly limited. Relative compactness does not allow us to proceed with the Tychonov procedure: if ψ_0 is on the boundary and thus does not belong to $\mathcal{M}(B_\delta)$, ϕ_0, if it exists at all, will not be in B_δ. Compactness and not just relative compactness, must be the goal.[4] Fortunately we can call on another theorem, that of Alaoglu and Banach, which, provided the appropriate conditions are satisfied, will allow us to show that $\mathcal{M}(B_\delta)$ is indeed compact.

Before stating the theorem we recall the distinction between strong, weak, and weak-* convergence. Consider some sequence $\{u_n\}$ in a Banach space U; convergence is defined as follows:

(i) strong (norm) convergence

$$u_n \to u \Leftrightarrow \|u_n - u\| \to 0 \tag{3}$$

(ii) weak convergence

$$u_n \rightharpoonup u \Leftrightarrow < l, u_n > \to < l, u >, \forall < l, . > \in U' \tag{4}$$

(U' is the dual space of U and we have used the notation $< l, . >$ for the elements of U')

[3]There need not exist a function ϕ_0 in terms of which ψ_0 could be represented as in eq. (2).

[4]To appreciate how much weaker a condition relative compactness as opposed to compactness actually is it may be helpful to appreciate that the boundary set can be large in comparison with the initial set. Consider, for example, the set of rational numbers.

(iii) weak-* convergence - this requires that U is itself the dual of some Banach space V, i.e. $U = V'$

$$u_n \xrightarrow{*} u \Leftrightarrow < l, u_n > \rightarrow < l, u >, \forall < l, . > \in V \tag{5}$$

($V \subset V'' = U'$. If $V = V''$ then U is reflexive and weak-* and weak are equivalent.)

We also need the concepts of weak, and weak-*, compactness. A set is said to be weak, or weak-*, compact if any infinite sequence $\{u_n\}$ within the set has a sub-sequence $\{u_{n'}\}$ which is weak, or weak-*, convergent to a point u also belonging to the set.

The statement of the Alaoglu-Banach theorem is as follows: *if U is the dual of some Banach space V ($U = V'$) then any ball B_δ (of finite radius δ) in U is weak-* compact.* To explore the implications of this we return to the map $\mathcal{M}$ defined by eq.(2). Any L^p space, with the sole exception of L^1, is the dual of another Banach space L^q, where $p^{-1} + q^{-1} = 1$. So provided that the $\phi(t)$ are L^p, for $1 < p \leq \infty$, the conditions of the theorem apply and B_δ will be weak-* compact. Further, if $K(s,t)$ is L^q with respect to t, for any value of s, which will be the case (for any q) if $K(s,t)$ is bounded, then the weak-* convergence of a sequence $\{\phi_{n'}(t)\}$ in B_δ to $\phi_0(t) \in B_\delta$ implies that for any individual value of s, the sequence $\{\psi_{n'}(s)\}$ of numbers converges to the number $\psi_0(s)$. In other words the sequence of functions $\{\psi_{n'}(s)\}$ converges point-wise to the function $\psi_0(s)$.

So, if the space C is L^p, for $1 < p \leq \infty$ (but $p \neq 1$), and if $K(s,t)$ is continuous in s and bounded, then the conditions for both theorems apply, giving the following two results, where $\{\phi_n\} \in B_\delta$ and $\{\psi_{n'}\}$, $\{\psi_{n''}\}$ are sub-sequences of $\{\psi_n\}$:

A. Arzelà-Ascoli

$$\psi_{n'} \rightarrow \psi_0. \tag{6}$$

ψ_0 need not lie in $\mathcal{M}(B_\delta)$. The convergence is uniform, i.e. it is with respect to the L^∞ norm (and, a fortiori, any other L^p norm).

B. Alaoglu-Banach

$$\psi_{n''}(s) \rightarrow \tilde{\psi}_0(s) \text{ for each } s \in \gamma. \tag{7}$$

In this case the convergence is pointwise in s, and $\tilde{\psi}_0(s)$ does belong to $\mathcal{M}(B_\delta)$. Combining these two results it is not difficult to show /Ciulli et al. 1990/ that $\psi_0 \equiv \tilde{\psi}_0$ and hence that $\psi_0 \in \mathcal{M}(B_\delta)$. It follows that $\mathcal{M}(B_\delta)$ is compact with respect to the L^∞ norm and, a fortiori, with respect to any other L^p norm. This is the result which we set out to obtain.

It is worth noting that if we are willing to restrict both C and S to be L^2 then the conditions imposed on $K(s,t)$ may be relaxed. To prove, in this case, that $\mathcal{M}(B_\delta)$ is compact, it is sufficient that $K(s,t)$ be Hilbert-Schmidt (i.e. that $\int_\gamma ds \int_\Gamma dt |K(s,t)|^2 < M$). Continuity in s is not required. More generally, if C is reflexive ($C'' = C$) and if $\mathcal{M}$ is a compact map (this is, by definition one which maps a ball B_δ into a relatively compact set) then $\mathcal{M}(B_\delta)$ will be compact.

3. Implications for the inverse problem

Where do the results of the previous section leave us as regards the inverse problem? We have shown that, subject to certain conditions, $\mathcal{M}(B_\delta)$ is compact. A linear map of the ball B_δ will itself certainly be a convex set. Hence there will be a uniquely defined nearest point ψ_0 which is continuous with respect to D. The corresponding point ϕ_0 in B_δ will be uniquely specified. Does this mean that we have solved the inverse problem $D \rightarrow \psi_0 \rightarrow \phi_0$? The difficulty is that ϕ_0 is extremely unstable with respect to small changes in ψ_0. This is illustrated in Figure 2 where we see that an arbitrarily small region about ψ_0 in $\mathcal{M}(B_\delta)$ maps into an extended *nebula* in B_δ. Although the point ϕ_0 is uniquely specified, and an algorithm [5] may be devised to determine it starting from D, given the instability of ϕ_0 with respect to ψ_0 it would be difficult to argue that ϕ_0 should be regarded as a physically meaningful solution. A potentially misleading, indeed treacherous, aspect of the situation is that the solution ϕ_0 may be continuous with respect to changes in D. This may be shown to be the case when C and S are both L^2/Ciulli et al, 1990/. So what we have to say is that the result of restricting C to the ball B_δ is to yield an extended *nebula* of solutions, all of which map into a small region

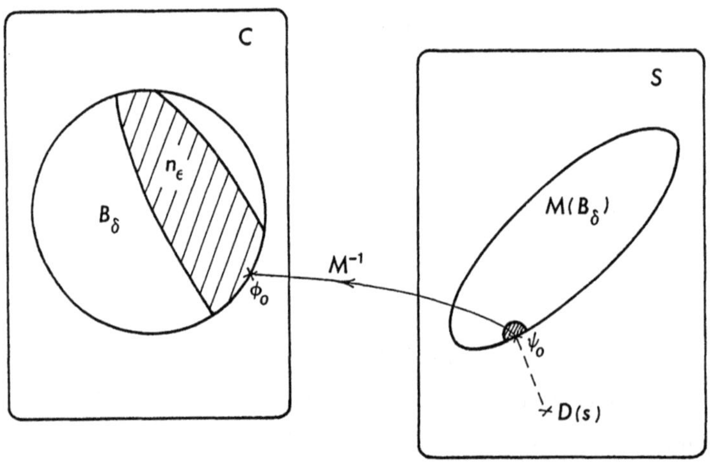

Figure 2: *The inverse map $D \rightarrow \psi_0 \rightarrow \phi_0$. Although ϕ_0 is uniquely specifed, it is quite unstable in the sense that an arbitrarily small region about ψ_0 in $\mathcal{M}(B_\delta)$ will map into an extended nebula in B_δ.*

[5] For an example of such an algorithm using the L^2-norm see Ciulli et al. /1984/, Ciulli /1988/. A procedure based on the L^∞ norm has been developed by Auberson et al. /1989/.

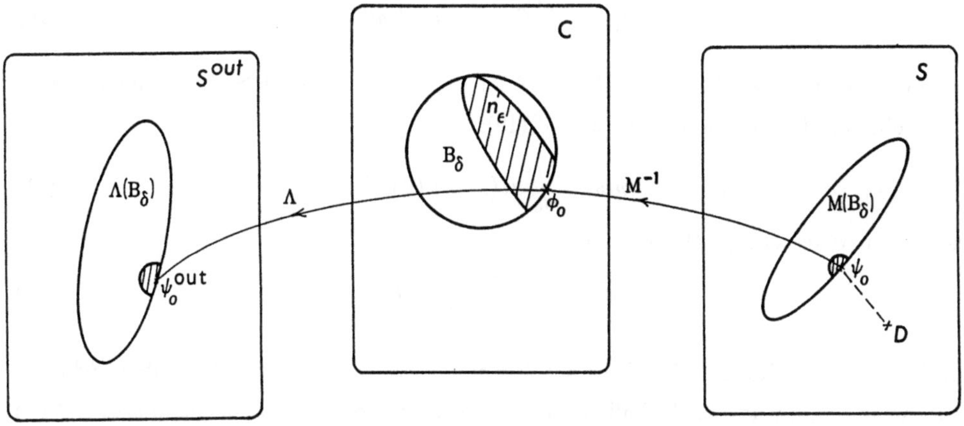

Figure 3: *The map* $D \to \psi_0 \to \phi_0 \to \psi_0{}^{\text{out}}$.

about ψ_0. The imposition of the ball constraint on C does not stabilize the inverse problem from S to C. It is important that this should be clearly stated and well understood.

Is this the end of the story? There are two ways in which we can usefully respond to this situation. The first is to ask if we really need to solve for ϕ itself or could instead settle for determining a function $\psi^{\text{ out}} \in S^{\text{ out}}$, defined by

$$\psi^{\text{ out}}(t) = \int_\Gamma \bigwedge(t,t')\phi(t')dt' \tag{8}$$

for any bounded $\bigwedge, \bigwedge \in V$ where $V' = U$. If $\bigwedge$ is continuous in t then any infinite sequence. $\{\psi_n{}^{\text{out}}\}$ corresponding to $\{\phi_n\} \in B_\delta$ has a subsequence $\{\psi_{n'}{}^{\text{out}}\}$ which converges in the L^∞ norm to a limit $\psi_0{}^{\text{out}}$. Hence the following map

$$D \to \psi_0 \to \phi_0 \to \psi_0{}^{\text{out}}$$

yields a solution $\psi_0{}^{\text{out}}$ which is stable. This is illustrated in Figure 3. Note that if the ϕ's are L^2, it is sufficient that $\bigwedge$ be Hilbert-Schmidt, $\int\int |\bigwedge(t',t)|^2 dt dt' < M$, for $\{\psi_{n'}\}$ to converge in the L^2 norm.

If this is not acceptable, and a solution ϕ_0 in C itself is what is sought, then a possible procedure is to define a new control space C' with respect to which C has the form of an $S_{\text{ out}}$ as defined in eq. (8). For example, if we were prepared to restrict the space C to differentiable functions then we could define C' to be the space of derivative functions $\phi'(t)$. The ball condition will now be imposed on C'; to be specific let us define B_δ using the L^2 norm

$$\|\phi'\|^2 = \int_0^1 |\phi'(t)|^2 \sigma(t)dt \tag{9}$$

where for convenience we have taken Γ to be $[0,1]$. $\phi(t)$ is given in terms of $\phi'(t)$ by

$$\phi(t) = \int_0^1 \Theta(t-t')\phi'(t')dt' \tag{10}$$

where $\Theta(t)$ is the usual Θ-function ($\Theta(t) = 0$ for $t < 0, \Theta(t) = 1$ for $t \geq 0$). The map from C' to S is obtained by combining eqs. (2) and (10):

$$\psi(s) = \int_0^1 \tilde{K}(s,t')\phi'(t')dt' \tag{11}$$

where the new kernel $\tilde{K}(s,t)$ is given by

$$\tilde{K}(s,t) = \int_t^1 K(s,\bar{t})d\bar{t} \ . \tag{12}$$

Now if $K(s,t)$ is a Hilbert-Schmidt kernel then so is $\tilde{K}(s,t)$; $\Theta(t-t')$ is also Hilbert-Schmidt. Hence weak compactness of the ball B_δ in C' implies L^2 - compactness of the images of B_δ in both C and S. It follows that the ball condition, where imposed on the space of the derivatives ϕ', leads to a stable map from S to C. This may be seen from Figure 3 if we read S_{out} as C and ${\psi_0}^{\text{out}}$ as ϕ_0, and if we replace C by C' and ϕ_0 by ϕ'_0.

Those readers who would wish to see the above methods applied to a specific inverse problem are referred to Ciulli et al (1990) where one such application is described in detail.

ACKNOWLEDGEMENTS

We are grateful to G. Auberson for many helpful discussions, and in particular for first suggesting to us the possible relevance of the Alaoglu-Banach theorem. We have also benefitted from many discussions with G. Mennessier. The on going contact with P.C. Sabatier and his group in Montpellier, with their wide experience of inverse problems, has been most valuable. One of us, T.D.S., acknowledges support from the CNRS and from the University of Montpellier which enabled him to make several visits to Montpellier during the course of which most of this work was done.

REFERENCES

Auberson, G., Mennessier, G. (1989): Commun. Math. Phys. **121**, 49

Ciulli, S., Spearman, T.D. (1984): Nuovo Cimento **83A**, 352

Ciulli, M., (1988): Ph.D. thesis, University of Dublin (unpublished)

Ciulli, M., Ciulli, S., Spearman, T.D. (1990): *Festschrift for André Martin*, Springer Tracts in Modern Physics (Springer, Berlin, Heidelberg, New York).

Sabatier, P.C. (1978): *Applied Inverse Problems*, Lecture Notes in Physics No. 85 (Springer, Berlin, Heidelberg, New York)

Tykhonov, A.N. (1963): Sov. Math. Dokl. **4**, 1035.

Part III

Spectral Inverse Problems

Algorithms for Solving Inverse Eigenvalue Problems for Sturm-Liouville Equations

L.-E. Andersson

Department of Mathematics, University of Linköping, S-58183 Linköping, Sweden

1. Introduction

In the present paper we will consider the inverse eigenvalue problem for a Sturm-Liouville equation in so called impedance form,

$$(p^2(x)u'(x))' + \omega^2 p^2(x)u(x) = 0, \quad 0 \le x \le L, \quad p(x) > 0 \tag{1}$$

and with the following boundary conditions and notations for the eigenfrequencies.

$$\begin{aligned} u'(0) = u'(L) = 0, &\quad \{\pm\omega_{2k}\}_{k=0}^{\infty} \\ u'(0) = u(L) = 0, &\quad \{\pm\omega_{2k+1}\}_{k=0}^{\infty}. \end{aligned} \tag{2}$$

It is well known that the eigenfrequencies are interlacing, so that

$$0 = \omega_0 < \omega_1 < \ldots < \omega_k < \omega_{k+1}.$$

By the aid of the substitution $v(x) = p^2(x)u'(x)/i\omega$ the second order equation (1) may be rewritten as the following first order system,

$$\begin{pmatrix} u'(x) \\ v'(x) \end{pmatrix} = i\omega \begin{pmatrix} 0 & 1/p^2(x) \\ p^2(x) & 0 \end{pmatrix} \begin{pmatrix} u(x) \\ v(x) \end{pmatrix} \tag{3}$$

or as an equation for the fundamental matrix U,

$$\begin{aligned} U'(x,\omega) &= i\omega \begin{pmatrix} 0 & 1/p^2(x) \\ p^2(x) & 0 \end{pmatrix} U(x,\omega) \\ U(0,\omega) &= I, \end{aligned} \tag{4}$$

I denoting the unit matrix. Let us also note that, if p is sufficiently regular then the substitution $y(x) = p(x)u(x)$ transforms (1), (2) into

$$\begin{aligned} &y''(x) + (\omega^2 - q(x))y(x) = 0 \\ &y'(0) - hy(0) = y'(L) + Hy(L) = 0 \\ &y'(0) - hy(0) = y(L) = 0, \end{aligned} \tag{5}$$

where $h = p'(0)/p(0)$ and $H = -p'(L)/p(L)$. Further, for

$$U(L,\omega) = \begin{pmatrix} u_{11}(L,\omega) & u_{12}(L,\omega) \\ u_{21}(L,\omega) & u_{22}(L,\omega) \end{pmatrix} \tag{6}$$

we have the canonical product formulas,

$$u_{11}(L,\omega) = \prod_{k=0}^{\infty}\left(1-\frac{\omega^2}{\omega_{2k+1}^2}\right), \quad u_{21}(L,\omega) = ic\omega\prod_{k=1}^{\infty}\left(1-\frac{\omega^2}{\omega_{2k}^2}\right) \tag{7}$$

with $c = \int_0^L p^2(s)\,ds/p^2(0)$. Here $u(L) = u_{11}(L,\omega)$ and $u'(L) = -p^2(0)\omega u_{21}(L,\omega)/p^2(L)$ if $u(0) = 1$, and $u'(0) = 0$. If $p \equiv 1$ then $u_{11}(L,\omega) = \cos\omega\delta$ and $u_{21}(L,\omega) = \sin\omega\delta$.

Previous work on the inverse eigenvalue problem for Sturm-Liouville equations has mostly been concerned with the equation in the form (5). The literature is extensive and only a few references will be given here.

The first one to study the inverse eigenvalue problem systematically was G. Borg, 1946, [3]. He established that, in general, two sets of eigenvalues were needed to obtain uniqueness for the potential q. In 1951 the famous Gelfand-Levitan equations, [4], appeared, providing a constructive procedure for the recovery of q. A different inversion algorithm was presented by H. Hochstadt [6], in 1973. O. Hald, [5], has used the Gelfand-Levitan equations to design numerical inversion algorithms. In two recent articles, [1,2], the present author has investigated the inverse eigenvalue problem for the impedance form, (1), of the Sturm-Liouville equations. The inversion methods given in this paper are based on ideas from these articles.

The point of view taken in this article, is to use the functions $u_{11}(L,\omega)$ and $u_{21}(L,\omega)$, or rather their inverse Fourier transforms, for the reconstruction of p. The rest of the paper is organized as follows.

In section 2 we consider a discretization of the problem obtained by requiring that the function p should be piecewise constant over intervals of equal length. A theorem of existence and uniqueness is given with an elementary proof. In section 3 we present an explicit inversion algorithm based on the theory of section 2. Apart from rounding off errors, the algorithm is exact. In section 4 we will indicate a closely related algorithm valid for the case when $p'/p \in L^2(0,L)$. In section 5, finally, a numerical result is given.

2. An existence and uniqueness result

Let $x_k = kL/N$, $-1 \le k \le N$ and assume that $p(x)$ is constant over each interval $I_k = (x_{k-1}, x_k)$, i.e. so that $p(x) = p_k$ for $x \in I_k$, $k = 0, 1, \ldots, N$. Also let $q_k = p_k/p_{k-1}$. The solution $U(x,\omega)$ of (3) is a continuous function of x and using that (1) takes the form $u''(x) + \omega^2 u(x) = 0$ on every subinterval I_k it is straightforward to verify that, with $\delta = L/N$,

$$U(x_k,\omega) = \begin{pmatrix} \cos\omega\delta & ip_k^{-2}\sin\omega\delta \\ ip_k^2\sin\omega\delta & \cos\omega\delta \end{pmatrix} U(x_{k-1},\omega). \tag{8}$$

Substituting $Y(x,\omega) = \begin{pmatrix} p(x) & 0 \\ 0 & 1/p(x) \end{pmatrix} U(x,\omega)$ it follows that

$$Y(x_k-,\omega) = \begin{pmatrix} q_k\cos\omega\delta & iq_k^{-1}\sin\omega\delta \\ iq_k\sin\omega\delta & q_k^{-1}\cos\omega\delta \end{pmatrix} Y(x_{k-1}-,\omega) \tag{9}$$

and that

$$Y(x_k+,\omega) = \begin{pmatrix} q_{k+1} & 0 \\ 0 & q_{k+1}^{-1} \end{pmatrix} Y(x_k-,\omega) \tag{10}$$

Let us introduce the notation $z = e^{i\omega\delta}$ and $C_k = C(q_k, z)$ where

$$C(q,z) = \frac{1}{2}\begin{pmatrix} q(z+1/z) & q^{-1}(z-1/z) \\ q(z-1/z) & q^{-1}(z+1/z) \end{pmatrix}. \tag{11}$$

Consequently

$$Y(L-,\omega) = C_N C_{N-1} \dots C_1 Y(0-,\omega) \tag{12}$$

and

$$U(L,\omega) = \begin{pmatrix} p_N^{-1} & 0 \\ 0 & p_N \end{pmatrix} C_N C_{N-1} \dots C_1 \begin{pmatrix} p_0 & 0 \\ 0 & p_0^{-1} \end{pmatrix} U(0,\omega). \tag{13}$$

or

$$U(L,\omega) = \begin{pmatrix} p_N^{-1} & 0 \\ 0 & p_N \end{pmatrix} Z_N \begin{pmatrix} p_0 & 0 \\ 0 & p_0^{-1} \end{pmatrix} U(0,\omega) \tag{14}$$

where $Z_N = C_N C_{N-1} \dots C_1$. Further, from (13) it follows that the elements of the matrix $U(L,\omega)$ are proportional to the elements of $Z_N(\omega)$ and therefore have the same zeroes.

Now, using a proof of induction, we may obtain the following lemma.

Lemma 1.

$$Z_N(z) = \begin{pmatrix} A_N(z) & D_N(z) \\ B_N(z) & E_N(z) \end{pmatrix}$$

where $A_N(z) = \sum_{k=0}^N a_k z^{N-2k}$ *and* $B_N(z) = \sum_{k=0}^N b_k z^{N-2k}$, *with* $A_N(z) = A_N(1/z)$, $B_N(z) = -B_N(1/z)$, *i.e.* $a_k = a_{N-k}$ *and* $b_k = -b_{N-k}$. *Further* $a_0 = b_0$.

Besides the interlacing properties the eigenfrequencies have the following periodicity properties.

Lemma 2. *For the zeroes* $\{\omega_{2k}\}_{k=0}^{\infty}$ *and* $\{\omega_{2k+1}\}_{k=0}^{\infty}$ *of* $B_N(e^{i\omega\delta})$ *and* $A_N(e^{i\omega\delta})$ *we have that* $\omega_{k+2N} = \omega_k + \pi N/L$ *and* $\omega_k = \pi N/L - \omega_{2N-k}$ *for all* k.

Proof: We have $A_N(z) = z^{-N}\tilde{A}(z^2)$ and $B_N(z) = z^{-N}\tilde{B}(z^2)$ wilth $\tilde{A}$ and $\tilde{B}$ polynomials. Now $\tilde{A}(e^{2i\omega\delta})$ and $\tilde{B}(e^{2i\omega\delta})$ have a period $\pi/\delta = \pi N/L$ whence the first statement follows. Further, the coefficients of $\tilde{A}$ and $\tilde{B}$ are real and therefore $e^{2i\omega\delta} = e^{2i\omega L/N}$ is a zero if and only iff $e^{-2i\omega\delta} = e^{2iL/N(\pi N/L-\omega)}$ is a zero whence the second statement follows. ∎

The next lemma solves the inverse eigenvalue problem for (1), (2) when p is piecewise constant over the intervals I_k. Moreover, the proof will provide an explicit algorithm for the determination of q_k and p_k.

Lemma 3. *Let there be given a set* $\{\omega_k\}_{k=0}^N$ *with the property* $0 = \omega_0 < \omega_1 < \omega_2 < \ldots < \omega_N = \pi N/2L$, *extended for* $k > N$ *so that* $\omega_{k+2N} = \omega_k + \pi N/L$ *and* $\omega_k = \pi N/L - \omega_{2N-k}$. *Take* $z_\nu = e^{i\omega_\nu\delta}$ *and form the two functions*

$$\begin{aligned} A_N(z) &= a_{N,0} \prod_{\nu=0}^{N-1} [z + z^{-1} - (z_{2\nu+1} + z_{2\nu+1}^{-1})] = \sum_{k=0}^{N} a_{N,k} z^{N-2k} \\ B_N(z) &= b_{N,0} \prod_{\nu=0}^{N-1} [z - z^{-1} - (z_{2\nu} - z_{2\nu}^{-1})] = \sum_{k=0}^{N} b_{N,k} z^{N-2k} \end{aligned} \tag{15}$$

with $a_{N,0} = b_{N,0} > 0$. *Then there exist uniquely defined positive numbers* $q_1, q_2, \dots q_N$ *so that*

$$Z_N = C(q_N, z)C(q_{N-1}, z)\dots C(q_1, z) = \begin{pmatrix} A_N(z) & D_N(z) \\ B_N(z) & E_N(z) \end{pmatrix}.$$

Here D_N *and* E_N *are trigonometric polynomials of the same structure as* A_N *and* B_N. *Moreover if* $a_{N,0} = b_{N,0}$ *are multiplied with a factor* c *then* q_1 *is replaced with* cq_1 *while* $q_2, q_3, \dots, q_N$ *are unchanged.*

Proof: For $N = 1$ the argument is trivial. Next assume that the statement of the theorem has been proved when the number of intervals is less than or equal to $N-1$. Consider the product

$$\begin{pmatrix} A_{N-1}(z) \\ B_{N-1}(z) \end{pmatrix} = C_N^{-1}(q,z) \begin{pmatrix} A_N(z) \\ B_N(z) \end{pmatrix} = \frac{1}{2} \begin{pmatrix} q^{-1}\left[(z+1/z)\,A_N(z) - (z-1/z)\,B_N(z)\right] \\ q\left[-(z-1/z)\,A_N(z) + (z+1/z)\,B_N(z)\right] \end{pmatrix}. \tag{16}$$

We get,

$$\begin{aligned}
A_{N-1}(z) &= \frac{q^{-1}}{2}\left[(z+1/z)\sum_{k=0}^{N} a_{N,k} z^{N-2k} - (z-1/z)\sum_{k=0}^{N} b_{N,k} z^{N-2k}\right] \\
&= \frac{q^{-1}}{2}(z^{N+1} + z^{-N-1})(a_{N,0} - b_{N,0}) \\
&+ \frac{q^{-1}}{2}\sum_{k=0}^{N-1}(a_{N,k+1} - b_{N,k+1} + a_{N,k} + b_{N,k}) z^{N-1-2k}, \\
B_{N-1}(z) &= -\frac{q}{2}\left[(z-1/z)\sum_{k=0}^{N} a_{N,k} z^{N-2k} - (z+1/z)\sum_{k=0}^{N} b_{N,k} z^{N-2k}\right] \\
&= -\frac{q}{2}(z^{N+1} - z^{-N-1})(a_{N,0} - b_{N,0}) \\
&+ \frac{q}{2}\sum_{k=0}^{N-1}(-a_{N,k+1} + b_{N,k+1} + a_{N,k} + b_{N,k}) z^{N-1-2k}.
\end{aligned} \tag{17}$$

Now, by (15), $a_{N,0} = b_{N,0}$, so that the highest degree terms in (17) vanish. Assume, to simplify notation, that $a_{N,0} = b_{N,0} = 1$. In order to make the coefficients of $z^{(N-1)}$ in these new expressions equal, we have to choose $q = q_N > 0$, so that

$$q = \sqrt{(2 + a_{N,1} - b_{N,1})/(2 - (a_{N,1} - b_{N,1}))}. \tag{18}$$

This is possible if $|a_{N,1} - b_{N,1}| < 2$. To verify this we use that, by lemma 2, $z_{\nu+2N} = -z_\nu$, $z_{2N-\nu} = -z_\nu^{-1}$, $z_0 = 1$, $z_{2N} = -1$ and $z_N = i$. A_N and B_N may then be rewritten

$$z^N A_N(z) = \prod_{\nu=0}^{N-1}(z^2 - z_{2\nu+1}^2), \quad z^N B_N(z) = \prod_{\nu=0}^{N-1}(z^2 - z_{2\nu}^2). \tag{19}$$

Hence

$$\begin{aligned}
-a_{N,1} &= \sum_{\nu=0}^{N-1} z_{2\nu+1}^2 = \frac{1}{2}\sum_{\nu=0}^{N-1}(z_{2\nu+1}^2 + z_{2N-1-2\nu}^2) \\
-b_{N,1} &= \sum_{\nu=0}^{N-1} z_{2\nu}^2 = \frac{1}{2}\sum_{\nu=0}^{N-1}(z_{2\nu}^2 + z_{2N-2\nu}^2)
\end{aligned} \tag{20}$$

Suppose first that N is odd. Then, with $\Re$ denoting the real part,

$$\begin{aligned}
-a_{N,1} &= 2\sum_{\nu=0}^{(N-3)/2} \Re z_{2\nu+1}^2 - 1 = 2\sum_{\nu=1}^{(N-1)/2} \Re z_{2\nu-1}^2 - 1 \\
-b_{N,1} &= 2\sum_{\nu=1}^{(N-1)/2} \Re z_{2\nu}^2 + 1.
\end{aligned} \tag{21}$$

Consequently,

$$a_{N,1} - b_{N,1} = 2 - 2\sum_{\nu=1}^{(N-1)/2}(\Re z_{2\nu}^2 - \Re z_{2\nu+1}^2).$$

By the interlacing property we have, for $1 \leq \nu \leq (N-1)/2$,

$$-1 < \Re z_{2\nu+1}^2 < \Re z_{2\nu}^2 < 1$$

whence we conclude that

$$0 < \sum_{\nu=1}^{(N-1)/2} (\Re z_{2\nu}^2 - \Re z_{2\nu+1}^2) < 2$$

and that $|a_{N,1} - b_{N,1}| < 2$. The argument for the case that N is even is similar.

In order to complete our induction argument it suffices to prove the the zeroes of $A_{N-1}(e^{i\omega\delta})$ and $B_{N-1}(e^{i\omega\delta})$ are interlacing when $0 \leq \omega\delta \leq \pi/2$. First consider two consecutive zeroes

$$0 < \omega_{2k-1} < \omega_{2k+1} < \pi/(2\delta)$$

of $A_N(e^{i\omega\delta})$. Then $B_N(e^{i\omega\delta})$ has opposite signs at these zeroes and, since

$$(z - 1/z)/i = 2\sin\omega\delta > 0 \quad 0 < \omega\delta < \pi/2$$

it follows, by (16), that $A_{N-1}(e^{i\omega\delta})$ also has opposite signs att these zeroes. Therefore every interval $0 < \omega_{2k-1} < \omega_{2k+1} < \pi/2$ contains at least one zero $\omega'_{2\nu+1}$ of $A_{N-1}(e^{i\omega\delta})$. Since A_N has exactly one zero more that A_{N-1} on $(0, \pi/(2\delta))$, we conclude that every interval $0 < \omega'_{2\nu-1} < \omega'_{2\nu+1} < \pi/2$ contains exactly one zero ω_{2k+1} of A_N and vice versa. Moreover, $A_N(e^{i\omega\delta})$ must have opposite signs at these zeroes. Further, if $z = e^{i\omega'_{2k+1}\delta}$ or $z = e^{i\omega'_{2k-1}\delta}$ then, by (16),

$$(z + 1/z)A_N(z) = (z - 1/z)B_N(z)$$

and therefore

$$B_N(z) = A_N(z)\left\{-(z - 1/z) + (z + 1/z)^2/(z - 1/z)\right\} q/2 = 2qA_N(z)/(z - 1/z).$$

Since $(z - 1/z)/i = 2\sin\omega\delta > 0$ for $0 < \omega\delta < \pi/2$ we conclude that $B_{N-1}(e^{i\omega\delta})$ has opposite signs at ω'_{2k+1} and ω'_{2k-1}. Now the number of zeroes of $B_{N-1}(e^{i\omega\delta})$ in $(0, \pi/(2\delta))$ is exactly one less than the number of zeroes of A_{N-1} and it follows that the zeroes of A_{N-1} and B_{N-1} are interlacing. This completes the proof. ∎

The next theorem, which is our main result, follows easily from lemma 3.

Theorem 1. *Given any set $\{\omega_k\}_{k=0}^N$ with the property $0 = \omega_0 < \omega_1 < \omega_2 < \ldots < \omega_N = \pi N/2L$ and extended for $k > N$ so that $\omega_{k+2N} = \omega_k + \pi N/L$ and $\omega_k = \pi N/L - \omega_{2N-k}$, there exists a unique function p which is piecewise constant over the intervals I_k, $1 \leq k \leq N$, equal to 1 on I_1 and so that equation (1) with the boundary conditions (2) has the eigenfrequencies $\{\pm\omega_{2k}\}_{k=0}^{\infty}$ and $\{\pm\omega_{2k+1}\}_{k=0}^{\infty}$.*

Proof: Take

$$\begin{aligned} A_N(z) &= a_0 \prod_{\nu=0}^{N-1} [z + z^{-1} - (z_{2\nu+1} + z_{2\nu+1}^{-1})] = \sum_{k=0}^{N} a_k z^{N-2k} \\ B_N(z) &= b_0 \prod_{\nu=0}^{N-1} [z - z^{-1} - (z_{2\nu} - z_{2\nu}^{-1})] = \sum_{k=0}^{N} b_k z^{N-2k} \end{aligned} \tag{22}$$

with $a_0 = b_0 > 0$ arbitrary. Define $q_1, q_2, \ldots, q_N$ by lemma 3. Then

$$U(L, \omega) = \begin{pmatrix} p_N^{-1} & 0 \\ 0 & p_N \end{pmatrix} Z_N \begin{pmatrix} p_0 & 0 \\ 0 & p_0^{-1} \end{pmatrix}.$$

$q_1 = p_1/p_0$ can be taken arbitrary, and therefore p_0 and p_1 are arbitrary. $p_k = q_k p_{k-1}$ for $2 \leq k \leq N$. If we take $p_1 = 1$ then $p_2, p_3, \ldots, p_N$ are uniquely defined. ∎

3. An inversion algorithm

Assume that $\{\omega_k\}_{k=0}^{\infty}$ is given satisfying the assumptions of theorem 1. Define the trigonometric polynomials $A_N(z)$ and $B_N(z)$ according to (15) with $a_{N,0} = b_{N,0} = 1$. The coefficients of these polynomials may be computed with good accuracy and with a number of operations being approximately $2N^2$ and requiring a storage memory of $4N \log_2 N$.

For $n = N, N-1, \ldots, 2, 1$ the numbers q_n and the coefficients $a_{n,r}$ and $b_{n,r}$ are computed according to (18) and (17) i.e. so that

$$q_n := \sqrt{(2 + a_{n,1} - b_{n,1})/(2 - (a_{n,1} - b_{n,1})}, \tag{23}$$

for $r = 0, 1, \ldots, n-1$,

$$\begin{aligned} a_{n-1,r} &:= (q_n^{-1}/2)\{a_{n,r+1} + a_{n,r} - b_{n,r+1} + b_{n,r}\} \\ b_{n-1,r} &:= (q_n/2)\{-a_{n,r+1} + a_{n,r} + b_{n,r+1} + b_{n,r}\}, \end{aligned} \tag{24}$$

for $r = 0, 1, \ldots, n-1$,

$$a_{n-1,r} := a_{n-1,r}/a_{n-1,0} \quad b_{n-1,r} = b_{n-1,r}/b_{n-1,0}. \tag{25}$$

Finally we take p_1 arbitrary, e.g. $p_1 = 1$ and $p_k = q_k p_{k-1}$ for $k = 2, 3, \ldots, N$. By lemma 3 $|a_{n,1} - b_{n,1}| < 2$ for all n.

The number of operations required in the equations (23)–(25) is approximately $3N^2$ and the required memory $4N$.

4. An Inversion algorithm for the case $p'/p \in L^2(0, L)$

First let us introduce some notation.

$\mathcal{K}$ denotes the set of sequences $s = \{s_k\}_{k=1}^{\infty} \in l^2$ with the property that the sequence $\{\omega_k\}_{k=0}^{\infty}$, with $\omega_0 = 0$ and $\omega_k = k\pi/(2L) + s_k$ for $k \geq 1$, is interlacing. $\mathcal{K}$ is an open convex subset of l^2. We introduce the notation $\alpha = p'/p$ and let the mapping

$$L^2(0, L) \ni \alpha \mapsto s = \mathcal{F}(\alpha) \in \mathcal{K}$$

be defined by the requirement that $\omega_0 = 0$ and $\omega_k = k\pi/(2l) + s_k$ for $k \geq 1$ are the eigenfrequencies for the problem (1)–(2).

In [2] the following result was proved.

Theorem 2. *The mapping $\mathcal{F}$ is one-to-one, onto and locally Lipschitz bicontinuous.*

For the calculation of $\mathcal{F}$ and $\mathcal{F}^{-1}$ we also have the following result.

Theorem 3. *The eigenfrequencies $\{\omega_{2k+1}\}_{k=0}^{\infty}$ and $\{\omega_{2k}\}_{k=0}^{\infty}$ respectively are obtained as the zeroes of two functions $\cos\omega L + \hat{a}(L, \omega)$ and $i \sin\omega + \hat{b}(L, \omega)$ where $\hat{a}$ and $\hat{b}$ are Fourier transforms of odd and even functions $a(L, \cdot), b(L, \cdot) \in L^2(\mathbb{R})$ with support in $[-L, L]$. Moreover the relation between α and $\hat{a}(L, \omega)$, $\hat{b}(L, \omega)$ is given by the integral equation*

$$f(x,t) = \int_0^x f(s, s - x - t)\alpha(s)\,ds + \frac{1}{2}\alpha((x+t)/2)H(x - |t|), \quad 0 \leq x \leq L. \tag{26}$$

This equation is equivalent to the following system.

$$\begin{aligned}\alpha(x) &= 2f(L,2x-L) - 2\int_x^L f(s,s-2x)\alpha(s)\,ds, \quad 0 \le x \le L \\ f(x,t) &= f(L,t+x-L) - \int_x^L f(s,s-x-t)\alpha(s)\,ds, \quad |t| \le x \le L.\end{aligned} \tag{27}$$

Here we have introduced the notation $f(x,t) = a(x,t) + b(x,t)$, $0 \le |t| \le x \le L$ with $a(x,\cdot)$ and $b(x,\cdot) \in L^2(R)$ even and odd functions respectively with their supports in $[-x,x]$.

Now let, $\tilde{\omega}_k = k\pi/(2L)$ and assume that the eigenfrequencies $\omega_k = \tilde{\omega}_k + s_k$, $\{s_k\}_{k=1}^{\infty} \in l^2$ are given, satisfying the interlacing conditions. Then

$$\cos\omega L + \hat{a}(L,\omega) = \cos\omega L \prod_{k=0}^{\infty} \frac{\omega^2 - \omega_{2k+1}^2}{\omega^2 - \tilde{\omega}_{2k+1}^2} = \cos\omega L \left[1 + \sum_{k=0}^{\infty} \frac{\alpha_k}{\omega^2 - \tilde{\omega}_{2k+1}^2}\right], \tag{28}$$

with

$$\alpha_k = (\tilde{\omega}_{2k+1}^2 - \omega_{2k+1}^2) \prod_{\substack{l=0 \\ l\neq k}}^{\infty} \frac{\tilde{\omega}_{2l+1}^2 - \omega_{2k+1}^2}{\tilde{\omega}_{2l+1}^2 - \tilde{\omega}_{2k+1}^2} \tag{29}$$

Consequently,

$$\hat{a}(L,\omega) = \sum_{k=0}^{\infty} \alpha_k \frac{\cos\omega L}{\omega^2 - \tilde{\omega}_{2k+1}^2}, \quad \text{i.e.} \quad a(L,t) = \textstyle\sum_{k=0}^{\infty} \alpha_k a_k(t)$$

where $\operatorname{supp} a_k \subset [-L,L]$ and

$$(\omega^2 - \tilde{\omega}_{2k+1}^2)\hat{a}_k(\omega) = \cos\omega L, \quad \text{i.e.} \quad a_k''(t) + \tilde{\omega}_{2k+1}^2 a_k(t) = -\tfrac{1}{2}(\delta(t-L) + \delta(t+L)).$$

Solving this differential equation we get,

$$a_k(t) = \begin{cases} 1/(2\tilde{\omega}_{2k+1})(-1)^{k-1}\cos(\tilde{\omega}_{2k+1}t) & \text{for } |t| \le L \\ 0 & \text{else.} \end{cases}$$

Therefore

$$a(L,t) = \frac{1}{2}\sum_{k=0}^{\infty} (\alpha_k/\tilde{\omega}_{2k+1})(-1)^{k-1}\cos(\tilde{\omega}_{2k+1}t). \tag{30}$$

In the same way we get,

$$i\sin\omega L + \hat{b}(L,\omega) = i\sin\omega L\left[1 + \sum_{k=1}^{\infty} \frac{\beta_k}{\omega^2 - \tilde{\omega}_{2k}^2}\right], \tag{31}$$

with

$$\beta_k = (\tilde{\omega}_{2k}^2 - \omega_{2k}^2) \prod_{\substack{l=1 \\ l\neq k}}^{\infty} \frac{\tilde{\omega}_{2l}^2 - \omega_{2k}^2}{\tilde{\omega}_{2l}^2 - \tilde{\omega}_{2k}^2} \tag{31}$$

and

$$b(L,t) = \frac{1}{2}\sum_{k=1}^{\infty} (\beta_k/\tilde{\omega}_{2k})(-1)^{k-1}\sin(\tilde{\omega}_{2k}t). \tag{32}$$

Assuming for example that only finitely many frequencies ω_k are different from $\tilde{\omega}_k = k\pi/(2L)$, we obtain the following inversion algorithm.

Compute α and β_k by (29) and (31).

Calculate, e.g. using FFT, $a(L,t)$ and $b(L,t)$ according to (30) and (32). Take $f(L,t) = a(L,t) + b(L,t)$.

Finally, solve the system (27) for $\alpha(x)$ and $f(x,t)$, $0 \le |t| \le x \le L$.

This inversion procedure is closely related to the previous one. The calculation of $a(L,t)$ and $b(L,t)$ corresponds to the calculation of the coefficients of $A_N(z)$ and $B_N(z)$ in section 3. Solving (27) from $x = L$ to $x = 0$ corresponds to the algorithm given by (23)–(25).

5. A numerical result

The numerical evaluation of the preceding algorithms are so far not very extensive. Only one example, for the discrete version of the problem, will be shown.

We have considered the equations (1)–(2), for the case that $p'(x)/p(x) = 4$ for $0 \le x \le \pi/2 = L$ and calculated the eigenfrequencies $\{\omega_k\}_{k=0}^{256}$ accurately. These have been fed into the inversion algorithm given by (23)–(25)and the piecewise constant function $p(x)$ has been plotted in the figure. The relative difference $(p(x) - e^{4x})e^{-4x}$ is very small, of the order of magnitude 1/2 % . On may however note that the q_k–values exhibit small oscillations so that every second value is a little too large and every second a little too small. this is most clearly seen when k is small, i.e. for those q_k–values that have been calculated last.

One may also note that for this case, $N = 256$, the use of double or single precision in the calculations makes little difference.

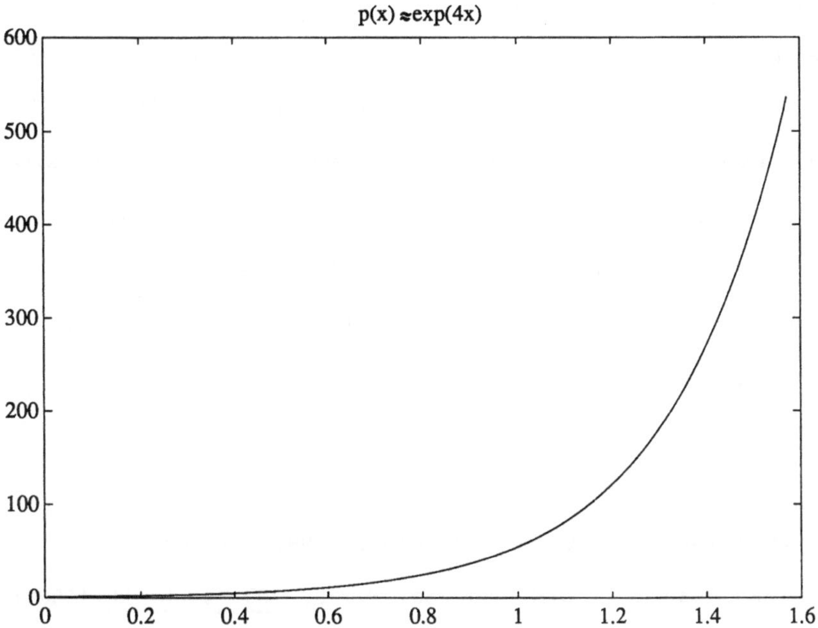

6. References

[1] Andersson L-E: *Inverse eigenvalue problems with discontinuous coefficients.* Inverse Problems **4**, (1988), 353-397.

[2] Andersson, L-E: *Inverse eigenvalue problems for a Sturm-Liouville equation in impedance form.* Inverse Problems **4**, (1988), 929-971.

[3] Borg, G: *Eine Umkehrung der Sturm-Liouvilleschen Eigenwertaufgabe. Bestimmung der Differentialgleichung durch die Eigenwerte.* Acta Math.,**78** (1946), 1-96.

[4] Gelfand,I.M., Levitan, B.M: *On the determination of a differential equation from its spectral function.* Izv. Akad. Nauk, SSSR, Ser. Mat. **15**, (1951), 309-60. (Engl. transl. 1955 Am. Math. Soc. Transl. (2),1., (1955), 253-304.

[5] Hald, O: *Numerical solution of the Gelfand-Levitan equation* . Linear Alg. and its Appl., **28**, (1979), 99-111.

[6] Hochstadt, H: *The inverse Sturm-Liouville problem* . Comm. of Pure and Appl. Math., **26**, (1973), 715-729.

Inverse Problem in the Coupling Constant

K. Chadan

Laboratoire de Physique Théorique et Hautes Energies,
Université de Paris XI, F-91405 Orsay, France

ABSTRACT

We show how to solve the inverse problem in the coupling constant by reducing it, with the help of Liouville transformation, to the classical case of Gel'fand-Levitan theory for discrete spectra. Once the Gel'fand-Levitan integral equation is solved, one has three more steps to get the potential, the first two being integrations, and the third the inversion of a function and differentiation. By generalizing to the case of two potentials, we show that one can solve the inverse problem in the coupling constant in the general case, at any (fixed) energy, and for finite or infinite intervals. Several soluble examples are presented.

K. Chadan & M. Musette : Inverse Problems 5 (1989) 257-268.

K. Chadan & M. Musette : C.R. Ac. Sc. Paris, 309, série II (1989) 1535-1540, with English abridged version.

Examples of Inverse Nodal Problems

O.H. Hald[1] *and J.R. McLaughlin*[2]

[1]Department of Mathematics, University of California, Berkeley, CA 94720, USA
[2]Department of Mathematical Sciences, Rensselaer Polytechnic Institute, Troy, NY 12180, USA

In this talk we will consider the following problem: What can you say about a vibrating rod, if you know the position of the nodes. A node is a point where an eigenfunction vanishes. We will assume that the mass per unit length is constant and try to determine the elasticity of the rod from the nodes. Instead of presenting general theories, (see [1,2,3]), we will consider two simple examples. In the first example the elasticity varies linearly and in the second the elastic modulus has a discontinuity. We will find that the presence of a discontinuity does not destroy the accuracy of the reconstruction.

To study the inverse problem we need a model for the vibrations of the rod. Here we consider infinitesimal oscillations and assume that

$$(p(x)y'(x))' + \lambda y(x) = 0$$

$$y(0) = y(1) = 0 .$$

We assume that $p(x)$ is a variable function of x and, for simplicity, that the rod has length 1. We will consider two choices of p. The first choice is $p_1(x) = \frac{1}{2} + x$ and the second is

$$p_2(x) = \begin{cases} \frac{1}{2} + x & \text{for } 0<x<\frac{1}{2} \\ \frac{3}{2} + x & \text{for } \frac{1}{2}<x<1 . \end{cases}$$

In both cases we require that $y(x)$ and $p(x)y'(x)$ are continuous. This means that when $p = p_2$ the function y' will have a discontinuity at $x = \frac{1}{2}$. In practice we get the data for the inverse problem from experiments. Here we get the data by solving the direct problem. Thus we calculate the eigenvalue λ_n and the points $x_1, \cdots, x_{n-1}$ where the eigenfunction $y_n(x)$ vanishes. To be specific, we pick the 10'th eigenfunction and the results are given in Table 1. As the elastic modulus increases, i.e. $p_1 \le p_2$ the eigenvalue increases, and the nodes move to the left.

We will now attack the inverse nodal problem for the two cases, and begin with an analysis of the differential equation. We claim that

Table 1. The nodes for our two test problems.

	p_1	p_2
x_1	0.0759	0.0674
x_2	0.1571	0.1391
x_3	0.2437	0.2151
x_4	0.3357	0.2953
x_5	0.4330	0.3798
x_6	0.5357	0.4685
x_7	0.6437	0.5812
x_8	0.7571	0.7165
x_9	0.8759	0.8561
λ_{10}	920.7745	1157.3107

Table 2. Reconstruction of $p_1(x)$.

$\bar{x}_i$	computed p_1	exact p_1	error
0.0379	0.5372	0.5379	-0.0007
0.1165	0.6158	0.6165	-0.0007
0.2004	0.6997	0.7004	-0.0007
0.2897	0.7890	0.7897	-0.0007
0.3843	0.8836	0.8843	-0.0007
0.4843	0.9836	0.9843	-0.0007
0.5897	1.0890	1.0897	-0.0007
0.7004	1.1997	1.2004	-0.0007
0.8165	1.3158	1.3165	-0.0007
0.9379	1.4372	1.4379	-0.0007

$$p\left(\frac{x_i+x_{i+1}}{2}\right) = \frac{\lambda_n (x_{i+1}-x_i)^2}{\pi^2} + \cdots$$

Here $\cdots$ indicates lower order terms. A similar result can be found in [1] and [2] for a rod with free boundary conditions. To derive such results we use asymptotic expansions of the eigenvalues and the nodes, but we need some smoothness to obtain these expansions. In our case it is enough that $p(x)$ is twice continuously differentiable. It is then natural to define

$$\text{computed } p\left(\frac{x_i+x_{i+1}}{2}\right) = \frac{\lambda_n (x_{i+1}-x_i)^2}{\pi^2}, \tag{1}$$

for $i = 0, 1, \cdots, n-1$ with $x_0 = 0$ and $x_n = 1$. Although this algorithms is suggested by looking at the smooth case, we will also use it when the elastic modulus has a discontinuity. Let us see how the algorithm works in practice. We start with a smooth function $p_1(x)$. The results are given in Table 2. Here $\bar{x}_i = \frac{1}{2}(x_i + x_{i+1})$. We note that the errors in the reconstruction are uniformly small. Experiments with different functions $p(x)$ show that the errors in the inverse problem are comparable to the errors in linear interpolation, and often smaller.

Next we look at the discontinuous case. The reconstruction of $p_2(x)$ is given in Table 3. The error is large when we use the interval that contains the discontinuity. This is to be expected. It is more interesting to observe that the errors in the remaining intervals are uniformly small. Thus the discontinuity does not destroy the accuracy of the algorithm in the intervals adjacent to the jump.

Our formula for computed p can also be derived by heuristic arguments. Pick two consecutive nodes, $x_i < x_{i+1}$ and assume that $p(x)$ is a constant between x_i and x_{i+1}. In this interval the eigenfunction

Table 3. Reconstruction of $p_2(x)$.

$\bar{x}_i$	computed p_2	exact p_2	error
0.0337	0.5331	0.5337	-0.0006
0.1033	0.6027	0.6033	-0.0006
0.1771	0.6765	0.6771	-0.0006
0.2552	0.7546	0.7552	-0.0006
0.3375	0.8370	0.8375	-0.0005
0.4242	0.9236	0.9242	-0.0006
0.5248	1.4873	2.0248	-0.5375
0.6488	2.1482	2.1488	-0.0006
0.7863	2.2857	2.2863	-0.0006
0.9281	2.4275	2.4281	-0.0006

$y(x)$ must be proportional to $\sin(\pi(x-x_i)/(x_{i+1}-x_i))$. Using differentiation we get

$$-(p\,y')' = \frac{\pi^2 p}{(x_{i+1}-x_i)^2}\, y = \lambda_n\, y\ .$$

After dividing by y and reordering the terms we obtain equation (1).

What is the rate of convergence of our algorithm. We claim that if $p(x)$ is twice continuously differentiable then

$$\left|\ \text{computed}\, p(\bar{x}_i) - \text{exact}\, p(\bar{x}_i)\ \right| \le \frac{\text{const.}}{n^2}\ ,$$

at the points $\bar{x}_i = \frac{1}{2}(x_i + x_{i+1})$. The constant can be given explicitly and depends on bounds for p, p' and p''. A proof for a similar result can be found in [2] for a vibrating rod with free boundary conditions. Our estimate shows that the algorithm converges quadratically, just like linear interpolation. Now, what happens if we have a discontinuity? In this case our analysis is no longer valid, and we must perform numerical experiments. To measure the error we use a discrete L^1 norm

$$\|\ \text{error}\ \| \;=\; \sum_{i=0}^{n-1} (x_{i+1}-x_i)\left|\ \text{computed}\, p(\bar{x}_i) - \text{exact}\, p(\bar{x}_i)\ \right|\ ,$$

where $\bar{x}_i = \frac{1}{2}(x_i + x_{i+1})$. The reason we have chosen this particular norm is that it is quite insensitive to the error at the discontinuity. However, to study the rate of convergence away from the discontinuity, we prefer the truncated L^1 norm, in which we delete the contribution from the interval that contains the discontinuity. The results of our numerical experiments are given in Table 4. In these inverse problems we have used the nodes from $y_{10}(x)$, $y_{20}(x)$ and $y_{40}(x)$. Based on these calculations and others like them, we believe that the rate of convergence is quadratic for functions that are piecewise smooth, provided we stay away from the discontinuities.

Table 4. More nodes give smaller errors in $p_2(x)$.

	$n = 10$	$n = 20$	$n = 40$
L^1 error	0.061	0.020	0.015
Truncated L^1 error	0.000521	0.000137	0.000035

Now, what can be said about the inverse problem from a mathematical point of view. The simplest statement is this: All the eigenvalues and all the nodes determine the elastic modulus uniquely. This is a lot of data. What happens if we don't know the eigenvalues? Then all the nodes determine $p(x)$ up to a multiplicative constant. But do we need all the nodes? The answer is no. If $p(x)$ is twice continuously differentiable then it is uniquely determined by any dense subset of the nodes. A similar result can be found in [1] and [2] for a rod with free boundary conditions. The proof is based on asymptotic expansions of the eigenvalues, the eigenfunctions and the nodes. Such expansions are not available for discontinuous problems. Here a new technique is needed. In a forthcoming paper we will show that if $p(x)$ is of bounded variation then it is uniquely determined (up to a multiplicative constant and at all points of continuity), by any dense set of pairs of nodes. Thus for any node in the set, one of its neighbors must also be in the set.

Our final question is this: What is the general principle underlying our algorithm. Let $\Omega = [0, 1]$ and $\Omega_i = [x_i, x_{i+1}]$. The heuristic derivation of (1) shows that $\pi^2/(x_{i+1} - x_i)^2$ is the first eigenvalue of $(1 \cdot y')' + \lambda y = 0$ on Ω_i. We can therefore rewrite equation (1) as

$$\text{computed}\, p = \frac{\lambda_n(\Omega, p)}{\lambda_1(\Omega_i, 1)} .$$

This formula can be used to solve the inverse nodal problem for the elliptic equation $div(p(x)\nabla u) + \lambda u = 0$ in, say, two dimensions. In this case Ω_i is a nodal domain, and the boundary of Ω_i consists of nodal lines. By using Rayleigh's principle we can even prove that the computed $p(x)$ converges towards the exact $p(x)$, provided p is smooth and the nodal domains shrink in size. The problem is that the diameters of the nodal domains may not decrease as the eigenvalues increase.

One possible application of our theory is to determine regions of fatigue in metals. An example, that is important in practice, is fatigue in turbine blades. One effect of fatigue is a dramatic shift of the nodal lines. We plan to analyze how the nodal line patterns are influenced by the regions where the material is weakened.

Acknowledgments

The work of OHH has been supported by NSF Grant no DMS-8701585 and by the Applied Mathematics Subprogram of the Office of Energy Research under contract DE-AC03-76F00098. The work by JRM has been supported by ONR Grant no N00014-84-K-516 and by NSF grant no DMS-8713722.

References

[1] O.H. Hald and J.R. McLaughlin, *Inverse problems using nodal position data - uniqueness results, algorithms and bounds*, in R.S. Anderssen and G.N. Newsam (eds.), *Special program on inverse problems*, Proc. of the Centre for Math. Anal., Australian National University, Canberra, 1988, pp. 32-59.

[2] O.H. Hald and J.R. McLaughlin, *Solutions of inverse nodal problems*, Inverse Problems, 5 (1989), pp. 307-347.

[3] J.R. McLaughlin, *Inverse spectral theory using nodal points as data - a uniqueness result*, J. Diff. Eq. 73 (1988), pp. 354-362.

Conditions on the Data in Inverse Problems in Vibration

G.M.L. Gladwell

Solid Mechanics Division, University of Waterloo,
Waterloo, Ontario, Canada N2L3G1

1. Introduction

This review concerns inverse problems for discrete or continuous elastic bodies of finite extent, executing linear, undamped, free vibration. The data for such inverse problems is *spectral*, i.e., it relates to the natural frequencies and/or normal modes of the system. The basic question is this: *Suppose some data is given, and it purports to be spectral data for a system of specified type, what tests can be carried out on the data to determine whether or not it is* acceptable, *i.e., whether or not it corresponds to a* realizable *system?*

This question is only one of those that must be asked, and answered, in Inverse Problems. Two others are: *Does the data specify a unique system?* and *How can a/ the system be reconstructed?* A general review of answers to these questions, as they relate to the particular systems described in the preceding paragraph, is given in Gladwell [13].

For this reason, we shall make no attempt here to document research in the field. Instead we will provide a largely personal account of our research into the basic question, starting with the simplest systems and proceeding to the more difficult.

2. Chain-like discrete, second-order systems

Figure 1 shows the simplest system of this type - 2 masses connected by linear springs.

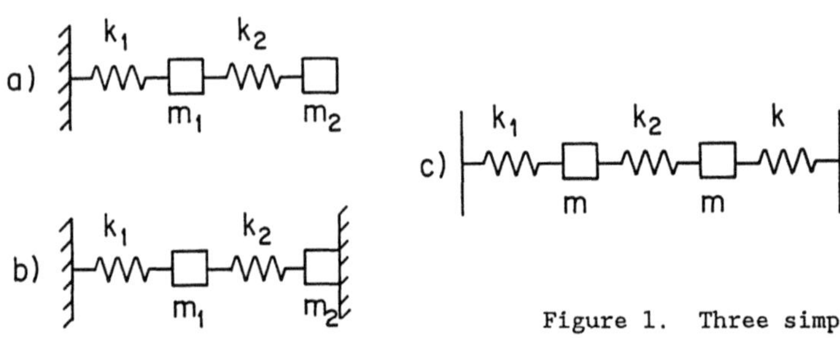

Figure 1. Three simple spring-mass systems

The (natural frequencies)^2 of the system of Figure 1a are the eigenvalues of the problem

$$(K - \lambda M)\,u = \begin{bmatrix} k_1+k_2-\lambda m_1 & -k_2 \\ -k_2 & k_2-\lambda m_2 \end{bmatrix} \begin{bmatrix} u_1 \\ u_2 \end{bmatrix} = 0,$$

i.e., they are the zeros of the quadratic

$$m_1 m_2 \lambda^2 - [m_2(k_1 + k_2) + m_1 k_2]\lambda + k_1 k_2 = 0, \tag{2.1}$$

so that, in particular

$$\lambda_1 + \lambda_2 = \frac{k_1+k_2}{m_1} + \frac{k_2}{m_2}, \quad \lambda_1\lambda_2 = \frac{k_1 k_2}{m_1 m_2}. \tag{2.2}$$

Suppose now that the right-hand end is fixed, as in Figure 1b. The system will have one (natural frequency)^2, given by

$$\mu_1 = \frac{k_1+k_2}{m_1}.$$

The inverse problem consists of reconstructing the system, i.e., the set of ratios $m_1 : m_2 : k_1 : k_2$ from the eigenvalues λ_1, λ_2, μ_1. Note that natural frequency data alone cannot determine the absolute magnitudes m_i, k_i $(i = 1,2)$ since the systems made from am_1, am_2, ak_1, ak_2 have the same eigenvalues for all $a > 0$. For the system of Figure 1, the solution of the inverse problem is provided by noting that

$$\frac{k_2}{m_2} = \lambda_1 + \lambda_2 - \mu_1, \quad \frac{k_1}{m_1} = \frac{\lambda_1 \lambda_2}{\lambda_1 + \lambda_2 - \mu_1}, \quad \frac{k_2}{m_1} = \frac{(\mu_1-\lambda_1)(\lambda_2-\mu_1)}{\lambda_1 + \lambda_2 - \mu_1}.$$

These equations show that the solution is meaningful, i.e., k_i, m_i are all of one sign (positive!) iff

$$0 < \lambda_1 < \mu_1 < \lambda_2.$$

If instead of the single eigenvalue of the system in Figure 1b, we are given the eigenvalues μ_1, μ_2 of 1c, in which the right-hand mass is attached to the support by an unknown spring, then

$$\mu_1 + \mu_2 = \frac{k_1 + k_2}{m_1} + \frac{k + k_2}{m_2}, \quad \mu_1\mu_2 = \frac{k_1 k_2}{m_1 m_2} + \frac{k(k_1 + k_2)}{m_1 m_2};$$

from this and (2.2) we may find $k_1 : k_2 : k : m_1 : m_2$. The data will be acceptable iff

$$0 < \lambda_1 < \mu_1 < \lambda_2 < \mu_2$$

The more general result is that if there are two systems of type 1c, having unknown springs k, k^*, and they have data λ_1, λ_2 and μ_1, μ_2, then the data is consistent iff *either* $\lambda_1 < \mu_1 < \lambda_2 < \mu_2$, *or* $\mu_1 < \lambda_1 < \mu_2 < \lambda_2$. In the former case $k < k^*$, in the latter $k^* > k$. Now k_1: k_2: k-k^*: m_1; m_2 may be found.

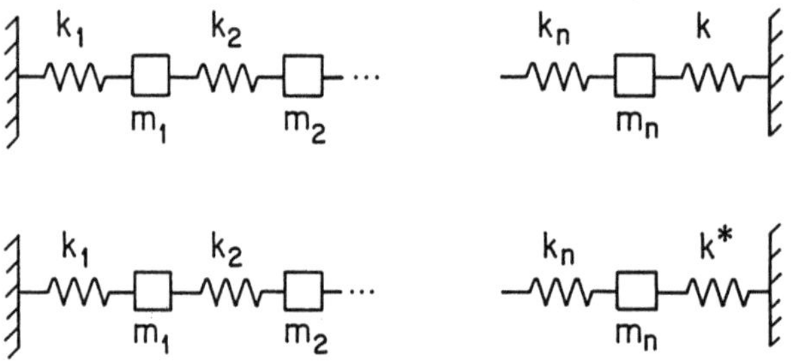

Figure 2. Two spring-mass systems with n degrees of freedom

These conditions generalise for systems made up of n masses and springs joined by springs k, k^* at the right hand end, as in Figure 2. These are second-order systems in the sense that there are only *nearest neighbour* interactions between elements, and the governing stiffness matrix is tridiagonal. Discrete masses on a taut string provide another, mathematically equivalent, second-order system. Now the data sets $(\lambda_i)_1^n$, $(\mu_1)_1^n$ are consistent iff either

$$\lambda_1 < \mu_1 < \lambda_2 < \ldots < \lambda_n < \mu_n \quad \text{or} \quad \mu_1 < \lambda_1 < \mu_2 < \ldots < \mu_n < \lambda_n \tag{2.3}$$

The data is sufficient for the determination of

$$k_1: k_2: \ldots : k_n: k - k^*: m_1: m_2: \ldots : m_n.$$

References to the literature on the general inverse problems for systems such as that shown in Figure 2 may be found in [11], but see also Boley and Golub [4], Gladwell and Willms [17], [19].

In the foregoing problems we have reconstructed the masses and the stiffnesses from *two* spectra, corresponding to two different conditions at one end. It is tempting to suppose that it would be easier to use just one spectrum to find *either* the masses when the stiffnesses were known *or* the stiffnesses when the masses were known. Consider the system in Figure 1a. The governing equation is (2.1), or equivalently (2.2). Suppose λ_1, λ_1, m_1, m_2 were known. Then

$$k_2 = m_1 m_2 \lambda_1 \lambda_2 / k_1,$$

so that k_1 is a root of the quadratic

$$k_1^2 - m_1(\lambda_1 + \lambda_2)k_1 + (m_1 + m_2)\, m_1 \lambda_1\lambda_2 = 0. \tag{2.4}$$

This has positive roots iff

$$\left(\frac{\lambda_2}{\lambda_1}\right)^{\frac{1}{2}} - \left(\frac{\lambda_1}{\lambda_2}\right)^{\frac{1}{2}} \geq 2\left(\frac{m_2}{m_1}\right)^{\frac{1}{2}}.$$

When this condition is satisfied, then there is a pair of isospectral systems with stiffnesses $k_1^{(i)}$, $k_2^{(i)}$, $i = 1,2$. If λ_1, λ_2, k_1, k_2 were known, then there will be a pair of isospectral systems with masses $m_1^{(i)}$, $m_2^{(i)}$, $i = 1,2$ iff

$$\left(\frac{\lambda_2}{\lambda_1}\right)^{\frac{1}{2}} - \left(\frac{\lambda_1}{\lambda_2}\right)^{\frac{1}{2}} \geq 2\left(\frac{k_2}{k_1}\right)^{\frac{1}{2}}.$$

These two problems relating to one spectrum are mathematically similar to the real versions of the so-called inverse *additive* and *multiplicative* eigenvalue problems (Friedland [5], Morel [24]): given **A**, a (real, symmetric) matrix of order n, find diagonal **D** such that *either* **A** + **D** *or* **AD** has given real eigenvalues $(\lambda_i)_1^n$, respectively. It is remarkable that, for $n \geq 3$, even though there are many conditions known to be either necessary or sufficient, no one has found explicit conditions on **A** and $(\lambda_i)_1^n$ which are *at once necessary and sufficient* for the existence of a *real* **D**. The problem revolves around an algebraic equation (like (2.4)) of degree n! which, for a solution, must have a real root.

Instead of using *eigenvalues*, we may use eigenvectors (i.e., principal modes) as data. Gladwell [14] considered this problem for the system of Figure 2a, with $k = 0$. In order to characterize acceptable modes it is necessary to introduce the indices S_u^+, S_u^- relating to a sequence of n real numbers making up a vector $\mathbf{u} = (u_1, u_2, \ldots u_n)$. If none of the u_r is zero, then we can unambiguously count the number of sign changes in $u_1, u_2, \ldots u_n$. If one or more u_r are zero we may arbitrarily assign signs to them; S_u^+, S_u^- are the maximum and minimum numbers of sign changes in the resulting sequences.

Let u_r be the displacement of mass r, and let $w_r = u_r - u_{r-1}$, then: *The necessary and sufficient conditions for* **u** *to be the ith mode* ($i = 1,2,\ldots n$) *of the spring mass system are that*

$$S_u^+ = S_u^- = S_w^+ = S_w^- = i-1. \tag{2.5}$$

Two modes $\mathbf{u}^{(i)}$, $\mathbf{u}^{(j)}$ may *separately* be acceptable modes of a system, but not *jointly*. In [13] we derived necessary and sufficient conditions for $\mathbf{u}^{(i)}$, $\mathbf{u}^{(j)}$ to be jointly acceptable, i.e., to be the ith and jth mode of the same system, and showed how, in that case, we could reconstruct $(m_r, k_r)_1^n$, uniquely, apart from a multiplicative factor.

The basis for the fundamental results (2.3), (2.5) is the nature of the matrices K, M appearing in the governing eigenvalue problem

$$(K - \lambda M)u = 0. \tag{2.6}$$

The matrix M is diagonal and positive, so that (2.6) is equivalent to

$$(B - \lambda I)v = 0, \quad B = M^{-\frac{1}{2}}KM^{-\frac{1}{2}}, \quad v = M^{-\frac{1}{2}}u. \tag{2.7}$$

The matrix B, like K, is tridiagonal, with *negative* off-diagonal elements. It is known ([11], Chapter 3) that such matrices have *distinct* eigenvalues, and their modes $v^{(i)}$ oscillate increasingly, with increasing i, as in (2.5). The *inverse* A $= B^{-1}$ of such a tridiagonal B is a full matrix which belongs to the class of *oscillating* (or *oscillatory*) matrices, studied by Gantmakher and Krein [7], see also Gantmakher [6] and Karlin [21]. A (real, square) matrix A is oscillatory if there is an index m such that A^m is *completely positive*, i.e., has all its minors positive. A is oscillatory iff i) A is non-singular ii) *all* of its minors are non-negative iii) $(a_{i,i+1}, a_{i+1,i}) > 0$ $i = 1,2, \ldots n-1$. Such matrices have special spectral properties. Write (2.7) as $(A - \mu I)v = 0$, $A = B^{-1}$, $\mu = \lambda^{-1}$, and label the μ in decreasing order. Then A has the following spectral properties, among others: i) its eigenvalues are distinct, i.e., $\mu_1 > \mu_2 > \ldots \mu_n > 0$ ii) the ith eigenvector $v = v^{(i)}$ has exactly i-1 changes of sign, i.e., $S_v^+ = S_v^- = i-1$. (Note that $v^{(1)}$ corresponds to μ_1 (the largest μ), i.e., to the smallest λ, i.e., the *lowest* natural frequency) iii) the *nodes* of successive modes interlace, i.e., the i nodes of the (i+1)th mode, interpreted as the broken line constructed by joining $v^{(i+1)}, v^{(i+1)}, \ldots v^{(i+1)}$, separate the i-1 nodes of the ith mode, the broken line constructed by joining $v^{(i)}, v^{(i)}, \ldots v^{(i)}$. Note that ii) shows that at a node, the mode always *crosses* the axis, it can never *touch* it.

When a second-order system is analysed by using the finite-element method, then the resulting matrices K, M are *both* tridiagonal, K with negative, M with positive, off-diagonal terms. Gladwell ([11], p. 117) showed that the resulting $A = K^{-1}M$ is still oscillatory, so that, in particular, the modes have properties ii) - iii) detailed above.

3. Chain-like, discrete, fourth-order systems

The representative system is the model of a beam in flexural vibration, shown in Figure 3. This consists of massless rigid rods of lengths $(l_r)_1^n$ linked by torsional springs of stiffness $(k_r)_1^n$ and having masses $(m_r)_1^n$ at the junctions. This is quite an accurate model of a continuous (Euler-Bernoulli) beam in

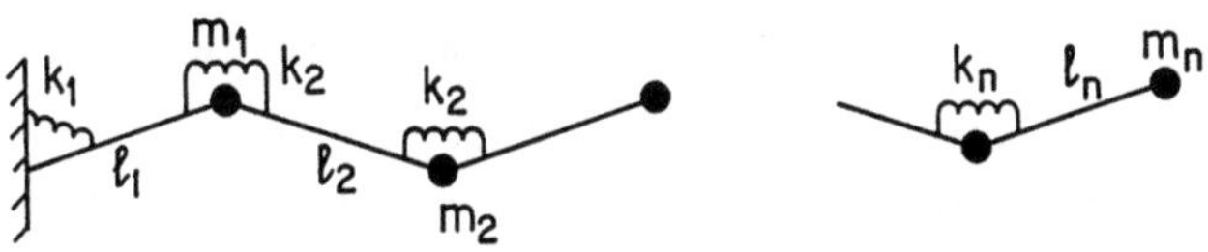

Figure 3. A discrete model of a cantilever beam

flexural vibration, provided that max l_r is small compared to the wavelength of the vibration. This model was analysed first by Barcilon e.g., [1], [2], and definitively by Gladwell [9]. For the inverse problem, the crucial questions centre around the spectral data needed for reconstruction. While the end of a second-order system, like a rod, or the system in Figure 2, may be either free or fixed, the end of a beam model, being specified by two quantities, displacement u and slope θ, may be free, sliding, pinned or clamped. Thus $\theta = 0$ at a sliding end; $u = 0$ at a pinned end; $u = 0 = \theta$ at a clamped end. Each end condition gives rise to a different spectrum and two questions immediately arise. Does each spectrum consist of distinct values? How are the spectra related? It is shown in [9] (see also [10]) that if the four spectra are labelled $(\lambda_i)_1^n$, $(\sigma_i)_1^{n-1}$, $(\mu_i)_1^{n-1}$, $(\gamma_i)_1^{n-2}$ respectively, then

$$\lambda_1 < \sigma_1 < \mu_1 < (\gamma_1, \lambda_2) < \sigma_2 < \mu_2 < \dots \qquad < \mu_{n-1} < \lambda_n \tag{3.1}$$

where the brackets around γ_1,λ_2, etc. indicate that these pairs have no definite ordering. One of the *consequences* of the ordering (3.1) is that *pinning*, giving rise to higher eigenvalues, is more constraining than *sliding*. The *basis* for the ordering, and the other properties of the beam-model modes, is that the dynamic flexibility matrix $\mathbf{A} = M^{\frac{1}{2}}K^{-1}M^{\frac{1}{2}}$ of the model is *oscillatory*.

It is shown in [9] that, in order to reconstruct the beam model, apart from *two* arbitrary factors linking the masses m_r, stiffnesses k_r and lengths l_r, we need *three* spectra, preferably $(\lambda_i)_1^n$, $(\sigma_i, \mu_i)_1^{n-1}$. However, now, the inequalities (3.1), while being necessary, are not like their second-order counterparts (2.3), sufficient, to ensure that the system will have positive m_r, k_r, l_r. Much more stringent conditions must be imposed on the spectra. Suppose u_i, θ_i, are abbreviations for the normalized $u_n^{(i)}$, $\theta_n^{(i)}$, the displacement and slope of the end mass, m_n, in the ith mode. The spectra $(\sigma_i, \mu_i)_1^{n-1}$ are the roots of the equations

$$\sum_{i=1}^{n} \frac{\theta_i^2}{\lambda_i - \lambda} = 0 , \qquad \sum_{i=1}^{n} \frac{u_i^2}{\lambda_i - \lambda} = 0$$

respectively. Suppose we know $(\lambda_i)_1^n$. Then if we know the $(\theta_i)_1^n$ we may find the

$(\sigma_i)_1^{n-1}$, and *vice versa*. So with $(u_i)_1^n$ and $(\mu_i)_1^{n-1}$. This means that the given spectral data $(\lambda_i)_1^n$, $(\sigma_i, \mu_i)_1^{n-1}$ is equivalent to $(\lambda_i, u_i, \theta_i)_1^n$. Now we find that the necessary and sufficient condition for this data to be acceptable is that the $(n+1) \times n$ matrix

$$P = \begin{bmatrix} u_1 & u_2 & \dots & u_n \\ \theta_1 & \theta_2 & \dots & \theta_n \\ \lambda_1 u_1 & \lambda_2 u_2 & \dots & \lambda_n u_n \\ \lambda_1 \theta_1 & \lambda_2 \theta_2 & \dots & \lambda_n \theta_n \\ \lambda_1^2 u_1 & \lambda_2^2 u_2 & \dots & \lambda_n^2 u_n \\ \vdots & \vdots & & \vdots \end{bmatrix} \tag{3.2}$$

be *completely positive*, i.e., *all* its minors must be positive. Note that the last row of **P** is

$$\lambda_1^s u_1 \quad \lambda_2^s u_2 \; \dots \; \lambda_n^s u_n \quad or \quad \lambda_1^s \theta_1 \quad \lambda_2^s \theta_2 \; \dots \; \lambda_n^s \theta_n ,$$

where $s = [n/2]$, according to whether n is even or odd, respectively. The proof of the sufficiency of this condition depends on a beautiful application of Perron's theorem for positive matrices, extended by the use of the Binet-Cauchy theorem (**AB** = **C** implies $A_p B_p = C_p$) for associated matrices, to oscillatory matrices.

The inverse *mode* problem for the beam model has not been fully solved. However, Gladwell et al [18] have precisely characterised an acceptable mode, as follows. It is natural to assume that the lengths $(l_r)_1^n$ are known: let $L = \text{diag}(l_i)$. Let **E** be the difference matrix $(e_{ij} = \delta_{ij} - \delta_{i,j-1})$ and $\mathbf{w} = \mathbf{E}^T \mathbf{L}^{-1} \mathbf{E}^T \mathbf{u}$, then the necessary and sufficient condition for **u** to be the ith mode of the beam model is (2.5). Note that the w's are now second-difference quantities. They are the differences $(w_r = \theta_r - \theta_{r-1})$ of the rotations θ_r $(= (u_r - u_{r-1})/l_r)$ of successive rods; in the second-order system of Figure 2, they were simply differences of the displacements.

An Euler-Bernoulli beam may be modelled by using the Finite Element Method based on consistent stiffness and inertia matrices derived from piece-wise cubic displacement modes. In that case there is a pair of generalised displacements, displacement u_r and slope θ_r at each element junction. Now these quantities are *independent* of each other; they are not related by $\theta_r = (u_r - u_{r-1})/l_r$ as in the model of Figure 3. Now the matrices **K**, **M** are pentadiagonal and the matrix $\mathbf{K}^{-1}\mathbf{M}$

is not oscillatory. However we may prove that, at least for a large class of beams, including the uniform beam, the finite element Green's function G(x,y), has oscillatory properties. Note that G(x,y) does not give the displacement at x due to a unit force at y, but due to the system of forces and moments at the extremities of the element in which y lies, and which, in the finite element sense, are equivalent to the force at y. For a finite element model, the Green's function is degenerate. We have established that, under certain conditions, it may be written in the form

$$G(x,y) = f^T(x)\, Bf(y)$$

where the vectors f(x), f(y) are of order (2n+2). The terms $f_1(x), \ldots f_{2n+2}(x)$, which are polynomials, possess positivity properties of the form

$$\begin{vmatrix} f_i(x_1) & f_j(x_1) & \ldots & f_m(x_2) \\ f_i(x_2) & f_j(x_2) & \ldots & f_m(x_2) \\ f_i(x_r) & f_j(x_r) & & f_m(x_r) \end{vmatrix} \geq 0$$

when $i < j < \ldots < m$ and $x_1 < x_2 < \ldots < x_r$. The matrix B has all non-negative minors. These results lead to the oscillatory properties for G(x,y), and to the properties of the finite-element eigensolutions of the equation

$$u(x) = \lambda \int_0^L G(x,y)A(y)u(y)dy$$

which is equivalent to the matrix eigenvalue problem $(K-\lambda M)u = 0$. In particular, the eigenvalues are distinct and the eigenmodes have properties i) - iii) listed above. Note that the nodes of modes must now be found as the points where the piecewise cubics cross the axis. These results will appear in a forthcoming paper.

4. Continuous chain-like systems

The simplest systems of this type are the taut string with distributed mass executing transverse vibration, and the thin straight rod executing longitudinal or torsional vibration. The free undamped vibrations of these systems are governed by second-order equations of Sturm-Liouville type, on a finite line, [0,L], the string by

$$y''(x) + \lambda\rho(x)y(x) = 0 \tag{4.1}$$

and the rods by

$$(A(x)y'(x)' + \lambda A(x)y(x) = 0 \quad . \tag{4.2}$$

Most of the vast literature on the inverse problem for these equations stems from Gel'fand and Levitan [8]. References may be found in [11], [13] or Mclaughlin [23]. The spectral properties of these systems are well known, viz. under the self-adjoint two-point end conditions appropriate (for instance) to the string or rod, i) the eigenvalues are positive and distinct; ii) the ith eigenmode ($i = 0,1, \ldots$) has exactly i nodes in $(0, L)$; iii) the nodes of successive modes interlace.

The Sturm-Liouville equation arises from the application of Newton's second law to an element of the system, or alternatively from Lagrange's equations. We may formulate the problem another way, by finding the static response $G(x,s)$ of the system, at position x, due to a unit load at s. In the free-vibration problem for the string, the system is excited by the reversed inertia forces

$$-\rho(s)\ddot{y}(s,t)ds = \lambda\rho(s)y(s,t)ds$$

distributed along the system with the density $\rho(s)$. This leads to the integral equation formulation

$$y(x) = \lambda \int_0^L G(x,s)\rho(s)y(s)ds \tag{4.3}$$

The Green's function $G(x,s)$ is the counterpart of the flexibility matrix K^{-1} in discrete problems.

Equation (4.3) is a Fredholm integral equation which is equivalent to the Sturm-Liouville equation and associated end conditions. There is a well developed spectral theory for such equations. However, the particular $G(x,s)$ in (4.3) is *special*: it must be, because its eigenvalues and eigenmodes, being those of the Sturm-Liouville equation, have properties i) - iii) listed above. If $G(x,s)$ were arbitrary, or even only symmetric and positive definite, the solutions of (4.3) would not necessarily have these properties.

It was Gantmakher and Krein who introduced the concept of an oscillating or oscillatory *kernel* $G(x,s)$. Thus they considered the equation (4.3) in itself, i.e., not necessarily derived from a Sturm-Liouville equation, and stated properties that $G(x,s)$ should satisfy in order that its solutions should have the spectral properties i) - iii). The defining properties of an oscillating kernel are generalisations of those of an oscillatory matrix and relate to the positivity of the minors of the matrix with elements $G(x_i, x_j)$, where $x_1, x_2 \ldots x_n$ are points in $(0,L)$ satisfying $0 < x_1 < x_2 < \ldots < x_n < L$.

Another important continuous chain-like system is the Euler-Bernoulli beam. The equation governing the free vibration modes is

$$(EI(x)y''(x))'' = \lambda A(x)\rho y(x) \qquad (4.4)$$

Here A(x), I(x) are the area and second moment of area about the neutral axis; E, ρ, assumed to be constant, are the Young's modulus and density of the beam. $\lambda = \omega^2$ is the square of a natural frequency. Gantmakher and Krein [7], (or see Karlin [21], Chapter 10, or Gladwell [11], Chapter 10) prove that the Green's function for an Euler-Bernoulli beam, under the usual free, sliding, pinned or clamped end-conditions, or indeed under the more general self-adjoint end conditions corresponding to spring supports, is oscillatory. (Provided that the beam is restrained against rigid body motion). From this we may deduce that the spectrum corresponding to a given set of self-adjoint end conditions is simple. Moreover, the four spectra $(\lambda_i, \sigma_i, \mu_i, \gamma_i)_0^\infty$ are related just as those for the discrete model, i.e.,

$$\lambda_0 < \sigma_0 < \mu_0 < (\gamma_0, \lambda_1) < \sigma_1 < \mu_1 < (\gamma_1, \lambda_2) < \ldots$$

Again, as we might expect from our experience with the discrete model of the beam, this ordering is insufficient to ensure that the spectra, or even the three spectra $(\lambda_i, \sigma_i, \mu_i)_0^\infty$ needed for reconstruction, are actually acceptable. It was shown by Gladwell [12] that the necessary and sufficient condition for acceptability is that the infinite matrix corresponding to P of (3.2) is completely positive. It is easy to show that the complete positivity of P is necessary. To prove that it is sufficient we show that if P is completely positive then the Gel'fand Levitan reconstruction procedure for the beam, derived by McLaughlin [22], cannot fail. Barcilon [3] gave an alternative proof of the sufficiency condition.

We have made no mention of asymptotic properties of the spectra. These have been discussed fully elsewhere. Hochstadt [20] made the most careful study of the asymptotic spectral properties of the Sturm-Liouville equation, and Barcilon [2] studied those for the Euler-Bernoulli beam. In a somewhat loose terminology we may say that the asymptotic conditions require that, for high frequencies, a non-uniform rod or beam should behave like a related uniform rod or beam having the same length. Indeed, in the Gel'fand-Levitan like reconstruction procedure derived by McLaughlin [22], [23], the finite set of spectral data corresponding to the lower eigenvalues is supplemented by spectral data corresponding to a known system, say a uniform rod or beam. In a pair of companion papers, Gladwell et al [15], [16], we carried out numerical investigations to determine how far data corresponding to the lower eigenvalues could depart from the corresponding eigenvalues for a uniform rod or beam and still yield a rod or beam that was actually slender; it is only for such systems that the rod equation (4.2) or the beam equation (4.4) actually hold. The general conclusion that we reached was that only very small departures are allowable; otherwise the system may vary so much that the ratio of the maximum to the minimum of a typical dimension, e.g., the depth of the beam, may be so large that the system cannot be regarded as slender.

5. Conclusions

Precise qualitative criteria have been found for the acceptibility of spectral data in the inverse problems related to second and fourth order chain-like vibrating systems.

References

1. Barcilon, V., "On the multiplicity of solutions of the inverse problem for a vibrating beam". SIAM J. Appl. Math. 37, 605-613, (1979).

2. Barcilon, V., "Inverse problems for the vibrating beam in the free-clamped configuration", Phil. Trans. Roy. Soc. Lond. A, 304, 211-252, (1982).

3. Barcilon, V., "Sufficient conditions for the solution of the inverse problem for a vibrating beam", Inverse Problems, 3, 181-193, (1987).

4. Boley, D. and Golub, G.H., "A survey of matrix inverse eigenvalue problems", Inverse Problems, 3, 595-622, (1987).

5. Friedland, S., "Inverse eigenvalue problems", Lin. Algebra Appl., 17, 15-51, (1977).

6. Gantmakher, F.R., The Theory of Matrices, New York: Chelsea Publishing Company, (1960), Vol II, Chapter 13.

7. Gantmakher, F.R. and Krein, M.G., Oscillation Matrices and Kernels and Small Vibrations of Mechanical Systems, Moscow-Leningrad: State Publishing House of Technical-Theoretical Literature. Translation, Washington, D.C.: U.S. Atomic Energy Commission, (1961).

8. Gel'fand, I.M. and Levitan, B.M., "On the determination of a differential equation from its spectral function" (in Russian), Izv. Akad. Nauk SSSR. Ser. Mat. 15, 309-362, (1951). English Translation, Amer. Math. Soc. Transl. Ser. 2,1, 253-304.

9. Gladwell, G.M.L., "The inverse problem for the vibrating beam", Proc. Roy. Soc. London A, 393, 277-295, (1984).

10. Gladwell, G.M.L., "Qualitative properties of vibrating systems", Proc. Roy. Soc. London A, 401, 299-315, (1985).

11. Gladwell, G.M.L., Inverse Problems in Vibration, Dordrecht, etc: Martinus Nijhoff, (1986).

12. Gladwell, G.M.L., "The inverse problem for the Euler-Bernoulli beam", Proc. Roy. Soc. London A, 407, 199-218, (1986).

13. Gladwell, G.M.L., "Inverse problems in vibration", Appl. Mech. Rev., 39, 1013-1018, (1986).

14. Gladwell, G.M.L., "The inverse mode problem for lumped-mass Systems", Q. Jl. Mech. Appl. Math., 39, 297-307, (1986).

15. Gladwell, G.M.L. and Dods, S.R.A., "Examples of reconstruction of vibrating rods from spectral data", J. Sound Vib., 119, 267-276, (1987).

16. Gladwell, G.M.L., England, A.H. and Wang, D., "Examples of reconstruction of an Euler-Bernoulli beam from spectral data", J. Sound Vib., 119, 81-94, (1987).

17. Gladwell, G.M.L. and Willms, N.B., "The reconstruction of a tridiagonal system from its frequency response at an interior point", Inverse Problems, 4, 1013-1024, (1988).

18. Gladwell, G.M.L., Willms, N.B., He B. and Wang, D., "How can we recognise an acceptable mode shape for a vibrating beam?", Q. Jl. Mech. Appl. Math., 42, 303-316, (1989).

19. Gladwell, G.M.L. and Willms, N.B., "A discrete Gel'fand-Levitan method for band-matrix inverse eigenvalue problems", Inverse Problems, 5, 165-179, (1989).

20. Hochstadt, H., "Asymptotic estimates of the Sturm Liouville spectrum", Comm. Pure Appl. Math., 14, 749-764, (1961).

21. Karlin, S., Total Positivity, Vol 1. Stanford: Stanford University Press, (1968).

22. McLaughlin, J.R., "On constructing solutions to an inverse Euler-Bernoulli beam problem", in Santosa, F., Pao, Y.-H., Symes, W.W. and Holland, C., Editors, Inverse Problems of Acoustic and Elastic Waves. Philadelphia: SIAM, 341-345, (1984).

23. McLaughlin, J.R., "Analytical methods for recovering coefficients in Sturm-Liouville equations", SIAM Rev., 28, 53-72, (1986).

24. Morel, P., "Algorithmes pour un probleme inverse discret de Sturm-Liouville", in Sabatier, P.C., Editor, Applied Inverse Problems. Berlin, etc: Springer Verlag, 371-387, (1978).

An Inverse Eigenvalue Problem and an Extremal Eigenvalue Problem

J.R. McLaughlin

Department of Mathematical Sciences, Rensselaer Polytechnic Institute, Troy, NY 12180, USA

This talk presents results for two inverse problems which arise in the study of vibrating systems. The first problem (Part I) extends the theory of second order inverse eigenvalue problems in one dimension and is joint work with Carol Coleman. The second problem (Part II) solves an identification problem for composite membranes in n-dimensions; this work is joint work with Steven Cox.

Part I:

We will present a selection of results from a complete solution for the inverse spectral problem for the eigenvalue problem for a vibrating rod

(1) $\quad (p^2 u_x)_x + \lambda\, p^2 u = 0, \quad 0 < x < 1, \quad p > 0$

(2) $\quad u(0) = u(1) = 0$

for the case where $p_x/p \in L^2(0,1)$ and $p(0) = 1$. We assume that the given data is comprised of two sequences plus the value of $p(1)$. The first data sequence is the set of eigenvalues $0 < \lambda_1 < \lambda_2 < \ldots$ If g_n is the normalized eigenfunction, $\int_0^1 p g_n^2 = 1$, corresponding to λ_n, the second data sequence is

$$k_n = \ell n \left| \frac{p(1) g_{n,x}(1)}{p(0) g_{n,x}(0)} \right| , \quad n = 1, 2, \ldots$$

With these two sequences and the value of $p(1)$ we will recover the function $p(x)$ for $0 < x < 1$.

The results contained in Poeschel and Trubowitz [8] could be directly applied to this problem if $p_{xx} \in L^2(0,1)$. Our assumption, $p_x/p \in L^2(0,1)$ requires new analysis and is of interest in applications where one seeks coefficients which model "rough" material parameters. We will state global existence and uniqueness results and establish that the above data set is a global coordinate system for all p with $p(0) = 1$, and $p_x/p \in L^2(0,1)$. We also establish the range of the map from p to the data set and that this map and its inverse are analytic.

Our analysis follows the model set in [8] for the smoother case. A different approach is taken in Andersson [1], [2] where the data set is two sequences of eigenvalues but one also assumes $p_x/p \in L^2(0,1)$. Andersson shows the map from p to the data set is bicontinuous, locally Lipschitz and one-to-one. He establishes the range of the map from p to the data set which is comprised of two sets of eigenvalues.

To present our results we require the alternative form of (1) - (2)

(3) $\quad u_{xx} + \alpha\, u_x + \omega^2\, u = 0 \qquad 0 < x < 1, \qquad \alpha \in L^2(0,1)$

(1) $\quad u(0) = u(1) = 0$

where $\alpha = \frac{p_x}{p}$ and $\lambda = \omega^2$. We write our data set as $\mu \times [\cdot] \times \mathbf{k}$ where

$$\mu = \left\{\ln \omega_n\right\}_{n=1}^{\infty}, \quad \mathbf{k} = \left\{k_n\right\}_{n=1}^{\infty} \text{ and } [\alpha] = \int_0^1 \alpha dx = \ell n[p(1)].$$

The first step in solving this problem is to establish the asymptotics for $\omega_n = \sqrt{\lambda_n}$ and k_n and to prove uniqueness. The asymptotics are established using the modified Prüfer transformation. The uniqueness proof is patterned after the uniqueness proof given in Borg [4] for the potential problem. We state

Theorem 1: Given $\alpha \in L^2(0,1)$, then

$$\left\{\omega_n - n\pi\right\}_{n=1}^{\infty} \in \ell^2 ,$$

$$\left\{k_n\right\}_{n=1}^{\infty} \in \ell^2 .$$

and

Theorem 2: Let

$$S = \left\{(\ln(n\pi + \sigma_n))_{n=1}^{\infty} \;\middle|\; (\sigma_n)\in\ell^2 \text{ and } n\pi + \sigma_n < (n+1)\pi + \sigma_{n+1},\ n = 1,2,\ldots\right\}.$$

Let R be the set of real numbers.

Let $L^2(0,1) = \left\{\alpha \;\middle|\; \alpha \text{ real}, \int_0^1 [\alpha(x)]^2 dx < \infty\right\}$.

Then $\mu \times [\cdot] \times \mathbf{k}$ maps $L^2(0,1)$ one-to-one into $S \times R \times \ell^2$.

A principle goal is to establish the range of the $\mu \times [\cdot] \times \mathbf{k}$ map and determine how smoothe the map is. Uniform complex extensions of each λ_n, k_n are shown to exist. This establishes the very strong local result that

Theorem 3: $\mu \times [\cdot] \times \mathbf{k}$ is a local real analytic isomorphism at every point of $L^2(0,1)$. The derivatives are

$$d_\alpha \mu(\nu) = \left\{\int_0^1 \left(-\frac{p^2 g_n g_n'}{(\omega_n)^2}\right)\nu(x)dx\right\}_{n=1}^{\infty}$$

$$d_\alpha \mathbf{k}(\nu) = \left\{\int_0^1 \left[-p^2[y_1 y_2]' - \frac{\partial\lambda_n}{\partial\alpha} < y_1, y_2 >_{p^2}\right]\nu(x)dx\right\}_{n=1}^{\infty}$$

$$d_\alpha [\cdot](\nu) = \int_0^1 \nu(x)dx$$

where y_1, y_2 satisfy (3) with

$$y_1(0) = y_2'(0) = 1 \quad , \quad y_1'(0) = y_2(0) = 0,$$

and

$$< y_1, y_2 >_{p^2} = \int_0^1 y_1(x)\, y_2(x) p^2 dx ,$$

and $$\frac{\partial\lambda_n}{\partial\alpha} = -2\, p^2 g_n\, g_n' .$$

The analyticity of the map given above enables us to establish flows on isospectral sets and to apply the inverse function theorem. Exact solutions are also exhibited to establish that the range of the $\mu \times [\cdot] \times \mathbf{k}$ map is $S \times R \times \ell^2$. We state here only the result which shows that for fixed $\mu = \mu(\alpha_0) = \mu^0$ and $[\alpha_0] = 0$ the map $\mathbf{k}$ is onto ℓ^2. We first establish some notation

Let $L_0^2(0,1) = \left\{\alpha\epsilon L^2(0,1) \;\middle|\; [\alpha]=0,\ \alpha \text{ real}\right\}$.

Let $M(\alpha_0) = \left\{\alpha \;\middle|\; \mu(\alpha)=\mu(\alpha_0),\ \alpha\epsilon L_0^2(0,1)\right\}$.

Let $V_n(x,\alpha) = (pg_n)^2 - \left(\frac{pg_n'}{\omega_n}\right)^2$, $n = 1,2,\ldots$

It is shown that

<u>Theorem 4</u>: $M(\alpha_0)$ is a real analytic submanifold of $L_0^2(0,1)$. For all $\alpha \in M(\alpha_0)$ the tangent space is

$$T_\alpha M(\alpha_0) = \left\{ V_\zeta(\alpha) = \sum_{n=1}^{\infty} \zeta_n V_n \;\middle|\; \zeta \in \ell^2 \right\} .$$

Let $\zeta \in \ell^2$ be fixed. Then the solution $\Phi^t(x)$ of

$$\frac{d}{dt}\Phi^t = \sum_{n=1}^{\infty} \zeta_n V_n$$

$$\Phi^0 = \alpha_0$$

exists for all time and $k_n(\Phi^t) = t\zeta_n + k_n(\alpha_0)$, $n = 1,2,\ldots$

See [5] for proofs and complementary results.

We remark that the inverse problem for (1) - (2) when $p_{xx} \in L^2(0,1)$ has a long history. We refer the reader to [7] for an extensive list of references.

Part II:

We consider here the eigenvalue problem for a vibrating membrane

$$\Delta u + \lambda \rho u = 0, \quad x \in \Omega \subset R^n,$$
$$u = 0, \quad x \in \Omega \subset R^n,$$

where Ω is a bounded, open, connected set in Euclidean n-space, R^n. We will show how to place the given materials for a composite membrane to solve extremal eigenvalue problems. For the composite membrane let $0 < \alpha < \beta$ and $\gamma > 0$ be given constants; then the density $\rho = \alpha\chi_D + \beta\chi_{\Omega-D}$ where $D \subset \Omega$ and the measure of D, $|D| = \gamma$. We seek to identify $D \subset \Omega$, and hence ρ, to solve either the extremal problem max λ_1 or min λ_1. This work is motivated by applications where it is desired to construct a material so that the eigenvalues (or natural frequencies) are constrained away from given intervals.

The mathematics is motivated by the one-dimensional work of Krein [17] where the eigenvalue problem is

$$u_{xx} + \lambda\rho u = 0, \qquad 0 < x < L$$
$$u(0) = u(L) = 0.$$

In Krein's work $\Omega = \{x \mid 0 < x < L\}$ and he shows max λ_1 can be achieved by choosing

$$D_{max} = \left\{x \;\middle|\; \frac{L-\gamma}{2} < x < \frac{L+\gamma}{2}\right\}$$

and min λ_1 can be achieved by choosing

$$D_{min} = \left\{x \;\middle|\; 0 < x < \gamma/2 \quad , \quad L - \gamma/2 < x < L\right\}.$$

Note that for the densities which solve the extremal eigenvalue problem, the first eigenfunction is positive, symmetric and has one maximum at $x = L/2$. To achieve max λ_1 then we place the light α-material in the center of the interval where the first eigenfunction u_1 is largest and the heavy β-material where u_1 is small. For min λ_1 the heavy β-material is placed where u_1 is largest.

We seek to generalize Krein's results to higher dimensions. The results will be stated in terms of the Rayleigh quotient formulation of the eigenvalue problem although the proofs use a functional developed by Auchmuty [10]. To begin we observe that the extremal problem min λ_1 can be rephrased as

$$\sup_{\rho} \left[\frac{1}{\lambda_1}\right] = \sup_{\rho} \sup_{u} \left[\int_\Omega \rho u^2\right]$$

and the extremal problem max λ_1 can be written as

$$\inf_{\rho} \left[\frac{1}{\lambda_1}\right] = \inf_{\rho} \sup_{u} \left[\int_\Omega \rho u^2\right] .$$

Of course we must choose the correct spaces for ρ and u. Here we choose $u \in H_0^1(\Omega)$ and normalize u so that $\int_\Omega |\nabla u|^2 = 1$. The density ρ is in the weak-* completion ad_γ^* of

$$ad_\gamma = \left\{\rho \mid \rho = \alpha\, \chi_D + \beta\, \chi_{\Omega-D},\ |D| = \gamma\right\}.$$

An aim of the present work is to show that the extreme values of λ_1 are taken on for specific choices of ρ in ad_γ. That is, ρ is one of the extreme points in ad_γ^*.

The extremal problem min λ_1 is the most straight forward to solve. In this case the 'sup's' are swapped to yield

$$\sup_{\rho} \left[\frac{1}{\lambda_1}\right] = \sup_{u} \sup_{\rho} \left[\int_\Omega \rho u^2\right].$$

If min λ_1 is thus achieved at $\tilde{\rho}$ with corresponding eigenfunction $\tilde{u}_1$ we must have the optimality criteria

$$\int_\Omega (\tilde{\rho} - \rho)\ (\tilde{u}_1)^2 \geq 0 \text{ for all } \rho \epsilon ad_\gamma^*.$$

It can be shown by applying a geometric argument similar to that found in [12] that this can (only) be achieved if for some $\bar{\ell} > 0$ we have

$$\begin{aligned} \tilde{\rho} &= \alpha && \text{when } |\tilde{u}_1| > \tilde{\ell} \\ \alpha \leq \tilde{\rho} &\leq \beta && \text{when } |\tilde{u}_1| = \tilde{\ell} \\ \tilde{\rho} &= \beta && \text{when } |\tilde{u}_1| < \tilde{\ell} \end{aligned}$$

If the measure $\left|\{x \mid |\tilde{u}| = \tilde{\ell}\}\right| = 0$, and this occurs, e.g., when $\partial\Omega \epsilon C^2$, then $\tilde{\rho}$ is completely determined by $\tilde{u}_1$. The optimality criteria then yields that $\tilde{\rho} \epsilon ad_\gamma$ and provides the n-dimensional generalization of Krein's result.

The extremal problem max λ_1 is more difficult. Here we must swap the 'sup' and the 'inf'. This requires a saddle point result. It is accomplished by generalizing the saddle point theorem, Theorem 3.7, in Barbu and Precupanu [11]. We show

$$\inf_\rho \sup_u \left[\int_\Omega \rho u^2\right] = \sup_u \inf_\rho \left[\int_\Omega \rho u^2\right]$$

Now if max λ_1 is achieved by $\bar{\rho}$ with corresponding first eigenfunction $\bar{u}_1$ then

$$\int_\Omega (\rho - \bar{\rho})\ (\bar{u}_1)^2 \geq 0 \text{ for all } \rho \ \epsilon\ ad_\gamma^*.$$

Again it is shown that this can only be achieved if for some $\bar{\ell} > 0$ we have

$$\begin{aligned} \bar{\rho} &= \alpha && \text{when } |\bar{u}_1| < \bar{\ell}, \\ \alpha \leq \bar{\rho} &\leq \beta && \text{when } |\bar{u}_1| = \bar{\ell}, \\ \bar{\rho} &= \beta && \text{when } |\bar{u}_1| > \bar{\ell}. \end{aligned}$$

The density $\bar{\rho}$ is then completely determined by $\bar{u}_1$ when we establish that the measure, $|\bar{u}_1^{-1}\ (\bar{\ell})| = 0$. We can also exploit the saddle point structure of the problem to prove that $\bar{\rho}$ is unique.

For the extremal problem max λ_1 it is possible to apply the work of Kawohl [16], Gidas, Ni, and Nirenberg [15], and Attouch [9], to exhibt how the geometric properties of Ω are preserved by $\bar{u}_1$ and therefore by $\bar{\rho}$; see [13], [14].

Acknowledgement: The work of JRM has been supported by ONR Grant No. N00014-84-K-516 and by NSF grant No. DMS-8713722.

References

[1] L. Andersson, Inverse problems with discontinuous coefficients, Inverse Problems 4 (1988) 353-397.

[2] L. Andersson, Inverse eigenvalue problems for a Sturm-Liouville equation in impedance form, Inverse Problems (4) (1988) 929-971.

[3] F. Atkinson, Discrete and Continuous Boundary Problems, Academic Press, New York, 1974.

[4] G.Borg, Eine Umkehrung der Sturm-Liouvilleschen Eigenwertaufgabe, Acta Math 78 (1946) 1-96.

[5] C. Coleman, An Inverse Spectral Problem with a Rough Coefficient, Ph.D. thesis, Rensselaer Polytechnic Institute, Troy, New York.

[6] I.M. Gel'fand and B.M. Levitan, On the determination of a differential equation from its spectral function, Izv. Akad. Nauk SSSR Ser. Mat. 15 (1951) 309-360; English transl. Amer. Math. Soc. Transl., 1 (1955) 253-304.

[7] J.R. McLaughlin, Analytical methods for recovering coefficients in differential equations from spectral data, SIAM Review 28 (1986) 53-72.

[8] J. Poeschel and E. Trubowitz, Inverse Spectral Theory, Academic Press, 1987.

[9] H. Attouch, Variational Convergence for Functions and Operators, Pitman, Boston, 1984.

[10] G. Auchmuty, Dual Variational Principles for Eigenvalue Problems, in Nonlinear Functional Analysis and its Applications, F. Browder, ed., AMS 1986, 55-71.

[11] V. Barbu and T. Precupanu, Convexity and Optimization in Banach Spaces, Reidel, Boston, 1986.

[12] J. Cea and K. Malanowski, An example of a max-min problem in partial differential equations, SIAM J Control, (3) 305-316.

[13] S.J. Cox and J.R. McLaughlin, Extremal Eigenvalue Problems for Composite Membranes I, Appl Math and Optimization (to appear).

[14] S.J. Cox and J.R. McLaughlin, Extremal Eigenvalue Problems for Composite Membranes II, Appl Math and Optimization (to appear).

[15] B. Gidas, W. Ni, and L. Nirenberg, Symmetry and related properties via the maximum principle, Comm. Math. Phys. 68 (1979) 209-243.

[16] B. Kawohl, Rearrangements and Convexity of Level Sets in PDE, Springer-Verlag, New York, 1985.

[17] M.G. Krein, On certain problems on the maximum and minimum of characteristic values and on the Lyapunov zones of stability, AMS Translations Ser. 2(1), (1955) 163-187.

Part IV

Scattering Inverse Problems (Exact Methods)

Scattering by Black-Hole for Electromagnetic Fields

A. Bachelot

Department of Applied Mathematics, University of Bordeaux I,
F-33405 Talence Cédex, France

I - Maxwell equations in Schwarzschild universe

We investigate the electromagnetic field outside spherical Black-Hole of radius $r_o > 0$, described by Schwarzschild metric

(1) $$ds^2 = \alpha^2\, dt^2 - \alpha^{-2}\, dr^2 - r^2(d\theta^2 + sin^2\,\theta\; d\varphi^2) \quad , \quad r_o < r\,,$$

and lapse function α is given by

(2) $$\alpha = (1 - r_o\, r^{-1})^{1/2}\,.$$

This metric is singular on the "Horizon" $\Gamma = \mathbb{R}_t \times \{r = r_o\} \times S^2$ and no radial null geodesic reaches Γ at finite time t. With Wheeler coordinate r_*, the equation of such geodesics is

(3) $$t = \pm r_* + C \quad , \quad r_* = r + r_o\, \ell n\,(r - r_o)\,.$$

In Schwarzschild vacuum, Maxwell's tensor F verifies equations :

(4) $$dF = 0 \quad , \quad d * F = 0\,,$$

where $*$ is the Hodge operator related to metric (1). We split F into electric and magnetic fields measured by an observer with four velocity u :

(5) $$E_\mu = F_{\mu,\nu}\; u^\nu \;, \quad B_\mu = -(* F)_{\mu,\nu}\; u^\nu\,.$$

Since we are concerned by scattering theory, we consider the Black-Hole as a perturbation and we choose an observer at rest by respect to the Black-Hole (Fiducial observer of [6]), and then

(6) $$u = \alpha^{-1}\, \partial_t\,.$$

By putting

(7) $${}^tU = (E^{\hat{r}}, E^{\hat{\theta}}, E^{\hat{\varphi}}, B^{\hat{r}}, B^{\hat{\theta}}, B^{\hat{\varphi}}) = (e, b)\,,$$

where

$$X = X^{\hat{r}}\,\alpha\,\partial_r + X^{\hat{\theta}}\,r^{-1}\,\partial_\theta + X^{\hat{\varphi}}\,(r\sin\theta)^{-1}\,\partial_\varphi \quad , \quad X = E, B \,,$$

Maxwell's equations (4) take a familiar form

$$(8) \qquad \partial_t U = -iHU \quad , \quad \nabla_S . E = \nabla_S . B = 0 \,,$$

where

$$(9) \qquad H = i\begin{pmatrix} 0 & \nabla_S \times \\ -\nabla_S \times & 0 \end{pmatrix} , \quad \nabla_S \times = \begin{pmatrix} 0 & -\frac{\alpha}{r\sin\theta}\partial_\varphi & \frac{\alpha}{r\sin\theta}\partial_\theta \sin\theta \\ \frac{\alpha}{r\sin\theta}\partial_\varphi & 0 & -\frac{\alpha}{r}\partial_r r\alpha \\ -\frac{\alpha}{r}\partial_\theta & \frac{\alpha}{r}\partial_r r\alpha & 0 \end{pmatrix}$$

$$(10) \qquad \nabla_S . X = \alpha\, r^{-2}\,\partial_r (r^2 X^{\hat{r}}) + (r\sin\theta)^{-1}\,[\partial_\theta(\sin\theta\, X^{\hat{\theta}}) + \partial_\varphi X^{\hat{\varphi}}]$$

If there is no Black-Hole, $\alpha = 1$ and we find the free dynamic in Minkowski space-time with spherical coordinates. We introduce the Hilbert space of finite redshifted energy :

$$\tilde{\mathscr{H}} = [L^2(\,]r_o\,, +\infty[_r \times S^2_\omega\,,\, r^2\,dr\,d\omega)]^6 \,,$$

and the subspace of free divergence :

$$\mathscr{H} = \{U \in \tilde{\mathscr{H}} \;;\; \nabla_S . E = \nabla_S . B = 0\} \,.$$

THEOREM I.1 - *H is a selfadjoint operator with dense domain on $\tilde{\mathscr{H}}$ and on $\mathscr{H}$.*

Then we solve the Cauchy problem for (8) by Stone's theorem.

REMARK : We are not concerned by a mixed problem : we do not need any boundary condition on horizon Γ which is not time like.

We have a result of finite velocity dependance :

THEOREM I.2 - *Let's be U in $\tilde{\mathscr{H}}$ such that*

$$supp\; U \subset \{r_*^1 \le r_* \le r_*^2\} \times S^2 \,;$$

then we have

$$supp\; e^{-itH}\, U \subset \{r_*^1 - |t| \le r_* \le r_*^2 + |t|\} \times S^2 \,.$$

Schwarzschild metric is trapping : all great circles of sphere with radius $3r_o/2$, so called "Photons-sphere", are null geodesics ; there exist so null geodesics asymptotic to the Photons-sphere. Therefore singularities of field can be trapped and do not escape at infinity. Despite of this difficulty, there is no time-periodic solution in Schwarzschild universe, unlike the euclidian case with an obstacle, for which, the second space of cohomology yields non trivial stationary solutions :

THEOREM I.3 - *The ponctual spectrum of* H *on* $\mathcal{H}$ *is empty.*

We can deduct from this result, the decay of local energy ; but we developp here a complete scattering theory for the electromagnetic field and in particular, we find the result of Damour [3] on the behaviour of fields near the horizon. The study of scalar case was treated by Dimock and Kay [4] [5] .

II - Wave operators at infinity

Schwarzschild universe is asymptotically flat and far from the Black-Hole we compare hamiltonian H with classical electromagnetic hamiltonian H_o :

(9) $$H_o = i \begin{pmatrix} 0 & curl \\ -curl & 0 \end{pmatrix},$$

in Minkowski space-time with metric

(10) $$ds^2 = dt^2 - d\rho^2 - \rho^2\,(d\theta^2 + sin^2\theta\ d\varphi^2)\ ,\ 0 \leq \rho\ .$$

For any choice of $\rho = \rho(r)$, the difference $H - H_o$ is a long-range type perturbation but because the radial null geodesics (3) are straight like their flat analogs, we avoid long range interaction between gravitational and electromagnetic fields by choosing :

(11) $$\rho = r_* \geq 0\ .$$

We introduce the usual finite energy Hilbert spaces :

$$\tilde{\mathcal{H}}_o = \{U_o = {}^t(E_o^{\hat{r}}, E_o^{\hat{\theta}}, E_o^{\hat{\varphi}}, B_o^{\hat{r}}, B_o^{\hat{\theta}}, B_o^{\hat{\varphi}}) \in [L^2(\mathbb{R}^+_{r_*} \times S^2_\omega, r_*^2\, dr_*\, d\omega)]^6\}\ ,$$
$$\mathcal{H}_o = \{U_o = {}^t(E_o, B_o) \in \tilde{\mathcal{H}}_o\ ; div\, E_o = div\, B_o = 0\ \}\ .$$

Given a cut-off function $\chi_o \in C^\infty(\mathbb{R}^+_{r_*})$ satisfying, $\chi_o(r_*) = 0$ for $0 \leq r_* < a$, and $\chi_o(r_*) = 1$ for $r_* > b$, for some $0 < a < b$, we construct an identification operator $\mathcal{J}_o : \tilde{\mathcal{H}}_o \rightarrow \tilde{\mathcal{H}}$ by putting :

$$\mathcal{J}_o U_o = \chi_o U_o \quad for \quad r_* \geq 0 \ , \quad \mathcal{J}_o U_o = 0 \quad for \quad r_* \leq 0 \, .$$

We define classical wave operators without Dollard's modification :

$$W_o^{\pm} U_o = s - \lim_{t \to \pm\infty} e^{itH} \mathcal{J}_o e^{-itH_o} U_o \quad in \quad \tilde{\mathcal{H}} \, .$$

The spherical invariance of Maxwell equations - that implies a t^{-2}decay of radial components - and our choice (11), cancel the long range effects and by Cook's method we prove the

THEOREM II.1 - $W_o^{\pm} : \mathcal{H}_o \to \mathcal{H}$ *exist, are independent of* χ_o *and* $\|W_o^{\pm}\| \leq 1$.

We deduct from this result, the existence of outgoing fields :

THEOREM II.2 - *If* $U_o \in \mathcal{H}_o$ *verifies* :

$$e^{-itH_o} U_o = 0 \quad for \quad 0 \leq r_* \leq \pm t + C \, ,$$

then we have

$$e^{-itH_o} W_o^{\pm} U_o = 0 \quad for \quad r_* \leq \pm t + C \, .$$

III - Wave operators near the black-hole

Hamiltonian H degenerates as $r \to r_o$, but $r\alpha H(r\alpha)^{-1}$ admits a formal limit H_1

$$(13) \qquad H_1 = i \begin{pmatrix} 0 & h_1 \\ -h_1 & 0 \end{pmatrix} , \quad h_1 = \begin{pmatrix} 0 & 0 & 0 \\ 0 & 0 & -\partial_{r_*} \\ 0 & \partial_{r_*} & 0 \end{pmatrix} .$$

H_1 is essentially the dynamic in Rindler metric that approximates Schwarzschild metric near the horizon. We introduce Hilbert spaces :

$$\tilde{\mathcal{H}}_1 = \{U_1 = {}^t(E_1^{\hat{r}}, E_1^{\hat{\theta}}, E_1^{\hat{\varphi}}, B_1^{\hat{r}}, B_1^{\hat{\theta}}, B_1^{\hat{\varphi}}) \in [L^2(\mathbb{R}_{r_*} \times S_\omega^2, dr_* \, d\omega)]^6\} ,$$

$$\mathcal{H}_1^{\pm} = \{U_1 \in \tilde{\mathcal{H}}_1 \, ; E_1^{\hat{r}} = B_1^{\hat{r}} = \pm E_1^{\hat{\theta}} + B_1^{\hat{\varphi}} = \pm E_1^{\hat{\varphi}} - B_1^{\hat{\theta}} = 0\} .$$

The fields in $\mathcal{H}_1^{+(-)}$ have an left (right) polarization and behave like a plane wave, falling into the future (coming out of the past) horizon :

$$U_1 \in \mathcal{H}_1^{\pm} \Rightarrow [e^{-itH_1} U_1](r_* , \omega) = U_1(\pm t + r_* , \omega) .$$

Given a cut-off function $\chi_1 \in C^\infty(\mathbb{R}_{r*})$ satisfying $\chi_1(r_*) = 1$ for $r_* < c$, $\chi_1(r_*) = 0$ for $r_* > d$, for some $c < d < 0$, we construct an identification operator

$\mathcal{J}_1 : \tilde{\mathcal{H}}_1 \to \tilde{\mathcal{H}}$ by putting

$$\mathcal{J}_1 U_1 = (r\alpha)^{-1} \chi_1 U_1 .$$

We define classical wave operators

$$(14) \qquad W_1^{\pm} U_1 = s - \lim_{t \to \pm\infty} e^{itH} \mathcal{J}_1 e^{-itH_1} U_1 \quad in \quad \tilde{\mathcal{H}} .$$

Because the Schwarzschild potential is exponentially decreasing as $r_* \to -\infty$, we prove easily by Cook's method the :

THEOREM III.1- *$W_1^{\pm} : \mathcal{H}_1^{\pm} \to \mathcal{H}$ exist, are independent of χ_1 and $\|W_1^{\pm}\| \le 1$.*

We deduct from this result, the existence of infalling fields, similar to the disappearing solutions in dissipative scattering :

THEOREM III.2 - *If $U_1 \in \mathcal{H}_1^{\pm}$ verifies*

$$U_1(r_*, \omega) = 0 \quad for \quad r_* \ge c$$

then we have

$$e^{-itH} W_1^{\pm} U_1 = 0 \quad for \quad r_* \ge \pm t + c .$$

IV - Asymptotic completeness

To study the asymptotic behaviour far from the Black-Hole we introduce

$$(15) \qquad W_o U = s - \lim_{t \to +\infty} e^{itH_o} \mathcal{J}_o^* e^{-itH} U \quad in \quad \tilde{\mathcal{H}}_o .$$

At infinity of Schwarzschild universe, the electromagnetic field is asymptotic to a free field in Minkowski space-time :

THEOREM IV.1 - *$W_o : \mathcal{H} \to \mathcal{H}_o$ exists, is independent of χ_o and $\|W_o\| \le 1$.*

To describe the field near the horizon as $t \to +\infty$ we define

$$(16) \qquad W_1 U = s - \lim_{t \to +\infty} e^{itH_1} \mathcal{J}_1^* e^{-itH} U \quad in \quad \tilde{\mathcal{H}}_1 .$$

THEOREM IV.2 - *$W_1 : \mathcal{H} \to \mathcal{H}_1^+$ exists, is independent of χ_1 and $\|W_1\| \le 1$.*

The physical meaning of this result of completeness is the famous "impedance condition" of Damour and Znajeck [3]. More precisely the asymptotic profile of regular fields satisfies a dissipative condition or infalling left-polarization :

THEOREM IV.3 - *Let's be* U *in* $\mathscr{H}$ *such that*

(17) $$U = Hf \quad , \quad f \in [C_0^\infty(]r_o, +\infty[\times S^2)]^6 .$$

We note $e^{-itH} U = {}^t(E^{\hat{r}}, \ldots, B^{\hat{\varphi}})$. *Then, for any* $s \in \mathbb{R}$, *there exist* $e^{\hat{r}}, \ldots, b^{\hat{\varphi}}$ *in* $L^2(S^2)$ *such that, as*

(18) $$r \to r_o \quad , \quad t + r_* = s ,$$

the following limits exist in $L^2(S^2)$:

(19) $$E^{\hat{r}} \to e^{\hat{r}}, B^{\hat{r}} \to b^{\hat{r}}, \alpha E^{\hat{\theta}} \to e^{\hat{\theta}}, \alpha E^{\hat{\varphi}} \to e^{\hat{\varphi}}, \alpha B^{\hat{\theta}} \to b^{\hat{\theta}}, \alpha B^{\hat{\varphi}} \to b^{\hat{\varphi}} .$$

Moreover, we have

(20) $$e^{\hat{\theta}} = -b^{\hat{\varphi}} \quad , \quad e^{\hat{\varphi}} = b^{\hat{\theta}} ,$$

(21) $$\partial_s e^{\hat{r}} + (\sin\theta)^{-1} [\partial_\theta(\sin\theta \, b^{\hat{\theta}}) + \partial_\varphi b^{\hat{\varphi}}] = 0 .$$

Remark by Theorem I.3, the set of data satisfying (17) is dense in $\mathscr{H}$.
So, the horizon is rather similar to a dissipative membrane in euclidian space with surface resistivity 377 ohms : (20) is formally the impedance condition and (21) the charge conservation law ; but we emphasize that, unlike the euclidian case for which the dissipative condition is posed at each time and is necessary to solve the mixed problem, impedance property (20) is a consequence of Maxwell equations verified at infinity of infalling null geodesics.
Now, we can introduce scattering operator S by putting

$$W^- : \mathscr{H}_1^- \times \mathscr{H}_o \to \mathscr{H} \quad , \quad W^-(U_1, U_o) = W_1^- U_1 + W_o^- U_o ,$$

$$W : \mathscr{H} \to \mathscr{H}_1^+ \times \mathscr{H}_o \quad , \quad WU = (W_1 U, W_o U), S = WW^- : \mathscr{H}_1^- \times \mathscr{H}_o \to \mathscr{H}_1^+ \times \mathscr{H}_o .$$

THEOREM IV.4 - W^- *is isometric from* $\mathscr{H}_1^- \times \mathscr{H}_o$ *onto* $\mathscr{H}$; W *is isometric from* $\mathscr{H}$ *onto* $\mathscr{H}_1^+ \times \mathscr{H}_o$, S *is isometric from* $\mathscr{H}_1^- \times \mathscr{H}_o$ *onto* $\mathscr{H}_1^+ \times \mathscr{H}_o$.

V - Membrane paradigm

The Membrane Paradigm [6] states that if we are concerned only by the behaviour, far from the Black-Hole, of an initially incoming field, we may approximate the Black-Hole by a dissipative spherical membrane of radius $r_o + \varepsilon$, $0 < \varepsilon$, called "streched horizon".

We consider the mixed problem for Maxwell equations (8) in $]r_o+\varepsilon,+\infty[\times S^2$ and on streched horizon $\Gamma_\varepsilon = \mathbb{R}_t \times \{r = r_o + \varepsilon\} \times S^2$ which is time like, we impose impedance condition

(22) $$E^{\hat{\theta}} = -B^{\hat{\varphi}} \quad , \quad E^{\hat{\varphi}} = B^{\hat{\theta}} \; .$$

It is a classical dissipative hyperbolic problem of which the solution is given by a semigroup $V_\varepsilon(t)$ on Hilbert space $\tilde{\mathcal{H}}_\varepsilon = \{L^2(]r_o+\varepsilon, +\infty[_r \times S^2_\omega, r^2\,dr\,d\omega)]^6$. For $0 < \varepsilon < a$ we define scattering operator

$$S_\varepsilon U_o = s - \lim_{t\to+\infty} e^{itH_o} \mathcal{J}_o^* V_\varepsilon(2t) \mathcal{J}_o e^{itH_o} U_o \quad in \quad \tilde{\mathcal{H}}_o .$$

THEOREM V.1 - *$S_\varepsilon : \mathcal{H}_o \to \mathcal{H}_o$ exists, is independant of χ_o and $\|S_\varepsilon\| \leq 1$.*

Now, in Schwarzschild universe, the asymptotic behaviour at infinity of an initially incoming field is described by operator S_{oo} defined by

(23) $$\forall U_o \in \mathcal{H}_o \, , \quad S_{oo} U_o = \Pi_o S(0, U_o)$$

where Π_o is the projector from $\mathcal{H}_1^+ \times \mathcal{H}_o$ onto $\mathcal{H}_o$. The following result is the mathematical foundation of Membrane Paradigm :

THEOREM V.2 - *For any $U_o \in \mathcal{H}_o$, $S_\varepsilon U_o$ tends to $S_{oo} U_o$ in $\mathcal{H}_o$ as $\varepsilon \to 0$.*

Of numerical analysis view point, impedance condition (22) is an absorbing boundary condition on artificial boundary Γ_ε, so called Silver-Müller radiation condition in euclidian case [2]. So, Theorem V.2 gives a method of numerical approximation, already used in [6] .

[1] A. Bachelot, *Scattering of electromagnetic field by Schwarzschild Black–Hole*, to appear.

[2] A. Bendali, L. Halpern, *Conditions aux limites absorbantes pour le système de Maxwell dans le vide en dimension* 3, C.R. Acad. Sci. Paris, t.307, série I, 1988, p.1011-1013.

[3] Th. Damour, *Black–Hole eddy currents*, Phys. Rev. D 18, 10, 1978, p.3598, 3604.

[4] J. Dimock, *Scattering for the wave equation on the Schwarzschild metric*, Gen. Relativ. Gravitation, 17, n° 4, 1985, p.353-369.

[5] J. Dimock, B.S. Kay, *Classical and quantum scattering theory for linear scalar fields on the Schwarzschild metric* I, Ann. Phys. 175, 1987, p.366-426.

[6] D.A. Mac Donald, W.M. Suen, *Membrane viewpoint on Black–Holes : Dynamical electromagnetic fields near the horizon*, Phys. Rev. D 32, n° 4, 1985, p.848-871.

Phase Space and Functional Integral Methods in Direct and Inverse Scattering

L. Fishman

Department of Mathematics, Colorado School of Mines, Golden, CO 80401, USA

The analysis and fast, accurate numerical computation of the wave equations of classical physics are often quite difficult for rapidly changing, multidimensional environments extending over many wavelengths. This is particularly so for environments characterized by a refractive index field with a compact region of arbitrary (n-dimensional) variability superimposed upon a transversely inhomogeneous ($(n-1)$-dimensional) background profile. For such environments, the entire domain is in the scattering regime, with the subsequent absence of an "asymptotically free", or homogeneous, region. For the most part, classical, "macroscopic" methods have resulted in direct wave field approximations (perturbation theory, ray-theory asymptotics, transmutation theory, modal analysis, hybrid ray-mode methods), derivations of approximate wave equations (scaling analysis, field-splitting techniques, formal operator expansions, approximation theory), and discrete numerical approximations (finite differences, finite elements, spectral methods). In the last several decades, however, mathematicians studying linear partial differential equations have developed, in the language of physicists, a sophisticated, "microscopic" phase space analysis (pseudo-differential and Fourier integral operators) [1]. In conjunction with the global functional integral techniques pioneered by Wiener (Brownian motion) and Feynman (quantum mechanics), and so successfully applied today in quantum field theory and statistical physics, the n-dimensional classical physics propagators can be both represented explicitly and computed directly. The phase space, or "microscopic," methods and path (functional) integral representations provide the appropriate framework to extend homogeneous Fourier methods to inhomogeneous environments, in addition to suggesting the basis for the formulation and solution of corresponding arbitrary-dimensional nonlinear inverse problems.

So, why bring all of the machinery of path integrals and pseudo-differential and generalized Fourier integral operators to bear on the problems of direct and inverse wave propagation? The answer lies in the difficult nature of the physical environments often encountered. The ocean/bottom environment provides an excellent example. This environment is inhomogeneous everywhere — there may be no accessible asymptotically homogeneous regime where simplified wave fields can be constructed and utilized or appropriate data collected. This impacts on both the direct and inverse problems. Methods based on approximations to the wave field, such as the Born, Rytov, and supereikonal, will generally fail due to the extended nature of the inhomogeneities. Even for frequency regimes where ray theory is appropriate, the resulting ray structure can be quite complex, exhibiting numerous caustic structures. Moreover, in many applications, the frequency is sufficiently low that the geometrical-acoustics limit is in serious quantitative error. Moreover, coupled-mode methods and purely numerical techniques can tend to obscure the underlying physics. These problems are typical for strong channeling (or focusing) media.

The mathematical modelling for such environments is provided by the n-dimensional scalar Helmholtz equation,

$$(\nabla^2 + \bar{k}^2 K^2(\underline{x}))\phi(\underline{x}) = 0 \ , \tag{1}$$

where $K(\underline{x})$ is the refractive index field and $\bar{k}$ is a reference wave number. The frequency-domain problem is chosen because there are corresponding physical experiments and, in the pseudo-differential operator (ΨDO) context, it presents novel problems. Sound propagation in the ocean, certain guided-wave electromagnetic propagation problems, and borehole-to-borehole seismic modelling are near one-way propagation problems in extended environments. To zeroth-order, these propagation problems can be viewed as transversely inhomogeneous, suggesting a weak-backscatter perturbation approach to the general direct and inverse wave propagation problems. In this context, the transversely inhomogeneous problem is first solved and then used to attack the more general formulation. Mathematically, this is expressed through the approximate factorization of the n-dimensional Helmholtz operator in the form

$$\partial_x^2 + \left[\bar{k}^2 K^2(\underline{x}) + \nabla_t^2\right] \approx \left[\partial_x - \lambda(\underline{x}, -i\nabla_t)\right]\left[\partial_x + \lambda(\underline{x}, -i\nabla_t)\right] + (\text{smooth error}) \quad , \qquad (2)$$

where the operator $\lambda(\underline{x}, -i\nabla_t)$ is constructed in an appropriate asymptotic fashion [2]. In the context of wave field splitting [3], this approximate operator factorization corresponds to an approximate diagonalization of the first-order Helmholtz system into forward $(\phi^+(\underline{x}))$ and backward $(\phi^-(\underline{x}))$ evolution equations [2-8]. For a transversely inhomogeneous environment, $K^2(\underline{x}) = K^2(\underline{x}_t)$, a normal-mode (spectral) analysis readily illustrates that this factorization (or diagonalization) is, in principle, exact, resulting in separate, physical forward and backward, one-way wave equations. The forward evolution (one-way) equation,

$$(i/\bar{k})\partial_x\phi^+(x,\underline{x}_t) + (K^2(\underline{x}_t) + (1/\bar{k}^2)\nabla_t^2)^{1/2}\,\phi^+(x,\underline{x}_t) = 0 \quad , \qquad (3)$$

is the formally exact wave equation for propagation in a transversely inhomogeneous half-space supplemented with appropriate outgoing wave radiation and initial-value conditions [2-7].

In the context of one-way modelling, the introduction and widespread application of the parabolic (paraxial) approximation to the Helmholtz equation,

$$\left[(i/\bar{k})\partial_x + (1/2\bar{k}^2)\nabla_t^2 + (1/2)(K^2(\underline{x})+1)\right]\phi(\underline{x}) = 0 \quad , \qquad (4)$$

marked a significant advance in wave propagation modelling for ocean, atmospheric, optical, and seismic problems of the type described above [3,9]. In many respects this is a most natural motivation for the phase space and path integral machinery. The parabolic is, after all, an approximation to the full square root (pseudo-differential) operator, while the associated split-step FFT computational algorithm [9] follows from a direct integration of the phase space Feynman path integral representation of the parabolic (Schrödinger) propagator [4,10]. It is then quite natural, in attempting to analyze and compute the one-way Helmholtz equation (3), to attempt to explicitly construct the formal square root operator and the more general path integral representation associated with its corresponding propagator (fundamental solution). Given the success of the parabolic approximation, it might be reasonable to expect that such constructions would have both analytical and computational utility.

For the transversely inhomogeneous half-space, the square root of two noncommuting operators must be constructed. While functions of a finite set of commuting self-adjoint operators can be defined through spectral theory, nontrivial functions of noncommuting operators are represented by pseudo-differential operators [2-8,11]. The formal wave equation (3) can be written explicitly as a Weyl pseudo-differential equation in the form [2-7]

$$(i/\bar{k})\partial_x\phi^+(x,\underline{x}_t) + (\bar{k}/2\pi)^{n-1}\int_{R^{2n-2}} d\underline{x}_t'\,d\underline{p}_t$$

$$\cdot\,\Omega_{\mathbf{B}}(\underline{p}_t, (\underline{x}_t+\underline{x}_t')/2)\exp(i\bar{k}\,\underline{p}_t\cdot(\underline{x}_t-\underline{x}_t'))\phi^+(x,\underline{x}_t') = 0 \quad . \qquad (5)$$

where $\Omega_{\mathbf{B}}(\underline{p},\underline{q})$ is the symbol associated with the square root Helmholtz operator $\mathbf{B} = (K^2(\underline{q}) + (1/\bar{k}^2)\nabla_{\underline{q}}^2)^{1/2}$. The principal thrust of pseudo-differential operator theory is to extend homogeneous Fourier analysis to inhomogeneous environments. The analysis can be viewed as a natural extension of the theory of partial differential operators with nonconstant coefficients. The focus is on the operator symbol, which contains the complete spectral information in just the appropriate manner to lead immediately to the infinitesimal propagator. Essentially, the singular operator (kernel) calculus is replaced by a calculus for smooth, well-behaved functions (symbols). Quantum mechanics provides an excellent illustration of the theory.

Of course, the operator symbol must be constructed. The square root of an operator is constructed in terms of its square. In the Weyl pseudo-differential operator calculus this means that the operator symbol $\Omega_{\mathbf{B}}(\underline{p},\underline{q})$ is defined through the Weyl composition equation

$$\Omega_{\mathbf{B}^2}(\underline{p},\underline{q}) = K^2(\underline{q}) - \underline{p}^2 = (\bar{k}/\pi)^{2n-2} \int_{R^{4n-4}} d\underline{t}\,d\underline{x}\,d\underline{y}\,d\underline{z}\; \Omega_{\mathbf{B}}(\underline{t}+\underline{p},\,\underline{x}+\underline{q})$$

$$\cdot\, \Omega_{\mathbf{B}}(\underline{y}+\underline{p},\,\underline{z}+\underline{q})\, exp(2i\bar{k}(\underline{x}\cdot\underline{y} - \underline{t}\cdot\underline{z})) \tag{6}$$

with $\Omega_{\mathbf{B}^2}(\underline{p},\underline{q})$ the symbol associated with the square of $\mathbf{B}$, $\mathbf{B}^2 = (K^2(\underline{q}) + (1/\bar{k}^2)\nabla_{\underline{q}}^2)$ [2-7,11]. For a given $K^2(\underline{q})$, Eq.(6) must be solved (exactly or approximately) consistent with the outgoing wave radiation condition.

The exact pseudo-differential evolution equation (5) and, in general, the wide-angle extended parabolic approximate equations derived from the analysis of the composition equation (6) [3,5,6] are singular integro-differential wave equations. Solution representations for such pseudo-differential equations can be directly expressed in terms of infinite-dimensional functional, or path, integrals [4,10], following from the Markov, or semigroup, property of the propagator. The path integral representation for the propagator takes the form [4,6]

$$G^+(x,\underline{x}_t\,|\,0,\underline{x}_t') = \lim_{N\to\infty} \int_{R^{(n-1)(2N-1)}} \prod_{j=1}^{N-1} d\underline{x}_{jt} \prod_{j=1}^{N} (\bar{k}/2\pi)^{n-1} d\underline{p}_{jt}$$

$$\cdot\, exp\left\{ i\bar{k} \sum_{j=1}^{N} \left[\underline{p}_{jt}\cdot(\underline{x}_{jt} - \underline{x}_{j-1t}) + (x/N)\,H(\underline{p}_{jt},\underline{x}_{jt},\underline{x}_{j-1t}) \right] \right\} \tag{7}$$

where $H(\underline{p},\underline{q}'',\underline{q}')$ is related to the operator symbol.

The one-way marching computational algorithm is based on (1) the marching range step (following from the path integral),

$$\phi^+(x+\Delta x,\, \underline{x}_t) \approx \int_{R^{n-1}} d\underline{p}_t\, exp(i\bar{k}\,\underline{p}_t\cdot\underline{x}_t)\, exp\left[i\bar{k}\,\Delta x\, h_{\mathbf{B}}^s(\underline{p}_t,\underline{x}_t)\right] \hat{\phi}^+(x,\underline{p}_t) \quad . \tag{8}$$

where

$$h_{\mathbf{B}}^s(\underline{p},\underline{q}) = (\bar{k}/\pi)^{n-1} \int_{R^{2n-2}} d\underline{s}\,d\underline{t}\; \Omega_{\mathbf{B}}(\underline{s},\underline{t})\, exp\left[-2i\bar{k}(\underline{q}-\underline{t})\cdot(\underline{p}-\underline{s})\right] \tag{9}$$

is the standard pseudo-differential operator symbol [2,3,6,11], (2) a sophisticated symbol analysis (reflecting the detailed study of the (Helmholtz) Weyl composition equation (6)), and (3) Fourier component, or wave number, filtering in phase space (for increased efficiency, decreased computational time, and reduced error). While the Weyl calculus is often most appropriate for analysis, the standard calculus leads to more efficient numerical algorithms [6]. For a two-dimensional model ocean/bottom propagation environment with a pressure release ocean surface, the propagated wave field can be expressed in the matrix form [6]

$$\phi^{+}(x+\Delta x, z_n) = \sum_m A_{nm}\,\hat{\phi}^{+}(x, p_m) \tag{10}$$

for each depth point z_n. In Eq.(10), ϕ^+ and the discrete fast sine transformed wave field $\hat{\phi}^+$ are column vectors and the matrix $\underline{\underline{A}}$ is defined by its matrix elements

$$A_{nm} = \eta \sin(\bar{k} p_m z_n + \bar{k}\,\Delta x h_{\mathbf{B}}^{s(o)}(p_m, z_n))\, exp(i\bar{k}\,\Delta x h_{\mathbf{B}}^{s(e)}(p_m, z_n)) \tag{11}$$

where $h_{\mathbf{B}}^{s(e)}$ and $h_{\mathbf{B}}^{s(o)}$ are the even and odd parts with respect to p of the standard pseudo-differential operator symbol of Eq.(9) and η is an appropriate transform normalization constant.

The computational algorithm is based on the fast Fourier transform and matrix multiplication, and as such, is ideally suited for vector and parallel computer architectures. Moreover, the recent advent of the fast wavelet transform [12] should favorably impact on this phase space marching algorithm. The algorithm provides the generalization of the Tappert/Hardin split-step FFT algorithm [9] to the full one-way Helmholtz equation. It is a full-wave algorithm. The manner of marching the wave field is independent of the specific properties of the medium and the symbol approximation employed. Further, the propagation models constructed and computed correspond to singular integro-differential equation as well as partial differential equation approximations to the one-way wave equation [3,6,13]. The principal idea is to construct several approximate symbols, which when taken together, will span a wide range of applied problems. Sufficiently accurate approximations of the square root ΨDO symbol over the relevant region of phase space should result in very accurate numerical wave field calculations. This is supported by extensive numerical computations [6,7].

Focusing on the symbol, it appears possible to develop inverse algorithms for the nonlinear generalized acoustic tomography problem. The inverse analysis is initially concerned with the reconstruction of a transversely inhomogeneous ($(n-1)$-dimensional) refractive index profile. Both the sources and receivers are located in the scattering region along parallel "planes" transverse to the x, or range, axis. Clearly, this is the first step in the more general inversion of environments characterized by a compact region of general inhomogeneity superimposed on a transversely inhomogeneous background profile.

The basic outline of the one-way inversion method is the following.

(1) There are two basic properties associated with the macroscopic and microscopic propagators. First, both G^{+}_{macro} and G^{+}_{micro} are Fourier integral operators (FIO) with complex phase [14] characterized by appropriate symbols. Second, the semigroup property allows for G^{+}_{macro} to be expressed in terms of the repeated composition of G^{+}_{micro} [3-7]. (This is just the basis for the path integral representation.)

(2) The macroscopic symbols are deduced from appropriate wave field data.

(3) The repeated composition of the generalized FIO [15] is used to construct a relationship between the macroscopic and microscopic symbols.

(4) The relationship in (3) is solved for the microscopic symbols.

(5) Direct application of the (Helmholtz) Weyl composition equation (6) for the microscopic symbol $\Omega_{\mathbf{B}}(\underline{p}, \underline{q})$ allows for the recovery of $K^2(\underline{q})$.

For applied inverse problems, approximate inversions may prove adequate. Approximate inversion algorithms follow readily from the perturbative treatments of the (Helmholtz) Weyl composition equation (6) [3,5]. $K^2(\underline{q})$ is related to $\Omega_{\mathbf{B}}(0, \underline{q})$ in a quadratic fashion and through a linear integral relationship, respectively, in the high-frequency ($\bar{k} \to \infty$) and weak-inhomogeneity (Born) limits. In particular, the high-frequency approximation replaces Eq.(6) in step (5) with

$$\Omega_{B^2}(\underline{p},\underline{q}) = K^2(\underline{q}) - \underline{p}^2 \approx \Omega_B^2(\underline{p},\underline{q}) \quad . \tag{12}$$

and is based upon choosing, in practice, a $|\underline{p}|$ such that the symbol approaches its asymptotic form, $\Omega_B(\underline{p},\underline{q}) \sim (K^2(\underline{q}) - \underline{p}^2)^{1/2}$. The approach to the asymptotic regime in phase space is governed both by the magnitude of $K^2(\underline{q}) - \underline{p}^2$ (large) and the variation of the refractive index field on the wavelength scale (small).

It is clear from the preceding outline of the phase space and path integral analysis that the square root Helmholtz operator construction, the phase space path integral representation and algorithm, and the inverse analysis all depend crucially upon a detailed symbol analysis. For the one-way Helmholtz equation, the symbol analysis is particularly interesting. Since the Helmholtz equation is elliptic, the one-way propagator will have smoothing properties — high-frequency components will be appropriately damped out. This manifests itself in the spectral analysis through the evanescent modes, while in the ΨDO analysis, the symbol corresponding to the square root of the indefinite Helmholtz operator will have both real and imaginary parts with characteristic oscillatory behavior [3,6-8]. The usual ΨDO constructions will lead to approximations which are not uniform over phase space, and, indeed, are singular [3,5,8]. This is the principal difficulty in the symbol analysis, and what ultimately necessitates new approximation methods. (Of course, what generates the difficulties in the symbol analysis leads to the computational feasibility of the numerical algorithm. The damping of the high-frequency components in the wave field makes a phase space integration algorithm stable and efficient.) Moreover, the symbol asymptotics has a dual nature. The expansions are in orders of smoothness [2,16] and in terms of a small parameter (formally $(1/\bar{k})$, but in reality, the variation in the refractive index field on the wavelength scale) [16]. For the purposes of approximation, the constructions are not primarily viewed in terms of smoothing operators, but rather, demand that the neglected terms are sufficiently small over the relevant region of phase space. Further, as the goal is to approximate the symbol over a wide-ranging parameter space, even a uniform high-frequency result is not, by itself, sufficient.

A general solution of the (Helmholtz) Weyl composition equation (6) would mark a significant achievement in pseudo-differential operator theory, and, at present, appears quite unlikely. Nonetheless, the Weyl symbol $\Omega_B(\underline{p},\underline{q})$ can be deduced from the Helmholtz Green's function G for the transversely inhomogeneous problem supplemented with an outgoing wave radiation condition and the relations [3,5,11]

$$\Omega_B(\underline{p},\underline{q}) = \int_{R^{n-1}} d\underline{u}\, exp(i\bar{k}\underline{p}\cdot\underline{u})\, B\left(\underline{q} - \frac{\underline{u}}{2}, \underline{q} + \frac{\underline{u}}{2}\right) \tag{13}$$

and

$$B(\underline{x}_t, \underline{x}_t') = \left(\frac{2i}{\bar{k}}\right) \lim_{x \to 0} \left(\partial_x^2 G(x,\underline{x}_t \,|\, 0,\underline{x}_t')\right) \quad . \tag{14}$$

In Eqs. (13) and (14), $B(\underline{x}_t, \underline{x}_t')$ is the integral operator kernel associated with the square root Helmholtz operator **B**. Further, G can be expressed in terms of the corresponding Schrödinger (parabolic) propagator Φ in the form [4,8]

$$G(x,\underline{x}_t \,|\, 0,\underline{x}_t') = 2^{-3/2} \left(\frac{i}{\pi\bar{k}}\right)^{1/2} \int_0^\infty d\tau\, \tau^{-1/2}\, exp\left(\left(\frac{i\bar{k}}{2}\right)\left(\tau + \frac{x^2}{\tau}\right)\right) \Phi(\tau,\underline{x}_t \,|\, 0,\underline{x}_t') \tag{15}$$

where Φ satisfies an equation analogous to Eq.(4). Thus, for all cases where the Schrödinger (parabolic) propagator Φ is known in closed form [10,17], the exact construction procedure given in Eqs.(13) – (15) can, in principle, be carried out. However, the integral operator kernel is singular [6], necessitating that the Weyl transform relation (13) and any resulting integral representation for the symbol be understood and subsequently computed in the sense of generalized functions [18]. The n-dimensional Helmholtz Green's function G in Eq.(15) can also be represented as an inverse Fourier transform of the transformed $(n-1)$-dimensional Green's function $\hat{G}$ [8]. Detailed symbol constructions have been presented [8].

Since the number of exact solutions is limited, it is desirable to develop a numerical construction procedure for the solutions to the composition equation. The first numerical construction method for $\Omega_B(p,q)$ follows from the spectral representation of the corresponding Helmholtz Green's function G and the two-dimensional $(n=2)$ forms of Eqs.(13) and (14). The computational algorithm takes the final form

$$\Omega_B(p,q) = (K^2(q)-p^2)^{1/2} + \int_{R^1} du\, exp(i\bar{k}pu) \left[\int_{R^1} d\xi\, \xi^2 \left[-2i\bar{k}\,\hat{G}(\xi, q-\frac{u}{2} \mid q+\frac{u}{2}) - \left(\frac{\bar{k}}{2\pi}\right) \frac{exp(i\bar{k}\,(K^2(q)-\xi^2)^{1/2} \mid u \mid)}{(K^2(q)-\xi^2)^{1/2}} \right] \right] . \tag{16}$$

where the transformed depth-dependent Green's function $\hat{G}$ is constructed by an appropriately modified form of the Scooter FFP algorithm [19]. The second numerical construction method is based on directly solving the (Helmholtz) Weyl composition equation (6). The equation is reformulated in terms of a difference symbol to yield a well-defined numerical problem in terms of a relatively small and slowly-varying unknown function. The resulting quadratically nonlinear integral equation on the infinite domain is reduced to a corresponding algebraic problem in two ways, and solved [19].

Beyond the exact and numerical symbol constructions, there are the approximate analytical constructions which provide the core of the marching algorithm [6,7]. The symbol constructions are initially based on the well-known pseudo-differential operator asymptotic (in smoothness) treatment of the composition equation [3,11]. This is, in effect, a high-frequency perturbation theory, ordered by increasing powers of $(1/\bar{k}^2)$. This nonuniform perturbation theory recovers well-known approximate wave theories such as the ordinary parabolic, range-refraction parabolic, Grandvuillemin-extended parabolic, half-space Born, and Thomson-Chapman (Feit-Fleck) extended parabolic [3,5-7]. The first term in the expansion provides for a full-wave, wide-angle high-frequency approximate symbol [3-8]. In constructing the square root of the indefinite Helmholtz operator, the usual asymptotic symbol analysis results in only an outer-scale algebraic expansion [3,6-8]. A scaling analysis, appropriate change of variables, and the construction of a leading-order canonical inner-scale solution provide for a uniform high-frequency approximation which properly represents the symbol in the high-propagating-angle, or near-evanescent, phase space regime. Moreover, the formal operator rational approximations to the square root Helmholtz operator [6-8,13,20] can also be reformulated in terms of operator symbols, allowing for the resulting high-angle-parabolic-equation (HAPE) implicit-finite-difference (IFD) algorithms [6,7,13,20,21] to be analyzed and computed by phase space and path integral methods. For every case where an exact symbol can be constructed, the symbols exactly corresponding to the operator rational approximations can be written in closed form in a manner analogous to that presented in Eqs.(13) – (15) [8].

The analysis of the one-way problem sets the stage for the full two-way Helmholtz equation. Rather than starting from a transversely inhomogeneous formulation and, subsequently, building in backscatter effects, the generalization of Fourier methods to arbitrary inhomogeneous environments can proceed from the construction of truly global configuration space path integrals. These elliptic-based (two-way) constructions, originating from the Fourier transform relationship between the Helmholtz and Schrödinger (parabolic) propagators, result in the approximate Feynman/Garrod path integral representation for the full-space propagator [4-7].

$$G(\underline{x}\,|\,\underline{x}') \approx (-1/2\bar{k}^2) \lim_{N\to\infty} \int_{R^{n(2N-1)}} \prod_{j=1}^{N-1} d\underline{x}_j \prod_{j=1}^{N} (\bar{k}/2\pi)^n d\underline{p}_j \, \frac{exp(i\bar{k}S_N)}{(1/2-\Sigma)} \quad . \tag{17}$$

where S_N corresponds to an appropriate discretized action and Σ plays a role analogous to an average energy. For a transversely inhomogeneous half-space, partial integration of Eq.(17) in conjunction with the reflection principle (or method of images) results in [4-7]

$$G^+(x,\underline{x}_t\,|\,0,\underline{x}_t') \approx \lim_{N\to\infty} \int_{R^{(n-1)(2N-1)}} \prod_{j=1}^{N-1} d\underline{x}_{jt} \prod_{j=1}^{N} (\bar{k}/2\pi)^{n-1} d\underline{p}_{jt}$$

$$\cdot\, exp\left[i\bar{k}\left(S_N + 2^{1/2}\,x(1/2-\Sigma)^{1/2}\right)\right] \tag{18}$$

with S_N and Σ taking on their appropriate forms in one-lower dimension. Equation (17) provides, in closed form, an approximation for the general (two-way) Helmholtz propagator, while the zeroth-order path functional construction in Eq.(18) is distinct from the high-frequency (locally-homogeneous) approximate one-way path integral. The approximate Feynman/Garrod path integral is exact in the homogeneous medium limit, incorporates significant backscatter information, and contains both the geometrical (ray) acoustic and ordinary parabolic approximations. These properties suggest that the Feynman/Garrod path integral representations in Eqs.(17) and (18) provide a viable approximation for a useful range of propagation experiments. It is important to get an accurate assessment of this approximate propagator [7].

In addition to providing a "zeroth-order" computational model, the Feynman/Garrod path integral representation motivates the symbolic Feynman/DeWitt-Morette representation and the notion of an underlying stochastic process [4-7,22]. The hope here is that the physics can be made more transparent by constructing a true path functional and that recent advances in massively parallel computing might eventually make a direct sum-over-paths calculation feasible. Clearly, this approach is distinct from the usual FIO treatments and from Taylor's use of Fourier-Airy integral operators in obstacle scattering [2,14]. Moreover, the use of FIO with complex phase in the one-way path integral construction is generally different from the usual treatments [2,14].

Acknowledgments

This work was supported by grants from the Office of Naval Research (N00014-85-K-0307 and N00014-88-K-0047) and a joint grant from the Air Force Office of Scientific Research and the National Science Foundation (DMS 8512295).

References

[1] A. Unterberger: Bull. Am. Math. Soc. **12(1)**, 149 (1985).

[2] M.E. Taylor: **Pseudodifferential Operators** (Princeton University Press, Princeton 1981).

[3] L. Fishman, J.J. McCoy: J. Math. Phys. **25(2)**, 285 (1984).

[4] L. Fishman, J.J. McCoy: J. Math. Phys. **25(2)**, 297 (1984).

[5] L. Fishman, J.J. McCoy: IEEE Trans. Geosc. Rem. Sens. **GE-22(6)**, 682 (1984).

[6] L. Fishman, J.J. McCoy, S.C. Wales: J. Acoust. Soc. Am. **81(5)**, 1355 (1987).

[7] L. Fishman, S.C. Wales: J. Comp. Appl. Math. **20**, 219 (1987).

[8] L. Fishman: Exact and operator rational approximate solutions of the Helmholtz, Weyl composition equation in underwater acoustics — the quadratic profile, submitted to Wave Motion (1989).

[9] F.D. Tappert: in **Wave Propagation in Underwater Acoustics**, edited by J.B. Keller and J.S. Papadakis (Springer-Verlag, New York 1977), p. 224.

[10] L.S. Schulman: **Techniques and Applications of Path Integration** (Wiley, New York 1981).

[11] L. Hörmander: Comm. Pure Appl. Math. **32**, 359 (1979).

[12] G. Beylkin, R. Coifman, V. Rokhlin: **Fast Wavelet Transforms and Numerical Algorithms I** (Research Report YALEU/DCS/RR-696, August, 1989).

[13] A. Bamberger, B. Engquist, L. Halpern, P. Joly: SIAM J. Appl. Math. **48(1)**, 129 (1988).

[14] F. Treves: **Introduction to Pseudodifferential and Fourier Integral Operators. Volume 2. Fourier Integral Operators** (Plenum Press, New York 1980).

[15] J. Duistermaat: **Fourier Integral Operators** (Courant Institute Lecture Notes, New York 1974).

[16] V.P. Maslov, V.E. Nazaikinskii: **Asymptotics of Operator and Pseudo-Differential Equations** (Consultants Bureau, New York 1988).

[17] B. Gaveau, L.S. Schulman: J. Phys. A. Math. Gen. **19**, 1833 (1986).

[18] I.M. Gelfand, G.E. Shilov: **Generalized Functions, Properties and Operations, Volume 1** (Academic Press, New York 1964).

[19] L. Fishman: in **13th International Congress on Acoustics, Vol. 5**, edited by P. Pravica (Sava Centar, Belgrade 1989), p. 149.

[20] L. Halpern, L.N. Trefethen: J. Acoust. Soc. Am. **84(4)**, 1397 (1988).

[21] D. Lee : **The State-of-the-Art Parabolic Equation Approximation as Applied to Underwater Acoustic Propagation with Discussions on Intensive Computations** (Naval Underwater Systems Center Technical Report TD 7247, 1984).

[22] C. DeWitt-Morette, A. Maheshwari, B. Nelson: Phys. Rep. **50(5)**, (March, 1979).

New Results on 3D Schrödinger Inverse Scattering

R.G. Newton

Physics Department, Indiana University, Bloomington, IN 47405, USA

The generalized Marchenko and Gel'fand-Levitan methods are based on the solution of a Riemann-Hilbert problem. This paper relates the solution of this problem to the standard Wiener-Hopf factorization of an operator-valued function. It also states a new theorem that equates the number of bound states of the underlying potential for an admissible S matrix to the dimensions of the eigenspaces, at the eigenvalues ± 1, of a relevant compact operator obtained from the Fourier transform of the scattering amplitude.

1 Introduction

In this paper I will tell you about some recent results in the three-dimensional inverse scattering problem for the Schrödinger equation. Some of them, and their proofs, are contained in my new book [2]; some of them are too new to have been included there (or even at the Arcata conference last July) and their proofs will be published elsewhere.

The starting point of the Marchenko method of solving the inverse scattering problem of the Schrödinger equation in $\mathbb{R}^3$ [1,2,3] is the relation

$$\psi^+(-k,\theta,x) = \int_{\mathbb{S}^2} d\theta'\, S(-k,-\theta',\theta)\psi^+(k,\theta',x)$$

for the scattering solution. Here and in the following k is the wave number, θ is a unit vector in $\mathbb{R}^3$ that indicates the direction of the beam sent toward the scattering center, ψ^+ is the solution of the Schrödinger equation that satisfies an outgoing-wave boundary condition at infinity, and $S(k,\theta,\theta')$ is the unitary S matrix. It is related to the scattering amplitude $A(k,\theta,\theta')$ by

$$S(k,\theta,\theta') = \delta(\theta,\theta') - \frac{k}{2\pi i}A(k,\theta,\theta').$$

If the potential V is in the Rollnik class then ψ^+, regarded as a function of k, is the boundary value of an analytic function that is meromorphic in $\mathbb{C}^+$ with simple poles at the points $i\kappa_m$ whenever $-\kappa_m^2$ is an eigenvalue (bound state) of the Schrödinger equation and with residues that lie in prescribed finite-dimensional subspaces $\mathcal{H}_m$ of $L^2(\mathbb{S}^2)$ [1,2]. Each of these subspaces is spanned by the *characters* $Y^b_{\kappa_m}(\theta)$ of the corresponding bound states (more than one in the case of degeneracy), i.e., functions that describe the directional dependence of the asymptotic form of the bound-state eigenfunctions. These are the generalizations of the specification of the angular momenta and norming constants for the bound states in the case of central potentials.

We shall collect these bound-state data (which can be obtained from the scattering amplitude by a procedure that is given in detail in [2]) in the following definition.

Definition 1.1 *The set $\mathcal{B}$ consists of all finite sets σ of p_σ pairs consisting of a positive number κ_m and a d_m-dimensional subspace $\mathcal{H}_m$ of $L^2(\mathrm{s}^2)$ $(d_m < \infty)$. The set $\{d_m\}$ will be denoted by $\mathbf{N}_\sigma$, $\Sigma\, d_m := n_\sigma$, and the set $\{\kappa_m\}$ will be called $\mathbf{P}_\sigma$.*

Furthermore, as $|k| \to \infty$ in C^+ the function $\zeta(k,\theta,x) := \psi^+(k,\theta,x)e^{-ik\theta\cdot x}$ approaches unity. Therefore, the equation

$$\zeta(-k,\theta,x) = \int_{\mathrm{S}^2} d\theta'\, S_x(-k,-\theta',\theta)\zeta(k,\theta',x),$$

where

$$S_x(k,\theta,\theta') := S(k,\theta,\theta')e^{ikx\cdot(\theta-\theta')},$$

together with the analyticity and asymptotic conditions, constitutes a generalized Riemann-Hilbert problem with respect to the real line.

It will simplify matters to regard ζ, for fixed values of k and x, as a vector in $L^2(\mathrm{s}^2)$ and $S_x(k,\theta,\theta')$ as the distribution kernel of an operator $L^2(\mathrm{s}^2) \mapsto L^2(\mathrm{s}^2)$, so that we may write more simply

$$\zeta(-k,x) = QS_x(-k)\zeta(k,x)$$

with some abuse of notation, in which operators and vectors are denoted by the same letters as the corresponding kernels and functions. Here Q denotes the operator defined by $(Qf)(\theta) := f(-\theta)$. Since the x-dependence of this equation is going to play no role in the following, we shall no longer explicitly indicate it.

To be more precise, we collect the properties that the scattering amplitude is known to have if the potential is in the class $\mathcal{V}_0$:

Definition 1.2 $\mathcal{V}_0 = \{V \mid V \in \mathrm{R},\ \lim_{|x|\to\infty} V(x) = 0,\ \text{and}\ \exists a, C, \epsilon > 0,\ \text{such that for all}\ x \in \mathrm{R}^3,\ |\nabla V(x)| < C(a+|x|)^{-4-\epsilon}\}$.

This class is smaller than it needs to be, but it is easy to define.

If $V \in \mathcal{V}_0$ then $A \in \mathcal{A}$:

Definition 1.3 $A \in \mathcal{A}$ *if*

(i) the kernel $A(k,\theta,\theta')$ *that defines the operator family* $A(k)$ *is a continuous, uniformly bounded, differentiable function* $\mathbf{R} \times \mathbf{s}^2 \times \mathbf{s}^2 \mapsto \mathbf{C}$,

(ii) $QAQ = \widetilde{A}$ *[*reciprocity*; the tilde here means the operator whose kernel is the transpose and* $(Qf)(\theta) = f(-\theta)$*]*

(iii) $A(-k) = \overline{A(k)}$,

(iv) $S := \mathbb{1} - \frac{k}{2\pi i}A$, $S^\dagger S = SS^\dagger = \mathbb{1}$, *(*unitarity; † *denotes the adjoint)*,

(v) $\|S - \mathbb{1}\| \in L^2(\mathbf{R})$ *[*$\|\cdot\|$ *here is the operator norm* $L^2(\mathbf{s}^2) \mapsto L^2(\mathbf{s}^2)$*]*,

(vi) the operators $\mathcal{G}$ *and* $\mathcal{G}^*$ *defined by (1), (2), and (3) in terms of A are compact and* $\mathcal{G}^2$ *and* $\mathcal{G}^{*2}$ *are Hilbert-Schmidt.*

We need the following functions:

$$\mathsf{G}(\alpha,\theta,\theta') := \frac{i}{(2\pi)^2}\int_{-\infty}^{\infty} dk\, kA(k,-\theta,\theta')e^{-ik\alpha} \tag{1}$$

$$\mathcal{G}(\alpha,\theta;\beta,\theta') := \mathsf{G}(\alpha+\beta,\theta,\theta'), \qquad \alpha,\beta \in \mathbf{R}_+,\ \theta,\theta' \in \mathbf{s}^2, \tag{2}$$

$$\mathcal{G}^*(\alpha,\theta;\beta,\theta') := \mathsf{G}(-\alpha-\beta,\theta,\theta'), \qquad \alpha,\beta \in \mathbf{R}_+,\ \theta,\theta' \in \mathbf{s}^2, \tag{3}$$

$$\mathcal{H}(\alpha,\theta;\beta,\theta') := \mathsf{G}(\alpha-\beta,\theta,\theta'), \qquad \alpha,\beta \in \mathbf{R}_+,\ \theta,\theta' \in \mathbf{s}^2. \tag{4}$$

The operators $\mathcal{G}$, $\mathcal{G}^*$, $\mathcal{H}$ are defined by these integral kernels. The generalized Marchenko equation can be written in terms of them:

$$\eta = \mathsf{g} + \tfrac{1}{2}(Q+\mathcal{H})\tilde{y}s^{-1}c + \mathcal{G}\eta, \tag{5}$$

where s is the square matrix and c is the column matrix whose elements are

$$s_{n,mb} = \langle z_n, y^b_m\rangle_+, \tag{6}$$

$$c_n = \langle z_n, \mathsf{g}^*\rangle_+, \tag{7}$$

$\mathsf{g} = G\hat{1}$, (where $\hat{1}$ is the vector in $L^2(\mathbf{s}^2)$ that is $f(\theta) = 1$), $\langle\cdot,\cdot\rangle_+$ is the inner product on $L^2(\mathbf{R}_+ \times \mathbf{s}^2)$, the row matrix $\tilde{y}$ has the elements

$$y^b_m(\alpha,\theta,x) = Y^b_{\kappa_m}(-\theta)e^{\kappa_m(\theta\cdot x-\alpha)}, \tag{8}$$

where the $Y^b_{\kappa_m}$ are the characters of the bound states, the z_n are a set of linearly independent eigenfunctions of $\mathcal{G}^*$ with the eigenvalue -1, and η is the inverse Fourier transform of $\zeta - \hat{1}$. The role of equation (5) will emerge below. It is known (Theorem 2.3.1 of [2]) that if $A \in \mathcal{A}$ then $\mathcal{G}$ and $\mathcal{G}^*$ are selfadjoint and $\|\mathcal{G}\| \leq 1$, $\|\mathcal{G}^*\| \leq 1$.

2 The Riemann-Hilbert Problem and Wiener-Hopf Factorization

The Riemann-Hilbert problem may now be stated more precisely as follows:

Problem $H^1_\sigma(S_x)$: *Let* $(S(k) - \mathbb{1}) \in \mathcal{A}$ *and* $\sigma \in \mathcal{B}$ *be given. Find a function* f, $\mathbf{R} \times \mathbf{S}^2 \mapsto \mathbf{C}$, *which satisfies the following conditions:*

(*i*) $f - \hat{1} \in L^2(\mathbf{R} \times \mathbf{S}^2)$, $\quad \hat{1}(k, \theta) := 1$;

(*ii*) $f(k)$, $\mathbf{R} \mapsto L^2(\mathbf{S}^2)$, *is the boundary value of an analytic function meromorphic in* $\mathbf{C}^+$ *such that* $\lim_{|k|\to\infty} \|f(k) - \hat{1}\| = 0$; *[Here* $\|\cdot\|$ *is the norm in* $L^2(\mathbf{S}^2)$.*]*

(*iii*) *In* $\mathbf{C}^+$ f *has simple poles at the points* $i\kappa_m$, $\kappa_m \in \mathbf{P}_\sigma$, *and nowhere else, and its residue at* $i\kappa_m$ *lies in the space* $\mathcal{H}_m$.

(*iv*) *On* $\mathbf{R}$, f *satisfies the equation*

$$f^* = QS^*f, \tag{9}$$

where $f^*(k) := f(-k)$.

The solvability of this problem is the content of the following theorem, which is proved in [2].

Theorem 2.1 *The Riemann-Hilbert problem* $H^1_\sigma(S)$ *has a unique solution if and only if the following three conditions hold:* (*i*) $\mathcal{G}$ *does not have the eigenvalue 1,* (*ii*) *the operator* $\mathcal{G}^*$ *has the eigenvalue* -1 *with an* n_σ*-dimensional eigenspace, and* (*iii*) *the matrix* s *defined by (6) (in which the functions* $Y^b_{\kappa_m}$ *span the space* $\mathcal{H}_m$*) is invertible. This solution is obtained by Fourier transformation from the solution of the generalized Marchenko equation (5).*

An immediate consequence of this theorem is:

Corollary 2.2 *If the two problems* $H^1_\sigma(S)$ *and* $H^1_\mu(S)$ *have unique solutions, then* $n_\sigma = n_\mu$.

The method based on a generalization of the Gel'fand-Levitan equation [1,2], on the other hand, requires the computation of a generalized operator-valued Jost function, whose inverse satisfies the equation $F^* = QS^*FQ$. The function $F(k)$ is the boundary value of an analytic function that is meromorphic in $\mathbf{C}^+$ with simple poles there at the points $i\kappa_m$ (see above) and residues that are operators whose ranges are prescribed finite-dimensional subspaces of $L^2(\mathbf{S}^2)$. As $|k| \to \infty$ in $\mathbf{C}^+$, $\|F(k) - \mathbb{1}\| \to 0$, where $\mathbb{1}$ is identity operator. Therefore this also constitutes a Riemann-Hilbert problem, but with an operator-valued solution. To be precise, we use the following definition:

Definition 2.3 *$\mathcal{M}^+$ is the set of all operator-valued ($L^2(\mathrm{s}^2) \mapsto L^2(\mathrm{s}^2)$) functions on $\mathbf{R}$ in L^2 that are boundary values of analytic functions, meromorphic in C^+, and whose operator norm approaches zero at infinity there. Similarly, $\mathcal{N}^+$ is the set of functions in $\mathcal{M}^+$ that are holomorphic in C^+.*

We then pose this problem:

Problem $W^1_\sigma(S)$: *Let* $(S(k) - \mathbb{1}) \in \mathcal{A}$ and $\sigma \in \mathcal{B}$ *be given. Find a function* F *on $\mathbf{R}$ such that*

(*i*) $F - \mathbb{1} \in \mathcal{M}^+$, *with simple poles at the points* $i\kappa_m$, $\kappa_m \in \mathbf{P}_\sigma$, and *residues there whose ranges equal* $\mathcal{H}_m$;

(*ii*) on $\mathbf{R}$, F *satisfies the equation*

$$F^* = QS^*FQ, \tag{10}$$

where $F^*(k) := F(-k)$.

This Riemann-Hilbert problem, in which the solution sought is operator-valued, subsumes the one whose solution is vector-valued because if F solves the first then $F\hat{1}$ solves the second. (Remember that $\hat{1}(k,\theta) = 1$ so that $Q\hat{1} = \hat{1}$.) For its solution one solves two generalized Marchenko equations in which, because of the appearance of Q on the right of the right-hand side of (10), both $+\mathcal{G}$ and $-\mathcal{G}$ appear as kernels.

If the inverse of F exists and is holomorphic in C^+ then the operator Riemann-Hilbert problem becomes a factorization problem, because it implies that

$$S = QFQF^{*-1},$$

where QFQ is meromorphic in C^+ with a holomorphic inverse, and F^{*-1} is holomorphic in C^- with a meromorphic inverse. It should be compared with the form in which a standard Wiener-Hopf factorization problem is stated.

A left standard factorization (also called *proper*) [4,5,6,7] is a decomposition of the form

$$S = W_+DW_-,$$

where

$$D = P_0 + \sum_{j\geq 1} P_j \left(\frac{k+i}{k-i}\right)^{\rho_j}, \tag{11}$$

$P_j = P_j^2 = P_j^\dagger$, $\mathrm{tr}P_j = 1$, $j \geq 1$, $P_0 + \Sigma_{j\geq 1} P_j = \mathbb{1}$, $P_jP_i = 0$ if $i \neq j$, $W_\pm$ is holomorphic and invertible everywhere in $\mathrm{C}^\pm$, $\lim_{|k|\to\infty} \|W_\pm - \mathbb{1}\| = 0$, and the *partial indices* ρ_j are nonzero integers; D is called the *diagonal factor*. If $D = \mathbb{1}$ then the factorization is called *canonical* (or *regular*). Whereas the partial indices are uniquely determined by S, the factors $W_\pm$ and D are not. (But if a canonical factorization exists, it is unique.)

The factorization defined by W^1_σ is also not necessarily unique. In the first place, for given S and σ the problem $W^1_\sigma(S)$ may have more than one solution;

in the second place, more than one set σ may be such that $W^1_\sigma(S)$ has a solution for a given S. The following lemma holds:

Lemma 2.4 *Suppose that the problem $W^1_\sigma(S)$ has a holomorphically invertible solution. Then for every choice of $\mathbf{P}_\mu$ with $p_\mu \leq n_\sigma$ there exist sets $\mu \in \mathcal{B}$ with $n_\mu = n_\sigma$ such that $W^1_\mu(S)$ also has a holomorphically invertible solution.*

In other words, for a given S there generally exists more than one problem $W^1_\sigma(S)$ (if there is one) that has a holomorphically invertible solution. The pole positions can be shifted around at will, and to a certain extent, so can the ranges of the residues. The latter, however, cannot be changed completely freely. (This corresponds to the well-known fact that for a given S matrix for a central potential the bound-state zeros of the Jost function can be shifted freely, but they cannot be shifted from one angular momentum to another.) The sum of the dimensions of the ranges of the residues, on the other hand, is fixed, as the following proposition asserts:

Proposition 2.5 *Suppose that the problem $W^1_\sigma(S)$ has a holomorphically invertible solution. Then the following two propositions hold:*

(i) If the problem $W^1_\mu(S)$ has a holomorphically invertible solution then $n_\sigma = n_\mu$.

(ii) There exists another set $\mu \in \mathcal{B}$ with $\mathbf{P}_\mu$ and $n_\mu = n_\sigma + m$, (where m is a nonnegative integer) arbitrarily given, such that $W^1_\mu(S)$ has a solution F whose inverse F^{-1} has m simple poles in $\mathbb{C}^+$.

The second part of this theorem tells us that if we are willing to allow zeros in the solution, in the sense that its inverse has poles, then we can arbitrarily increase the number of poles in the solution sought.

We have the analogue of Theorem 2.1, which is also proved in [2].

Theorem 2.6 *The problem $W^1_\sigma(S)$ has a unique solution if and only if the following three conditions hold: (i) the operator $\mathcal{G}^2$ does not have the eigenvalue 1, (ii) $\dim \mathrm{nul}(\mathbb{1}+\mathcal{G}^*) = n_\sigma$, and (iii) the matrix s of (6) (in which the functions $Y^b_{\kappa_m}$ span the space $\mathcal{H}_m$) is invertible. The solution is obtained by Fourier transformation from the solution of the generalized Marchenko equation (5).*

Note that if S is given, then $\mathcal{G}$ and $\mathcal{G}^*$ are given, and hence, so is the nullspace of $\mathbb{1} + \mathcal{G}^*$. Therefore the number n_σ of the problem $W^1_\sigma(S)$ that has a unique solution, if it exists, is determined. The question is, does such a problem $W^1_\sigma(S)$ always exist if $\mathcal{G}^2$ does not have the eigenvalue 1? First, we have the following.

Lemma 2.7 *Given S and $\mathbf{P}_\sigma$ (whose membership must be no greater than n_σ) in a set $\sigma \in \mathcal{B}$ with $n_\sigma = \dim \mathrm{nul}(\mathbb{1} + \mathcal{G}^*)$, there always exist functions $Y^b_{\kappa_m}$ so that the matrix s in (6) is invertible.*

As a result we obtain this proposition:

Proposition 2.8 *Suppose that $(S - \mathbb{1}) \in \mathcal{A}$ is such that the operator $\mathcal{G}^2$ does not have the eigenvalue 1. Then there exists a set $\sigma \in \mathcal{B}$ with $n_\sigma = \dim \mathrm{nul}(\mathbb{1} + \mathcal{G}^*)$ such that the problem $W^1_\sigma(S)$ has a unique solution. Moreover, the set $\mathbf{P}_\sigma$ can be chosen at will, so long as its membership does not exceed n_σ.*

3 The Indices

Even though S "approaches the unit operator" in the sense that $\|S - \mathbb{1}\| \in L^2(\mathbf{R})$, the Fredholm determinant of the S matrix of the Schrödinger equation with a potential in the class $\mathcal{V}_0$ does not approach unity as $k \to \infty$. It is, however, in the class of functions $\mathfrak{U}$ defined as follows [2]:

Definition 3.1 *Let $\mathcal{U}$ be the set of unitary operators $L^2(\mathbf{s}^2) \mapsto L^2(\mathbf{s}^2)$. The function $S(k) : \mathbf{R} \mapsto \mathcal{U}$ is in $\mathfrak{U}$ if and only if it has the following properties:*
(i) $S(k)$ is continuous;
(ii) $S(-k) = \overline{S(k)}$;
(iii) for each $k \in \mathbf{R}$ the Fredholm determinant $\det S$ *exists;*
(iv) $\exists c_1, c_2$ such that as $k \to \infty$

$$\eta(k) = c_1 k + c_2 + o(1),$$

where $\eta(k)$ is defined to be continuous and such that

$$\det S(k) = e^{2i\eta(k)}.$$

Three different kinds of indices, each of which is known to be an integer [1,4,5,6,7], have been defined. We will denote them as follows.

Definition 3.2 *(1) Suppose that $S \in \mathfrak{U}$. Then*

$$\mathrm{ind_L} S := \frac{1}{\pi}[\eta(0) - c_2].$$

(2) Suppose that S has a standard left Wiener-Hopf factorization, as defined earlier. Then

$$\mathrm{ind_{WH}} S := \sum_j \rho_j.$$

This is the standard definition of the total left Wiener-Hopf index (or "sum index").

(3) Suppose that $W_\sigma^1(S)$ has a unique solution F. Then

$$\mathrm{ind_J} S := n_\sigma.$$

In other words, $\mathrm{ind_J} S$ *is the sum of the dimensions of the ranges of the residues of F at all its (simple) poles in* $\mathbb{C}^+$.

Definition (3) is justified by part (i) of Theorem 2.5. According to Theorem 2.8 $\dim \mathrm{nul}(\mathbb{1} + \mathcal{G}^*) = \mathrm{ind_J} S$ if $\mathcal{G}^2$ does not have the eigenvalue 1.

In this language a three-dimensional generalization of Levinson's theorem [1] can be stated in the following form:

Levinson's theorem: *If S is the S matrix of a potential $V \in \mathcal{V}_0$ that produces n bound states (counting their degeneracies) and there is no half-bound state, then* $\mathrm{ind_L} S = n$.

The following theorem relates the canonical factorization to the problem W_0^1 (which is W_σ^1 with $\sigma = \emptyset$). Note that if $(S - \mathbb{1}) \in \mathcal{A}$ then $S^{-1} = QS^*Q$.

Lemma 3.3 *Suppose that $S^{-1} = QS^*Q$. Then S has a left canonical factorization $S = W_+W_-$ if and only if $W_0^1(S)$ has a solution F that is holomorphically invertible. We then have $F = QW_+Q$, $W_- = QW_+^{*-1}Q$, and F is the unique solution of $W_0^1(S)$.*

A convenient tool in the study of the more general case with poles is the reduction method [1,2]. It utilizes the rational function

$$\Pi_\sigma := \left(\mathbb{1} - B_1 + B_1 \frac{k + i\kappa_1}{k - i\kappa_1}\right) \left(\mathbb{1} - B_2 + B_2 \frac{k + i\kappa_2}{k - i\kappa_2}\right) \cdots . \tag{12}$$

Here the B_j are self-adjoint projections that are successively constructed so that the function $F^{\mathrm{red}} := \Pi_\sigma^{-1} F$ is free of poles. It allows us to reduce the solvability of $W_\sigma^1(S)$ to the existence of a canonical factorization of the reduced S matrix.

Lemma 3.4 *If $W_\sigma^1(S)$ has a holomorphically invertible solution F then $S^{\mathrm{red}} = Q\Pi_\sigma^{-1} QS\Pi_\sigma^*$ has a left canonical factorization; conversely, if S^{red} has a left canonical factorization, then $W_\sigma^1(S)$ has a unique solution F and the inverse F^{-1} is holomorphic in* $\mathbb{C}^+$. *Here Π_σ is the factor of the form (12) appropriate for W_σ^1.*

As an immediate consequence we have the following.

Corollary 3.5 *If $W_\sigma^1(S)$ has a holomorphically invertible solution then that solution is the only solution of $W_\sigma^1(S)$.*

The most powerful method, however, is the reduction of the solvability of $W_\sigma^1(S)$ to a similar problem for the diagonal factor, if a left standard factorization exists. This, of course, implies that it is reduced to a problem for a rational function. It was proved by Aktosun and van der Mee [8] that if S is admissible and the underlying potential is in a specified class, then S has a left standard factorization. They also showed that if $S = QS^{*-1}Q$ and S has a left standard factorization $S = W_+DW_-$ then it is always possible to choose the factorization so that $W_- = QW_+^{*-1}Q$. It follows that in that case $DQ = QD$, because $D^* = D^{-1}$. (Remember that $W^*(k) := W(-k)$.)

Proposition 3.6 *Suppose that $S = QS^{*-1}Q$ has the left standard factorization $S = W_+DQW_+^{*-1}Q$; then the following holds: If $W_\sigma^1(S)$ has a unique solution then there exists a set $\sigma' \in \mathcal{B}$ with $\mathbf{P}_{\sigma'} = \mathbf{P}_\sigma$ and $\mathbf{N}_{\sigma'} = \mathbf{N}_\sigma$ such that $W_{\sigma'}^1(D)$ has a unique solution, and vice versa: if $W_\sigma^1(D)$ has a unique solution then $\exists \sigma' \in \mathcal{B}$ with $\mathbf{P}_{\sigma'} = \mathbf{P}_\sigma$ and $\mathbf{N}_{\sigma'} = \mathbf{N}_\sigma$ such that $W_{\sigma'}^1(S)$ has a unique solution. The same holds if "unique" is everywhere replaced by "holomorphically invertible."*

The problem $W_\sigma^1(D)$ for the rational function D, on the other hand, is solved by this result:

Lemma 3.7 *Suppose that*

$$D = P_0 + \sum_{j\geq 1} P_j \left(\frac{k+i}{k-i}\right)^{\rho_j} \neq \mathbb{1},$$

$P_j = P_j^2 = P_j^\dagger$, $\mathrm{tr} P_j = 1$, $j \geq 1$, $P_0 + \Sigma_{j\geq 1} P_j = \mathbb{1}$, $P_j P_i = 0$ if $i \neq j$, *all the ρ_j are integers, and $QD = DQ$. A necessary condition for the problem $W_\sigma^1(D)$ to have a holomorphically invertible solution is that 1) each ρ_j is even and positive, and 2) $n_\sigma = \frac{1}{2}\Sigma_j \rho_j$. Conversely, if each ρ_j is even and positive, then there exists a set $\sigma \in \mathcal{B}$ with $n_\sigma = \frac{1}{2}\Sigma_j \rho_j$ such that the problem $W_\sigma^1(D)$ has a holomorphically invertible solution.*

The combination of Proposition 3.6 and Lemma 3.7 allows us to conclude the following theorem.

Theorem 3.8 *Suppose that $S = QS^{*-1}Q$ has a left standard factorization. Then a necessary condition for the problem $W_\sigma^1(S)$ to have a holomorphically invertible solution is that 1) either S has no left partial indices, or each left partial index of S is even and positive, and 2) $n_\sigma = \frac{1}{2}\mathrm{ind}_{\mathrm{WH}} S$. Conversely, if 1) holds, then there exists a set $\sigma \in \mathcal{B}$ with $n_\sigma = \frac{1}{2}\mathrm{ind}_{\mathrm{WH}} S$ and arbitrarily prescribed $\mathbf{P}_\sigma$ such that the problem $W_\sigma^1(S)$ has a holomorphically invertible solution.*

At this point the following questions need to be answered:

If S is *admissible*, i.e., it is the S matrix of some underlying potential (in $\mathcal{V}_0$, say), does there exist a set $\sigma \in \mathcal{B}$ such that $W_\sigma^1(S_x)$ has a unique solution for each $x \in \mathbf{R}^3$? If so, is $\text{ind}_\text{J} S_x = \text{ind}_\text{L} S$? We know, of course, that if S is admissible then $H_\sigma^1(S_x)$ has a solution, where $n_\sigma = \text{ind}_\text{L} S_x = \text{ind}_\text{L} S$. But it is not known that this solution is unique or that it is obtainable from the unique solution of $W_\sigma^1(S_x)$. This uniqueness is a sufficient condition for the Fourier transform of the solution of the generalized Marchenko equation to solve the Schrödinger equation if the *miracle* happens [1,2]. It would simplify matters very much if it were also a necessary condition (as is the miracle). It is conceivable that even for an admissible S $\text{ind}_\text{L} S \neq \text{ind}_\text{J} S$, in which case some poles in the unique solution of $W_{\sigma'}^1(S)$ with $\text{ind}_\text{L} S = n_{\sigma'}$ do not appear in the solution of $H_\sigma^1(S)$, or the solution of $W_{\sigma'}^1(S)$ has zeros.

The following result is a step in that direction.

Theorem 3.9 *Suppose that S is admissible and the underlying potential causes N negative-energy bound states (counting their degeneracies). Then* $\dim \text{nul}(\mathbb{1} + \mathcal{G}^*) = \dim \text{nul}(\mathbb{1} - \mathcal{G}^*) = N$.

Note that $\mathcal{G}^*$ here refers to S with $x = 0$. However, of $x \neq 0$ then S_x corresponds to a potential that is shifted by x, and the number of bound states is invariant under such a shift. Therefore we have the following

Corollary 3.10 *If $\mathcal{G}^*$ corresponds to an admissible S_x then* $\dim \text{nul}(\mathbb{1} + \mathcal{G}^*)$ *and* $\dim \text{nul}(\mathbb{1} - \mathcal{G}^*)$ *are independent of x.*

It also follows that if S is admissible and $\|\mathcal{G}\| < 1$ then $\text{ind}_\text{L} S = \text{ind}_\text{J} S$. However, I do not have a complete proof yet that if S is admissible then $\|\mathcal{G}\| < 1$.

References

[1] R. G. Newton. *Scattering Theory of Waves and Particles.* Springer Verlag, New York, 2nd edition, 1982.

[2] R. G. Newton. *Inverse Schrödinger Scattering in Three Dimensions.* Springer Verlag, New York, 1989. To be published.

[3] K. Chadan and P. C. Sabatier. *Inverse Problems in Quantum Scattering Theory.* Springer Verlag, New York, second edition edition, 1989.

[4] I. C. Gohberg. The factorization problem for operator functions. *Izv. Akad. Nauk SSSR Ser. Mat.*, 28:1055–1082, 1964. English translation: American Mathematical Society Translations 49, 130-161 (1966).

[5] K. Clancey and I. C. Gohberg. *Factorization of Matrix Functions and Singular Integral Operators.* Birkhäuser Verlag, Basel, 1981.

[6] I. C. Gohberg and M. A. Kaashoek. *Constructive Methods in Wiener-Hopf Factorization.* Birkhäuser Verlag, Basel, 1986.

[7] W. Greenberg, C. van der Mee, and V. Protopopescu. *Boundary Value Problems in Abstract Kinetic Theory.* Birkhäuser Verlag, Basel, 1987.

[8] T. Aktosun and C. van der Mee. Solution of the inverse scattering problem for the 3-d Schrödinger equation by Wiener-Hopf factorization of the scattering operator. Preprint, University of Delaware, 1989.

Electromagnetic Inverse Problems at Low Frequencies

A.G. Ramm[1] *and E.J. Somersalo*[2]

[1]Mathematics Department, Kansas State University, Manhattan, KS 66506, USA
[2]Department of Mathematics, University of Helsinki, Hallituskatu 15, SF-00100 Helsinki, Finland

Consider the following electromagnetic inverse problem arising for example in geophysical exploration: A small time-harmonic current loop is moved along the surface of the earth, modelled as a half space. The earth is assumed to contain a bounded regular region of inhomogeneity where the electromagnetic parameters ε, μ and σ (permittivity, permeability and conductivity) vary; outside this region they are assumed to be constants. The scattered magnetic field is then measured everywhere on the surface of the earth with low transmitter frequencies. The problem is to find the parameters describing the inhomogeneity from these measurements.

Below is an overview of the results obtained in the article [5], where the details and proofs as well as some further references can be found.

The scattering configuration is described by Maxwell's equations

$$\begin{cases} \nabla \times H + i\omega\varepsilon E - \sigma E = j \\ \nabla \times E - i\omega\mu H = 0, \end{cases} \tag{1}$$

where the external current density j is given as

$$j(x, y) = \nabla \times (\delta(x - y)e_3),$$

x being an arbitrary field point, y the location of the transmitter and e_3 a unit vector along the x_3-axis. (See Figure below).

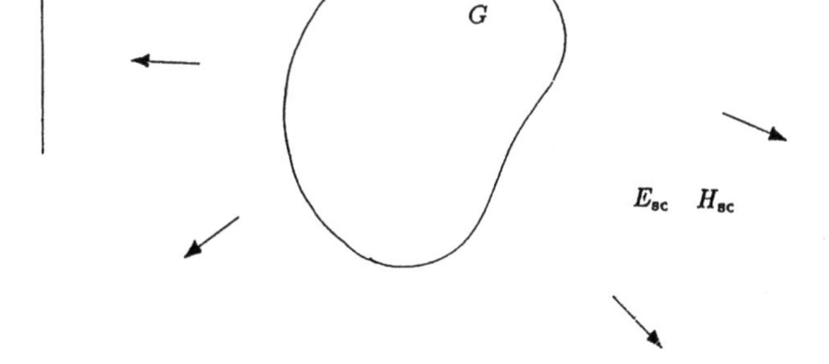

We assume that $y \in P = \{x = (x_1, x_2, x_3) | x_3 = 0\}$. To guarantee certain degree of regularity of the solutions we have to assume that ε, σ and μ are in the space $C^2(\mathrm{R}^3)$ of twice continuously differentiable functions. Furthermore, we shall assume that $\mu > 0$ and $\varepsilon > 0$. Regarding the conductivity σ it is assumed that either $\sigma > 0$ everywhere or σ vanishes identically. Let G denote a bounded regular domain, $G \subset \mathrm{R}^3_- = \{(x_1, x_2, x_3) \in \mathrm{R} \mid x_3 < 0\}$, such that for some constants ε_0, σ_0 and μ_0 $\varepsilon = \varepsilon_0$, $\sigma = \sigma_0$ and $\mu = \mu_0$ outside the domain G. To single out the physically meaningful solution of the system (1) it is assumed that the fields E and H satisfy the Silver-Müller type radiation condition at infinity: Denoting $k^2 = \omega^2 \mu_0 \gamma_0$, $0 \leq \arg k < \pi$, where $\gamma_0 = \varepsilon_0 + i\sigma_0/\omega$, we assume that for $\omega > 0$

$$\begin{aligned} \omega\mu_0 x^0 \times H + kE &= O\left(\frac{1}{|x|^2}\right), \\ \omega\gamma_0 x^0 \times E - kH &= O\left(\frac{1}{|x|^2}\right), \end{aligned} \tag{2}$$

where $x^0 = x/|x|$.

As a model problem of the low frequency inversion consider first the acoustic scattering, where Maxwell's equations are replaced by the reduced wave equation

$$(\Delta + k^2 n(x)^2) u(x, y, k) = \delta(x - y),$$

the function $x \mapsto u(x, y, k)$ satisfying the Sommerfeld radiation condition at infinity. Here n is the refractive index of the medium, and it is assumed that $n(x) = 1$ outside a bounded region G. The unique solution of this equation with the outgoing radiation at infinity is found by solving the Lippmann-Schwinger equation

$$u(x, y, k) = \Phi(x - y, k) + k^2 \int_G \Phi(x - z, k) V(z) u(z, y, k) dz,$$

where

$$\Phi(x, k) = \frac{e^{ik|x|}}{4\pi|x|} \tag{3}$$

and $V(z) = 1 - n(z)^2$. The inverse problem of recovering V from the low frequency measurements of $u(x, y, k)$ as $x, y \in P$ is solved in the reference [4] in the following way. First, one can prove that as k tends to zero, the the field u approaches the primary field at zero frequency, i.e.,

$$\lim_{k \to 0} u(x, y, k) = \Phi_0(x - y) = \frac{1}{4\pi|x - y|}.$$

Therefore we get immediately

$$16\pi^2 \lim_{k \to 0} \frac{1}{k^2}(u(x, y, k) - \Phi(x - y, k)) = \int_G \frac{V(z)}{|x - z||z - y|} dz,$$

which can be considerded as data when $x, y \in P$. At low frequencies the acoustic inverse problem reduces thus asymptotically to a linear one that admits a unique, and in fact constructive solution.

One can ask how much of this structure of the inverse problem is inherent when the scalar waves are replaced by electromagnetic ones.

By transforming Maxwell's equations into a second order system one can show that the system (1) with the additional condition (2) is equivalent to a pair of integral equations

$$E(x,y,\omega) = E_{\mathrm{i}}(x,y,\omega) + E_{\mathrm{sc}}(x,y,\omega) = E_{\mathrm{i}}(x,y,\omega) \tag{4a}$$

$$+ \int_G (-\frac{\nabla\gamma(z,\omega)}{\gamma(z,\omega)} \cdot E(z,y,\omega)\nabla\Phi(x-z,\omega) + \omega^2\mu_0\tilde{\gamma}(z,\omega)E(z,y,\omega)\Phi(x-z,\omega)$$

$$+ i\omega\tilde{\mu}(z)H(z,y,\omega) \times \nabla\Phi(x-z,\omega))dz,$$

and

$$H(x,y,\omega) = H_{\mathrm{i}}(x,y,\omega) \tag{4b}$$

$$+ \int_G (-\frac{\nabla\mu(z)}{\mu(z)} \cdot H(z,y,\omega)\nabla\Phi(x-z,\omega) + \omega^2\gamma_0\tilde{\mu}(z)H(z,y,\omega)\Phi(x-z,\omega)$$

$$- i\omega\tilde{\gamma}(z,\omega)E(z,y,\omega) \times \nabla\Phi(x-z,\omega))dz,$$

where $\nabla = \nabla_z$, $\gamma = \varepsilon + i\sigma/\omega$ and Φ is the fundamental solution of the scalar Helmholtz equation given by (3), $\tilde{\mu}(x) = \mu(x) - \mu_0$ and $\tilde{\gamma}(x) = \gamma(x) - \gamma_0$. The incident fields E_{i} and H_{i} correspond to the situation when no inhomogeneity is present, i.e.,

$$E_{\mathrm{i}}(x,y,\omega) = -i\omega\mu_0\nabla \times (e_3\Phi(x-y,\omega)),$$
$$H_{\mathrm{i}}(x,y,\omega) = \nabla \times (\nabla \times (e_3\Phi(x-y,\omega))), \quad x \neq y,$$

(See e.g. [1]).

The low frequency asymptotics of the solutions of Maxwell's equations at least in the non-dissipative case are fairly well known (see e.g. [3] and [6]). By using techniques similar to the classical limiting absorption principle (regularity results for solutions of elliptic equations and compact embedding theorems) one obtains the following result ([5]).

Proposition. The solution (E,H) of equations (4 a–b) has a low frequency limit

$$\lim_{\omega\to 0}(E(x,y,\omega), H(x,y,\omega)) = (0, H_0(x,y))$$

in $H^2_{\mathrm{loc}}(\mathbb{R}^3 \setminus \{y\})$, and H_0 satisfy the equations

$$\begin{cases} \nabla \times H_0 = j \\ \nabla \cdot (\mu H_0) = 0 \end{cases}$$

with the asymptotic behavior

$$|H_0| = O\left(\frac{1}{|x|}\right)$$

This low frequency limit can be used in linearizing the integral equations (4 a–b), in general cases approximately and in a special case exactly as expained below.

(i) A special case: Assume that $\mu = \mu_0 > 0$ and $\sigma = \sigma_0 > 0$ everywhere in $\mathbb{R}^3$. Then we have

$$\lim_{\omega\to 0}\frac{1}{\omega^2}(H(x,y,\omega) - H_i(x,y,\omega)) = -\mu_0 \int \tilde{\varepsilon}(z)\nabla \times (e_3\Phi_0(z-y)) \times \nabla\Phi_0(x-z)dz,$$

the convergence taking place in H^2_{loc}. Here,

$$\Phi_0(x) = \frac{1}{4\pi|x|},$$

and $\tilde{\varepsilon}(x) = \varepsilon(x)/\varepsilon_0$.

Denoting by $I(x,y)$ the right hand side of the above equation we can consider $\{I(x,y)|x,y\in P\}$ as being the data of the inverse problem. Let $x = (x',0), y = (y',0)\in P$ and $z = (z',z_3)\in\mathbb{R}^3$. Then, for $\xi,\eta\in\mathbb{R}^2$ it turns out that the Fourier transform of $I(x,y)$ over $P \times P$ yields

$$\hat{I}(\xi,\eta) \equiv \int_{\mathbb{R}^2} dx' \int_{\mathbb{R}^2} dy' e^{i(x'\cdot\xi+y'\cdot\eta)} I((x',0),(y',0))$$

$$= -\frac{1}{4}(i\eta^o + (\eta^o\cdot\xi^o)e_3)\int_G \tilde{\varepsilon}(z)e^{i(\xi+\eta)\cdot z'}e^{(|\xi|+|\eta|)z_3}dz$$

where $\eta^o = \eta/|\eta|$ and $\xi^o = \xi/|\xi|$. Let $p = \xi+\eta \in \mathbb{R}^2$, $r = |\xi|+|\eta| \in \mathbb{R}_+$. Considering η^o as a fixed parameter, the change of variables $(|\eta|,\xi) \to (r,p)$ is non–singular, and in terms of (p,r) we have

$$\frac{2}{\eta^o\cdot\xi^o}(i\xi^o - e_3)\cdot\hat{I}(\xi,\eta) = \int_G \tilde{\varepsilon}(z)e^{ip\cdot z'+rz_3}dz.$$

Since this integral is known for all $p \in \mathbb{R}^2$, $r \in \mathbb{R}_+$ and $\eta^o \in S^1$ can be an arbitrary unit vector, the function $\tilde{\varepsilon}$ can be recovered analytically in a unique way.

(ii) The general case: If μ and σ are no longer constants, one cannot hope to get a linear dependence between the data and the unknown parameters, since the low frequency limits of the fields are not independent of the parameters as the Proposition demonstrates. However, if the scattering potentials in the integral equations for E and H are weak, one can use the Born approximations

$$\lim_{\omega\to 0} H(x,y,\omega) \approx \lim_{\omega\to 0} H_i(x,y,\omega) = \nabla\times(\nabla\times(e_3\Phi_0(x-y))),$$

$$\lim_{\omega\to 0}\frac{1}{i\omega}E(x,y,\omega) \approx \lim_{\omega\to 0}\frac{1}{i\omega}E_i(x,y,\omega) = -\mu_0\nabla\times(e_3\Phi_0(x-y))$$

at low frequencies. Rearranging the terms according to the asymptotic behavior at low frequencies we have

$$H(x,y,\omega) = H_i(x,y,\omega) - \int_G \frac{\nabla\mu(z)}{\mu(z)}\cdot H(z,y,\omega)\nabla\Phi(x-z,\omega)dz$$

$$+ i\omega\left(\sigma_0\int_G \tilde{\mu}(z)H(z,y,\omega)\Phi(x-z,\omega)dz + \int_G \tilde{\sigma}(z)\frac{E(z,y,\omega)}{i\omega}\times\nabla\Phi(x-z,\omega)dz\right)$$

$$+\omega^2\left(\varepsilon_0\int\limits_G \tilde{\mu}(z)H(z,y,\omega)\Phi(x-z,\omega)dz+\int\limits_G \tilde{\varepsilon}(z)\frac{E(z,y,\omega)}{i\omega}\times\nabla\Phi(x-z,\omega)dz\right)$$

$$= H_{\mathrm{i}}(x,y,\omega)-I_0(x,y,\omega)+i\omega I_1(x,y,\omega)+\omega^2 I_2(x,y,\omega).$$

Using the Born linearizations, the Taylor coefficients of H_{sc} at $\omega=0$ are written approximately as

$$\lim_{\omega\to o} I_0(x,y,\omega)\approx I_0^B(x,y)$$

$$:=\int\limits_G \frac{\nabla\mu(z)}{\mu(z)}\cdot\nabla\times(\nabla\times(e_3\Phi_0(z-y)))\nabla\Phi_0(x-z)dz,$$

$$\lim_{\omega\to o} I_1(x,y,\omega)\approx I_1^B(x,y)$$

$$:=\sigma_0\int\limits_G \tilde{\mu}(z)\nabla\times(\nabla\times(e_3\Phi_0(z-y)))\Phi(x-z)dz-\mu_0\int\limits_G \tilde{\sigma}(z)(\nabla\times(e_3\Phi_0(z-y))\times\nabla\Phi_0(x-z)dz,$$

and

$$\lim_{\omega\to o} I_2(x,y,\omega)\approx I_2^B(x,y)$$

$$:=\varepsilon_0\int\limits_G \tilde{\mu}(z)\nabla\times(\nabla\times(e_3\Phi_0(z-y)))\Phi(x-z)dz-\mu_0\int\limits_G \tilde{\varepsilon}(z)(\nabla\times(e_3\Phi_0(z-y))\times\nabla\Phi_0(x-z)dz.$$

In the same way as in case (i), we can now take the Fourier transforms over $P\times P$ of the approximately known functions $I_j^B(x,y)$, $j=0,1,2$ to obtain the following formulas. If $\eta^o\neq\xi^o$, we have

$$\frac{4|\xi|}{\varepsilon_0|\eta|(1-\xi^o\cdot\eta^o)}(i\xi^o+e_3)\cdot\hat{I}_2^B(\xi,\eta)=\int\limits_G \tilde{\mu}(z)e^{i(\xi+\eta)\cdot z'+(|\xi|+|\eta|)z_3}dz$$

and

$$-\frac{4|\xi|}{\mu_0(1-\xi^o\cdot\eta^o)}(i\eta^o+e_3)\cdot\hat{I}_2^B(\xi,\eta)=\int\limits_G \tilde{\varepsilon}(z)e^{i(\xi+\eta)\cdot z'+(|\xi|+|\eta|)z_3}dz.$$

After the recovery of $\tilde{\mu}$ from the above equation we get

$$\frac{4}{\mu_0}i\eta^o\cdot\left(\hat{I}_1^B(\xi,\eta)-\sigma_0\frac{|\eta|}{4|\xi|}(i\eta^o+e_3)\int\limits_G \tilde{\mu}(z)e^{i(\xi+\eta)\cdot z'+(|\xi|+|\eta|)z_3}dz\right)$$

$$=\int\limits_G \tilde{\sigma}(z)e^{i(\xi+\eta)\cdot z'+(|\xi|+|\eta|)z_3}dz$$

and $\tilde{\sigma}(x)=\sigma(x)-\sigma_0$ is finally recovered.

[1] D. Colton and R. Kress: Integral equation methods in scattering theory. John Wiley&Sons, New York, 1983.

[2] C. Müller: Foundations of mathematical theory of electromagnetic waves. Springer-Verlag, Berlin, 1969.

[3] R. Picard: On the low frequency asymptotics in electromagnetic theory. J. reine angew. Math. 354 (1984) 50–73.

[4] A. G. Ramm: Inverse scattering for geophysical problems. Phys. lett. 99A (1983) 258–60.

[5] A. G. Ramm and Erkki Somersalo: Electromagnetic inverse problems with surface measurements at low frequencies. Inverse Problems 1989 (to appear).

[6] P. Werner: Über das Verhalten elektromagnetischer Felder für kleine Frequenzen in mehrfach zusammenhangenden Gebieten I–II. J. reine angew. Math. 278/9 (1975), 365–397, 280 (1976) 98–125.

Retrieval of a Time-Dependent Source in an Acoustic Propagation Problem

G. Saracco[1] *and Ph. Tchamitchian*[2]

[1]L.M.A., Equipe Ultra-Sons, CNRS 31, Chemin Joseph Aiguier, F-13009 Marseille, France
[2]Centre de Physique Théorique, CNRS-Luminy, Case 907, F-13288 Marseille Cédex 9, France

Abstract : Consider two media separated by a plane interface, a time-dependent source F(t) at a point S in the first medium (of the lower celerity), receivers in the second medium, located at a large radial distance from the source. Thanks to a diffusion-like behaviour of the transmitted wave, we are able to retrieve the term F(t), under high frequencies hypothesis in an elementary way.

1. About homogeneity and cancellation

We begin with the Poisson semi-group of operators

$$P_y = e^{-y\Lambda},$$

where $y > 0$ and $\Lambda = \sqrt{-\Delta}$, acting on $L^2(\mathbb{R}^n)$. Then, we define

$$Q_y = -y \frac{\partial}{\partial y} P_y .$$

Because of the fact that P_y is an approximation of the identity, which means that

$$I = \lim_{y\to 0} P_y \quad \text{and} \quad 0 = \lim_{y\to +\infty} P_y$$

in the strong topology of $L^2(\mathbb{R}^n)$, we deduce the following decomposition of the identity :

$$I = \int_0^{+\infty} Q_y \frac{dy}{y} . \tag{1}$$

This formula is known as the Littlewood-Paley decomposition, and has extensively been studied in harmonic analysis since the 30's. See for example the classical treatise by Zygmund [4]. For the amateurs, we mention that is in fact the ancestor of the so-called wavelet transform. But we do not want to go further into this matter.

What we would like to point out here is that analogous formulas to (1) can be obtained, and may be useful, in some physical situations which a priori have nothing to do with diffusion.

Before stating what we have in mind, let us have a closer look at (1), on the Fourier side.

(1) may be interpreted as a way of analysing signals with a continuous chain of filters associated to the operators Q_y. More precisely, if we denote these filters by $Q_y(\omega)$, then :

$$Q_y(\omega) = y\,|\,\omega\,|\,e^{-y|\omega|}\,.$$

Now, it becomes apparent why (1) is true. It is because of the following two basic properties of $Q_y(\omega)$:

a. an homogeneity property : $Q_y(\omega)$ depends effectively on the product $y\,|\omega|$. It is centered around the sphere in $\mathbb{R}^n$ of radius $\frac{1}{y}$, with a bandwidth $\Delta Q_y = \frac{c_n}{y}$, where c_n is a constant (varying with your normalisation). Notice that $\frac{dy}{y}$ is the dilation-invariant measure on $\mathbb{R}^+$.

b. a cancellation property : $Q_y(\omega)$ cancels at the origin. Such a property is necessary in order to insure the convergence at $y = 0$ of the integral in (1). A contrario, the same integral with P_y instead of Q_y would strongly diverge.

We now turn our attention to a seemingly very different problem : the propagation through a plane interface.

2. An asymptotic formula for retrieving a time-dependent source

The physical situation is the classical one : two infinite homogeneous and isotropic media are separated by a plane interface. We think to air and water for numerical and experimental purposes [3]. An acoustical point source is emitting in the first medium (air), at height h, and receivers are located in the second one (water), at a large radial distance r from the source, and on a vertical straight line starting from the interface at the point R_o :

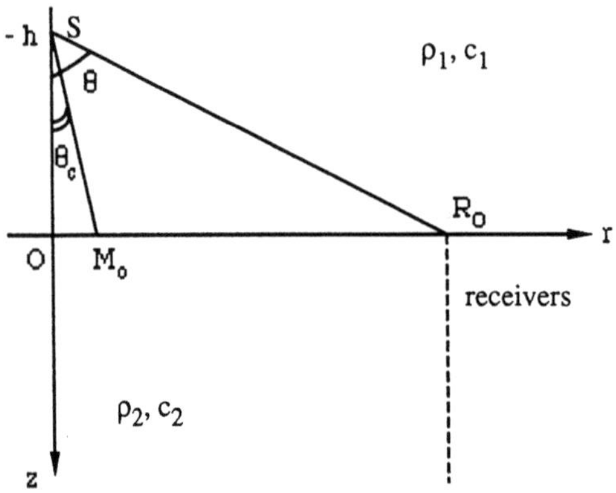

$\frac{\rho_2}{\rho_1} = m \sim 800$ (densities)

$\frac{c_1}{c_2} = n \sim 0.23$ (celerities)

$\theta_c = \arcsin n \sim 13°$ (critical angle)

Fig. 1

Here, "large radial distance" means that $\theta = \mathrm{arctg}\,\frac{r}{h}$ is greater than the critical angle θ_c (that is to say, $\frac{r}{\sqrt{r^2+h^2}} > n$).

If F(t) is the acoustic potential of the source, and if $\psi_2\,(r,z,t)$ is the transmitted potential at a point (r,z), $z > 0$, and at time t, the associated pressure $P_2\,(r,z,t)$ is

$$P_2(r,z,t) = \rho_2 \frac{\partial}{\partial_y} \psi_2(r,z,t) .$$

We recall that (see Landau and Lifshitz [2]) :

$$\psi_2(r,z,t) = \int_{\mathbb{R}} e^{-i\omega t} \hat{F}(\omega)\ i \int_0^{+\infty} e^{i(k_1 h + k_2 z)} J_o(kr) \frac{1}{m\ k_1 + k_2}\ k\ dk\ d\omega ,$$

where $k_j = \sqrt{\left(\frac{\omega}{c_j}\right)^2 - k^2}$.

We suppose that the source is of a high frequency nature : $\hat{F}(\omega) \neq 0$ only if $|\omega| \geq \omega_{min}$, and $\frac{r\ \omega_{min}}{c_1} >> 1$.

Then, we have the following asymptotic formula :

$$\int_0^{+\infty} P_2(r,z,t)\ dz \sim C_1(r,h)\ F(t-t_1) + C_2(r,h)\ F(t-t_2) + C_3(r,h)\ \mathcal{H}F(t-t_2) \qquad (2)$$

where

$\mathcal{H}$ is the Hilbert transform ,

$t_1 = \frac{h\sqrt{1-n^2}+nr}{c_1}$,

$t_2 = \frac{\sqrt{r^2+h^2}}{c_1}$,

$C_j(r,h)$ are constants depending only on the positions of the source and the receivers.

t_1 is the time for a way to travel over the path S M_o R_o with velocity c_1 on S M_o and c_2 on M_o R_o, and t_2 corresponds to the path S R_o with velocity c_1 (see fig. 1).

The presence of the Hilbert transform is due to the condition $\frac{r}{\sqrt{r^2+h^2}} > n$. It disappears in the opposite case, where a similar (and much less interesting, at least for an air-water interface) formula holds :

$$\int_0^{+\infty} P_2(r,z,t)\ dt \sim C(r,h)\ F(t - t_2). \qquad (2')$$

If we take into consideration the fact that $m = \frac{P_2}{P_1} \approx 800$, we have for the constants in (2) the following simplified expressions :

$$C_1(r,h) = \frac{\rho_1}{(1-n^2)^{1/4}\ [r(r\sqrt{1-n^2} - hn)]^{1/2}} + 0\left(\frac{1}{m}\right) , \qquad C_2(r,h) = 0\left(\frac{1}{m}\right) ,$$

$$C_3(r,h) = \frac{\rho_1}{\left[\frac{r^2}{r^2+h^2} - n^2\right]^{1/2} r} + 0\left(\frac{1}{m}\right) .$$

We are not going to enter in the details of the proof of the formula (2), which is straightforward and typical of asymptotic arguments (one could ask whether such a kind of argument is necessary or not : the answer is yes, for the result is simply false for arbitrary sources).

We prefer to conclude with a comparison between formulas (1) and (2).

Let us remark however that, once (2) is valid and the position of the source is known, it is easy to retrieve the source term F(t). Maybe it is also possible to find the coordinates of the point source from (2), but this is still under investigations.

3. Link between the formulas (1) and (2)

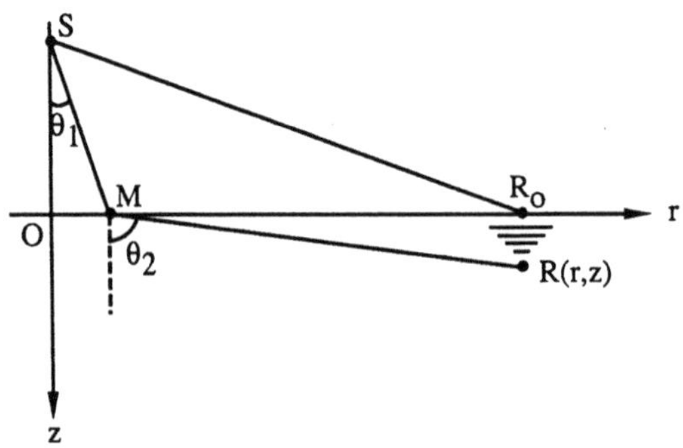

$$\sin\theta_1 = n \sin\theta_2$$

Fig. 2

Looking at P_2 (r,z,t) as the result of a filtering process applied on F(t), we denote by $\Gamma(r,z,\omega)$ the associated impulse response.

$\Gamma(r,z,\omega)$ is nothing but the transmitted pressure at the point R(r,z) when the source is harmonic, with frequency ω. It is usual, following Brekhovskikh [1], to decompose it in two contributions, under asymptotic conditions :

$$\Gamma(r,z,\omega) = \Gamma_{geo}(r,z,\omega) + \Gamma_{lat}(r,z,\omega).$$

Γ_{geo} is given by the approximation of the geometrical acoustics (see fig. 2) :

$$\Gamma_{geo}(r,z,\omega) = \gamma_{geo}(r,z,h)\, i\,\omega\, e^{i\omega\left(\frac{SM}{c_1}+\frac{MR}{c_2}\right)}, \qquad (3)$$

while Γ_{lat} is specific of the case when $n = \frac{c_1}{c_2} < 1$:

$$\Gamma_{lat}(r,z,\omega) = \gamma_{lat}(r,h)\, e^{i\omega\frac{\sqrt{r^2+h^2}}{c_1}}\, i\omega\, e^{-\alpha|\omega|\frac{z}{c_2}}, \qquad (4)$$

$$\alpha = \sqrt{\frac{r^2}{r^2+h^2} - n^2}\,.$$

Looking at these expressions, we may understand why (2) is valid.

Consider at first Γ_{lat}, which is the main term, at least near the interface (see Saracco [3]). It is usually said that Γ_{lat} corresponds to a wave propagating in the air along the path S R_0 and then diffusing in the water (fig. 2). The term $e^{-\alpha|\omega|\frac{z}{c_1}}$ in (4) is typical of a diffusion process, while $e^{i\omega\frac{\sqrt{r^2+h^2}}{c_1}}$ is only a delay term. $\Gamma_{lat}(r,z,\omega)$ appears to be very much like $Q_y(\omega)$ in §1, with $\alpha\frac{z}{c_1}$ playing the role of y.

What about Γ_{geo} ? It is different in nature from Γ_{lat}, and associated to a wave propagating in air and in water along the path SMR (fig. 2). But it still has the properties of homogeneity and cancellation we described in § 1, at least again near the interface. In fact, if $\frac{z}{r}$ is small, we have

$$\Gamma_{geo}(r,z,\omega) \sim \gamma'_{geo}(r,h)\; e^{i\omega\frac{h\sqrt{1-n^2}+rn}{c_1}}\; i\omega z\, e^{i\frac{\omega z^2}{c_1\ell}}, \qquad (5)$$

where $\ell = \frac{r}{n} - \frac{h}{\sqrt{1-n^2}}$. The cancellation property is already clear in (3), and the homogeneity is apparent in (5).

These remarks explain why the formulas (1) and (2) are analogous.

REFERENCES

[1] L.N. Brekhovskikh, Waves in Layered Media, 1960 Wiley, New-York.

[2] L.D. Landau, E.M. Lifshitz, Fluid mechanics, Pergamon Press 1959.

[3] G. Saracco, Etude de la transmission acoustique à travers le dioptre air-eau, Note-Laboratoire de Mécanique et d'Acoustique n° 101, CNRS.

[4] A. Zygmund, Trigonometric series, Cambridge University Press 1968.

Inverse Scattering Problems for the Wave Equation with Time Dependent Impurities

P.D. Stefanov

Institute of Mathematics, Bulgarian Academy of Sciences, 1090 Sofia, Bulgaria

1. Statement of the results

In this paper we present some uniqueness theorems in the inverse scattering for the wave equation with a time dependent potential and for the wave equation in the exterior of a moving obstacle. The complete proofs can be found in author's articles [23-27].

In what follows we assume that the space dimension n is odd, $n \geq 3$. First, consider the wave equation

$$u_{tt} - \Delta u + q(t,x)u = 0, \qquad t \in \mathbb{R},\ x \in \mathbb{R}^n \tag{1}$$

with a potential $q(t,x)$ depending on t. We assume that

(A1) $q \in C^\infty$, $|q| \leq C(1+|t|)^N$ with some $C>0$, $N>0$;

(A2) there exists $\varrho>0$ such that $q(t,x) = 0$ for $|x|>\varrho$, all t.

Our aim is to prove that $q(t,x)$ can be determined uniquely by the scattering data corresponding to (1).

Secondly, consider the wave equation in the exterior of a moving obstacle. Let $Q \subset \mathbb{R}^{n+1}$ be an open connected set with a smooth boundary ∂Q. Given $t \in \mathbb{R}$, we denote by $\Omega(t) = \{x \in \mathbb{R}^n;\ (t,x) \in Q\}$ the exterior domain and by $\mathcal{O}(t) = \mathbb{R}^n \setminus \Omega(t)$ the obstacle at the time t. We impose the following conditions.

(B1) There exists $\varrho>0$ such that $\mathcal{O}(t) \subset B_\varrho = \{x;\ |x|<\varrho\}$ for each t.

(B2) If $\nu = (\nu_t, \nu_x)$ is the inner unit normal to ∂Q, then $|\nu_t| < |\nu_x|$.

Conditions (B1), (B2) mean that the obstacle remains within a fixed compact set and that the boundary moves with a speed less than 1. Consider the wave equation in Q with Dirichlet boundary conditions

$$\begin{cases} u_{tt} - \Delta u = 0 & \text{in } Q, \\ u = 0 & \text{on } \partial Q. \end{cases} \tag{2}$$

We are interested in the problem of recovering Q from the scattering data related to (2).

The case of stationary (time independent) potentials and stationary obstacles has been treated by many authors (see for example [12,13,22]). Then the variables can be separated and we can reduce (1) and (2) to stationary elliptic problems. The uniqueness of the inverse scattering problem for (1) with time independent $q(x)$ follows from the Born approximation of the scattering amplitude (see e.g. [22]). In [12] there are two proofs showing that a stationary obstacle can be uniquely recovered from the scattering operator.

In the non-stationary case the situation changes considerably. The separation of variables is no longer available. A time dependent scattering theory for (2) has been developed by Cooper and Strauss (see [3-7] and the references herein), generalizing the Lax-Phillips theory (see also [2,18-20,29]). Equation (1) has been studied in [1,9,16-18,30]. Since the local energy for (1) or (2) may increase [20], the scattering operator S does not exist in general. The main object in the scattering theory for (1), (2) is the *generalized scattering (echo) kernel* $K^{\#}$ introduced by Cooper and Strauss [5]. Below we shall give a brief definition of $K^{\#}$.

Given $f=(f_1,f_2)$ define the energy norms

$$\|f\|_0^2 = \int_{\mathbb{R}^n} (|\nabla f_1|^2 + |f_2|^2)\,dx, \qquad \|f\|_{(t)}^2 = \int_{\Omega(t)} (|\nabla f_1|^2 + |f_2|^2)\,dx.$$

Denote by $\mathcal{H}_0$ the completition of $C_0^\infty(\mathbb{R}^n) \times C_0^\infty(\mathbb{R}^n)$ with respect to the norm $\|\cdot\|_0$ and by $\mathcal{H}(t)$ the completition of $C_0^\infty(\Omega(t)) \times C_0^\infty(\Omega(t))$ with respect to the norm $\|\cdot\|_{(t)}$. Denote by $U_q(t,s)$ and $U_Q(t,s)$ the propagators related to (1) and (2) respectively. In other words the solutions to (1) and (2) with initial data $u(s,\cdot) = f_1$, $u_t(s,\cdot) = f_2$ are given by $(u(t,\cdot),u_t(t,\cdot)) = U_q(t,s)f$ and $(u(t,\cdot),u_t(t,\cdot)) = U_Q(t,s)f$, respectively. The operators $U_q(t,s):\mathcal{H}_0 \to \mathcal{H}_0$ and $U_Q(t,s): \mathcal{H}(s) \to \mathcal{H}(t)$ are bounded ones for fixed t, s but their norms may increase exponentially as $|t-s|\to\infty$. Using the finite speed of propagations we can define $U_q(t,s)f$, $U_Q(t,s)f$ for functions f which belong locally to $\mathcal{H}_0$ and $\mathcal{H}(t)$, respectively.

Set $h_n(\xi) = \xi^n/n!$ for $\xi\ge 0$, $h_n(\xi) = 0$ otherwise, $n = 0,1,\ldots$ Thus $h_n' = h_{n-1}$, h_0 is the Heaviside function. Given $(s,\omega) \in \mathbb{R}\times S^{n-1}$ let $\Gamma_q(t,x;s,\omega)$ be the solution of the problem

$$\begin{cases} (\Box + q(t,x))\Gamma_q = 0, \\ \Gamma_q\big|_{t\ll 0} = h_1(t+s-x\cdot\omega), \end{cases}$$

and let $\Gamma_Q(t,x;s,\omega)$ solve the problem

$$\begin{cases} \Box\Gamma_Q = 0 & \text{in } Q, \\ \Gamma_Q = 0 & \text{on } \partial Q, \\ \Gamma_Q\Big|_{t\ll 0} = h_1(t+s-x\cdot\omega). \end{cases}$$

Here $\Box = \partial_t^2 - \Delta_x$. Since the pair $(h_1(t+s-x\cdot\omega), h_0(t+s-x\cdot\omega))$ considered as a function of x has locally finite energy, Γ_q and Γ_Q are well defined. The functions $\Gamma_{q,sc} = \Gamma_q - h_1(t+s-x\cdot\omega)$, $\Gamma_{Q,sc} = \Gamma_Q - h_1(t+s-x\cdot\omega)$ possess *asymptotic wave profiles* [5], i.e. the limit

$$\Gamma^{\#}_{sc}(s',\omega';s,\omega) = \lim_{t\to\infty} (t+s')^{(n-1)/2}\partial_t\Gamma_{sc}(t,(t+s')\omega';s,\omega)$$

exists in $L^2_{loc}(\mathbb{R}_{s'}\times S^{n-1}_{\omega'})$, where $\Gamma^{\#}_{sc} = \Gamma^{\#}_{q,sc}$, $\Gamma_{sc} = \Gamma_{q,sc}$ and $\Gamma^{\#}_{sc} = \Gamma^{\#}_{Q,sc}$, $\Gamma_{sc} = \Gamma_{Q,sc}$, respectively.

We define the *generalized scattering kernels* $K^{\#}_q$, $K^{\#}_Q$ by the following relations

$$K^{\#}_q(s',\omega';s,\omega) = (-2\pi)^{(1-n)/2}\partial_s^{(n+1)/2}\Gamma^{\#}_{q,sc}(s',\omega';s,\omega),$$

$$K^{\#}_Q(s',\omega';s,\omega) = (-2\pi)^{(1-n)/2}\partial_s^{(n+1)/2}\Gamma^{\#}_{Q,sc}(s',\omega';s,\omega).$$

$K^{\#}_q$ and $K^{\#}_Q$ are distributions with respect to s' smoothly depending on ω', s, ω. It should be noted that if the scattering operator S exists, then $K^{\#}_q$ (respectively $K^{\#}_Q$) is the Schwartz kernel of $S\text{-}Id$ in the Lax-Phillips translation representation.

We are interested in the following inverse problems:

(IP1) Does $K^{\#}_q$ determine q uniquely?

(IP2) Does $K^{\#}_Q$ determine Q uniquely?

As mentioned above if $q(x)$ does not depend on t, then (IP1) has affirmative answer. The first result concerning time dependent q has been obtained by Ferreira and Perla Menzala [9]. They have shown that if $q_1 \ge 0$ and $q_2 \ge 0$ are small in a suitable sense, $q_1 \ge q_2$ for all t, x and $\partial_t q_1$, $\partial_t q_2$ have appropriate decay as $|t| \to \infty$, then the equality of the scattering operators for q_1 and q_2 implies $q_1 = q_2$.

Our first result gives an affirmative answer to (IP1).

Theorem 1. *Let q_1 and q_2 satisfy* (A1), (A2). *Suppose there exist $\varepsilon > 0$ and $\omega_0 \in S^{n-1}$ so that $K^{\#}_{q_1}(s',\omega';s,\omega) = K^{\#}_{q_2}(s',\omega';s,\omega)$ for $|\omega-\omega_0| < \varepsilon$, $|\omega'-\omega_0| < \varepsilon$, $|s'-s| < \varepsilon$. Then $q_1(t,x) = q_2(t,x)$ for all t, x.*

The kernel $K_q^{\#}$ will be shown to be smooth off the diagonal $(s',\omega') = (s,\omega)$ and to have a singularity at $(s',\omega') = (s,\omega)$. Theorem 1 states that all the information about q is contained in the behavior of $K_q^{\#}$ near the singularity.

Next we turn our attention to the inverse back-scattering problem for (1), i.e. the problem of recovering $q(t,x)$ from knowledge of $K_q^{\#}$ for $\omega'=-\omega$. This is a still unsolved problem even for stationary q. Recently Eskin and Ralston [8] proved that the map $q(x) \longrightarrow a(k,-\omega,\omega)$ is a local homeomorphism at least for an open dense set of potentials q. Here $a(k,\omega',\omega)$ is the scattering amplitude related to $q(x)$. However the (global) uniqueness problem is still open. In Theorem 2 we give a partial answer to the inverse back-scattering problem.

Let us assume $n=3$ and the following condition

(A1') $q \in C(\mathbb{R}_t;W^{1,\infty}(\mathbb{R}_x^3))$ satisfied instead of (A1). Then we have the following.

Theorem 2. *Let q_i, $i=1,2$ satisfy (A1'), (A2) with $n=3$. Let $q_1(t,x) \geq q_2(t,x)$ for $t_1 \leq t \leq t_2$, all x, where $t_2-t_1 > 4\varrho$. Suppose there exist $\omega_0 \in S^2$, $d > 2\varrho$, such that $K_{q_1}^{\#}(s',-\omega_0;s,\omega_0) = K_{q_2}^{\#}(s',-\omega_0;s,\omega_0)$ for $|s'-s| \leq d$, $t_1+\varrho < -s < t_2+\varrho$. Then $q_1(t,x) = q_2(t,x)$ for $t_1+2\varrho \leq t \leq t_2-2\varrho$ and for all x.*

If q does not depend on t, then $K_q^{\#}$ depends merely on $s'-s$ and if in addition $q(x)$ has no bound states, then the scattering amplitude $a(k,\omega',\omega)$ related to q coincides with the Fourier transform of $K_q^{\#}(s',\omega';0,\omega)$ with respect to s' up to some multiplication factors. This leads to the following.

Corollary 3. *Let $q_i(x) \in L^{\infty}(\mathbb{R}^3)$, q_i have compact support and q_i have no bound states, $i=1,2$. Let a_i be the scattering amplitude related to q_i, $i=1,2$. Suppose $q_1 \geq q_2$ and let for some ω_0 we have $a_1(k,-\omega_0,\omega_0) = a_2(k,-\omega_0,\omega_0)$ for all k. Then $q_1 = q_2$.*

Let $\mu(\omega) = \inf\{x\cdot\omega;\ x \in \operatorname{supp}V\}$ be the support function of $\operatorname{supp}V$. The technique developed for the proof of Theorem 2 enables us to get the following.

Corollary 4. *Let q be a non-negative function in $L^{\infty}(\mathbb{R}^3)$ with compact support and let $a(k,\omega',\omega)$ be the scattering amplitude associated with q. Then $-2\mu(\omega) = \sup_{s'} \operatorname{supp} K_q^{\#}(s',-\omega;0,\omega)$. Therefore, $K_q^{\#}(s',-\omega;0,\omega)$ (respectively $a(k,-\omega,\omega)$) determines uniquely the convex hull of the support of q.*

Next we turn our attention to an inverse boundary value problem for (1). Let D be a bounded domain with smooth boundary ∂D, and let $q(t,x)$ satisfy (A1) and $q=0$ for $x \notin D$. Here the space dimension $n \geq 2$ may be arbitrary. Given $f \in C^{\infty}(\mathbb{R}\times\partial D)$ with $f=0$ for $t \ll 0$, we consider the mixed problem

$$\begin{cases} (\Box+q(t,x))u = 0 & \text{in } \mathbb{R}\times D, \\ u = f & \text{on } \mathbb{R}\times\partial D, \\ u = 0 & \text{for } t \ll 0. \end{cases}$$

We are interested in the following inverse problem:

(IP3) Does the map $\Lambda: f \longrightarrow \partial_N u\big|_{\mathbb{R}\times\partial D}$ determine q uniquely?

Here N is the outer normal to ∂D. A similar problem for time-independent q has been considered in [21]. In this case the results of Nachman, Sylvester and Uhlmann [28,14,15] could also be applied. Although such kind of problems are not scattering problems, there is a close relation between them and the scattering theory. Using the tools developed for the proof of Theorem 1, we can easily prove the following.

Theorem 5. *The map* Λ *determines* q *uniquely.*

Now consider (IP2). As mentioned above it is well-known that for stationary bodies the scattering operator S always exists and the answer to (IP2) is affirmative [12]. In the case of moving obstacles it is known that $K_Q^{\#}$ determines uniquely the convex hull of $\mathcal{O}(t)$ for each t [6]. The proof of this result is based on the analysis of the leading singularity of $K_Q^{\#}$ (see also [2,18,19]) This solves (IP2) for convex moving bodies. Nevertheless, nothing is known about (IP2) for moving obstacles of arbitrary geometry. It looks a little surprising that the answer to (IP2) is negative in general. This will be proved in Theorem 7 below. For this reason in order to investigate (IP2) we need to impose some additional restrictions.

Definition. *We say that* $Q \in \mathcal{Q}$ *if* Q *satisfies* (B1), (B2) *and the following condition*

(B3) *There exists some* $T > 0$ *such that* $\nu_t = 0$ *for* $|t| > T$.

In other words the class $\mathcal{Q}$ consists of all obstacles which are stationary in the far past and in the far future. Let $U_0(t)$ be the unitary group related to the Cauchy problem for the wave equation [12]. It is not hard to prove that for $Q \in \mathcal{Q}$ the scattering operator

$$S_Q = \operatorname*{s-lim}_{t\to\infty} U_0(-t)U_Q(t,-t)\chi U_0(-t)$$

exists. Here χ is a multiplication of a smooth function $\chi(x)=1$ for $|x|>3\varrho$, $\chi(x)=0$ for $|x|<2\varrho$ and we assume $\mathcal{H}(t) \subset \mathcal{H}_0$. In the next theorem we show that (IP2) has an affirmative answer if $Q \in \mathcal{Q}$.

Theorem 6. *Let $Q_i \in \mathcal{Q}$, $i=1,2$ and let the scattering operators S_{Q_i}, $i=1,2$, associated with Q_i, coincide. Then $Q_1=Q_2$.*

In contrast to Theorem 1 we note that the proof of Theorem 5 is not based on the analysis of the singularities of $K_Q^{\#}$ but on a specific Holmgren's type theorem (see section 3).

Finally we show that in general the uniqueness of (IP2) fails even for periodically moving obstacles. For this reason we construct in sect. 2 an explicit counter-example.

Theorem 7. *There exists a family of infinitely many distinct periodically moving obstacles satisfying* (B1), (B2) *with the same generalized scattering (echo) kernel.*

2. Outline of the proofs. IP1

First we are going to sketch the proof of Theorem 1. Let $u = \partial_s^2\Gamma_q$ be the solution of the Cauchy problem

$$\begin{cases} (\Box + q(t,x))u = 0, \\ u\Big|_{t\ll 0} = \delta(t+s-x\cdot\omega). \end{cases} \tag{3}$$

First we derive the following representation of $K_q^{\#}$:

$$K_q^{\#}(s',\omega';s,\omega) = -2^{-1}(2\pi)^{1-n}\,\partial_{s'}^{(n-1)/2}\partial_s^{(n-3)/2}\iint q(t,x) \times u(t,x;s,\omega)\delta(t+s'-x\cdot\omega')\,dt\,dx, \tag{4}$$

where the integral is to be considered in distribution sense. Secondly, we show that $u(t,x;s,\omega)$ vanishes for $t+s<x\cdot\omega$, u is smooth for $t+s>x\cdot\omega$ and near the plane $t+s=x\cdot\omega$ we have the following singular expansion

$$u(t,x;s,\omega) \sim \delta(t+s-x\cdot\omega) + \sum_{j=0}^{\infty} A_j(t,x,\omega)h_j(t+s-x\cdot\omega), \tag{5}$$

where $A_0(t,x,\omega) = -\frac{1}{2}\int_{-\infty}^{0} q(t+\tau,x+\tau\omega)\,d\tau$, $A_j \in C^{\infty}$. Therefore $K_q^{\#}(s',\omega';s,\omega)$ is C^{∞} off the diagonal. Moreover, all the information about q is contained in the behavior of $K_q^{\#}$ near the singularity $(s',\omega') = (s,\omega)$. More precisely, choosing a unit vector α orthogonal to ω and setting

$$\omega'(\mu) = \omega.\cos\mu + \alpha.\sin\mu,$$

$$s'(\mu) = s + \alpha.\sin\mu, \qquad (\alpha,\mu)\in\mathbb{R}^2,$$

we get

$$\lim_{\mu\to 0} |\omega'(\mu)-\omega| M(s'(\mu),\omega'(\mu);s,\omega) = \int_{x\cdot\alpha=\alpha} q(x\cdot\omega-s,x)\,dS_x,$$

where M is the integral in (4). Now, suppose we know $K_q^{\#}$ for $|\omega-\omega_0|<\varepsilon$, $|\omega'-\omega_0|<\varepsilon$, $|s'-s|<\varepsilon$. Then we can find the integral above for all s, α, $\alpha \perp \omega$, $|\omega-\omega_0|<\varepsilon$. It is easy to deduce that this enables us to determine the integrals

$$\int_{-\infty}^{\infty} q(t+\beta,x+\beta\omega)\,d\beta \tag{6}$$

for all t, x, and $\omega\in S^{n-1}$ such that $|\omega-\omega_0|<\varepsilon$. So we arrive at the following problem of integral geometry: prove that if (6) vanishes for all t, x, $|\omega-\omega_0|<\varepsilon$, then $q = 0$. Note that (6) is the "X"-ray transform of q, but the integral is taken only over characteristic (for $\Box$) lines. In case q has compact support in t the solution to the above problem is known [10] and it is based on the analyticity of the Fourier transform $\hat{q}(\tau,\xi)$ of q. In the general case the proof is more complicated and uses the fact that $\hat{q}(\tau,\xi)$ is an analytic function of ξ with values in $\mathcal{S}'(\mathbb{R}_\tau)$.

Next, we are going to sketch the proof of Theorem 2. We shall make some changes in the notations. Denote Γ_q by Γ_q^{+} and let Γ_q^{-} be the solution of (1) with initial data $h_1(-t-s+x\cdot\omega)$ for $t\gg 0$. In other words, we have

$$\begin{cases} (\Box + q(t,x))\Gamma_q^{+} = 0, \\ \Gamma_q^{+}\big|_{t\ll 0} = h_1(t+s-x\cdot\omega), \end{cases} \qquad \begin{cases} (\Box + q(t,x))\Gamma_q^{-} = 0, \\ \Gamma_q^{-}\big|_{t\gg 0} = h_1(-t-s+x\cdot\omega). \end{cases}$$

Let q_1 and q_2 be two potentials satisfying the regularity assumptions of Theorem 2. Since q_i are not necessarily smooth, $K_i^{\#}$ are distributions in s',s depending continuously on ω',ω. We show that the following formula takes place:

$$K^{\#}_{q_1}(s',\omega';s,\omega) - K^{\#}_{q_2}(s',\omega';s,\omega) =$$

$$-\frac{1}{8\pi^2}\partial^3_{s'}\partial^2_s\iint\left(q_1(t,x)-q_2(t,x)\right)\Gamma^+_{q_1}(t,x;s,\omega)\Gamma^-_{q_2}(t,x;s',\omega')dtdx. \tag{7}$$

Since we have $u = \partial^2_s\Gamma^+_q$ in (4) and $\partial^2_{s'}\Gamma^-_{q_2} = \delta(t+s'-x\cdot\omega')$ provided $q_2=0$, the above relation can be considered as a generalization of (4). According to the finite speed of propagation we have $\mathrm{supp}\Gamma^+_{q_1} \subset \{(t,x);\ t+s\geq x\cdot\omega\}$, $\mathrm{supp}\Gamma^-_{q_2} \subset \{(t,x);\ t+s'\leq x\cdot\omega'\}$. Moreover, if t,x,s,s' run over compact sets, we can find a constant $C > 0$ such that

$$|\Gamma^+_{q_1}(t,x;s,\omega) - h_1(t+s-x\cdot\omega)| \leq Ch_2(t+s-x\cdot\omega),$$

$$|\Gamma^-_{q_2}(t,x;s',\omega') - h_1(-t-s'+x\cdot\omega')| \leq Ch_2(-t-s'+x\cdot\omega').$$

As a corollary there exists $\delta > 0$ such that $\Gamma^+_{q_1} > 0$ and $\Gamma^-_{q_2} > 0$ for $0 < t+s-x\cdot\omega < \delta$ and $-\delta < t+s'-x\cdot\omega' < 0$, respectively. Now set

$$A_\mu = \left\{(t,x);\ x\cdot\omega_0+t_1+\varrho \leq t \leq -x\cdot\omega_0+t_2-\varrho,\ |x|\leq\varrho,\ x\cdot\omega_0 \leq \mu\right\}.$$

Let $\mu_0 = \sup\{\mu;\ q(t,x)=0$ for $(t,x) \in A_\mu\}$, where $q = q_1-q_2$. If Theorem 2 fails, then $\mu_0<\varrho$. Put $s' = s-2\mu_0-\varepsilon$, where $0 < \varepsilon < \varrho-\mu_0$. Assume in what follows that $t_1+\varrho < -s < t_2-\varrho-2\mu_0-\varepsilon$, $(t,x) \in A_\varrho$. Then t,x,s,s' belong to compact sets and let $\delta>0$ be the corresponding constant. We may assume that $\varepsilon < \delta$. Then by (7) we get

$$\int_{\Omega(s)} q(t,x)\Gamma^+_{q_1}(t,x;s,\omega_0)\Gamma^-_{q_2}(t,x;s-2\mu_0-\varepsilon,-\omega_0)dtdx = 0. \tag{8}$$

Here the integration is taken over the set

$$\Omega(s) = \left\{(t,x);\ \mu_0 < x\cdot\omega_0 < \min(t+s,-t-s+2\mu_0+\varepsilon),\ |x|\leq\varrho\right\}.$$

Since $\varepsilon<\delta$, we get $\Gamma^+_{q_1}>0$, $\Gamma^-_{q_2}>0$ in $\Omega(s)$. Further, by the assumptions of Theorem 2, $q\geq0$ thus each integrand in (8) is nonnegative and moreover $\Gamma^+_{q_1}$ and $\Gamma^-_{q_2}$ are positive in $\Omega(s)$. Therefore $q(t,x) = 0$ for $(t,x) \in \Omega(s)$. Letting s run over the interval $t_1+\varrho < -s < t_2-\varrho-2\mu_0-\varepsilon$, we get $\overline{\bigcup_s\Omega(s)} \cup A_{\mu_0} = A_{\mu_0+\varepsilon}$ hence $q = 0$ on $A_{\mu_0+\varepsilon}$ which contradicts the choice of μ_0.

Corollary 3 follows immediately from Theorem 2. Indeed, if q is stationary, then we have $\Gamma_q^{\pm}(t,x;s,\omega) = \Gamma_q^{\pm}(t+s,x;0,\omega)$ and $K_q^{\#}(s',\omega';s,\omega) = K_q^{\#}(s'-s,\omega';0,\omega)$. Moreover, if $-\Delta+q$ has no bound states, then

$$-\frac{k}{2\pi i}a(k,\omega',\omega) = \int e^{-iks'}K_q^{\#}(s',\omega';0,\omega)\,ds',$$

where a is the scattering amplitude related to q. Therefore, the knowledge of $a(k,-\omega,\omega)$ enables us to find $K_q^{\#}(s',-\omega;0,\omega)$ and to apply Theorem 2.

Remark. Let q be stationary, $-\Delta+q$ has no bound states. Then, after performing the indicated differentiations in (7), setting $s=0$ and making Fourier transform with respect to s', we get the following formula

$$\begin{aligned} &a_1(k,\omega',\omega) - a_2(k,\omega',\omega) \\ &= -\frac{1}{4\pi}\int (q_1(x)-q_2(x))\,\psi_1(k,x,\omega)\,\psi_2(k,x,-\omega')\,dx, \end{aligned} \tag{9}$$

where a_j are the scattering amplitudes related to q_j and ψ_j are the solutions of the problems

$$\begin{cases} (-\Delta-k^2+q_j(x))\psi_j = 0, \\ \psi_j = \exp(ik\omega\cdot x) + O(|x|^{-1}) \quad \text{as } |x|\to\infty, \end{cases}$$

$j=1,2$. If $q_2=0$, then (9) reduces to the well-known representation of the scattering amplitude. We hope that (9) might be useful for solving other inverse problems in the potential scattering.

Let us sketch the proof of Corollary 4. Let q satisfy the assumptions of the corollary and denote by $v(t,x,\omega)$ the solution of the problem

$$(\Box+q)v = 0, \qquad v = h_0(t-x\cdot\omega) \text{ for } t<-\varrho.$$

Recall that h_0 is the Heaviside function. We have the following obvious relations between v, Γ_q^+ and u (see (3)): $u(t,x;s,\omega) = \partial_s^2\Gamma_q^+(t,x;s,\omega) = \partial_s v(t+s,x,\omega)$. From (4) we deduce

$$K_q^{\#}(s',\omega;0,\omega) = -\frac{1}{8\pi^2}\partial_{s'}^2\int q(x)\,v(-s'-x\cdot\omega,x,\omega)\,dx.$$

Taking into account that $v(t,s,\omega) = h_0(t-x\cdot\omega) + v_{sc}(t,s,\omega)$, where $|v_{sc}| \le Ch_1(t-x\cdot\omega)$, we prove the following

Proposition 8. *Let* $q \in L^{\infty}(\mathbb{R}^3)$, *q has compact support. Let* μ *be such that* $q(x) = 0$ *for* $x\cdot\omega \le \mu$. *Then* $K_q^{\#}(s',-\omega;0,\omega) = 0$ *for* $s' > -2\mu$ *and if q is non-negative near the plane* $x\cdot\omega = \mu$ *we have*

$$K_q^{\#}(-2\mu-\varepsilon,-\omega;0,\omega) = -\frac{1}{8\pi^2}\partial_\varepsilon^2\Big[\int_{\mu\le x\cdot\omega\le\mu+\varepsilon/2} q(x)\,dx\,(1+O(\varepsilon))\Big] \quad \text{as } \varepsilon \to +0.$$

This proposition shows that if $\mu = \mu(\omega) = \inf\{x\cdot\omega;\ x \in \operatorname{supp} V\}$, then $K_q^{\#}(s',-\omega;0,\omega) \ne 0$ for $s' = -2\mu-\varepsilon$, $\varepsilon>0$ sufficiently small. This proves Corollary 4.

Finally, we give the proof of Theorem 5. Let q_1, q_2 be such that $\Lambda_1=\Lambda_2$. Given $\varphi \in C^{\infty}(\mathbb{R}\times S^{n-1})$, let $\Phi_i = \iint u_i(t,x;s,\omega)\varphi(s,\omega)\,ds\,d\omega \in C^{\infty}$, where u_i is the solution of (3) with $q=q_i$, $i=1,2$. The functions Φ_i solve (1) with $q=q_i$. Denote by w the solution of the problem

$$\begin{cases} (\Box+q_1)w = 0 & \text{in } \mathbb{R}\times D,\\ w = \Phi_2 & \text{on } \mathbb{R}\times\partial D,\\ w = 0 & \text{for } t\ll 0,\ x\in D. \end{cases}$$

Set $\tilde w = w$ for $x\in D$, $\tilde w = \Phi_2$ for $x \notin D$. We have $w = \Phi_2$ on ∂D and $\Lambda_1=\Lambda_2$ implies $\partial_N w = \partial_N\Phi_2$ on ∂D. Therefore, $\tilde w$ solves $(\Box+q_1)\tilde w = 0$ in $\mathbb{R}_t\times\mathbb{R}_x^n$. Since $\tilde w=\Phi_1$ for $t\ll 0$, we conclude that $\tilde w \equiv \Phi_1$ thus $\Phi_1=\Phi_2$ for $x\notin D$, all t. Since φ was arbitrary, we get $(u_1-u_2)(t,x;s,\omega) = 0$ for all t,s,ω and $x\notin D$. Combining this with (5), we get that the integrals (6) for q_1 and q_2 are equal for all t,x,ω, which leads to $q_1=q_2$. Notice that the equality $u_1=u_2$ for $x\notin D$ implies $K_1^{\#} = K_2^{\#}$, which yields another proof of Theorem 5. We note that the requirement $q \in C^{\infty}$ can be relaxed to $q \in C^k$ for some finite k.

Remark. Prof. A.G.Ramm informed the author that he and Prof. J.Sjöstrand have a proof of Theorem 5 different from ours which also leads to the inversion of the light-ray transform (6).

3. Outline of the proofs. IP2

Let us sketch the proof of Theorem 6. Let Q_1 and Q_2 be two distinct domains in $\mathcal{Q}$ whose scattering operators S_{Q_i}, $i=1,2$ coincide. Fix some $s \in \mathbb{R}$ and fix $\varphi \in [C_0^{\infty}(\mathbb{R}^n) \times C_0^{\infty}(\mathbb{R}^n)] \cap D_-^{\varrho}$, D_-^{ϱ} being the incoming space of Lax and Phillips [12]. We denote by u_i the solution of (2) related to Q_i having initial data φ for $t=s$. First we prove that $S_{Q_1} = S_{Q_2}$ implies that $u = u_1-u_2$ vanishes for

$|x|>\varrho$ and all $t \in \mathbb{R}$. In the stationary case it is not hard to show that, in fact, u vanishes in the unbounded connected component of $\Omega_1 \cap \Omega_2$ which easily yields $\Omega_1 = \Omega_2$ [12]. Indeed, in this case the function $v(k,x) = \int \exp(-itk)u(t,x)dt$ is a solution to the Helmholtz equation, consequently v is a real-analytic function with respect to x. Clearly, in the case of moving obstacles this argument is no more available. Nevertheless, assuming for definiteness that $\partial Q_2 \cap Q_1 \neq \emptyset$, we aim to prove that there exists $(t_0,x_0) \in \partial Q_2 \cap Q_1$ such that $u(t_0,\cdot)$ vanishes in a neighborhood of x_0 in $\partial\Omega_2(t_0)$. This fact makes possible to obtain a contradiction.

To this end we apply two new ideas. First, we construct a two-parameter family $F_{t,s} : \Omega(s) \longrightarrow \Omega(t)$ of diffeomorphisms related to a domain Q satisfying (B1), (B2). We choose $F_{t,s}$ to be the flow associated with a (time-dependent) vector field $v \in C^{\infty}(\overline{Q};\mathbb{R}^n)$ which is chosen so that $|v|<1$, $v(t,x)=0$ for $|x|>\varrho$ and $(1,v)$ is tangent to ∂Q. In some sense $v(t,x)$ can be considered as the velocity of $x \in \Omega(t)$ at the moment t. Thus we treat all points in the exterior of the obstacle as moving ones. Note that if we introduce the coordinates $(t,y) = (t,F_{t_0,t}(x))$, t_0-fixed, then Q becomes cylindric.

Secondly we formulate and prove a specific Holmgren's type theorem. In order to state it we need some preparation. Using $F_{t,s}$ we can easily show that $\Omega(t)$ is connected for each t. Fix t_0 and $x_0 \in \Omega(t_0)$ and let

$$\gamma = \Big\{ x;\ x = x(\sigma),\ 0 \leq \sigma \leq l \Big\} \tag{10}$$

be a smooth curve in $\Omega(t_0)$ which is not self-intersecting, such that $x(0)=x_0$, $x(l)=x_1$ for some $x_1 \in \Omega(t_0)$ and $|x'(\sigma)|=1$. Consider the two-dimensional surface $\Gamma = \bigcup_{t\in\mathbb{R}} \big(t,F_{t,t_0}(\gamma)\big) \subset Q$. Obviously, σ and t are coordinates on Γ. We define vector fields $A^{\pm} \in T\Gamma$ in a following way. Set $A^{\pm} = \frac{\partial}{\partial\sigma} \pm a^{\pm}(t,\sigma)\frac{\partial}{\partial t}$, where $a^{\pm}>0$ is chosen so that for each (t,σ) the vector $A^{\pm}$ considered as a vector in $\mathbb{R}^{n+1}$, is characteristic, i.e. $|A^{\pm}_t|=|A^{\pm}_x|$. Let $\sigma \longrightarrow (t,x) = \big(t_{\pm}(\sigma), F_{t_{\pm}(\sigma),t_0}(x(\sigma))\big)$ be the integral curve of $A^{\pm}$ such that $t_{\pm}(0)=t_0$. Condition (B3) guarantees that $a^{\pm}$ is uniformly bounded, thus the function $t_{\pm}(\sigma)$ is defined for all $\sigma \in [0,l]$. We define a subset X of Γ by the relation

$$X = \Big\{(t,x);\ x = F_{t,t_0}(x(\sigma)),\ 0\leq\sigma\leq l,\ t_-(\sigma)\leq t\leq t_+(\sigma)\Big\}. \tag{11}$$

X depends on t_0, x_0, x_1 and on the choice of γ joining x_0 and x_1.

We have the following Holmgren's type theorem.

Theorem 9. *Let u be a distribution satisfying the wave equation $\Box u = 0$ in an open subset of $\mathbb{R}^{n+1}$ containing X. Suppose that u vanishes in a neighborhood of the curve $\{(t,x);\ t_-(l) \le t \le t_+(l),\ x = F_{t,t_0}(x(l))\}$. Then* $\operatorname{supp} u \cap X = \emptyset$.

The proof of Theorem 9 is based on the local Holmgren theorem (see e.g. [11], Theorem 8.6.5). In the partial case when $\mathcal{O}(t)$ is stationary and $F_{t,s} = I$, Theorem 9 follows easily from Theorem 1.4 of Ch. IV in [12].

Now, let $Q_i \in \mathcal{Q}$ and $S_{Q_1} = S_{Q_2}$. Let $u = u_1-u_2$ be the function defined above. As mentioned above, $u = 0$ for $|x|>\varrho$ and all $t \in \mathbb{R}$. Suppose $Q_1 \ne Q_2$. Then we can assume that $Q_1 \cap \partial Q_2 \ne \emptyset$. Let $(t_0,x_0) \in Q_1 \cap \partial Q_2$ and choose some $x_1 \notin B_\varrho$ and a curve $\gamma \subset \Omega_1(t_0)$ of the kind (10), joining x_0 and x_1. Denote by $X \subset Q_1$ the set (11) associated with t_0, x_0, x_1 and the flow $F^1_{t,s}$ related to Q_1. Without loss of generality we can assume that $X \subset \overline{Q}_2$, $X \cap \partial Q_2 = (t_0,x_0)$. Given $\varepsilon > 0$ denote by $X_\varepsilon \subset X$ the set (11) related to t_0, $x(\varepsilon)$, x_1. Since $u=0$ in a neighborhood of $\{(t,x);\ t_-(l) \le t \le t_+(l),\ x=x_1\}$, and $\Box u=0$ in $Q_1 \cap Q_2 \supset X_\varepsilon$, then u vanishes at the point $(t_0,x(\varepsilon))$, in virtue of Theorem 9. Letting $\varepsilon \to 0$, we find $u(t_0,x_0) = 0$. Moreover, we can prove that $u(t_0,x) = 0$ for $x \in V \cap \partial\Omega_2(t_0)$, $V \subset \Omega_1(t_0)$ being a small neighborhood of x_0. Recall that $u_2=0$ on ∂Q_2. Consequently, $u_1(t_0,x) = [U_{Q_1}(t_0,s)\varphi]_1$ vanishes on $V \cap \partial\Omega_2(t_0)$ for all $s \in \mathbb{R}$, $\varphi \in D^\varrho_- \cap C_0^\infty \times C_0^\infty$. Since such solutions are dense in $\mathcal{H}_1(t_0)$, we conclude that the first component of every $f \in \mathcal{H}_1(t_0)$ vanishes on the surface $V \cap \partial\Omega_2(t_0)$. Thus we obtain a contradiction, which completes the proof of Theorem 6.

Finally, we shall construct explicitly a family of periodically moving obstacles with the same generalized scattering kernel. For the sake of simplicity we assume $n=3$.

It is not hard to see that there exists a function $\psi \in C_0^\infty(\mathbb{R})$ with the properties: $\psi(\sigma) = 1$ for $|\sigma|\le 2$, $\psi(\sigma) = 0$ for $|\sigma|\ge 11/4$, $|\psi|\le 1$, $|\psi'|\le 3/2$. Next set

$$\phi(t,\sigma) = \psi(\sigma)\sin(k(\sigma-t/2)).$$

Here $k \ge 2\pi$ is a large parameter. Consider the hypersurface $\Sigma \subset \mathbb{R}^4$ and the domain $Q_0 \subset \mathbb{R}^4$ given by

$$\Sigma = \left\{(t,x) \in \mathbb{R}^4;\ (x_2-\phi(t,x_1))^2 + x_3^2 = r^2,\ t \in \mathbb{R},\ |x_1| < 3\right\},$$

$$Q_0 = \left\{(t,x) \in \mathbb{R}^4;\ (x_2-\phi(t,x_1))^2 + x_3^2 < r^2,\ t \in \mathbb{R},\ |x_1| < 3\right\}.$$

Here $r \in (0,1/2)$ is a small parameter. Clearly, Σ is a smooth manifold and $\Sigma \subset \partial Q_0$. We set $\Omega_0(t) = \{x;\ (t,x) \in Q_0\}$. The domain $\Omega_0(t)$ can be considered as a small neighborhood of the curve

$$\gamma_t = \left\{x;\ x_2 = \phi(t,x_1),\ x_3 = 0,\ |x_1| \le 3\right\}.$$

It is not hard to see that Σ is time-like. Let us define a velocity field v associated with Σ. Further we denote by K_α the cube $K_\alpha = \{x \in \mathbb{R}^3;\ |x_i| \le \alpha,\ i = 1,2,3\}$. Pick a function $\chi \in C_0^\infty(\mathbb{R}^3)$, such that $|\chi| \le 1$, $\chi(x) = 1$ for $x \in K_{11/4}$ and $\chi(x) = 0$ for $x \in K_3$. Given $(t,x) \in \mathbb{R}^4$ we set

$$v(t,x) = \tfrac{1}{2}\chi(x)\left[\ 1\ ,\ \psi'(x_1)\sin(k(x_1-t/2))\ ,\ 0\ \right].$$

Clearly, $v(t,x) = (1/2,0,0)$ for $x \in K_2$ while $v = 0$ for $x \notin K_3$. The properties of ψ imply $|v|<1$. A straightforward calculation shows that $(1,v)$ is tangent to Σ. Thus $v(t,x)$ generates a flow $F_{t,s}:\mathbb{R}^3 \longrightarrow \mathbb{R}^3$ that maps $\Omega_0(s)$ onto $\Omega_0(t)$. Note that $F_{t,s} = Id$ on $\mathbb{R}^3\setminus K_3$.

Let $Q \subset \mathbb{R}^4$ be a domain with smooth boundary such that $\Omega(t) = \{x;\ (t,x) \in Q\}$ satisfies the conditions

C1) $\Omega(t) \cap K_3 = \Omega_0(t)$;

C2) $\Omega(t) \cap (\mathbb{R}^3\setminus K_3)$ is stationary (time-independent);

C3) if $K' = \{\ x;\ -4 \le x_1 \le -3,\ |x_2| \le 1,\ |x_3| \le 1\ \}$, then $\partial\Omega(t) \cap \partial K' = \partial\Omega_0(t) \cap \partial K' = \{\ x;\ x_1 = -3,\ x_2^2 + x_3^2 = r^2\}$;

C4) $\Omega(t) \subset K_5$;

(see Figure 1). Clearly, Q satisfies (B1), (B2) with $\varrho > 5\sqrt{3} = \frac{1}{2}\mathrm{diam}K_5$. Moreover, $\Omega(t+4\pi/k) = \Omega(t)$ thus the motion is periodic with

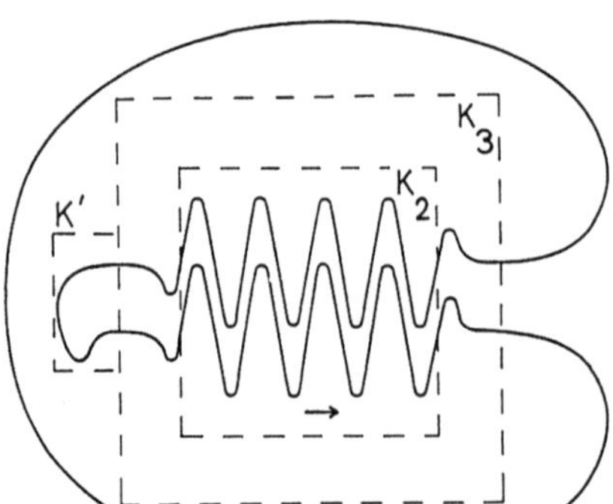

Fig. 1

period $4\pi/k$. Below we denote $\Omega' = \Omega(t) \cap K'$, which does not depend on t by virtue of (C2). We have the following property.

Proposition 10. *There is no piecewise smooth curve* $\{x = x(t),\ a \le t \le b\}$ *such that* $x(a) \notin B_\varrho$, $x(t) \in \overline{\Omega(t)}$ *for all* $t \in [a,b]$, $x(b) \in \Omega'$ *and* $|x'| \le 1$.

In other words, if we travel in the exterior of the obstacle with a speed not greater than 1, starting from $\mathbb{R}^3 \setminus B_\varrho$, we shall never reach Ω'. Using the principle of causality we get the following.

Proposition 11. *For any* $b \ge a$ *and for any* $f \in \mathscr{H}(a)$ *with* $\operatorname{supp} f \cap B_\varrho = \emptyset$ *we have* $U_Q(b,a)f = 0$ *in* Ω'.

Now consider a family $\mathscr{F}$ of domains Q such that each $Q \in \mathscr{F}$ satisfies (C1)-(C4) and $\Omega_1(t) \cap (\mathbb{R}^3 \setminus K') = \Omega_2(t) \cap (\mathbb{R}^3 \setminus K')$ for any $Q_1 \in \mathscr{F}$, $Q_2 \in \mathscr{F}$. Note that we do not impose any restriction on $\Omega(t) \cap K'$, thus the geometry of $\partial\Omega(t) \cap K'$ may be arbitrary, provided that (C3) holds. Hence $\mathscr{F}$ consists of infinitely many distinct domains Q and the corresponding obstacles $\mathcal{O}(t)$ move with the same period $4\pi/k$. Using Proposition 11 we prove that $K_Q^{\#}$ does not "feel" the shape of $\partial\Omega(t) \cap K'$ therefore all obstacles in $\mathscr{F}$ have the same generalized scattering kernel.

References

[1] A.Bachelot, V.Petkov, *Existence des opérateurs d'ondes pour les systèmes hyperboliques avec un potentiel périodique en temps*, Ann.Inst.Henry Poincaré.Phys.Theor. **47**(1987), 383-428.

[2] F.Cardoso and V.Petkov, *Leading singularity of the scattering kernel for moving obstacles with dissipative boundary condition*, preprint, Univ. de Pernambuco, Recife, Brazil, 1988.

[3] J.Cooper, *Scattering of plane waves by a moving obstacle*, Arch. Rat. Mech. Anal. **71**(1979), 113-149.

[4] J.Cooper and W.Strauss, *Representation of the scattering operator for moving obstacles*, Indiana Univ. Math. J. **28**(1979), 643-671.

[5] J.Cooper and W.Strauss, *Scattering of waves by periodically moving bodies*, J. Funct. Anal. **47**(1982), 180-229.

[6] J.Cooper and W.Strauss, *The leading singularity of a wave reflected by a moving boundary*, J. Differential Equations **52**(1984), 175-203.

[7] J.Cooper and W.Strauss, *Time-periodic scattering of symmetric hyperbolic systems*, J.Math.Anal.Appl. **122**(1987),444-452.

[8] G.Eskin, J.Ralston, *The inverse back-scattering problem in three dimensions*, preprint UCLA.

[9] J.A.Ferreira, G.Perla Menzala, *Time dependent approach to the inverse scattering problem for wave equations with time dependent coefficients*, Appl.Anal. **26**(1988), 223-254.

[10] S.Helgason, *The Radon Transform*, Birkhäuser, 1980.

[11] L. Hörmander, *The Analysis of Linear Partial Differential Operators*, vol.1, Springer-Verlag, New York Heidelberg Berlin, 1983.

[12] P. Lax, R. Phillips, *Scattering Theory*, Academic Press, New York, 1967.

[13] P. Lax, R. Phillips, *The acoustic equation with an indefinite energy form and the Schrödinger equation*, J. Funct. Anal. **1**(1967), 37-83.

[14] A. Nachman, A. Sylvester, J. Uhlmann, *An n-dimensional Borg-Levinson theorem*, Commun. Math. Phys. **115**(1988), 595-605.

[15] A. Nachman, *Reconstruction from boundary measurements*, Ann. Math. **128**(1988), 531-576.

[16] G. Perla Menzala, *Sur l'opérateur de diffusion pour l'équation des ondes avec des potentiels dependant du temps*, C. R. Acad. Sci. Paris **300**(1985), 621-624.

[17] G. Perla Menzala, *Scattering properties of wave equations with time dependent potentials*, Comput. Math. Appl. **12A**(1986), 457-475.

[18] V. Petkov, *Scattering Theory for Hyperbolic Operators*, North Holland, in press.

[19] V. Petkov and Tz. Rangelov, *Leading singularity of the scattering kernel for moving obstacles*, C. R. Acad. Bulg. Sci. **40**, No. 12(1987), 5-8.

[20] G. Popov and Tz. Rangelov, *On the exponential growth of the local energy for periodically moving obstacles*, C. R. Acad. Bulg. Sci. **40**, No. 8(1987), 21-23.

[21] Rakesh, W. Symes, *Uniqueness for an inverse problem for the wave equation*, Commun. PDE, **13**(1988), 87-96.

[22] Y. Saitō, *Some properties of the scattering amplitude and the inverse scattering problem*, Osaka J. Math. **19**(1982), 527-547.

[23] P. Stefanov, *Uniqueness of the inverse scattering problem for the wave equation with a potential depending on time*, Inverse Problems **4**(1988), 913-920.

[24] P. Stefanov, *Uniqueness of the multi-dimensional inverse scattering problem for time-dependent potentials*, Math. Z. **201** (1989), 541-559.

[25] P. Stefanov, *Inverse scattering problem for moving obstacles*, preprint, April, 1989.

[26] P. Stefanov, *On the inverse back-scattering problem*, preprint, November, 1989.

[27] P. Stefanov, *On an inverse problem for the wave equation*, preprint, June, 1989.

[28] J. Sylvester, G. Uhlmann, *A global uniqueness theorem for an inverse boundary value problem*, Ann. Math. **125**(1987), 541-560.

[29] H. Tamura, *On the decay of the local energy for wave equations with a moving obstacle*, Nagoya Math. J. **71**(1978), 125-147.

[30] H. Tamura, *On the decay of local energy for wave equations with time dependent potentials*, J. Math. Soc. Japan **33**(1981), 605-618.

Wave Splitting and the Reflection Operator for the Wave Equation

V.H. Weston

Department of Mathematics, Purdue University, West Lafayette, IN 47907, USA

The problem of wave splitting in a non-homogeneous medium (with sufficiently smooth velocity) is $\mathbb{R}^3$ is considered. The wave equation is factorized into an up and down-going wave system using certain integral and integral-differential operators. The equation for the reflection operator (which relates the up-going wave to a down-going wave) is obtained, and certain properties of the reflection operator are deduced, including the ideal set of measurements needed to determine the kernel of the reflection operator. The possible application to inverse problems is considered.

Introduction

Invariant imbedding and wave splitting [**1-8**] techniques have been used successfully to formulate transient inverse problems in one spatial dimension for physical phenomena represented by a class of hyperbolic equations such as the wave, telegraph equations and the equation of viscoelasticity. These techniques have led to robust numerical methods for the inverse problem. The techniques have recently been extended [**9,10**] to planar stratified medium in three dimensions. This in turn has led to developments in the generalization of wave splitting and invariant imbedding in three dimensions for the wave equation where the medium is no longer stratified [**11**]. A review of these results are presented here.

We will consider the wave equation

$$\frac{1}{c^2}u_{tt} - \nabla^2 u = 0, \qquad x \in \mathbb{R}^3 \tag{1}$$

where the velocity c is a smooth function (at least twice differentiable [**12**]). Assume that the scatterer is below the plane $x_3 = 0$, $c = c_0$ (constant) for $x_3 \geq 0$ and the sources that produce the scattered field lie in the half-space $x_3 \geq 0$. The invariant imbedding geometry (see Fig. 1) that we are concerned with here is obtained by introducing a one-parameter family of planes $x_3 = \alpha$, with the velocity c being the actual physical velocity $c = c(x_1, x_2, x_3)$ in the half-space $x_3 \leq \alpha$. In the region $x_3 > \alpha$ the velocity will depend upon transverse variables x_1, x_2 only, and to have continuity of c at $x_3 = \alpha$, we take $c = c(x_1, x_2, \alpha)$ for $x_3 \geq \alpha$. We thus have a

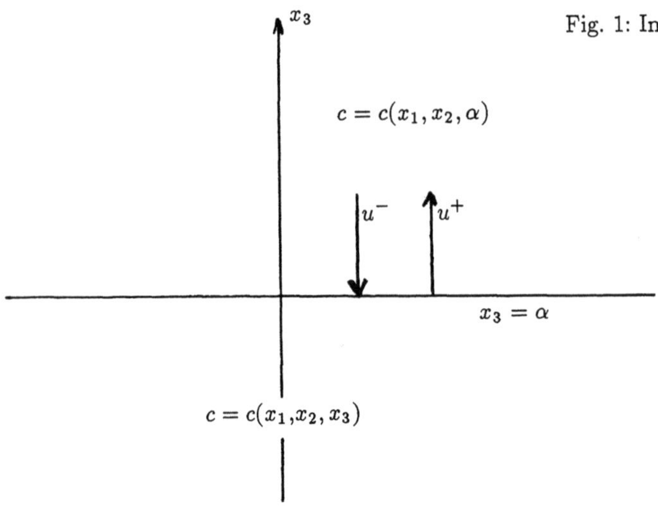

Fig. 1: Invariant Imbedding Geometry

one-parameter family of scattering media. When $\alpha \geq 0$, the invariant imbedding geometry becomes identical to the physical geometry.

To generate a one–parameter family of scattering problems, we introduce a down-going wave u^- in the medium $c = c(x_1, x_2, \alpha)$ incident upon the interface $x_3 = \alpha$. This will produce an up-going wave u^+. The two will be related by a scattering operator $\boldsymbol{R}$, $u^+ = \boldsymbol{R}u^-$ at $x_3 = \alpha$. We thus have a one-parameter family of scattering operators. When $\alpha \geq 0$, we can relate u^- to an actual incident wave and u^+ to the scattered wave.

Wave Splitting

We need to define precisely what is meant by an up-and down-going wave in a medium where $c = c(x_1, x_2, \alpha)$. This is done by considering the mixed (initial-value Neumann boundary value) problem in the half-space $x_3 > x_3^0$, for a finite time interval $0 < t < T$

$$\text{(i)}\ \frac{1}{c^2}u - \nabla^2 u = 0, \qquad 0 < t < T, \qquad x_3 > x_3^0$$

$$\text{(ii)}\ u(x,0) = u_t(x,0) = 0, \qquad x > x_3^0$$

$$\text{(iii)}\ \frac{\partial u}{\partial x_3}(x,t) = v(x_1, x_2, t), \quad x_3 = x_3^0, \quad 0 < t < T$$

where v has compact support in $\mathbf{R}^2$, v, v_t satisfy certain integrability conditions (in reference **[11]**, smoother conditions were placed upon v to get a classical solution), and $c = c(x_1, x_2, \alpha)$.

In reference **[11]** T was restricted so that there were no caustics or intersections of rays that originated from a point, but in reference **[12]** the only restriction is that T be finite. Physically, the initial-value boundary-value problem formulation implies that all sources that generate up-going waves lie below the plane $x_3 = x_3^0$.

Taking the limit as $x_3 \to x_3^0+$ in the solution $u(x,t)$ of the mixed problem, one obtains a condition of the form

$$u + \boldsymbol{K}_\alpha u_n = 0, \; x_3 = x_3^0 \tag{2}$$

which represents a mapping of Neumann to Dirichlet data. The operator $\boldsymbol{K}_\alpha$ (where [**12**] $\boldsymbol{K}_\alpha u_n \in \mathcal{L}_2(\mathsf{R}^2 \times (0,T))$ has the precise form

$$\boldsymbol{K}_\alpha u = c(\alpha) \int_0^t u(x_3, s) ds$$

in one dimension. In three dimensions for the special case where $c = c(\alpha)$, it has the form [**11**]

$$\boldsymbol{K}_\alpha u = \iint_{\mathsf{R}^2} \frac{u(\boldsymbol{y}, t-r)/c)H(t-r/c)d\boldsymbol{y}}{2\pi r}$$

where $H(\eta)$ is the Heaviside step function, $r = |\boldsymbol{x} - \boldsymbol{y}|$, and we use the notation that the transverse components of points $x, y \in \mathsf{R}^3$ are given by

$$\boldsymbol{x} = (x_1, x_2), \qquad \boldsymbol{y} = (y_1, y_2). \tag{3}$$

Condition (2) represents the up-going wave condition at $x_3 = x_3^0$. Since the kernel of the operator $\boldsymbol{K}_\alpha$ does not depend upon the variable x_3, condition (2) gives the up-going wave condition on any plane $x_3 =$ constant in a medium where $c = c(x_1, x_2, \alpha)$.

By considering the mixed problem for the lower half-space $x_3 < x_3^0$, one obtains the down-going wave condition

$$u - \boldsymbol{K}_\alpha u_n = 0, \quad x_3 = \text{ constant} \tag{4}$$

in a similar manner.

Another up-going wave condition can be obtained by investigating the mixed (initial-value Dirichlet boundary-value) problem. This yields the up-going wave condition

$$u_n + \Box^T \boldsymbol{K}_\alpha u = 0 \tag{5}$$

where $\Box^T$ is the transverse d'Alembertian

$$\Box^T = \frac{\partial^2}{c^2 \partial t^2} - \frac{\partial^2}{\partial x_1^2} - \frac{\partial^2}{\partial x_2^2}. \tag{6}$$

The two up-going wave conditions yield the result (for a suitable domain)

$$\boldsymbol{K}_\alpha^{-1} = \boldsymbol{K}_\alpha \Box^T = \Box^T \boldsymbol{K}_\alpha \tag{7}$$

and this leads to the important identity

$$f - \boldsymbol{K}_\alpha^2 \Box^T f = 0, \qquad x_3 = \text{ constant}, \quad 0 < t < T \tag{8}$$

for functions $f(\boldsymbol{x}, t) \in C^2(\mathrm{R}^2 \times (0, T))$ such that $f(\boldsymbol{x}, 0) = f_t(\boldsymbol{x}, 0) = 0$. Identity (8) is generalized in a rigorous manner to apply to a general class of second-order differential hyperbolic operators [**12**]. This is the key identity that allows one to do the wave-splitting.

Let $u(x, t)$ be a solution of the wave equation (1) with velocity $c = c(x_1, x_2, \alpha)$ and satisfy the initial conditions $u(x, 0) = u_t(x, 0)$ in the slab $a < x_3 < b$, then define

$$u^{\pm} = \frac{1}{2}[u \mp \boldsymbol{K}_\alpha u_n]. \tag{9}$$

Identity [**8**] can then be employed to show that u^+ and u^- satisfy the up-going and down-going wave conditions (2) and (4) respectively on any plane $x_3 =$ constant in the slab $a < x_3 < b$. Since

$$u = u^+ + u^- \tag{10}$$

the solution to the wave-equation has been decomposed into the two wave components.

To get the generalization of the "Ricatti" equation for the reflection operator, the wave splitting has to be generalized to the medium where $c = c(x_1, x_2, x_3)$, c smooth. On each plane $x_3 = x_3^0$, define the up and down-going wave components as follows

$$u^{\pm} = \frac{1}{2}(u \mp \boldsymbol{K}_{x_3^0} \frac{\partial u}{\partial x_3}), \qquad x_3 = x_3^0 \tag{11}$$

where the parameter $\alpha = x_3^0$. This means that $\boldsymbol{K}_{\alpha = x_3^0}$ corresponds to the mapping of Neumann to Dirichlet data for the mixed-problem in the half-space $x_3 > x_3^0$ with velocity $c = c(x_1, x_2, x_3^0)$. The decomposition (11) will be expressed in matrix form

$$\begin{bmatrix} u^+ \\ u^- \end{bmatrix} = T \begin{bmatrix} u \\ \frac{\partial u}{\partial x_3^0} \end{bmatrix} = \frac{1}{2} \begin{bmatrix} 1 & -\boldsymbol{K}_\alpha \\ 1 & \boldsymbol{K}_\alpha \end{bmatrix} \begin{bmatrix} u \\ \frac{\partial u}{\partial x_3^0} \end{bmatrix} \tag{12}$$

where $\alpha = x_3^0$.

For simplicity of notation set

$$\boldsymbol{K} = \boldsymbol{K}_{\alpha = x_3^0}. \tag{13}$$

The system of equations for u^+ and u^- is obtained in a similar manner as the one-dimensional case [**3,4**], by employing the identity

$$\frac{\partial u}{\partial x_3^0} = \frac{\partial u}{\partial x_3^0}$$

together with the wave equation written in the form

$$\frac{\partial^2 u}{\partial (x_3^0)^2} = \Box^T u$$

where the velocity $c = c(x_1, x_2, x_3^0)$ in the expression [6] for $\Box^T$.

These equations are placed in matrix form

$$\frac{\partial}{\partial x_3^0}\begin{bmatrix} u \\ \frac{\partial u}{\partial x_3^0} \end{bmatrix} = \begin{bmatrix} 0 & 1 \\ \Box^T & 0 \end{bmatrix}\begin{bmatrix} u \\ \frac{\partial u}{\partial x_3^0} \end{bmatrix}$$

and relationship (12) is employed to express the variable u, $\frac{\partial u}{\partial x_3^0}$ in terms of u^+ and u^-. The resulting system is given by

$$\frac{\partial}{\partial x_3^0}\begin{bmatrix} u^+ \\ u^- \end{bmatrix} = \begin{bmatrix} -\boldsymbol{K}^{-1} & 0 \\ 0 & \boldsymbol{K}^{-1} \end{bmatrix}\begin{bmatrix} u^+ \\ u^- \end{bmatrix} + \frac{1}{2}\begin{bmatrix} 1 & -1 \\ -1 & 1 \end{bmatrix}\boldsymbol{K}'\boldsymbol{K}^{-1}\begin{bmatrix} u^+ \\ u^- \end{bmatrix} \tag{14}$$

where

$$\boldsymbol{K}' = \left(\frac{\partial}{\partial \alpha}\boldsymbol{K}_\alpha\right)_{\alpha = x_3^0}. \tag{15}$$

For the case where the velocity $c = c(x_3)$ (continuously stratified medium) the operator in the last term of Eq. (14) has the explicit form

$$\boldsymbol{K}'\boldsymbol{K}^{-1}u = \frac{1}{c^3}\frac{\partial c}{\partial x_3}\frac{\partial^2}{\partial t^2}\boldsymbol{K}^2 u.$$

Reflection Operator

Returning to the invariant imbedding geometry given in Figure 1 (with the parameter α replaced by x_3^0) we will consider the reflection operator

$$u^+ = \boldsymbol{R}u^- \tag{16}$$

associated with the up-and down-going wave at the interface $x_3 = x_3^0$.

An equation for the reflection operator that yields its form and some of its properties is obtained as follows. Let $\boldsymbol{G}$ be the operator that maps Neumann data to Dirichlet data on the surface $x_3 = x_3^0$

$$u + \boldsymbol{G}\frac{\partial u}{\partial x_3} = 0, \qquad x_3 = x_3^0, \quad 0 < t < T \tag{17}$$

arising from the solution of the mixed problem in the lower half-space $x_3 < x_3^0$

(i) $\frac{1}{c^2}u_{tt} - \nabla^2 u = 0, \qquad x_3 < x_3^0, \quad 0 < t < T$

(ii) $u(x,0) = u_t(x,0) = 0, \qquad x_3 < x_3^0$

(iii) $\dfrac{\partial u}{\partial x_3} = v(x_1, x_2, t), \qquad x_3 = x_3^0, \quad 0 < t < T$

where the velocity $c = c(x_1, x_2, x_3)$. The Neumann data v has compact support in $\mathbf{R}^2$, and satisfies certain smoothness conditions. Using the fact that u, $\dfrac{\partial u}{\partial x_3}$ are continuous at the interface $x_3 = x_3^0$, one can combine Eq. (17) with the wave splitting $u^{\pm} = \dfrac{1}{2}\left(u - \boldsymbol{K}\dfrac{\partial u}{\partial x_3}\right)$ at $x_3 = x_3^0$. Eliminating $\dfrac{\partial u}{\partial x_3}$ from these equations, and replacing u^+ by $\boldsymbol{R}u^-$, one obtains

$$(\boldsymbol{I} + \boldsymbol{A})\boldsymbol{R}u^- = \boldsymbol{A}u^-, \qquad x_3 = x_3^0 \tag{18}$$

where

$$\boldsymbol{A} = \frac{1}{2}(\boldsymbol{G} - \boldsymbol{K})\boldsymbol{K}^{-1}.$$

Equation (18) is not useful from the inverse problem standpoint since it requires knowledge of the fundamental solution associated with the wave equation, hence knowledge of the direct scattering problem. However Eq. (18) is useful to develop certain properties of $\boldsymbol{R}$.

It can be shown under certain smoothness conditions, that the $C^2(\mathbf{R}^2 \times (0,T))$ solution of $(\boldsymbol{G} + \boldsymbol{K})\nu = 0$ such that $\nu(\boldsymbol{x}, 0) = 0$, is the solution $\nu \equiv 0$. From this one can deduce $u^+ = (I + \boldsymbol{A})^{-1}\boldsymbol{A}u^- = \boldsymbol{R}u^-$.

Using the properties of the fundamental solution one can show that one concrete form of $\boldsymbol{R}$ is given by

$$\boldsymbol{R}u^- = \int_0^t \int_{\mathbf{R}^2} R_N(\boldsymbol{x}, \boldsymbol{y}, x_3^0; t - s) u(\boldsymbol{y}, x_3^0, s) d\boldsymbol{y}\, ds$$

where the kernel R_N belongs to the space of distributions. (R_N is related to the kernel R in reference [11] in the following manner, $R_N = \dfrac{\partial^2}{\partial t^2}R$.) It can be shown that R_N has compact support as indicated by the relation

$$R_N(\boldsymbol{x}, \boldsymbol{y}, x_3; t) = 0 \text{ for } |\boldsymbol{x} - \boldsymbol{y}| > c_M t \tag{19}$$

where c_M = maximum value of $c(x_1, x_2, x_3)$, and its initial value is given by

$$R_N(\boldsymbol{x}, \boldsymbol{y}, x_3; 0) = -\frac{1}{4}\frac{\partial c}{\partial x_3}\delta(\boldsymbol{x} - \boldsymbol{y}). \tag{20}$$

For a restricted time T (such that in the interval $0 \le t < T$ rays originating from a point do not intersect, or form a caustic) the support of R_N is contained in the set $\boldsymbol{x}, \boldsymbol{y}, x_3^0, t$ such that

$$\tau(x, y) \le t$$

where τ is the time of arrival of a ray from the point $x = (\boldsymbol{x}, x_3^0)$ to the point $y = (\boldsymbol{y}, x_3^0)$ in a medium where $c = c(x_1, x_2, x_3^0)$, i.e. c depends on the transverse variables only. In this case it can be shown that R_N has a singularity on the support boundary given by

$$R_N(\boldsymbol{x}, \boldsymbol{y}, x_3; t) \sim \frac{\partial}{\partial t}\left\{\frac{H(t-\tau(x,y))}{\sqrt{t-\tau(x,y)}} r_0(x,y)\right\}.$$

Another form of the reflection operator is given by

$$\boldsymbol{R}u^- = \int_0^t \int_{\mathbb{R}^2} R_D(\boldsymbol{x}, \boldsymbol{y}, x_3^0; t-s) v^-(\boldsymbol{y}, x_3^0, s) d\boldsymbol{y}\, ds \tag{21}$$

where

$$v^- = \boldsymbol{K}^{-1} u^-. \tag{22}$$

Whereas R_N is not symmetric in the transverse variables, R_D is symmetric, the symmetry condition given by

$$R_D(\boldsymbol{x}, \boldsymbol{y}, x_3^0; t) = R_D(\boldsymbol{y}, \boldsymbol{x}, x_3^0; t). \tag{23}$$

The initial conditions for this kernel are given by

$$\left.\begin{aligned} R_D(\boldsymbol{x}, \boldsymbol{y}, x_3^0; 0) &= 0 \\ \frac{\partial}{\partial t} R_D(\boldsymbol{x}, \boldsymbol{y}, x_3^0; 0) &= -\frac{1}{8}\frac{\partial c^2}{\partial x_3}\delta(\boldsymbol{x}-\boldsymbol{y}) \end{aligned}\right\} \tag{24}$$

and its singularity on the support boundary is given by

$$R_D(\boldsymbol{x}, \boldsymbol{y}, x_3^0; t) \sim \delta(t-\tau(x,y)) r_D(\boldsymbol{x}, \boldsymbol{y}, x_3^0).$$

Note that the two reflection kernels are related by

$$\boldsymbol{K}(\boldsymbol{y}) R_N(\boldsymbol{x}, \cdot, x_3^0, \cdot) = R_D(\boldsymbol{x}, \boldsymbol{y}, x_3^0; t). \tag{25}$$

Next we come to the measurements of the reflection operator kernels. Since in the actual physical problem $c = c_0$ for $x_3 \geq 0$, we can identify in region $x_3 \geq 0$, u^- and u^+ with the incident wave (produced by sources in the region $x_3 \geq 0$) and the scattered wave respectively.

For an *ideal* set of measurements, consider a point pulse source located at $y = (y_1, y_2, 0)$ producing an incident wave of the form

$$u^i(x,t) = \frac{\delta(t - |x-y|/c_0)}{4\pi|x-y|}, \qquad x_3 \geq 0.$$

This gives rise to the scattered wave $u^s(\boldsymbol{x}, x_3, \boldsymbol{y}, t)$ at a point $x = (\boldsymbol{x}, x_3)$. It can be shown that

$$R_D(\boldsymbol{x},\boldsymbol{y},0;t) = u^s(\boldsymbol{x},0,\boldsymbol{y},t), \tag{26}$$

$$R_N(\boldsymbol{x},\boldsymbol{y},0;t) = -\frac{\partial u^s(\boldsymbol{x},0,\boldsymbol{y},t)}{\partial x_3}. \tag{27}$$

Thus the reflection kernels are directly measured in the ideal set-up with the kernel R_D being obtained from Dirichlet measurements of the scattered field, and R_N from Neumann measurements.

"Ricatti" Equation for the Scattering Kernels

The imbedding or generalized "Ricatti" equation (so named, since in one dimensions in the frequency domain, it is an ordinary differential Ricatti-type equation) for the kernel $R(\boldsymbol{x},\boldsymbol{y},x_3;t)$ will be presented. Using combinatorics involving the scattering matrices associated with slabs $\beta < x_3 < \alpha$ of material $c = c(x_1,x_2,x_3)$ imbedded in a medium where c depends upon the transverse variables only, one can show that the expression for the derivative of the reflection operator $\frac{\partial}{\partial x_3^0}R$ (where the parameter $\alpha = x_3^0$) can be obtained directly from system (14). To obtain the corresponding equation for the kernel R_D of the reflection operator first set $u^- = \boldsymbol{K}^{-1}v^-$, and $u^+ = \boldsymbol{R}u^- = R_D \circ v^-$

$$R_D \circ v^- = \int_0^t \int_{\mathbf{R}^2} R_D(\boldsymbol{x},\boldsymbol{y},x_3^0;t-s)v^-(\boldsymbol{y},x_3,s)d\boldsymbol{y}\,ds \tag{28}$$

into system (14). This yields the equations

$$\frac{\partial}{\partial x_3^0}(R_D \circ v^-) + (\boldsymbol{K}^{-1} - \frac{1}{2}\boldsymbol{K}'\boldsymbol{K}^{-1})(R_D \circ v^-) = -\frac{1}{2}\boldsymbol{K}'v^-,$$

$$\frac{\partial v^-}{\partial x_3^0} - (\boldsymbol{K}^{-1} - \frac{1}{2}\boldsymbol{K}^{-1}\boldsymbol{K}')v^- = -\frac{1}{2}\boldsymbol{K}^{-1}\boldsymbol{K}'\boldsymbol{K}^{-1}\frac{\partial}{\partial t}(R_D \circ v^-).$$

Elimination of $\frac{\partial v^-}{\partial x_3^0}$ results in a single non-linear equation involving both R_D and v^-. Letting v^- be an arbitrary C^2 function such that $v^- = 0$ at $t = 0$, and using the initial condition that $R_D = 0$ at $t = 0$, the "Ricatti" or imbedding equation for the kernel R_D is obtained (after dropping the superscript o on x_3)

$$\begin{aligned}\frac{\partial R_D(\boldsymbol{x},\boldsymbol{y},x_3;t)}{\partial x_3} &+ (\boldsymbol{K}^{-1} - \frac{1}{2}\boldsymbol{K}'\boldsymbol{K}^{-1})R_D(\cdot,\boldsymbol{y},x_3;\cdot) + (\boldsymbol{K}^{-1} - \frac{1}{2}\boldsymbol{K}'\boldsymbol{K}^{-1})R_D(\boldsymbol{x},\cdot,x_3;\cdot)\\ &- \frac{1}{2}(\boldsymbol{K}'\boldsymbol{K}^{-1}R_D(\boldsymbol{x},\cdot,x_3;\cdot) \circ \boldsymbol{K}^{-1}R_D(\cdot,\boldsymbol{y},x_3;\cdot))\\ &= -\frac{1}{2}k'(\boldsymbol{x},\boldsymbol{y},x_3;t)\end{aligned} \tag{29}$$

where k' is the kernel of the operator $\boldsymbol{K}'$, i.e.

$$\boldsymbol{K}'u = \int_0^{t+} \iint_{\mathbf{R}^2} k'(\boldsymbol{x},\boldsymbol{y},x_3;t-s)u(\boldsymbol{y},s)d\boldsymbol{y}\,ds. \tag{30}$$

The corresponding equation for the kernel R_N can be obtained in a similar manner and is indirectly given in reference **[11]** where the corresponding equation involves the kernel R with R_N related to R by

$$R_N = \frac{\partial^2}{\partial t^2} R.$$

The "Ricatti" equation for R_N will not be given here except for the special case when the medium is stratified, depending continuously on the single variable x_3, i.e. $c = c(x_3)$. In this case due to rotational and translational invariance the reflection operator kernel depends upon the transverse variables $\boldsymbol{x}$ and $\boldsymbol{y}$ in the following manner $R_N = R_N(|\boldsymbol{x} - \boldsymbol{y}|, x_3; t)$. If we let $\phi(\boldsymbol{x})$ be a smooth function such that $\phi(\boldsymbol{x}) \equiv 1$ in the neighbourhood of the support of R_N, and set

$$\widetilde{R}(x_3, t) = \iint_{\mathbf{R}^2} R_N(|\boldsymbol{x} - \boldsymbol{y}|, x_3; t)\phi(\boldsymbol{x})d\boldsymbol{x} \tag{31}$$

then the "Ricatti" or imbedding equation for R_N reduces to

$$\frac{\partial}{\partial x_3}\widetilde{R} + \frac{2}{c(x_3)}\frac{\partial \widetilde{R}}{\partial t} - \frac{c'(x_3)}{2c(x_3)}\int_0^t \widetilde{R}(x_3, t-s)\widetilde{R}(x_3, s)ds = 0 \tag{32}$$

for $0 \leq t < T$. The corresponding initial condition for $\widetilde{R}(x_3, t)$ is given by

$$\widetilde{R}(x_3, 0) = -\frac{1}{4}c'(x_3). \tag{33}$$

Equation (32) is the one-dimensional equation developed previously.

The Inverse Problem

The inverse problem for the three-dimensional case can be stated as follows:

Given (i) values of $\frac{\partial c}{\partial x_3}$, c on the plane $x_3 = 0$,

(ii) suitable data set (knowledge of $R_D(\boldsymbol{x}, \boldsymbol{y}, 0; t)$ or $R_N(\boldsymbol{x}, \boldsymbol{y}, 0; t)$ for a set of values of $\boldsymbol{x}, \boldsymbol{y}$ and t where $0 \leq t < T$, or certain functionals of R_D or R_N)

to *determine* $c(x_1, x_2, x_3)$ for $-\ell < x_3 < 0$, where $\ell = \ell(T)$ using

(i) the "Ricatti" or imbedding equation for R_D or R_N

(ii) the initial conditions for R_D or R_N.

Remark: Complete knowledge of $R_D(\boldsymbol{x}, \boldsymbol{y}, 0; t)$ or $R_N(\boldsymbol{x}, \boldsymbol{y}, 0, t)$ for $\boldsymbol{x}, \boldsymbol{y} \in \mathbf{R}^2, 0 \leq t < T$ is not needed. For the stratified medium, where $c = c(x_3)$, one needs knowledge only of $\widetilde{R}(0, t)$ for $0 \leq t < T$ where $\widetilde{R}(x_3, t)$ is the average value of R_N as indicated by equation (31) and definition of ϕ. The corresponding one-dimensional problem given by Eqs. (32, 33) has a solution for finite T.

This leads to the key problem that needs to be further investigated:

(i) what is the suitable data set or sets

(ii) what is the optimum procedure to use the Ricatti equation and inverse condition in the inverse problem.

Note: For the numerical computation standpoint the operators $\boldsymbol{K}, \boldsymbol{K}'\boldsymbol{K}^{-1}$ in Eq. (29) can be easily handled. Numerical algorithms for computing $\boldsymbol{K}u$, etc., where u is sufficiently smooth, have been obtained. The problem of computing $\boldsymbol{K}R_D$ etc., comes from the singularity in the kernel of the reflection operator.

Non-planar Splitting

Non-planar wave splitting [**13**] of solutions of the wave equation and the invariant imbedding process has been applied to the case where the physical medium is a non-planar stratified medium, i.e.: where the velocity c depends upon a single parameter. Special cases of such geometry are spherical and cylindrical symmetric media. Incoming and out-going waves across a smooth closed surface S are defined using the associated initial-value boundary-value problem in the appropriate region of space either exterior or interior to S. In the invariant imbedding geometry one takes such surfaces to coincide with the level surfaces $c =$ constant. Thus the initial imbedding geometry corresponding to the planar geometry of Fig. 1 would have $c = c(x)$ in the interior to the level surface $c = \alpha$ where α is a parameter, and $c = \alpha$ (constant) in the exterior region. The condition for an incoming (lower sign) and out-going wave (upper sign) across a surface S in a medium where $c =$ constant is given by the equation

$$(1 \pm M)u \pm Ku_n = 0$$

where K and M are given by

$$\begin{aligned} Ku(x,t) &= \frac{1}{2\pi}\int_s \frac{1}{R}u(y,t-R/c)H(t-R/c)d\sigma, \\ Mu(x,t) &= \frac{1}{2\pi}\int_s \frac{1}{R^2}\partial_{n_y}R\left[u(y,t-R/c)+\frac{R}{c}u_t(y,t-R/c)\right]H(t-R/c)d\sigma, \end{aligned}$$

with $R = |x-y|$, and the unit normal to S is directed into the exterior domain.

For the case of a cylindrical symmetric medium where $c = c(\rho)$, with (ρ,θ,z) being cylindrical polar coordinates, the system of equations for out-going u^+ and in-coming waves u^- corresponding to Eq. (14) for the planar case is given by

$$\frac{\partial}{\partial\rho}\begin{bmatrix} u^+ \\ u^- \end{bmatrix} = \begin{bmatrix} \Gamma^+ & -\frac{1}{2}c_\rho B^- \\ -\frac{1}{2}c_\rho B^+ & \Gamma^- \end{bmatrix}\begin{bmatrix} u^+ \\ u^- \end{bmatrix} \tag{34}$$

where

$$\Gamma^{\pm} = \mp K^{-1}(I \pm M) + \frac{1}{2}c_\rho B^{\pm}$$
$$B^{\pm} = \mp M_c + K_c K^{-1}(I \pm M)$$
$$K_c u = \frac{t}{c}Ku_t - \frac{1}{c}K(tu_t)$$

with a similar definition for M_c and where $c_\rho = \frac{\partial c}{\partial \rho}$. By taking the average of the field quantities over a cylindrical surface,

$$\hat{u}(\rho, t) = \int_S u(\rho, \theta, z; t)\rho d\theta dz$$

system (34) can be reduced to a system involving the variables ρ and t only. This can be used to get the appropriate equation for the kernel of the reflection operator given by

$$\hat{u}^+(\rho, t) = R\hat{u}^- = \int_0^t R(t - s; \rho)\hat{u}^-(\rho, s)ds.$$

This was used by Kreider **[14]** to get numerical solutions of the inverse problem associated with a cylindrical symmetric medium.

Acknowledgement: The research in this paper was supported by O.N.R. on Contract No. 00014-89-K-0003.

REFERENCES

[1] BELLMAN R. AND WING G. N., (1975), "An Introduction to Invariant Imbedding," Wiley, New York.

[2] REDHEFFER R., (1962), *On the relation of transmission-line theory to scattering and transfer*, J. Math. Phys. **41**, p. 1, Cambridge, MA.

[3] CORONES J. P., DAVISON M. E., AND KRUEGER R. J., (1983), *Wave splittings, invariant imbedding and inverse scattering*, edited by A. J. Devaney, Proc. SPIE, in "Inverse Optics," SPIE, Bellingham, WA, pp. 102–106.

[4] CORONES J. P., KRUEGER R. J., AND WESTON V. H., (1984), *Some recent results in inverse scattering theory*, edited by Santosa, Poa, Symes and Holland, in "Inverse Problems of Acoustic and Elastic Waves," S.I.A.M., Philadelphia, pp. 65–81.

[5] CORONES J. P., DAVISON, M. E., AND KRUEGER, R. J. (1983), *Direct and inverse scattering in the time domain by invariant imbedding techniques*, J. Acoust. Soc. Am. **74**, p. 1535.

[6] CORONES J. P. AND KRUEGER, R. J., (1983), *Obtaining scattering kernels using invariant imbedding*, J. Math. Anal. Appl. **95**, p. 393.

[7] BEEZLEY R. A. AND KRUEGER R. J., (1985), *An electromagnetic inverse problem for dispersive media*, J. Math. Phys. **26**, p. 317.

[8] KRISTENSSON G. AND KRUEGER R. J., (1986), *Direct and inverse scattering in the time domains for a dissipative wave operation equation I. Scattering Operators*, J. Math. Phys. **27**, p. 1667.

[9] WESTON V. H., (1987), *Factorization of the wave equation in higher dimensions*, J. Math. Phys. **28**, p. 1061.

[10] WESTON V. H., (1988), *Factorization of the dissipative wave equations and inverse scattering*, J. Math. Phys **29**, p. 2205.

[11] WESTON V. H., (1989), *Wave splitting and the reflection operator for the wave equation in* $\mathbb{R}^3$, J. Math. Phys. **30**, p. 2545.

[12] WESTON V. H., (1990), *Square root of a second order hyperbolic differential operator and wave-splitting*, to be published.

[13] WESTON V. H., (1988), *Factorization of the wave equation in a nonplanar medium*, J. Math. Phys. **29**, p. 36.

[14] KREIDER K. L., (1989), *A wave-splitting approach to the time dependent inverse scattering for the stratified cylinder*, S.I.A.M. J. App. Math. **49**, p. 932.

Multidimensional Inverse Quantum Scattering Problem and Wiener-Hopf Factorization*

T. Aktosun[1] *and C. van der Mee*[2]

[1]Department of Mathematical Sciences, University of Texas at Dallas, Richardson, TX75083, USA

[2]Department of Mathematical Sciences, University of Delaware, Newark, DE19716, USA

*Research supported in part by the NSF grant DMS-8823102

Abstract: We consider the direct and inverse scattering for the n-dimensional Schrödinger equation, $n \geq 2$, with a potential having no spherical symmetry. Sufficient conditions are given for the existence of a Wiener-Hopf factorization of the corresponding scattering operator. This factorization leads to the solution of a related Riemann-Hilbert problem, which plays a key role in inverse scattering.

1. INTRODUCTION

Consider the n-dimensional ($n \geq 2$) Schrödinger equation

$$\Delta_x \psi + k^2 \psi = V(x)\psi \tag{1}$$

where $x \in \mathbf{R}^n$, Δ_x is the Laplacian, and $V(x)$ is the potential. In quantum mechanics the behavior of a particle in the force field of $V(x)$ is governed by (1). We assume that $V(x) \to 0$ as $|x| \to \infty$, but we do not assume any spherical symmetry for $V(x)$. All of our results hold for real and locally square-integrable potentials $V(x) \in L^2_{\text{loc}}(\mathbf{R}^n)$ belonging to the class $\mathbf{B}_\alpha$ with $0 \leq \alpha < 2$. Here $\mathbf{B}_\alpha$ denotes the class of potentials such that for some $s > 1/2$, $(1+|x|^2)^s V(x)$ is a bounded linear operator from $H^\alpha(\mathbf{R}^n)$ into $L^2(\mathbf{R}^n)$, where $H^\alpha(\mathbf{R}^n)$ denotes the Sobolev space of order α. For the reader who is solely interested in the case $n = 3$, the following conditions on the potential will suffice:

1. $\int_{\mathbf{R}^3} dx\, |V(x)| \left(\frac{|x|+|y|+1}{|x-y|}\right)^2 < \infty, \quad \forall y \in \mathbf{R}^3.$

2. There exist some $c > 0$ and some $s > 1/2$ such that $|V(x)| \leq \frac{c}{(1+|x|^2)^s}, \quad \forall x \in \mathbf{R}^3.$

3. $\int_{\mathbf{R}^3} dx\, |x|^\beta\, |V(x)| < \infty$ for some $\beta \in (0,1]$.

4. $k = 0$ is not an exceptional point. This condition is satisfied if there are neither bound states nor half-bound states at zero energy.

As $|x| \to \infty$, the wavefunction ψ behaves as

$$\psi(k,x,\theta) = e^{ik\theta\cdot x} + ie^{-\frac{\pi}{4}i(n-1)}\frac{e^{ik|x|}}{|x|^{\frac{n-1}{2}}}A(k,\frac{x}{|x|},\theta) + o(|x|^{\frac{1-n}{2}}) \tag{2}$$

where $\theta \in S^{n-1}$ is a unit vector in $\mathbf{R}^n$ and $A(k,\theta,\theta')$ is the scattering amplitude. The scattering operator $S(k)$ is defined as

$$S(k,\theta,\theta') = \delta(\theta-\theta') + i(\frac{k}{2\pi})^{\frac{n-1}{2}}A(k,\theta,\theta'), \tag{3}$$

where δ is the Dirac delta distribution.

The inverse quantum scattering problem consists of recovering $V(x)$ for all x from the knowledge of $S(k)$ for all k. Most information about molecular, atomic, and subatomic particles comes from scattering experiments, and one of the most important problems in physics is to understand the forces between these particles. Solving the inverse scattering problem accomplishes this task, and it can be described as the determination of the force from the scattering data. A comprehensive review of the methods and open problems for the 3-D inverse scattering prior to 1989 can be found in [**Ne89**] and in [**CS89**]. None of the methods developed to solve the multidimensional inverse problem have yet led to a complete and satisfactory solution, but there has been a lot of progress made in this research area. The methods to solve the multidimensional inverse scattering problem include the Newton-Marchenko method [**Ne80, Ne81, Ne82**], the generalized Gel'fand-Levitan method [**Ne74, Ne80, Ne81, Ne82**], the $\bar{\partial}$ method [**NA84, BC85, BC86, NH87**], the generalized Jost-Kohn method [**Pr69, Pr76, Pr80, Pr82**], and a method that uses the Green's function of Faddeev [**Fa65, Fa74, Ne85**].

Here we discuss the Wiener-Hopf factorization of the scattering operator and give sufficient conditions for the existence of such a factorization. The Wiener-Hopf factorization of a unitary transform of the scattering operator, which we also establish, leads to the solution of a Riemann-Hilbert problem which plays a major role in inverse quantum scattering. We only state our results here, and for proofs we refer the reader to [**AV89a,**

AV89b]. To keep the discussion short, we only consider the case where the potentials do not have any bound states.

2. RIEMANN-HILBERT PROBLEM

It is possible to pose the inverse scattering problem as a vector-valued Riemann-Hilbert problem (Hilbert's twenty-first problem in a generalized sense). The solution of the Schrödinger equation satisfies the functional equation

$$\psi(k,x,\theta) = \int_{S^{n-1}} S(k,-\theta,\theta')\,\psi(-k,x,\theta')\,d\theta' \tag{4}$$

or equivalently

$$f_+(k,x,\theta) = \int_{S^{n-1}} G(k,x,\theta,\theta')\,f_-(k,x,\theta')\,d\theta', \quad k \in \mathbf{R} \tag{5}$$

where

$$f_\pm(k,x,\theta) = e^{\mp ik\theta\cdot x}\psi(\pm k,x,\pm\theta) \tag{6}$$

and

$$G(k,x,\theta,\theta') = e^{-ik(\theta-\theta')\cdot x}S(k,-\theta,-\theta'). \tag{7}$$

For potentials as specified in Section 1, $f_\pm$ has an analytic extension in $k \in \mathbf{C}^\pm$. Here $\mathbf{C}^+$ denotes the upper half complex plane and $\mathbf{C}^-$ denotes the lower half complex plane. Let us write (5) in vector form and suppress the x-dependence as

$$f_+(k) = G(k)f_-(k), \quad k \in \mathbf{R}, \tag{8}$$

or equivalently as

$$X_+(k) = G(k)\,X_-(k) + [G(k) - \mathbf{I}]\hat{1}, \tag{9}$$

where

$$X_\pm(k) = f_\pm(k) - \hat{1}. \tag{10}$$

In $L^2(S^{n-1})$, the Hilbert space of square integrable functions on S^{n-1}, the strong limit of $f_\pm$ is $\hat{1}$ as $k \to \infty$ in $\mathbf{C}^\pm$. Note that in our notation $\mathbf{I}$ denotes the identity operator on $L^2(S^{n-1})$ and $\hat{1}$ denotes the vector in $L^2(S^{n-1})$ such that $\hat{1}(\theta) = 1$, $\forall\theta \in S^{n-1}$. Hence, (9) constitutes a Riemann-Hilbert problem: From the knowledge of $G(k)$, determine $X_\pm(k)$.

3. WIENER-HOPF FACTORIZATION

We have the following result.

Theorem 1 For potentials as specified in Section 1, $G(k)$ defined in (7) has a (left) Wiener-Hopf factorization; i.e., there exist operators $G_+(k)$, $G_-(k)$, and $D(k)$ such that $G(k) = G_+(k)\, D(k)\, G_-(k)$ where

1. $G_+(k)$ is continuous in $\mathbf{C}^+$ in the operator norm of $\mathcal{L}(L^2(S^{n-1}))$ and is boundedly invertible there. Here $\mathcal{L}(L^2(S^{n-1}))$ denotes the Banach space of linear operators acting on $L^2(S^{n-1})$. Similarly, $G_-(k)$ is continuous in $\mathbf{C}^-$ in the operator norm of $\mathcal{L}(L^2(S^{n-1}))$ and is boundedly invertible there.

2. $G_+(k)$ is analytic in $\mathbf{C}^+$ and $G_-(k)$ is analytic in $\mathbf{C}^-$.

3. $G_+(\pm\infty) = G_-(\pm\infty) = \mathbf{I}$.

4. $D(k) = P_0 + \sum_{j=1}^{m} \left(\frac{k-i}{k+i}\right)^{\rho_j} P_j$, where $P_1, \ldots, P_m$ are mutually disjoint, rank-one projections, and $P_0 = \mathbf{I} - \sum_{j=1}^{m} P_j$. The (left) partial indices $\rho_1, \ldots, \rho_m$ are nonzero integers. In case there are no partial indices; i.e., when $D(k) = \mathbf{I}$, the resulting Wiener-Hopf factorization is called canonical.

Note that, as seen from (7), $G(k)$ is a unitary transform of the scattering operator $S(k)$. In particular, when $x = 0$, $G(k)$ reduces to $S(k)$. The proof of Theorem 1, using some results of Gohberg and Leiterer regarding factorization of operator functions on contours in the complex plane [**GL73**], is essentially based on the Hölder continuity of $S(k)$ and can be found in [**AV89a, AV89b**].

The solution of the Riemann-Hilbert problem (9) is obtained in terms of the Wiener-Hopf factors of $G(k)$ and can be written as

$$X_+(k) = [G_+(k) - \mathbf{I}]\hat{1} + G_+(k) \sum_{\rho_j > 0} \frac{\phi_j(k)}{(k+i)^{\rho_j}} \pi_j \tag{11}$$

and

$$X_-(k) = [G_-(k)^{-1} - \mathbf{I}]\hat{1} + G_-(k)^{-1} \sum_{\rho_j > 0} \frac{\phi_j(k)\pi_j + [(k+i)^{\rho_j} - (k-i)^{\rho_j}]P_j\hat{1}}{(k-i)^{\rho_j}}, \tag{12}$$

provided $P_j\hat{1} = 0$ whenever $\rho_j < 0$. Here π_j is a fixed nonzero vector in the range of P_j, and $\phi_j(k)$ is an arbitrary polynomial of degree less than ρ_j associated with each $\rho_j > 0$. We can state our result as follows.

Theorem 2 For potentials as specified in Section 1, the Riemann-Hilbert problem (9) has a solution if and only if $P_j\hat{1} = 0$ whenever $\rho_j < 0$. When this happens, the solution is given by (11) and (12).

A simple condition that assures the unique solvability of the Riemann-Hilbert problem (9) is given by $\sup_{k\in\mathbf{R}} ||S(k)-\mathbf{I}|| < 1$, where the norm is the operator norm in $\mathcal{L}(L^2(S^{n-1}))$. If this holds, neither the scattering operator $S(k)$ nor its unitary transform $G(k)$ has any partial indices. As a result, in this case, (9) is uniquely solvable.

REFERENCES

[**AV89a**] T. Aktosun and C. van der Mee, *Inverse Scattering Problem for the 3-D Schrödinger Equation and Wiener-Hopf Factorization of the Scattering Operator*, J. Math. Phys., to appear.

[**AV89b**] T. Aktosun and C. van der Mee, *Wiener-Hopf Factorization in the Inverse Scattering Theory for the n-D Schrödinger Equation*, preprint.

[**BC85**] R. Beals and R. R. Coifman, *Multidimensional Inverse Scattering and Nonlinear P.D.E.'s*, Proc. Symp. Pure Math. **43**, 45-70 (1985).

[**BC86**] R. Beals and R. R. Coifman, *The D-bar Approach to Inverse Scattering and Nonlinear Evolutions*, Physica D **18**, 242-249 (1986).

[**CS89**] K. Chadan and P. C. Sabatier, *Inverse Problems in Quantum Scattering Theory*, 2nd ed., Springer, New York, 1989.

[**Fa65**] L. D. Faddeev, *Increasing Solutions of the Schrödinger Equation*, Sov. Phys. Dokl. **10**, 1033-1035 (1965) [Dokl. Akad. Nauk SSSR **165**, 514-517 (1965) (Russian)].

[**Fa74**] L. D. Faddeev, *Three-dimensional Inverse Problem in the Quantum Theory of Scattering*, J. Sov. Math. **5**, 334-396 (1976) [Itogi Nauki i Tekhniki **3**, 93-180 (1974) (Russian)].

[**GL73**] I. C. Gohberg and J. Leiterer, *Factorization of Operator Functions with Respect to a Contour. III. Factorization in Algebras*, Math. Nachrichten **55**, 33-61 (1973) (Russian).

[**NA84**] A. I. Nachman and M. J. Ablowitz, *A Multidimensional Inverse Scattering Method*, Studies in Appl. Math. **71**, 243-250 (1984).

[**Ne74**] R. G. Newton, *The Gel'fand-Levitan Method in the Inverse Scattering Problem in Quantum Mechanics*. In: J.A. Lavita and J.-P. Marchand (Eds.), *Scattering Theory in Mathematical Physics*, Reidel, Dordrecht, 1974, pp. 193-225.

[**Ne80**] R. G. Newton, *Inverse Scattering. II. Three Dimensions*, J. Math. Phys. **21**, 1698-1715 (1980); **22**, 631 (1981); **23**, 693 (1982).

[**Ne81**] R. G. Newton, *Inverse Scattering. III. Three Dimensions, Continued*, J. Math. Phys. **22**, 2191-2200 (1981); **23**, 693 (1982).

[**Ne82**] R. G. Newton, *Inverse Scattering. IV. Three Dimensions: Generalized Marchenko Construction with Bound States*, J. Math. Phys. **23**, 2257-2265 (1982).

[**Ne85**] R. G. Newton, *A Faddeev-Marchenko Method for Inverse Scattering in Three Dimensions*, Inverse Problems **1**, 371-380 (1985).

[**Ne89**] R. G. Newton, *Inverse Schrödinger Scattering in Three Dimensions*, Springer, New York, 1990.

[**NH87**] R. G. Novikov and G.M. Henkin, *Solution of a Multidimensional Inverse Scattering Problem on the Basis of Generalized Dispersion Relations*, Sov. Math. Dokl. **35**, 153-157 (1987) [Dokl. Akad. Nauk SSSR **292**, 814-818 (1987) (Russian)].

[**Pr69**] R. T. Prosser, *Formal Solution of Inverse Scattering Problems*, J. Math. Phys. **10**, 1819-1822 (1969).

[**Pr76**] R. T. Prosser, *Formal Solution of Inverse Scattering Problems. II*, J. Math. Phys. **17**, 1775-1779 (1976).

[**Pr80**] R. T. Prosser, *Formal Solution of Inverse Scattering Problems. III*, J. Math. Phys. **21**, 2648-2653 (1980).

[**Pr82**] R. T. Prosser, *Formal Solution of Inverse Scattering Problems. IV*, J. Math. Phys. **23**, 2127-2130 (1982).

The Direct and Inverse Scattering Transforms for a Sturm-Liouville Equation on Finite Interval as Smooth Maps. The Geometry of a Set of Finitely-Zoned Potentials

I.S. Zakharevich

The Institute for Problems in Mechanics, pr. Vernadskogo 101, SU-117526 Moscow, USSR

Editor's note: Professor I.S. Zakharevich was an invited lecturer but unfortunately he could not attend the meeting.

Here we consider from the slightly new point of view the very old problem of finding the potential u from the spectral data of Sturm–Liouville operator $Q = -\partial^2/\partial x^2 + u(x)$ on $[0,1]$ with some boundary conditions. Two potentials u_1 and u_2, which are near in some sense should have the spectral data, which are near in some other sense. Here we will discuss these notions of neighborhood of spectral data for different notions of neighborhood for potentials and three classical sets of spectral data:

i) the tied spectrum (i.e., with $\psi(0) = \psi(1) = 0$) and the norming numbers (i.e., the L_2-norms of the tied eigenfunctions with $\psi'(0) = 1$);

ii) the tied and semitied (i.e., with $\psi(0) = \psi'(1) = 0$) spectrum;

iii) the periodic and antiperiodic spectrum, the tied spectrum and some sequence of discrete parameters (signs).

In more explicit terms, we consider a potential u, that depends on some parameter $t \in M$, where M is a smooth variety, and study the t-dependence of the spectral data from *i*), *ii*), *iii*) sets. We introduce the smoothness structures on these data in such a way, that their t-dependence becomes smooth. First, we may restrict ourselves to the universal problem, where m is the space of all potentials with given smoothness (and u depends on t tautologically). In this case we can also consider the inverse problem map and study its smoothness. In cases *i*) and *ii*) we get an answer quite easily. Since the image of direct problem map is well known and it is the open set in some functional space, the reminiscence of the Open Map Theorem from functional analysis says that in inherited from this functional space smoothness structure the direct problem map can be diffeomorfism. So it is in these cases, and the proof is absolutely straightforward.

But in case *iii*) the situation is more complicated, since there are discrete parameters in the set of spectral data. So we may first restrict ourselves to the stratum in the space of potentials with fixed signs in the spectral data. In their articles McKean and Trubowitz have studied the tori, which join the $\pm$-strata, so this suggest the way of gluing together $+/-$-sign strata. Then we may (hypothetically) study only the more coarse stratification on strata with 0/non-0 signs. The potential u with only finite number of non-0 signs appears to be finitely zoned. The tori of McKean–Trubowitz (isospectral families) form the fibration on these strata. Since the smoothness structure on these tori is well known, we may conjecture the kind of the smoothness structure on an entire stratum. Under these hypothesis the only remaining question is the way of gluing together the strata. The answer to this question is the simplest possible: the intersections of closures of strata are of maximal possible transversality. That means that there is the coordinate system (ξ_i, η_i) in the space of potentials, such that the closure of every stratum has in this coordinate system the equations $\xi_i = \eta_i = 0, i \in I$, with some set of indices I. This set I is the set of indices of 0-signs in the sequence of signs.

This information allows us to build the smoothness structure on the set of possible spectral data for the *iii*) problem (this set is known for a long time), such that the direct and inverse problem maps are diffeomorfisms. It is easy to see that as a byproduct we get the (global) diffeomorfism of the space of potentials of given degree of smoothness, which maps the set of finitely zoned potentials to the set of trigonometric polynoms (it is sufficient to apply the inverse Fourier transform to the sequence (ξ_i, η_i) .)

We obtain all these results only for the Sobolev spaces of potentials, but it is clear that with some technical improvement they can be extended to other spaces, such that the Hoelder spaces and spaces of analytical functions.

Part V

Theoretical Imaging

Application of Singular Systems to Some Data Reduction Problems in Modern Optics

M. Bertero[1], *C. De Mol*[2], *and E.R. Pike*[3]

[1]Dipartimento di Fisica dell'Università di Genova and Istituto Nazionale di Fisica Nucleare, Via Dodecaneso 33, I-16146 Genova, Italy
[2]Département de Mathématique, Université Libre de Bruxelles, Campus Plaine C.P. 217, Bd. du Triomphe, B-1050 Bruxelles, Belgium
[3]Department of Physics, King's College, Strand, London WC2R 2LS and RSRE, Great Malvern, Worcester WR14 3PS, UK

INTRODUCTION

Many data reduction or "inverse" problems in optics originate from scattering or imaging experiments, where radiation is incident on a sample as a probe of some unknown characteristic of that sample. The scattered radiation is measured after interaction with the sample by means of a detector or a set of detectors. Since, in most problems, the property of interest is not directly accessible to measurement, the output of the detectors has to be processed and interpreted in order to provide as much relevant information as possible about the required structural function characterizing the sample. As usual in optics, we will call this unknown function the object and denote it by $f(x)$, where x represents a one- or multi-dimensional space or time (or both space and time) variable, according to the problem under study. When considering an inverse problem, we shall assume that the direct problem has been solved, i. e. that we have a good mathematical model describing the interaction between the radiation and the sample and associating the corresponding data function $g(x)$, also called the image, to a given object $f(x)$. Moreover, we will restrict ourselves to situations where a linear or linearized model holds true or is good enough in practice, so that the problem of recovering the object $f(x)$ from its image $g(x)$ is a linear inverse problem. In the present paper, we review the basic features of some data reduction problems in which we have been particularly interested during the last few years. They relate to several practical applications, namely various situations in confocal scanning microscopy and various techniques for particle sizing by laser light scattering. More generally, the problems discussed hereafter could be viewed as case studies for many other applied linear inverse problems encountered in optics and other fields.

In particle sizing, the unknown function f represents the distribution of sizes of small scattering particles in air or fluid suspension. We discuss first a suspension of particles of the same size (i. e. a so-called monodisperse suspension) sufficiently large with respect to the wavelength of light so that when illuminated by a monochromatic laser source they behave like opaque spheres. The scattered intensity at angle x in the far field or Fraunhofer region is then proportional to the well-known Airy diffraction pattern $y^2[J_1(xy)/x]^2$, where $y=ka=2\pi a/\lambda$, λ being the wavelength of the incident light, a the radius of the particles and J_1 the first-kind Bessel function of order 1. This Fraunhofer-diffraction model is valid in the forward direction. It can also be used for smaller values of the radius a (but still greater than λ) if so-called "Mie corrections" for each value of radius are made after inversion. For a polydisperse suspension, the scattered intensity will be a superposition of these elementary patterns for different values of the size parameter y. Hence in the case of a continuous distribution $p(y)$ of sizes, the measured intensity is given by

$$g(x) = \int_0^{+\infty} k(xy) f(y) dy \tag{1}$$

with

$$k(xy) = [J_1(xy)/xy]^2 \tag{2}$$

and $f(y)=y^4 p(y)$. In order to recover the size distribution, we have to solve the Fredholm integral equation of the first kind (1).

In another type of experiment for dilute scatterers, one measures instead the attenuation (or "extinction") coefficient $\tau(k)$ in the forward direction at zero angle as a function of the wavenumber k of the incident light beam. This is usually referred to as particle sizing from "multispectral extinction data" or else from "spectral turbidity" measurements. The equation relating the volume size distribution $f(y)$ to the data function $g(k)=\tau(k)/k$ is again of the form (1), with a kernel given by

$$k(xy) = \frac{3}{4xy}\left[2 + \frac{4}{x^2y^2} - 4\frac{\sin xy}{xy} - 4\frac{\cos xy}{x^2y^2}\right] \tag{3}$$

in the so-called anomalous diffraction approximation. Note that both the kernels (2) and (3) are approximations of the Mie scattering kernel, which describes exact scattering by spheres and can be computed and used for a more accurate modeling of the diffraction process [1].

For submicron particle sizing, the previous "static" light scattering techniques are no longer sufficient and one has to resort to dynamic methods, which make use of the phase shifts induced by the Brownian motion of the macromolecules. By the Wiener-Khinchine theorem, the frequency power spectrum of the time-fluctuating scattered field is the Fourier transform of its autocorrelation function. This function can be measured by photon-counting techniques and digital hardware. The spectrum of the light scattered from a monochromatic laser beam by a monodisperse suspension of spherical particles has the shape of a Lorenzian function with half-width y inversely proportional to the hydrodynamic radius of the particles. The field autocorrelation function $g(x)$ is then just an exponential function $g(x)=\exp(-xy)$, where x is the correlation delay time. Note that the output of the correlator is usually the intensity autocorrelation function, from which $g(x)$ has first to be recovered. In the case of a polydisperse sample, the data function is a superposition of such exponential functions, so that the recovery of the size distribution requires the inversion of a Laplace transform, i. e. again the solution of the integral equation (1), with the following kernel

$$k(xy) = \exp(-xy) \quad . \tag{4}$$

This technique for particle sizing is known as photon correlation spectroscopy (PCS) [2]. Let us mention that particle-velocity distributions can also be recovered from correlation functions by photon correlation velocimetry also called Laser Doppler Anemometry (LDA) in aerodynamics, using yet another kernel function.

Concerning the second group of problems reviewed here, let us recall that an imaging system can be viewed in Fourier optics [3] as a linear system entirely characterized by its impulse response, also called the "point spread function" (PSF). For isoplanatic systems, the point spread function $s(x)$ is shift invariant and the image amplitude $g(x)$ is then just the convolution of $s(x)$ by the amplitude transmission function of the object $f(x)$:

$$g(x) = \int_{-\infty}^{+\infty} s(x-y)f(y)dy \quad . \tag{5}$$

The inverse problem of object restoration consists in recovering the object f from its image g. Although x and y are here clearly at least two-dimensional space variables, we will for simplicity only deal here with one-dimensional imaging models, which can be easily generalized. Note that by the convolution theorem, equation (5) becomes in the conjugate space a simple product of Fourier transforms: $\hat{g}(\omega)=\hat{s}(\omega)\hat{f}(\omega)$. The function $\hat{s}(\omega)$ is called the transfer function. All optical systems are limited by diffraction, i. e. the transfer function has a finite support, say the frequency band $(-\Omega,+\Omega)$. The corresponding spatial resolution limit $R=\pi/\Omega$ is called the Rayleigh limit and represents the distance between the closest resolvable details in the image. The transfer function of an ideal (aberration-free) diffraction-limited system is the characteristic function of the frequency band $(-\Omega,+\Omega)$ and the corresponding point spread function is given by $s(x)=\sin(\Omega x)/\pi x$. In the language of electrical engineering, we have to recover the input signal of a perfect low-pass filter from its output, which is a classical problem occurring in many other instances. In writing (5), we have implicitly assumed the object to be uniformly illuminated by coherent light. When the illumination is not uniform, but instead weighted according to a certain known profile function $P(y)$, such as the Gaussian profile of a laser beam for example, the transmission function $f(y)$ has to be multiplied by $P(y)$, so that the equation to solve is then

$$g(x) = \int_{-\infty}^{+\infty} s(x-y)P(y)f(y)dy \quad . \tag{6}$$

In confocal scanning microscopy [4], which will be discussed in the last section, the laser beam is focussed into the sample by means of a lens with the same impulse response as the imaging lens, so that $P(y)$ in (6) should be replaced by $s(y)$ itself. In incoherent light imaging, the impulse response to be used in (5) is the squared modulus of the coherent point spread function, while f and g represent respectively the field intensities of the object and of the image (instead of the amplitudes).

THEORETICAL FRAMEWORK

Let us first recast the linear inverse problems we wish to solve in a more concise operator form, viz, find f belonging to a suitable functional solution space F, from the data g (belonging to a certain data space G), such that

$$g = Kf \tag{7}$$

where K is a bounded linear operator from F to G. In the following, we will refer principally to integral operators K of the form given by equations (1) or (5). For the uses we have in mind, it is convenient to assume that F and G are Hilbert spaces of square integrable functions (L^2-spaces). We consider first the rather academic situation, where we have access to the data function on its full range, i. e. on the whole real line for equation (5) and on the semi-axis $(0,+\infty)$ for equation (1). As for equation (5) which becomes diagonal in Fourier space, equation (1) can also be diagonalized by means of the Mellin transform, defined for the function g by

$$\tilde{g}(\omega) = \int_0^{+\infty} x^{-(1/2)+i\omega}\, g(x)dx \quad . \tag{8}$$

In conjugate space, equation (1) simply becomes $\tilde{g}(\omega)=\tilde{k}(\omega)\tilde{f}(-\omega)$ and it is now easy to see that its solution will be unique if the Mellin transform $\tilde{k}$ of the kernel k does not vanish on some interval. But then, if $|\tilde{k}(\omega)|$ tends to zero when $\omega \to \infty$, the inverse operator K^{-1}, however defined, is unbounded. As noticed in [5], this situation happens for the Laplace kernel (4), for which the squared modulus of the Mellin transform is given by $\pi/\cosh(\pi\omega)$, and also for the kernels (2) and (3), for which the explicit Mellin transforms are given in [6] and [7], respectively. Clearly, such a behaviour is the source of instabilities in the presence of noise on the data. Moreover, there is no reason for a generic noisy data function to belong to the range of the operator K (i. e. to the set of possible noise-free images), since the noise does not in general vanish for high frequencies ω. The usual way of coping with both the possible uniqueness problem and the existence (data compatibility) problem is to introduce the so-called generalized solution f^+ of equation $Kf=g$. This is defined as the pseudo-solution (a pseudo-solution minimizes the deviation $||Kf-g||$ measured in the norm of the data space G) of minimal norm (in space F) and exists for any g belonging to the direct sum of the range of K and of its orthogonal complement (see for example [8]). The generalized or Moore-Penrose inverse K^+ is by definition the operator transforming each such data function g to the corresponding (unique) generalized solution f^+. Note that the requirement that f^+ should have minimal norm implies that f^+ is orthogonal to the null-space of the operator K, i. e. to the set of "transparent" or "invisible" objects f such that $Kf=0$. Any object is the sum of a visible part and of an invisible part. Since the data obviously cannot provide any information about the latter one, a reasonable choice, in the absence of further information, is to put the invisible part of the object equal to zero. Let us remark that there are no transparent objects for equation (1) with the kernels (2), (3) or (4), since their Mellin transforms are different from zero everywhere. The concept of generalized solution does not cope however with the stability problem, which has to be overcome by means of regularization theory which we will consider shortly. In fact, the inverse problem is ill-posed, i. e. f^+ is unstable whenever the range of the operator K is not closed [8]. A simple way of looking at this stability problem now is to notice that the data noise will be amplified in the inversion essentially at those frequencies for which the Fourier or Mellin spectrum $\tilde{k}(\omega)$ of the operator K is too small. When the modulus of these transforms decreases gently to zero when $\omega \to \infty$, as in the examples quoted above, it appears quite natural to stabilize the inversion by suppressing the frequencies higher than some cut-off frequency Ω, such that $|\tilde{k}(\Omega)|$ is of the order of magnitude of the noise level. This implies that the object is not perfectly recovered, but only with a finite resolution given in the case of Fourier-type convolutions by the resolution limit $R=\pi/\Omega$. The analogue in the case of equation (1) is the following resolution ratio introduced in [9]

$$\delta = \exp(\pi/\Omega) \qquad (9)$$

which represents the ratio of the abscissae of closest resolvable elements or otherwise the ratio between any two successive zeroes of the real part of $\exp(i\Omega \log y)$. The analogy is obvious when making the change of variable $t=\log y$.

In practical problems, full-range data are not available, so that we will assume now that the data function is given only on a finite range, say the interval (c,d). This restriction makes of course the inverse problem even more hopeless, unless we have at our disposal a priori information about the support of the object f itself, which might compensate somehow for the lack of data. If we assume that $f(y)$ has a given finite support (a,b), the operator K becomes an operator from $L^2(a,b)$ into $L^2(c,d)$, which is compact if the kernel k is square integrable on the rectangle $(a,b)x(c,d)$. From the analyticity properties of the kernels (2), (3) and (4) we may see that uniqueness also holds in this case for the corresponding integral equations.

As will be discussed later, problems with spurious edge effects in reconstructions (Gibbs phenomenon) have led us to work instead with weighted L^2-spaces, i. e. choosing the norm in the solution space F to be

$$||f||^2 = \int dx \, \frac{|f(x)|^2}{|P(x)|^2} \quad . \qquad (10)$$

Regularization, as we shall see, amounts to the introduction of some penalization on the norm of the solution f. Hence, the role of the "profile" function $P(x)$, which is a sort of first guess for the solution, is to force $f(x)$ to be small where $P(x)$ is small. This is a way to express a priori information about the localization of f. The previous case fits into this framework if we take for $P(x)$ the characteristic function of the support (a,b), but it is often more convenient in practice to use a smoother profile function. For finite-range data, when $P(x)$ is square-integrable, the operator K is again compact.

In diffraction-limited imaging according to equation (5), all objects whose Fourier transform is zero on the band $(-\Omega,+\Omega)$ are transparent objects. However, the assumption that $f(x)$ has finite support, say $(-X,+X)$ ensures uniqueness of the restored object, since the Fourier transform of f is then an entire (bandlimited) function, whose extrapolation outside the band $(-\Omega,+\Omega)$ is unique. This is the cornerstone of the many attempts to perform "super-resolution", i. e. to go beyond the Rayleigh limit. This analytic extrapolation is unstable, however, and requires regularization. A careful study of the reconstruction errors on stabilized solutions vitiates then the hopes of achieving significant super-resolution for reasonable noise levels [10], except in the case of a very low space-bandwidth product $c=X\Omega$, as shown in [11]. When localization of the object is expressed by introducing a weight in the L^2-norm, as in (10), the compactness of the operator K is guaranteed (for full- or limited-range data) by the square-integrability of $s(x)$ and $P(x)$. Uniqueness, however, depends on the choice of $P(x)$. Let us remark that solving equation (5) in the corresponding weighted space is equivalent to solving equation (6) in the (unweighted) space $L^2(-\infty,+\infty)$ [12]. The solution is unique in the case of a Gaussian profile but is not in the case of the band-limited profile arising in confocal microscopy [12].

When the operator K is compact, its spectrum is discrete and a canonical tool for analyzing and solving the associated linear inverse problem is the so-called singular system of K, i. e. the set of solutions $\{\alpha_k; u_k, v_k\}$ of the coupled equations

$$K\, u_k = \alpha_k\, v_k \quad ; \quad K^*\, v_k = \alpha_k\, u_k \qquad (\alpha_k > 0) \tag{11}$$

where K^* is the adjoint operator. The α_k are called the singular values of K and the singular vectors (or singular functions) u_k and v_k are eigenvectors (with eigenvalues $\alpha_k{}^2$) of the selfadjoint operators K^*K and KK^*, respectively. The set $\{v_k\}$ constitutes an orthogonal basis in the closure of the range of K, while the set $\{u_k\}$ is a basis in the orthogonal complement of the null-space of K. Hence the generalized solution is easily shown to admit the following expansion:

$$f^+ = \sum_k \frac{1}{\alpha_k}(g, v_k) u_k \tag{12}$$

where the summation is made over all singular values (different from zero, by definition). We assume in the following the α_k to be arranged in decreasing order of magnitude ($\alpha_0 \geq \alpha_1 \geq \alpha_2$ ...), each value being repeated as many times as required by its multiplicity. The range of a compact operator cannot be closed (unless it is finite-dimensional) and the singular values accumulate to zero. Therefore, the corresponding generalized inverse K^+ is unbounded and the solution (12) unstable. A regularization algorithm can be defined as a family of bounded operators $\{R_\mu\}$, depending on a positive regularization parameter μ and which approximate K^+ in the sense that $R_\mu g$ tends, in the norm of F, to $K^+ g$ when μ tends to zero and for any g in the domain of K^+ [8]. For compact operators, a wide class of such regularization algorithms can be viewed as "spectral windows" of the form

$$\tilde{f} = \sum_k \frac{F_\mu(\alpha_k)}{\alpha_k}(g, v_k) u_k \tag{13}$$

where the $F_\mu(\alpha_k)$ $(0 \leq F_\mu(\alpha_k) \leq 1)$ are filtering or windowing coefficients tending to 1 for all k when μ tends to zero. A famous example of a regularization algorithm is the Tikhonov filter [8] corresponding to

$$F_\mu(\alpha_k) = \frac{\alpha_k{}^2}{\alpha_k{}^2 + \mu} \tag{14}$$

Another common choice, sometimes referred to as numerical filtering, is the truncated (or "top-hat") filter:

$$F_\mu(\alpha_k) = 1 \quad \text{if} \quad \alpha_k{}^2 \geq \mu \quad \text{and} \quad F_\mu(\alpha_k) = 0 \quad \text{if} \quad \alpha_k{}^2 < \mu \quad . \tag{15}$$

In order to ensure continuous dependence of the regularized solution $\tilde{f}$ on the data or in other words its stability in the presence of noise, a rule must still be given for the choice of the regularization parameter μ. This problem has been extensively discussed in the literature (see for instance [8] or [13] and the references quoted therein) and we will not do it here. Let us just say that a practical rule of thumb consists in taking $1/\mu$ roughly equal to the signal-to-noise ratio (SNR). Truncated solutions defined by (15) allow us to give a precise meaning to the concept of "number of degrees of freedom" (NDF) which has been widely used in the literature on optical inverse problems e.g. in [14] and is analogous to the Shannon number which measures the "information content" of a bandlimited signal in communication theory [15]. We can define indeed this NDF or generalized Shannon number as being the number of terms in the truncated expansion (15). This number represents for any given linear inverse problem the number of independent "pieces of information" (independent since we use an orthogonal expansion) about the object f, which can be reliably (stably) retrieved from the data in the presence of a given amount of noise. The sensitivity of this number to the actual noise level depends on the problem and on its "degree of ill-posedness": for "severely" ill-posed problems, the NDF increases very slowly with the signal-to-noise ratio [16]. This happens when the singular values α_k fall off to zero very rapidly, e. g. exponentially. In fact, the solution f of an ill-posed problem has not only components belonging to the null-space of the operator K applying f on the data which remain "invisible" even in the absence of noise, but also "quasi-invisible" components which cannot be recovered because they are drowned in the noise. As already said, this implies that we cannot achieve arbitrary resolution in the object recovery. Indeed, it is not difficult to show [13] that the regularized solution (13) is a blurred or mollified version of the true object f

$$\tilde{f}(x) = \int a(x,y)f(y)dy \tag{16}$$

where the averaging kernel $a(x,y)$ is given by

$$a(x,y) = \sum_k F_\mu(\alpha_k)u_k(x)u_k(y) \tag{17}$$

and represents the reconstruction of a Dirac delta peak in y for a given choice of the regularization algorithm, i. e. of the set of filters. Therefore we can call $a(x,y)$ the impulse response of the inversion method, taking into account the response of the experimental device (the operator K) as well as the data processing by means of a regularization algorithm to control the noise. The width of this impulse response is a measure of the resolution that can thus be achieved in the reconstruction. As clearly seen from expression (17), regularization provides a trade-off between resolution and instability.

True experimental data are of course sampled data. The data g then belong to an N-dimensional data space G and the operator K to be inverted becomes a finite-range compact operator K_N. Its singular system will be denoted by $\{\alpha_{N,k}; u_{N,k}, v_{N,k}\}$. The generalized inverse K_N^+ is then a bounded operator, but f^+ can still be numerically unstable because of the ill-conditioning arising when the ratio between the largest and the smallest singular values, i. e. the condition number, becomes too large. This is most likely to happen whenever the finite-range operator results from a discretization of a Fredholm integral equation of the first kind. Hence, regularized expansions in the singular system, similar to (13), are still to be used in the case of discrete data. Because of the noise, the NDF will in general be smaller than the dimension of the range of K_N, i. e. the number of its (non-zero) singular values. Let us finally mention that the regularized inversion of linear operators with continuous spectrum, such as the operators (1) or (5) with full-range data, can also be achieved by windowing the spectrum with an appropriate filter $F_\mu(\omega)$. "Quasi-positivity" of the regularized solution, i. e. positivity in the absence of noise, is ensured by special tailoring of that filter, as shown in [17]. The price to be paid is a widening of the impulse response $a(x,y)$, i. e. a loss in resolution. The use of similar filter factors in the case of limited-range or discrete data, allows also to reduce the negative side-lobes in the response (17) and hence to attenuate the negative parts in the reconstructed objects. This is useful when positivity of the solution is required and is preferred to very sharp resolution.

PARTICLE SIZING

In this section, we will report a few results relative to the theoretical problems of recovering particle-size distributions, using equation (1) with kernels (2), (3) or (4), from full-range, limited-range and sampled data. We will emphasize here some generic features, which are common to all three applications. A detailed discussion of more specific properties can be found in [9,18] for the Laplace kernel (4), in [6] for Fraunhofer diffraction (2) and in [7] for the extinction kernel (3), and also in the references quoted therein.

A first interesting feature to note is that a priori knowledge of the support of the object can somehow compensate for the lack of information due to truncation and sampling of the data. Let us assume that the support of f is the interval (a,b) ($a>0$), which by a change of variables can be reduced to $(1,\gamma)$, $\gamma=b/a$ representing then the ratio of size between the largest and the smallest particles of the polydisperse suspension. For the three examples considered in this paper, the operator K is then invertible ($Kf=0$ implies $f=0$). Moreover the corresponding singular value spectrum presents some nice properties that are established in [9,6,7]. For fixed γ, each singular value $\alpha_k(\gamma,J)$ increases when J tends to cover the whole semi-axis and tends to the singular value $\alpha_k(\gamma)$ corresponding to full-range data. On the other hand, for fixed J, $\alpha_k(\gamma,J)$ increases with γ and $\alpha_k(\gamma)$ tends for $\gamma \rightarrow \infty$ to the value $|\tilde{k}(\omega_k)|$ of the modulus of the Mellin transform of the kernel k at the point $\omega_k=\pi k/\log\gamma$. Moreover, it appears from numerical experiments that the corresponding singular functions are regularly oscillating: as a rule $u_k(x)$ has k zeroes in the interval $(1,\gamma)$, which are roughly equidistant in the variable $t=\log x$. Hence the resolution achieved when using the regularized solution (13), with top-hat filter (15), is essentially determined by the spacing between the zeroes of the largest-order singular function involved, i. e. by the following resolution ratio, introduced in [9]

$$\delta_o = \gamma^{1/K_o} \tag{18}$$

where α_{K_o} is the smallest singular value which is still greater than or equal to $\sqrt{\mu}$ ($\mu \approx 1/\text{SNR}$). In other words, $K_o + 1$ is the number of degrees of freedom. In many instances, the resolution ratio obtained in this way is comparable to the ratio defined by (9) for full-range data, so that a lack of data is compensated by the finiteness of the support of the object. For example, in the case of the Fraunhofer-diffraction kernel (2), with $\gamma=5$ and data support (0.16, 3.84), $\delta_o=1.6$, 1.26 for a SNR of 10^4 and 10^6, respectively, whereas the corresponding values of the resolution ratio (9) are 1.72 and 1.27, respectively.

Another common feature of these problems is their severe ill-posedness. Because of the analyticity properties of the integral kernels, the singular values α_k decrease exponentially with k. Therefore, in practice, the NDF is small and only weakly sensitive to an improvement of the signal-to-noise ratio. For example, when the SNR passes from 10^4 to 10^6, the NDF passes from 3 to 5 in the case of the Laplace kernel (4) and from 4 to 8 in the case of kernel (2), for $\gamma=5$. Figure 1a shows a

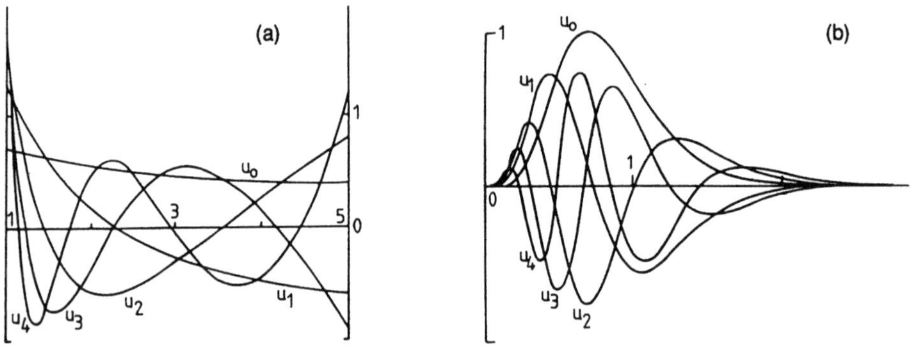

Figure 1. Singular functions corresponding to the Laplace kernel (4): (a) with P(x) = 1 on the interval (1,5) *and* 0 *elsewhere; (b) with* $P(x) = (128/3)x^3\exp(-4x)$.

numerical computation of the first five singular functions for the Laplace kernel (4), again for γ=5. Let us mention that those singular functions are also the eigenfunctions of a second-order differential operator as shown in [19], and therefore can be shown to have very useful properties (e. g. the fact that u_k has exactly k zeroes in the interval $(1,\gamma)$) . We notice that the singular functions are rather large at the edges of the support of the object. This explains why troublesome edge effects arise in the reconstructions made through formula (13). As already noted above, these undesirable effects can be avoided by introducing a smooth weight function $P(x)$ in the norm (10) of the object space F. To understand this, it is enough to look at Figure 1b, which reproduces the same singular functions as of Figure 1a, using a smooth profile, namely a gamma distribution. Figure 1 is also quite representative for the singular functions relative to the two other kernels.

The true experimental situation of sampled data is linked to the previous case of data in a finite or infinite interval by means of perturbation lemmas for compact operators. It can be shown indeed that when the number of data points $N \to \infty$ (under appropriate assumptions that they cover well the whole data interval), the singular values $\alpha_{N,k}$ of the operator K_N tend to the singular values α_k of the operator K corresponding to an L^2-data function. The associated singular functions $u_{N,k}$ also tend in the norm of F to the corresponding functions u_k (the precise statements are given in [6] and [9]). The numerical computation of the singular system $\{\alpha_{N,k};u_{N,k},v_{N,k}\}$ reduces essentially to the diagonalization of a NxN matrix and can be carried out by means of standard algorithms. For the problems discussed here, it appears that good approximations of the first singular values and functions (i. e. of the only relevant ones for stable inversions) can already be obtained by relatively small numbers of data points. This observation agrees with common sense, since we know that, in the presence of noise, only a finite number (the NDF) of independent coefficients in the expansion of the object can be retrieved from the data. Therefore, we can expect to determine those coefficients from a number of data points which is roughly equal to the NDF. This happens to be quite true in the problems we discuss, but the location of the data points has to be chosen in some optimal way. For integral kernels depending on the product of the variables, geometric distributions of data points (the so-called "exponential sampling" method) perform better than equidistant points. Moreover, the position of the first point and the ratio of the geometric progression can be optimized, for instance, in order to minimize the condition number. This problem is discussed in [18] where examples are given for the

Laplace kernel. This shows how helpful inverse problem theory can be for optimal design of experiments and better data collection (using a minimum number of detectors for example). Expansions in the singular system like (13) can be used for experimental data inversion actually to recover the particle-size distribution. The corresponding algorithms can be easily implemented on personal computers coupled with the data acquisition system. They perform very fast since for a given experiment, i. e. for a given operator K_N, the singular system can be computed once for all and prestored in the computer. The inversion requires then, for each point in the object, only N multiplications and additions to compute the scalar products (g,v_k), followed by K_o multiplications by the values of u_k at that point, plus K_o additions. Good recoveries of size distributions are obtained in this way, provided that one does not attempt to exceed the resolution limits and that, of course, the a priori information about the support of the object contained in the smooth profile function $P(x)$ is correct. For PCS, a method for determining the mean value and the width of P from the data themselves has been proposed in [20].

CONFOCAL SCANNING MICROSCOPY

As we have already explained in the introduction, the coherent and ideal (aberration-free) imaging process in a confocal scanning microscope is described, in a one-dimensional situation, by the equation $g=Kf$, with K given by

$$(Kf)(x) = \int_{-\infty}^{+\infty} \mathrm{sinc}(x-y)\mathrm{sinc}(y)f(y)dy \tag{19}$$

where $\mathrm{sinc}(x)=\sin(\pi x)/\pi x$ (by means of a change of variables, the Rayleigh distance $R=\pi/\Omega$ has been taken here to be equal to 1). In a conventional scanning microscope, only the central point of the image is recorded, so that it is easy to see that the recorded image is just the convolution of the object with the square of the PSF, $\mathrm{sinc}(x)$. The corresponding transfer function is a triangle, with support $(-2\pi,+2\pi)$, which is to be compared to the characteristic function of the band $(-\pi,+\pi)$ corresponding to imaging by just one lens of bandwidth π. Hence, in a noise-free situation, the effect of the scanning would lead to an improvement in resolution by a factor of 2. In the presence of noise, however, the frequencies near the edges of the triangle will be lost, so that the practical gain in resolution provided by the usual confocal scanning light microscope (CSLM) is estimated to be less than a factor of 1.4, depending on the size of the detector. An alternative proposal was made in [11] to record the full diffraction image at a few points for each scanning position and to invert the data thus obtained. As shown in [12], this allows in principle the recovery of object frequencies up to 2π or equivalently leads to a true resolution improvement of a factor of 2 with respect to an ordinary microscope. Let us see now why this happens. The operator K defined by (19) is compact in $L^2(-\infty,+\infty)$. The closure of its range is the set of bandlimited functions on the band $(-\pi,+\pi)$ and therefore, by the Shannon sampling theorem, the image $g(x)$ can be expanded as follows:

$$g(x) = \sum_n g(x_n)\mathrm{sinc}(x-x_n) \qquad \text{where } x_n=n \quad (n=0,\pm1,\pm2,\ldots) \tag{20}$$

and needs only to be measured at the integer points. Moreover, it is not difficult to see that the set of visible objects is the set of bandlimited functions on the band $(-2\pi,+2\pi)$, which vanish on all integer points except zero. Hence, the generalized solution admits the following expansion:

$$f^{+}(y) = \sum_{m} f(y_m)\mathrm{sinc}[2(y-y_m)] \qquad (m = 0,\pm1,\pm2,\ldots)$$

$$y_0 = 0\ ;\ y_m = \mathrm{sign}(m)\ (|m| - \tfrac{1}{2}) \qquad (m = \pm1,\pm2,\ldots) \tag{21}$$

Substituting these two expansions in the equation $g=Kf$, we see that we have to solve the infinite matrix equation

$$g(x_n) = \sum_{m} A_{nm}\, f(y_m) \qquad (n,m = 0,\pm1,\pm2,\ldots) \tag{22}$$

where the matrix A is given by

$$A_{no} = \frac{1}{2}\delta_{no} \quad ; \quad A_{nm} = \frac{1}{2\pi^2}\,\frac{(-1)^{n+1}}{y_m(n-y_m)} \quad (m\neq0) \tag{23}$$

By using the explicit form of the singular system given in [21], we have been able to determine the inverse matrix analytically, which is given by [22]

$$(A^{-1})_{on} = 2(-1)^n \quad ; \quad (A^{-1})_{mn} = \frac{2n(-1)^{n+1}}{(n-y_m)} \quad (m\neq0) \tag{24}$$

This leads to a very simple reconstruction formula for the central value of the object (more precisely of its visible part f^{+}), namely,

$$f^{+}(0) = \sum_{n} t_n\ g(n) \qquad \text{with} \quad t_n = 2(-1)^n \tag{25}$$

The formulas (21) and (24) provide the reconstruction of the object $f^{+}(y)$ at other points. The impulse response (17) of the inversion in the absence of noise (i.e. when $F_\mu(\alpha_k)=1$ for all k) was shown in [21] to be

$$a(x,y) = 2\ [\ \mathrm{sinc}(2x)\mathrm{sinc}(2y) + \mathrm{sinc}(x-y)\sin(\pi x)\sin(\pi y)\ ] \tag{26}$$

which reduces for the central point $x=0$ to $a(0,y)=2\mathrm{sinc}(2y)$. When scanning through the object one gets then the convolution of the true object by this function $a(0,y)$, which has twice the bandwidth relative to a single lens of bandwidth $\Omega=\pi$. Hence the implementation in a confocal scanning microscope of the inversion formula given above should lead to a resolution improvement of a factor of 2 with respect to the ordinary microscope. Note that in practice only a finite number of sampled values $g(n)$ of the image can be recorded. The singular system of finite-dimensional sections of the matrix A can be computed numerically. It appears then that good approximations of the infinite-dimensional case are already obtained with a small number of data points. Indeed, using only five data points (and 17 object points), one finds for α_0 the value 0.821805 instead of the exact value 0.821898, for α_2 the value 0.206304 instead of 0.206417 and for α_4 the value 0.109697 instead

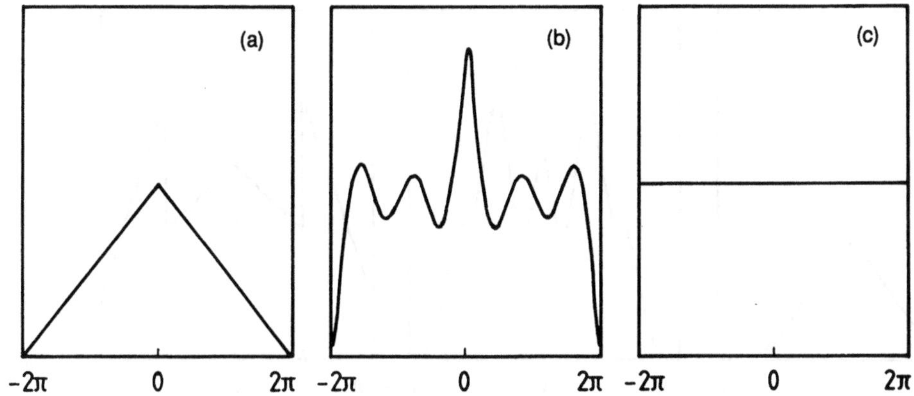

Figure 2. Transfer functions for the coherent CSLM: (a) conventional; (b) with 5 data points; (c) with an infinite number of data points.

of 0.109845. We report only the even singular values since it can be seen that the odd ones do not contribute to the central value of the restored object. The "mask vector" t_n appearing in (25) is replaced then by the approximate value $t_n=1.85(-1)^n$. Note that the condition number with this small number of data points is only $\alpha_0/\alpha_4 = 7.5$, so that regularization is not required for signal-to-noise ratios greater than the square of this condition number. Figure 2 shows the Fourier transform of the corresponding impulse response compared to the transfer function relative to an infinite number of data points. Other numerical examples are discussed in [23] as well as simulations of reconstructions which confirm the expected gain in resolution. Experimental confirmation is given in [24].

When the microscope is working in the incoherent mode and in particular in fluorescence microscopy, formula (19) has to be replaced by

$$(Kf)(x) = \int_{-\infty}^{+\infty} \mathrm{sinc}^2(x-y)\mathrm{sinc}^2(y)f(y)dy \tag{27}$$

The operator K is again compact in $L^2(-\infty,+\infty)$ and its null-space is described in [25]. The image is now a bandlimited function with bandwidth 2π, so that we have by the sampling theorem:

$$g(x) = \sum_n g(x_n)\mathrm{sinc}[2(x-x_n)] \qquad \text{where} \quad x_n=n/2 \ (n=0,\pm1,\pm2,\ldots) \tag{28}$$

The visible part of the object is a bandlimited function of bandwidth 4π, vanishing at the integer points except at zero, so that the following expansion holds

$$f^+(y) = \sum_m f(y_m)\mathrm{sinc}[4(y-y_m)] \qquad \text{with}$$

$$y_0 = 0\ ;\ y_m = \pm(k+j/4)\ ,\ m = \pm(3k+j) \quad (k=0,1,2,\ldots\ ;\ j=1,2,3) \tag{29}$$

Using these expansions, the object restoration problem is equivalent to the solution of the infinite-dimensional linear system (22), with the matrix A given here by

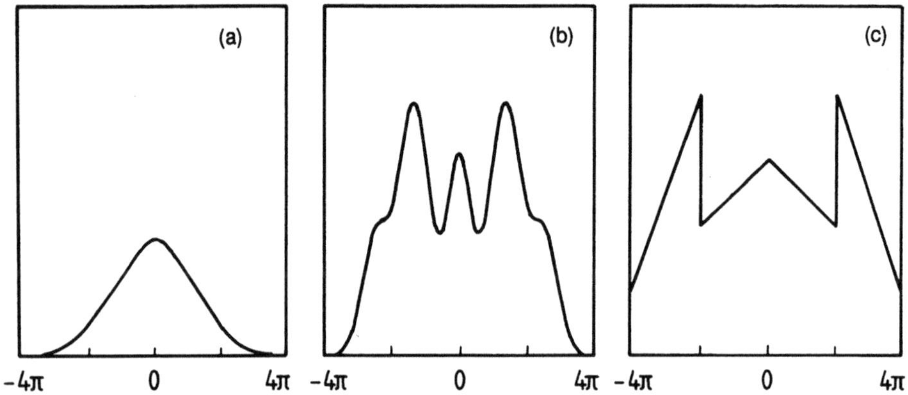

Figure 3. Transfer functions for the incoherent CSLM: (a) conventional; (b) with 5 data points; (c) with an infinite number of data points.

$$A_{nm} = \frac{1}{4}\,\mathrm{sinc}^2(x_n - y_m)\mathrm{sinc}^2(y_m) \tag{30}$$

We have not been able to find an explicit inverse of this matrix, so that the singular system has to be computed numerically. Nevertheless, thanks to a characterization of the null-space, we could derive, in the noise-free case the following expression for the impulse response (17) at the central point $x=0$ [25]

$$a(0,y) = \frac{9}{2}\,\mathrm{sinc}^2(2y) - \frac{3}{4}\,\mathrm{sinc}^2(y) - 3\,\mathrm{sinc}(2y)\sin^2(\pi y) \tag{31}$$

Its Fourier transform is given by

$$\begin{aligned} \hat{a}(\omega) &= (3/8)[3-(1/2\pi)|\omega|] && \text{for } |\omega| < 2\pi \\ \hat{a}(\omega) &= (3/8)[7-(3/2\pi)|\omega|] && \text{for } 2\pi < |\omega| < 4\pi \end{aligned} \tag{32}$$

The function (31) has to be compared with the impulse response $\mathrm{sinc}^4(y)$ of the conventional confocal scanning microscope. The corresponding transfer functions are shown in Figure 3, together with the transfer function obtained with only five data points. This demonstrates that a significant improvement in resolution can be expected even in the presence of noise, because the frequency band $(-4\pi,+4\pi)$ is better filled. This question is discussed in more detail in reference [25], and experimental confirmation is given in [26].

The one-dimensional model described here is a good mathematical, numerical and experimental laboratory for assessing the performance and the resolution power of this new type of confocal scanning microscope, which requires the inversion of several data points at each scanning position. The results given here can be extended straightforwardly to two-dimensional imaging with square pupils. The case of circular pupils considered in [27] and the case of aberrated system have to be treated numerically. The full three-dimensional imaging problem is investigated in [28] and work is still in progress concerning the practical implementation of these theoretical results. An important recent experimental development has been the implementation of the masking operation (25) by simple optical means [29].

ACKNOWLEDGEMENTS

C. De Mol is "Chercheur qualifié" of the Belgian National Fund for Scientific Research.

REFERENCES

[1] H. C. van de Hulst: "Light Scattering by Small Particles" (Dover, 1957)
[2] H. Z. Cummins and E. R. Pike (eds): "Photon Correlation and Light Beating Spectroscopy" (Plenum Press, 1974)
[3] J. W. Goodman: "Introduction to Fourier Optics" (McGraw-Hill, 1968)
[4] T. Wilson and C. Sheppard: "Theory and Practice of Scanning Microscopy" (Academic Press, 1984)
[5] J. G. McWhirter and E. R. Pike: J. Phys. **A11** (1978): 1729
[6] M. Bertero and E. R. Pike: Optica Acta **30** (1983): 1043 and
M. Bertero, P. Boccacci and E. R. Pike: Inverse Problems **1** (1985): 111
[7] M. Bertero, C. De Mol and E. R. Pike: Inverse Problems **2** (1986): 247
[8] C. W. Groetsch: "The theory of Tikhonov regularization for Fredholm integral equations of the first kind" (Pitman, 1984)
[9] M. Bertero, P. Boccacci and E. R. Pike: Proc. R. Soc. Lond. **A383** (1982): 15 and **A393** (1984): 51
[10] M. Bertero, C. De Mol and G. A. Viano: Optica Acta **27** (1980): 307
[11] M. Bertero and E. R. Pike: Optica Acta **29** (1982): 727 and 1599
[12] M. Bertero, C. De Mol, E. R. Pike and J. G. Walker: Optica Acta **31** (1984): 923
[13] M. Bertero, C. De Mol and E. R. Pike: Inverse Problems **1** (1985): 301 and **4** (1988): 573
[14] G. Toraldo di Francia: J. Opt. Soc. Am. **59** (1969): 799
[15] C. E. Shannon and W. Weaver: "The Mathematical Theory of Communication" (Univ. of Illinois Press, 1949)
[16] M. Bertero and C. De Mol: Atti della Fondazione Giorgio Ronchi, Anno **XXXVI** (1981): 619
[17] M. Bertero, P. Brianzi, E. R. Pike and L. Rebolia: Proc. R. Soc. Lond. **A415** (1988): 257
[18] M. Bertero, P. Brianzi, E. R. Pike: Proc. R. Soc. Lond. **A398** (1985): 23 and Inverse Problems **1** (1985): 1
[19] M. Bertero, F. A. Grünbaum: Inverse Problems **1** (1985): 181
M. Bertero, F. A. Grünbaum and L. Rebolia: Inverse Problems **2** (1986): 131
[20] M. Bertero, P. Brianzi, E. R. Pike, G. De Villiers, K. H. Lan and N. Ostrowsky: J. Chem. Phys. **82** (1985): 1551
[21] F. Gori and G. Guattari: Inverse Problems **1** (1985): 67
[22] M. Bertero, C. De Mol and E. R. Pike: J. Opt. Soc. Am. **A4** (1987): 1748
[23] M. Bertero, P. Brianzi and E. R. Pike: Inverse Problems **3** (1987): 195
[24] M. Young, R. E. Davies, J. Walker, E. R. Pike and M. Bertero: Europhysics Letters **9** (1989): 773
[25] M. Bertero, P. Boccacci, M. Defrise, C. De Mol and E. R. Pike: Inverse Problems **5** (1989): 441
[26] M. Young, R. E. Davies, J. Walker and E. R. Pike: in preparation
[27] M. Bertero, P. Boccacci: "Computation of the singular system for a class of integral operators related to data inversion in confocal microscopy", to appear in Inverse Problems
[28] M. Bertero, P. Boccacci, G. J. Brakenhoff, F. Malfanti and H. T. M. van der Voort: "Three-dimensional image restoration and super-resolution in fluorescence confocal microscopy", to appear in J. Microscopy
[29] J. Walker, E. R. Pike and M. Bertero: U. K. Patent Application Nº 89 13, 129-6

Temporal Homogenization

J. Corones and R. Winther

Applied Mathematical Sciences, Ames Laboratory-USDOE,
Iowa State University, Ames, IA 50011, USA

Introduction

Most, if not all, work on inverse scattering problems assumes a well defined model for the scattering processes within the target medium. This model is normally assumed fixed for the purposes of the investigation. So, for example, it might be assumed that the target medium is an elastic material characterized by its density and elastic constants. With this given efforts are made to find ways of reconstructing these material attributes from knowledge of incident and scattered waves. If loss mechanisms are also present in the material these are incorporated in the model and an attempt is made to recover both the elastic and loss properties from scattering data, now in the context of the extended model.

If a constraint is placed on the available information e.g. attention is restricted to reflected data, it is sometimes evident that, given the constraint, there is not sufficient data given to reconstruct the desired information. In the case of electrodynamics it is clear that given only reflection data from a normally incident signal it is not possible to reconstruct both the speed of light and the DC conductivity of a (by *assumption)* layered material. For problems of this sort it is necessary either to attempt to incorporate more data or to make a priori assumptions about the target material e.g. assume knowledge of the DC conductivity. It is to be expected that in the context of limited data more than one model for the material will reproduce the available data e.g. the reflected field.

In the case treated in this note the ambiguity in the modeling of a layered material that yields a given reflected field is discussed for the case of a normally incident electromagnetic wave. Two types of models are discussed, an inhomogeneous, possibly lossey dielectric material [1] and a homogeneous but dispersive material [2]. It is shown that given the first an equivalent medium of the second type can always be found. Since the equivalent medium is homogeneous this procedure represents of sort of homogenization procedure for the first (inhomogeneous) medium. Since the equivalent medium is dispersive its constitutive relation is time dependent and thus the process is called temporal homogenization.

This study was motivated by several applications. Among the most interesting is the study of low reflective coatings of one form or another. Often such coatings consist of layers, sometimes many layers of dielectric materials, which in many cases are difficult to manufacture. Likewise layered 'synthetic dielectrics', sometimes consisting of several hundred layers are of technological interest. In both these cases the layered material is designed to have specified reflective properties. This note shows that it is possible to replace this complex layered structures with simple homogeneous coatings that yields the same reflectivity.

Two Model Problems

Consider a half-space, illuminated by an electromagnetic field at normal incidence. Suppose the half-space is located in $x \geq 0$ and that free space occupies $x<0$. Two models for the half-space $x \geq 0$ will be considered. In the first the constigutive relations are given by

$$\vec{D}(x,t) = E(x)\ \vec{E}(x,t)$$

$$\vec{J}(x,t) = \sigma(x)\ \vec{E}(x,t)$$

$$\mu = \mu o$$

The wave equation for the electric field is

$$E_{xx}(x,t) - c^{-2}(x)\ E_{tt}(x,t) = b(x)\ E_t(x,t) \tag{1}$$

with

$$c^{-2}(x) = \epsilon(x)\mu o; \qquad b(z) = \sigma(z)\ \mu o \tag{2}$$

Suppose a right going field, $E^+(x,t)$ is incident on the half-space and that $E^+(0,t) = 0\ t < 0$. Assume further

$$E(x,t) = 0 \qquad x \leq 0 \qquad t < 0. \tag{3}$$

so that the medium is quiescent for $t < 0$. It is of interest here to find the reflected field in $x < 0$, $E^-(x,t)$. It has been shown previously that the left going field $E^-(x,t)$ is related to $E^+(x,t)$

$$E^-(x,t) = (\hat{R}(x_1\cdot)*E^+(x,\ \cdot))(t) \qquad x < 0 \tag{4}$$

In addition it has been shown that [1]

$$\frac{\partial R(x,t)}{\partial x} - \frac{2}{c(x)}\frac{\partial R(x,t)}{\partial t} = b(x)\ c(x)\ R(x,t) \tag{5}$$

$$+ \frac{1}{2}\left[b(x)\ c(x) + \frac{c'(x)}{c(x)}\right]\int_0^t R(x,t-s)R(x,s)ds$$

where R(x,t) is the kernel of the convolutional reflection operator $\hat{R}(x,t)$. In addition

$$R(x,0^+) = -\frac{1}{4} (b(x)\ c^2(x) - c'(x)) \tag{6}$$

In the second model the constitutive relations are given by

$$\vec{D}(x,t) = \epsilon_o\ \vec{E}(x,t) + \int_0^\infty G(t-s)\ E(x,s)\ ds \tag{7}$$

$$\mu = \mu_o$$

where G(t) is the susceptibility kernel. The relevant wave equation is

$$E_{xx}(x,t) - \frac{1}{c^2}\left\{E_{tt}(x,t) + \partial_t^2 \int_0^\infty G(s)E(z,t-s)ds\right\} = 0 \tag{8}$$

with $c^{-2} = \epsilon o \mu o$. Defining the kernel of the transient reflectivity as before, it can be shown that [2]

$$2cR_x(x,t) - 4R_t(x,t) = G'(t) + G(0)[2R(x,t) + R(x,\cdot)R(x\cdot)(t)]$$

$$+ G'(\cdot)\alpha[2R(x,\cdot) + R(x_1\cdot) - R(x,\cdot)](t) \tag{9}$$

In addition

$$R(x,0^+) = -\frac{1}{4} G(0^+) \tag{10}$$

Since the half-space x≥0 is homogeneous (here ϵ_o, μ_o, and G(t) do not depend upon α) $R_x(x,t) = 0\ x \geq 0$ and (9) can easily be integrated to yield

$$4R(t) + G(t) + G(\cdot) * [2R(\cdot) + R(\cdot)*R(\cdot)J(t) = 0 \tag{11}$$

The Direct and Inverse Scattering Problem

For each of the models discussed the direct or inverse problem can be solved using (5) and (6) or (9) and (10). Detailed discussions of these can be found in [1,2]. In particular given $\epsilon(x)$ and $\sigma(x)$ (5) and (6) provide an effective method for computing the transient reflectivity. With respect to the reverse problem note the (11), the equation for a homogeneous half-space is a Volterra equation of the second kind for G(t) given R(t). Thus the inverse problem is well posed and quite amendable to numerical computation.

Clearly if R(t) for an inhomogeneous half-space is given it is always possible to compute the susceptibility kernel of an associated homogeneous medium which yields the same transient reflectivity. The numerical results presented in the next section suggest that the G(t) are realistic from a physical point of view. It should be noted that by construction G(t) is causal, a constriant that is often difficult to implement in the frequency domain. Finally, it should be emphasized that given any incident signal the reflected signal is found by convolving the incident signal with the transient reflectivity. Since the procedure discussed here shows how to obtain equivalent transient reflectivities the construction yields the same reflected field for either medium given any incident field.

Numerical Examples

In this section several numerical examples will be presented that illustrate the method of constructing a homogeneous dispersive medium equivalent, from the point of view reflectivity, to an inhomogeneos dielectric medium with loss. In each example the inhomogeneous medium has a periodic dielectic function $\epsilon(x)$ with relative permativity ranging from 1 to 3 with a spatial period of .00002 meters. In each case the medium has constant loss, $\sigma(x)$ constant. In case one $\sigma = 40$ mho/meter, in case two $\sigma = 10$ mho/meter and in case three $\sigma = 4$ mho/meter.

In each case the first step is to compute R(t) by integrating equation (5). The resulting R(t) for case one (40 mho/m, a strongly attenuating medium) is given in Figure (1). Note that the medium is strongly scattering so that the resulting reflectivity has little resemblance to the material profile due to the presence of many multiple reflections. This time trace gets more complicated as time goes on due to an ever increasing number of multiple reflections. Of course R(t) goes to zero for large t. The reflectivity for the other cases are qualitatively similar to Figure (1) though as expected they tend to zero more slowly than R(t) for case one.

Once R(t) has been computed the equivalent G(t) is found by solving equation (11) for G(t) given R(t) as input. Consider first the results in the strongly attenuating case. The resulting G(t) is displayed in Figure 2. The most striking aspect of this result is the clear effect of the periodic nature of the original inhomogeneous medium in the nearly monochromatic "carrier frequency" of G(t). This frequency gives the short time scale of the problem. It is particularly interesting that the spacial length scale set by the periodicity of the original medium, which is equivalent to a time scale (time of flight across a period) is first greatly obscured in the transient reflectivity, observe Figure 1, due to the strongly nonlinear nature of the scattering, but then recovered through the strongly nonlinear process of solving equation ( ) for G(t).

The second important feature of G(t) is the existence of the long time scale. that is the scale on which the medium equilibrates. This time scale is related to the decay rate of the transient reflectivity and clearly increases as the attenuation decreases as can be seen in Figures 3 and 4. It should be pointed out that these later figures are the envelopes only of the resulting susceptibility kernels. The high frequency (monochromatic) component has been

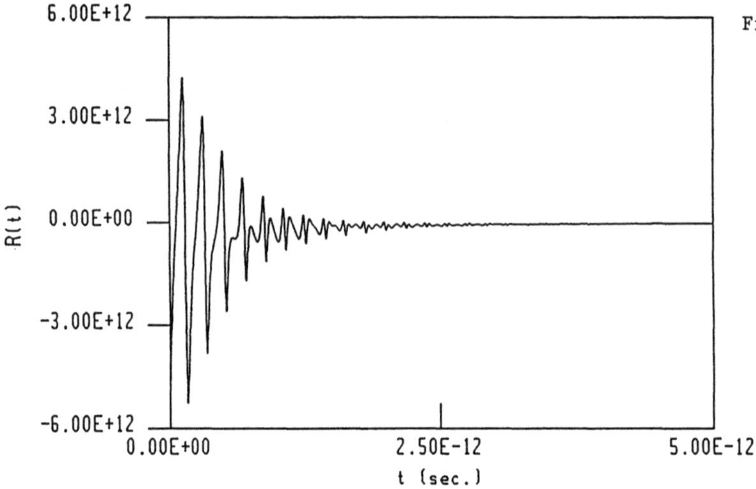

Figure 1

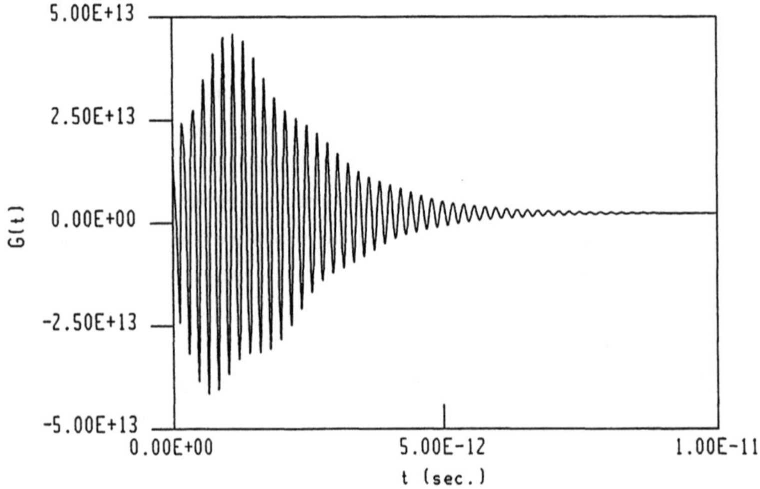

Figure 2

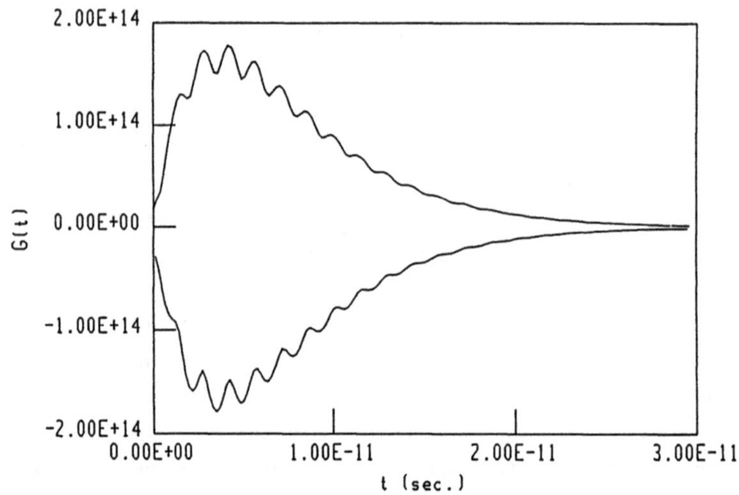

Figure 3

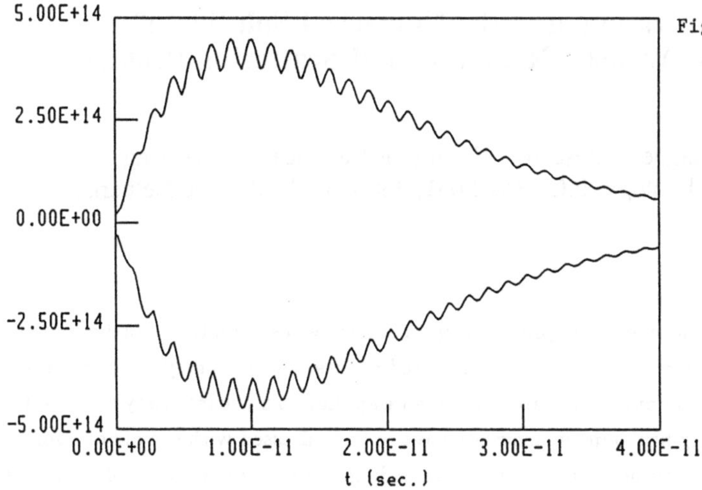

Figure 4

suppressed for ease of graphic display. Note the lengthening of the times from Figure 1 to 2 to 3.

Acknowledgment

This work was supported by the Applied Mathematical Sciences subprogram of the Office of Energy Research, U.S. Department of Energy under contract No. W-7405-ENG-82 and by the Office of Naval Research contract No. N0014-83-K-0038.

References

1. J.P. Corones, M.E. Davison and R.J. Krueger, "Dissipative inverse problems in the time domain, " in Inverse Methods in Electromagnetic Imaging, NATO ASI series, Series C, Vol. 143, edited by W-M Boerner, Riedel, Dordrecht, Holland, p. 121-130, (1985).

2. R.S. Beezley and R.J. Krueger, "An electromagnetic inverse problem for dispersive media," J. Math. Phys. 26, 317-325 (1985).

Reciprocity, Discretization, and the Numerical Solution of Direct and Inverse Acoustic Radiation and Scattering Problems

A.T. de Hoop

Laboratory of Electromagnetic Research, Faculty of Electrical Engineering, Delft University of Technology, P.O. Box 5031, 2600 GA Delft, The Netherlands

Abstract. Except for the canonical problems in acoustics whose solution can be expressed in terms of analytic functions of a not too complicated nature, and for analytic approximation techniques (usually of an asymptotic nature) that can be applied to a wider variety of cases, wave propagation and scattering problems in acoustics have to be addressed with the aid of numerical techniques. Many of these methods can be envisaged as being discretized versions of appropriate 'weak' formulations of the pertinent operator (differential or integral) equations. For the relevant problems as formulated in the time Laplace-transform domain it is shown that the Rayleigh reciprocity theorem encompasses all known weak formulations, while its discretization leads to the discretized forms of the corresponding operator equations, in particular to their finite-element and integral-equation modeling schemes. Both direct (forward) and inverse problems are discussed.

1. Introduction

Except for the canonical problems in acoustics whose solution can be expressed in terms of analytic functions of a not too complicated nature (examples of which can be found in Bowman, Senior and Uslenghi 1969) and for analytic approximation techniques (usually of an asymptotic nature), both in the long and the short wavelength regimes, that can be applied to a wider variety of cases (see, Felsen and Marcuvitz 1973), radiation and scattering problems in acoustics have to be addressed with the aid of numerical methods. Many of these methods can be envisaged as being discretized versions of appropriate 'weak' formulations of the pertinent operator (differential or integral) equations. Specifically, the finite- element method, based on the method of weighted residuals (see, for example, Zienkiewicz and Taylor 1989), applied to the first-order acoustic wave equations, and the collocation method (method of point matching) applied to the source-type acoustic integral relations (see, for example, Herman 1981) can be grouped in this category. The present contribution is an attempt to systematize the different approaches and to bring, to a certain extent, consistency and coherence in the procedures.

To this end, the time Laplace-transform domain (s-domain or complex frequency domain) acoustic Rayleigh reciprocity theorem for time-invariant configurations is taken as point of departure. By taking the transform parameter s to be positive and real (as is done in the Cagniard method for calculating impulsive waves in stratified media (see, for example, Van der Hijden 1987), or complex in the right half $\mathrm{Re}(s) > 0$ of the complex s-plane, the causality of the wave motion is ensured by requiring the time Laplace-transform domain wave-field quantities to be bounded functions of position in space, especially at infinitely large distances from the sources (of bounded extent) that generate the wave field. Also, arbitrarily inhomogeneous and anisotropic materials with arbitrary relaxation behavior can in this way be incorporated in the analysis in an easy manner. From the s-domain formulations the time-domain counterparts easily follow upon using some standard rules of the one-sided Laplace transformation, while the results for the limiting case of sinusoidally in time varying field quantities follow upon replacing s by $j\omega$, where j is the imaginary unit and ω the angular frequency of the oscillations, on the condition that imaginary values of s are approached via the right half of the complex s-plane.

Taking one of the two states in the reciprocity theorem to be the acoustic state to be actually computed, and the other to be an appropriately chosen 'auxiliary' or 'computational' one, it is shown that the reciprocity theorem encompasses all known 'weak' formulations of the acoustic-wave differential equations and source-type integral relations. For this reason, the Rayleigh reciprocity theorem is considered to best serve as the point of departure for the computational modeling of acoustic wave fields. (Note that this theorem is global rather than local in nature.)

Turning to numerics, first a geometrical discretization procedure is applied, with the tetrahedron (simplex in three-dimensional space) as the basic building block. The mesh size (supremum of the maximum diameters of the tetrahedra) of the discretization should be consistent with both the geometry of the configuration and the (inhomogeneous) distribution of matter in it, in the sense that it guarantees that the discretized configuration and the actual one differ relatively not more than a given fractional number according to some agreed-upon relative error criterion (for example, the normalized root-mean-square error of the difference in the material parameters). Subsequently, the wave-field quantities are expanded in terms of a base in an appropriately chosen linear function space, and a weighting procedure with an appropriately chosen 'computational state' is carried out. This procedure leads to a system of linear algebraic equations in the expansion coefficients of the wave-field quantities. To solve the system of equations in the expansion coefficients, again, an error criterion is needed to define what the 'best approximation' to their (iterative) solution is (cf. Van den Berg 1985). In this manner, several versions of the finite-element method can be understood as well as certain discretized versions of the integral equations describing scattering phenomena. Finally, it is remarked that both direct (forward) and inverse modeling are natural consequences of the reciprocity theorem (cf. De Hoop 1988).

Through an analysis of the type presented here, it is, at least theoretically, feasible that the whole operation of arriving at numerical results of a predetermined accuracy is computer controlled.

2. The Acoustic Wave Field in the Configuration

The configuration in which the acoustic wave field is present consists of a fluid material that occupies three-dimensional space R^3. In the bounded subdomain $D^s \subset R^3$ the fluid is, in general, inhomogeneous and anisotropic. The boundary surface ∂D^s of D^s is assumed to be piecewise smooth. In the unbounded domain $D_0 = \bar{D}^s$ that is the complement of the closure of D^s in R^3, the fluid is homogeneous and isotropic.

The position of observation in the configuration is specified by the coordinates $\{x_1, x_2, x_3\}$ with respect to a fixed, orthogonal, Cartesian reference frame with origin O and the three mutually perpendicular base vectors $\{\mathbf{i}_1, \mathbf{i}_2, \mathbf{i}_3\}$ of unit length each. In the indicated order, the base vectors form a right-handed system. The subscript notation for Cartesian vectors and tensors is used and the summation convention applies. The corresponding lowercase Latin subscripts are to be assigned the values $\{1, 2, 3\}$. Whenever appropriate, the position vector will be denoted by $\mathbf{x} = x_m \mathbf{i}_m$. The time coordinate is denoted by t. Partial differentiation is denoted by ∂; ∂_m denotes differentiation with respect to x_m, ∂_t is a reserved symbol for differentiation with respect to t.

For any causal space-time function $u = u(\mathbf{x}, t)$ the one-sided Laplace transform is introduced as

$$\hat{u}(\mathbf{x}, s) = \int_{t=0}^{\infty} \exp(-st) u(\mathbf{x}, t) dt, \tag{1}$$

where the instant $t = 0$ marks the onset of the events. Obviously, for bounded $| u(\mathbf{x}, t) |$, $\hat{u}(\mathbf{x}, s)$ is an analytic function of the complex transform parameter s in the right half $\mathrm{Re}(s) > 0$ of the complex s-plane. For ease of notation the circumflex over a symbol denoting its s-domain counterpart will be omitted in the remainder of the paper. The acoustic wave motion is started from a configuration at rest; then, under the one-sided Laplace transformation the operator ∂_t is replaced by an algebraic factor s.

In each subdomain of the configuration where the fluid's acoustic properties vary continuously with position, the acoustic wave field quantities are continuously differentiable and satisfy the s-domain equations

$$\partial_k p + \varsigma_{k,r} v_r = f_k, \tag{2}$$

$$\partial_r v_r + \eta p = q, \tag{3}$$

in which p is the acoustic pressure, v_r is the particle velocity, $\varsigma_{k,r}$ is the fluid's longitudinal acoustic impedance per length ($\varsigma_{k,r} = s\mu_{k,r}$, where $\mu_{k,r}$ is the time Laplace transform of the fluid's inertia relaxation function; for an instantaneously reacting, i.e. lossless, fluid, $\mu_{k,r} = \rho_{k,r}$, where $\rho_{k,r}$ is the fluid's s-independent volume density of mass), η is the fluid's transverse acoustic admittance per length ($\eta = s\chi$, where χ is the time Laplace transform of the fluid's compliance relaxation function; for an instantaneously reacting, i.e. lossless, fluid, $\chi = \kappa$, where κ is the fluid's s-independent compressibility), f_k is the volume source density of force, and q is the volume source density of injection rate. Full anisotropy in the inertia properties of the fluid is taken into account. For lossy fluids (i.e., fluids with relaxation), $\mu_{k,r}$ and χ are s-dependent, subject to the condition of causality which entails analyticity in $\mathrm{Re}(s) > 0$. Across interfaces between different kinds of fluids (which can slip with respect to each other), p and $\nu_r v_r$ are to be continuous, where ν_r is the unit vector along the normal to such an interface. In the configuration, objects can be present that are impenetrable to the acoustic wave motion. On the boundary surface of such an object explicit boundary conditions hold; these are: either $p \to 0$ (void) or $\nu_r v_r \to 0$ (immovable rigid object).

In the domain D_0, which is denoted as the embedding, the fluid is assumed to be homogeneous, isotropic and lossless, and we have $\varsigma_{k,r} = s\rho_0 \delta_{k,r}$ and $\eta = s\kappa_0$, in which ρ_0 and κ_0 are position- and s-independent positive constants, and $\delta_{k,r}$ is the unit tensor of rank two (Kronecker tensor: $\delta_{k,r} = 1$ if $k = r$, $\delta_{k,r} = 0$ if $k \neq r$). The acoustic wave field quantities that are generated by known sources in such a fluid are analytically known (cf. De Hoop 1985). In deriving the relevant representations causality plays, again, an essential role.

3. The Contrast-Source or Scattering Formulation of the Acoustic Wave Problem

The assumed simple acoustic properties of the embedding enable us to reformulate the acoustic wave problem as a contrast-source or scattering problem. To this end, it is assumed that in some bounded domain $D^i \subset D_0$ external sources with given volume distributions $\{f_k^i, q^i\}$ generate the acoustic wave field. The wave field $\{p^i, v_r^i\}$ that would be generated by these source distributions if also the domain D^s had the acoustic properties of the embedding D_0, is denoted as the incident wave field. Obviously, it satisfies the acoustic wave equations

$$\partial_k p^i + s\rho_0 v_k^i = \{f_k^i, 0\} \quad \text{for } \mathbf{x} \in \{D^i, \bar{D}^i\}, \tag{4}$$

$$\partial_r v_r^i + s\kappa_0 p^i = \{q^i, 0\} \quad \text{for } \mathbf{x} \in \{D^i, \bar{D}^i\}, \tag{5}$$

where $\bar{D}^i$ is the complement of the closure of D^i in R^3. The total acoustic wave field $\{p, v_r\}$ in the configuration satisfies on account of Eqs.(2) and (3) the acoustic wave equations

$$\partial_k p + \zeta_{k,r} v_r = \{f_k^i, 0\} \quad \text{for } \mathbf{x} \in \{D^i, \bar{D}^i\}, \tag{6}$$

$$\partial_r v_r + \eta p = \{q^i, 0\} \quad \text{for } \mathbf{x} \in \{D^i, \bar{D}^i\}. \tag{7}$$

Upon introducing the scattered wave field $\{p^s, v_r^s\}$ as the difference between the total wave field and the incident wave field, i.e.,

$$\{p^s, v_r^s\} = \{p - p^i, v_r - v_r^i\}, \tag{8}$$

it follows from Eqs.(4)-(5) and (6)-(7) that this wave field satisfies the acoustic wave equations

$$\partial_k p^s + s\rho_0 v_k^s = \{f_k^s, 0\} \quad \text{for } \mathbf{x} \in \{D^s, \bar{D}^s\}, \tag{9}$$

$$\partial_r v_r^s + s\kappa_0 p^s = \{q^s, 0\} \quad \text{for } \mathbf{x} \in \{D^s, \bar{D}^s\}, \tag{10}$$

where the contrast volume source densities $\{f_k^s, q^s\}$ of force and injection rate, respectively, are given by

$$f_k^s = (s\rho_0\delta_{k,r} - \zeta_{k,r})v_r, \tag{11}$$

$$q^s = (s\kappa_0 - \eta)p. \tag{12}$$

Of course, the values of the contrast volume source densities are not known as long as the values of the total acoustic wave field in the contrasting domain D^s have not been determined.

4. The Acoustic Reciprocity Relation

For our further analysis, the s-domain reciprocity relation that is associated with Eqs.(2) and (3) will serve as the point of departure. A general wave-field reciprocity theorem interrelates, in a specific manner, the quantities that characterize two different physical states that could occur in one and the same domain in space-time. For time-invariant configurations the application of the one-sided Laplace transformation of Eq.(1) to the convolution-type reciprocity theorem leads to an equivalent s-domain result. For this to be applicable to the fluid configuration under investigation, the fluids in the two states should be present in one and the same time-invariant subdomain D of the configuration. The two states will be distinguished by the superscripts A and B, respectively. The reciprocity relation will be arrived at by combining certain weighted forms, over the domain D, of the acoustic wave equations pertaining to the two states.

First, the acoustic wave equations (2) and (3) pertaining to the State A are multiplied through by the particle velocity and the acoustic pressure, respectively, pertaining to the State B, and integrated over the domain D. The result is

$$\int_{\mathbf{x}\in D} v_k^B \partial_k p^A dV + \int_{\mathbf{x}\in D} v_k^B \varsigma_{k,r}^A v_r^A dV = \int_{\mathbf{x}\in D} v_k^B f_k^A dV, \tag{13}$$

$$\int_{\mathbf{x}\in D} p^B \partial_r v_r^A dV + \int_{\mathbf{x}\in D} p^B \eta^A p^A dV = \int_{\mathbf{x}\in D} p^B q^A dV. \tag{14}$$

Secondly, the acoustic wave equations (2) and (3) pertaining to the State B are multiplied through by the particle velocity and the acoustic pressure, respectively, pertaining to the State A and integrated over the domain D. The result is

$$\int_{\mathbf{x}\in D} v_k^A \partial_k p^B dV + \int_{\mathbf{x}\in D} v_k^A \varsigma_{k,r}^B v_r^B dV = \int_{\mathbf{x}\in D} v_k^A f_k^B dV, \tag{15}$$

$$\int_{\mathbf{x}\in D} p^A \partial_r v_r^B dV + \int_{\mathbf{x}\in D} p^A \eta^B p^B dV = \int_{\mathbf{x}\in D} p^A q^B dV. \tag{16}$$

Equations (13)-(14) and (15)-(16) can be regarded as weighted forms of the acoustic wave equations. Assuming that in D the acoustic properties of the fluid vary continuously with position (under which condition the acoustic wave quantities are continuously differentiable in D), taking the sum of Eqs.(13) and (16), subtracting from the result the sum of Eqs.(14) and (15), and applying Gauss' divergence theorem to the terms containing the spatial derivatives, it follows that

$$\begin{aligned}\int_{\mathbf{x}\in\partial D} \nu_m (p^A v_m^B - p^B v_m^A) dA &= \int_{\mathbf{x}\in D} [(\varsigma_{r,k}^B - \varsigma_{k,r}^A) v_r^A v_k^B - (\eta^B - \eta^A) p^A p^B] dV \\ &+ \int_{\mathbf{x}\in D} (f_k^A v_k^B - f_r^B v_r^A + q^B p^A - q^A p^B) dV,\end{aligned} \tag{17}$$

in which ∂D is the boundary surface of the domain D, which is assumed to be piecewise smooth and ν_m is the unit vector along its normal, pointing away from D. The validity of Eq.(17) is extended to a domain in which the acoustic fluid properties are only piecewise continuously differentiable by adding the results of the (finite number of) subdomains to which Eq.(17) applies. In this procedure, the contributions from interfaces cancel in view of the pertaining boundary conditions of the continuity type, while the contributions from the boundary surfaces of impenetrable objects vanish in view of the pertaining boundary conditons of the explicit type. The thus obtained Eq.(17) is the global Rayleigh reciprocity relation that will be used in the considerations that follow. Since, in it, spatial differentiations no longer occur, it can be regarded as the 'weakest' form of the acoustic wave equations (2) and (3).

Note that in the first integral on the right-hand side the differences in medium properties in the States A and B occur. This term vanishes if we take $\varsigma^A_{k,r} = \varsigma^B_{r,k}$ and $\eta^A = \eta^B$ at all points of D. If these conditions hold, the media in the States A and B are denoted as each others' adjoints. If the conditions hold for one and the same medium, the medium is denoted as self-adjoint. In the second integral on the right-hand side the volume source distributions in the States A and B occur. This term vanishes if the domain D is sourcefree. From the derivation it is clear that, due to the arbitrariness of the States A and B, subject of course to Eqs.(2) and (3), Eq.(17) and Eqs.(13)-(16) are equivalent as long as the pertaining boundary conditions are satisfied, both in State A and in State B.

In some of the applications, D will be the entire three-dimensional space. To address this situation, Eq.(17) is first applied to the domain interior to the sphere S_Δ of radius $\Delta > 0$ and with center at the origin of the chosen reference frame, after which the limit $\Delta \to \infty$ is taken. From some Δ onward, S_Δ will be entirely situated in the homogeneous, isotropic, lossless (ideal) fluid of the embedding. In view of this, the far-field representations for the acoustic wave-field quantities can on S_Δ, for sufficiently large values of Δ, be used. From the latter, the contribution from S_Δ can be shown to vanish in the limit $\Delta \to \infty$ (cf. De Hoop 1988).

5. Discretization Procedure

The numerical handling of acoustic wave problems always implies that some discretized version of the wave problem is used to 'approximate' the actual analytic one. In the following, the global reciprocity relation Eq.(17) will serve as the point of departure to define in what sense such an approximation holds. First, it is observed that each quantity $Q = Q(\mathbf{x}, s)$ occurring in the acoustic wave problem (which can be a scalar, a vector, or a tensor of arbitrary rank) and defined on some domain D has, after discretization, a discretrized counterpart $[Q] = [Q](\mathbf{x}, \mathbf{s})$ defined on the discretized version $[D]$ of D. Further, the actual machine computations are finite in number and can therefore only be carried out for some bounded computational domain $D \subset R^3$. In the scheme to be presented, the inhomogeneous part D^s of the configuration has to be entirely incorporated in D, so we take D^s to be a proper subset of D. Note that the acoustic wave motion is defined in the entire R^3, which implies that D also contains some part of D_0, where the fluid is homogeneous, isotropic and lossless. Without loss of generality we can therefore take the discretized version of the domain of computation identical to the actual one, i.e., $[D] = D$.

Discretization of the Computational Domain

The domain of computation D is discretized by taking it to be the union of a finite number of tetrahedra (simplices in R^3) that all have vertices, edges and faces in common (see, Naber 1980). The vertices of the tetrahedra will also be denoted as the nodes of the (geometrical) mesh and the supremum h of the maximum diameters of the tetrahedra will be denoted as the mesh size.

Except for the Green's functions to be introduced in Section 9, each quantity $Q = Q(\mathbf{x}, s)$ occurring in the acoustic wave-field problem will, in the interior of each tetrahedron, be approximated by the linear interpolation of its values at the vertices. Let $\{\mathbf{x}(0), \mathbf{x}(1), \mathbf{x}(2), \mathbf{x}(3)\}$ denote the position vectors of the vertices of the tetrahedron (simplex) *SMPLX*, then the corresponding linear interpolation is given by

$$[Q](\mathbf{x}, s) = \sum_{I^V=0}^{3} A^Q(I^V, s)\lambda(I^V; \mathbf{x}) \quad \text{for } \mathbf{x} \in SMPLX, \tag{18}$$

in which

$$A^Q(I^V, s) = Q(\mathbf{x}(I^V), s) \tag{19}$$

is the value of Q at the vertex with ordinal number I^V, and $\{\lambda(I^V; \mathbf{x}); I^V = 0, 1, 2, 3\}$ are the barycentric coordinates of $\mathbf{x}$ in *SMPLX*. The latter have the property

$$\lambda(I^V, \mathbf{x}(J^V)) = \{1, 0\} \quad \text{if } \{I^V = J^V, I^V \neq J^V\}, \tag{20}$$

and are expressed in terms of the vectorial areas of the faces of *SMPLX* through

$$\lambda(I^V; \mathbf{x}) = 1/4 - (1/3V)(x_m - b_m)A_m(I^V) \quad \text{with } I^V = 0, 1, 2, 3, \tag{21}$$

where $A_m(I^V)$ is the outwardly oriented vectorial area of,the face opposite to the vertex with ordinal number I^V.

Discretization of the Medium Parameters

In the discretization of the medium parameters a distinction must be made between subdomains in which these parameters vary continuously with position and subdomains in which surfaces of discontinuity in these parameters occur. It is assumed that across such surfaces of discontinuity in medium parameters, the parameter values jump by finite amounts. Especially in applications where accurate values of the wave-field quantities up to these surfaces are needed (such as, for example, in the modeling of borehole measurement situations in exploration geophysics, and in the non-destructive evaluation of mechanical structures), special measures have to be taken to model the behavior of these quantities accurately. In principle, the medium properties can jump across any face of any tetrahedron of the discretized geometry. To accomodate this feature, all nodes of the geometrical mesh are considered as multiple nodes, where the multiplicity of each node is equal to the number of tetrahedra that meet at that node. The values of the constitutive parameters at the vertices follow either from user-supplied input expressions that are spatially sampled in the interior of each tetrahedron close to each of its vertices (as is the case in direct or forward scattering problems) or from computationally derived values (as is the case in inverse scattering problems). Out of the thus constructed local expansions of the medium parameters, their global expansions over the domain of computation are composed. If in the latter procedure, at a particular node no discontinuity turns up, the multiple node is replaced by a simple one, with an associated single value of the relevant constitutive parameter. For a non-scalar constitutive parameter (such as the longitudinal impedance per length for a fluid with anisotropic inertia properties) the components with respect to the background Cartesian reference frame are used in the discretization procedure. The relevant global expansions for the medium properties are written as

$$[\varsigma_{k,r}(\mathbf{x},s)] = \sum_{I^\varsigma=1}^{N^\varsigma} A^\varsigma(I^\varsigma,s)\Phi^\varsigma_{k,r}(I^\varsigma,\mathbf{x}) \quad \text{for } \mathbf{x} \in [D], \tag{22}$$

where $\{\Phi^\varsigma_{k,r}(I^\varsigma,\mathbf{x}); I^\varsigma = 1,\ldots,N^\varsigma\}$ is the sequence of global expansion functions for the fluid's longitudinal impedance per length and $\{A^\varsigma(I^\varsigma,s); I^\varsigma = 1,\ldots,N^\varsigma\}$ is the sequence of its global expansion coefficients, and

$$[\eta(\mathbf{x},s)] = \sum_{I^\eta=1}^{N^\eta} A^\eta(I^\eta,s)\Phi^\eta(I^\eta,\mathbf{x}) \quad \text{for } \mathbf{x} \in [D], \tag{23}$$

where $\{\Phi^\eta(I^\eta\mathbf{x}); I^\eta = 1,\ldots,N^\eta\}$ is the sequence of global expansion functions for the fluid's transverse admittance per length and $\{A^\eta(I^\eta,s); I^\eta = 1,\ldots,N^\eta\}$ is the sequence of its global expansion coefficients.

Discretization of the Volume Source Densities

For the discretization of the volume source densities the same procedure as for the discretization of the medium parameters is followed. In the case of direct or forward modeling, the values of the volume source densities at the vertices of the tetrahedra out of which the discretized geometry is composed, follow from user-supplied input expressions that are sampled in the interior of each tetrahedron close to each of its vertices. In the modeling of inverse source problems their values are computationally constructed. Out of the thus constructed local expansions of the volume source densities, their global expansions over the domain of computation are composed. If in this procedure at a particular node no discontinuity shows up, the multiple node is replaced by a simple one, with an associated single value of the relevant volume source density. Here, too, for the non-scalar volume source densities the components with respect to the background Cartesian reference frame are used in the discretization procedure. The relevant global expansions for the volume source densities are written as

$$[f_k(\mathbf{x},s)] = \sum_{I^f=1}^{N^f} A^f(I^f,s)\Phi^f_k(I^f,\mathbf{x}) \quad \text{for } \mathbf{x} \in [D], \tag{24}$$

where $\{\Phi^f_k(I^f,\mathbf{x}); I^f = 1,\ldots,N^f\}$ is the sequence of global expansion functions for the volume source density of force and $\{A^f(I^f,s); I^f = 1,\ldots,N^f\}$ is the sequence of its global expansion coefficients, and

$$[q(\mathbf{x},s)] = \sum_{I^q=1}^{N^q} A^q(I^q,s)\Phi^q(I^q,\mathbf{x}) \quad \text{for } \mathbf{x} \in [D], \tag{25}$$

where $\{\Phi^q(I^q,\mathbf{x}); I^q = 1,\ldots,N^q\}$ is the sequence of global expansion functions for the volume source density of injection rate and $\{A^q(I^q,s); I^q = 1,\ldots,N^q\}$ is the sequence of its global expansion coefficients.

Discretization of the Wave-Field Quantities

In the discretization of the acoustic wave field quantities the situation is more complicated. Here, some components are by necessity continuous across an interface of discontinuity in material properties, while other components show a finite jump across such a discontinuity surface. To preserve accuracy in the computational results, it is necessary, both in the modeling of direct or forward problems and in the modeling of inverse problems, to take computational measures that enforce the continuity conditions across an interface (in machine precision) and leave the non-continuous components free to jump by finite amounts. For this purpose, local expansions of the type of Eq.(18) have been developed where a non-scalar quantity at a vertex is expressed in terms of those of its tensor components that are continuous across an interface of discontinuity in material properties. These components are, in general, not the components with respect to the background Cartesian reference frame (see, Mur and De Hoop 1985, and Stam and De Hoop 1990). For the acoustic wave field quantities this difficulty does not arise in connection with the acoustic pressure: this quantity is continuous across any interface of discontinuity in material properties. For the particle velocity, however, we have to take special measures. Its component normal to an interface is continuous across the interface, while the remaining components show a jump discontinuity. To guarantee the continuity of the normal component of the particle velocity across each face of adjoining tetrahedra, we consider each node as a multiple node and construct at each vertex the particle velocity out of its three values on the three faces that meet at that vertex and use the relevant values in the local expansion for the particle velocity. (In the terminology of the finite-element method we have denoted such elements as 'face' elements (cf. Stam and De Hoop 1990)). Out of the thus constructed local expansions, the global expansions over the domain of computation are composed. If in this procedure simple nodes are met, the particle velocity is just expressed in terms of its components in the background Cartesian reference frame. The face-element representation for the global expansion of the particle velocity could also be used at simple nodes, but this leads to an unnecessarily larger number of expansion coefficients to be computed (without yielding an increased accuracy) since the number of tetrahedral faces that meet at a particular node is larger than three. The relevant global expansions for the wave-field quantities are written as

$$[p(\mathbf{x},s)] = \sum_{I^p=1}^{N^p} A^p(I^p,s)\Phi^p(I^p,\mathbf{x}) \quad \text{for } \mathbf{x} \in [D], \tag{26}$$

where $\{\Phi^p(I^p,\mathbf{x}); I^p = 1,\ldots,N^p\}$ is the sequence of global expansion functions for the acoustic pressure and $\{A^p(I^p,s); I^p = 1,\ldots,N^p\}$ is the sequence of its global expansion coefficients, and

$$[v_r(\mathbf{x},s)] = \sum_{I^v=1}^{N^v} A^v(I^v,s)\Phi^v_r(I^v,\mathbf{x}) \quad \text{for } \mathbf{x} \in [D], \tag{27}$$

where $\{\Phi^v_r(I^v,\mathbf{x}); I^v = 1,\ldots,N^v\}$ is the sequence of global expansion functions for the particle velocity and $\{A^v(I^v,s); I^v = 1,\ldots,N^v\}$ is the sequence of its global expansion coefficients.

6. Source-Type Wave-Field Integral Representations for the Embedding

In an unbounded, homogeneous, isotropic, lossless fluid with volume density of mass ρ_0 and compressibility κ_0 (like the one in the embedding D_0) the s-domain source-type wave-field representations are analytically known. They are given by (cf. De Hoop 1985)

$$p = s\rho_0\Psi - \partial_k W_k, \tag{28}$$

$$v_r = (s\rho_0)^{-1} f_r - \partial_r\Psi + (s\rho_0)^{-1}\partial_r\partial_k W_k, \tag{29}$$

where

$$\{W_k, \Psi\}(\mathbf{x}, s) = \int_{\mathbf{x}'\in D^T} G(\mathbf{x}-\mathbf{x}', s)\{f_k^T, q^T\}(\mathbf{x}', s)dV \tag{30}$$

are the acoustic-wave vector and scalar potentials associated with the volume source distributions f_k^T of force and q^T of injection rate, respectively,

$$G(\mathbf{x}) = \exp(-s \mid \mathbf{x} \mid /c_0)/4\pi \mid \mathbf{x} \mid \quad \text{for } \mathbf{x} \neq \mathbf{0} \tag{31}$$

is the infinite-medium Green's function of the modified Helmholtz equation

$$\partial_m\partial_m G - (s^2/c_0^2)G = -\delta(\mathbf{x}), \tag{32}$$

the quantity

$$c_0 = (\rho_0\kappa_0)^{-1/2} \tag{33}$$

is the acoustic wave speed in the fluid and D^T is the bounded spatial support of the volume source distributions. Representations of the type of Eqs.(28) and (29) play a vital role in the hybrid finite-element modeling to be discussed in Section 8. In that kind of modeling, also the acoustic wave equations (cf. Eqs.(4)-(8))

$$\partial_k p^s + \varsigma_{k,r} v_r^s = \{(s\rho_0\delta_{k,r} - \varsigma_{k,r})v_r^i, 0\} \quad \text{for } \mathbf{x} \in \{D^s, \bar{D}^s\}, \tag{34}$$

$$\partial_r v_r^s + \eta p^s = \{(s\kappa_0 - \eta)p^i, 0\} \quad \text{for } \mathbf{x} \in \{D^s, \bar{D}^s\}, \tag{35}$$

satisfied by the scattered wave-field, play a role.

7. The different direct (forward) and inverse problems

In the following we shall distinguish between the direct source, inverse source, direct scattering, and inverse scattering problems. The characteristics of these problems are summarized below.

Direct Source Problem

In the direct source problem, the constitutive coefficients (medium parameters) and the volume source distributions are known, and the values of the acoustic wave-field quantities are to be found, at least in principle, in all space. The support of the volume source distributions that generate the acoustic wave field has the discretized version $[D^i]$. Depending on the computational modeling scheme to be employed, we have either $[D^i] \subset [D]$ or $[D^i] \subset [\bar{D}]$, where $[\bar{D}]$ is the complement of $[D] \cup \partial[D]$ in R^3.

Inverse Source Problem

In the inverse source problem, the constitutive coefficients of the medium are known, while in some domain $[D^{\Omega}] \subset [D]$ the values of the acoustic wave-field quantities are assumed to be known from measurements, the acoustic radiation being assumed to be due to radiating sources with the bounded support $[D^T] \subset [D]$. The distribution of the volume source densities over $[D^T]$ is to be determined from the measured values of the acoustic wave-field quantities in $[D^{\Omega}]$.

Direct Scattering Problem

In the direct scattering problem, the constitutive coefficients of the medium and the volume source densities that generate the incident wave field are known, while the scattered wave field is to be found in all space. In view of Eqs.(9) and (10) and the integral representations (28)-(29), the latter problem can be reduced to computing either the distribution of the quantities of the total acoustic wave field or the contrast source volume source densities (11) and (12) over the domain $[D^s]$, which is always chosen to be a proper subdomain of $[D]$.

Inverse Scattering Problem

In the inverse scattering problem the constitutive coefficients of the medium in the bounded domain whose discretized version is $[D^s]$, where $[D^s] \subset [D]$, are unknown, while this domain is embedded in a medium whose Green's functions (point-source acoustic wave fields) are analytically known. In particular, the latter applies to the chosen embedding D_0 of Section 2. Further, in some domain $[D^{\Omega}] \subset [D]$ the values of the acoustic wave-field quantities are assumed to be known from measurements.The constitutive coefficients in the domain $[D^s]$ are now to be reconstructed from the measured values of the acoustic wave-field quantities in $[D^{\Omega}]$.

In what follows we shall, to save on notation, collectively denote the medium parameters by M, the volume source densities by S and the wave-field quantities by F, and add these indications to the symbol Φ for the expansion functions and the symbol A for the expansion coefficients, while the indication of the ranks of the different tensors involved is suppressed.

8. Hybrid Finite-Element Modeling

To construct the system of equations in the expansion coefficients of the pertaining unknown functions that result from the finite-element modeling of acoustic wave problems, we apply either the weighted forms of the acoustic wave equations (13)-(16) or the global Rayleigh reciprocity relation Eq.(17) to the discretized computational domain $[D]$ of which D^s is taken to be a proper subdomain. For State A we take the representations of Section 5 applying to the volume source densities, the medium parameters and the (total or scattered, depending on whether $[D^i] \subset [D]$ or $[D^i] \subset [\bar{D}]$), acoustic wave field. State B is identified with successive, appropriately chosen, computational states. A feature of the finite-element modeling is that in all computations the medium parameters applying to the State B are taken to be zero, while the chosen wave fields in the State B, which are taken to be the successive elements of the expansion sequences of Section 5, and the volume source densities in this state are (exactly) made to correspond to each other in accordance with Eqs.(2) and (3). Further, on the boundary $[\partial D]=\partial[D]$ of the discretized computational domain $[D]$ 'absorbing' boundary conditions are invoked that account for the fact that across it either the total acoustic the wave field (in case the embedding is sourcefree) or the scattered acoustic wave field (in case the sources that generate the incident wave field are situated in the embedding) radiates (causally) into the embedding. In the so-called hybrid formulation of the finite-element method, the corresponding boundary values on $\partial[D]$ are taken to follow from the source-type integral representations (28) and (29), applied to the contrast-source acoustic wave equations (9) and (10), discretizing these representations, taking the point of observation at the successive nodes of the boundary $\partial[D]$ and thus establishing a relationship between the values of p and v_r (or p^s and v_r^s) on $\partial[D]$ and the values of p and v_k in D^s that occur in the contrast volume source densities given by Eqs.(11) and (12). This idea has been put forward by Lee, Pridmore and Morrison (1981).

Direct Source Problem

At the interior nodes of the domain of computation $[D]$ we employ for the known constitutive parameters, the known volume source distributions and the unknown acoustic wave-field quantities in State A the expansions that have been discussed in Section 5. On the boundary $[\partial D] = \partial[D]$ of the domain of computation 'absorbing' boundary conditions of the type indicated above are invoked. On the nodes coinciding with the boundary surfaces of impenetrable objects, the pertaining explicit boundary conditions are substituted. Depending on whether the source domain $[D^i]$, where the sources that generate the field are located, is part of $[D]$ or of $[\bar{D}]$, we proceed differently. The case $[D^i] \subset [D]$ is the simpler of the two. In this case the total wave field in $[D]$ is expanded according to Section 5, i.e.,

$$[F^A] = [F] = \sum_{I^F=1}^{N^F} A^F(I^F,s)\Phi^F(I^F,\mathbf{x}) \quad \text{for } \mathbf{x} \in [D]. \tag{36}$$

If $[D^i]$ is, however, located at a large distance from $[D^s]$, this procedure is impractical, since the domain of computation is then exceedingly large with an associated large demand for computer storage capacity and number of unknown expansion coefficients. In this case, we apply the scattering formulation of Section 3 in that we first compute the incident wave field from the integral representations (28) and (29) of Section 6, and next apply the finite-element formulation to the wave-field equations (34) and (35) for the scattered wave field. Correspondingly, the known right-hand sides are expanded according to Section 5 and for the scattered wave field in $[D]$ the expansion is written as

$$[F^A] = [F^s] = \sum_{I^F=1}^{N^F} A^{F;s}(I^F,s)\Phi^F(I^F,\mathbf{x}) \quad \text{for } \mathbf{x} \in [D]. \tag{37}$$

Subsequently, for State B we put $[M_B] = 0$, take successively $[F^B] = \Phi(J^F,\mathbf{x})$, with $J^F = 1,\ldots,N^F$, and match $[S^B]$ to these choices. Substituting everything in Eqs.(13)-(16) or Eq.(17), a square system of linear, algebraic equations in $\{A^F(I^F,s); I^F = 1,\ldots,N^F\}$ or $\{A^{F;s}(I^F,s); I^F = 1,\ldots,N^F\}$ results, from which the relevant field expansion coefficients can be solved.

Inverse Source Problem

At the interior nodes of the domain of computation $[D]$ we employ for the known constitutive parameters, the unknown volume source distributions $[S^T]$ in their known support $[D^T] \subset [D]$, the known acoustic wave-field quantities $[F^\Omega]$ in the domain of observation $[D^\Omega]$ and the unknown acoustic wave-field quantities $[F]$ in $[D]\backslash[D^\Omega]$ in State A the expansions that have been discussed in Section 5. On the boundary $[\partial D] = \partial[D]$ of the domain of computation 'absorbing' boundary conditions of the type indicated above are invoked. On the nodes coinciding with the boundary surfaces of impenetrable objects, the pertaining explicit boundary conditions are substituted. In accordance with Section 5 the following expansions are used:

$$[S^A] = [S^T] = \sum_{I^S=1}^{N^S} A^S(I^S,s)\Phi^S(I^S,\mathbf{x}) \quad \text{for } \mathbf{x} \in [D^T], \tag{38}$$

$$[F^A] = [F^\Omega] = \sum_{I^{F;\Omega}=1}^{N^{F;\Omega}} A^{F;\Omega}(I^{F;\Omega},s)\Phi^{F;\Omega}(I^{F;\Omega},\mathbf{x}) \quad \text{for } \mathbf{x} \in [D^\Omega], \tag{39}$$

$$[F^A] = [F] = \sum_{I^F=1}^{N^F} A^F(I^F,s)\Phi^F(I^F,\mathbf{x}) \quad \text{for } \mathbf{x} \in [D]\backslash[D^\Omega]. \tag{40}$$

Next, for State B we put $[M_B] = 0$, take successively $[F^B] = \Phi^{F;\Omega}(J^{F;\Omega},\mathbf{x})$, with $J^{F;\Omega} = 1,\ldots,N^{F;\Omega}$, with the correspondingly matched $[S^B]$ for $\mathbf{x} \in D^\Omega$ to produce known terms in Eqs.(13)-(16) or Eq.(17), and $[F^B] = \Phi^F(J^F,\mathbf{x})$, with $J^F = 1,\ldots,N^F$, with the correspondingly matched $[S^B]$ for $\mathbf{x} \in [D]\backslash D^\Omega$. Substituting everything in Eqs.(13)-(16) or Eq.(17), a square system of linear, algebraic equations in $\{A^S(I^S,s); I^S = 1,\ldots,N^S\}$ and $\{A^F(I^F,s); I^F = 1,\ldots,N^F\}$ results if $N^{F;\Omega} = N^S + N^F$, from which system the relevant source expansion coefficients as well as the expansion coefficients for the wave field in $[D]\backslash[D^\Omega]$ can be solved, while an overdetermined system of linear, algebraic equations in $\{A^S(I^S,s); I^S = 1,\ldots,N^S\}$ and $\{A^F(I^F,s); I^F = 1,\ldots,N^F\}$ results if $N^{F;\Omega} > N^S + N^F$, from which system the 'best' values of the source expansion coefficients and the expansion coefficients for the wave field in $[D]\backslash[D^\Omega]$ can be determined by minimizing the 'error' in the satisfaction of the equality signs according to some error criterion.

Direct Scattering Problem

This problem is encompassed in the finite-element modeling of the direct source problem. To emphasize the scattering aspect of the problem, the computed acoustic wave-field quantities in $[D^s] \subset [D]$ can be used to compute the contrast volume source densitites of Eqs.(11)-(12). From the latter, the scattered acoustic wave field everywhere in the configuration can be computed by using the integral representations (28)-(29).

Inverse Scattering Problem

This problem is most consistently dealt with by considering it as an inverse source problem, in the embedding, for the contrast volume source densities of Eqs.(11)-(12), that have the support $[D^s]$ and generate the scattered acoustic wave field. After the sequences of expansion coefficients of these volume source densities have been computed, the resulting field computation problem can be formulated as a direct source problem whose solution leads to the values of the expansion coefficients of the scattered acoustic wave field in $[D^s]$, where the constitutive coefficients are still unknown. Subsequently, the reciprocity relation Eq.(17) is applied to the domain $[D^s]$ and Eqs.(34)-(35). For the State A we take $[F^A] = [F^s]$, substitute for $[M^A]$ an expansion with N^M coefficients for the unknown constitutive coefficients of the type discussed in Section 5, and identify $[S^A]$ with the discretized versions of the right-hand sides of Eqs.(34)- (35) in which the incident wave field is known. For the State B we take successively a number of N^F local field states $[F^B] = \Phi^F(J^F, \mathbf{x})$, with $J^F = 1, \ldots, N^F$, $[M^B] = 0$, and the correspondingly matched $[S^B]$. Then, a square system of linear, algebraic equations in the expansion coefficients of the medium parameters results if $N^F = N^M$, from which system the expansion coefficients of the medium parameters can be solved, while an overdetermined system of linear, algebraic equations in the expansion coefficients of the medium parameters results if $N^F > N^M$, from which the 'best' values of the expansion coefficients of the medium parameters can be determined by minimizing the 'error' in the satisfaction of the equality signs according to some error criterion.

9. Integral-Equation Modeling

To construct the system of equations in the expansion coefficients of the pertaining unknown functions that result from the integral-equation modeling of acoustic wave problems, we apply in a particular number of steps the global Rayleigh reciprocity relation Eq.(17) to the discretized computational domain $[D]$ of which D^s is taken to be a proper subdomain. The integral-equation formulation is invariably based on the contrast-source description discussed in Section 3 and on source-type wave-field integral representations of the type given in Section 6 for the case of a homogeneous, isotropic, lossless medium. In fact, the formulation can be extended to any case where the acoustic wave-field Green's functions (point-source solutions to the acoustic wave equations) can be constructed analytically.

The first step in the integral-equation formulation consists of constructing the radiated wave fields $[F^G] = G(\mathbf{x}, s; I^G)$ with $\{G(\mathbf{x}, s; I^G); \mathbf{x} \in R^3, I^G = 1, \ldots, N^G\}$ that correspond to the localized source distributions $[S^G] = \Phi^S(I^G, \mathbf{x}')$ for $\mathbf{x}' \in [D^G(I^G)]$, where $[D^G(I^G)]$ is some localized source domain of the geometrical discretization as discussed in Section 5. It is assumed that these radiated wave fields are analytically evaluated from the expressions at the right-hand sides of Eqs.(28)-(29). Next, the contrast volume-source densities $[S^s]$ are written as the expansions

$$[S^s(\mathbf{x}, s)] = \sum_{I^{S;s}=1}^{N^{S;s}} A^{S;s}(I^{S;s}, s)\Phi^{S;s}(I^{S;s}, \mathbf{x}) \quad \text{for } \mathbf{x} \in [D^s], \tag{41}$$

with the unknown coefficients $\{A^{S;s}(I^{S;s}, s); I^{S;s} = 1, \ldots, N^{S;s}\}$. The corresponding expansion for the scattered wave field is given by

$$[F^s(\mathbf{x}, s)] = \sum_{I^{S;s}=1}^{N^{S;s}} A^{S;s}(I^{S;s}, s)G(\mathbf{x}, s; I^{S;s}) \quad \text{for } \mathbf{x} \in R^3. \tag{42}$$

A similar expansion holds for the incident wave field. Let the volume source densities that generate the incident wave field be expanded as

$$[S^i(\mathbf{x},s)] = \sum_{I^{S;i}=1}^{N^{S;i}} A^{S;i}(I^{S;i},s)\Phi^{S;i}(I^{S;i},\mathbf{x}) \quad \text{for } \mathbf{x} \in [D^i], \tag{43}$$

with the known coefficients $\{A^{S;i}(I^{S;i},s); I^{S;i} = 1,\ldots,N^{S;i}\}$. Then, the corresponding expansion for the incident wave field is given by

$$[F^i(\mathbf{x},s)] = \sum_{I^{S;i}=1}^{N^{S;i}} A^{S;i}(I^{S;i},s)G(\mathbf{x},s;I^{S;i}) \quad \text{for } \mathbf{x} \in R^3. \tag{44}$$

The different problems enumerated in Section 7 will now be discussed separately for the case of penetrable scatterers. The case of impenetrable scatterers runs along similar lines, be it that the contrast source representations of the volume type that follow from Section 6 have to be replaced by contrast source representations of the surface type, in which the contrast sources are located at the boundary surfaces of the impenetrable objects.

Direct Source Problem

In the direct source problem the sources in $[D^i]$ radiate in the known embedding, and the incident wave field as given by Eq.(44) is already the total wave field, a scattered wave field being absent.

Inverse Source Problem

In the inverse source problem the unknown volume source distributions $[S^T]$ are expanded as

$$[S^T(\mathbf{x},s)] = \sum_{I^{S;T}=1}^{N^{S;T}} A^{S;T}(I^{S;T},s)\Phi^{S;T}(I^{S;T},\mathbf{x}) \quad \text{for } \mathbf{x} \in [D^T], \tag{45}$$

in which $\Phi^{S;T}(I^{S;T},\mathbf{x})$ is a local volume source expansion function of the discretized source domain $[D^T]$ and the coefficients $\{A^{S;T}(I^{S;T},s); I^{S;T} = 1,\ldots,N^{S;T}\}$ are to be determined. The corresponding radiated wave field, which is measured in $[D^\Omega]$, is then given by

$$[F^T(\mathbf{x},s)] = \sum_{I^{S;T}=1}^{N^{S;T}} A^{S;T}(I^{S;T},s)G(\mathbf{x},s,I^{S;T}) \quad \text{for } \mathbf{x} \in [R^3], \tag{46}$$

where $\{G(\mathbf{x},s;I^{S;T}); \mathbf{x} \in R^3, I^{S;T} = 1,\ldots,N^{S;T}\}$ are the radiated wave fields that correspond to the localized source distributions $\Phi^{S;T}(I^{S;T},\mathbf{x})$ with supports $[D^T(I^{S;T})] \subset [D]$. Next, the reciprocity relation Eq.(17) is applied to the domain $[D^T]\cup[D^\Omega]$, while the medium parameters in the States A and B are both chosen to be the ones of the embedding, i.e., $[M^A] = [M^B] = [M_0]$. Further, we take $[S^A] = [S^T]$ and $[F^A] = [F^T]$. For State B we successively take $[F^B] = G(\mathbf{x},s;J^{S;\Omega})$, where $\{G(\mathbf{x},s;J^{S;\Omega}); \mathbf{x} \in R^3, J^{S;\Omega} = 1,\ldots,N^{S;\Omega}\}$ are the radiated wave fields that correspond to the localized source distributions $[S^\Omega] = \Phi^{S;\Omega}(I^{S;\Omega},\mathbf{x}')$ with supports $[D^\Omega(I^{S;\Omega})] \subset [D^\Omega]$. Substituting everything in Eq.(17), a square system of linear, algebraic equations in $\{A^{S;T}(I^{S;T},s); I^{S;T} = 1,\ldots,N^{S;T}\}$ results if $N^{S;\Omega} = N^{S;T}$, from which system the relevant source expansion coefficients can be solved, while an overdetermined system of linear, algebraic equations in $\{A^{S;T}(I^{S;T},s); I^{S;T}=1,\ldots,N^{S;T}\}$ results if $N^{S;\Omega} > N^{S;T}$, from which system the 'best' values of the source expansion coefficients can be determined by minimizing the 'error' in the satisfaction of the equality signs according to some error criterion.

Direct Scattering Problem

In the direct scattering problem the reciprocity relation Eq.(17) is applied to the domain $[D^s]$. For State A we take the wave field to be the scattered wave field as it follows from the contrast-source equations (9)-(10), i.e., $[F^A]=[F^s]$, $[M^A]=[M_0]$ and $[S^A]=[S^s]$. For State B we take the volume source distributions to be successively one of the local discretized source distributions $[S^B]=[S^G]=\Phi^{S;s}(J^{S;s},\mathbf{x})$ for $J^{S;s}=1,\ldots,N^{S;s}$, whose corresponding wave fields in the medium $[M^B]=[M_0]$ are $[F^B]=G(\mathbf{x},s,J^{S,s})$. Substituting everything in Eq.(17), using for $[S^s]$ and $[F^s]$ the expansions given by Eqs.(41) and (42), respectively, and substituting in the right-hand sides of Eqs.(11)-(12) the expansions applying to the known constitutive coefficients, the known incident wave field and the unknown scattered wave field, a square system of linear algebraic equations results from which the expansion coefficients $\{A^{S;s}(I^{S;s},s); I^{S;s}=1,\ldots,N^{S;s}\}$ of the contrast source distributions can be solved. Substituting these values in Eq.(42), the scattered wave field follows anywhere in space.

Inverse Scattering Problem

This problem is most consistently dealt with by considering it as an inverse source problem for the contrast volume source densities of Eqs.(11)-(12), that have the support $[D^s]$ and generate the scattered acoustic wave field. After the sequences of expansion coefficients of these volume source densities have been computed, the scattered wave field follows from Eq.(42) anywhere in space. Next, the reciprocity relation Eq.(17) is applied to the domain $[D^s]$ with $[F^A]=[F^s]$, $[S^A]=[S^s]$, $[M^A]=[M_0]$, while for the unknown contrasts in the constitutive coefficients in the right-hand sides of Eqs.(11)-(12) an expansion, with N^X coefficients, of the type discussed in Section 5 are taken. For the State B we successively take a number N^G of discretized Green's states as introduced in the beginning of the present section: $[F^B]=[F^G]$, $[S^B]=[S^G]$, $[M^B]=[M_0]$. Subsequently applying the reciprocity relation Eq.(17) to the domain $[D^s]$, a system of linear, algebraic equations in the expansion coefficients of the contrasts in the constitutive coefficients results. For $N^G=N^X$ this is a square system from which the expansion coefficients for the contrasts in the constitutive coefficents can be solved, for $N^G>N^X$ this system is overdetermined and the 'best' values of the expansion coefficients for the contrasts in the constitutive coefficents can be determined by minimizing the 'error' in the satisfaction of the equality signs according to some error criterion. From the obtained values, the values of the expansion coefficients of the constitutive coefficients themselves follow directly.

10. Conclusion

It has been shown that the Rayleigh reciprocity relation can be used as the point of departure for the numerical modeling of acoustic direct (or forward) and inverse wave propagation and scattering problems. In fact, all known numerical procedures (finite-element method, integral-equation method) are shown to be consequences of it.

Acknowledgements

The research reported in this paper has been financially supported through a Research Grant from the Stichting Fund for Science, Technology and Research (a companion organization to the Schlumberger Foundation in the U.S.A.), a Research Grant from Schlumberger-Doll Research, Ridgefield, CT, U.S.A., and a Research Grant from Etudes et Productions Schlumberger, Clamart, France. This support is gratefully acknowledged.

References

Bowman, J.J., T.B.A. Senior and P.L.E. Uslenghi, 1969. *Electromagnetic and acoustic scattering by simple shapes*, North-Holland Publishing Company, Amsterdam.

De Hoop, A.T., 1985. "A time-domain energy theorem for scattering of plane acoustic waves in fluids", Journal of the Acoustical Sopciety of America, Vol. **77**(1), January, pp. 11-14.

De Hoop, A.T., 1988. "Time-domain reciprocity theorems for acoustic wave fields in fluids with relaxation", Journal of the Acoustical Society of America, Vol. 84, No. 5, November, pp. 1877-1822.

Felsen, L.B. and N. Marcuvitz, 1973. *Radiation and scattering of waves*, Englewood Cliffs, NJ, Prentice-Hall.

Herman, G.C., 1981. *Scattering of transient acoustic waves in fluids and solids*, Report Nr. 1981-13, Laboratory of Electromagnetic Research, Department of Electrical Engineering, Delft University of Technology, Delft, the Netherlands, 183 pp.

Lee, K.H., D.F. Pridmore, and H.F. Morrison, 1981. "A hybrid three-dimensional electromagnetic modeling scheme", Geophysics, Vol. 46, No. 3, May, pp. 796-805.

Mur, G. and A.T. de Hoop, 1985. "A finite-element method for computing three-dimensional electromagnetic fields in inhomogeneous media", IEEE Transactions on Magnetics, Vol. MAG-21, No. 6, November, pp. 2188-2191.

Naber, 1980. *Topological methods in Euclidean space*, Cambridge University Press, Cambridge.

Stam, H.J. and A.T. de Hoop, 1990. "Theoretical considerations on a finite- element method for the computation of three-dimensional space-time elastodynamic wave fields", accepted for publication in Wave Motion, **12**.

Van den Berg, P.M., 1985. "Iterative schemes based on the minimization of the error in field problems", Electromagnetics, **5** , 237-262.

Van der Hijden, J.H.M.T., 1987. *Propagation of transient elastic waves in stratified anisotropic media*, North-Holland Publishing Company, Amsterdam.

Zienkiewicz, O.C. and R.L. Taylor, 1989. *The finite element method*, McGraw-Hill, London, 4th. ed.

A Quadratic Inverse Problem: The Phase Retrieval

T. Isernia[1], *G. Leone*[2], *and R. Pierri*[1]

[1]Dip. di Ing. Elettronica, Univ. di Napoli, Via Claudio 21, I-80125 Napoli, Italy
[2]Ist. di Ing. Elettronica, Univ. di Salerno, I-84081 Baronissi, Salerno, Italy

1 . Introduction

The phase retrieval from the knowledge of the amplitude of a complex signals is an interesting problem encountered in many fields of the applied sciences as : astrophysics, electron microscopy, numerical holography, antenna testing, inverse optics, quantum mechanics, etc.. A list of references on such subjects can be found in [1-2]. In this paper we refer to the phase of radiated fields.

The phase retrieval has been studied from different points of view, but the key points of the problem are: i) the existence of a solution; ii) the uniqueness of the solution; iii) an effective algorithm to find the solution.

As far as the question of the existence of a solution is concerned, it can be seen that the answer depends, essentially, on the presence of noise in the data. For the sake of semplicity in this paper we suppose, first, that exact data are available and successively the presence of noise will be explicitly taken into account.

Now, supposing exact data, let us discuss briefly the problem of the uniqueness. This problem has been examined with reference to the following situations: a) one amplitude distribution; b) two amplitude distributions.

In the case of *one* amplitude distribution a large work has been done in the case of the far field of a planar space limited source which is a bandlimited function. In two-dimensional geometry the uniqueness can be estabilished when some a priori information on the source are known, f.i. whenever the source is a part of a band-limited function [1-2]. Moreover in the near field zone the knowledge of the position, the dimensions and the structure of the source can allow to retrieve more easily the phase [3].

The three-dimensional geometry has been studied first in [4]; successively many authors have shown that the only fact that a two-variables band-limited function cannot be in general factorized with the exception of a "zero-measure set", ensures that, apart *trivial symmetries*, the phase corresponding to a given amplitude must be unique [5].

When *two* amplitude distributions are given, the demonstration of the uniqueness is easier. In fact in the *one-dimensional* case, considering two circular scanning curves, the uniqueness has been shown in [6], from just the knowledge of the dimension of the source. It can be argued that the same results could be extended to the other scanning geometries, mainly when some information on the structure of the source are available.

Let us now discuss briefly the algorithms introduced for the phase retrieval. To this end we recall that the first algorithms were introduced, essentially, in a heuristic way; only in a second moment they were justified. In the case of a single amplitude distribution (far field) an iterative procedure was introduced in [7]. This algorithm is basically of a first order, or gradient type, which ensure the minimization of a *non convex* functional by an iterative procedure [8]. It can be shown that it can be interpreted in terms of a topological problem: find the intersection point of a number of non convex sets. This last approach has suggested the use of a projection procedure to find the intersection points. The main drawback of the method is the existence of "*traps*", or local minima of the functional involved; therefore it is a *local method* which is useful when a priori information on the source are available or a "rough" phase measurement is possible.

In this paper we propose a new method in the phase retrieval: it is based on the inversion of a *quadratic* operator, which can be done by minimization of an adequate functional, that, at variance of previous ones, seems not to present local minima, but only *one* minimum corresponding to the solution.

2. Theoretical analysis

Let us consider a scalar case referring to a *two-dimensional* geometry; if E_1 and E_2 are the radiated fields in the near zone on two circles of radius r_1 and r_2 respectively, then

$$E_1(\theta) \cong \sum_{-N}^{N} c_n \exp(jn\theta) \tag{1}$$

$$E_2(\theta) \cong \sum_{-N}^{N} c_n t_n \exp(jn\theta) \tag{2}$$

wherein $t_n = H_n(\beta r_2)/H_n(\beta r_1), H_n(\cdot)$ are Hankel functions, $N \cong \beta a$, $\beta = 2\pi/\lambda, \lambda$ is the wavelenght, $2a$ is the dimension of the source and θ is the usual angular variable. As pointed out in [9], it can be said that N is the number of "degrees of freedom" of the field, which means that the truncation error made in the series (1) and (2) decays exponentially to zero with N.

Let us suppose, now, that we know the amplitude distribution on the two circles; i.e. $|E_1|^2 = M_1^2(\theta), |E_2|^2 = M_1^2(\theta)$. As well known, the phase retrieval problem consists in the recovering of E_1 or E_2 from M_1 and M_2. As already pointed out in [6], from the series (1) and (2) respectively, it is easy to obtain the following "convolution type" equations

$$\sum_{-N_n}^{N} c_n c_{n-m}^* = a_m \tag{3}$$

$$\sum_{-N_n}^{N} c_n t_n c_{n-m}^* t_{n-m}^* = b_m \tag{4}$$

for $m = -2N, \cdots, 2N$ wherein a_m and b_m are the Fourier coefficients of $M_1^2(\theta)$ and $M_2^2(\theta)$ respectively.

As it can be seen, the system of equations (3) and (4) is non linear, but, fortunately, it can be solved uniquely by a *recursive* procedure, which is shown in the appendix.

It can be observed that the uniqueness can be, in principle, established for every value of N, but, because a_{2N} tends to zero exponentially [9], it can be said that, in any real measurement situation a finite value for N can be considered. However, the above cited recursive procedure is not useful as an effective algorithm in the real situations due to the unavoidable computational (round-off) errors which does not allow to obtain the solution when the number N is large. Thus, in order to obtain an effective algorithm, let us define

$$\underline{x} = (c_{-N}, \cdots, c_0, \cdots, c_N); \qquad \underline{y} = \begin{pmatrix} \underline{y}_1 \\ \underline{y}_2 \end{pmatrix} \tag{5}$$

wherein

$$\underline{y}_{1,2} = \{M_{1,2}^2(\theta_i)\}, \qquad \theta_i = \frac{i\pi}{2N+1} \tag{6}$$

for $i = -2N+1, \cdots, 2N$ and if A is the quadratic operator relating $\underline{x}$ to $\underline{y}$, from the relations (1),(2) and (6)

$$\underline{y} = A(\underline{x}) \tag{7}$$

The eq. (7) defines a one-to-one mapping of $R^{4N} \times R^+$ (where R^+ denotes the non-negative part of the real axis and c_0 has been chosen real and positive), onto a manifold Y_o of $R^{+(8N+2)}$, which, as it can be seen from the Appendix, is an omeomorphism. Only the points $\underline{y}$ belonging to such manifold Y_o (see fig. 1) allow to obtain a solution to our problem. It can be noted, on the other side, that the knowledge from eq. (3) of the Euclidean norm of $\underline{x}$, i.e. $||\underline{x}|| = a_0$ determines the emisphere, where the solution is: we define X_o a compact set containing such an emisphere. Let us consider now the analytical properties of the mapping. It is easy to see that the operator A is differentiable, it can be defined on a compact set, it defines a one to one mapping. Therefore it is invertible and its inverse is continuous [10,11],which means that when we move on the manifold around a data point $\underline{y}_0$, to a small change in the data a small change corresponds around the solution $\underline{x}_0$. In the literature these properties correspond to the fact that the inversion of the operator A is a "well posed problem", in the Tickonov sense [11].

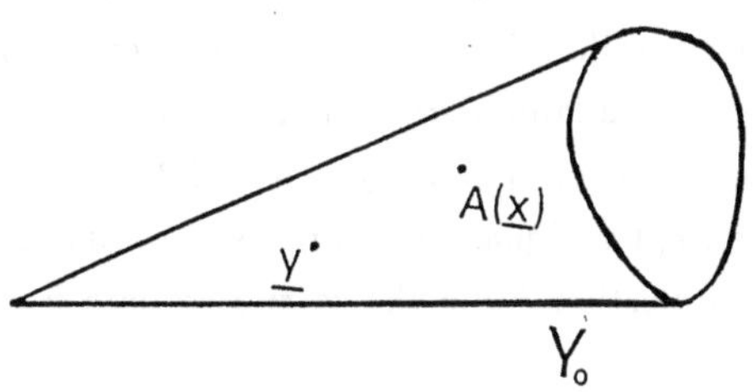

Fig. 1

In order to find an effective algorithm, different approaches can be followed [12,13]. The most useful, perhaps, is the variational approach based on the minimization of the functional

$$\Phi(\underline{x}) = ||\underline{y} - A(\underline{x})||^2 \tag{8}$$

wherein $||\cdot||$ denotes the Euclidean norm.

This functional has some interesting properties: 1) it is semidefinite positive; 2) it is analytical; 3) it can be argued that it has one minimum, which corresponds to its zero.

It is easy to see that the equation (7) defines a cone [13] (or a part of a cone when $\underline{x} \in X_0$). In fact if we consider

$$\underline{y}' = \alpha \underline{y} \qquad \alpha > 0 \tag{9}$$

to $\underline{y}'$ corresponds $\underline{x}' = \sqrt{\alpha}\underline{x}$, i.e. $\underline{y}' \in Y_0$, see fig. 1.

As far as the minimization of the functional Φ is concerned, it can be performed, in principle, by different approaches [13] of first order (gradient type), and second order (Newton-type). The possible presence of "flat regions" can render the minimization of the functional very slow, or even impossible when the number of the dimensions of the space involved is very large. Therefore, in general, a second order algorithm should be considered. It is based, as well known, on successive quadratic approximation of the functional Φ, i.e.

$$\Phi(\underline{x} + \Delta\underline{x}) \simeq \Phi(\underline{x}) + \underline{\nabla\Phi} \cdot \Delta\underline{x} + \frac{1}{2}\Delta\underline{x} \cdot \underline{\underline{\Phi}}'' \cdot \Delta\underline{x} \tag{10}$$

wherein $\underline{\nabla\Phi}$, $\underline{\underline{\Phi}}''$ are respectively the gradient and the Hessian of Φ evaluated in $\underline{x}$. Searching for the stationary points of $\Phi(\underline{x} + \Delta\underline{x})$ defines an iterative scheme where at each step $\Delta\underline{x}$ is the solution of

$$\underline{\underline{\Phi}}'' \cdot \Delta\underline{x} = -\underline{\nabla\Phi} \tag{11}$$

If $\underline{\underline{\Phi}}''$ presents ill conditioning problems, two, essentially equivalent, method can be suggested : the first one is based on the inversion of eq. (11) through the minimization of the functional $||\underline{\underline{\Phi}}'' \cdot \Delta\underline{x} + \underline{\nabla\Phi}||^2 + \alpha||\underline{x}||^2$ wherein α is a regularization parameter [11]. The second one is related to the inversion of the equation (11), at every step, through a filtered inversion of the Hessian [8]. Around the solution $\underline{x}_o$, it can be argued that the filtering process can determine the maximum information that we can extract from the data, i.e. we can say that a crucial role is played by the eigenvalues of $\underline{\underline{\Phi}}''(\underline{x}_o)$.

Let us now consider the case when the noise is explicitely taken into account. The data vector $\underline{y}$,in general, does not belong to the manifold Y_0, therefore the inversion of the operator A *is not a well posed* problem in the Hadamard meaning, i.e. to a small change in the data a solution to the problem does not correspond any more [11,13]. Thus, in order to consider a well posed problem the concept of "quasi solution" [10,11] must be introduced, as follows. Let us consider first the projection $P\underline{\tilde{y}}$ of the corrupted data $\underline{\tilde{y}}$ on the manifold $\underline{Y}_0$, i.e. $P\underline{\tilde{y}}$ is such that the distance

$$||P\underline{\tilde{y}} - \underline{\tilde{y}}|| \qquad P\underline{\tilde{y}} \in Y_0 \tag{12}$$

is minimum. Therefore a quasi solution, in the Tikhonov sense [11], can be introduced considering the corresponding element of $\underline{\tilde{x}}$, such that

$$A(\underline{\tilde{x}}) = P\underline{\tilde{y}} \tag{13}$$

To this element $P\underline{\tilde{y}}$ the above expressed consideration for the exact data can be made, about the recovering of the solution. Thanks to the properties of A, the element $\underline{\tilde{x}}$ (or "quasi solution") can be obtained minimizing the functional

$$\tilde{\Phi} = ||\underline{\tilde{y}} - A(\underline{x})|| \tag{14}$$

and $\underline{\tilde{x}} \to \underline{x}_o$ whenever measurement error tends to zero [11].

3. Test example

In order to validate the method and discuss the actual convergency properties of the method, a numerical analysis has been performed using a first order (Fletcher - Reeves type) algorithm. The field of a parabolic cylinder of diameter $2a = 6\lambda$ and focal lenght 3λ has been considered over two circle in near zone of radii $r_1 = 9\lambda, r_2 = 18\lambda$.

The continuous line of fig. 2 displays the behaviour of Φ normalized to its maximum value versus the number of iterations n starting from a random point. When the algorithm tends to stagnate, it results convenient to modify the iterative algorithm considering the weighted functional

$$\Psi = ||\underline{\underline{w}} \cdot (A(\underline{x}) - \underline{y})||^2 \tag{15}$$

wherein

$$\underline{\underline{w}} = \begin{bmatrix} \underline{\underline{w}}^1 & 0 \\ 0 & \underline{\underline{w}}^2 \end{bmatrix} \tag{16}$$

and $\underline{\underline{w}}^{1,2}$ are diagonal blocks with elements $w_{ii}^{1,2} = 1/M_{1,2}^2(\theta_i)$. Such a functional has the

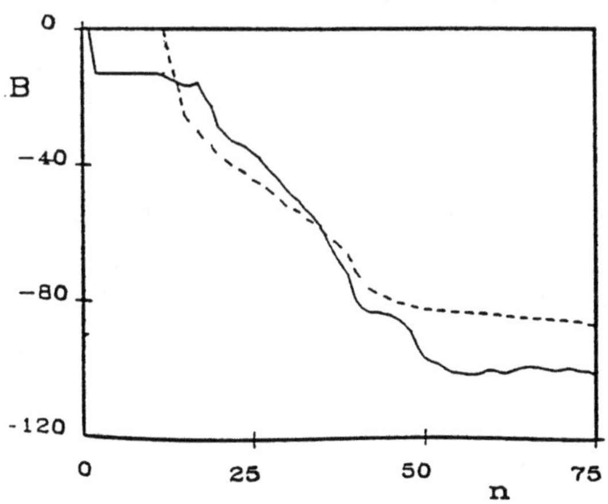

Fig. 2

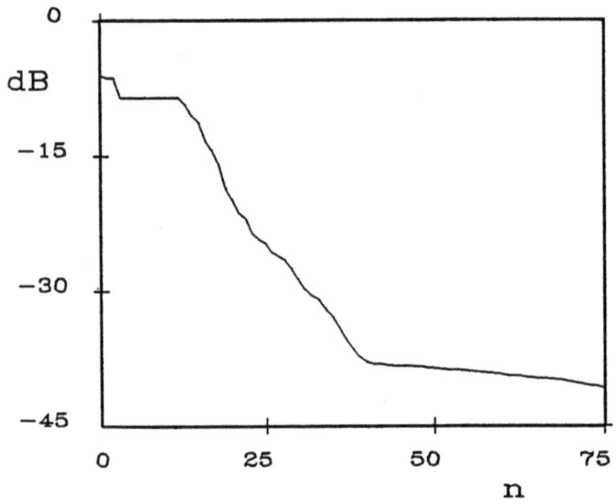

Fig. 3

same properties of Φ, but it emphasizes the small values of $M_1^2(\theta)$ and $M_2^2(\theta)$. The dashed line of fig. 2 represents the behaviour of the functional Ψ normalized to its maximum value.

Fig. 3 furnishes the euclidean distance between $\underline{x}$ and $\underline{x}_o$ normalized to the energy of $\underline{x}_o$. It can be seen that a good agreement is obtained between $\underline{x}_o$ and the final result of the iteration scheme.

4. Appendix

The uniqueness was already shown in [6] by a slightly different approach. It is here reported for the sake of completeness. Let us write explicitly the eq. (3) and (4)

$$\begin{cases} c^*_{-N}c_N & = a_{2N} \\ c^*_{-N}c_N t^*_{-N}t_N & = b_{2N} \end{cases} \qquad (A1)$$

$$\begin{cases} c^*_{-N}c_{N-1} + c^*_{-N+1}c_N & = a_{2N-1} \\ t^*_{-N}t_{N-1}c^*_{-N}c_{N-1} + t^*_{-N+1}t_N c^*_{-N+1}c_N & = b_{2N-1} \end{cases} \qquad (A2)$$

which can be solved obtaining the products $c^*_{-N}c_{N-1}$ and $c_N c^*_{-N+1}$.

Now, following,

$$\begin{cases} c^*_{-N}c_{N-2} + c^*_{-N+2}c_N & = a_{2N-2} - c^*_{-N+1}c_{N-1} \\ t^*_{-N}t_{N-2}c^*_{-N}c_{N-2} + t^*_{-N+2}t_N c^*_{-N+2}c_N & = b_{2N-2} - t^*_{-N+1}t_{N-1}c^*_{-N+1}c_{N-1} \end{cases} \qquad (A3)$$

from which , trough the knowledge of the second members from (A1) and (A2), we derive the quantities $c^*_{-N}c_{N-2}$ and $c^*_{-N+2}c_N$.

In the same way we obtain all the products $c^*_n c^*_{-N}$ and $c^*_n c_N$, for $n = 0, \cdots, N-1$.

In particular, supposing c_0 real, we know $c_0 c^*_{-N}$ and $c_0 c_N$ from which, together with (A_1) we derive c_0. It is obvious that from the knowledge of c_0, c^*_{-N} and c_N can be derived and consequently all the c_n.

This work has been sponsored by MPI and CNR.

REFERENCES

[1] L.S. Taylor: "The phase retrieval problem", *IEEE Trans. on Antennas and Propagat.*, vol. 29, n.2, pp. 386-391,1981.

[2] H.A. Ferwerda: "The phase reconstruction problem for wave amplitudes and coherence functions", *Inverse Source Problems in Optics*, H.P. Baltes ed., Springer Verlag 1978.

[3] T. Isernia, G. Leone, R. Pierri: "Numerical experiments on phase retrieval in near zone", *Proc. of JINA 88*, Nice, France, 1988.

[4] Y. Bruck, L. Sodin: "On the ambiguity of the image reconstruction problem", *Opt. Comm.*,30, 1979.

[5] I. Sabba Stefanescu: "On the phase retrieval problem in two dimensions", *J. Math Phys.*, 26 (9), pp. 2114-2160, 1985.

[6] T. Isernia, G. Leone, R. Pierri: "On the uniqueness and stability of phase retrieval in near zone", *Proc. of '89 URSI Symposium on Electromagnetic Theory*, Stockholm, Sweden, 1989.

[7] R.W. Gerchberg, W.O. Saxton: "Phase determination from image and diffraction plane pictures in the electron microscope", *Optik*, 33, pp. 275-284, 1971.

[8] Barakat, Newsam: "Algorithms for reconstruction of partially known, bandlimited Fourier transform pairs from noisy data", *J. Int. Eq.*, vol. 9, 1985.

[9] O.M. Bucci, G. Franceschetti: "On the degrees of freedom of scattered fields", *IEEE Trans. on Antennas and Propagat.*, vol. 37, n. 7, pp. 918-926,1989.

[10] V.A. Morozov: *Method for solving incorrectly posed problems*, Springer Verlag.

[11] A.Tickhonov, V.Arsenine: *Methodes de resolution de problemes mal posée*, MIR.

[12] M. Bertero, T.A. Poggio, V. Torre: "Ill posed problems in early vision", *Proc of IEEE*, vol. 76, n. 8, pp. 869-889, 1988.

[13] J.M. Ortega, W.C. Rheinboldt: *Iterative solution of non linear equations in several variables*, Academic Press, 1970.

[14] D.G. Luenberger: *Optimization by vector space methods*, John Wiley, 1969.

Part VI

Wave Propagation and Scattering Problems (Imaging and Numerical Methods)

Numerical Investigation of the Probing of Sea-Bottoms Using Exact and Approximate Methods

R. Bohbot[1], *D. Lesselier*[1], *and W. Tabbara*[1;2]

[1]Equipe Electromagnétisme, Laboratoire des Signaux et Systèmes, CNRS-ESE, F-91192 Gif-sur-Yvette Cédex, France
[2]Laboratoire d'Electronique Générale, ESE, F-91192 Gif-sur-Yvette Cédex, France

Probing of the ocean floor by means of acoustical waves is addressed through the analysis of four methods. The first one based on a geometrical acoustics approximation and the second one which uses the step response of the medium are applied to configurations of a homogeneous sediment lying on a homogeneous solid medium. The third one, an exact Gel'fand-Levitan-like technique, and the last one, known as impediography, are used to probe an inhomogeneous fluid sediment on the top of a fluid subbottom. Synthetic data are employed to assess the effectiveness of these four methods.

Introduction

Acoustical parameters of the ocean floor play a prominent role in the propagation of sound waves in shallow water (at depths less than about 10 wave-lengths). Direct measurements of these parameters are often difficult to make. These last years, many theoretical investigations have been conducted, but numerical works based on realistic models are much less frequent (see, e. g., discussion in [1]).

The ocean is modeled as a three-layer structure : water/sediment/subbottom, with the first layer of density ρ_1 = 1030 kg/m^3, velocity c_1 = 1500 m/s and of large enough thickness such that plane waves can be considered as the insonification sources and the reflection at the air-water interface negligible (Fig. 1). The sediment is a fluid of thickness *d*, density ρ_2 where P-waves are propagated with velocity c_2 and attenuation α_2. The subbottom is either a fluid or a solid of density ρ_3 and of infinite thickness supporting P-waves (velocity c_P, attenuation α_P) and, if a solid, S-waves (velocity c_S, attenuation α_S). Wave numbers are $k_i = \omega/c_i + j\omega\alpha_i$, $i = 1, 2, P, S$, with time-dependence of the acoustic field $e^{-j\omega t}$. The incident angle of the plane-wave is θ.

We aim at reconstructing the profiles of the acoustic parameters of the structure (velocities, densities, attenuations) from synthetic values of the reflection coefficient $R(\theta, \omega)$. We consider four methods, exact and approximate ones, and we assess their effectiveness using two structures : one with high contrasts in

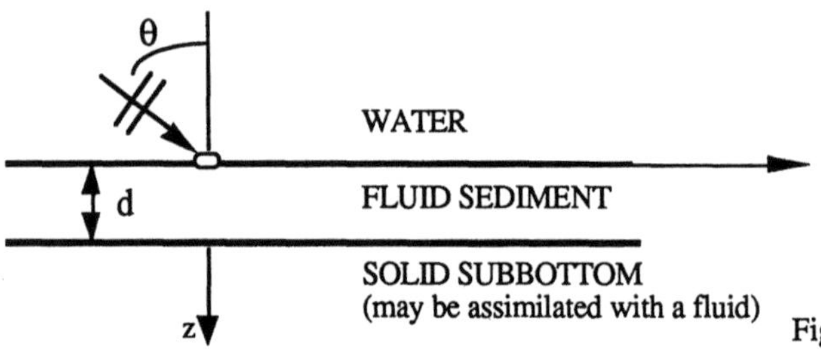

Fig. 1 : The scattering configuration

densities and velocities (HC), e.g. a sandy sediment above a basaltic subbottom, the other with low constrasts (LC), e.g. a clay layer with a subbottom whose parameters are not very different.

The first two methods are applied to a fluid/fluid/solid structure made of homogeneous layers. The first one is based on a geometrical acoustics approximation and takes into account the first-order reflections at each interface. A study of the direct scattering problem [2] shows that this approach is suitable for weakly inhomogeneous media even at very low frequencies. The second one is based on the analysis of the step response of the structure. The waveform is made of an infinite sequence of plateaux whose first two, which are linked to the reflection coefficients at the interfaces, provide an exact solution for the parameters. The remaining two methods are devised for a fluid/fluid/fluid lossless structure. The sediment is an inhomogeneous medium with smoothly varying density and velocity profiles. The first method [3] reduces the wave equation to a Schrödinger equation which is solved using a Gel'fand-Levitan approach [4]. The transient impulse response (the Fourier transform of $R(\theta, \omega)$) provides the synthetic data. The second one (impediography, e.g. [5]) is based on a Born's approximation and allows the retrieval of the impedance profile as a function of travel-time. The next sections describe the fundamentals of each method and give an analysis of the results they allowed to obtain as a function of θ, ω, contrast and noise. Part of the analysis and some results are found in a preliminary paper [6].

1. The homogeneous structure

1.1 Geometrical Acoustics

When multiple reflections inside the sediment slab are neglected, the reflection coefficient $R(\theta, \omega)$ of the structure is made of the contribution of the waves reflected once at each interface :

$$R(\theta, \omega) \approx R_{12} + T_{12}R_{23}T_{21}e^{2jk_{2z}d} \tag{1}$$

where R_{12} and R_{23} are the Fresnel reflection coefficients at the interfaces, $T_{21} = 1 + R_{12}$ and $T_{21} = k_{2z} T_{12} / k_{1z}$ the Fresnel transmission coefficients at the upper interface, and k_{1z}, k_{2z} the z-component of the wave vectors in water and in sediment respectively. Retrieving the sediment parameters is done using three frequencies and two angles of incidences by first solving for K^i_{2z}, $i = 1, 2$, non-linear equations

$$\chi(\theta_i)\left[1 - \exp 2jK^i_{2z}(\omega_3 - \omega_1)\right] - \left[1 - \exp 2jK^i_{2z}(\omega_2 - \omega_1)\right] = 0 \;\Big|_{i=1,2},$$

$$\text{where : } \chi(\theta) = \frac{R(\omega_1,\theta) - R(\omega_2,\theta)}{R(\omega_1,\theta) - R(\omega_3,\theta)}, \quad K^i_{2z} = d\sqrt{\left[\frac{1}{c_2} + j\alpha_2\right]^2 - \left[\frac{\sin\theta_i}{c_1}\right]^2}\;\Bigg|_{i=1,2}. \tag{2}$$

The solution is obtained using a complex root search technique in a domain bounded by extremal values (chosen from physical consideration) [7]. Knowledege of K^i_{2z} yields d, c_2 and α_2. Density ρ_2 follows :

$$\rho_2 = \rho_1 \frac{k_{2z}(\theta_1)}{k_{1z}(\theta_1)} \frac{1 + R_{12}(\theta_1)}{1 - R_{12}(\theta_1)}, \quad R_{12}(\theta_1) = R(\theta_1, \omega_1) - \frac{R(\theta_1, \omega_1) - R(\theta_1, \omega_2)}{1 - \exp 2jK^1_{2z}(\omega_2 - \omega_1)}. \tag{3}$$

As the sediment is now determined one can calculate R_{12}, T_{12}, T_{21} and R_{23}. Using three more angles of incidences at "any" given frequency ω and R_{23} we are able to retrieve the parameters of the subbottom from a set of four non linear equations (two if the subbottom is lossless). A Newton method is used, but

Table 1 : Geometrical Acoustics method (1.1). How many solutions ? / How do they behave ? Results are excellent : *** (with less than 5% error), good : ** (less than 10% error), * : so-so (up to 20% error), 0 : bad (up to 50% error) and 00 : non-existent. (SED = sediment, SUB = subbottom.)

		LOW θ	HIGH θ
LC LF	SED	***	***
	SUB fluid	*	**
	SUB solid	many, but good $\rho_3 c_P$	
LC HF	SED	*** save density	several (one with less than 1% error), but good $\rho_2 c_2$
	SUB fluid	00	one for each SED (one with less than 5% error)
	SUB solid	0 (several solutions)	

		LOW θ	HIGH θ
HC LF	SED	00	***
	SUB		many
HC HF	SED	00	several (one with less than 1% error), but good $\rho_2 c_2$
	SUB		many for each SED

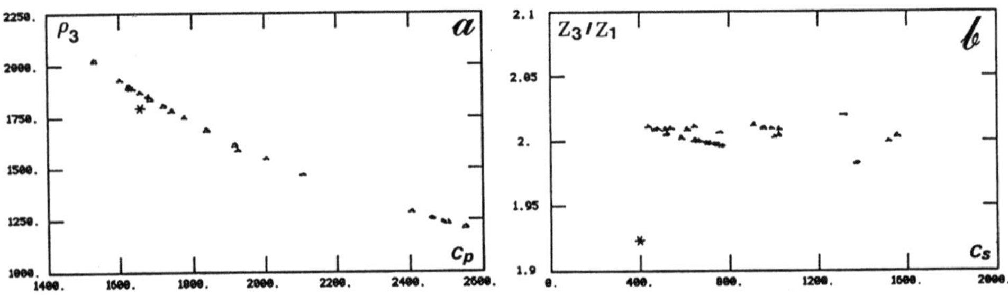

Fig. 2 : Geometrical Acoustics method (1.1). Low-Contrast LC, Low-Frequency LF, Low-θ. Probing of a unique and almost exact sediment (ρ_2 = 1445 kg/m^3, c_2 = 1495 m/s, d = 9.6 m) but of 29 subbottoms distributed as follows : a) ρ_3 vs. c_P, b) Z_3/Z_1 vs. c_S. where $Z_3 = \rho_3 c_P$ and $Z_1 = \rho_1 c_1$. The symbol * shows the exact solution.

different sets of initial values may lead to different equivalent subbottoms. (On the contrary, neglecting shear waves lead to a unique fluid subbottom.) We have probed two types of structures,

Low-contrast (LC) : wave speeds (m/s) : $c_1 = 1500$, $c_2 = 1550$, $c_P = 1650$, $c_S = 400$
densities (kg/m^3) : $\rho_1 = 1030$, $\rho_2 = 1400$, $\rho_3 = 1800$
attenuations (dB/m/kHz) : $\alpha_2 = 0.1$, $\alpha_P = 0.3$, $\alpha_S = 2$

High-contrast (HC) : wave speeds (m/s) : $c_1 = 1500$, $c_2 = 1700$, $c_P = 4000$, $c_S = 2000$
densities (kg/m^3) : $\rho_1 = 1030$, $\rho_2 = 1920$, $\rho_3 = 2200$
attenuations (dB/m/kHz) : $\alpha_2 = 0.1$, $\alpha_P = 0.01$, $\alpha_S = 0.3$

using low frequencies (LF) with $d/\lambda_{sed.} \approx 0.5$ and high frequencies (HF) with $d/\lambda_{sed.} \approx 3.$, and at low$\theta$ ($0 \leq \theta \leq 20°$) or high θ ($30° \leq \theta \leq 45°$). The efficiency of this method is summarized in Table 1 and some examples displayed in figures 2 and 3. Fig. 2 shows "a good case" although non uniqueness is obvious, 29 possible solutions for the subbottom being obtained. It must be noted that the product $\rho_3 c_p$ is constant and slightly different from the true values of the P-wave impedance. On the other hand Fig. 3 shows a typical "bad reconstruction". Some improvement can be obtained when the layers are lossless.

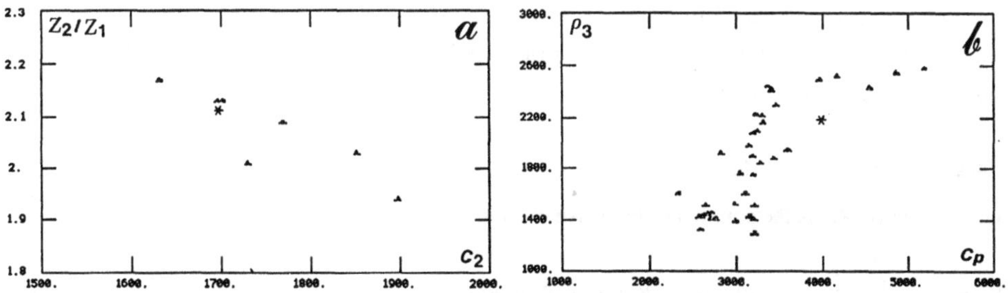

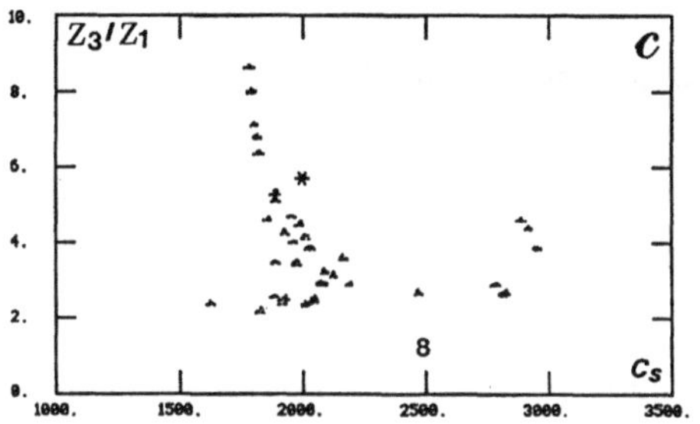

Fig. 3 : Geometrical Acoustics method (1.1). High-Contrast HC, High-Frequency HF, High-θ. Probing of 7 sediments and 41 subbottoms distributed as follows :
a) Z_2/Z_1 vs. c_2 where $Z_2 = \rho_2 c_2$,
b) ρ_3 vs. c_P,
c) Z_3/Z_1 vs. c_S,
The symbol * shows the exact solution.

1.2 Step response

In this approach only lossless layers are considered. The transient step response is the Fourier Transform of $jR(\theta, \omega) / \omega$ with respect to ω and is made of a sequence of plateaux whose heights are related to the amplitude of simply and multiply reflected waves as can be seen in Fig. 4 for the first two levels. Perfect knowledge of $h(\theta, t)$ obtained if $R(\theta, \omega)$ can be measured at all frequencies, allows the retrieval of the exact values of the unknowns using few incidences. (In practice we only have band-limited data.)

Progressing along similar lines as in geometrical acoustics, one can retrieve the values of the unknown parameters from R_{12} and R_{23}. A summary of the results is given in Table 2, where $\theta = 0°$, 10° and 15°, and low-pass filtered reflection coefficients are used. Fig. 4 shows an example of reconstruction. A main drawback of this approach is the importance of the low frequency data in $R(\theta, \omega)$, being known that a real source lacks this part in its spectrum.

Table 2

Relative errors (in %) of the parameters reconstructed from the low-pass filtered step response of the Fig.4.

	R_{12}	R_{23}	ρ_2	C_2	d	ρ_3	C_P	C_S
LC	5.4	0.1	3.5	-1.5	-0.3	1.5 -32	-3 45	35 148
HC	4.6	1.5	3	0.8	0.6	23 -20	-15 31	-27 44

Fig. 4

Step response method (1.2). Exact : - - - and low-pass filtered [0, 5 kHz] :-□-□- responses for the HC configuration used in order to calculate R_{12} and R_{23}.

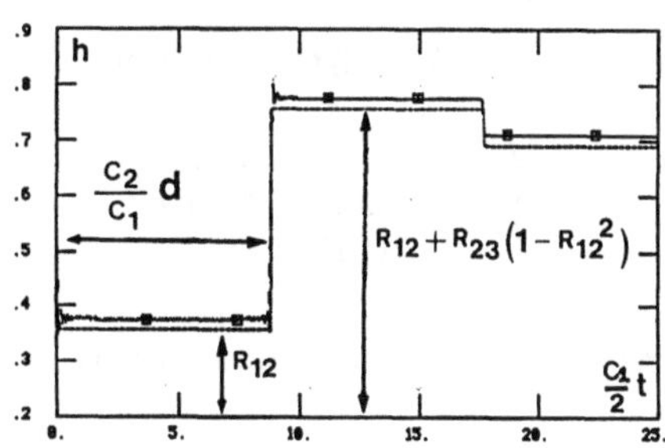

2. The inhomogeneous structure

There we consider a fluid filling the entire half-space z > 0 and described by the two profiles $\rho(z)$ and $c(z)$ (losses can also be accounted for).

2.1 The Gel'fand-Levitan like method

The pressure wave $p(z)$ is the solution of the wave equation :

$$\frac{d^2}{dz^2} p(z) + k_1^2 \left[\frac{c_1^2}{c(z)^2} - \sin^2\theta \right] p(z) = \frac{d}{dz} \operatorname{Log} \rho(z) \frac{d}{dz} p(z). \tag{4}$$

The method used here was introduced in [3] and is based on a transformation defined by apparent depth :

$$s(z) = \int_0^z \frac{\rho(z')}{\rho_1} dz', \tag{5}$$

followed by Liouville transformation :

$$\frac{ds}{dx} = \frac{1}{N_\theta(x)}, \quad g(x) = \sqrt{N_\theta(x)}\, p(s), \quad N_\theta(x) = \frac{\rho_1}{\rho(s)} \sqrt{\frac{c_1^2}{c^2(s)} - \sin^2\theta}\,. \tag{6}$$

These transformations lead to the 1-D Schrödinger equation :

$$\left[\frac{d^2}{dx^2} + k_1^2 \right] g(x) = \frac{1}{\sqrt{N_\theta(x)}} \frac{d^2\sqrt{N_\theta(x)}}{dx^2} g(x). \tag{7}$$

The well-known Gel'fand-Levitan method [4] is then applied to (7). Using the spatial impulse response of the inhomogeneous half-space $r(\theta, x)$ [= Fourier Transform of $R(k_1 = \omega/c_1, \theta)$ with respect to k_1] at two angles of incidences, we are able to retrieve ρ and c as functions of x first, then by inverting the two previous transforms as functions of z.

The results show on the average a good efficiency of this approach. Typical reconstructions are displayed in Fig. 5. Two points can be highlighted. First, the stability of the reconstructed velocity profile with respect to noise (on the contrary, errors on the density profile increase sharply with noise), and of both profiles with respect to the losses. Second, the fact that a discontinuity in the profiles seems to oppose the perturbation due to these losses. Although, as seen in (7), this method requires the profiles to be twice differentiable, but still gives reliable results when this condition is not satisfied.

2.2 Impediography

Starting again from (4) to which the transformation

$$t(z) = \int_0^z \sqrt{\left[\frac{1}{c(z')^2} - \frac{\sin^2\theta}{c_1^2} \right]}\, dz', \tag{8}$$

$t(z)$ travel-time, is applied we are led to :

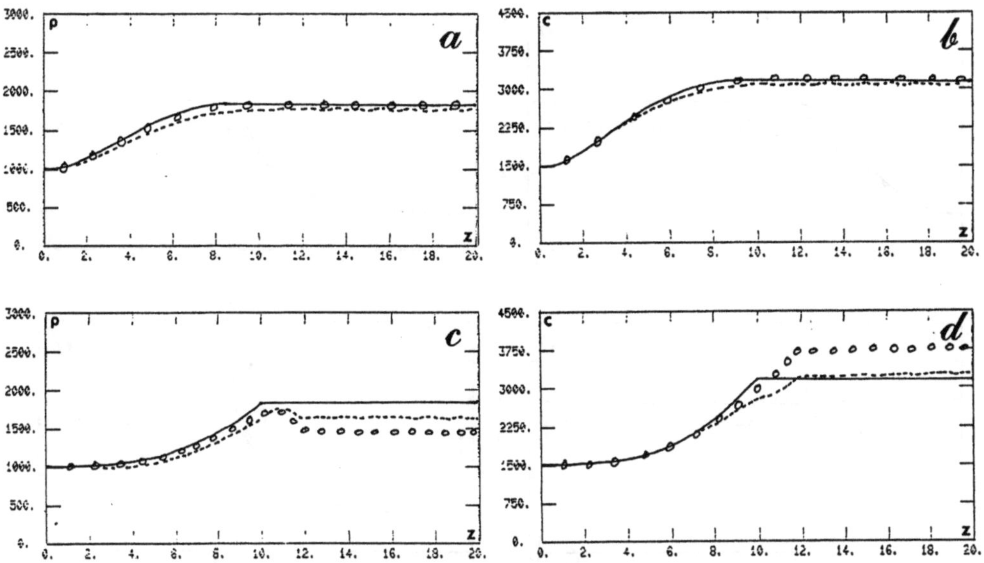

Figs. 5 a-d: Gel'fand-Levitan like method (2.1). Probing two inhomogeneous fluids with either twice-differentiable (a, b) density and velocity profiles or only once (c, d). Exact profiles are ——. Influence of losses on the reconstruction : without o o o and with (0.5 dB/m/kHz) - - - . *Note that the "profiles in lossy media" depicted in Fig. 7 in ref. [6] are erroneous.*

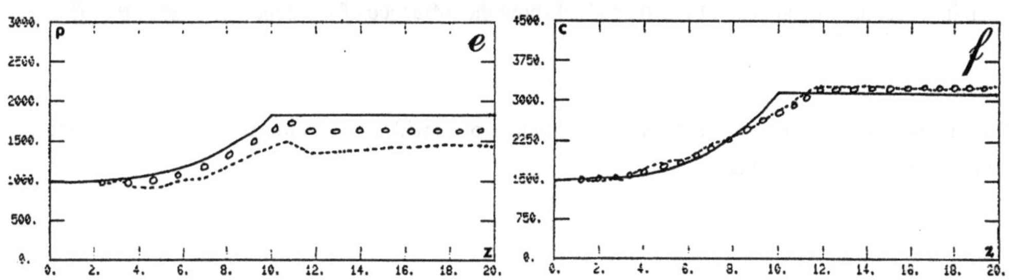

Figs. 5 e-f : Gel'fand-Levitan like method (2.1). Influence of noise obtained by truncating the spatial impulse response. The Normalized Mean-Square Error (NMSE) is about 10^{-4} : o o o and 10^{-2} : - - - (losses have been taken into account in both cases). Density (e) and velocity (f) reconstructions of a once-differentiable profile.

$$\frac{d^2}{dt^2} p(t) + \omega^2 p(t) = \frac{d}{dt} \mathrm{Log}\, Z(t) \frac{d}{dt} p(t), \quad Z(t) = \rho(t)\, c(t) \Big/ \sqrt{1 - \sin^2\theta \frac{c^2(t)}{c_1^2}}, \tag{9}$$

Z impedance of the medium. Under the Born approximation it can be shown from (9) that :

$$Z(t) = Z_1 \exp 2 \int_0^{2t} \mathrm{h}(\theta, \mathrm{t}')\mathrm{dt}', \tag{10}$$

where Z_1 is the impedance of the water and $h(\theta, t')$ the impulse response of the medium [5]. Attempt to solve for both $\rho(t)$ and $c(t)$ from the knowledge of $Z(t)$ at two different angles has not been successful. Examples of the usefulness of this method, which requires differentiable profiles, are shown in Fig. 6.

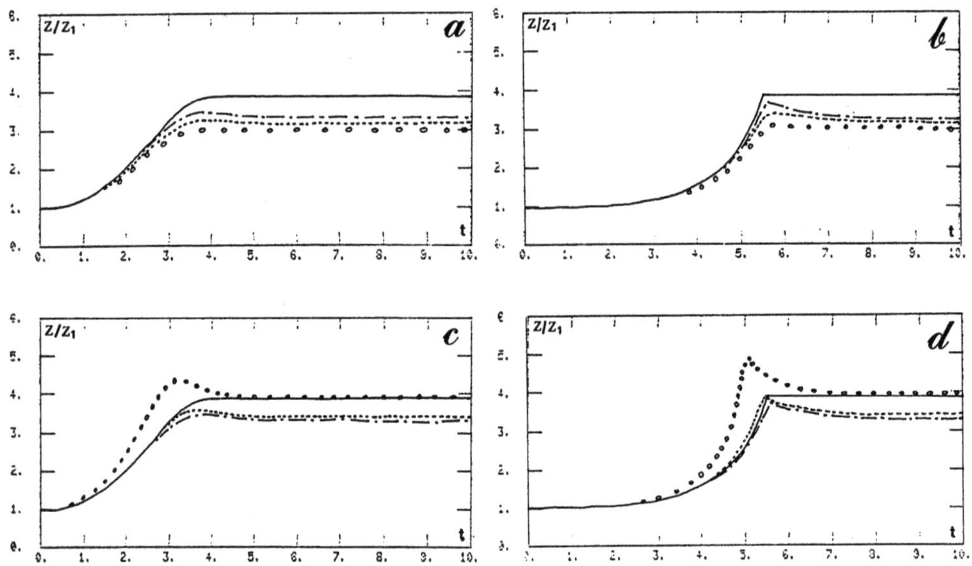

Fig. 6 : Impediography (2.2). Probing the impedance of the two media considered in Fig. 5. Exact profiles are ——. Figs. 6a and 6b show the influence of losses : without – · –, with (0.5 dB/m/kHz) - - -, and of noise obtained by truncating the time impulse response with NMSE about 10^{-2} : o o o (losses are taken into account in that case). Figs. 6c and 6d show the influence of incidence (no loss) : $\theta = 0°$ –.–.–, 10° _ _ _ _, 20° o o o.

This approach gives fair results at least for low values of θ, with much better reliability vs. noise than with the previous exact method.

Conclusion

The extensive comparative study presented here shows that all four methods provide useful information about fairly realistic profiles. Improvement of their efficiency is one of our future goals along with their application to the case of a water layer of finite thickness. As a general conclusion one can say that the 1-D problem is still "to be solved". This means that for real world applications where the medium is highly idealized in the modeling process, the methods presented here and any others are still to be checked against in situ experimental data. Unless this is done it is not possible to draw conclusions about the practical usefulness of one method or another.

References

[1] D. Lesselier and W. Tabbara, "Probing one-dimensional inhomogeneous media : how can it be done ?", in *Electromagnetic and Acoustic Scattering : Detection and Inverse Problem*, C. Bourrely *et al.* Eds., World Scientific, Singapore, 303-316, 1989.

[2] P. Grassin *et al.*, "Analyse de quelques approximations utiles à la caractérisation acoustique du fond marin", *J. Acoust.*, **2**, 241-254, 1989.

[3] S. Coen, "Density and compressibility profiles of a layered acoustic medium from precritical incidence data", *Geophys.,* **46**, 1244-1246, 1981.
[4] I. Kay, "The inverse scattering problem for transmission line", in *Mathematics of Profile Inversion*, Ed.Colin, NASA TM X-62, **150**, 6-2 - 6-17, 1972.
[5] J. P. Jones, "Impediography : a new ultrasonic technique for non-destructive testing and medical diagnosis", in *1973 Ultrason. Int. Conf. Proc.*, IPC Sci. Techn. Press, Guilford, 214-218, 1974.
[6] R. Bohbot *et al.*, "Exact and Approximate probing of sea-bottom", in *Proc. 1989 IEEE Ultrasonics Symp.*, Montréal, 1989, to appear.
[7] A. G. Tijhuis, *Electromagnetic Inverse Profiling : Theory and Numerical Implementation*, VNU Science Press, Utrecht, 1987.

Quantitative Reconstruction of Complex Permittivity Distributions by Means of Microwave Tomography

J.P. Hugonin[1], *N. Joachimowicz*[2], *and Ch. Pichot*[2]

[1]Institut d'Optique, Bâtiment 503, F-91405 Orsay, France
[2]Groupe d'Electromagnétisme, Laboratoire des Signaux et Systèmes, F-91192 Gif-sur-Yvette Cédex, France

INTRODUCTION

This paper concerns an iterative method for solving an inverse electromagnetic scattering problem : the quantitative reconstruction of complex permittivity distribution of inhomogeneous objects. During the last decade, intensive research and interest have been focussed on reconstruction algorithms based on the so-called Diffraction Tomography with applications in microwaves or ultrasound [1-16]. The main motivations for using such approaches are to obtain explicit formulas for solving the imaging problem, to take advantage of existing numerical algorithms (Fast Fourier Transform) and to implement the algorithms on PC's or minicomputers for imaging system purposes (planar microwave camera [5], experiments with ultrasounds and microwaves [11][13][14][16], cylindrical microwave system [15]). However, Diffraction Tomography is subjected to various limitations which include artefacts due to diffraction effects in strongly inhomogeneous media [7][10][16][17][18].The final objective of microwave imagery is to determine the complex permittivity profile (permittivity and conductivity profiles) of the object under investigation. An increasing number of papers [17-25] have been devoted to a such non-linear inverse problem. Different solutions based on Moment Methods [19-25] have been explored, but, convergence depends on contrasting objects. Furthermore, stability is very sensitive to the observation point locations and measurement accuracy (due to the ill-conditioning of the matrix which has to be inverted). Stability and sensivity are why the use of an iterative scheme is important : effects of ill-conditioning can be significantly reduced by enforcing the convergence with "a priori" information (object external shape, upper and lower bounds of complex permittivity, presence of different media,...). Concerning the uniqueness of object reconstruction, it has been proved [24] that if the background medium is dissipative, then the inverse scattering problem has a unique solution.

In the first section, the derivation approach is described where 2D-TE and 2D-TM polarization as well as 3D cases are considered. The authors are obviously aware of the controversy concerning the computations of the electromagnetic field in the 2D-TE polarization and 3D cases [30-33], and are presently considering this problem [34]. Any way, by starting from exact integral formulations, the Moment Method could be utilized to generate matrix equations for the forward scattering problem. Then, an iterative variation scheme for the inverse problem is proposed, based on the relation between a small change in complex permittivity and the measured scattered field. A regularization procedure is applied on matrices which have to be inverted. The iterative scheme has been constructed in order to incorporate a multiple view configuration (the illuminating source rotates with the probing surface). Numerical programs involving the implementation of forward and inverse problems for 2D and 3D cases have been developed. In the second section, the method is illustrated with numerical simulations performed for biomedical applications of microwave imaging. Reconstruction in presence of noise is presented in order to test the robustness of the algorithm.

THEORETICAL ANALYSIS

Let $\mathbf{e_x}$, $\mathbf{e_y}$, $\mathbf{e_z}$ be the unitary vectors of the rectangular coordinate system 0xyz. For ease of representation, the basic ideas are described for the 2D electromagnetic problem in the simple case of polarization TM, i.e. where $\mathbf{E} = E(x,y)\,\mathbf{e_z}$. A cylindrical object (invariant along oz) of arbitrary inhomogeneous cross-section D_1, characterized by a complex permittivity $\varepsilon_1^*(x,y)$, is immersed in a homogeneous medium D_0 of complex permittivity ε_0^*. This object is illuminated by a time-harmonic plane wave whose direction of propagation $\mathbf{k_0}$ lies in the (x,y) plane. The field scattered by the object is collected by M detectors located (on a circle in our simulations) around the latter (Figure 1).

Let us sample the object at N points and let s be the vector (dimension N) which contains the values of the contrast $(k_1^2(x,y) - k_0^2)$ (k_j being the propagation constant of the medium j) at the N sampling points. We define a diagonal matrix S whose diagonal terms are the components of s. We realize V different experiments of diffraction upon the object, by varying the direction of propagation of the incident wave. In the experiment number v, the scattered field measured at M measurement points are the components of a vector Ψ_v of dimension M. The incident and total field at the sampling points in the object during experiment number v, form the N components vectors E^i_v and E_v. By applying Moment Method with pulse function basis and point matching, we obtain from the Electric Field Integral Equation, the two following matrix equations :

$$E_v - E^i_v = G\,S\,E_v \qquad (1)$$

$$\Psi_v = H\,S\,E_v \qquad (2)$$

where G and H are NxN matrices which depend on the geometry of the object and on detector positions. Solving (1) gives the total field everywhere and particulary inside the object :

$$E_v = (I - GS)^{-1}\,E^i_v \qquad (3)$$

Once the internal field is obtained, Ψ_v can be computed from (2). Let Ψ be the vector (dimension VM) whose components are those of Ψ_v for different experiments. When the vector of contrast s is given, Ψ

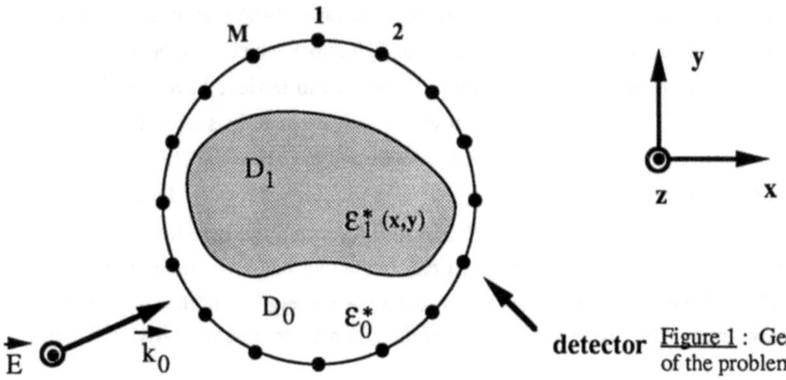

Figure 1 : Geometrical configuration of the problem

can be computed by using relations (3) and (2). This defines an operator Ω such that :

$$\Psi = \Omega\,(s) \tag{4}$$

This operator is not linear and consequently cannot be easily inverted. The variation $\Delta\,\Psi_v$ of Ψ_v induced by a small variation Δs of s can be written after some easy calculations, under the form :

$$\Delta\,\Psi_v\ = [H\,(\,I - SG\,)^{-1}\,e_v\ ]\,\Delta s \tag{5}$$

where e_v is the diagonal matrix whose elements are the components of E_v.
Collecting $\Delta\,\Psi_v$ for different experiments leads to a linear relation between Δs and $\Delta\Psi$:

$$\Delta\Psi = D\,\Delta s \tag{6}$$

where D is a VM x N matrix. The inverse problem consists of the inversion of the operator Ω. This inversion is made by using an iterative technique similar to the method developped by Roger and al [35] in a different configuration. Data are now the measurement vector Ψ and we seek for a vector s which satisfies (4). Let s_o be an "a priori" estimate of s and s_k the approximation obtained after k iteration steps. The next approximation of s is given by : $s_{k+1} = s_k + \Delta\,s_k$, where Δs is the quantity which minimizes the norm of $\{D\,\Delta s - (\Psi - \Omega\,(s_k))\}$. This problem is equivalent to solve the matrix equation $\Delta\,s_k\ = (D^*D)^{-1}\ [D^*\ \ (\Psi\ -\ \Omega\ (s_k))]$, where D^* is the conjugate transpose of D. Due to the ill-conditioning of the matrix D^*D, a Tikhonov procedure of regularization is used. This iterative process is stopped when $\Omega\,(s_k)$ is closed to the measurement vector Ψ.

The above method can be easily extended to other geometries where the electric field is described by two or three components (2D-TE and 3D cases). Let d denotes the number of dimensions for the electric field intensity vectors. In these cases the vectors E_v and E^i_v have now Nd components and in the equations (1) and (2) S is the diagonal matrix of dimension Nd, whose diagonal elements are obtained by repeating d times the N components of s.

NUMERICAL RESULTS

1) General considerations

This section is devoted to some illustrative results obtained by means of the previously described reconstruction algorithm. These results concern some basic configurations which have been studied in view of biomedical applications of microwave imaging. The objects to retrieve have the dielectric properties of bones, muscles, etc... and are immersed in water, as it is usual for such applications in order to improve the penetration of microwaves in living tissues and to reduce the effect of external parasitic paths. At the 3 GHz operating frequency, the complex permittivity of such media are $\varepsilon^* = (8., 1.2)$ for bone, $\varepsilon^* = (46., 12)$ for muscle and $\varepsilon^* = (76., 14.4)$ for water. Both two-dimensional and three-dimensional cases have been considered. The numerical simulations have been mainly conducted in order to assess the effect of "a priori" information and noise on the convergence of the reconstruction process. For this purpose, the relative mean square error of the reconstructed permittivity profile has been defined as :

$$ERR_S = \left[\sum_{i=1}^{N} |\Delta s^k (i)|^2 / \sum_{i=1}^{N} | s (i)|^2 \right]^{1/2} \qquad (7)$$

where i, s and Δs^k denote the number of the cell, the exact value of the contrast and the difference between the reconstructed and the exact value at the k-th iteration step respectively.

The noise is introduced by adding to the real and imaginary parts of the simulated scattered field a random signal with uniform repartition function between two opposite values ± w. The signal over noise ratio S/N is defined as follows :

$$S/N = 20 \log \{ \text{Max scattered field} / \text{Mean square noise} \} \qquad (8)$$

where Max scattered field is the maximum "measured" value of the scattered field over all the projections and Mean square noise is obtained from all the noise samples introduced in the data.

Unless it is explicitely indicated, the convergence criterion ERR_ψ on the scattered field is 10^{-6}. However the iterative process is also stopped if the error ERR_ψ remains stable and no longer decreases after a few additional iterations.

2) Two-dimensional case

The object consists of a circular cylindrical shell, whose inner and outer dielectric properties are those of bone and muscle respectively, and whose inner and outer radii are 3 cm and 6 cm respectively. This object is enclosed within a square divided into 11 x 11 elementary cells. It is illuminated by TM or TE polarized plane waves. The measurements are assumed to be performed on a circle whose center is located at the center of the object and whose radius is equal to 3.5 λ (λ being the wavelength in the surrounding water). For TM cases, 8 plane waves are used. For each incidence, the scattered field is measured by means of 64 receivers. For TE cases, the number of views is 16 while the number of receivers is 32, maintaining the same number of measured data as for TM cases.

Two different configurations have been investigated in order to study the effect of "a priori" knowledge on the convergence rate. In the first configuration, the initial guess corresponds to a homogeneous 6 cm square of water (no "a priori" information), while in the second one, the initial guess is a homogeneous 6 cm radius muscle cylinder surrounded by water (approximate knowledge of the external boundary). Figure 2 shows the convergence of the complex permittivity in TM polarization cases. Even if no "a priori" information is introduced, the object is accurately retrieved. Figure 3 gives the relative mean square error as a function of the number of iterations. White and black arrows indicate the value of this error calculated from the initial guess. It appears that, for the same accuracy, the second configuration needs only half of the number of iterations of the first configuration. The CPU time required per iteration is around 10 minutes, on a VAX 8300 computer.

Figure 4 shows similar results for TE polarization cases. The same comment apply as for TM cases except that, due to the fact that the calculation involves the two components of the electric field, the CPU time is now 20 minutes per iteration.

The noise effects, have been studied with the second configuration in the TM polarization case. The S/N ratio was chosen equal to 31dB, which constitutes a realistic value for real experimental

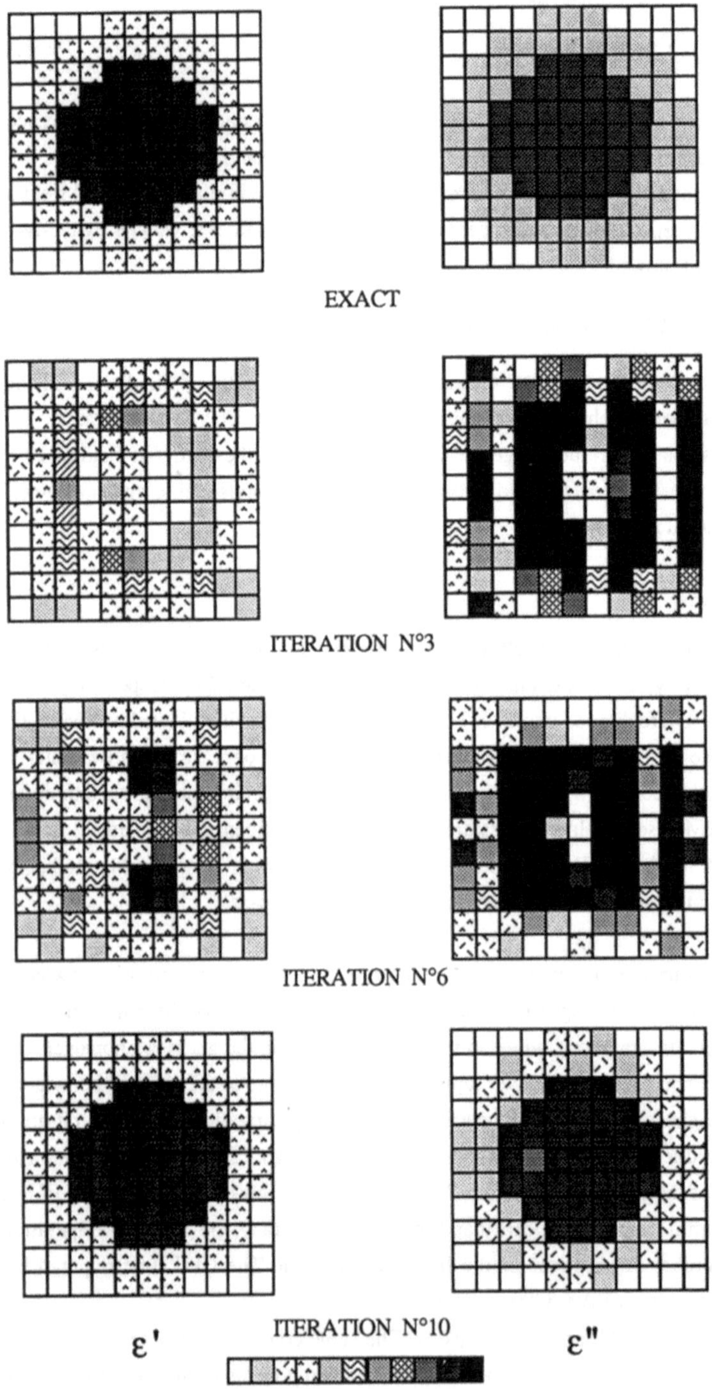

Figure 2 : Reconstruction of complex permittivity (ε'+ j ε") with 8 plane waves in the TM polarization case (initial configuration n°1).

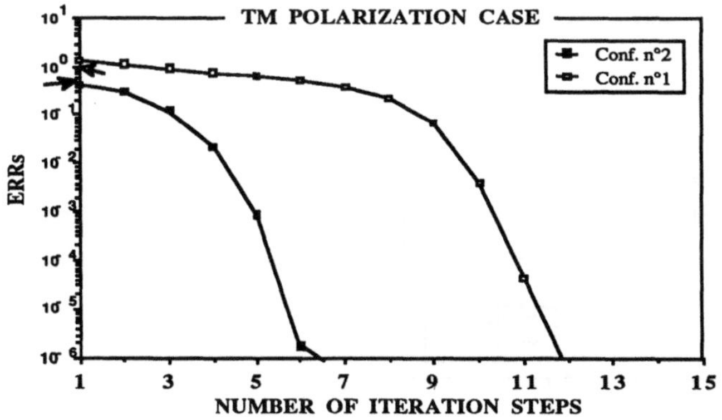

Figure 3 : The relative mean squared error of the reconstructed complex permittivity distribution (ERRs) as a function of the iteration steps in the TM polarization case, for each initial configuration

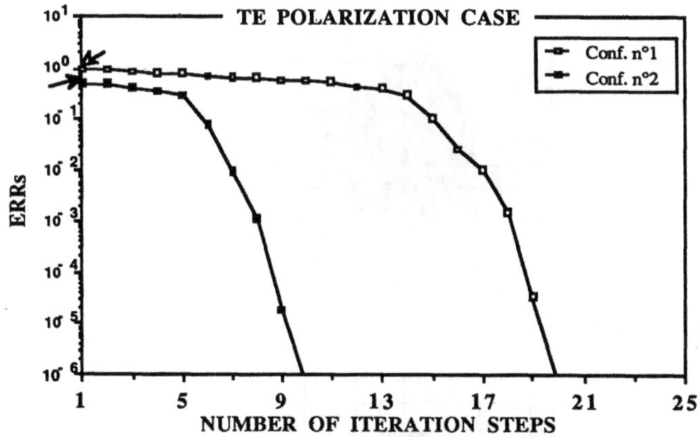

Figure 4 : The relative mean squared error of the reconstructed complex permittivity distribution (ERRs) as a function of the iteration steps in the TE polarization case, for each initial configuration

measurements. It is worth noting that, according to the convergence criterion defined on ERR_ψ, the iterative scheme stops after 7 steps, while the mean square error on the permittivity is not negligible (0.18 instead of 10^{-6} in the absence of noise). In order to improve the accuracy of the reconstruction process, the number of views was increased from 8 to 16. As previously, the iterative scheme stops after 7 steps, but the mean square error on the permittivity is now decreased to 0.08. Figure 5 shows, step by step, how the accuracy is improved by increasing the number of plane waves. Figure 6 gives the reconstructed permittivity at the last iteration step for 8 and 16 views.

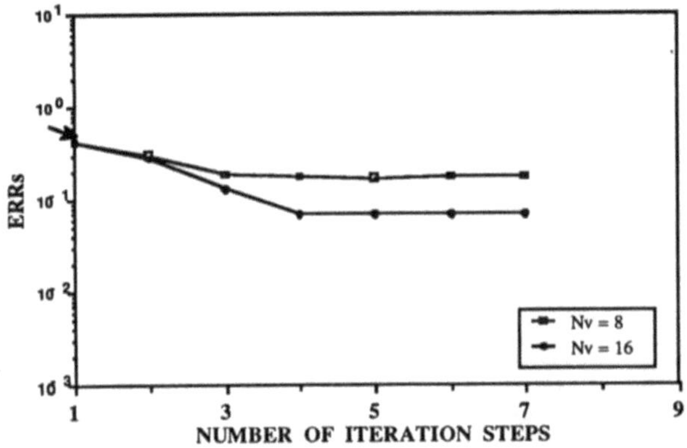

Figure 5 : Effect of the noise on the convergence of ERR_S and influence of the number of views (S/N = 31dB).

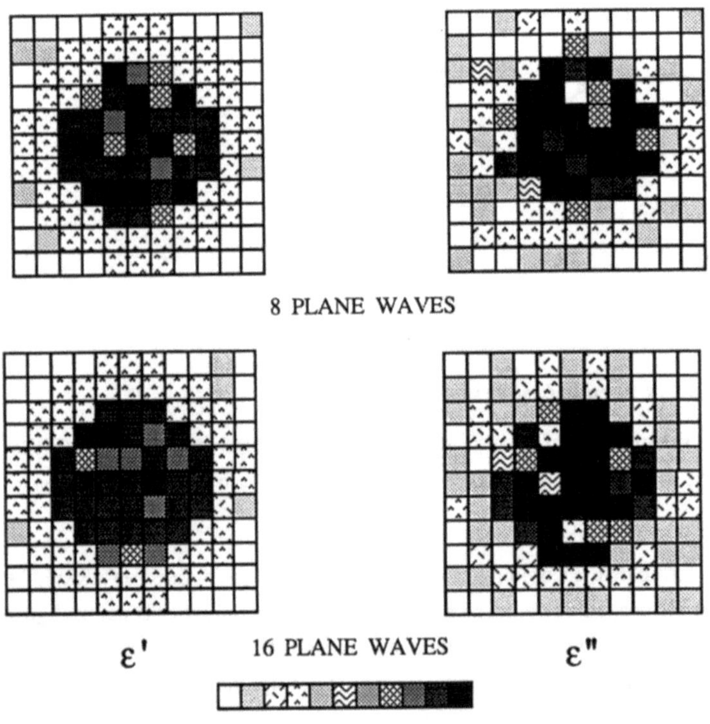

Figure 6: Profile of the reconstructed complexe permittivity at the last iteration step for 8 and 16 views.

3) Three-dimensional case

The object is a dielectric cube, whose each side is equal to the wavelength λ in the surrounding water. The cube consists of 27 cells. 26 cells correspond to muscle and 1 to bone. The object is illuminated by 8 plane waves and 6 receivers are located on 2 parallel lines located on both sides of the dielectric object, at λ from the center. The distance between two receivers on each lines has been set to λ/3. The initial guess is a homogeneous muscle cube surrounded by water. The convergence is achieved in 4 iterations, with the same order of magnitude for ERRψ and ERRs. As for two-dimensional cases, the noise effect has been studied. Typically, for a 25 dB S/N ratio, the mean square errors on the permittivity is about 2.2% and 0.7%, for 8 and 32 views, respectively. The same trends can be observed than for TM cases.

By means of this example, only a small part of the wide possibilities offered by the algorithm has been examined. In the future, it should be tested for more realistic objects and more convenient measurement configurations.

CONCLUSION

An algorithm for solving both 2D and 3D electromagnetic nonlinear inverse scattering problems has been proposed. Sactifactory results have been obtained for realistic levels of noise even in the case of high contrast, due to the possibility of including in the iterative procedure multiple view configuration and "a priori" information. Such results still remains to be validated through further experiments, by using existing microwave imaging equipment.

REFERENCES

[1] R.K. Mueller, M. Kaveh, G. Wade, "Reconstructive tomography and applications to ultrasonics," Proc. IEEE, 67 (4), pp. 567-587, 1979.
[2] F.M. Adams, A.P. Anderson, Synthetic aperture tomography (SAT) imaging for microwave diagnostics," IEE Proc.-H, Micr.,Opt and Ant., 129 (2), pp. 83-88, 1982.
[3] J.Ch. Bolomey, A. Izadnegahdar, L. Jofre, Ch.Pichot, G. Peronnet, M. Solaimani, "Microwave diffraction tomography for biomedical applications," IEEE Trans. Microwave Theory Tech., MTT-30, pp. 1998-2000, 1982.
[4] M. Baribaud, F. Dubois, R. Floyrac, M. Kom, S. Wong, "Tomographic image reconstruction of biological objects from coherent microwave diffraction data," IEE Proc.-H, Micr., Opt and Ant., 129 (2), pp. 356-359, 1982.
[5] G. Peronnet, Ch. Pichot, J. Ch Bolomey, L. Joffre, A. Izadnegahdar, C. Szeles, Y. Michel, "A microwave diffraction tomography system for biomedical applications," Proc. XIIIth European Microwave Conference (Nuremberg, FRG), pp.529-533, 1983.
[6] A.J. Devaney, "A computer simulation study of diffraction tomography," IEEE Trans. Biomed. Eng., BME-30, pp 377-386, 1983.
[7] M. Slaney, A. Kak, L. Larsen, "Limitations of imaging with first order diffraction tomography, " IEEE Trans. Microwave Theory Tech., MTT-32, pp. 860-873, 1984.
[8] Ch. Pichot, L. Joffre, J. Ch Bolomey, G. Peronnet, " Active Microwave Imaging of Inhomogenous bodies," IEEE Trans. Ant. Propagat., AP-33, pp. 416-425, 1985.
[9] *Inverse methods in Electromagnetic Imaging*, W.-M. Boerner et al. Eds, NATO ASI Series, Reidel Publ. Comp., Dordrecht, 1985.
[10] H. Ermert, M. Dohlus, "Microwave-diffraction-tomography of cylindrical objects using 3-dimensional wave-fields," ntzArchiv, Bd.8 H.5, pp.111-117, 1986.
[11] F.J. Paolini, "Implementation of microwave diffraction tomography for measurement of dielectric constant distribution," IEE Proc.-H, Micr., Opt and Ant., 134 (1), pp. 25-29, 1987.
[12] J.M. Rius, M.Ferrando, L. Joffre, A. Broquetas, " Microwave tomography : an algorithm for cylindrical geometries," Electr. Lett., 23 (11), pp. 564-565, 1987.
[13] W. Tabbara, B. Duchêne, Ch. Pichot, D. Lesselier, L. Chommeloux, N. Joachimowicz, "Diffraction tomography : contribution to the analysis of some applications in microwaves and ultrasonics," Inverse Problems, 4, pp. 305-331, 1988.
[14] R. Aitmehdi, A.P. Anderson, S. Sali, M. Ferrando, "The determination of dielectric loss tangent by microwave phase tomography," Inverse Problems, 4, pp. 333-345, 1988

[15] L. Joffre, M.S. Hawley, A. Broquetas, E. De Los Reyes, M. Ferrando, "Medical imaging with a microwave tomographic scanner, " Submitted for publication in IEEE Trans. Biomed. Eng.
[16] J.Ch. Bolomey, Ch. Pichot, G. Gaboriaud, "Critical and prospective analysis of reconstruction algorithms devoted to a planar microwave camera for biomedical applications, " URSI Int. Symp. Proc.(Stockholm, Sueden), pp. 144-146, 1989.
[17] B. Duchêne, D. Lesselier, W. Tabbara, " Diffraction tomography approach to accoustical omaging and media characterization," J. Opt. Soc. Am., 2 (11), pp. 1943-1953, 1985.
[18] M.A. El Khalifa, "Amélioration de la qualité des images en tomographie microonde. Contribution a une imagerie quantitative, " Thèse de doctorat en Sciences Physiques, Université de Paris-Sud (Orsay), 1989.
[19] D.K. Ghodgaonkar, O.P. Gandhi, M.J. Hagmann, "Estimation of complex permittivities of three-dimensional biological bodies, " IEEE Trans. Microwave Theory Tech., MTT-31, pp. 442-446,1983.
[20] S.A. Johnson, M.L. Tracy, "Inverse scattering solutions by a sinc basis, multiple source, Moment Method- Part I : Theory, Part II : Numerical Evaluations," Ultrasonic Imaging, 5, pp.361-392, 1983.
[21] S.A. Johnson, M.L. Tracy, "Inverse scattering solutions by a sinc basis, multiple source, Moment Method- Part III : Fast Algorithms, " Ultrasonic Imaging, 6, pp.103-116, 1984.
[22] M.M. Ney, A.M. Smith, S.S. Stuchly, "A solution of electromagnetic imaging using pseudoinverse transformation, " IEEE Trans. Medical Imaging, MI-3, pp.155-162, 1984.
[23] A.N. Datta, B. Bandyopadhyay, "Nonlinear extension to a moment method iterative reconstruction algorithm for microwave tomography, " Proc. IEEE,74 (4), pp.604-606, 1986.
[24] T.C. Guo, W.W. Guo, "Physics of image formation by microwave scattering, " Medical Imaging, SPIE Proc., 767, pp.1-10, 1987.
[25] S. Caorsi, G.L. Gagnani, M. Pastorino, "Electromagnetic vision-oriented numerical solution to three-dimensional inverse scattering, " Radio Science, 23 (6), pp.1094-1106, 1988.
[26] Y.M. Wang, W.C. Chew, "An iterative solution of two-dimensional electromagnetic inverse scattering problem, " Int. J. Imag. Syst., 1, pp.100-108, 1989.
[27] D.G.H. Tan, R.D. Murch, R.H.T. Bates, "Inverse scattering for penetrable obstacles, " URSI Int. Symp. Proc.(Stockholm, Sueden), pp.160-162, 1989.
[28] T.M. Habashy, D.G. Dudley, "Simultaneous inversion of permittivity and conductivity profiles using a renormalized source-type integral equation approach," URSI Int. Symp. Proc.(Stockholm, Sueden), pp.578-580, 1989.
[29] M.Schueller, H. Chaloupka, "Improve reconstruction in difraction tomography by means of a nonlinear scattering model, " URSI Int. Symp. Proc.(Stockholm, Sueden), pp.584-586, 1989.
[30] H. Massoudi, C. H. Durney and M. F. Iskander, "Limitations of the cubical block model of man in calculating SAR distributions, " IEEE Trans. Microwave Theory Techn., MTT-32, pp.746-752, 1984.
[31] M. J. Hagmann, " Comments on 'Limitations of the cubical block model of man in calculating SAR distributions', " IEEE Trans. Microwave Theory Techn., MTT-33, pp.347-350, 1985.
[32] D. T. Borup, D. M. Sullivan, and O.P.Ghandhi, "Comparison of the FFT Conjugate Gradient Method and the Finite-Difference Time-Domain Method for the 2-D absorption problem," IEEE Trans. Microwave Theory Tech., MTT-35, pp.383-395, 1987.
[33] N.Joachimowicz, Ch. Pichot, " Comparison of Three Integral Formulations for the 2D-TE Scattering Problem', " to be published in IEEE Trans. Microwave Theory Techn., Feb. 1990.
[34] A. Roger, D. Maystre, M. Cadilhac, "On a problem of inverse scattering in optics : The dielectric inhomogeneous medium, " J. Optics (Paris), 9 (2) , pp. 83-90, 1978.

Some Asymptotic Results on Coherent-State Matrix Elements

O. Brander

Chalmers University of Technology, S-41296 Göteborg, Sweden

Abstract: It is proved that the coherent-state matrix elements of a multiplication operator $V(x)$ are uniformly bounded, provided V belongs to the weighted space $L^{p,\sigma}(R^1)$, where (p, σ) is restricted to a certain region Δ.

A coherent-state path-integral formulation of inverse scattering theory was introduced in Ref. [1]. In such a formulation a fundamental role is played by the symbol r of an observable R, defined as the matrix element between coherent states,

$$r(a,b) = <\xi_a|R|\xi_b> \quad . \tag{1}$$

Here, for simplicity restricting ourselves to one dimension, the coherent state is given by[2)]

$$|\xi_b> = \exp\left(-\frac{1}{2}|b|^2\right) \sum_{j=0}^{\infty} (j!)^{-\frac{1}{2}} b^j |j> \quad . \tag{2}$$

The state is labeled by a point b in the two-dimensional phase space Γ, and $|j>$ is a harmonic oscillator number state.

In the present study we shall restrict ourselves to a particular problem, namely that of characterizing for which multiplication operators $R = V(x)$ the quadratic form $v(a,b)$ is bounded. This is of fundamental importance for the theory of Ref. [1], but is also of more general interest. It involves the use of uniform asymptotic analysis[3,4)].

We shall assume that $V(x)$ is a momentum-independent multiplication operator with a finite weighted L^p-norm:

$$||V||_{p,\sigma} := \{\int_{-\infty}^{\infty} dx (1+x^2)^{\sigma/2} |V(x)|^p\}^{1/p} \quad . \tag{3}$$

The proof goes in two steps. First we prove a Lemma about the asymptotic properties of $L^{p,\sigma}(R^1)$-norms of high harmonic oscillator states, then a Theorem about the finiteness of the quadratic form.

Lemma: Let $\varphi_n(x) =< x|n >$ be a harmonic oscillator number state,

$$\left(\frac{d^2}{dx^2} + 2n + 1 - x^2\right)\varphi_n(x) = 0 \quad , \tag{4}$$

and let $\rho \leq q^{-1} - \epsilon$ for $1 \leq q \leq 4$ or $\rho \leq -(1-q^{-1})/3 - \epsilon$ for $4 \leq q \leq \infty$, $\epsilon > 0$. Then the $L^{q,\rho}$-norm of φ_n is finite and bounded as $n \to \infty$ by a constant times $n^{-1/4}$.

Proof: As is well-known,

$$\varphi_n(x) = 2^{-n/2}\pi^{-1/4}(n!)^{-1/2}\exp\left(-\frac{1}{2}x^2\right)H_n(x) \quad , \tag{5}$$

where H_n are Hermite polynomials. For those Olver (Ref. [3], p. 403) gives the asymptotic estimate

$$H_n(\nu z) = (2\pi)^{1/2}e^{-\nu^2/4}\nu^{(3\nu^2-1)/6}e^{\nu^2 z^2/2}\left(\frac{\zeta}{z^2-1}\right)^{1/4}\left\{Ai\left(\nu^{4/3}\zeta\right) + \epsilon(z)\right\} \quad , \tag{6}$$

valid for $z > -1$, with

$$\begin{cases} \nu = (2n+1)^{1/2}, z = x\nu^{-1} \\ \zeta = \begin{cases} \left\{\frac{3}{4}z(z^2-1)^{1/2} - \frac{3}{4}\cosh^{-1}z\right\}^{2/3} & \text{for } z \geq 1 \\ -\left\{\frac{3}{4}\cos^{-1}z - \frac{3}{4}z(1-z^2)^{1/3}\right\}^{2/3} & \text{for } -1 < z \leq 1. \end{cases} \end{cases} \tag{7}$$

The error term $\epsilon(z)$ is bounded by

$$|\epsilon(z)| \leq \begin{cases} 1.36\{\exp(0.09\nu^{-2}) - 1\}Ai(\nu^{4/3}\zeta) & \text{for } z \geq 1 \\ 0.97\{\exp(0.28\nu^{-2}) - 1\}E^{-1}(\nu^{4/3}\zeta)M(\nu^{4/3}\zeta) & \text{for } 0 \leq z \leq 1, \end{cases} \tag{8}$$

where M and E are modulus and weight functions, respectively, for the Airy functions (Ref. [3], p. 394).

The relation between z and ζ can also be written[4)]

$$\zeta = (z^2-1)f^{2/3}(z), \; f(z) := \frac{1}{2}F\left(\frac{3}{2},\frac{1}{2};\frac{5}{2};1-z^2\right) \tag{9}$$

Here, F is the hypergeometric function, and it is easy to prove that the function $f(z)$ is finite and positive for all $z \epsilon (0, \infty)$, slowly varying and has the asymptotic forms

$$f(z) \sim \begin{cases} \frac{3\pi}{8} \text{ for } z \to 0 \\ \frac{1}{2} \text{ for } z \to 1 \\ \frac{3}{4} z^{-1} \text{ for } z \to \infty \end{cases} . \tag{10}$$

Now, using Eq. (6) in the expression for the norm, one obtains

$$||\varphi_n||_{q,\rho} = \text{const} . \times \; n^{-\frac{1}{12}+\frac{1}{2q}} \left[J_{q,\rho}(n) \right]^{\frac{1}{q}} \quad , \tag{11}$$

where

$$J_{q,\rho}(n) := \int_0^\infty dz (1+\nu^2 z^2)^{q\rho/2} \left(\frac{\zeta}{z^2-1} \right)^{q/4} |Ai\left(\nu^{4/3}\zeta\right) + \epsilon(z)|^q . \tag{12}$$

To within the error bounds of Eq. (8), these formulae give a faithful estimate of the norm integral. We now turn to upper bounds for the integrals. Since $q \geq 1$ and ρ are arbitrary, we have to use a good bound for the Airy function.

As is well-known, the Airy function oscillates for $\xi < 0$ with an amplitude proportional to $|\xi|^{-1/4}$, reaches a maximum

$$\sup \; Ai(\xi) = C_1 \; \approx \; 0.536 \tag{13}$$

at $\xi \approx -1.019$ and goes to zero exponentially for larger ξ. More details are given, e.g., in Ref. [3], pp. 392-7. As a simplified bound, sufficient for our purpose, we take

$$|Ai(\xi)| \leq \begin{cases} \pi^{-1/2}(-\xi)^{-1/4} \text{ for } -\infty < \xi < -c \\ C_1 \text{ for } -c < \xi < c \\ C_2 \xi^{-1/4} \exp\left(-\frac{2}{3}\xi^{3/2}\right) \text{ for } c < \xi < \infty \end{cases} , \tag{14}$$

where C_1 is the supremum, c is chosen such that $\pi^{-1/2}c^{-1/4} = C_1 \Rightarrow c \approx 1.228$ and the constant $C_2 < 1.4$.

We now divide the integration interval of Eq. (12) into four parts and write

$$J_{q,\rho}(n) = \sum_{i=1}^{4} J_{q,\rho}^{(i)}(n) \quad , \tag{15}$$

with the correspondance

$$\begin{cases} i=1 & : \quad 0<z<\frac{1}{2} \\ i=2 & : \quad \frac{1}{2}<z<z(-c) \\ i=3 & : \quad z(-c)<z<z(c) \\ i=4 & : \quad z(c)<z<\infty . \end{cases} \tag{16}$$

Here,

$$\xi = \pm c \Rightarrow \zeta = \pm \nu^{-4/3} c \Rightarrow z^2 - 1 = \zeta f^{-2/3}(z(\pm c)) \cong \pm 2^{2/3} \nu^{-4/3} c \quad , \tag{17}$$

such that

$$z(\pm c) = 1 \pm 2^{-1/3} \nu^{-4/3} c + 0(\nu^{-8/3} c^2) \quad . \tag{18}$$

To estimate the integrals in Eq. (15) is now rather straightforward, so we just give the result:

$$|J_{q,\rho}(n)| \leq \text{const} \times \begin{cases} \nu^{-\frac{q}{3}-1} \begin{cases} \text{for } q<4 \text{ and } \rho<-q^{-1} \\ \text{or } q=4 \text{ and } \rho \leq -a^{-1}-\epsilon \\ \text{or } q>4 \text{ and } \rho<-3^{-1}+3^{-1}q^{-1} \end{cases} \\ \nu^{-\frac{q}{3}+q\rho+\epsilon} \text{ for } q \leq 4 \text{ and } \rho \geq -q^{-1} \\ \nu^{-\frac{4}{3}+q\rho+\epsilon} \text{ for } q>4 \text{ and } \rho \geq -3^{-1}+3^{-1}q^{-1} \end{cases} \tag{19}$$

Here, an arbitrarily small $\epsilon > 0$ is introduced to account for logarithmic cases. Insertion of this result into Eq. (11) now proves the Lemma, except for the supremum norm case $q = \infty$. However, that the Lemma is true also in this case follows directly from Eqs. (5), (6) and (14).

We now proceed to the Theorem on the phase space behaviour of the symbol $v(a,b)$.

Theorem: Let the multiplication operator $V(x)(x \epsilon R^1)$ have a finite $L^{p,\sigma}$-norm, with $(p,\sigma)\epsilon\Delta$, where

$$\Delta := \left\{ p,\sigma \;\middle|\; \begin{array}{l} 1 \leq p \leq 2 \text{ and } \sigma \geq (1+p^{-1})/3+\epsilon, \text{ or} \\ 2 \leq p \leq \infty \text{ and } \sigma \geq 1-p^{-1}+\epsilon, \ \epsilon>0. \end{array} \right\} \quad . \tag{20}$$

Then its symbol $v(a,b)$,

$$v(a,b) = \langle \xi_a | V | \xi_b \rangle \quad , \tag{21}$$

is a uniformly bounded quadratic form in $\Gamma \times \Gamma$.

Proof: Expand in the harmonic oscillator wave functions of Eq. (5):

$$v(a,b) = \sum_{j,m=0}^{\infty} \langle \xi_a | j \rangle \langle j | V | m \rangle \langle m | \xi_b \rangle \tag{22}$$

Here, Hölder's inequality can be used to estimate the matrix elements of V:

$$|\langle j|V|m\rangle| \leq ||V||_{p,\sigma} \times ||\langle j|x\rangle\langle x|m\rangle||_{q,-\sigma} \leq ||V||_{p,\sigma} \times |||j > ||_{2q,-\frac{1}{2}\sigma} \times |||m > ||_{2q,-\frac{1}{2}\sigma} \quad , \tag{23}$$

where p and q are conjugate, $p^{-1} + q^{-1} = 1$.

Now, applying the Lemma gives

$$|\langle j|V|m\rangle| \leq \text{ const. } \times j^{-\frac{1}{4}} m^{-\frac{1}{4}} ||V||_{p,\sigma} \tag{24}$$

provided $\sigma \geq q^{-1} + \epsilon$ for $1 \leq q \leq 2$ or $\sigma \geq (2 - q^{-1})/3 + \epsilon$ for $2 \leq q \leq \infty$, which amounts to $(p, \sigma) \epsilon \Delta$.

Eq. (24) implies that under the assumed condition on V each term of the sum (22) is finite and the sum is bounded by

$$|v(a,b)| \leq g_{q,-\sigma}(|a|) g_{q,-\sigma}(|b|) ||V||_{p,\sigma} \quad , \tag{25}$$

where

$$g_{q,-\sigma}(a) := \exp\left(-\frac{1}{2}a^2\right) \sum_{m=0}^{\infty} (m!)^{-\frac{1}{2}} a^m |||m > ||_{2q,-\frac{1}{2}\sigma} \quad . \tag{26}$$

Using here Laplace's method for the asymptotic evaluation of sums (Ref. [5], p. 304), we find that for $(p, \sigma) \epsilon \Delta$ the function $g_{q,-\sigma}(a)$ tends to a constant as $a \to \infty$. This concludes the proof of the theorem.

Corollary: The weighted trace

$$Tr_\tau(V) := \int_\Gamma da \; (1 + a^2)^{\tau/2} |v(a,a)| \tag{27}$$

and the Bargmann space norm

$$||V||_{F_\rho} := \left\{ \int_{\Gamma\times\Gamma} dadb \; (1 + a^2 + b^2)^\rho |v(a,b)|^2 \right\}^{\frac{1}{2}} \tag{28}$$

are finite for $\tau \leq -2-\epsilon,\ \rho \leq -2-\epsilon,\ \epsilon > 0$, provided the conditions of the Theorem are fulfilled.

Proof: Since the symbol is finite, the corollary follows from an elementary integral estimate.

Acknowledgement: This work is part of a larger study together with prof. B. DeFacio on a coherent state formulation of inverse scattering problems. I want to thank him for valuable, inspiring discussions. A grant from the Swedish Board for Technical Development is gratefully acknowledged.

References

1. DeFacio, B., and Brander, O., Suppl. Rend. Circ. Mat. Palermo, Series II, 17 (1987) 185-208 and in Nonlinear Evolutions, J. Leon editor (World Scientific Press, Singapore 1988) 750-766

2. Klauder, J.R., and Sudarshan, E.C.G., Fundamentals of Quantum Optics (Benjamin, New York 1968)
   Perelomov, A., Generalized Coherent States and Their Applications (Springer, Berlin and New York 1986)

3. Olver, F.W.J., Asymptotics and Special Functions (Academic Press, New York 1974)

4. Brander, O., Phys. Lett. 75A (1980) 471-474

5. Bender, C.M., and Orzag, S.A., Advanced Mathematical Methods for Scientists and Engineers (Mc Graw Hill, New York 1978)

Inverse Problems for Dispersive Biological Media

B. De Facio

Department of Physics and Astronomy, University of Missouri,
Columbia, MO 65211, USA

1. Introduction

Recently electromagnetic pulses with widths of picoseconds, 10^{-12} s, and even femtoseconds, 10^{-15} s, have become available. This opens a new field of inquiry called ultrafast spectroscopy. This new family of tools also provides new ways to look at old problems. This note will present a tutorial review of how these pulses together with inverse problems may enable scientists to address certain biological problems.

It is fitting and proper that at this celebration of the *700th* birthday of the University of Montpellier, this paper will confront some issues which were first raised in the Journal of the Medical School here over a century ago.

In 1861 G.E. Stahl [1] published a paper in the Journal of the University of Montpellier postulating *vitalistic forces* in living cells. Stahl was a major scientist of his day, for example, Prigogine and Stengers [2] credit him with the invention of chemical systematics. Louis Pasteur attacked Stahl's ideas and showed experimentally that bacteria and germs cause diseases and can be killed with heat; i.e., Pasteurization.

This, of course, does not disprove *vitalistic forces* but merely shifts them into other areas: Recently, Prigogine[2] has interpreted the nonlinearity and irreversible entropy which he introduced into physical biochemistry as precisely *vitalistic forces*, in the sense of Stahl! There is still another sense in which Stahl may have been correct; *solitons*, but these experiments are doubtful. Thus, Stahl's ideas are still very much alive today.

There are a number of other reasons to study biological systems without subscribing to the bold idea of Dyson [3] that physics is finished. A number of health and safety issues for humans can be improved by research in this area. For example, today's pulse dosimetry standards neglect transients, envelope effects, dispersion and Doppler shifts [4]. It is necessary to determine whether any of these effects is serious and, if so, how serious. In addition, the dispersion of these short pulses may provide new diagnostic tools either in the Laboratory or the Clinic.

2. Very Elementary Biology

The organization of a biological entity [5] is carried by its *genome*, which is the collection of information on the structure and function of the unit. During growth the genome arranges all added material into the proper subunits for correct operation. Both the atomic and molecular constituents and their spatial and temporal arrangements play a role. The genome plays the same role as the blueprint in von Neumann's self-reproducing automata [6].

The biological materials of interest include water, lipids (which are fat cells consisting of sixteen to eighteen carbon atoms), proteins and nucleic acids. Lipid bilayers form both boundary and internal membranes. The membranes asymmetrically pass nutrients into the cell and expell waste.

The proteins and nucleic acids are macromolecular biopolymers. Most are heteropolymers consisting of different groups, i.e., M-K-P-V-T... in DNA 1-5 from E Coli. The nuclei acids are formed according to

$$\left(\begin{array}{cc} & Base_1 \\ P0_4 & | \\ & Sugar_1 \end{array} \right) - \left(\begin{array}{cc} & Base_2 \\ P0_4 & | \\ & Sugar_2 \end{array} \right) - \circ\circ\circ - \left(\begin{array}{cc} & Base_N \\ P0_4 & | \\ & Sugar_N \end{array} \right) .$$

Proteins are other heteropolymers made of the 20 amino acids. Most, though not all, biological molecules are called polar (having non-zero permanent electric dipole moments) and being mostly hydrophilic. It is the dynamics of the dipole moments which give the dielectric response of biosolutions, especially their strong dispersion.

3. Direct Problems, Dielectric Response of Biological Solutions

The permanent electric dipole moment, $\vec{p}(t) = e\vec{d}(t)$, can only be non-zero if the parity symmetry

$$\Pi(0) = diag(-1, -1, -1)$$

is broken. Only non-centrosymmetric molecules and crystals which are non-degenerate can have non-zero permanent electric dipole moments, $\vec{p}$. Because the external applied fields can be weak or strong, the dielectric response of the medium may be:

1) *Linear*[7-9] for weak applied fields, in which case far more is known.

2) *Weakly non-linear*[10-12] for intermediate strong applied fields.

3) *Nonlinear*[13,14] for strong applied fields, in which case little is known.

The sharp rise-time of the ultrafast pulses will necessitate the study of both *memory* and *non-linearity*[13]. It is useful to begin with the linear regime when possible, however.

One useful property possessed by classical, linear, causal systems is half-plane analyticity. Considering the dielectric response $\epsilon(\omega)$ with temporal but not spatial dispersion the Fourier transform is given by

$$\epsilon(\omega) = \epsilon'(\omega) + i\epsilon''(\omega) \quad , \tag{1}$$

where ϵ' is the real part and ϵ'' is the imaginary part. The imaginary part is also called the attenuation or the extinction of the medium. Causality together with linearity implies that ϵ' and ϵ'' are a (possibly subtracted) Hilbert transform pair. If spatial dispersion is absent either $\epsilon(\omega)$ or $1/\epsilon(\omega)$ have causal responses but if the system has spatial dispersion then only $1/\epsilon(\omega, \vec{k})$ is causal. This is easily seen from the fact that the linear response of the free charges can be controlled in the finite volume capacitor where the dielectric response is measured, whereas the total charge cannot. This requires that the response function $R(\omega, \vec{k}) = 1/\epsilon(\omega, \vec{k})$

The static arrangement of the individual sub-molecule and the shape of a biopolymer can be determined from X-ray crystallography, if single crystals are available. This phase problem can easily be solved chemically by isotopic substitution. This is a familiar lesson in inverse problems, that if one data set is insufficient for inversion it may be possible to get additional information and solve a different problem. Although crystallographic information has proven extremely useful, it has two drawbacks. One is that many interesting biological systems do not form single crystals. The other is that crystallography is a static measurement of the system under unnatural conditions. When the molecular weight is less than 75,000 proton masses, Wüthrich and collaborators [15] have developed an NMR technique to observe the nuclear quadrupoles of proteins in solution. This technique uses the nuclear Overhauser effect, where two spins $\vec{S}_1$ and $\vec{S}_2$ are coupled through the lattice via a $1/r_{12}^3$ dipolar interaction. Upon plotting ω_1 of $\vec{S}_1$vs. ω_2 of $\vec{S}_2$ for all of the resonant peaks observed, the diagonal elements correspond to spins $\vec{S}_1^{(0)}$ and $\vec{S}_2^{(0)}$ which are separated by less than 5 Å. All other peaks are out of resonance due to the fall-off of the lattice coupling between them. This technique is a tremendous probe for many proteins and light biomolecules, giving the spatial structure of the molecule, called its conformation. The pH, concentration, ions and other environmental factors can then be freely varied. This allows the biophysicists to obtain the dependence of the conformation on the environment. For example, in sites 1-12 in DNA from E. Coli are MKPVTLLYDAEY loops are found at the L and A sites. A side project of the author, which RCP participants might find interesting is the search for a simple relation between the conformation loops in direct $\vec{x}$- space and the C^*-algebra of the equilibrium configurations in the β, V variable $\beta = (k_B T)^{-1}$ of the free-energy. Since this idea will directly use knots, loop spaces and homotopy [16] it should be mathematically beautiful, independently of whether it is useful to biologists in the forseeable future.

To summarize the physical properties of the biological systems of interest. They are: large complex polymers in bulk water with boundary and internal cell membranes, ions, water in strange states called water of hydration, clusters, etc., and trace elements. Liquid crystals have most of the features of membranes and bio-macromolecules. It may be necessary to also investigate rough surfaces, classical localization, bending and torsion modes and fractals. Thus, the medium is a teaming fluid of large molecules in a complicated water solution at temperature $T \sim 290$ K. Many of these molecules have a number of electric dipoles and possibly quadrupoles. The fundamental question is now, **Exactly what knowledge is to be sought on these complicated materials?** The answer for this work is that the *dielectric response* will be the property to be studied. There are a number of measurements by Grant, Sheppard and South who have also collected reliable data available ten years ago in their monograph [17]. The theory was pioneered by Debye [18]. The extension and clarification of the Debye theory of dielectric response by Langevin suffices for electromagnetic frequencies up to about 100. G Hz (10^9 Hz). At frequencies below 1 G Hz the response is constant leading to the misnomer "dielectric constant." Above 100 G Hz Debye and Langevin's theory is frequency independent in complete conflict with experiment. A schematic of the frequency dependence of $\epsilon'(\omega)$ from Eq. (1) is given in Fig. 1. The dotted horizontal line indicates the frequency independence of Debye theory.

As a crucial ingredient of life (as we know it) water plays a special role. The real and imaginary parts of $\epsilon(\omega)$ for water, at 20°C taken from Ref. [16] are shown in Fig. 2.
Grant *et al.* have also measured the temperature dependence of the real part $\epsilon'(\omega_0 T)$ at $\omega_0 \sim 1.$ G Hz and $0\text{C} < T < 100$ C and found it to be given by

$$\epsilon'(1 \text{ G Hz}, T) \cong 87.74 - .40008T + .00094T^2 - .0000014T^3 \quad . \tag{2}$$

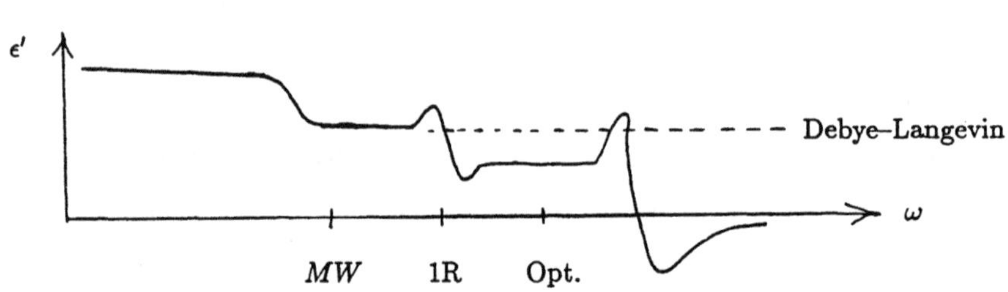

Fig. 1. Schematic of the real part of the dielectric response ϵ' of a heterogenous dielectric as a function of frequency ω.

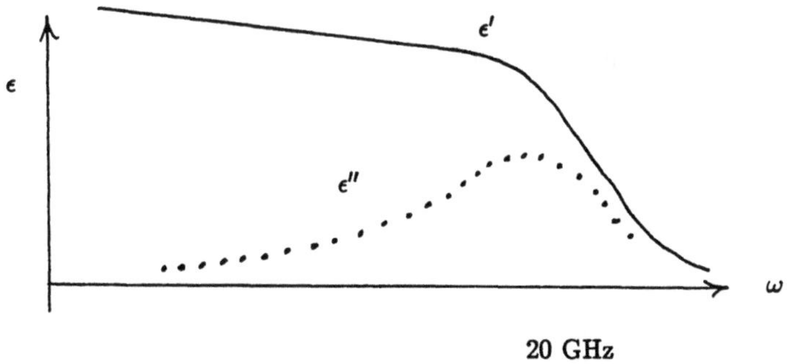

Fig. 2. A schematic form of the frequency dependence of the real ϵ' and imaginary parts of the dielectric response of water. Debye theory is satisfactory at these frequencies but fails at higher frequencies.

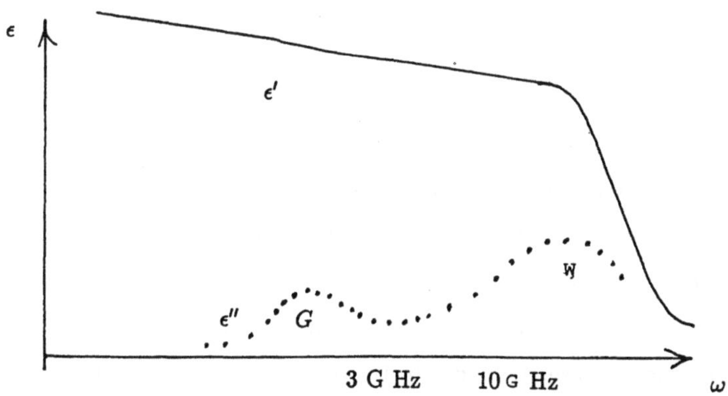

Fig. 3. A schematic of the dielectric response of a 10% solution of Glycine in water. The peaks in ϵ'' are labeled as G for Glycine relaxation and W for the water relaxation times.

In Figs. 3 and 4 the dielectric response for a 10Glycine in water and of muscle tissue are shown. Glycine is the simplest amino acid and muscle is a complex biological system of considerable interest. Observe that the glycine solution exhibits a new peak in ϵ'' at about 3. G Hz which was absent in water. It is identified with a second Debye relaxation of the 10% Glycine in the solution. The muscle tissue is more complex and far more interesting. First of all, it has a real dielectric response value of 10^8 at low frequency compared to 87.7 for water! This could indicate a near ferroelectric phase transition or a Lorentz catastrophe (or an experimental mistake). There are four apparent Debye relaxations in the real part $\epsilon'(\omega)$ for this sample. The fact that many biological solutions have these various dispersions was the reason for "Highly Dispersive" in the title. It is also one of the reasons that these biological systems are physically interesting . To understand the mechanisms consider the following facts. At UHF to microwave frequencies it is the motion of the electric dipole moments on the biopolymers which interact

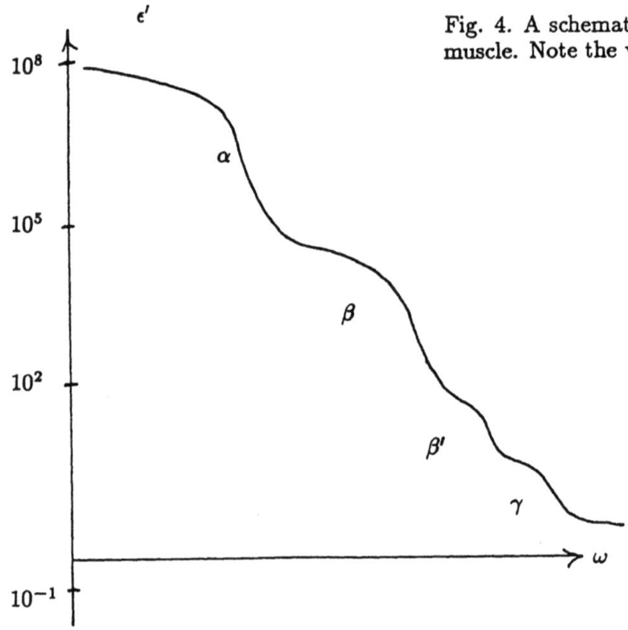

Fig. 4. A schematic of the dielectric response of muscle. Note the value of $\epsilon'(0^+)$ as 10^8.

with the applied external field, $\vec{E}$. As infrared vibrational bands in the molecules are driven and stretch, bending, torsion, and vibration modes begin to respond and increases as the near IR is reached. At optical frequencies electronic excitations and possibly molecular disassociations contribute and at higher frequencies, disassociation of the individual atoms occur.

The pioneering work on dielectric response is due to Debye [18] early in this century. His theory described experiments at frequencies up to the microwave region, but it was frequency independent at all higher frequencies in contrast to later experiments. This necessitated the extensions by McConnell [19] who added inertia I and Reichl [9], [20], Sparling, Reichl and Sedlak [21] who added memory and work in progress by the author [22]. The idea is to use rotational Brownian motion of the dipolar molecule in a solvent fluid with viscosity and to study memory in the interaction, following refs. [18-22].

Consider a single spherical, polar Brownian solute molecule of radius R, density ρ and electric dipole moment $\vec{p}(t)$ in a solvent fluid with mean density ρ_0 and Stokes viscosity N. At the surface of the particle *stick* boundary conditions are the correct choice. For simplicity of exposition, axial symmetry with $\theta(t)$ the orientation angle of the dipole at time t will be assumed. For hindered rotation where the electric dipole occurs in the interior of a long chain of non-polar groups this assumption is exact. When spatial dispersion is absent the dielectric response $\epsilon(\omega)$ is given by

$$\frac{\epsilon(\omega)-\epsilon_\infty}{\epsilon_0-\epsilon_\infty}=1+i\omega\int_0^\infty dt\; e^{i\omega t}\frac{\langle\vec{p}(t)\cdot\vec{p}(0)\rangle_E}{\langle p^2(0)\rangle_E}\quad, \tag{3}$$

where $\epsilon_0, \epsilon_\infty$ denote the response as frequency $\omega \to 0$ and $\omega \to \infty$, respectively, $\vec{p}(t)$ is the electric dipole moment at time t and $\langle\cdot\rangle_E$ is the equilibrium thermal average. Let $P(\delta\theta, t)$ be the probability that the dipole has rotated through $\delta\theta$ in time t, the angular velocity at t is

$$\Omega_t = \left(\frac{d\theta}{dt}\right)(t) \tag{4}$$

and

$$\left(\frac{d\Omega_t}{dt}\right) = \frac{\tau(\theta)}{I} - \int_{-\infty}^{t} \beta(t-t')\Omega(t')dt' \tag{5}$$

with $\tau(\theta)$ is the random torque, $I = 8\pi\rho_0 R^5/15$ is the moment of enertia of the particle and $\beta(t-t')$ is the friction. The variance σ_E^2 is given by

$$\begin{aligned}\sigma_t^2 &= \int_{-\infty}^{\infty} d(\delta\theta)(\delta\theta)^2 P(\delta\theta, t) \\ &= 2\int_0^t d\tau(t-\tau)\langle\Omega_t\Omega_0\rangle_E \quad .\end{aligned} \tag{6}$$

This last equation shows that Debye's idea of a single relaxation-time has been replaced by a continuum of relaxations in Eq. (6). At equilibrium, the time dependent dielectric susceptibility $\chi(t)$ is related to the equilibrium pair correlation function of Ω by

$$\langle\Omega_t\Omega_0\rangle_E = \frac{k_BT}{I}\chi(|\, t\, |) \tag{7}$$

where k_BT is the Boltzman constant times the (absolute) temperature T. For $t > 0$, one can differentiate the second part of Eq. (6) twice and use Eq. (7) to obtain

$$\frac{1}{2}\frac{d\sigma_t^2}{dt^2} = \frac{k_BT}{I}\chi(t) \tag{8}$$

If $\chi(\cdot)$ is in L^2 its Fourier transform is given by

$$\hat{\chi}(\omega) = \frac{1}{\hat{\beta}(\omega) - i\omega} \quad . \tag{9}$$

The choice of a Gaussian probability density for the probability distribution P yields a Fokker–Planck equation

$$\frac{\partial P}{\partial t} = D(t)\frac{\partial^2 P}{\partial(\delta\theta)^2} \tag{10}$$

which is local in time t with a time dependent diffusion coefficient $D(t)$. As $t \to \infty$,

$$D = \lim_{t\to\infty}[D(t)] = \frac{k_BT}{\xi} = \int_0^{\infty} \langle\Omega_s\Omega_0\rangle_E ds \quad . \tag{11}$$

The simplest models which show the relevant physics of the biological systems described in Section 2 are given by Reichl and collaborators in refs. [9,20,21]. In [21] the frictional response

$\beta(t)$ from an earlier paper is studied. It has Laplace transform in $\beta_L(\omega) = \mathcal{L}[\beta(t)]$ given by

$$\beta_L(\omega) = \frac{1}{\tau_I}\left[1 + \frac{\omega\tau_v/3}{(i + \sqrt{i\omega\tau_v}}\right] \quad , \tag{12}$$

where τ_v is the hydrodynamic relaxation time given by $\tau_v = R^2\rho_0/\eta$ and τ_I is the inertial relaxation time given by $\tau_I = I/\xi$. The variance for this model was carefully shown to be

$$\frac{\sigma_t^2}{2} = Dt - D\tau_I\, g(t) \tag{13}$$

where

$$g(t) = \frac{1}{2\pi}\int_{-\infty}^{\infty} d\omega\; e^{-i\omega t}\hat{g}(\omega) \tag{14}$$

with

$$\hat{g}(\omega) = \frac{i\sqrt{i\omega\tau_v} - (1 + \tau_v/3\tau)}{\omega[i + \sqrt{i\omega\tau_v} + (1 + \tau_v/3\tau_I)\omega\tau_I - i\omega\tau_I\sqrt{i\omega\tau_v}]} \quad . \tag{15}$$

As $\tau_v \to 0$, the variance reduces to

$$\frac{\sigma_t^2}{2} = Dt - D\tau_I(i - e^{-t/\tau_I}) \tag{16}$$

and the dielectric response is given by

$$\frac{\epsilon(\omega) - \epsilon_\infty}{\epsilon_0 - \epsilon_\infty} = \frac{1 + D\tau_I}{(1 - i\omega\tau_D)(1 + D\tau_I - i\omega\tau_I)} \quad . \tag{17}$$

As $\tau_I \to 0$, the Debye theory is recovered. When τ_I and τ_v are both non-zero, the results for the variance include the nonlocal term I_t with

$$\frac{\sigma_t^2}{2} = Dt - D\tau_I\left[1 + \frac{\tau_v}{3\tau_I} - \frac{\tau_v}{3\pi\tau_I}I_t\right] \tag{18}$$

with

$$I_t = \int_0^{\infty} \frac{e^{-rt/\tau_v}}{\sqrt{r}P_3(r)}dr \tag{19}$$

where $P_3(r)$ is the cubic polynomial

$$P_3(r) = [1 - (\alpha + 1/3)r]^2 + r(1 - \alpha r)^2 \quad , \tag{20}$$

where α is a constant given in ref. [21]. The dielectric response must be calculated numerically in this case and the Cole-Cole plots for $Im(\epsilon(\omega))$ vs. $Re(\epsilon(\omega))$ are very different at higher frequencies from the simpler models. All of these models have a $t^{-5/2}$ decay at large positive times in contrast to the exponential time decay of a Debye model.

These direct problems all have a time dependent dielectric response of the form

$$\epsilon(t) = \epsilon_\infty + (\epsilon_0 - \epsilon_\infty)f(t) \quad . \tag{21}$$

For the special case of inertial relaxation without hydrodynamic memory, $f(t)$ is simply the right hand side of Eq. (17). Even a non-Markovian response function can be calculated this way (21), at least numerically. In this latter case, a non-local function of the form

$$f(t,t_0) = f(t) - g(t)I_{t,t_0} \tag{22}$$

will occur where I_{t,t_0} is an integral over a t' time variable which includes at least one distinguished starting time t_0, in addition to the continuum infinity of relaxation times inside I_{t,t_0} . Using Eq. (21)

$$\epsilon_{,t}(t) := \left(\frac{\partial \epsilon}{\partial t}\right)(t) = \left(\frac{\partial f}{\partial t}\right)(t) \tag{23}$$

and

$$\epsilon_{,tt} := \left(\frac{\partial^2 \epsilon}{\partial t^2}\right)(t) = \left(\frac{\partial^2 f}{n^2}\right)(t) \tag{24}$$

are non-zero in general and any one of the three $\{\epsilon, \epsilon_{,t}, \epsilon_{,tt}\}$ may be the largest. It is possible to use either the time-domain or the frequency-domain for an electromagnetic pulse or wave in the medium. The usual frequency-domain approach has poles which move about as the medium (i.e., they move with the parameters inside $\hat{\epsilon}(\omega)$) changes. Using a non-dispersive reference medium with speed of light c_1 related to the vacuum value by

$$c_1^2 = \frac{c^2}{\mu_1 \epsilon_1}$$

with $\mu_1 \epsilon_1$ the constant constitutive parameters, it is straightforward to solve Maxwell's equations for the electric field $\vec{E}(\vec{x}, t)$ in the form

$$\left(\Delta - \frac{1}{c_1^2}\frac{\partial^2}{\partial t^2}\right)\vec{E} = \vec{\nabla}(\vec{\nabla}\cdot\vec{E}) + \left(\frac{\mu_1\epsilon_1}{c^2} - \frac{\mu\epsilon}{c^2}\right)\vec{E}$$
$$+ \mu\frac{\epsilon_{,tt}}{c^2}\vec{E} + 2\mu\frac{\epsilon_{,t}}{c^2}\vec{E}_{,t} \quad . \tag{26}$$

In Eq. (26), Δ is the Laplacian $\Delta = \vec{\nabla}\cdot\vec{\nabla}$ where $\vec{\nabla}$ is the del operator on functions in $C^2(\Omega) \subset R^3$ and the Markovian case has been assumed.

4. Inverse Problems

These studies are just beginning the simplest approach seems to be to first assume transversely stratified media. By then choosing a finite number of angles of incidence $\eta_i = \sin(\theta_i), i = 1, 2, \cdots, N$ a splitting method by Corones and Karlsson [23] can be used to unify the direct and inverse problems. Given $\epsilon, \epsilon_{,t}$ and $\epsilon_{,tt}$ which is equivalent to an initial reflection operator $R(0,x)$ a Ricatti equation was derived from which the full reflection operator $R(t,x)$ in the direct problem. The inverse problem maps $R(t,0)$ into $R(t,x)$ using a marching technique along the characteristic. The variable x is a travel-time which straightens out the characteristics in the slab. Karlsson [24] has shown how to reformulate this approach using only transmission data.

5. Conclusions

This review should have shown that biological solutions, including human tissue, furnish an interesting class of direct and inverse problems. Perhaps Stahl's idea published here in Montpellier fully 128 years ago, may find a secure place in science after all. Whatever is learned will be reported at a future RCP 264 meeting.

Acknowledgements

This work was partially supported by the University of Missouri Faculty Foreign Travel Fund, NATO Travel Grant 095/87, AFOSR Grant 89-0311 and the French CNRS for a visiting professor position at the Laboratory of Physical Mathematics at USTL in Montpellier.

References

1. G.E. Stahl, J. Univ. Montpellier Med. Sch., 1861.
2. I. Progigine and I. Stengers *Order Out of Chaos* (Bantam, New York, 1984.
3. F. Dyson, Rev. Mod. Phys. **51**, 447 (1979).
4. R. Albanese, Private Communication (June, 1989).
5. B.S. Guttman and John W. Hopkins III, *Understanding Biology* (Harcourt Brace Jovanovich Inc., New York, 1983).
6. J. von Neumann, *Theory of Self-Reproducing Automata*, Edited by A.W. Burks (University of Illinois press, Urbana, 1966).
7. F. Hynne and R.K. Bullough, Phil. Trans. Roc. London **A321**, 305-360 (1987); *ibid* **A312**, 251-293 (1984).
8. J. McConnell, *Rotational Brownian Motion and Dielectric Theory* (Academic, New York, 1980).
9. L.E. Reichl, Phys. Rev. Lett. **49**, 85 (1982).
10. J. Kerr, Phil. Mag. **50**, 337 (1875); *ibid* (157 (1880).
11. C.T. O'Konski, Biophys. J. **5**, 607 (1965).
12. S.H. Fine and W.P. Hansen, Appl. Optics **10**, 2350 (1971).
13. D.J. Kaup, Rocky Mount. Math. J. **8**, 71 (1978); Physica **6D**, 143 (1983).

14. A.C. Scott, Rev. Mod. Phys. **47**, 487 (1975).

15. K. Wüthrich, Adv. Chem. Res. **22**, 36-44 (1989).

16. L.H. Kaufman, *On Knots*, Ann Math Studies, Vol. **115** (Princeton University Press, Princeton, NJ, 1987), 480 pp.

17. E.H. Grant, R.H. Sheppard and G.P. South, *Dielectric Behavior of Molecules in Solution* (Oxford University Press, Oxford, 1978).

18. P. Debye, *Polar Molecules* (The Chemical Catelog Company, New York, 1929).

19. J. McConnell, *Rotational Brownian Motion and Dielectric Theory* (Academic Press, New York, 1980.

20. L. Reichl, Phys. Rev. **A24**, 1617 (1981).

21. L.C. Sparling, L. Reichl and J.E. Sedlak, Phys. Rev. **A33**, 699 (1986).

22. B. DeFacio (in preparation), and work in progress with Mr. C.R. Dick.

23. J.P. Corones and A. Karlsson, Inv. Probs **4**, 643 (1988).

24. A. Karlsson, Inv. Probs **3**, 691 (1987).

Image Estimation from Scattered Field Data (at 10 GHz) Using the Distorted Wave Born Approximation and Other Prior Knowledge

A. Alavi [1], *F.C. Lin* [1], *M.A. Fiddy* [1], *and C.L. Byrne* [2]

[1]Department of Electrical Engineering, University of Lowell, Lowell, MA 01854, USA
[2]Department of Mathematics, University of Lowell, Lowell, MA 01854, USA

A practical inversion algorithm based on the Fourier relation between scattered field data and the scatterer is presented; this procedure attempts to go beyond the limitations of conventional diffraction tomographic techniques which are largely limited to scatterers that satisfy the first Born or the Rytov approximation. A distorted wave approximation is used and reconstructions based on this have been compared with (first) Born inverted images and those based on the Rytov approximation. The new procedure is evaluated using microwave scattered field data taken from sand filled cylinders with diameters of 15.24cm and 30.48cm using 10GHz radiation. The approach adopted, being Fourier based, lends itself to further improvement through the application of well known signal processing (spectral estimation) techniques. The latter can improve the resolution of the reconstructed data by interpolating and extrapolating the available Fourier data, which lie on semicicular arcs in the Fourier domain, through the minimization of a cost function which incorporates prior knowledge about the scatterer, such as its support.

Introduction

Since the papers by Wolf [1], Devaney [2] and others, there has been considerable interest in the reconstruction of a scattering potential (i.e. a function or object which generates the scattered field) from measurements of the scattered field, the objective being to obtain a quantitative image of the scattering structure from measurements of the field scattered by it. This has particular relevance to acoustic and microwave scattering, in the context of non-destructive testing and medical imaging. The success of computerized (projection) tomography, where the high-frequency of the illuminating radiation allows straight line propagation to be assumed, motivated much of the work on the development of analagous tomographic algorithms for a wider class of objects. Of particular interest are scatterers having structures on the scale of the order of the wavelength for which scattering and diffraction effects can no longer be ignored; consequently, such algorithms are known generically as diffraction tomography.

The majority of these algorithms are based upon weak scattering approximations to linearise the wave equation and render it invertable. Within these approximations the measured field may, as in projection tomography, be related to the Fourier transform of the scattering potential. The precise domains of validity of the approximations depend upon the physical interpretation of strong inequalities and hence have been the source of much controversy over the years, [3]. The "quality" of any reconstruction will depend upon how well any scatterer being imaged will satisfy these inequalities and few real objects of interest will do so. If these algorithms are used regardless of their applicability one can consider the distortions in the resulting reconstructions [4, 5].

Due to these limitations of 'conventional' diffraction tomography, there has been recent interest in trying to extend the applicability and interpretation of these methods to include stronger scattering effects. One approach, for example, bases reconstruction algorithms upon distorted-wave approximations in which it is assumed that a prior estimate for all but a weakly scattering component of the potential is available, [6, 7, 8].

We present an approach based on the determination of an estimate for a scattering potential under the assumption that the Born approximation is not valid. An algorithm for the inversion of sampled noisy Fourier data is evaluated using microwave scattered field data from constructed (and therefore known) targets. We have focused on the use of real data from known simple targets, because of the difficulty in testing these procedures with simulated data, due to the inherent difficulty in computing such data sets reliably. These experiments have been carried out at the RADC Radar Cross Section Facility in Ipswich, MA. Based on this, an estimate of the scattering potential can be determined which is optimal in a certain specified sense. It is data consistent, i.e. consistent with the scattered field data, has minimum (weighted) energy and minimizes the distance from the true potential to a given subspace of estimators.

Theoretical background

Let us consider a scattering potential $V(\underline{r}) = V_1(\underline{r}) + V_2(\underline{r})$, where $V_1(\underline{r})$ is an assumed known background scattering potential. The integral equation to be inverted is given by

$$f(k\hat{\underline{r}},k\hat{\underline{r}}_o) = \int_D \exp(-ik\hat{\underline{r}}.\underline{r}_1)V(\underline{r}_1)\psi(\underline{r}_1,k\hat{\underline{r}}_o)\, d\underline{r}_1$$

or, in terms of the scattering amplitudes, f, associated with V_1 and V_2

$$f_1(k\hat{\underline{r}},k\hat{\underline{r}}_o) = \int_D \exp(-ik\hat{\underline{r}}.\underline{r}_1)\, V_1(\underline{r}_1)\psi_1(\underline{r}_1,k\hat{\underline{r}}_o)\, d\underline{r}_1$$

$$f_2(k\hat{\underline{r}},k\hat{\underline{r}}_o) = \int_D \psi_1(\underline{r}_1, -k\hat{\underline{r}})V_2(\underline{r}_1)\psi(\underline{r}_1,k\hat{\underline{r}}_o)\, d\underline{r}_1$$

The distorted wave technique assumes that the scattering amplitude f_2 associated with the unknown part of the potential $V_2(\underline{r})$ is

$$f_2(k\hat{\underline{r}},k\hat{\underline{r}}_o) = \int_D \psi_1(\underline{r}_1, -k\hat{\underline{r}})V_2(\underline{r}_1)\psi_1(\underline{r}_1,k\hat{\underline{r}}_o)\, d\underline{r}_1$$
$$= \int_D \exp(-ik(\hat{\underline{r}} - \hat{\underline{r}}_o).\underline{r}_1)U(\underline{r}_1)d\underline{r}_1$$

where f_2 is the scattering amplitude deduced from the data f and the known or calculated f_1; this allows one to estimate V_2 from an estimate of U using

$$U(\underline{r}) = \left(\frac{B_1(\underline{r})}{V_1(\underline{r})}\right)^2 V_2(\underline{r}) \qquad (1)$$

(see [8]). This approach is limited by several approximations, but offers a fast method for an improvement over the 1st Born approximation. The two main problems are limited k space coverage rendering B_1 a poor (low pass filtered) estimate of the 1st Born object-estimate of V_1 and secondly the fact that it is assumed that the approximation to the field within D is satisfactory. The latter really assumes that the product of the potential and the internal field differs only by a rotation from view to view.

Using experimental data we have studied the distorted wave method described above (equation (1)) and compared reconstructions using these with those obtained using a simple first Born and Rytov inversion scheme. For the Born inversion case, the restricted spatial frequency content of the scattering potential, suggests the need for using a spectral estimation procedure that exploits prior knowledge about the background V_1. Assuming that we have a prior estimate P(r) of the broad features of V(r), eg $V_1(r)$, we can incorporate this spectral estimation method into the inversion scheme. It has been shown [9] that solving a set of equations of the form:

$$f(m) = \sum_{n=-M}^{M} a_n \, p(m-n) , \quad -M \leq m \leq M$$

(where p denotes the Fourier transform of P(r)) for the coefficients a_n, allows one to define an estimator that minimises the approximation error

$$\int_{-\pi}^{\pi} | V(r) - P(r) \sum_{n=-M}^{M} a_n \exp(inr)|^2 / P(r) \, dr$$

The resulting estimator of V(r) is termed the PDFT estimator, because of its form, namely PDFT(r) = P(r)A(r), where A(r) is the trigonometric polynomial with coefficients a_k. If the prior estimate P(r) is a constant for $|r| \leq \pi$, then $a_n = f(n)$ and the PDFT reduces to the discrete Fourier transform (DFT) estimator, that is usually calculated. This estimator is both continuous and data consistent, and the algorithm is easily extended to a two-dimensional (or higher) form (see [9]).

Incorporation of the prior knowledge about V_1 in the method neglects the fact that the wavenumber associated with the field within the scatterer is not the free space wavenumber k. Alternatively consider V(r) weak but not weak enough to satisfy first Born approximation:

$$V \ (r) = - \beta^2 / k^2$$

$$\psi(r_1, k\hat{r}_o) \approx \exp(-i\gamma\hat{r}_o . r_1) \text{ where } \gamma^2 = \beta^2 + k^2$$

$$\psi_s(r, k\hat{r}_o) \propto \int_D \exp(-i \, (k\hat{r} - \gamma\hat{r}_o) . r_1 \, dr_1 \tag{2}$$

i.e. the scattered field corresponds to Fourier data about the scatterer on semi-circular arcs of radius k but center $\gamma\hat{r}_o$. Thus based on the assumption of a known homogeneous background, V_1, with *refractive index* n, one can assign the more appropriate wavenumber, nk. Using this, one is forced to reinterpret the relationship of the scattered field data to the Fourier data associated with the scattering object. The semi-circular arcs on which Fourier data can be assigned remain of radius k but now have center nk leading to a region around the k origin without data. The consequences of this are an apparent high pass filtering of Born inverted scatterered field data, which is well known (e.g. [5]).

The assumption underlying this Fourier arc-shifting is that it assumes a plane wave approximation to the internal field for each illumination direction; this is in the spirit of the distorted wave approach, however. For a sufficiently large weakly scattering object, this is not an unreasonable assumption. The reconstructions obtained in this way can be compared directly with those obtained using the Rytov approximation, which assumes that the scattering potential is slowly varying on the scale of a wavelength. We relate the scattered field data to Fourier data for V rather than V_2 in the case of equation (2).

Experimental and numerical results

We assume a scattering potential for which multiple scattering is assumed, i.e. either from $V = V_1+V_2$ or from the known stronger scattering background component V_1. We have constructed several scattering targets and collected data from them at 10 GHz (or 3 cm wavelength). Examples are a 15.24cm diameter cylinder and a 30.48 diameter cylinder, each with a rigid base and a height of 5 feet to avoid end effects in the scattering experiments. They were both either filled with sand or partially filled with sand and empty smaller diameter cylinders acting as "voids". These objects are very heavy when filled with sand, and they have to be firmly mounted onto the turntable in the chamber but in such a way that they can be turned by hand for each new incident field direction. The two cylinders described have diameters of approximately 6 and 12 wavelengths respectively. The product of the wavenumber (k) with the diameter of the target (d) and the magnitude of the scattering potential (V) has to be $kVd \ll 1$ for the first Born approximation to hold and < 1 for the Born series to converge. For these scatterers, kVd is 37 and 75 respectively, when the cylinders are filled with dry sand. The incident beam has a half-power full-beam divergence angle of approximately 60°.

For the geometry of the experimental arrangement, one can assume that the scattered field data are measured in the far field, provided $kd^2/2 \ll z$ (see e.g. [10]). Here, z represents the distance from the center of the target to the detector, which is 3.78 m in this case. The quantity $kd^2/2$ is 2.35m and 9.4m respectively, and hence the approximation that we have far field (Fourier) data for the larger cylinder is not a good one. As a result, interpretation of such scattered field data as Fourier data requires the incorporation of the appropriate quadratic phase factor for the larger cylinder data, in principle. Also, the 30cm diameter cylinder is sufficiently large that some attenuation is seen in the forward direction.

In principle, one could use the estimation technique to overcome the problem of having only a limited number of illumination directions. However, this estimation method requires a matrix inversion which would be computationally costly and somewhat ill-conditioned for the large size and sparse data sets we have here. For this reason, it is preferable to take the Fourier data available and interpolate first onto a cartesian grid and then apply the extrapolation procedure if necessary. With Fourier data on a square grid, the method can run much faster and more reliably. This of course puts more significance onto the fidelity of the results of the interpolation routine; to date, we have been using IQHSCV in the IMSL library. The IMSL algorithm can be very unreliable in areas of sparse data coverage; it was from a series of experiments with this algorithm that we have now been able to conclude that approximately 40 illumination directions are the minimum necessary to achieve a reliable interpolation using this routine. Also, for some data, the incident beam was not incident normal to the cylinder but at a squint angle of 20°. This required a further correction to the interpretation of the data onto semi-circular arcs that are rotated by 20° their respective centers.

We show below some reconstructions based on these data.

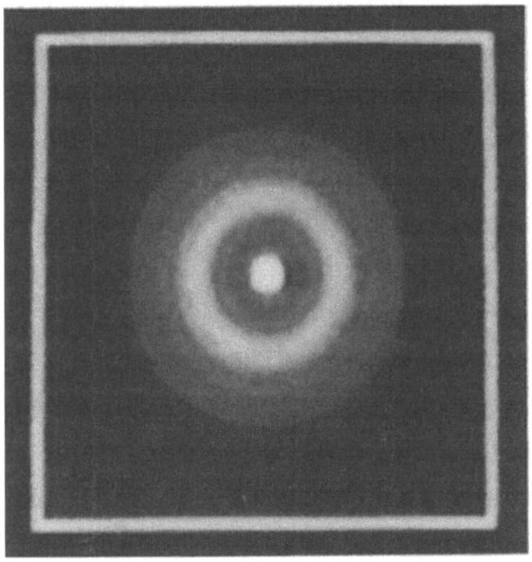

Figure 1

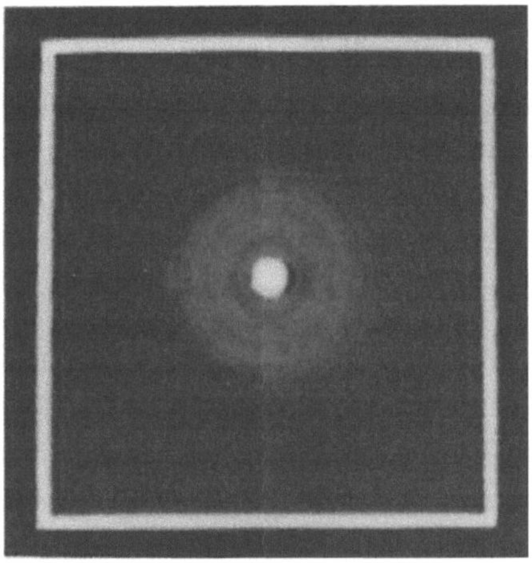

Figure 2

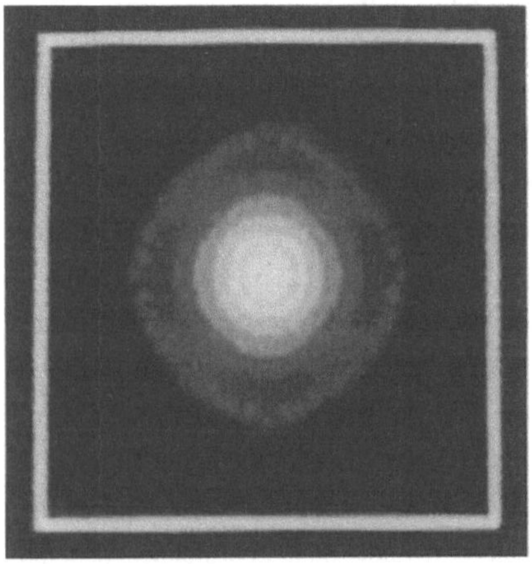

Figure 3

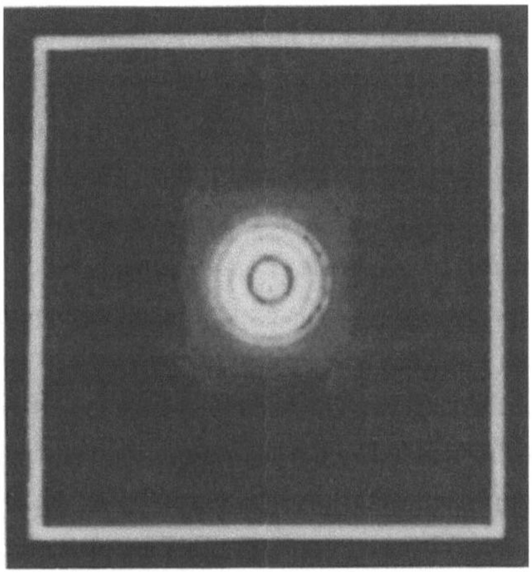

Figure 4

Figure (1) shows the reconstructed image of the 30cm sand filled cylinder, obtained from the scattered field data prior to the subtraction of the background field. Removal of this background is difficult, because it is uncertain how to accurately scale it with respect to the scattered field measured with the cylinder present.This difficulty arises because of attenuation and shadowing effects arising from the cylinder. Inverting the background field with no scatterer present, leads to an image of the source distribution. As a result, the reconstruction from an estimate of the scattered field from the cylinder alone, shown in figure (2), still contains evidence of the source distribution at the center of the reconstructed image. Figure (3) shows the reconstruction of the 15.24cm cylinder filled with styrofoam granules. In this case, the field measured with the cylinder empty was subtracted from the field when the styrofoam was present. In this way, one should obtain the scattered field just from the styrofoam alone. The reconstruction shows the characteristic size of the cylinder, as expected from an inversion procedure based on the first Born approximation. There is some distortion present even for such a weak scatterer, with $kVd \approx 10$. Moreover, there is still some evidence of the source distribution in the reconstruction. In figure (4), the reconstruction shown is that obtained after shifting the Fourier data from arcs centered at k to arcs centered at $\gamma = 1.1k$. The reconstruction shows a marked improvement, especially in the definition of the edge of the cylinder.

Conclusions

Our motivation is that of developing a procedure which remains computationally straightforward in the sense of conventional diffraction tomography algorithms. We wish to retain a Fourier based inversion procedure, in order to easily incorporate spectral interpolation and extrapolation procedures. We described a method which does this and we have begun to evaluate it using data collected from simple targets. It transpires that using microwave illumination in an anechoic chamber, there are difficulties associted with beam squint, phase errors and accurate background field removal. Despite these problems, which continue to be addressed, we expect to identify a class of scatterers, more stronly scattering than those satisfying the first Born approximation, for which the new procedure yields a reliable image.

Acknowledgements

The authors acknowledge the support of the Office of Naval Research under grant number N00014-89-J-1158. We would also like to thank R. McGahan for permission to use the RADC RCS facility at Ipswich, MA, and to J. Quinn for assistance with the experiments.

References

1. E. Wolf: Optics Communications1,153 (1969).
2. A.J. Devaney: Optics Letters 7, 111 (1982).
3. M.A. Fiddy: J. Phys. D, 19, 357 (1986).
4. L. Zapalowski and M.A. Fiddy: Inverse Problems 1, L5 (1985).
5. L. Zapalowski, S. Leeman and M.A. Fiddy: Acoustical Imaging, 14, 295 (1985) (Plenum Press).
6. A.J. Devaney and M.L. Oristaglio: Phys. Rev. Lett. 51, 237 (1983).
7. K.I. Schultz and D.L. Jaggard: J. Opt. Soc. Am. A 4, 1773 (1987).
8. R.J. Wombell and M.A. Fiddy: Inverse Problems 4, L23-27, (1988).
9. C.L. Byrne and M.A. Fiddy, J. Opt. Soc. Amer., A4, 112-117, (1987).
10. J.W. Goodman, Introduction to Fourier Optics, McGraw Hill, (1968), p61.

Phase Equivalence Between Nonlocal and Local Potentials Using Inversion and Supersymmetry*

H. Fiedeldey[1], *S.A. Sofianos*[1], *R. Lipperheide*[2], *and H. Leeb*[3]

[1]Department of Physics, University of South Africa, Pretoria, South Africa
[2]Bereich Physik, Hahn-Meitner-Institut Berlin, D-1000 Berlin 39, Germany
[3]Institut für Kernphysik, Technische Universität Wien, Vienna, Austria

We briefly review recent work on the construction of phase equivalent local potentials to given nonlocal interactions of the Resonating Group Model type and the exact three-body Faddeev type. Direct, supersymmetric as well as inversion methods are discussed. We report on a method for constructing nonlocal potentials and wave functions from their phase shifts by fixed-E inversion.

1. Introduction

Microscopic theories of the interaction between two clusters result in effective nonlocal two-body potentials, which often have a very complicated structure. As a consequence they provide far less insight than the more familiar local potentials. Our understanding of nuclear reactions, for example, in particular heavy-ion reactions, is largely based on their intuitive interpretation in terms of features of local optical potentials. All classical and semiclassical approaches employ local potentials. They are also widely used in calculations of nuclear structure and even of elementary-particle spectra. Local effective potentials provide a meeting ground for the data on cross sections and spectra, and microscopic theory.

2. Equivalent Local Potentials by Direct and Supersymmetric Methods

The interaction between composite nuclei is commonly described by means of an optical potential. One of the most successful and widely used microscopic theories for determining this potential is the Resonating Group Model (RGM)[1]. Here the wavefunction $\chi(\mathbf{r})$ for the relative motion of the two nuclei is obtained as the solution of the equation

$$<\phi(C_1)\phi(C_2)\mid \Big[H-(E+E_1+E_2)\Big] \mid \mathcal{A}\Big\{\chi(\mathbf{r})\,\phi(C_1)\,\phi(C_2)\Big\}> = 0\,, \tag{2.1}$$

where H is the total Hamiltonian, E_i the binding energy of nucleus C_i (with wave function ϕ (C_i)) and E the energy of the relative motion. $\mathcal{A}$ is the antisymmetrizer. Equation (2.1) becomes an integro-differential equation for the wave function of the relative motion, $\chi(\mathbf{r})$:

$$\left(-\frac{\hbar^2}{2\mu}\nabla^2 + V_D(r)\right)\chi(\mathbf{r}) + \int d\mathbf{r}'\, G_{ex}(\mathbf{r},\mathbf{r}')\,\chi(\mathbf{r}') = E\,\chi(\mathbf{r})\,, \tag{2.2}$$

where $V_D(r)$ is the local, and $G_{ex}(\mathbf{r},\mathbf{r}')$ the nonlocal part of the RGM interaction.

---

* Dedicated to Professor J.P.F. Sellschop on the occasion of his 60th birthday.

The phase equivalent local potential (ELP), $V_L(r)$, associated with the nonlocal RGM interaction satisfies

$$\left(-\frac{\hbar^2}{2\mu}\nabla^2 + V_L(r)\right)\chi_L(\mathbf{r}) = E\,\chi_L(\mathbf{r})\,, \tag{2.3}$$

with

$$\chi(\mathbf{r}) = F(r)\,\chi_L(\mathbf{r})\,,\quad where\quad F(r)\rightarrow 1\quad for\quad r\rightarrow\infty\,. \tag{2.4}$$

The factor F(r) is called the Perey factor[2]. A mathematically rigorous construction of $V_L(r)$ and F(r) is possible by the Wronskian method[3]. Horiuchi[4] employed a WKB approximation for the construction of $V_L(r)$. The loss of exactness is more than compensated by the ease of application.

Applying the WKB approximation to Eqs. (2.2) and (2.3) we obtain two equations for the local wave number vector **k**(r):

$$\frac{\hbar^2}{2\mu}\,\mathbf{k}^2(r) + V_D(r) + G^W_{ex}\left(\mathbf{r},\mathbf{k}(r)\right) = E \tag{2.5}$$

and

$$\frac{\hbar^2}{2\mu}\,\mathbf{k}^2(r) + V_L(r) = E\,. \tag{2.6}$$

In Eq. (2.5) $G_{ex}(\mathbf{r},\mathbf{k})$ is the Wigner transform of $G_{ex}(\mathbf{r},\mathbf{r}')$,

$$G^W_{ex}\left(\mathbf{r},\mathbf{k}\right) \equiv \int ds\, e^{i\mathbf{k}\cdot\mathbf{s}}\, G_{ex}\left(\mathbf{r}-\mathbf{s}/2\,,\mathbf{r}+\mathbf{s}/2\right). \tag{2.7}$$

Equations (2.5) and (2.6) together yield

$$V_L(r) = E - \frac{\hbar^2}{2\mu}\,\mathbf{k}^2(r) = V_D(r) + G^W_{ex}\left(\mathbf{r},\mathbf{k}(r)\right), \tag{2.8}$$

with **k**(r) a solution of Eq. (2.5). To solve for **k**(r) in Eq. (2.5) implies to solve for the radial part $k_r(r)$ of **k**(r) for each value of the angular momentum ℓ, $\mathbf{k}(r) = k_r(r)\,\mathbf{r}/r + (\ell+1/2)/r\,\mathbf{n}$, $\mathbf{r}\perp\mathbf{n}$. The ELP $V_L(r)$ is therefore not only E- but also ℓ-dependent. The Perey factor in the WKB approximation is given by

$$F(r) = 1/\sqrt{\partial H^W/\partial k_r^2} = \sqrt{(1-\partial V_L/\partial E)}\,. \tag{2.9}$$

For light-ion scattering, x + ^{4}He, x + ^{16}O and x + ^{40}Ca, where x is a 0s shell nucleus (n,d,t, ^{3}He or α), and for light heavy-ion scattering, ^{16}O + ^{16}O and ^{16}O + ^{40}Ca, Horiuchi and collaborators have shown that the $V_L(r)$ closely reproduce the phase shifts from the original RGM nonlocal potentials[5].

Their calculations also showed that $V_L(r)$ is *quite deep* and relatively close to the double folding potential $V_D(r)$. There is a moderate dependence of $V_L(r)$ on the energy leading to a decreasing depth with increasing energy. However, $V_L(r)$ depends only weakly on ℓ.

The reason why $V_L(r)$ is deep is the existence of a Pauli forbidden region in the phase space of the relative motion of the two nuclei. This is closely related to the fact that the nonlocal RGM interaction supports not only the physical bound states of the two-cluster system, but also the redundant states (these correspond to the so-called Pauli Forbidden States (PFS) in the ELP). Consequently it produces phase shifts which satisfy a generalized Levinson theorem[6],

$$\delta_\ell(0) - \delta_\ell(\infty) = n_P \pi + n_{FS} \pi , \tag{2.10}$$

where n_P = number of physical bound states and n_{FS} = number of PFS. These PFS lie below the physical bound states, and, although not physically present, show up in the high-energy behaviour of the physical phase shifts. The ELP associated with the RGM interaction, being a local potential, has to support the PFS as if they were ordinary bound states, and consequently has to be deep.

On the other hand, equally good fits to the same phase shift "data" were obtained by using *shallow* energy independent phenomenological potentials which are strongly ℓ-dependent. Potentials of this type are the well-known α-α potentials of Ali and Bodmer[7]. These potentials, however, have phase shifts which satisfy $\delta_\ell(0) - \delta_\ell(\infty) = n_P\pi$, and disagree with the high-energy data.

Baye[8] replaced the energy dependent ELP of the RGM interaction for α-α scattering by the energy independent deep local potential of Buck et al.[9] valid at low energies, which supports the same PFS as the RGM interaction. To remove the PFS while retaining the same phase shifts as the potential of Buck et al., he introduced a supersymmetric scheme (SS).

In this scheme the Hamiltonian is given by:

$$H_0 \equiv - \frac{d^2}{dr^2} + V_0 , \tag{2.11}$$

where V_0 incorporates the centrifugal, Coulomb and nuclear potentials. The Hamiltonian H_0 and its supersymmetric partner H_1 are factorized as

$$H_0 = A_0^+ A_0^- + \varepsilon_0, \quad H_1 = A_0^- A_0^+ + \varepsilon_0 , \tag{2.12}$$

where

$$A_0^- = \left(A_0^+ \right)^+ = - \frac{d}{dr} + \frac{d}{dr} \ell n \, \psi_0 (\varepsilon_0) . \tag{2.13}$$

The wave function $\psi_0(\varepsilon_0)$ is a nodeless, non-renormalizable solution of $H_0\psi_0 = \varepsilon_0\psi_0$. We must therefore have $\varepsilon_0 \leq E_0^{(0)}$, where $E_0^{(0)}$ is the ground state energy of H_0.The eigenfunctions $\psi_0(E)$ and $\psi_1(E)$ of H_0 and H_1 are related by

$$\psi_1(E) = \left(E - \varepsilon_0\right)^{-1/2} A_0^- \, \psi_0(E) . \tag{2.14}$$

The choice of $\varepsilon_0 = E_0^{(0)}$ makes the spectrum of H_1 identical to the spectrum of H_0, except for the absence of the ground state of H_1. The potential corresponding to H_1 is then given by

$$V_1 = V_0 - 2 \frac{d^2}{dr^2} \ell n \, \psi_0\left(E_0^{(0)}\right) . \tag{2.15}$$

For small values of r we have $V_1 \sim V_0 + 2(\ell+1)/r^2 \sim (\ell+1)(\ell+2)/r^2$, which behaves as a nonsingular potential with angular momentum $\ell+1$. The potentials V_1 and V_0 are, however, not phase equivalent. To impose phase equivalence the supersymmetric partner H_2 of H_1 with $\varepsilon_1 = E_0^{(0)}$ as factorization energy is required. Since $\varepsilon_1 < E_1^{(0)}$, the ground state energy of H_1, the function $\psi_1(\varepsilon_1)$ is not square integrable and behaves asymptotically as exp. $(\gamma_0 r)$, with $\gamma_0 = (-E_0^{(0)})^{1/2}$. However, it is possible to choose it to be regular at the origin. The bound state spectra of H_1 and H_2 are identical and to any physical state of H_1 with energy E corresponds a state of H_2 with the wave function

$$\psi_2(E) = \left(E_0^{(0)} - E\right)^{-1} A_1^- A_0^- \psi_0(E) , \tag{2.16}$$

where $\psi_2(E)$ and $\psi_0(E)$ have the same phase shift, $\delta_\ell^{(2)}(E) = \delta_\ell^{(0)}(E)$, and where $\psi_2(E)$ is normalizable for the bound states. The corresponding potential is given by

$$V_2 = V_0 - 2 \frac{d^2}{dr^2} \ell n \left[\psi_0\left(E_0^{(0)}\right) \psi_1\left(E_0^{(0)}\right)\right] , \tag{2.17}$$

which for small r behaves like $V_2 \sim V_0 + 2(2\ell+3)/r^2 \sim (\ell+2)(\ell+3)/r^2$. The singular potential V_2 is shallow and has one bound state less than V_0. It follows from the equality of the phase shifts that $\delta_\ell(0) - \delta_\ell(\infty)$ is identical for V_0 and V_2, although V_2 has one bound state less than V_0 as proved by Swan[10]. This two-step process can be iterated to remove all the nonphysical bound states. Baye[8] was able to show that if this SS procedure is applied to the deep and energy independent local potential of Buck et al.[9], the phase equivalent shallow ℓ-dependent potentials closely resemble the α-α potentials of Ali and Bodmer. However, to achieve this result the inherently energy dependent ELP of the α-α interaction in the RGM had to be approximated by a local energy independent potential, an approximation which is only valid at low energies. The strict phase equivalence between the RGM interaction and the shallow local potentials is also lost in this way.

To retain strict phase equivalence, inverse scattering methods have to be used. These will be discussed in the next Section.

3. Equivalent Local Potentials by Inverse Scattering Methods

To avoid approximations we extend Baye's supersymmetric scheme by including an intermediate step to remove the energy dependence of the deep ELP. In this way we guarantee that the shallow ℓ-dependent potentials are energy independent. This has been explained in detail in Ref. 11, for the case of two dineutrons. Although dineutrons do not exist in reality, the scattering of two dineutrons has often been used for a simple illustration of the methods of the RGM. In Ref. 11 the treatment of Kukulin et al.[12] of this case was followed. The internal spatial wave function of the dineutron is chosen to be a translationally invariant shell-model function of the $0sp^2$ configuration in a harmonic oscillator well of width parameter α. The (2+2) system then corresponds to the shell model configuration s^2p^2. In this system one s-state of the relative motion is forbidden by the Pauli principle. Kukulin et al. have shown that in this case the RGM equation can be written as

$$\left(T_r + V_0 + V_N - E\right)\phi(\mathbf{r}) = 0 \,, \tag{3.1}$$

where

$$V_N\,\phi(\mathbf{r}) \equiv \int d\mathbf{r}'\,V_N \cdot e^{-\alpha(\mathbf{r}+\mathbf{r}')^2/2} \cdot e^{-\beta(r-r')^2}\,\phi(\mathbf{r}') \,, \tag{3.2}$$

with the orthogonality condition $<\phi|\phi_0> = 0$, where $\phi_0(\mathbf{r})$ is the wave function of the PFS. In this case that condition is automatically fulfilled.

The ELP of the nonlocal interaction of Eq. (3.2) is obtained by the use of Horiuchi's "WKB + RGM" method, which in this case reduces to the old Perey-Buck procedure[13]. It results in an ELP which is ℓ-independent. It has, however, a substantial E-dependence which for $r=0$ and low energies can be represented by

$$V_0(E) \approx V_0(0)\cdot e^{-E/E_0}, \quad where \quad E_0 = 84\,MeV\, and\, V_0 = 20.5\,MeV\,. \tag{3.5}$$

To remove this E-dependence, which is necessary to apply Baye's SS scheme for the construction of shallow *energy independent ELP's*, we do not neglect it, as Baye did in the α-α case, but rather using the phase shifts of the ELP (or, for exact phase equivalence, those of the RGM) we perform a Marchenko inversion at fixed ℓ, to obtain deep ℓ-dependent but *energy-independent* local potentials $V^{(\ell)}(r)$. We then apply Baye's two-step procedure to remove the PFS for $\ell=0$.

For the Marchenko inversion we represent the $S_\ell(k)$ of the RGM in rational form by

$$S_\ell(k) = \Pi\;(k+q_n)/(k-q_n)\,. \tag{3.6}$$

This can be done to high accuracy and leads to Bargmann potentials. In Fig. 1 we show the potential $V_0(r)$, the potential $V_1(r)$ obtained after the first step of the SS-scheme, and finally the potential $V_2(r)$, which is the shallow singular potential strictly phase equivalent to the RGM interaction, satisfying the same Levinson relation $\delta_0(0) - \delta_0(\infty) = \pi$, but having no PFS. This is obvious from the figure since the potential $V_2(r)$ is purely repulsive.

The scheme described here [11] is a mathematically rigorous method to obtain shallow singular potentials which are ℓ-dependent but E-independent, and are strictly phase equivalent to the corresponding RGM interaction.

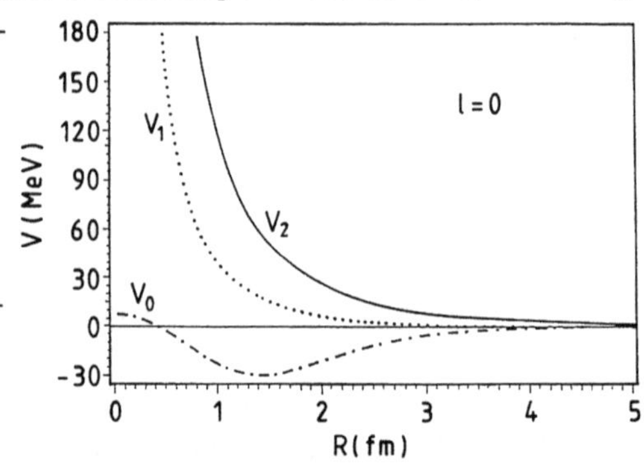

Fig. 1. ELP's obtained by the supersymmetric + inversion scheme.

Leaving the supersymmmetric scheme, we briefly consider recent work on the inversion of the neutron-deuteron phase shifts in the quartet (S = 3/2) state, to construct strictly phase equivalent ELP's. The phase shifts were calculated by means of the exact three-body Faddeev equation. We therefore deal here with a fully microscopic theory, which treats break-up exactly, unlike the semi-microscopic RGM theory, which is of limited applicability

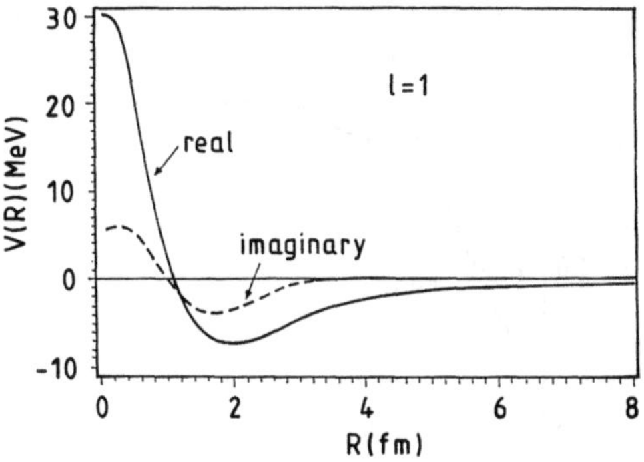

Fig. 2. The p-wave fixed-ℓ Marchenko ELP in the quartet state of the n-d system

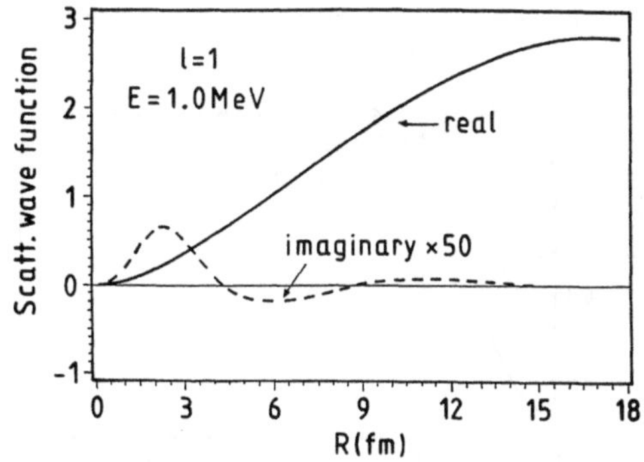

Fig. 3. The p-wave function corresponding to the potential of Fig. 2 for E = 1.0 MeV

in this case. It is therefore interesting to see if the qualitative features of ELP's for the RGM interaction are also present in the ELP's of the Faddeev theory. For this purpose both fixed-ℓ Marchenko and fixed-E inversion have been employed by Papastylianos et al.[14] and Alt et al. [15] in these investigations. The n-d quartet phase shifts are of course real below and complex above the break-up threshold. Marchenko inversion therefore results in ELP's $V^{(\ell)}(r)$ which are energy independent and complex. To perform the inversion of theFaddeev phase shifts $\delta_\ell^F(k)$ accurately, we again employ a rational Bargmann representation of $S_\ell(k)$, which reproduces the $\delta_\ell^F(k)$ to high accuracy up to 3 GeV. This is necessary to obtain a stable potential. In Fig. 2 the p-wave potential is shown. It is seen that the imaginary part is partly emissive, a well-known feature in ELP's from nonlocal interactions. The emissive part, however, is mainly concentrated at short distances, where the real part is already strongly repulsive and the wave function is relatively small. The phase shifts calculated with these potentials agree with the input Faddeev phase shifts within 0.004 degrees or less, even in the region below the break up threshold, where the $\delta_\ell^F(k)$ are real, while $\delta_0(0) = \pi$ like $\delta_0^F(0)$. What is even more surprising, however, is that even the wave functions $U^{(\ell)}(r)$ of the complex potential $V^{(\ell)}(r)$ are nearly real, for real boundary conditions, not only in the asymptotic region, but also in the interior. This is shown in Fig. 3 for an incident energy of 1.0 MeV. This is a remarkable feature of our inverted optical potentials. It is unlikely that it would have been possible to construct such optical potentials by phenomenological means. A point of major interest is that these optical potentials $V_\ell(r)$ clearly are the analogs in the Faddeev case of the shallow potentials. This is particularly clear for the s-wave, where the real part of the potential $V_0(r)$ is purely repulsive, but we have $\delta_0(0) = \pi$.

It is also interesting to apply fixed-E inversion to these Faddeev phase shifts for the construction of E-dependent but ℓ-independent optical potentials which are analogous to the deep ELP's of the RGM. The rational/nonrational mixed inversion scheme[16] at fixed E was employed for this purpose at 200, 150 and 100 MeV [15]. In Figs. 4 and 5 the energy dependence of the inversion potentials for odd and even parity waves is shown. The n-d quartet scattering shows a strong parity dependence (a feature which also occurs in the RGM description of cluster-cluster scattering). We therefore have to invert the odd and even partial

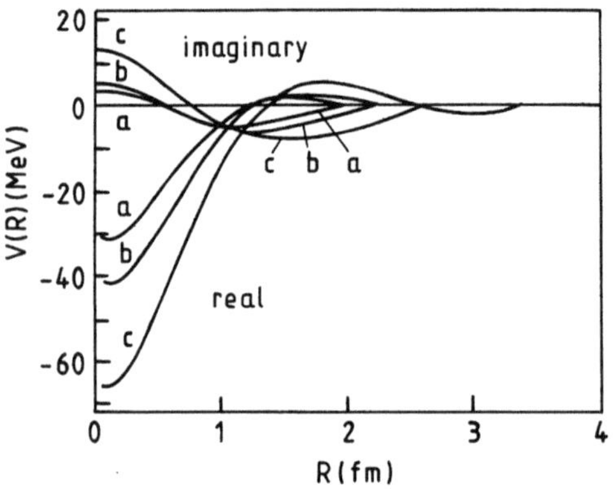

Fig. 4. The fixed-E ELP for even partial wave quartet states of the n-d system, for E = 200 (a), 150 (b) and 100 MeV (c)

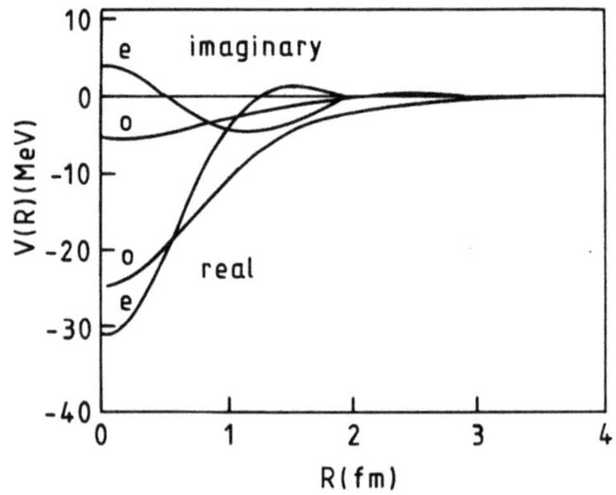

Fig. 5. Parity dependence of the fixed-E ELP for the quartet states at E = 200 MeV (e-even, o-odd partial waves)

waves separately. It is seen from these figures that the real parts of the potentials have similar energy dependence as the deep potentials of Horiuchi. Thus depth decreases with increasing energy. Their shapes are quite similar to those of Woods-Saxon potentials while the imaginary parts of the potentials for even waves, again have some emissive regions.

We can therefore conclude that the inversion of the n-d quartet scattering in the Faddeev formalism qualitatively shows a behaviour very similar to the deep ELP for cluster scattering in the "RGM + WKB" formalism of Horiuchi. Since the Faddeev theory is an exact theory, including all break-up effects, while the RGM theory is an approximate one mainly designed for the correct treatment of exchange effects, this result indicates that the RGM theory is indeed a good description of the qualitative features of cluster-cluster scattering.

Finally, we mention that fixed-E inversion methods have been used to determine the wave function of an energy independent nonlocal potential from its phase shifts[17]. If the phase shifts of such a potential are known for all ℓ and E, the nonlocal potential itself can also be determined by inversion[18].

As to the nonlocal wave function, one proceeds in the following way: The phase shifts of the energy independent nonlocal potential V_{NL} are used to determine an energy dependent ELP $V_L(E)$. From the wave function $\chi_L(r)$ of this local $V_L(E)$ the nonlocal wave function $\chi(r)$ is obtained with the help of Eqs. (2.4) and (2.9) in the WKB approximation. Evidently, the potential $V_L(E)$ has to be determined for a range of neighbouring energies in order to determine $\partial V_L/\partial E$. This method has been successfully applied to N-α scattering[7].

The determination of an energy independent nonlocal potential from its phase shifts makes use of the analogs of the WKB equations (2.5) and (2.6), which lead to

$$V_L(E,R) = V^W_{NL}\Big(R^2, 2\mu(E - V_L(E,R))\Big) . \qquad (3.7)$$

Instead of solving this equation for V_L, knowing V_{NL} (as done in Sect. 2), we now regard $V_L(E,R)$ as given from the inversion of the "nonlocal" phase shifts, and determine the nonlocal potential $V^W_{NL}(R^2,k^2)$, or rather $V_{NL}(\mathbf{r},\mathbf{r}')$, from it. For example, for a nonlocal potential of the Frahn-Lemmer type, $V_{NL}(\mathbf{rr'}) = V(R)\,H(s^2)$, $\mathbf{R} = (\mathbf{r}+\mathbf{r'})/2$, $\mathbf{s} = \mathbf{r}-\mathbf{r'}$, one finds

$$V(R)/V(0) = exp\left\{ \int_0^R dR'[V_L'(R')/V_L(R')] \cdot \; [1-\partial V_L(R')/\partial E]^{-1} \right\} , \tag{3.8}$$

and for the Fourier transform of $H(s^2)$,

$$V(0) \cdot \; H_{FT}(k^2) = V_L(E,R) \cdot \; [V(R)/V(0)]^{-1} \; , \;\; where \;\; k^2 = 2\mu(E - V_L(E,R)) \; . \tag{3.9}$$

Equations (3.8) and (3.9) solve the inverse problem for this type of nonlocal potential. In principle, $V_L(E,R)$ must be known for all E, that is the phase shifts of the nonlocal potential V_{NL} must be known for all ℓ and E. The procedure has been applied to n-^{40}Ca scattering[18].

This project has been supported in part by the Fonds zur Förderung der Wissenschaftlichen Forschung in Österreich, Project Number P6608.

References

1) K. Wildermuth and Y.C. Tang: A Unified Theory of the Nucleus (Nieweg, Braunschweig, 1977)
2) F.G. Perey: Direct Interactions and Nuclear Reaction Mechnanisms, eds. E. Clementel and C. Villi (Gordon and Breach Science Publishers, Inc., New York, 1963) p. 125
3) H. Fiedeldey: Nucl. Phys. A96 (1967) 463
4) H. Horiuchi: Prog. Theor. Phys. 69 (1980) 725; 64 (1980) 184
5) K. Aoki and H. Horiuchi: Prog. theor. Phys. 69 (1989) 1154
6) S. Sarto: Prog. Theor. Phys. Suppl. 62 (1977) Chap. II
7) S. Ali and A.R. Bodmer: Nucl. Phys. 80 (1966) 99
8) D. Baye: Phys. Rev. Lett. 58 (1987) 2738
9) B. Buck, H. Friedrich and C. Wheatly: Nucl. Phys. A275 (1977) 246
10) P. Swan, Am. Phys. (N.Y.) 48 (1968) 455
11) H. Fiedeldey, S.A. Sofianos, A. Papastylianos, K. Amos and L.J. Allen, submitted to Phys. Rev. C.
12) V.I. Kukulin, V.G. Neudatchin and Yu. F. Smirnov: Nucl. Phys. A245 (1975) 429
13) F.G. Perey and B.Buck:. Nucl. Phys. 32 (1962) 353
14) A. Papastylianos, S.A. Sofianos, H. Fiedeldey and E.O. Alt: Proc. of the 12th Int. Conf. on Few Body Problems in Physics, Vancouver, B.C., Canada, July 2-8, 1989, Contributions D28 and D50
15) E.O. Alt, H. Fiedeldey and S.A. Sofianos: Few-Body Systems, Suppl. 2 (1987) 544
16) R. Lipperheide and H. Fiedeldey: Z. Phys. A286 (1978) 45; A301 (1981) 81
17) H. Fiedeldey, S.A. Sofianos, L.J. Allen and R. Lipperheide: Phys. Rev. A32 (1985) 3095
18) H. Fiedeldey, S.A. Sofianos, L.J. Allen and R. Lipperheide: Phys. Rev. A33 (1986) 1581

Inverse Scattering Theory for Phase-Conjugate Reflection Coefficients: Application to Optical Switching

A.K. Jordan[1], *S. Lakshmanasamy*[2], *and J. Xia*[3]

[1]Research Laboratory of Electronics, Massachusetts Institute of Technology, Cambridge, MA 02139, USA
Permanent address: Space Science Division, Naval Research Laboratory, Washington, DC 20375, USA
[2]Jonsson School of Engineering, University of Texas at Dallas, Richardson, TX 75083, USA
[3]Department of Electrical Engineering and Computer Science, Massachusetts Institute of Technology, Cambridge, MA 02139, USA

Introduction. The importance of the phase function for the solution of inverse problems can be demonstrated by considering phase-conjugate reflection coefficients, i.e., two reflection coefficients with equal amplitudes but opposite phases. Propagation of light rays along a dielectric waveguide can be modelled as a progression of total internal reflections at the boundaries of the waveguide. Since it is known from direct scattering theory that total internal reflection at a dielectric interface is accompanied by a phase change of π radians, it will be useful for the synthesis of inhomogeneous optical waveguides to investigate the implications of this phenomenon for the corresponding inverse problem.

In this paper we use inverse scattering theory to design an optical device that can be switched between two states, depending on the amplitude of the incident wave. In one state the optical waveguide can support a single discrete guided mode (in addition to radiated modes); in the second state the waveguide can support only radiated modes. Thus the device acts as an exclusive-or (XOR) gate by allowing or not allowing the propagation of a guided mode. The device contains an active material whose permittivity depends nonlinearly upon the amplitude of the incident electromagnetic wave due to the Kerr effect. With appropriately designed permittivity profiles, such a device should be able to switch optical pulses at picosecond speeds.

Physical Model of a Planar Optical Waveguide. The propagation of the electric field $\overline{E}$ in an inhomogeneous planar wave-guide, shown in Fig. 1, is governed by the vector wave equation

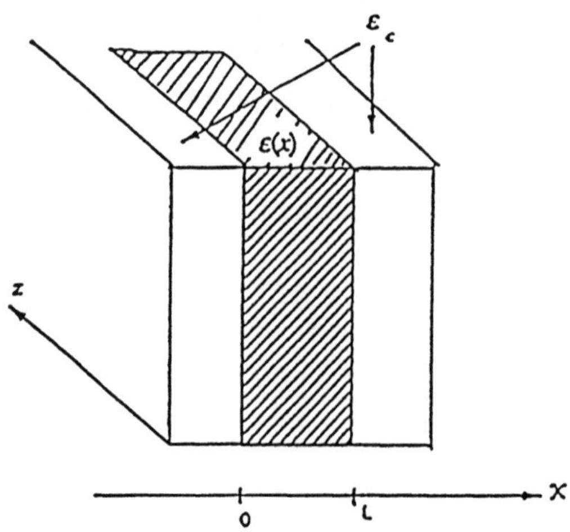

Figure 1: Physical model of planar optical waveguide.

$$\nabla^2\overline{E}+\nabla\left(\overline{E}\cdot\frac{1}{\epsilon_r}\nabla\epsilon_r\right)+\frac{\epsilon_r}{c^2}\frac{\partial^2}{\partial t^2}\overline{E}=0 \tag{1}$$

where $k_c = \omega\sqrt{\mu_0\epsilon_r}$ is the wavenumber in the external cladding region with a permittivity ϵ_c and $\epsilon_r = \epsilon_r(x) = \frac{\epsilon(x)}{\epsilon_c}$ is the relative permittivity of the inhomogeneous core region that supports the guided mode. We will be interested in time-harmonic transverse-electric (TE) waves propagating in the positive z-direction so that the electric field vector $\overline{E}$ has the form

$$\overline{E}=\hat{y}\psi(x,k)\ e^{i\beta z-i\omega t} \tag{2}$$

where β is the longitudinal wavenumber and k is the transverse wavenumber. Eq. (1) can now be written

$$\frac{d^2}{dx^2}\psi(k,x)+\left[k^2-q(x)\right]\psi(k,x)=0 \tag{3}$$

where the scattering potential q(x) is related to the relative permittivity by

$$q(x)=k_c^2\left[1-\epsilon_r(x)\right], \tag{4}$$

so that

$$q(x)\to 0 \text{ as } |x|\to\infty, \tag{5}$$

since $\epsilon_r(x) \to 1$ as $|x| \to \infty$. The transverse wave number k is found from the dispersion relation

$$k^2=k_c^2-\beta^2 \tag{6}$$

The general properties of the electromagnetic waves can be classified according to the values of the spectral parameter k:

$k^2 < 0$, bounded solutions for ψ exist only for a finite number of discrete values k_n of k; these are termed guided modes. For propagation in the positive z-direction, $\beta > 0$

$k^2 > 0$, the spectrum is continuous and is divided into two regions: $0 < k < k_c$, β is positive and the modes are radiated; $k > k_c$, the modes do not propagate and are termed evanescent.

Application of Inverse Scattering Theory. For coherent single-mode propagation in the $+z$-direction, the longitudinal wavenumber β can only have a single discrete value. Thus from Eq. (5) the transverse wavenumber k can only have one discrete value. Any realistic optical waveguide will always have continuous nonguided modes. Thus, in terms of the spectral variable k of eq. (4), both a discrete spectrum and the continuous spectrum are present. Information about the modal structure of the waveguide is contained in the analytic properties of the transverse reflection coefficient $r(k)$.

The synthesis of a waveguide device to be used for optical switching can be demonstrated with a three-pole approximation for $r(k)$,

$$r(k)=\frac{r_0}{(k-k_1)(k-k_2)(k-k_3)} \tag{7}$$

where $r_0 = k_1k_2k_3$ and $k_1 = c_1 - ic_2$ and $k_2 = -k_1^*$ are used to approximate the continuous portion of the spectrum, which represents the radiating and evanescent modes. The third pole k_3 is defined by $k_3 = i\delta a$, where $a = 1$ and

$$\delta=\begin{cases}+1 \text{ and } k_3 \text{ is in the discrete spectrum}\\ -1 \text{ and } k_3 \text{ is in the continuous spectrum with no discrete spectrum.}\end{cases}$$

Pole switching is similar to the Darboux transformation of inverse-scattering theory [3]. The magnitude and phase of $r(k)$ are

$$|r(k)|^2 = \frac{(c_1^2 + c_2^2)^2 a^2 \delta^2}{[(k+c_1)^2 + c_2^2]\,[(k-c_1)^2 + c_2^2]\,[k^2 + \delta^2 a^2]} \tag{8}$$

and

$$\phi(k) = -\delta\left[-\pi/2 + \tan^{-1}\left(\frac{a}{k}\right)\right] - \left[\tan^{-1}\left(\frac{c_2}{k+c_1}\right) + \tan^{-1}\left(\frac{c_2}{k-c_1}\right)\right] \tag{9}$$

There is no difference in the magnitude of $r(k)$ between the two cases. However there is a difference in the phase of

$$\Delta\Phi = \pi - 2\tan^{-1}\left(\frac{a}{k}\right). \tag{10}$$

In the optical limit $\Delta\Phi \to \pi$ as $k \to \infty$, i.e., the reflection coeffieients become phase conjugate.

This inverse scattering problem can now be stated: Given these prescribed three-pole transverse reflection coefficients $r(k)$, determine the corresponding scattering potentials $q(x)$. For this synthesis a previously developed inversion method [1], which was based upon the version of the Gelfand-Levitan-Marchenko inverse theory appropriate for rational reflection coefficients [2], is used.

It is necessary to examine the "allowed regions" for the poles k_1, k_2, k_3 in the complex $k-$plane. These regions are determined by the appropriate physical criteria–in this case the conservation of energy

$$|r(k)|^2 \leq 1 \text{ for all real } k \tag{11}$$

and normalization

$$r_0 = r(0) = -1 \tag{12}$$

The allowed regions for the three-pole example is shown in Fig. 2; analytic expressions for their boundaries are given in [1]. All poles used in these examples are in the allowed regions.

The Gelfand-Levitan-Marchenko integral equation of inverse-scattering theory is

$$R(x+t) + K(x,t) + \int_{-t}^{x} K(x,\zeta)R(\zeta+t)d\zeta = 0,\; t < x \tag{13}$$

where the scattering data are contained in the function $R(x+t)$

$$R(x+t) = \frac{1}{2\pi}\int_{-\infty}^{\infty} r(k)\, exp[-ik(x+t)]dk - i\sum_{n=1}^{N} r_n\, exp[-ik_n(x+t)] \tag{14}$$

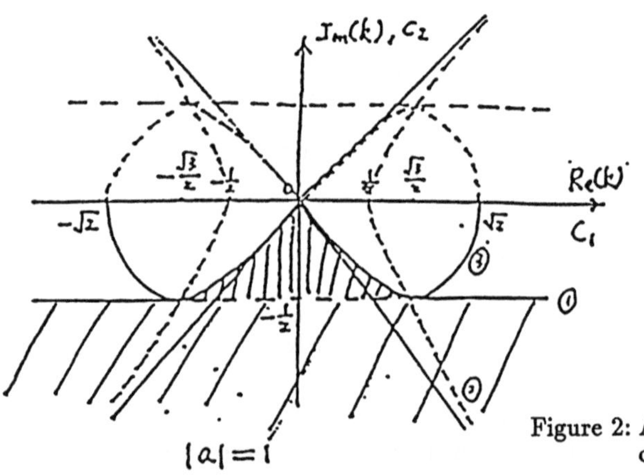

Figure 2: Allowed regions for poles of $r(k)$.

which is solved for the transformation kernel $K(x,y)$ to obtain the solution for the scattering potential $q(x)$

$$R(x+t) = -\frac{\delta a(c_1^2+c_2^2)}{c_1^2+(c_2+\delta a)^2}\{exp[-c_2(x+t)]\left[cos[c_1(x+t)]\right.$$
$$\left.+\frac{c_2-\delta a}{c_1}sin[c_1(x+t)]\right] - exp[\delta a(x+t)]\Big\} \tag{15}$$

In order to solve the GLM integral Eq.(13) a differential operator is constructed

$$\begin{aligned} f(p) &= (p+ik_1)(p+ik_2)(p+ik_3) \\ &= p^3 + (2c_2-\delta a)p^2 + \{(c_1^2+c_2^2) - 2\delta ac_2\}p - \delta a(c_1^2+c_2^2) \end{aligned} \tag{16}$$

so that $f(p)R(x+t)=0$, where $p=\frac{d}{dt}$. Applying $f(p)$ to the integral equation (13), yields

$$f(p)K(x,t) + \delta a(c_1^2+c_2^2)K(x,-t) = 0. \tag{17}$$

A second independent equation can be obtained by using the time-invarient property of the scattering data when $t \to -t$ and $p \to -p$.

$$f(-p)K(x,-t) + \delta a(c_1^2+c_2^2)\,K(x,t) = 0. \tag{18}$$

These last two equations can be combined into a sixth-order differential equation for $K(x,t)$ whose solution together with the appropriate boundary conditions gives the analytic solution for $K(x,t)$; details are given in [1]. The scattering potential $q(x)$ is found from

$$q(x) = 2\frac{d}{dx}K(x,x) \tag{19}$$

The GLM integral has been solved for the case of a single discrete mode with the two-pole approximation for the continuous spectrum[1]. This solution is easily generalized to include this case and the case of the three-pole approximation to the continuous spectrum. The solution for $q(x)$ is

$$q(x) = 2\left[\mathbf{a}'^T(x) - \mathbf{a}^T(x)\mathbf{A}^{-1}(x)\mathbf{A}'^{-1}\right]\mathbf{A}^{-1}(x)\mathbf{b} \tag{20}$$

where ^{T} denotes transpose, $'$ denotes $\frac{d}{dx}$,

$$\mathbf{a}^T(t) = [1 \quad t \quad exp(\eta_1 t) \quad exp(-\eta_1 t) \quad exp(\eta_2 t) \quad exp(-\eta_2 t)] \tag{21a}$$
$$\mathbf{b}^T = \left[0 \quad 0 \quad 0 \quad 0 \quad 0 \quad -a\delta\,(c_1^2+c_2^2)\right] \tag{21b}$$

and the matrix $\mathbf{A}$ is

$$\begin{bmatrix} 0 & 1 & 0 & 0 & 0 & 0 \\ 0 & 0 & f(\eta_1) & a\delta(c_1^2+c_2^2) & 0 & 0 \\ 0 & 0 & 0 & 0 & f(\eta_2) & a\delta(c_1^2+c_2^2) \\ 1 & -x & exp(-\eta_1 x) & exp(\eta_1 x) & exp(-\eta_2 x) & exp(\eta_2 x) \\ 0 & -1 & -\eta_1 exp(-\eta_1 x) & \eta_1 exp(\eta_1 x) & -\eta_2 exp(-\eta_2 x) & \eta_2 exp(\eta_2 x) \\ 0 & 0 & \eta_1^2 exp(-\eta_1 x) & \eta_1^2 exp(\eta_1 x) & \eta_2^2 exp(-\eta_2 x) & \eta_2^2 exp(\eta_2 x) \end{bmatrix} \tag{21c}$$

in which $f(\cdot)$ has the functional form of $f(p)$ and

$$\eta_1 = [(\sigma+\rho)/2]^{1/2}\ ,\ \eta_2 = [(\sigma-\rho)/2]^{1/2}\,,$$
$$\sigma = a^2 + 2c_2^2 - 2c_1^2\ ,\ \rho = \left[(a^2-4c_2^2)(a^2+4c_1^2)\right]^{1/2}.$$

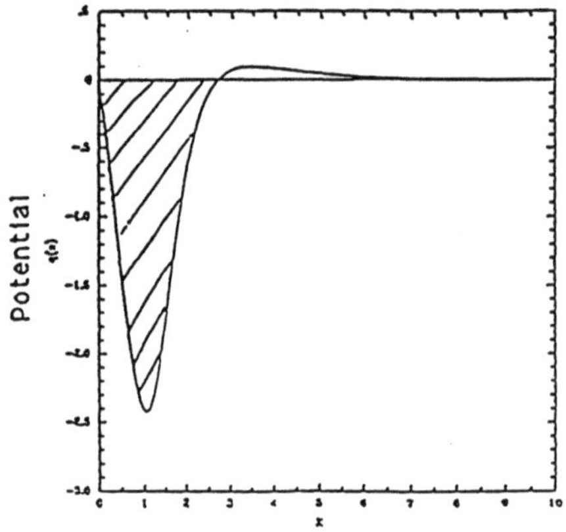

Figure 3: Example of scattering potential $q(k)$.

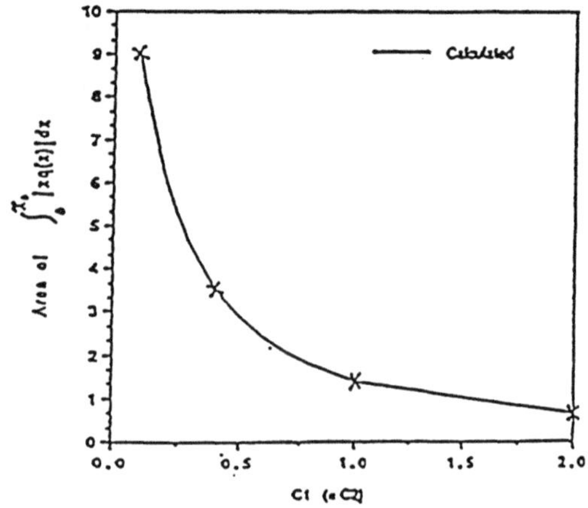

Figure 4: Dependence of area function A_- on pole positions $c_1 = c_2$.

The form of the scattering potential $q(x)$ is a function of the pole configuration of $r(k)$. For the present three-pole case we will vary c_1 and c_2 in the allowed region and calculate the variations in $q(x)$. A useful parameter to measure this variation is the first moment of the area under the negative $q(x)$ curve

$$A_- = \int_0^{x_0} |x|\,|q_-(x)|\,dx, \tag{22}$$

where $q_-(x)$ is the negative portion of $q(x)$ and x_0 is the zero crossing. The change in this area function is shown in Fig. 4. The magnitude of the area function A_- provides a measure of the switching speed possible for the "pole flip" of k_3 from the negative to the positive imaginary axis. The continuous spectrum determines, to a large extent, the magnitude of A_- and thus the switching speed. The appropriate profile for a particular switching application can be determined by optimizing the continous spectrum of $r(k)$. Experimental results have been obtained for reflection from uniform nonlinear interfaces. Switching speeds of the order of 2 picoseconds with 10^9 watt power densities have been observed [4] for CS_2 with ruby laser pulses. For $LiNiO_3$ the corresponding speed was 2 psec. [5]. It is anticipated that for single-mode switching from an inhomogeneous interface with a profile of Fig. 3 generated by using appropriately doped materials to optimize this profile area according to Fig. 4 can provide fast switching speeds at lower power densities.

References

1. Jordan, A. K., & Lakshmanasamy, S.,"Inverse scattering theory applied to the design of single-mode planar optical waveguides", *J. Opt. Soc. Am.* A6, 1206-1212 (1989).
2. Kay, I., "The inverse scattering problem", *Rep. EM-74* (New York University, New York, N.Y., 1955); and "The inverse scattering problem when the reflection coefficient is a rational function", *Comm. Pure & Applied Math.* 13, 371-393 (1960). See also I. Kay & H. Moses, *Inverse Scattering Papers: 1955-1963,* (Math Sciences, Brookline, Mass., 1982).
3. Chadan, K., & Sabatier, P.C., *Inverse Problems in Quantum Scattering Theory,* 2nd ed. (Springer-Verlag, New York, 1989).
4. Smith, P.W., et al, "Experimental studies of a nonlinear interface", *IEEE J. Quantum Electronics QE-17,* 340-346; 1981.
5. Lattes, A. L., MIT Ph.D. Thesis, 1982.

An Inverse Problem in Propellant Combustion

H.D. Ladouceur

Naval Research Laboratory, Code 6110, Washington, D.C.20375, USA

INTRODUCTION

If the temperature or heat flux at the surface of a solid are given as a function of time, then the temperature distribution in the solid's interior can be calculated. This is termed a direct problem. However, there are many practical situations which arise in engineering where the surface environment may be unsuitable for attaching a sensor such as a thermocouple. One example of such a situation is designing an experiment to determine the temperature distribution and surface heat flux history on an ablating heat shield undergoing chemical reactions with the atmosphere. Since the shield's surface is receding and located in a hostile environment, the sensor cannot monitor the temperature accurately. However, it is easier to measure the transient temperature at a known interior point of the shield. The mathematical problem of determining the temperature distribution and surface boundary conditions from such an interior measurement is known as the inverse heat conduction problem (IHCP) [1]. There are two general classes of such inverse problems: parameter estimation and function estimation. An example of parameter estimation is the determination of a solid's thermal diffusivity from time-dependent temperature measurements. The estimation of the surface heat-flux as a function of time is an example of a function estimation problem. There are many papers and some books on both classes of problems. Beck et al. [2] have written the only English language book on the IHCP. This book provides an excellent introduction to various aspects of the IHCP and contains numerous references up to 1985.

In developing models to describe propellant combustion, the thermal diffusivity, α, of the condensed phase is a necessary parameter for calculating the phase's thermal evolution. The determination of the surface heat flux from the gas-phase combustion zone is required as a boundary condition for the condensed phase and is useful in understanding the material's behavior and performance. A common experimental technique employed by the propellant community is based upon combustion wave analysis in a strand burner [3]. In strand-burner experiments, the propellant is prepared as a small circular cylinder called a grain which burns in the fashion of a cigarette and can be modeled as a one-dimensional heat-flow problem. Thermocouples on the order of several microns in diameter are embedded in the propellant at known positions which are determined by X-ray analysis. Upon ignition at constant pressure, the transient burning behavior is dependent upon the characteristic times associated with the thermal diffusion process in the condensed phase, t_c, in the surface region, t_s, and in the gas phase, t_g [4]. Typical magnitudes of these relaxation time scales have been calculated in reference [5] and are on the order of t_c= 0.01 sec, t_s=0.001 sec, and t_g=0.0001 sec. Thus, the burning gas phase above the propellant surface attains a steady-state 100 times faster than the condensed phase; and to a first approximation the gas phase can be treated as a steady-state source of the heat flux Q to the surface. The temperature distribution T(z,t) within the solid phase, subject to this heat flux, is governed by the standard heat-conduction equation [6] given by

$$\frac{\partial}{\partial z}\left[k\frac{\partial T}{\partial z}\right] + g(z,t) = \rho c\frac{\partial T}{\partial t} \qquad [1.1]$$

$$t > o \qquad S(t) < z < L_o,$$

where z is a fixed spatial coordinate oriented along the grain's axis, L_o is the grain's initial length, ρ denotes the density, c is the specific heat, k is the thermal conductivity and g(z,t) is a volumetric energy source or sink due to chemical reactions in the condensed phase. Since the solid phase is consumed during the combustion, the z domain is time dependent. The position of the gas-solid interface is denoted by S(t) and in general depends upon the heat flux Q(t) via the boundary condition

$$-k\frac{\partial T}{\partial z} + Q_p = Q(t) \quad \text{at } z = S(t), \qquad [1.2]$$

where Q_p is the heat flux absorbed by the condensed phase in passing to the gas phase.

The direct heat-transfer problem is nonlinear. The parabolic heat conduction equation (1.1) is to be solved over a region with a moving boundary S(t), the location of which is unknown. Heat-conduction problems of this type are encountered in melting, freezing and ablation processes. There are very few analytic solutions to this class of problems [7] which are called Stefan problems after J. Stefan, who in 1891 investigated the freezing problem in relation to polar ice formation.

By introducing a moving coordinate system

$$x(z,t) = z - S(t) \qquad [1.3]$$

for which the receding surface always remains at x=0, the heat-conduction equation can be written as

$$\frac{\partial}{\partial x}\left[k\frac{\partial T}{\partial x}\right] + g(x,t) = \rho c\left[\frac{\partial T}{\partial t} - V(t)\frac{\partial T}{\partial x}\right] \qquad [1.4]$$

$$t > o \qquad o < x < L_o - S(t),$$

where V(t) is the instantaneous burning rate. Equation (1.4) is to be solved subject to the following initial and boundary conditions

$$T(x,o) = T_o$$

$$-k\frac{\partial T}{\partial x} + Q_p = Q(t) \quad \text{or} \quad T = T_s \quad \text{at } x=0 \qquad [1.5]$$

$$Q_p = \rho \Delta Q\, V(t)$$

$$T = T_o \quad \text{as } x \to \infty.$$

If the source term g(x,t) and Q_p are negligible, the thermal conductivity, k, is a constant, and the system is in a steady-state, the temperature profile in the solid phase is given by

$$T(x) = T_o + Q\,\text{Exp}(-V_o x/\alpha)/V_o\,\rho\,c \qquad \text{or} \qquad [1.6]$$

$$T(x) = T_o + (T_s - T_o)\,\text{Exp}(-V_o x/\alpha)$$

where V_o denotes the steady-state burning rate and α is the thermal diffusivity ($k/\rho c$). Equation

(1.6) shows that the steady-state thermal wave which propagates into the solid is exponential in its spatial dependence. By plotting the natural logarithm of $T(x)-T_o$ versus x, the material's thermal diffusivity, α, can be determined from the slope and the surface heat flux, Q, or the surface temperature, T_s, estimated from the intercept. The steady-state burning velocity, V_o, is determined from the total combustion time and the grain length, L_o. Typical burning velocities are on the order of 1 to 5 mm/sec over a pressure range of 1 to 50 atm and α is on the order of 1.0×10^{-3} cm^2/sec.

Equation (1.6) has been derived upon the assumption that the thermal conductivity is a constant. Chen et al. [8] have refined the analysis such that this assumption is no longer required. Integrating Eqn. (1.4) over x and neglecting g(x,t) provides the following expression for the thermal conductivity k as a function of x or the measured temperature T(x)

$$k(T) = \int_{T_o}^{T(x)} (\rho\, V_o\, c dT)/\mathrm{grad}[T(x)], \qquad [1.7]$$

where V_o is the measured burning velocity and the local temperature gradient, grad[T(x)], is calculated from the measured subsurface temperature profile. If the density, ρ, and the specific heat, c, are evaluated from mass-averaged values, the thermal diffusivity is readily deduced from Eqn. (1.7). A distinct advantage of Eqn (1.7) is that it allows the estimation of the thermal properties as a function of temperature from <u>one</u> combustion wave experiment.

Equation (1.6) demonstrates that the combustion wave has an exponential spatial dependence. In materials where the thermal diffusivity, α, is small or the burning velocity is large, it may be difficult if not impossible to resolve the combustion wave structure. Moreover, the combustion wave experiment requires that several fine thermocouples be embedded in the grain and their locations be accurately determined. This can be tedious. If the propellant contains fine metal particles such as aluminum, then the thermocouple bead diameter must be smaller than these particles in order to obtain a meaningful temperature. The precise insertion of these fine thermocouples into a pyrotechnic can also be difficult.

PROPOSED EXPERIMENTAL TECHNIQUE

In this paper, we examine the theoretical feasibility of an alternative technique for determining the heat flux to a propellant or pyrotechnic surface based upon temperature-time measurements at one location. From an experimental point of view, the proposed technique is simpler. A fine wire of high thermal conductivity such as silver (α=1.70 cm^2/sec) is embedded along the grain's axis normal to the burning surface. Such samples could be prepared by pressing the pyrotechnic composition around the wire. The wire is longer than the pyrotechnic grain and is to be used as a thermal probe of the combustion front. The temperature measurements are taken at one known location X_1 on the wire, but <u>outside</u> of the grain. These temperature-time measurements are to be "inverted" in order to determine the heat flux to the surface. In a well-designed experiment, the wire should not perturb the normal combustion. This can be realized by using a wire of small cross sectional area, perhaps on the order of 1% of the grain's cross sectional area. Thus, the wire radius might be 1/10 of the grain's radius. The wire should also have a large contact resistance [9] with the surrounding material. This will minimize radial heat loss from the wire to the unreacted propellant. If the contact resistance is too small, the temperatures at the measurement station will be damped in time and preheating of the unreacted material may effect the burning velocity. The contact resistance could be regarded as a parameter to be estimated by inversion from the temperature-time measurements. Both the inverse and direct problem require a mathematical model of the proposed experiment.

MATHEMATICAL MODEL

Consider the one-dimensional transient heat conduction equation for a long slender metal wire of length L, cross sectional area A and perimeter P embedded in a propellant grain of length $L_o < L$

$$\frac{\partial^2 T}{\partial z^2} - hP/kA(T-T_o) = \frac{1}{\alpha}\frac{\partial T}{\partial t} \qquad [2.1]$$

$$t > o \qquad S(t) < z < L,$$

where α denotes the thermal diffusivity of the metal wire, k the thermal conductivity, and h denotes the reciprocal of the contact resistance [9] between the wire and the surrounding mixture. The z coordinate system is fixed in space. All physical properties of the metal are assumed to be independent of temperature. As indicated in the Introduction, the gas phase rapidly attains steady-state combustion and provides a heat flux back to the condensed phase. Since we are primarily interested in determining the sensitivity of the proposed experimental technique and estimating the order of magnitude of the parameters involved, the propellant combustion will not be modeled in detail at this time. Instead, the gas-phase propellant combustion will be modeled as a temporal heat flux, Q(t), of the form

$$Q(t) = Q_o\,(1-\exp[-\beta t]), \qquad [2.2]$$

where β is a gas-phase time constant. The unreacted portion of the propellant can act as a heat sink at temperature T_o. This interaction is modeled by the term $hP/kA\,(T-T_o)$ in Eqn. (2.1). If the contact resistance is large, this term can be neglected or it can be regarded as a parameter to be estimated from the temperature-time measurements.

At pressures below 10 atm, a typical propellant's surface temperature [10] will not be sufficient to melt the silver wire (MP 1235K). However, in the gas-phase combustion zone, the temperatures may exceed 2000 K. The burning rate is expected to be on the order of 1 to 2 mm/sec [10]. Therefore, there should be a range of parameters for L_o, L, X_1 and Q(t) over which the proposed technique can provide useful estimates of Q(t) from temperature measurements $T(X_1,t)$ without considering melting and locating the moving boundary. The exact details of the model formulation are dependent upon experimental observation of the system's behavior. One might also consider using a low melting point wire and including heat transfer through the liquid phase and solid phase to the sensor. Therefore, the subsequent mathematical development will include the possibility of phase transformations in the wire. By using wire materials of lower melting point, the movement of the interface, S(t), may follow the combustion wave.

Equation (2.1) is to be solved subject to the following conditions

$$T(z,0) = T_o \qquad [2.3]$$

$$-k\frac{\partial T}{\partial z} + \rho\,\Delta Q\,\frac{dS}{dt} = Q(t) \qquad \text{at } z = S(t)$$

$$\frac{\partial T}{\partial z} = o \qquad \text{at } z = L.$$

The middle equation expresses the fact that the heat flux Q(t) equals the rate of heat flow into the solid by conduction plus the rate of heat absorption by melting at the interface S(t). The conditions expressed in Eqn. (2.3) apply during both melting and nonmelting if at the heated surface z= S(t), we require that [11]

$$\frac{dS}{dt} > o \qquad T(S(t),t) = T_m \tag{2.4}$$

$$\frac{dS}{dt} = o \qquad T(S(t),t) < T_m,$$

where T_m is the wire's melting temperature. The moving boundary S(t) can be eliminated by the transformation [11]

$$x(z,t) = (L-z)/(L-S(t)). \tag{2.5}$$

The mathematical model now reduces to the following equations

$$f_2 \frac{\partial^2 T}{\partial x^2} - \gamma(T-T_o) = \frac{1}{\alpha}\left(\frac{\partial T}{\partial t} + f_1 \frac{\partial T}{\partial x}\right) \tag{2.6}$$

$$t > o \qquad o < x < 1$$

$$f_2 \equiv 1/[1-S(t)]^2 \qquad \gamma \equiv hP/kA \qquad f_1 = -\,xV(t)/[1-S(t)]$$

$$T(x,o) = T_o < T_m \qquad \frac{\partial T}{\partial x} = o \quad \text{at } x = o \ (z=L)$$

$$\frac{k}{[1-S(t)]}\frac{\partial T}{\partial x} + \rho\Delta Q \frac{dS}{dt} = Q(t) \ \text{ at } x=1 \ (z=S(t)) \qquad V(t) = \frac{dS}{dt}.$$

The nonlinearity of the model is evident from the fact that S(t) and its derivative occur as coefficients of the differential equation. A numerical technique was utilized to solve the direct problem.

NUMERICAL SOLUTION OF THE DIRECT PROBLEM

There are very few analytic solutions to Eqn. (2.6). Therefore, the model problem was solved by numerical integration. An algorithm based upon the method of lines [12] was developed and the spatial derivatives in Eqn. (2.6) were replaced by finite difference representations of the second order in Δx. The method of lines reduces the problem to a coupled system of fifty ordinary differential equations which are solved by a Gear method using IMSL software [13] on a VAX 6310 computer. This approach allows for rapid investigation of the effects of various Q(t) functions and input parameters such as α and Q_o on the temperature profile. The detailed numerical results of these calculations along with the computer code will be published and discussed in an NRL technical report [14].

However, for the benefit of the experimentalists, we have tabulated partial results of these calculations in Table I. The tabulation is intended to address three practical questions: 1) How large is the expected temperature $[T(X_1,t_j)]$ in the wire at location X_1 as a function of the heat flux parameter Q_o prior to melting ? 2) How much time is required for the onset of melting ? and 3) What is the propellant's expected temperature $[T_p(X_1,t_j)]$ if the sensor were embedded in it at the same

Table I

Q_o [cal/cm^2-sec]	300	400	500
$T(X_1,t_j)$ [K]	682	561	472
t_j [sec]	4.66	2.56	1.61
$T_p(X_1,t_j)$ [K]	300	300	300

location ? The tabulated results assumed a silver wire (α=1.70 cm^2/sec & MP=1235 K) of 10.0 cm length and a temperature sensor located at X_1 = 2.5 cm. The propellant's thermal diffusivity was taken as 1.0 x 10^{-3} cm2/sec and V(t) was roughly 1 x 10^{-1} cm/sec. The system's T_o = 300K.

THE INVERSE PROBLEM

In the inverse problem, we are expected to utilize the temperature-time data $T(X_1,t_j)$ measured at the known location X_1 and discrete times t_j to determine the heat flux $Q(t_j)$ to the propellant's surface. The thermal properties of the wire probe and the average burning velocity of the propellant are known.

This inverse problem is mathematically ill-posed [15]. An ill-posed problem is one in which the solution does not satisfy three criteria first discussed by Hadamard [16]. In order for the problem to be well-posed it is necessary that the solution (1) exist, (2) be unique, and (3) be continuously dependent on the data ie. stable. The existence of the solution has been proved by Weber [17]. Burggraf [18] and Langford [19] independently constructed exact solutions to the one-dimensional inverse problem. Their analysis decomposes the problem's spatial domain into two regions: an inverse region and a direct region. These regions are separated at X_1 the location of the temperature sensor. The solution is first determined in the direct region subject to the boundary condition at X_1. This solution may be constructed numerically or analytically. After this solution has been determined in the direct region, the heat flux at the sensor's location X_1 is computed from the direct solution's spatial derivative. The inverse region now has two boundary conditions specified at the same boundary X_1. The solution in the inverse region can be constructed as a Taylor series expansion about the sensor's location X_1. The coefficients in this series expansion are time derivatives of the temperature and heat flux at X_1. Thus the algorithm requires that all time derivatives of the sensor temperature and heat flux at X_1 be known. In applications with discrete data, this requirement limits the utility of the solution. However, the solution does provide insight into the IHCP and by uniqueness of power series provides a proof that the inverse problem considered here has a unique solution.

We now consider the question of stability [17]. By stability, we mean that the solution should be continuously dependent on the data. Qualitatively, this means that the differences between two possible solutions with slightly different data can be made arbitrarily small as the data sets approach each other. By using the exact Burggraf solution [18] with continuous data one can prove that the inverse solution is not always stable. The practical consequence of this observation is that small differences in the data set can produce arbitrarily large differences in the inverse solution. Therefore, the proposed experiment must incorporate appropriate statistics. For a more detailed discussion of these issues, consult chapter 4 of reference [2] or chapters 1 and 2 of reference [15].

There are several methods available for solving the inverse problem described here. However, due to space limitations, we can not discuss them in any detail. Beck et al. [2] describe and provide references to these methods, which include: (1) inverse convolution techniques based upon Duhamel's theorem (2) the function specification method in which a functional form for the unknown heat flux is assumed (3) the regularization method and (4) numerical methods based upon finite differences such as space marching.

Space marching is based upon the observation that the temperature and heat flux being both specified at the sensor location X_1 suggests that the IHCP can be related to an initial value problem in space. A number of numerical techniques have been published based upon finite difference representations of the spatial derivatives. The inverse problem is solved by space marching the known solution from the direct region into the inverse region. The approach is a numerical analogue of the Burggraf solution [18].

The approach taken here is based upon a function specification method. To solve the inverse problem, we assume functional forms of the surface heat flux Q(t) and S(t). Frank [20] suggested that the time-dependent heat flux be approximated by a polynomial series. The direct problem is solved using this flux model and the coefficients of the polynomial are determined by either matching the predicted model temperatures with the experimentally measured temperatures or using the method of least squares to estimate the coefficients which provide the best fit of the experimental temperature data at X_1.

As stated in the introduction, the gas phase rapidly attains a steady-state upon ignition ie. the time constant β in Eqn. (2.2) is large. Therefore, a reasonable functional form of the heat flux Q(t) is to approximate it by a constant Q_o for $t > 0$ over the observation time. Computations on the direct problem demonstrate that S(t) is very well approximated as a linear function of time. The numerical calculations on the direct problem also show that the thermal wave down the wire does not penetrate beyond 5.0 cm during the first phase of the heat transfer reported in Table I. To a good approximation, one can treat the finite wire as a semi-infinite medium. Under these conditions, the direct problem is linear and the solution can be obtained by Laplace transforms or Duhamel's theorem

$$T(x,t) - T_o = W(x,t)\ \mathrm{Exp}[-bx/2\alpha-(b^2/4\alpha + \gamma\alpha)t] \qquad [4.1]$$

$$W(x,t) = (Q_o-\rho\Delta Qb)\ \Phi(x,t)$$

$$\Phi(x,t) = \frac{2\sqrt{\alpha t}}{k}\left(\mathrm{Exp}(-y^2)/\sqrt{\pi} - y\ \mathrm{Erfc}(y)\right) + \ldots$$

$$y \equiv x/2\sqrt{\alpha t}$$

$$\mathrm{Erfc}(y) \equiv 2/\sqrt{\pi}\int_y^{\infty} \mathrm{Exp}[-\tau^2]d\tau$$

where $\gamma = hP/kA$, x is given by Eqn. (1.3), and b is the speed of the melting interface

$$S(t) = bt. \qquad [4.2]$$

Equation (4.1) shows that the radial heat transfer tends to attenuate the measured signal at X_1. The parameters γ and b can be estimated from the temperature-time measurements at X_1 by a least squares method. This would involve a nonlinear parameter estimation.

As a simple example of this calculation, we consider the estimation of Q_o from $T(X_1,t_j)$. Eqn. (4.1) provides a mathematical model of the direct solution. In the least squares optimization, we minimize the function

$$F(Q_o,b,\gamma) = \sum_{j=1}^{n} [T(X_1,t_j) - T_o - Q_o \Phi (X_1,t_j)]^2 \quad [4.3]$$

with respect to Q_o. For simplicity at the moment, we set γ and b to zero. Taking the first derivative of F with respect to Q_o and solving for the latter yields

$$Q_o = \frac{\sum_{j=1}^{n} \Delta T_j \Phi (X_1,t_j)}{\sum_{j=1}^{n} \Phi^2(X_1,t_j),} \quad [4.4]$$

Where ΔT_j denotes the temperature rise above T_o and the functions $\Phi(X_1,t_j)$ are evaluated at the experimental times and position. In general, to determine γ and b we would have to solve a nonlinear estimation problem. The general nonlinear problem and the effects of experimental noise will be considered in more detail in [14].

SUMMARY

A model to describe a proposed experimental technique for estimating the heat-flux to the condensed phase surface of a burning propellant or pyrotechnic of very low thermal diffusivity has been developed. Preliminary calculations (Table I) of the direct problem indicate an observable temperature rise is possible over the time scale of the proposed experiment. The design of an optimum experiment is still to be investigated. The proposed technique may work best for slowly burning (low pressure) pyrotechnic materials. The inverse problem is fairly simple. Based upon the assumption that the gas-phase combustion rapidly attains a steady-state, a function estimation procedure utilizing least-squares can provide an estimate of the heat-flux. Contact resistance and the combustion wave speed can also be estimated by the approach. Numerical details of the calculations will be reported elsewhere [14]. Finally, it should be noted that the time-dependent measurements of temperature at one location within a propellant of sufficient thermal diffusivity could also be use directly with the inversion algorithm proposed here to determine the heat flux and thermal diffusivity. Details of this calculation will also be given in [14].

ACKNOWLEDGEMENT

The support of the Office of Naval Research, Dr. A. Roberts is gratefully acknowledge.

REFERENCES

1. J. V. Beck, B. Blackwell: "Inverse Problems", in Handbook of Numerical Heat Transfer, ed. by W. J. Minkowycz, E. M. Sparrow, G. E. Schneider, R. H. Pletcher (Wiley, New York, 1988), pp. 787-834

2. J. V. Beck, B. Blackwell, C. R. St. Clair, Jr.: *Inverse Heat Conduction Ill-posed Problems* (Wiley, New York, 1985)

3. A. A. Zenin: Structure of Temperature Distribution in Steady-State Burning of a Ballistite Powder, Combustion, Explosion and Shock Waves (Fizika Goreniya i Vzryva), 2 , 67-76 (1966)

4. K. K. Kuo, J. P. Gore, M. Sumerfield: "Transient Burning of Solid Propellants", in *Progress in Astronautics and Aeronautics, v. 90, Fundamentals of Solid-Propellant Combustion*, ed. by K. K. Kuo, M. Sumerfield (AIAA, New York, 1984) pp. 599-651

5. M. Sumerfield, L. H. Caveny, R. A. Battista, N. Kubota, Gostintsev, A. Yu, H. Isoda: Theory of Dynamic Extinguishment of Solid Propellants with Special Reference to Nonsteady Heat Feedback Law, Journal of Spacecraft and Rockets, 8, 251-258 (1971)

6. M. N. Ozisik: *Heat Conduction* (Wiley, New York, 1980) pp. 4-6

7. H. S. Carslaw, J. C. Jaeger: *Conduction of Heat in Solids* (University Press, Oxford, 1980) pp. 282-296

8. D. M. Chen, W. H. Hsieh, T. S. Snyder, V. Yang, K. K. Kuo: Study of the Thermophysical Properties and Combustion Behavior of Metal-Based Solid Fuels, AIAA/ASME/SAE/ASEE 24th Joint Propulsion Conference, Paper No. AIAA-88-3041, Boston, MA, July 11-13 (1988); see also D. M. Chen: "Pyrolysis, Ignition, and Combustion of Solid Fuels for Ramjet Applications", Ph.D. Thesis, Department of Materials Science and Engineering, The Pennsylvania State University, August (1988)

9. P. J. Schneider: "Conduction" in *Handbook of Heat Transfer Fundamentals*, ed. by W. M. Rohsenow, J. P. Hartnett, E. N. Ganic (McGraw-Hill, New York, 1985) Chapter 4

10. G. Lengelle, A. Bizot, J. Duterque, J. F. Trubert: "Steady-State Burning of Homogeneous Propellants", reference 4, pp. 361-407

11. H. G. Landau: Heat Conduction in a Melting Solid, Quart. Appl. Math., 8, 81-94 (1950)

12. M. B. Carver: "The Choice of Algorithms in Automated Method of Lines Solutions of Partial Differential Equations", in *Numerical Methods in Differential Equations*, ed. by L. Lapidus, W. Schiesser (Academic Press, New York, 1976) pp. 243-265

13. IMSL routine DIVPAG, Version 1.0 (1987)

14. H. D. Ladouceur, NRL memorandum report in preparation

15. A. V. Goncharsky; "Ill-posed Problems and Their Solution" in *Ill-Posed Problems in the Natural Sciences*, ed. by A. N. Tikhonov, A. V. Goncharsky (Mir, Moscow, 1987) pp 21-52; available from Imported Publications, Inc. Chicago, IL

16. J. Hadamard: *Lectures on Cauchy's Problem in Linear Partial Differential Equations*, (Yale University Press, New Haven, 1923)

17. C. F. Weber: Analysis and Solution of the Ill-Posed Inverse Heat Conduction Problem, Int. J Heat Mass Transfer, 24, 1783-1792 (1981)

18. O. R. Burggraf: An Exact Solution of the Inverse Problem in Heat Conduction Theory and Applications, ASME J. Heat Transfer, 86C, 373-382 (1964)

19. D. Langford: New Analytic Solutions of the One-Dimensional Heat Equation for Temperature and Heat Flow Rate Both Prescribed at the Same Fixed Boundary (With Applications to the Phase Change Problem), Quart. Appl. Math., 24, 315-322 (1967)

20. I. Frank: An Application of Least Squares Methods to the Solution of the Inverse Problem of Heat Conduction, J. Heat Transfer, 85, 378-379 (1963)

Inversion of Three-Quark Spectral Data

H. Leeb[1], *H. Fiedeldey*[2], *R. Lipperheide*[3], *and W.A. Schnizer*[1]

[1]Institut für Kernphysik, Technische Universität Wien, Austria
[2]Physikalisches Institut, Universität Bonn, D-5300 Bonn, Fed. Rep. of Germany
[3]Hahn-Meitner-Institut, D-1000 Berlin, Germany

A method for the inversion of the spectrum of low-lying three-quark s-states within the hypercentral approximation is proposed. The procedure is a modification of the inverse spectrum method of Thacker, Quigg and Rosner and is tested on theoretically determined three-quark observables assuming a reasonable quark-quark interaction. The reconstruction of the hypercentral potential is sufficiently accurate in the physically relevant radial range 0.6 fm<r<1.5fm.

1. The Nonrelativistic Quark Model

Today it is commonly accepted that hadrons are composed of quarks, and that the new degree of freedom, color, plays an essential role in the physics of strong interactions. An important feature based on experimental experience is confinement which states that only color singlets can be observed. Thus mesons (quark-antiquark bound states) and baryons (bound states of three quarks) are the simplest quark states occurring in nature. There are strong indications that the underlying dynamics of color is given by QCD which should describe the properties of hadrons as well as the nuclear forces. Until now, however, no full solution of the QCD field equations is known, and consequently our understanding of the rich spectroscopy of hadrons is based on models and simulations.

The nonrelativistic quark model [1-3] has been widely and successfully applied. It is based on a quark-antiquark potential which can conceptually be justified in the limit of high quark masses. But contrary to expectations the model has also a surprisingly high predictive power for light-quark systems. Usually, the quark-antiquark interaction $V_{q\bar{q}}(r)$ is determined from the meson masses assuming rather simple parametrised potential shapes [1-3] and adjusting the parameters in order to reproduce the masses M_n and the leptonic widths Γ_n of the mesons. An indication for reasonable potential shapes is given by perturbation theory and lattice gauge calculations [4], which show a Coulombic term at small distances whereas the asymptotic behaviour is dominated by a monotonically increasing confinement term.

In order to calculate baryon properties the quark-quark interaction $V_{qq}(r)$ is commonly determined by the Lipkin rule,

$$V_{qq}(r) = \frac{1}{2} V_{q\bar{q}}(r) , \qquad (1)$$

which is based on group theoretical arguments. However, there are several reasons for a deviation from the relationship (1). i) The Lipkin rule is valid for the one-gluon exchange which is only one part of the interaction. ii) The nonrelativistic quark model deals with constitutent quarks which effectively contain relativistic and polarisation contributions. These contributions will be rather different in a meson from those in a baryon. iii) In a baryon three-quark forces can affect the bound states. All these effects together will lead to a violation of (1). From this point of view it is very important to determine $V_{qq}(r)$ independently.

It is the aim of this contribution to present a method which allows the determination of $V_{qq}(r)$ by inversion of baryon spectral data in a good approximation. Before we describe the method in detail in section 3, we discuss in section 2 the inversion of meson spectral data by the procedure of Thacker, Quigg and Rosner [5].

2. Inversion of Meson Spectral Data

The determination of the quark-antiquark potential from the meson spectra is in principle an inverse scattering problem of the Schrödinger operator which has been extensively studied [6]. Assuming a spherically symmetric quark-antiquark potential the eigenfunctions $u_n^{(\ell)}(r)$ of the radial Schrödinger equation,

$$\left\{ \frac{d^2}{dr^2} + \frac{2\mu_{q\bar{q}}}{\hbar^2} \left[E_n^{(\ell)} - V_{q\bar{q}}(r) - \frac{\ell(\ell+1)}{r^2} \frac{\hbar^2}{2\mu_{q\bar{q}}} \right] \right\} u_n^{(\ell)}(r) = 0 , \tag{2}$$

describe the mesons with orbital angular momentum quantum number ℓ. The eigenenergies $E_n^{(\ell)}$ are directly related to the meson masses M_n,

$$M_n c^2 = m_q c^2 + m_{\bar{q}} c^2 + E_n^{(\ell)} , \tag{3}$$

where n denotes the radial excitation, $\mu_{q\bar{q}}$ is the reduced mass of the quark-antiquark system and m_q, $m_{\bar{q}}$ are the masses of the quark and the antiquark, respectively. Because of the confinement property of $V_{q\bar{q}}$ the Schrödinger equation (2) has only a discrete spectrum. Hence all bound states together with their normalisation constants determine the underlying potential $V_{q\bar{q}}$ uniquely, which latter can be calculated by the Marchenko or the Gel´fand-Levitan equations. In practice, however, only the lowest-lying bound states are given. Because of this incomplete knowledge of the spectrum additional assumptions are required.

A very simple and elegant method for $\ell=0$ has been developed by Thacker et al. [5]. They truncate the spectrum to the known bound states by introducing a continuum threshold E_0 which corresponds to the asymptotic value of the potential $V^{TQR}(r=\infty)$. Furthermore, they consider a one-dimensional problem with a symmetric and transparent potential which can be directly related to the three-dimensional one for s-waves. Let us characterize the N bound states of the one-dimensional problem by the wave numbers κ_n, n=1,2,...,N. The odd-parity states in the one-dimensional problem correspond to s-state mesons with mass M_n,

$$M_n c^2 = E_0 - \frac{\kappa_{2n}^2}{2\mu_{q\bar{q}} c^2} , \quad 2n=2,4,...,N . \tag{4}$$

The wave numbers κ_n (n=1,3,...) of the even-parity states can be determined from the wave functions $\Psi_n(0)$ of the three-dimensional problem which can be evaluated for each s-state meson M(ns) from the leptonic width $\Gamma(M(ns) \rightarrow e^+e^-)$ via the Weisskopf-van Royen formula [7],

$$\Gamma(M(ns) \rightarrow e^+e^-) = 16\pi \frac{e_q^2\,\alpha^2}{M_n c^2}\, f_{QCD}\, |\Psi_n(0)|^2 . \tag{5}$$

Here e_q is the quark charge, α is the fine structure constant and f_{QCD} takes into account radiative chromodynamical corrections [8].

It has been pointed out by Thacker et al. [5] that the resulting potential depends strongly on the choice of E_0 and the quark masses m_q, which are open parameters in a constitutent quark model. In order to remove this deficiency we have embedded the inversion in a least square fit [9] by adjusting E_0 and the wave numbers κ_n of higher states which are usually poorly determined by experiment. Thus we can also include the information contained in the spectra of other partial waves and determine the potential more precisely in the relevant radial range. We have applied the method to a calculated spectrum which was obtained from a reasonable quark-antiquark potential (all quantities are given in MeV units),

$$V_{q\bar{q}} = -\frac{0.41}{r} + 100.0 + 40850.0\, r^{2/3}, \tag{6}$$

with a quark mass m_q=4550 MeV. In fact the inclusion of additional p-states fixes the truncation point E_0 and leads, apart from oscillations, to a good reproduction of the potential (Fig. 1).

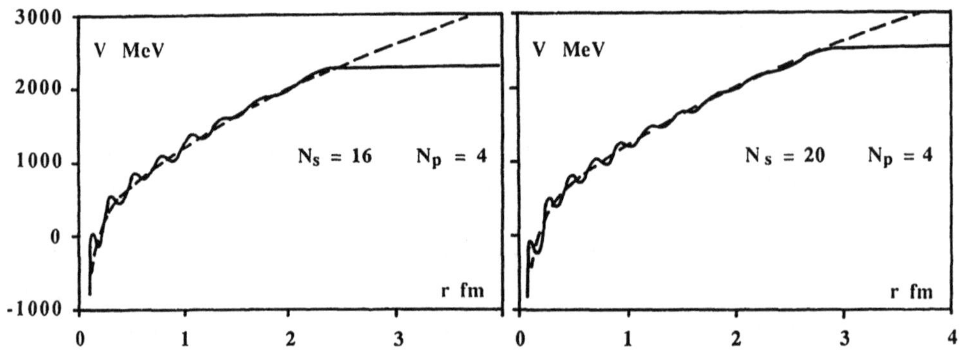

Fig. 1 Reproduction of the original potential from the spectral data by using the combined inversion and fitting procedure described in the text. The quantities N_s and N_p denote the number of the lowest-lying s- and p-states which have been included in the inversion. The dashed line shows the radial dependence of the original potential.

3. Inversion of Baryon Spectral Data

Another source of information on the interquark potential is provided by the spectroscopy of baryons which are composed of three quarks. The determination of the quark-quark interaction $V_{qq}(r)$ from the baryon spectra is therefore a three-body inverse spectrum problem. Recently, several procedures for the inversion of three-body scattering data have been proposed by Zakhariev [10]. These are, however, not suitable for our inverse spectrum problem. Instead we rely on the hypercentral approximation, which is valid in lowest order of the hyperspherical formalism [11], taking only that part of the interaction into account which is invariant under rotations in the 6-dimensional space of the relative coordinates. This hypercentral approximation has been shown to be excellent for the lower part of the baryon spectra [12]. The description of the three-body bound state problem reduces in this approximation to an effective Schrödinger equation for the hyperradial wave function $F_n^{(\mathcal{L})}(\rho)$. Restricting ourselves to baryons consisting of quarks with equal masses m_q, the effective Schrödinger equation in the hyperradius takes the form

$$\left\{ \frac{d^2}{d\rho^2} + \frac{m_q}{\hbar^2} \left[E_n^{(\mathcal{L})} - V_{HC}(\rho) - \frac{\mathcal{L}(\mathcal{L}+1)}{\rho^2} \frac{\hbar^2}{m_q} \right] \right\} F_n^{(\mathcal{L})}(\rho) = 0 , \tag{7}$$

where $\mathcal{L} = \ell + \frac{3}{2}$. The hypercentral potential $V_{HC}(\rho)$ is given by

$$V_{HC}(\rho) = \frac{6}{\pi} \int_0^{\pi} d\varphi \, \sin^2\varphi \, V_{qq}(\rho \cos\tfrac{\varphi}{2}) . \tag{8}$$

The hypercentral potential maintains the confinement property of $V_{qq}(r)$ and as in the meson case we have only a discrete spectrum of (6). Because of the smoothness of the quark-quark interaction the hypercentral approximation gives sufficiently accurate results for three-quark bound states [12-14].

In the following we restrict our considerations to s-state baryons, where the method works particularly well. In this case the hyperradial equation (6) is a Schrödinger equation with a centrifugal barrier corresponding to the effective angular momentum quantum number $\mathcal{L} = 3/2$. In principle, using the Marchenko or Gel'fand-Levitan integral equations, one can determine the potential uniquely from the knowledge of the whole spectrum. Again one needs the eigenenergies and the normalisation constants. Similar to the meson case the eigenenergies $E_n^{(\mathcal{L})}$ are directly related to the baryon masses M_n^B,

$$M_n^B c^2 = m_q c^2 + m_q c^2 + m_q c^2 + E_n^{(\mathcal{L})} . \tag{9}$$

For the normalisation constants no direct relationship with a physical observable is known today, because (6) is only an effective equation for the hyperradial part of the wave function. A physically relevant quantity is

$$J_n^B := \int d\rho \frac{1}{\rho^3} \left[F_n^{(\mathcal{L})}(\rho)\right]^2 \quad \text{with} \quad \int d\rho \left[F_n^{(\mathcal{L})}(\rho)\right]^2 = 1 \tag{10}$$

which can be determined from the mass splitting between $s = 3/2$ and $s = 1/2$ baryons [13]. The values J_n^B are not suitable for the construction of the spectral kernel of the Marchenko equation. Hence even with complete knowledge of all $E_n^{(\mathcal{L})}$ and J_n^B we must take recourse to a procedure which combines Marchenko inversion and fitting.

As in the meson case we have only limited knowledge of the baryon spectrum. In practice only two or three orbital excitations are known, and we require additional assumptions to fix the potential. Recognizing to this constrained uniqueness of the inversion of experimental baryon spectra we propose a simple approximation which is based on the procedure of Thacker et al. [5].

The main difference of the radial Schrödinger equation (2) for s-state mesons and the Schrödinger equation for the hyperradial wave function (6) for s-state baryons is the occurrence of a centrifugal barrier, which tends to infinity when ρ approaches 0. As a consequence the behaviour of the hyperradial wave function $F_n^{(\mathcal{L})}(\rho)$ near the origin is given by

$$F_n^{(\mathcal{L})}(\rho) \underset{r\to 0}{\longrightarrow} \alpha\, \rho^{5/2} \;, \mathcal{L} = 3/2 \;. \qquad (11)$$

The inversion method of Thacker et al.[5] uses the similarity of the radial Schrödinger equation for ℓ=0 with a one-dimensional problem. A straightforward extension of this concept to higher partial waves does not seem promising due to the infinite barrier at the origin. Therefore we use the original procedure treating the s-state baryon spectrum as an $\mathcal{L} = 0$ problem. In this case the resulting potential from the inversion procedure should correspond to the sum of hypercentral and centrifugal potentials (6). However, a complete reproduction of the centrifugal potential is not possible because the inversion of Thacker et al. leads to a finite value of the potential at the origin. The corresponding bound-state wave functions reflect the property of s-states being proportional to ρ near the origin. Using these bound-state wave functions in (9) would lead to completely wrong values of J_n^B because of the wrong behaviour of the integrand near the origin. In order to correct for this deficiency we must modify the wave function near the origin to obtain the required $\rho^{5/2}$ behaviour thus guaranteeing the correct value of J_n^B . In the formalism of Thacker at al. [5] we know the wave functions in terms of κ_n analytically. Changing the wave functions at a certain matching radius, say ρ_m, to a $\rho^{5/2}$ behaviour we are in principle able to renormalize the wave function and to evaluate the integral analytically. However, the equations will become so complicated that it will hardly be possible to extract the unknown κ-values of the even-parity states from these relations.

Actually we perform the inversion in a more economic way numerically. We consider a problem with N s-state baryons. Assuming that we know all 2N wave numbers κ_n of the associated one-dimensional problem we can calculate the potential $V^{TQR}(\rho)$ by the standard procedure of Thacker et al. [5]. In the next step we modify the potential near the origin assuming the hypercentral potential to be constant below a certain matching radius ρ_m and obtain an effective interaction $V_{eff}(\rho)$ for the hyperradial Schrödinger equation,

$$V_{eff}(\rho) = \begin{cases} V^{TQR}(\rho_m) + \dfrac{15\,\hbar^2}{4\,m_q}\left(\dfrac{1}{\rho^2} - \dfrac{1}{\rho_m^2}\right) & \rho < \rho_m \;, \\ V^{TQR}(\rho) & \rho > \rho_m \;. \end{cases} \qquad (12)$$

With this potential the eigenenergies and eigenfunctions of (6) are calculated numerically. Since $V_{eff}(r)$ differs from $V^{TQR}(\rho)$ only near the origin we expect the eigenenergies to remain essentially unchanged. In fact this is true as can be seen in our example (Tab. 1). In contrast to the wave function obtaind by

Table 1 Deviation ΔM_n^B of the masses M_n^B of the lowest-lying three-quark s-states due to the modification (11) of the potential near the origin. The exact values of M_n^B and J_n^B are determined by the hypercentral approximation using the potential (14). The calculations denoted by N=2 and N=3 are the deviations ΔM_n^B of the best fits assuming a matching radius ρ_m=0.2 fm and inverting the 2 and 3 lowest lying three-quark states with our approximation, respectively.

n	M_n^B MeV	J_n^B MeV	ΔM_n^B (N=2)	ΔM_n^B (N=3)
1	1617.56	0.2206 107	8.79	5.08
2	1892.65	0.2094 107	2.11	1.56
3	2130.49	0.2032 107		2.35

the inversion the new wave functions associated with $V_{eff}(\rho)$ have the correct behaviour at the origin and lead to reasonable values of the integrals J_n^B. Performing a least square fit on M_n^B and J_n^B we adjust the wave numbers κ_n of the associated one-dimensional problem. Finally we determine the hypercentral potential $V_{HC}(\rho)$ from the optimized potential $V_{eff}(\rho)$ by

$$V_{HC}(\rho) = V_{eff}(\rho) - \frac{15}{4}\frac{\hbar^2}{m_q}\frac{1}{\rho^2}. \qquad (13)$$

In order to test our procedure we construct three-quark data from a reasonable quark-antiquark potential

$$V_{q\bar{q}}(r) = -\frac{0.4}{r} + 100 + 0.4 \cdot 10^5 r \ , \quad r \text{ in MeV}^{-1}. \qquad (14)$$

Assuming the Lipkin rule (1) we obtain for the hypercentral potential

$$V_{HC}(\rho) = -\frac{16}{\pi}\frac{0.4}{\rho} + 300 + \frac{32}{5\pi} 0.4 \cdot 10^5 \rho \ , \quad \rho \text{ in MeV}^{-1}. \qquad (15)$$

With the quark mass m_q=300 MeV we calculate the three-quark s-state masses M_n^B and integral values J_n^B given in Tab. 1. As mentioned above only 2 to 3 orbital excitations are known in realistic systems. Therefore we perform the inversion with 2 and 3 states using different matching radii 0.1fm < ρ_m < 0.4fm. In all cases we obtain perfect agreement of the masses M_n^B and the integral values J_n^B. As can be seen from Fig. 2 the inverted potential (12) reproduces the original hypercentral potential in the radial range 0.65fm < ρ < 1.45 fm quite well. Due to the approximation (11) the deviation at smaller ρ-values is not surprising. Since we use a truncated spectrum it is also not unexpected that we cannot reproduce the hypercentral potential at large ρ-values. This result is very promising because we can determine the potential in the radial range where also meson data are sensitive, thus allowing a comparison of $V_{qq}(r)$ and $V_{q\bar{q}}(r)$, which have been obtained independently.

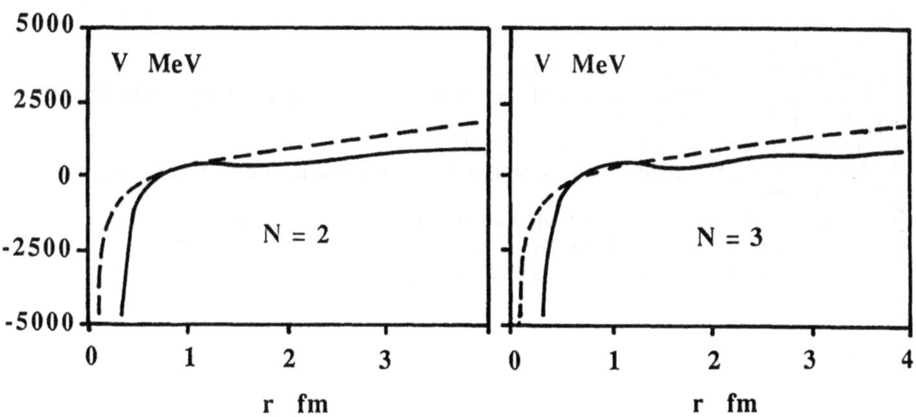

Figure 2 Comparison of the inverted hypercentral potenial (solid line) obtained from three-quark data M_n^B and J_n^B with the original hypercentral potential (dashed line). The matching radius is fixed at ρ_m=0.2 fm. The resulting potentials of the inversion of the N=2 and N=3 lowest-lying three-quark states are shown.

4. Summary

In this contribution we have developed and tested an approximation which allows the inversion of three-quark s-state observables. Our method is based on the hypercentral approximation and uses a numerical modification of the inversion formalism of Thacker et al. [5]. The modification corrects for the centrifugal barrier occuring in the hypercentral approximation but does not appreciably influence the three-quark bound states. Using three-quark data obtained from a reasonable quark-antiquark potential of the Cornell type [1] we succeeded in reproducing the hypercentral potential in a radial range of about 1 fm. This radial range coincides with the radial range where the quark-antiquark interaction is determined by meson data [9]. Three-quark states are based on the quark-quark potential, and thus we can directly test the Lipkin rule. Because of the more complicated structure of the three-quark observables we must embed the inversion procedure in a least square fit. Similar to the meson case this allows the inclusion of information from other partial waves. In particular, as we have shown for mesons in section 2, this additional information can be used to determine the truncation point E_0 of the formalism of Thacker et al. [5].

Acknowledgement

The work has been supported by Fonds zur Förderung der wissenschaftlichen Forschung in Österreich (Project Nr. P6608)

References

1. E. Eichten et al., Phys. Rev. **D17**, 3090 (1978); **D21**, 203 (1980).
2. A. Martin, Phys. Lett. **93B**, 338 (1980).
3. J. L. Richardson, Phys. Lett. **82B**, 272 (1979).
4. H.-Q. Ding, Phys. Lett. **200B**, 133 (1988).
5. H. B. Thacker et al., Phys. Rev. **D18**, 274 (1978); **D18**, 287 (1978).
6. K. Chadan, P. C. Sabatier, Inverse Problems in Quantum Scattering Theory, Springer, 1989.
7. V. van Royen, W. Weisskopf, Nuovo Cimento **50**, 617 (1967).
8. R. Barbieri, Phys. Lett. **B57**, 455 (1975).
9. H. Leeb et al., Proc. of the International Conference on Hadron Spectroscopy - HADRON 89, Ajaccio, 23 - 27. 9. 1989, France.
10. B. N. Zakhariev, Few-Body Systems **4**, 25 (1988).
11. M. Fabre de la Ripelle, Models and Methods in Few-Body Physics, Lecture Notes in Physics **273**, 283 (1987).
12. M. Fabre de la Ripelle et al., Few-Body Systems, Suppl. **2**, 493 (1987).
13. A. M. Badalyan, Phys. Lett. **B199**, 267 (1987).
14. M. Fabre de la Ripelle, Phys. Lett. **205B**, 97 (1988).

Stability and Convergence in the 1D Time-Dependent Problem

A. Roger

Laboratoire d'Electromagnétisme, Micro-ondes et Optoélectronique,
23 Avenue des Martyrs, B.P. 257, F-38016 Grenoble Cédex, France

I. INTRODUCTION

This paper deals with the one dimensional electromagnetic inverse problem for a stratified dielectric medium illuminated at normal incidence by an electromagnetic pulse. The reflected field is measured and hence one tries to retrieve the permittivity profile of the medium.
Many authors have studied this inverse problem and a very large number of papers is available on the subject. Without trying to be exhaustive, we will only mention approaches based on the similar quantum mechanics inverse problem [1], [2], direct marching-in-time reconstructions [3] [4], and optimization techniques [4] [5]. Much work has also be done in Geophysics for the inversion of sismograms [6] [7]. In this paper we first use the Liouville transform commonly used in Geophysics so as to replace the ordinary space variable x by the propagation time variable τ, which yields a simple representation of the problem. Then we establish an integral relation following the approach usual in Electromagnetics. This integral relation is examined in details and gives much insight into the physics of the problem. Then the numerical experiments are conducted with the so-called Goupillaud model [8] constituted of elementary homogeneous layers with constant thickness in the τ space. We show that with the help of our integral relation we are able to implement efficient numerical techniques and to build a fast optimization algorithm for the retrieval of the permittivity profile.

II. THE PHYSICAL PROBLEM

A plane electromagnetic wave is incident upon a dispersionless stratified medium, the permittivity $\varepsilon(x)$ of which depends only on the x coordinate (figure 1). For the sake of simplicity, the conductivity is assumed to be zero and the magnetic permeability equal to μ_0 like in vacuum. The incident electric field is parallel to the Oz axis, so that the same property holds for the total electric field $\vec{E}$:

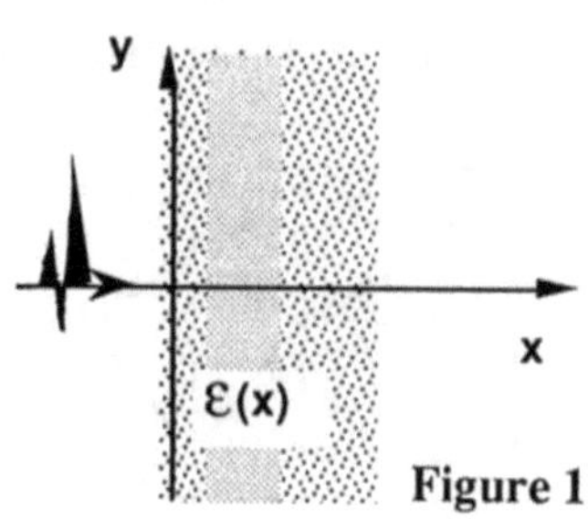

Figure 1

$$\vec{E}(x,t) = E(x,t)\,\vec{e}_z$$

From Maxwell's equations and the well-known continuity conditions, it is easily shown that E(x,t) obeys the following propagation equation :

$$\frac{\partial^2 E}{\partial t^2} - \frac{1}{\varepsilon(x)\,\mu_0}\frac{\partial^2 E}{\partial x^2} = 0$$

and that E(x,t) and $\partial E(x,t)/\partial x$ are continuous functions of x.
The direct problem is that of computing the total field E(x,t) given the incident field $E_i(x,t)$ and the permittivity profile $\varepsilon(x)$. The inverse problem is that of reconstructing $\varepsilon(x)$ from the field E(0,t) at the surface x = 0, or from related measurable quantities.
Following Kay [9], we now perform a Liouville transformation and use instead of x the propagation time τ defined by :

$$\tau := \int_0^x \sqrt{\mu_0\,\varepsilon(x')}\;dx'$$

The variable τ has a clear physical meaning : it is the optical path divided by the speed of the light in vacuum. The propagation equation for E now writes :

$$\frac{\partial^2 E}{\partial t^2} - \frac{\partial^2 E}{\partial \tau^2} = r(\tau)\,\frac{\partial E}{\partial \tau}$$

$$r(\tau) := -\frac{1}{2}\,\mu_0\,(\mu_0\,\varepsilon)^{-\frac{3}{2}}\,\frac{d\varepsilon}{dx}$$

In addition $E(\tau,t)$ satisfies some initial conditions :

$$E(\tau,0) = 0 \qquad \frac{\partial E}{\partial t}(\tau,0) = 0 \qquad \forall \tau \in [0,T]$$

$$\frac{\partial E}{\partial t} - \frac{\partial E}{\partial \tau} = \frac{\partial E_i}{\partial t} - \frac{\partial E_i}{\partial \tau} \qquad \tau=0,\ \forall t \in [0,T]$$

where E_i is the incident field. These equations suggest that the medium should be described by a reflection function $r(\tau)$ rather than by a permittivity profile. The model is valid even when $\varepsilon(\tau)$ is then a discontinous function : since Maxwell's equations are valid in the sense of distributions, the some property holds for the above equations and a discontinuous permittivity only implies the presence of Dirac distributions.

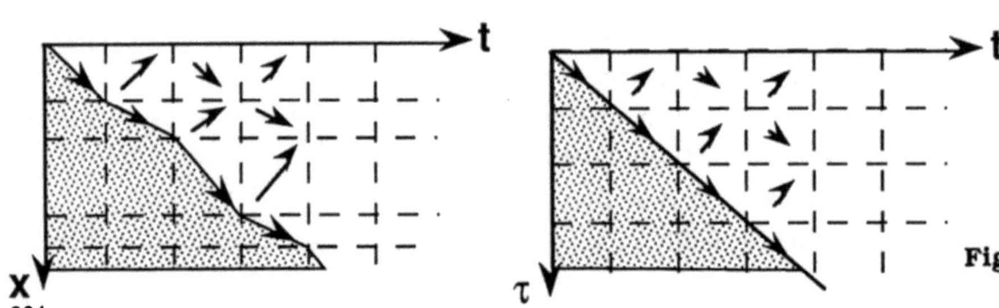

Figure 2

Not only the propagation time is convenient for the theoretical analysis, but moreover it makes the numerical implementation simpler. Indeed, in any numerical model the medium is represented as a stack of elementary homogeneous layers. In a (t,z) model with equally spaced points in t and z, each elementary layer is crossed with a different time interval and the computation of the field requires the solving of an integral equation or the using of a many-points differentiation scheme. On the opposite, in a model with equally spaced points in t and τ the propagation equation for the field can be very easily integrated because the waves propagate at speed one, the permittivity being taken into account by the reflection properties of the interfaces between layers. This model is known in Acoustics as "Goupillaud model" [8] and is widely used in Geophysics.

The propagation equation for E can be easily transformed with the help of the elementary solution of the free space wave equation. An integro-differential relation is found as :

$$E(\tau,t) = E_i(\tau-t) + \frac{1}{2}\int_{\tau-t}^{(t+\tau)/2} d\tau'\, r(\tau') \int_0^{t-|\tau-\tau'|} dt' \frac{\partial E}{\partial \tau}(t', \tau')$$

It is more interesting to work with a purely integral relation. Differentiating the above equation with respect to τ leads to :

$$\frac{\partial E}{\partial \tau}(\tau,t) = \frac{\partial E_i}{\partial \tau}(t-\tau) + \int_{(\tau-t)/2}^{(\tau+t)/2} d\tau'\, r(\tau') \frac{\partial E}{\partial \tau}(\tau', t-|\tau'-\tau|)$$

This relation is new in Electromagnetics at least to our knowledge. Several facts may be hence inferred. Let us first remark that $\partial E/\partial\tau$ is clearly identified and measurable since this quantity is proportional to the time derivative of the component B along the Oy axis of the magnetic field :

$$\frac{\partial E}{\partial \tau} = c\frac{\partial B}{\partial t}$$

As a consequence of this integral relation, a slowly varying medium (with $r(\tau) \ll 1$) cannot be distinguished from an homogeneous medium ($r(\tau) \equiv 0$) with any constant permittivity profile ; also, at first order in r, it gives rise to a field equal to the convolution of $r(\tau)$ by the incident field (just replace $\partial E/\partial\tau$ in the integrand by $\partial E_i/\partial\tau$). This approximation is very popular in Geophysics and allows one to implement linear inversion procedures.

It is worth noting that this integral relation states a new and important result : **the reflected field is rigorously equal to the convolution of the reflection function by the total field along certain caracteristic lines.**

III. THE OPTIMIZATION APPROACH

Following a widely used technique, the function $r(\tau)$ is searched for by performing the minimization of an objective function $J(r)$ defined as :

$$J(r) = \int_0^T \left| \frac{\partial E}{\partial \tau}(\tau = 0,t,r) - \frac{\partial E_i}{\partial \tau} - \frac{\partial E_0}{\partial \tau}(t) \right|^2 dt$$

where $\partial E_0/\partial t$ is the measured value of the reflected field at the surface, and $\partial E/\partial t$ is estimated using the given reflection function r. We used the classical Fletcher-Reeves conjugate gradient to minimize J(r). The gradient was computed using an adjoint state as usual [10] [11] [4].

The gradient may be viewed as the first order term of the Taylor expansion of the objective function. In order to perform a more subtle analysis, let us write down this expansion up to order two :

$$J(r + s) = J(r) + (G,s) + \frac{1}{2}(s,Hs) + \ldots$$

where s is a small perturbation of r, (.) denotes a scalar product and (s,Hs) is a bilinear symmetric form ; H is called the <u>Hessian</u>.

Most of optimization algorithms have been built using the example of **quadratic** objective functions, i.e. functions with constant Hessians ; moreover they are generally more efficient when the Hessian is close to identity. So one may try to perform a linear change in the space of parameters (here in the space of reflection functions) in order to change the features of the Hessian ; this mathematical trick is well known in Applied Mathematics and is called **preconditioning**. More precisely, if one defines the new variable r' by :

$$r = D\,r'$$

then the new objective function and its Hessian are :

$$J'(r') = J(D\,r') \quad , \quad H' = D^t H D$$

The point is now : how to choose D so that H' be close to identity without computing the eigensystem of H ? The theorems available in the mathematical literature do not give the answer. Fortunately our physical analysis does. In view of the definition of the objective function and of our integral relation, it is obvious that setting :

$$r'(t) := \int_{-t/2}^{t/2} d\tau'\, r(\tau') \frac{\partial E}{\partial \tau}(\tau', t-|\tau'|)$$

yields a new objective function perfectly quadratic in r' with diagonal Hessian. However things are not that simple : the actual value of $\partial E/\partial \tau$ is unknown since r(t) is also unknown. But we want to build an algorithm which will hopefully provide a sequence r_i converging towards r ; thus we may use the corresponding approximate value of $\partial E/\partial \tau$ to implement the above change of variable.

One may foresee that this transform will have several effects : first the new Hessian will be less varying during the iterative process and nearly diagonal (at least if the algorithm converges !) ; thus the speed of convergence should be much increased. Second, in order to obtain physical quantities, we need to go back from r' to r ; since the transformation from r' to r is a smoothing one, the backwards transformation will enhance the noise, if any is present on r'. This feature is all the more pronounced as the kernel of the integral transformation, namely $\partial E/\partial \tau$ along certain lines, is smooth, i.e. as the incident wave is narrow-banded. The numerical experiments come up to these expectations, as shown below.

V. NUMERICAL EXPERIMENTS

Figure 3 shows that the reconstructed profile with preconditioning is indeed more noisy (the initial data have been added a 20% white noise) than without preconditioning.

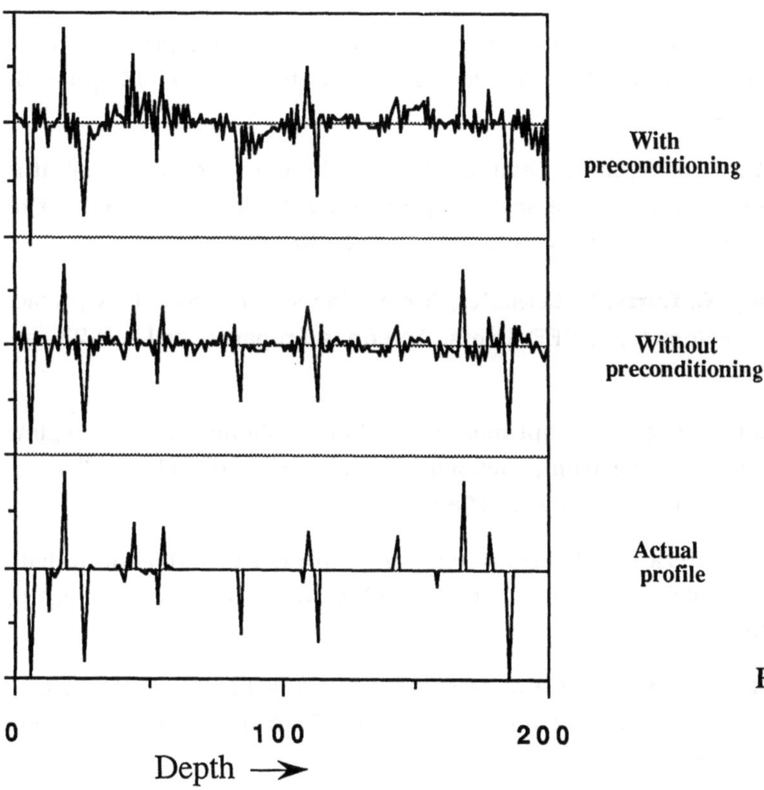

Figure 3

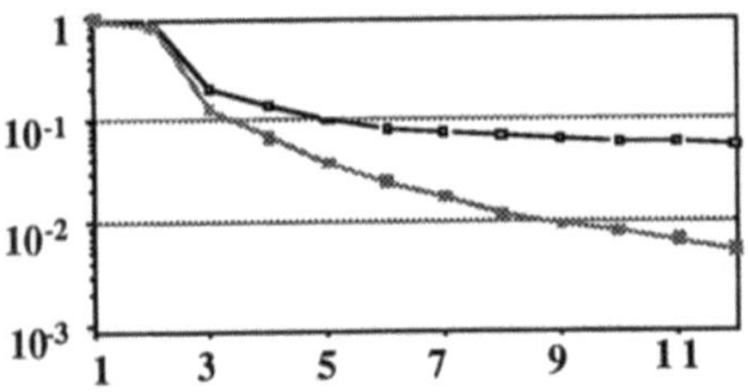

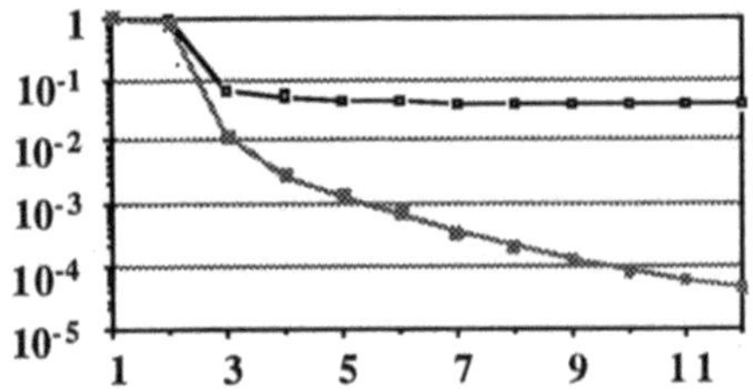

Figure 4

Figure 4 describes the decreasing of the objective function in four cases : at left without preconditioning using uncorrupted data and 30% noisy data, and at right with the same data but using preconditioning,. It is clear that preconditioning enhances considerably the convergence speed of the algorithm, even in presence of noise.
Thus one has to find a trade-off between the speed of convergence and the stability of the algorithm.

REFERENCES

[1] **Weston V.H., Krueger R.J.,** "On the inverse scattering problem for a hyperbolic dispersive partial differential equation",J. Math Phys, vol 14, pp 406-408, 1972.

[2] **Burridge R.,** "The Gelfan'd Levitan, the Marchenko, and the Gopinath-Sondhi integral equation of inverse scattering theory, regarded in the context of inverse impulse-response problems", Wave Motion 2, 305-323, 1980

[3] **J.C. Bolomey, C. Durix, D. Lesselier,** "Determination of conductivity profiles by time-domain reflectometry" ,IEEE Trans. Antennas Propagat., vol. AP-27, pp. 244-248, 1979

[4] **D. Lesselier,** "Diagnostic optimal de la lame inhomogène en régime temporel. Applications à l'électromagnétisme et à l'acoustique", Ph. D. Thesis, Université Pierre et Marie Curie, Paris, 1982

[5] **A.G. Tijhuis,** "Iterative determination of permittivity and conductivity profiles of a dielectric slab in the time domain", IEEE Trans. Antennas Propagat., vol. AP-29, pp. 239-245, 1981

[6] **Mendel J.M., Habibi-Ashrafi F.,** " A survey of approaches to solving inverse problems for lossless layered media systems", IEEE Trans. Geosci. Remote Sensing, vol. grs-18, pp. 320-330, 1980

[7] **P. Kolb, G. Canadas,** "Least squares inversion of prestack data : simultaneous identification of density and velocity of stratified media" ,56th annual SEG Meeting, Houston, 1986

[8] **P.L. Goupillaud,** "An approach to inverse filtering of near-surface layer effects from seismic records", Geophys., vol. 26, pp. 754-760, 1961

[9] **Kay I.,** "The inverse scattering problem", Research Rept. EM-74, New York University, New York, 1955

[10] **J.L. Lions,** "Contrôle optimal des systèmes gouvernés par des équations aux dérivés partielles", Dunod, Gauthier Villars, Paris 1968, (English Translation : Springer, New York, 1982)

[11] **A. Roger,** "Reciprocity theorem applied to the computation of functional derivatives", Electromagnetics, vol. 2, pp. 69-83, 1982

Inverse Profiling for an Inhomogeneous, Plane-Stratified Lossy Causal Medium

A.G. Tijhuis

Laboratory of Electromagnetic Research, Department of Electrical Engineering,
Delft University of Technology, P.O. Box 5031, 2600 GA Delft, The Netherlands

1 Introduction

In this contribution, we consider the inverse-profiling problem for a one-dimensionally inhomogeneous, lossy layer of given thickness embedded in free space. The reflected field caused by one or more plane waves incident from one or both sides of the layer is assumed to be known. The waves may be of an electromagnetic or acoustic nature. The aim of the computation is to reconstruct the spatial distribution of one or more material parameters of the layer (profiles) from these reflected-field data.

With present-day computer systems, the solution of multidimensional inverse-profiling problems has come within reach [1]. Nevertheless, one-dimensional problems are still of interest, especially for developing and analyzing inversion schemes. Furthermore, keeping the configuration simple allows us to take into account other aspects that are important in a "practical" approach:

- We consider a realistic *time signal.* In particular, we restrict the analysis to incident pulses with a high-frequency *band limitation*, and we add *noise.* Further, we systematically generate the "known" reflected-field data by a forward scheme that is independent from the one on which the inversion scheme is based, and we use only algorithms for which the error can be estimated.
- We consider a realistic *configuration*, where the material parameters are *discontinuous* at the *known* obstacle boundaries. In view of the band limitation mentioned above, the location of these discontinuities must be available as *a priori* information in the inversion. This means that we avoid Liouville transformations, and use the *actual* space coordinates. Thus, we also keep open the possibility of a generalization to more complicated geometries.
- To *validate* our results, we justify *all approximations*, and we compare the reconstructions from different inversion procedures. Further, we use the *natural frequencies* of the object, which can be obtained directly from the scattered field.

In the present contribution, we consider whether the inverse-profiling problem outlined above can be solved by using contrast-source integral representations that express the known reflected fields in terms of the unknown fields and the material parameters inside the layer. In these integral relations, we apply the *Born-type approximation* of making some educated guess about the unknown field quantity inside the layer. This leads to one or two integral equations of the first kind, whose solution yields an approximation to the unknown profiles. When the contrast between the unknown medium and free space is small, this procedure directly yields an acceptable reconstruction. For larger contrasts, an iterative procedure is employed, in which successive approximations for the unknown field inside the layer and the unknown material parameters are obtained by alternately solving an approximate direct-scattering problem and an approximate inverse-scattering problem.

In addition, we outline various techniques for solving the underlying direct-scattering problem in the time domain as well as in the frequency domain. Finally, we discuss the validation of our conclusions by using alternative inversion techniques such as Prony's algorithm and nonlinear optimization methods. Because of space limitations, it has not been possible to include an extensive literature review or numerical results. For a historical perspective on one-dimensional inverse profiling, the reader is referred to a [2,3]. Numerical results are presented in the references cited throughout the text.

2 Formulation of the Problem

In this contribution, we consider acoustic or electromagnetic plane waves, normally incident on a one-dimensionally inhomogeneous, lossy layer of given thickness embedded in a homogeneous, lossless "exterior" medium. According to the nature of the waves, different material parameters influence their propagation. In this section, we review the systems of differential equations that describe both types, and we reduce them to a single, generic form.

2.1 Acoustic Waves in a Viscoelastic Fluid

As explained in [4], the propagation of an acoustic plane wave in a one-dimensionally inhomogeneous, linearly reacting, isotropic *viscoeleastic fluid* is governed by the first-order space-time differential equations

$$\partial_z p(z,t) + \rho(z)\,\partial_t^2 u_z(z,t) = 0, \tag{2.1}$$

$$\partial_z u_z(z,t) - \theta(z,t) = 0, \tag{2.2}$$

and the constitutive relation

$$\theta(z,t) = -\int_{-\infty}^{t} J(z,t-t')\,p(z,t')\,dt', \tag{2.3}$$

where z is the space coordinate (m), t the time coordinate (s), $p(z,t)$ the acoustic pressure (Pa), $\rho(z)$ the volume density of mass (kg/m^3), $u_z(z,t)$ the longitudinal particle displacement (m), $\theta(z,t)$ the bulk strain (dimensionless), and $J(z,t)$ the linear creep compliance (Pa^{-1} s^{-1}).

In the present analysis, we restrict ourselves to the case where the volume density of mass is constant, i.e. $\rho(z) = \rho_0$, and where the creep compliance is given by the *Maxwell model*:

$$J(z,t) = \left[J_g(z) + \frac{t}{\eta(z)}\right] U(t), \tag{2.4}$$

where $J_g(z)$ denotes the instantaneous creep compliance ($\mathrm{Pa}^{-1}\mathrm{s}^{-1}$, g for glass), $\eta(z)$ the steady-state shearing viscosity (Pa s^2), and $U(t)$ a unit time-step function (dimensionless).

Substituting (2.3) and (2.4) in (2.2), and eliminating $u_z(z,t)$ from (2.1) and (2.2) result in the second-order differential equation

$$\partial_z^2 p(z,t) - \rho_0\, J_g(z)\, \partial_t^2 p(z,t) - \frac{\rho_0}{\eta(z)}\, \partial_t p(z,t) = 0. \tag{2.5}$$

The restriction to $\rho(z) = \rho_0$ is not essential, and is introduced to simplify the formulation.

2.2 Electromagnetic Waves in a Lossy Delectric

Electromagnetic waves in a plane-stratified, linearly and instantaneously reacting, isotropic, lossy *dielectric medium* are described by the first-order differential equations

$$\partial_z E_y(z,t) - \mu(z)\, \partial_t H_x(z,t) \;=\; 0, \tag{2.6}$$

$$\partial_z H_x(z,t) - \varepsilon(z)\, \partial_t E_y(z,t) - \sigma(z)\, E_y(z,t) \;=\; 0, \tag{2.7}$$

where $E_y(z,t)$ is the transverse electric-field strength (V/m), $\mu(z)$ the permeability (H/m), $H_x(z,t)$ the transverse magnetic-field strength (A/m), $\varepsilon(z)$ the permittivity (F/m), and $\sigma(z)$ the conductivity (S/m). In most practical dielectric materials, we have $\mu(z) = \mu_0$, with μ_0 being the permittivity of free space (vacuum). Restricting ourselves to that situation, we eliminate $H_x(z,t)$ from (2.6) and (2.7). This leads to the second-order differential equation

$$\partial_z^2 E_y(z,t) - \mu_0\, \varepsilon(z)\, \partial_t^2 E_y(z,t) - \mu_0\, \sigma(z)\, \partial_t E_y(z,t) = 0, \tag{2.8}$$

which has the same structure as (2.5).

2.3 Generic Formulation

In both the acoustic and the electromagnetic case, we consider the situation where the medium is homogeneous and lossless outside a finite domain $0 < z < d$. A normalization involving the wave speed in the "exterior" medium and the layer thickness d reduces each of the second-order differential equations (2.5) and (2.8) to the form

$$\partial_z^2 \varphi(z,t) - [1 + \chi(z)]\, \partial_t^2 \varphi(z,t) - A(z)\, \partial_t \varphi(z,t) = 0, \tag{2.9}$$

where the *contrast functions* $\chi(z)$ and $A(z)$ vanish outside the interval $0 \leq z \leq 1$. The field $\varphi(z,t)$ corresponds to either the pressure $p(z,t)$ or the electric-field strenght $E_y(z,t)$. The function $\chi(z)$ represents the contrast in velocity between the layer and the surrounding medium. The function $A(z)$ represents the local attenuation; for a passive medium, we have $A(z) \geq 0$.

For the incident field, we take one or both of the pulses

$$\varphi_+^i(z,t) = F(t - z), \quad \varphi_-^i(z,t) = F(t - (1 - z)), \tag{2.10}$$

where $F(t)$ vanishes outside the domain $0 < t < T$. The subscripts + and − refer to the direction of propagation of these fields. The corresponding total fields are denoted as $\varphi_\alpha(z,t)$, where α stands for either of the labels + and − .

For $\alpha = +$ and $\alpha = -$, the incident field originates from $z < 0$ and $z > 1$, respectively. These regions will be referred to as the *region of incidence* $\mathcal{D}^i_\alpha$. The half-space on the opposite side of the layer will be called the *region of transmission* $\mathcal{D}^t_\alpha$. Using these definitions, we can identify the reflected and transmitted fields as

$$\varphi^r_\alpha(z,t) = \varphi_\alpha(z,t) - \varphi^i_\alpha(z,t) \quad \text{in } \mathcal{D}^i_\alpha, \tag{2.11}$$

$$\varphi^t_\alpha(z,t) = \varphi_\alpha(z,t) \quad \text{in } \mathcal{D}^t_\alpha. \tag{2.12}$$

For the reflected and transmitted fields, we have the relations

$$\varphi^r_\alpha(z,t) = \varphi^r_\alpha(z^i_\alpha, t - |z - z^i_\alpha|) \quad \text{in } \mathcal{D}^i_\alpha, \tag{2.13}$$

$$\varphi^t_\alpha(z,t) = \varphi^t_\alpha(z^t_\alpha, t - |z - z^t_\alpha|) \quad \text{in } \mathcal{D}^t_\alpha, \tag{2.14}$$

with $\alpha = \pm$, where z^i_α and z^t_α are the coordinates of the interfaces of the layer with the regions $\mathcal{D}^i_\alpha$ and $\mathcal{D}^t_\alpha$, i.e. $z^i_+ = z^t_- = 0$ and $z^t_+ = z^i_- = 1$.

In addition to the time-domain fields specified above, we also consider the *Laplace transforms* of these field quantities, e.g.

$$\tilde{\varphi}_\alpha(z,s) \stackrel{\text{def}}{=} \int_{-\infty}^{\infty} \exp(-st)\, \varphi_\alpha(z,t)\, dt. \tag{2.15}$$

Since the incident field reaches the layer at $t = 0$, the time-domain signals are causal, and the Laplace transforms are regular for $\mathrm{Re}(s) \geq 0$. These Laplace transforms are related to the conventional, unit-amplitude frequency-domain solutions $\hat{\varphi}_\alpha(z,s)$ (with $\alpha = \pm$) via the equality

$$\tilde{\varphi}_\alpha(z,s) = \tilde{F}(s)\, \hat{\varphi}_\alpha(z,s), \tag{2.16}$$

where

$$\tilde{F}(s) = \int_0^T F(t) \exp(-st)\, dt \tag{2.17}$$

denotes the Laplace transform (spectrum) of the incident pulse. Outside the layer, the normalized solutions $\hat{\varphi}_\alpha(z,s)$ can be written in the form

$$\hat{\varphi}_\alpha(z,s) = \begin{cases} \exp(-s|z - z^i_\alpha|) + r_\alpha(s) \exp(s|z - z^i_\alpha|) & \text{in } \mathcal{D}^i_\alpha, \\ t_\alpha(s) \exp(-s|z - z^t_\alpha|) & \text{in } \mathcal{D}^t_\alpha, \end{cases} \tag{2.18}$$

where $r_\alpha(s)$ and $t_\alpha(s)$ are the reflection and transmission coefficients.

2.4 Direct versus Inverse Problems

In this contribution, we consider the direct-scattering problem as well as the inverse-scattering problem for the configuration defined above. In the *direct-scattering* problem, one or both of the incident fields $\varphi^i_\alpha(z,t)$ and the material parameters are known. The aim of the computation is to determine the corresponding total field(s) $\varphi_\alpha(z,t)$. From (2.13), (2.14), and (2.18), it is observed that it suffices to determine $\varphi_\alpha(z,t)$ for $0 \leq z \leq 1$. Over the last decade, several methods have been suggested for attacking this problem. In the present contribution, we confine ourselves to giving a comprehensive review, in Section 3.

In the *inverse-scattering* problem, we distinguish between *inverse profiling* and *identification*. In both problems, the reflected field(s) caused by one or more plane waves incident from one or both sides if the layer is assumed to be known. The aim of the *inverse-profiling* computation is to reconstruct one or both of the material parameters, i.e. $\chi(z)$ and $A(z)$, from these reflected-field data. This problem will be our main subject attention. In particular, we investigate whether it can be resolved by utilizing integral relations for these reflected fields. We will consider both the frequency-domain case, where the reflection coefficient for plane-wave incidence, as introduced in (2.18), is known for a specified range of real frequencies ($s = i\omega$), and the time-domain case, where the known data consists of the (sampled) reflected field caused by an incident pulse of finite duration T. In view of (2.13), (2.14), and (2.18), it suffices in both cases to consider the reflected fields at $z = z^i_\alpha$. Techniques for solving the inverse-profiling problem will be outlined in Sections 4–6.

The *identification* problem amounts to establishing whether the scattering configuration is or is not identical to some unknown reference configuration. For the geometry at hand, one approach is to extract the *natural frequencies* from the time-domain reflected field(s), with the aid of a Prony-type algorithm (see [5], [2, Chapter 4]). The natural frequencies can be used as well to assess the quality of the reconstructions obtained by the inverse-profiling algorithms mentioned above. This issue will be addressed briefly in Section 7.

3 Solution of the Direct-Scattering Problem

Most procedures for solving inverse-scattering problems are derived from specific techniques for solving the underlying *direct-scattering* problems. The present contribution, therefore, would not be complete without at least a comprehensive review of such techniques. One way classify the various schemes is to distinguish between time-domain and frequency-domain techniques.

3.1 Time-Domain Techniques

As such, the second-order differential equation (2.9) is not extremely well-suited for a numerical discretization. In general, alternative formulations are employed. As indicated by (2.1)–(2.2) and (2.6)–(2.7), one possibility is to break (2.9) up into two coupled *first-order differential equations*, according to

$$\partial_z\varphi(z,t) - \partial_t\psi(z,t) = 0, \tag{3.1}$$

$$\partial_z\psi(z,t) - [1+\chi(z)]\,\partial_t\varphi(z,t) - A(z)\,\varphi(z,t) = 0. \tag{3.2}$$

The field quantity $\psi(z,t)$ corresponds to the velocity $-\partial_t u_z(z,t)$ in the acoustic case, and to the magnetic-field component $H_x(z,t)$ in the electromagnetic case. In the *finite-difference time-domain technique*, the differentiations with respect to z and t figuring in (3.1) and (3.2) are approximated by two-point central differences on a staggered, regularly spaced space-time grid. Boundary conditions at the layer's interfaces are available from (2.12) and (2.13). When augmented with these boundary conditions, the system of difference equations can be solved by marching on in time. In the literature, this technique is known as the two-step Lax-Wendroff scheme or, more descriptively, as staggered leapfrogging [6,7].

A second possibility is to use a *contrast-source integral equation*. The most simple variant of such an equation is

$$\varphi(z,t) = \varphi^i(z,t) - \tfrac{1}{2} \int_0^1 [\chi(z')\partial_t + A(z')]\, \varphi(z', t - |z - z'|)\, dz'. \tag{3.3}$$

In the space-time discretization, the space integral is replaced by a repeated trapezoidal rule, and the time differentiation by a three-point backward interpolation formula with a double time step. For $\chi(z) \geq 0$, the resulting sytem of equations is solved by a step-by-step updating procedure in time. For $\chi(z) < 0$, an alternative integral equation is available. A considerable reduction in computation time can be achieved by evaluating the contrast term in the right-hand side of (3.3) recursively. Two different types of recurrence relations have been proposed in [8,9] and [2, Subsection 3.3.3].

The most significant features of time-marching methods are:

- Time-domain methods are computationally efficient, due to the fact that we use the same space discretization for *all* instants.
- The accuracy of the numerical solution can be controlled by reducing the mesh size h of the space-time discretization.
- In principle, both methods can be generalized to nonlinear scattering problems. This involves the solution of one nonlinear equation in one unknown for each discrete space-time point.
- In the time-stepping procedure, the errors caused by the space-time discretization will accumulate. As a consequence, the time signals obtained will be a factor of $\mathcal{O}(h^{\frac{1}{2}})$ less accurate than the approximation(s) used in that discretization.
- Finally, it would require additional effort to handle cases where (part of) the scatterer consists of dispersive material.

We can trade off the last disadvantage for the capability of handling nonlinear scattering problems. This leaves the fundamental problem of error accumulation.

3.2 Frequency-Domain Techniques

As in time-domain techniques, it is more convenient to convert the second-order differential equation (2.9) to either a system of coupled first-order differential equations or a contrast-source integral equation. Applying the Laplace transformation (2.15) to the system of equations (3.1)–(3.2) results in

$$\partial_z\tilde{\varphi}(z,s) - s\tilde{\psi}(z,s) = 0, \tag{3.4}$$

$$\partial_z\tilde{\psi}(z,s) - s\,[1 + C(z,s)]\tilde{\varphi}(z,t) = 0, \tag{3.5}$$

where the frequency-dependent contrast function $C(z,s)$ is defined as

$$C(z,s) \stackrel{\text{def}}{=} \chi(z) + A(z)/s. \tag{3.6}$$

In (3.4)–(3.6), the complex angular frequency s occurs as a parameter. The system (3.4)–(3.5) is solved by direct numerical integration from $z = z^t_\alpha$ to $z = z^i_\alpha$ with the aid of a standard fifth- and sixth-order *Runge-Kutta-Verner* procedure. The starting values follow by using the linearity of the equations and the fact that $\hat{\varphi}_\alpha(z,s)/t_\alpha(s)$ is known in closed form in region $\mathcal{D}^t_\alpha$ from (2.18). We terminate the numerical integration at any point where $\chi(z)$, $A(z)$, or one of their first few derivatives is discontinuous. At those points, we use the continuity of $\hat{\varphi}(z,s)$ and $\hat{\psi}(z,s)$ to restart the integration. Details can be found in [10] and [2, Subsection 2.4.2].

The frequency-domain counterpart of the *contrast-source integral equation* (3.3) is given by

$$\tilde{\varphi}(z,s) = \tilde{\varphi}^i(z,s) - s^2 \int_0^1 G(z,z';s)\, C(z',s)\, \tilde{\varphi}(z',s)\, dz', \tag{3.7}$$

where $G(z,z';s)$ denotes the conventional free-space Green's function

$$G(z,z';s) \stackrel{\text{def}}{=} (2s)^{-1} \exp(-s|z - z'|). \tag{3.8}$$

As in the time-domain case, the integral on the right-hand side of (3.7) is approximated by a repeated trapezoidal rule with constant space step h. Recently [11], it was discovered that the matrix equation thus obtained can be reduced analytically to a tridiagonal form, with nonvanishing elements on the diagonal and the first two superdiagonals only. This reduced system can then be solved recursively.

Frequency-domain techniques can be used as well for solving time-domain problems. In that case, the frequency-domain solutions $\tilde{\varphi}(z,s)$ are computed repeatedly for varying s. Next, the time-domain field $\varphi(z,t)$ is obtained by determining the Bromwich inversion integral

$$\varphi(z,t) = (2\pi)^{-1} \exp(\beta t) \int_{-\infty}^{\infty} \exp(i\omega t)\, \tilde{\varphi}(z, \beta + i\omega)\, d\omega, \tag{3.9}$$

with β being a real-valued, nonnegative parameter. Typically, the integral in (3.9) is evaluated numerically with the aid of a Fast Fourier Transformation (FFT).

The most characteristic features of frequency-domain methods are:

- The use of the temporal Laplace transformation is an implicit application of the superposition principle. This restricts the use of frequency-domain techniques to linear scattering problems.

- Because of the use of that transformation, solving for the case where (part of) the scatterer consists of a linearly reacting, dispersive material requires no extra effort.
- The accuracy of the numerical solution can be controlled by adjusting the step size in the local approximations employed.
- The computational effort required in determining $\tilde{\varphi}(z, \beta+i\omega)$ with a fixed *relative* accuracy increases linearly with $|\omega|$ as $|\omega| \to \infty$. This can be explained from the observation that both the numerical integration of (3.4)–(3.5) and the discretization of the integral in (3.6) take a fixed number of steps per wavelength.

When the frequency-domain algorithms are used in the context of solving time-domain scattering problems, we further observe:

- The results are stable for late times, because the Fourier transformation in (3.9) selects only the integrable part of the time signal corresponding to the numerically obtained frequency-domain solution $\tilde{\varphi}(z, \beta + i\omega)$.
- The FFT does not significantly increase the error in $\tilde{\varphi}(z, \beta + i\omega)$. Moreover, the absolute error in $\varphi(z, t)$ is usually almost constant in magnitude for $0 \le z \le 1$ and t in the time interval of interest.
- It suffices to determine $\tilde{\varphi}(z, \beta+i\omega)$ with a constant *absolute* accuracy [11]. This means that we can utilize the pulse spectrum $\tilde{F}(\beta + i\omega)$ introduced in (2.16) in the error estimation. Doing so shows that, for most practical pulse shapes $F(t)$, the same space-discretization can be chosen for all relevant values of ω. As a consequence, the computational effort in determining $\tilde{\varphi}(z, \beta + i\omega)$ becomes (almost) invariant with ω.

The last observation, in particular, has recently made frequency-domain techniques more attractive for solving linear time-domain scattering problems.

4 Born Approximation for Small Contrasts

The most simple example of an application of the Born-type method of solution is the reconstruction of a small velocity contrast $\chi(z)$ and a small attenuation function $A(z)$ from the known real-frequency behavior of the reflection coefficients $r_\alpha(s)$ for both directions of incidence. As such, this problem constitutes an ideal starting point for gaining insight into the type of results that we can expect in more complicated cases.

4.1 Single-Step Procedure

For small contrasts, we formally write the unknown unit-amplitude fields $\hat{\varphi}_\alpha(z, s)$ as a *Neumann series* of the form

$$\hat{\varphi}_\alpha(z,s) = \hat{\varphi}^i_\alpha(z,s) + \sum_{n=1}^{\infty} \hat{\varphi}^{(n)}_\alpha(z,s), \tag{4.1}$$

where $\hat{\varphi}_\alpha^{(n)}(z,s)$ is assumed to be "n-th order small" [12,13]. This series converges only for a small enough $|s|$ [13]. Substituting (4.1) in (3.7), and enforcing the equality sign for each order of magnitude separately result in the hierarchy of equations

$$\hat{\varphi}_\alpha^{(n)}(z,s) = -s^2 \int_0^1 G(z,z';s)\, C(z',s)\, \hat{\varphi}_\alpha^{(n-1)}(z',s)\, dz', \tag{4.2}$$

with $n = 1, 2, \ldots, \infty$, and with $\hat{\varphi}_\alpha^{(0)}(z,s) \stackrel{\text{def}}{=} \hat{\varphi}_\alpha^i(z,s)$. The first two terms of the expansion thus obtained are substituded in the integral relation

$$r_\alpha(s) = -\tfrac{1}{2}s \int_0^1 C(z,s)\, \hat{\varphi}_\alpha^i(z,s)\, \hat{\varphi}_\alpha(z,s)\, dz, \tag{4.3}$$

which is obtained directly from (2.18) and (3.7). This results in

$$r_\alpha(s) = -\tfrac{1}{2}s \int_0^1 C(z,s)\, \hat{\varphi}_\alpha^i(z,s)^2\, dz + \Delta r_\alpha(s), \quad \text{with} \tag{4.4}$$

$$\Delta r_\alpha(s) \approx \tfrac{1}{2}s^3 \int_0^1 C(z,s)\, \hat{\varphi}_\alpha^i(z,s) \int_0^1 G(z,z';s)\, C(z',s)\, \hat{\varphi}_\alpha^i(z',s)\, dz'\, dz. \tag{4.5}$$

The *Born approximation* can be phrased as neglecting the error $\Delta r_\alpha(s)$ in (4.4). In this approaximation, the right-hand side of (4.4) can be recognized as a Laplace transform. This allows us to derive closed-form approximations for the unknown profiles $\chi(z)$ and $A(z)$. We arrive at

$$\chi_B(z) = (2\pi)^{-1} \int_{-\infty}^{\infty} \exp(ikz)\, \hat{\chi}_B(k)\, dk, \tag{4.6}$$

$$A_B(z) = (2\pi)^{-1} \int_{-\infty}^{\infty} \exp(ikz)\, \hat{A}_B(k)\, dk. \tag{4.7}$$

where the Born-approximated Fourier coefficients $\hat{\chi}_B(k)$ and $\hat{A}_B(k)$ are given by

$$\hat{\chi}_B(k) = \frac{2i}{k}\left[r_+(ik/2) - \exp(-ik)\, r_-(-ik/2)\right], \tag{4.8}$$

$$\hat{A}_B(k) = -r_+(ik/2) - \exp(-ik)\, r_-(-ik/2). \tag{4.9}$$

The quality of these approximations can be assessed by estimating the right-hand side of (4.5). From (4.6)–(4.9), it is observed that we only need to estimate $\Delta r_\alpha(s)$ for $s = i\omega$, with ω being a conventional, real-valued frequency. For such frequencies, this term is nothing but a sum of single and double Fourier integrals, whose magnitude follows from standard Fourier theory (see e.g. [14]). Assuming that the unknown contrasts $\chi(z)$ and $A(z)$ are of the same order of magnitude, and that $\chi(z)$ exhibits at least one discontinuity in the interval $-\infty < z < \infty$, we find

$$\Delta r_\alpha(i\omega) = \begin{cases} O(\chi^2) & \text{for } |\omega| = O(1), \\ \omega\, O(\chi^2) & \text{as } |\omega| \to \infty. \end{cases} \tag{4.10}$$

With the aid of this estimate, we can determine the size of the error in the approximate Fourier coefficients $\hat{\chi}_B(k)$ and $\hat{A}_B(k)$. We end up with

$$\hat{\chi}(k) - \hat{\chi}_B(k) = O(\chi^2), \tag{4.11}$$

$$\hat{A}(k) - \hat{A}_B(k) = k\, O(\chi^2), \tag{4.12}$$

for $-\infty < k < \infty$.

The results listed in (4.6)–(4.9) and (4.11)–(4.12) lead to the following conclusions.

- For a complete determination of $\chi_B(z)$ *and* $A_B(z)$, we need to know the real-frequency behavior of *both* reflection coefficients over the *entire* interval $-\infty < \omega < \infty$. By a similar argument, it can be shown that a *single* unknown profile can be reconstructed from the frequency behavior of *one* of these coefficients, provided that the remaining profile is known.

- Any band limitation in the known reflected-field data will correspond with a band limitation in the spectral representation of the unknown profile(s).

- The error in $\hat{A}_B(k)$ increases linearly in magnitude with increasing $|k|$. Consequently, the approximate inversion formula (4.7) can at best produce a band-limited reconstruction, even when wide-band reflected-field information is available. In particular, the short-range variation of $A(z)$ will be recovered poorly.

- The error in $\hat{\chi}_B(k)$ remains approximately constant in magnitude over the entire range of k. In line with the observations made in Subsection 3.2, this indicates that, particularly for small contrasts, it may be possible to reconstruct $\chi(z)$ up to the band limitation imposed by the available information. On the other hand, the actual Fourier coefficient $\hat{\chi}(k)$ is of $\mathcal{O}(k^{-1})$ as $|k| \to \infty$. Even in this case, therefore, the possibility of an inherent high-frequency band-limitation cannot be excluded.

The main disadvantage of the estimation procedure described above is that the agreement between the reconstructed and the actual profiles may be confined to a limited range of the spatial wave number k. One way to remedy this situation is to supply additional *a priori* information about the short-range behavior of the unknown profiles, for example by representing them by piecewise-linear expansions of the type

$$\chi(z) = \sum_{m=0}^{M} \chi_m \, \phi_m(z), \qquad A(z) = \sum_{m=0}^{M} A_m \, \phi_m(z), \tag{4.13}$$

with the $\{\phi_m(z)\}$ being triangular expansion functions. The expansion coefficients $\{\chi_m\}$ and $\{A_m\}$ can be interpreted as being approximations to values of the corresponding profiles $\chi(z)$ and $A(z)$ at $z = m/M$. Taking into account the information that the unknown profiles can be represented by (4.13) requires a different inversion procedure. We now obtain the coefficients $\{\chi_m\}$ and $\{A_m\}$ from minimizing cost functions of the type

$$\int_0^{k_{max}} w(k)^2 \left| \sum_{m=0}^{M} \xi_m \, \hat{\phi}_m(k) - \hat{\xi}_B(k) \right|^2 dk$$
$$+ \delta \sum_{m=1}^{M-1} (\xi_{m-1} - 2\xi_m + \xi_{m+1})^2 + \delta(2\xi_0 - 3\xi_1 + \xi_3)^2 + \delta(2\xi_M - 3\xi_{M-1} + \xi_{M-3})^2, \tag{4.14}$$

where ξ stands for either χ or A. The weighting function $w(k)$ is chosen in conformity with the error estimations (4.11)–(4.12). The number of subintervals in the piecewise-linear expansions

(4.13) is chosen such that the additional information about the short-range variation of $\chi(z)$ and $A(z)$ links up with the information about the longer-range variation available from $\hat{\chi}_B(k)$ and $\hat{A}_B(k)$. The terms proportional to δ are *regularization terms* that can be envisaged as favoring the "most linear" solution in case of ambiguity, with δ being a small nonnegative *regularization parameter* (see also [2, Subsection 5.3.4], [15]).

4.2 Perturbation Approach

The Neumann-series formulation of (4.1)–(4.2) suggests that it should be possible to cast the Born inversion procedure proposed above into the framework of a *perturbation approach.* So far, only a few numerical results of such an approach have been obtained. Nevertheless, its close resemblance to the Born-type iterative approach that will be discussed in Section 5 warrants its inclusion in the present contribution.

To arrive at the relevant equations, we augment the series representation (4.1) by a similar representation for the contrast function $C(z,s)$, i.e.

$$C(z,s) = \sum_{n=1}^{\infty} C^{(n)}(z,s). \tag{4.15}$$

Substituting the representations (4.1) and (4.15) in the equations (3.7) and (4.3), and enforcing the equality signs in both these equations for each order of magnitude separately then again yield a hierarchy of equations. For the first-order terms $\chi^{(1)}(z)$ and $A^{(1)}(z)$, we again end up with (4.4). The higher-order terms can be obtained from

$$\hat{\varphi}_\alpha^{(n)}(z,s) = -s^2 \int_0^1 G(z,z';s) \sum_{m=0}^{n-1} C^{(n-m)}(z',s)\, \hat{\varphi}_\alpha^{(m)}(z',s)\, dz', \tag{4.16}$$

and

$$\int_0^1 C^{(n)}(z,s)\, \hat{\varphi}_\alpha^i(z,s)^2\, dz = -\int_0^1 \hat{\varphi}_\alpha^i(z,s) \sum_{m=1}^{n-1} C^{(n-m)}(z,s)\, \hat{\varphi}_\alpha^{(m)}(z,s)\, dz, \tag{4.17}$$

which hold for $n = 1,2,\ldots,\infty$ and for $n = 2,3,\ldots,\infty$, respectively. (4.16) can be resolved by simply evaluating the Fourier-type integral on the right-hand side. (4.17) is of the same form as (4.3), and, hence, can be resolved by either of the procedures proposed in Subsection 4.1.

As in the case of single-step procedure, it is to be expected that the perturbation approach will suffer from an inherent band limitation. This band limitation can possibly be overcome by using the minimization procedure to resolve (4.17), and keeping track of the the residual error in the equality sign in each iteration step.

5 Born-Type Iterative Procedure for Larger Contrasts

One possibility for retrieving larger contrasts is to invoke a *Born-type iterative scheme.* In such a scheme, we start from some initial guess, which should be available from *a priori* information, or be procured from some characteristic features of the known reflected fields. In each iteration step, we then consider the the approximate reconstruction found in the previous step as the reference medium in an inversion procedure similar to the "numerical" one explained in Subsection 4.1.

The essential difference with the perturbation approach proposed in Subsection 4.2 is that, in the field-computation part of each iteration step, we now compute *exact* solutions to an *approximated* integral equation of the type (3.7) instead of an n-th order *approximate* solution to the *exact* equation. In the past, several versions of the Born-type iterative scheme have been investigated [9,16]. In the present context, we restrict ourselves to the only one that does not suffer from inherent band limitations.

5.1 INTEGRAL RELATIONS

Before formulating the inversion procedure, we review the underlying contrast-source integral relations. We introduce a *reference medium* with *known* constitutive coefficients $\bar{\chi}(z)$ and $\bar{A}(z)$, and $\bar{\varphi}_\alpha(z,s)$ as the corresponding unit-amplitude solutions. Further, we introduce the Green's function of the reference medium as the product

$$\bar{G}(z,z';s) = \bar{\varphi}_+(z_>,s)\,\bar{\varphi}_-(z_<,s)/W(\bar{\varphi}_-,\bar{\varphi}_+), \tag{5.1}$$

where $z_< \stackrel{\text{def}}{=} \min\{z,z'\}$ and $z_> \stackrel{\text{def}}{=} \max\{z,z'\}$, and where $W(\bar{\varphi}_-,\bar{\varphi}_+)$ denotes the Wronskian

$$W(\bar{\varphi}_-,\bar{\varphi}_+) = \bar{\varphi}_+(z,s)\,\partial_z\bar{\varphi}_-(z,s) - \bar{\varphi}_-(z,s)\,\partial_z\bar{\varphi}_+(z,s) = 2s\bar{t}^-(s) = 2s\bar{t}^+(s), \tag{5.2}$$

and we define the *contrast function*

$$\bar{C}(z,s) \stackrel{\text{def}}{=} \chi(z) - \bar{\chi}(z) + [A(z) - \bar{A}(z)]/s, \tag{5.3}$$

which represents the difference in material properties between the actual and the reference configurations. In terms of these quantities, the integral relation can be written as

$$\hat{\varphi}_\alpha(z,s) = \bar{\varphi}_\alpha(z,s) - s^2\int_{-\infty}^{\infty}\bar{G}(z,z';s)\,\bar{C}(z',s)\,\hat{\varphi}_\alpha(z',s)\,dz'. \tag{5.4}$$

This integral relation assumes a special form when we take $z = z^i_\alpha$, and use the identity

$$\bar{\varphi}_\alpha(z^t_\alpha,s) = \bar{t}_\alpha(s), \tag{5.5}$$

which follows directly from (2.18). We obtain

$$r_\alpha(s) - \bar{r}_\alpha(s) = \hat{\varphi}_\alpha(z^i_\alpha,s) - \bar{\varphi}_\alpha(z^i_\alpha,s) = -\tfrac{1}{2}s\int_0^1\bar{C}(z,s)\,\bar{\varphi}_\alpha(z,s)\,\hat{\varphi}_\alpha(z,s)\,dz, \tag{5.6}$$

with $\alpha = \pm$. The pair of equations given in (5.6) can also be envisaged as a pair of *reciprocity relations* between the fields $\hat{\varphi}_\alpha(z,s)$ and $\bar{\varphi}_\alpha(z,s)$ [17].

5.2 THE BORN-TYPE APPROXIMATION

Substituting the integral relation (5.4) in the "additional" relation (5.5) yields

$$r_\alpha(s) - \bar{r}_\alpha(s) = -\tfrac{1}{2}s\int_0^1\bar{C}(z,s)\,\bar{E}_\varphi(z,s)^2\,dz + \Delta r_\alpha(s), \tag{5.7}$$

where the *Born error* $\Delta r_\alpha(s)$ is given by

$$\Delta r_\alpha(s) \stackrel{\text{def}}{=} \tfrac{1}{2}s^3 \int_0^1 \bar{C}(z,s)\, \check{\varphi}_\alpha(z,s) \int_0^1 \bar{G}(z,z';s)\, \bar{C}(z',s)\, \hat{\varphi}_\alpha(z',s)\, dz'\, dz. \tag{5.8}$$

Now, the *Born-type approximation* consists of neglecting $\Delta r_\alpha(s)$ in (5.7). This reduces (5.7), with $\alpha = \pm$, to a system of approximate integral equations of the first kind for the unknown profiles $\chi(z)$ and $A(z)$. The Born-type approximation is incorporated in an *iterative scheme* that leads to successive approximations for the unknown total field(s) as well as for the unknown constitutive parameter(s). Starting from some initial guess, we take, in each iteration step, the reference medium identical to the approximation obtained previously.

For a complete justification of this procedure, we would need closed-form expressions for the unit-amplitude fields $\check{\varphi}_\alpha(z,s)$ and $\hat{\varphi}_\alpha(z,s)$ and for the Green's function $\bar{G}(z,z';s)$ for media of arbitrary inhomogeneity. Since such expressions are not available, we have to content ourselves with WKB expressions as derived and analyzed in [2, Sections 2.3, 2.4], [10]. In the present context, we restrict ourselves to summarizing the results.

The differences $r_\alpha(s) - \bar{r}_\alpha(s)$ for $\alpha = \pm$ are each dominated by a term that can be identified with a linear combination of a distorted version of $d_z[\chi(z) - \bar{\chi}(z)]$ and a distorted version of $A(z) - \bar{A}(z)$. In the corresponding time signals, these terms show up for $0 < t < 2\bar{\tau}$, with τ being the one-way travel time across the layer. For the Born error $\Delta r_\alpha(s)$ defined in (5.8), we end up with the estimate

$$\Delta r_\alpha(i\omega) = \Delta r_\alpha^{(1)}(i\omega) + \Delta r_\alpha^{(2)}(i\omega), \tag{5.9}$$

with

$$\Delta r_\alpha^{(1)}(i\omega) = \omega\, \mathcal{O}[(\chi - \bar{\chi})^2], \quad \Delta r_\alpha^{(2)}(i\omega) = \omega \mathcal{O}(\chi - \bar{\chi})\, \mathcal{O}(\chi), \tag{5.10}$$

as $|\omega| \to \infty$. In the time signals corresponding to the Born errors, the first term takes effect from $t = 0$, and the second one from $t = 2\bar{\tau}$.

These results lead to the following conclusions for the feasibility of the Born-type approach:

- The errors $\Delta r_\alpha(i\omega)$ in (5.9) and $\Delta r_\alpha^{(2)}(i\omega)$ in (5.10) are of the same order of magnitude as the improvements $[\chi(z) - \bar{\chi}(z)]$ and $[A(z) - \bar{A}(z)]$ that one would like to obtain. Moreover, as $|\omega| \to \infty$, these errors dominate over $r_\alpha(i\omega) - \bar{r}_\alpha(i\omega)$. Hence, a computation neglecting them may suffer from an *inherent band limitation* in the reconstructed profile(s).

- To avoid such a band limitation, we must restrict the consideration to the *time-domain* signals corresponding to $r_\alpha(s) - \bar{r}^\alpha(s)$ in the interval $0 < t < 2\bar{\tau}$. In that case, the band limitation will gradually be removed as the computation progresses. Owing to this restriction, we can at best reconstruct a *single* unknown profile from *one-sided* reflection data. For a unique reconstruction of *both* profiles, *two-sided* reflection data must be available.

- From the correspondence between $r_\alpha(s) - \bar{r}_\alpha(s)$ and the configuration parameters $\chi(z) - \bar{\chi}(z)$ and $A(z) - \bar{A}(z)$ mentioned above, it follows as well that a band limitation in the incident field(s) will give rise to a band limitation in the reconstructed profile(s).

From a physical point of view, the conclusion reached above seems satisfactory since the version of the Born-type iterative scheme selected is the only one that directly relates the material properties inside the layer to the transient fields at those instants when the relevant reflected waves actually arrive at the interfaces $z = z^i_\alpha$.

5.3 Numerical Implementation

The numerical implementation of the Born-type iterative scheme runs along the same lines as the numerical inversion procedure discussed in Subsection 4.1. For details, the reader is referred to [2, Section 6.4]. The principal features of our implementation are:

- The "known" and the "reference" fields are generated by applying different algorithms. In this manner, we automatically avoid the possibility of the solution of the inverse-scattering problem being biased by numerical errors made in solving the underlying direct-scattering problems.

- In some of our numerical experiments, multiplicative or additive noise was superimposed on the "known" time-domain reflected fields. In all cases, a high-frequency band limitation was simulated by subjecting these time signals to a low-pass filtering.

- The cost function is formulated in the time domain, taking into account the travel time across the layer as well as the finite duration of the incident pulse.

- The number of cells in the piecewise-linear expansions (4.13), is adjusted to the cut-off frequency ω_{max} in the low-pass filtering as well as to the two-way travel time 2τ. According to Shannon's theorem, the shortest sampling period for which samples of the band-limited time signals are independent is $\Delta t = \pi/\omega_{max} = T/4$. Hence, we can at most recover $M \approx 8\tau/T$ independent parameters for each unknown profile. In view of the distortion effects observed in our WKB analysis, we choose M at least a factor of 2 larger to ensure that the maximum amount of profile information is drawn from the reflected fields.

- As in Subsection 4.1, we include regularization terms to eliminate any ambiguity due to this oversampling, and, in the first few iteration steps, due to the band limitations inherent in the Born-type approach.

For a representative numerical results, the reader is referred to [2, Section 6.4] or [3].

6 Nonlinear Optimization

An alternative way to resolve the inverse-profiling problem discussed in this paper is to minimize a cost function of the form

$$\sum_\alpha \int_0^\infty \left\{\varphi^r_\alpha(z^i_\alpha, t) - \bar{\varphi}^r_\alpha(z^i_\alpha, t)\right\}^2 dt$$
$$+ \delta \sum_{m=1}^{M-1} [\chi_{m-1} - 2\chi_m + \chi_{m+1}]^2 + \delta \sum_{m=1}^{M-1} [A_{m-1} - 2A_m + A_{m+1}]^2 \qquad (6.1)$$

as a function of the parameters $\{\chi_m\}$ and/or $\{A_m\}$ with the aid of a *nonlinear optimization method.* In (6.1), the summation runs over one or both of the labels $\alpha = \pm$, according to whether one or both of the contitutive parameters $\chi(z)$ and $A(z)$ are unknown. The reflected fields $\bar{\varphi}^r_\alpha(z^i_\alpha, t)$ pertain to some *trial medium*, in which the unknown profile(s) are given by one or both of the piecewise-linear expansions (4.13). The main advantage of the optimization approach is that we can now use *all* the reflected-field information in the time interval $0 < t < \infty$.

An approach of this type was first proposed by Lesselier [18]. The fundamental improvement in our version is that we purposely choose the expansion of the unknown profile(s) in conformity with the amount of information available from the known reflected field(s). We experimented with both quasi-Newton and conjugate-gradient procedures available from standard mathematical subroutine libraries. The most efficient reconstruction procedure was obtained by combining a conjugate-gradient method with an independent evaluation of the "gradient" (the vector containing the partial derivatives of the cost function with respect to the unknown profile parameters) as outlined in [18]. Even for this procedure, however, a typical computation took an effort equivalent to tens of iteration steps in the Born-type procedure.

The fact that a conjugate-gradient method converges faster than quasi-Newton methods can possibly be explained from the feature that, in the latter, the discretization error in the numerically obtained reference solutions $\bar{\varphi}_\alpha(z,t)$ are amplified in the estimation of the Hessian matrix. In conjugate-gradient methods, such an estimation is avoided.

7 Using the Natural Frequencies

Now that we can solve the inverse-scattering problem, we have two different ways of finding the *natural frequencies* of the unknown configuration from time-domain reflected-field data. In the first place, we can follow the indirect route of first reconstructing the configuration with the aid of the Born-type iterative procedure. Subsequently, we can then compute the natural frequencies of the reconstructed configuration as explained in [10]. In the second place, we can resample the filtered time signals mentioned in Subsection 5.3 with a suitably chosen sampling period, and directly apply Prony's method [5], [2, Section 4.3] to the resampled signal.

By comparing the two sets of results thus obtained with each other and with the actual natural freqencies, we can resolve two issues at once. In the first place, we can verify our conclusion that reconstructed configurations have the same reflection coefficients as the actual configuration over the real-frequency range excited. If this conclusion is correct, we should observe good agreement between the natural frequencies of the reconstructed configurations whose imaginary parts lie in that range and their actual counterparts. In the second place, we can decide whether the piecewise-linear expansions employed are flexible enough. If the deviation between the indirectly obtained natural frequencies and those obtained otherwise increases considerably with their increasing imaginary parts, it may be necessary to choose a larger number of subintervals, i.e. a larger M.

Numerical experimentation has shown that especially the natural frequencies nearest to the origin of the complex s-plane are recovered well by the Prony-type method. Hence, comparing those frequencies with those obtained indirectly via profile reconstruction may indeed provide an additional way of judging the quality of the reconstructed profiles.

References

[1] K.J. Langenberg, "Applied inverse problems for acoustic, electromagnetic and elastic waves". In: *Basic Methods of Tomography and Inverse Problems*, P.C. Sabatier (Ed.), Adam Hilger, Bristol, 1987, Part II, 125–453.

[2] A.G. Tijhuis, *Electromagnetic Inverse Profiling: Theory and Numerical Implementation*, VNU Science Press, Utrecht, The Netherlands, 1987.

[3] A.G. Tijhuis, "Born-type reconstruction of material parameters if an inhomogeneous, lossy dielectric slab from reflected-field data", *Wave Motion 11*, 151–173, 1989.

[4] A.C. Pipkin, *Lectures on viscoelastic theory*, Springer-Verlag, New York, 1972, Chapter 1.

[5] E.K. Miller, "Natural mode methods in frequency and time domain analysis". In: *Theoretical Methods for Determining the Interaction of Electromagnetic Waves with Structures*, J.K. Skwyrzinski (Ed.), Sijthoff and Noordhoff, Alphen aan den Rijn, The Netherlands, 1981, 173–212.

[6] W.H. Press, B.P. Flannery, S.A. Teukolsky, and W.T. Vetterling, *Numerical Recipes*, Cambridge University Press, Cambridge, 1986, Section 17.1.

[7] A. Taflove and K.R. Umashankar, "Review of FT-TD numerical modeling of electromagnetic wave scattering and radar cross section", *Proc. IEEE 77*, 682–699, 1989.

[8] J.-Ch. Bolomey, Ch. Durix and D. Lesselier, "Determination of conductivity profiles by time-domain reflectometry", *IEEE Transactions Antennas Propagat. 27*, 244–248, 1979.

[9] A.G. Tijhuis, "Iterative determination of permittivity and conductivity profiles of a dielectric slab in the time domain", *IEEE Trans. Antennas Propagat. 29*, 239–245, 1981.

[10] A.G. Tijhuis and H. Blok, "SEM approach to the transient scattering by an inhomogeneous, lossy dielectric slab; Part 2: the inhomogeneous case", *Wave Motion 6*, 167–182, 1984.

[11] A.G. Tijhuis, R. Wiemans, and E.F. Kuester, "A hybrid method for solving time-domain integral equations in transient scattering", *Journal of Electromagnetic Waves and Applications 3*, 485–511, 1989.

[12] D.S. Jones, *The Theory of Electromagnetism*, Pergamon Press, Oxford, 1964, 335–336.

[13] F.G. Tricomi, *Integral Equations*, Dover Publications, New York, 1985, 49–55.

[14] E.C. Titchmarsh, *The Theory of Functions*, 2nd ed., Oxford University Press, New York, 1950, Chapter 13.

[15] A.N. Tikhonov and V.Y. Arsenin, *Solution of Ill-Posed Problems*, V.H. Winston and Sons, Washington D.C., 1977, Chapters II, III.

[16] A.G. Tijhuis and C. van der Worm, "Iterative approach to the frequency-domain solution of the inverse-scattering problem for an inhomogeneous, lossy dielectric slab", *IEEE Transactions Antennas Propagat. 32*, 711–716, 1984.

[17] H. Blok and M.C.S. Zeijlmans, "Reciprocity and the formulation of inverse scattering problems", *Radio Science 22*, 1137–1147, 1987.

[18] D. Lesselier, "Optimization techniques and inverse problems: reconstruction of conductivity profiles in the time domain", *IEEE Transactions Antennas Propagat. 30*, 59–65, 1982.

Part VII

Miscellaneous Problems

Some Remarks on the Relationship Between Mathematics and Physics and on Inverse Problems

G. Anger

Department of Mathematics, Martin Luther University Halle-Wittenberg, DDR-4010 Halle a.d. Saale, German Democratic Republic

1. Some reflections on mathematical modelling and on the way of thinking of mathematicians

Any member of the animal world has the capacity to perceive and to digest certain physical and chemical information. Man, in addition, is capable of storing and, when needed, using stored information. It has to be pointed out that our sense organs on the whole are very complex and cannot totally be substituted for single devices (machines). This statement is of great importance when applied to medicine, as owing to the ever existing lack of information, practical experience of physicians, as well as from laboratory outcomes, is necessary for diagnosis (ars medica). Furthermore, it has to be mentioned that our sense organs furnish information on which inverse problems are subject to mathematical solution. For reasonable processing of the above information natural numbers are needed on a somewhat higher level - rational numbers, in science and technology - real numbers, such as π, e, etc. The set of rational numbers is denumerable while the set of real numbers is non-denumerable. Transition from rational numbers to real numbers is a qualitatively large jump.

For millenia, man has tried to improve living conditions, making use of physics and mathematics. It was at a rather early stage that algebraic and geometric problems were topical. Euclid's elements of geometry which were written in 325 B.C. in principle are plotted in an axiomatic way. In 1899, D. Hilbert corrected some inaccuracies of these elements, at the same time giving parts of geometry a strictly axiomatic structure.

Later on, these ideas were carried on systematically. The axiomatic (constructive) way of thinking is one of the great strength of mathematics. It leads to new outcomes, which cannot be found the purely experimental way. However, the formalism which is nowadays overstated in mathematics also has its drawbacks, which may have a detrimental effect particularly on junior scientists. Any suggestions, both from mathematics and from natural sciences, have to be considered in order to acquire additional knowlegde.

Physics, in its theoretical foundation, also has an axiomatic structure. In 1687, Newton in his famous "Principia" established theoretical physics [6], [8]. Axioms of physics are, among others, Newton's laws of motion, the law of conservation of mass, the law of conservation of energy and so on. The basic equations of mathematical physics are derived by assuming the knowledge of the material parameters at each point of the special system considered. V. I. Arnold [8] pointed out that later revisions of "Principia" omitted substantial ideas of Newton. With material parameters of a physical system well-known, both direct and inverse problems can be formulated and, provided the availability of a sufficient stock of information, solved [3], [4]. However, in geophysics and in medicine, one tries to determine interior parameters inaccessible to measurements by remote measurements. This class of problems are also inverse problems. It has to be pointed out that in complex systems there exists practically no theory of measurement [3]. Biologists stick to their style of working, which may be considered to be a model for all domains of science. Using laboratory outcomes they observe the course of biological processes in open nature. From a mathematical point of view, inverse problems related to a solid body, in most cases, cannot be solved due to the large information content. But the knowledge gained is eminently suitable for the solution of particular experimental and constructional tasks. The sensational outcomes of physics, such as in space technology, microelectronics, etc. are also related to constructional aspects.

I. Newton and G. Leibniz were the founders of differential and integral calculus. The methods which they developed continue, even today, to be the basis of equations in mathematical physics. In a word, as far as mathematical modelling is concerned, we still are living in Euler's intellectual world. For certain parameters, differential equations are derived making use of certain basic physical laws. This renders possible advance determination of the behaviour of a special system. In doing so, little attention is being paid to the complexity of both nature and the great experimental progress achieved. In deriving the differential equations, one uses mass distributions (charge distributions) μ mostly in the form $d\mu(y) = \rho(y)dy$, dy volume element. The Dirac measure δ_x and others do not belong to this class. So, from the very outset principal decisions are anticipated. Mass distribution of the earth and of other celestial bodies cannot be described by the relation $d\mu(y) = \rho(y)dy$. Owing to special assumptions, many solutions are lost, which could be of principal importance for many uses [3], [5]. Similar conditions are true for other material parameters.

In the 19th century, the theoretical basis of mathematics was developed. Fundamental notions, such as real numbers, algebraic structure, convergence, function, set etc. were created in this period. In an interview, V. I. Arnold said: "The 200-year interval from Huygens and Newton to Riemann and Poincaré seems to me to be a mathematical desert filled only with calculations". In order to apply the outcomes of mathematics one needs a complete space such as R^1. Such a space is always non-denumerable. For practical use, the spaces $L^p(\nu)$ of all functions f, such that $|f|^p$ is ν-integrable, are of central importance. In order to achieve these spaces, intense basic research in the domain of real functions was necessary.

In this regard, the first step consisted in the integral concept introduced by H. Lebesgue in 1904. In 1918, J. Radon transferred these ideas to general measures. For equations of the second kind $Af + \lambda f = g$, I. Fredholm (1903), D. Hilbert (1904), E. Schmidt (1905), F. Riesz (1918) and J. Schauder (1927) made deep descoveries. F. Riesz studied the continuous linear forms on the space of continuous functions C([a,b]). From these considerations resulted the functional analysis, which today is an essential basis for the study of equations of mathematical physics [3], [9]. In studying applied problems, one of the major difficulties arises in finding the natural coupling between applied problem and mathematical problem. In 1958, the author during a study visit to the Paris University found such a coupling for the gravitational field [1]. Problems between potentials are transformed to problems between continuous linear forms on the space of continuous functions. Such a transformation is equally possible with respect to other differential equations [3], [7].

V. I. Arnold estimates the present state of mathematics as follows: "It seems to me, however, that the greater part of the contemporary mathematical production does not satisfy the aesthetic requirements, on the one hand, and will never be of use, on the other. Perhaps that was always the case and is an indispensable condition for the emergence of the necessary part of mathematics ...".

In applied occupations, strict legal regulations have to be observed with regard to vocational training as a basis for the occupational practice. As far as mathematics - a pure intellectual domain of science - is concerned, such regulations are highly necessary in order to enable a leader to be such a one in the field of applied mathematics. At last, he has to master the specific process as well as relevant mathematics, which often is related to inverse problems.

2. Inverse problems in mathematical physics

For practical applications, the interpretation of experimental findings is of fundamental importance. Many physical parameters f do not lend themselves to direct measurement and the effect g which they induce is then the only known criterion; that is, often g is the only measureable value. Between f and g there exist relations based on physical laws which may be expressed as equation of the first kind: $Af = g$, with f to be determined by calculation. In doing so, an essential step consists in modelling the process resulting in a symbolic description of A. The basic equations of mathematical physics are derived by assuming the knowledge of the material parameters f at each point of the system considered. Such parameters are known, for instance, for the atmosphere, for interplanetary space and for materials for which experiments of high level exist, while in geophysics the earth's parameter are to be determined by measurements on of the surface of the earth. The calculation of Af with given f is called a direct problem. Typically, this does not offer any major difficulties, as it is possible to use the constructive methods of mathematics. The determination of f from the relation $Af = g$, with given g, is called an inverse problem.

The solution of inverse problems is the only way to completely investigate experimental findings. Equations of the first kind often have a great number (non-denumerably many) of solutions f with Af = g. Each equation of the first kind describes a particular process and a uniform theory for such equations cannot be expected.

Most mathematical problems in areas such as physics, chemistry, biology, medicine, technology, ecology, economy are often governed by a very complex system of nonlinear partial differential equations or other equations. The coefficients and the source terms of the system are internal parameters having a special meaning. The problem mathematics has to solve in inverse theory is to identify those external parameters of the system to be measured, together with those subsets on R^3, on which the measurements are to be made, so that several unknown coefficients or sources of the differential equations, or functionals of these quantities inaccessible to measurement and depending continuously on the measured values can be uniquely determined.

In the applied sciences the efficiency of experiments is determined relatively quickly. In contrast to this the efficiency of the equations of mathematical physics in the area of inverse problems has scarely been investigated. This process of clarification must take place with the aid of counterexamples. Such counterexamples are usually found only by mature mathematicians! By using the sense organs one can to certain degree determine the information content of electromagnetic waves, acoustic waves, chemical products, and so on. But it is not possible to look behind a hill or behind the moon. With the help of a scale the mass of a body, but not its physical composition or its geometric shape, can be determined. These examples raise the question of the <u>information content</u> of a physical process or of the relevant mathematical model: which values f do lend themselves to unique and stable determination using the mapping A from Af = g?

The following principle seems to be valid: if the functionals are measured on a surface S, then the coefficients or the source term of the differential equation can be uniquely determined only on a set which has the dimension of the surface S.

Most inverse problems in differential equations relative to a solid body cannot be solved uniquely. Large systems in fact, have intrinsic ambiguities: thus it happens that many theoretical investigations, while mathematically correct, do not lead to any "real world" solution, and hence have no consequences for applications. The following example illustrates the last remark. In the north of the USSR, a bore-hole was drilled to a depth of 12 kilometers. Geophysicists expected to find (once past shallow depths) different materials - granite, basalt, etc. But they found only granite. In addition, the temperature is greater than expected. In other words, from the set of possible solutions, the scientists involved did not select the right solution. Such negative results are well-known in geophysical exploration as well as medicine. In the case of an inverse source problem (gravimetry) only the set of all solutions can be introduced and studied [2], [3]. In this case the information content of the sources to be determined is too large. If, however, the information content of individual coefficients or of the source term is not too large, these parameters can be uniquely calculated on the basis of additional information. This is true for the diffusion coefficient $k(u)$ in the equation $u_t = \mathrm{div}(k(u)\,\mathrm{grad}\, u)$, for the density $\rho = \rho(|\mathrm{grad}\, u|)$ in the system $\rho(|\nabla u|)u_{x_1} = v_{x_2}$, $\rho(|\nabla u|)u_{x_2} = -v_{x_1}$, the right-hand side $f(u)$ in the equation $u_t - \Delta u = f(u)$ and for one-dimensional inverse problems, where the parameters to be determined depends only on one variable (layered media).

If an inverse problem $Af = g$ can be uniquely solved, i. e., $f = A^{-1}g$ exists and A is compact, the inverse mapping A^{-1} is discontinuous. Often, the mapping A^{-1}, like in computed tomography, contains derivatives. In practice, scientists often can measure only the function g and not its derivatives. From this fact it follows that the problem under consideration is ill-posed.

The present state of inverse theory may be described as follows: each particular physical or biological problem is of complex nature; the models used in mathematics comprise only a small number of material parameters. When faced with a multiplicity of inverse problems, one tries to determine just one parameter by measurement outside an open set. This approach leads to a limitation of the problem, which in general can be dealt with in laboratories only. In laboratories most parameters can be kept constant, and different experiments can be controlled. Every inverse problem requires a particular approach. In designing instruments one finds certain additional degrees of freedom as far as geometrical dimensions and materials are concerned. The great success in design engineering is due to this fact. Design perception corresponds to deductive perception in mathematics. On the other hand, interpretive sciences such as geophysics, soil mechanics, meteorology, space research, biology, medicine, etc., are facing particular difficulties. There, one tries to determine a great number of inner parameters from a small number of measurements. Lack of information often results in ambiguity, so different inner parameters produce the same measured values. This ambiguity can only be eliminated by means of additional information to be obtained from experiment.

For inverse problems we have the following principles: only in exceptional cases can an inverse problem be decided in a purely mathematical way. In such cases, the solution often depends discontinuously on measured values. In case of ambiguity, only the set of all solutions can be introduced and studied as a first step (information content of an inverse problem). One can try to choose a solution relevant for the given use on the basis of laboratory experiments. This result has important consequences for scientific work. Success may be achieved only through an interdisciplinary approach. For the complex systems of nature, the scientists largely depends on practical experience. In the future, intense fundamental research on inverse problems will be imperative.

One can find all results in the author's book on inverse problems. The book contains an extensive bibliography (340 titles).

References

[1] G. Anger, Sure le role des potentiels continues dans les fondements de la théorie du potentiels. In: Séminaire de théorie du potentiel, t. 2 (1958), 30 pp., Institut H. Poincaré, Université de Paris.

[2] G. Anger, The information content of an inverse problem: a theory of equivalence classes. Indianapolis J. Math. Phys. 3 (1984-88), 57 - 83 (ed. H. H. Snyder, Southern Illionois University at Carbondale).

[3] G. Anger, Inverse problems in differential equations. Plenum, London - New York 1989.

[4] J. V. Beck, B. Blackwell and C. R. S. Clair, Inverse heat conduction. John Wiley, Sussex 1985.

[5] W. Hansen and H. Hueber, Eigenvalues in potential theory. J. Differential Equations 73(1988), 133 - 152.

[6] G. Lassner, Mathematik und Physik. NTM-Schriftenr. Gesch. Naturwiss., Techik, Med. 25(1988), 1 - 19.

[7] B.-W. Schulze and G. Wildenhain, Methoden der Potentialtheorie für elliptische Differentialgleichungen beliebiger Ordnung. Birkhäuser-Verlag, Basel - Boston, Mass. 1977.

[8] S. Zdravkovska, Conversation with V. I. Arnold. The Mathematical Intelligencer 9(1987), no. 4, 28 - 38.

[9] E. Zeidler, Lectures on nonlinear functional analysis and its applications I-V. Springer-Verlag, Berlin - New York 1985 - 1990.

Variational Averaging of Problems with Stochastic Data

H. Attouch
Université de Montpellier, USTL – Mathématiques, F-34095 Montpellier, France

1. INTRODUCTION

We sketch the main lines of a new method recently introduced by Attouch & Wets [4] and Attouch & Valadier & Wets [5] in order to handle variational problems with stochastic data. Our initial motivation was the study of random media. The idea is to average the problem with respect to the stochastic parameter and to introduce an average problem (which is now a deterministic one) which solution is the average of the solutions of the stochastic problem. Clearly a naïve averaging procedure does not enjoy such property ! We focuss our attention on minimization problems

$$(1) \qquad \min\{F(v,\omega) \ : \ v \in X\}$$

where ω, the stochastic parameter, belongs to some probability space $(\Sigma,\mathcal{A},P)$ and for each ω belonging to Σ, the functional

$$F(.,\omega) \ : \ X \longrightarrow \mathbb{R} \cup \{+\infty\}$$

is the corresponding energy or criteria...

In that case we describe a "*variational*" averaging procedure directly on the stochastic functional. Its geometrical interpretation is the following :

For each $\omega \in \Sigma$ let us consider the epigraph of $F(.,\omega)$, that is

$$\operatorname{epi}F(.,\omega) = \{(v,\lambda) \in X \times \mathbb{R} \ : \ \lambda \geqslant F(v,\omega)\}.$$

Then the integral of the set-valued mapping $\omega \mapsto \operatorname{epi} F(.,\omega)$ (see Castaing & Valadier [7]) is the epigraph a new function, called the <u>epigraphical average</u> of F, and which we denote $\oint F$. It is given by the following formula :

$$(2) \qquad \left(\oint F\right)(u) = \overline{\operatorname{conv}} \left\{\inf \left[\int_\Sigma F(v(\omega),\omega)) \ P(d\omega) \ : \ \int_\Sigma v(\omega) \ P(d\omega) = u\right]\right\}$$

where $\overline{\operatorname{conv}}$ is the closed convex regularization. In the convex case, when considering the corresponding subdifferential operators this operation can be viewed as an <u>inverse averaging method</u> : take the inverse operators, then take classical averaging, and finally take again the inverse.

Indeed, the epiaverage functional can be obtained quite naturally by sampling the problem let

$$F(.,\omega_1) \ \dots \ F(.,\omega_n)$$

then computing the average

$$(3) \qquad \frac{1}{n} *_e \ [F(.,\omega_1) +_e \dots +_e F(.,\omega_n)]$$

where $+_e$ stands for the episum and $*_e$ for the scalar epimultiple, and let n go to $+\infty$ thanks to the epigraphical law of large numbers [5].

This is the key for a numerical approach to problem (1) : if the observations $F(.,\omega_i)$ $i = 1,2,\dots,n$ are (epi) independant, identically distributed then (3) gives a good approximation of $\oint_e F$ and hence to problem (1).

The above method is then applied to the stochastic homogenization of random media (like porous material, rough surface, soils and aggregate...). Numerically, its main advantage is that ϵ, the parameter which describes the tightness of the structure, has not to be taken "small", which is important where computing corresponding finite elements, and it still provides a good approximation.

Up to now, this method has been mostly developped as a direct method. It would be interesting to apply it to inverse problems, like scattering problems with noise to which, thanks to its nonlinear character, it seems to be well adapted. This first requires a good understanding of the "*variational averaging*" of general operators and evolution equations, a matter which is still in progress.

2. THE AVERAGE PROBLEM

In the sequel X is a normed linear space. Given F, G : $X \longrightarrow \mathbb{R} \cup \{+\infty\}$ two real extended valued functions, let us define the two basic operations from epigraphical analysis :

$$(F +_e G)(x) := \inf\{F(u) + G(v) : u + v = x\}$$

is the epigraphical sum of F and G, and for any $\lambda > 0$

$$(\lambda *_e F)(x) := \lambda F(\lambda^{-1}x)$$

is the λ-epimultiple of F.
The terminology is justified by the following geometrical interpretation of these operations : the strict epigraph of $F +_e G$ is the vectorial sum of the strict epi graphs of F and G, while the epigraph of $\lambda *_e F$ is the λ-homothetic of the epigrap of F.

When averaging the epigraphs of two functions, clearly, the solutions of the minimization problems as well as the corresponding infimal values are averaged. This, combined with the above interpretation of these operations, yields

Proposition 2.1. Let $F_1, F_2 : X \longrightarrow \mathbb{R} \cup \{+\infty\}$ be two real extended valued functions and u_1, u_2 some respective minimizers :

$$F_1(u_1) = \min\{F_1(u) : u \in X\}$$
$$F_2(u_2) = \min\{F_2(u) : u \in X\}$$

Then, $\dfrac{u_1 + u_2}{2}$ minimizes $\dfrac{1}{2} *_e (F_1 +_e F_2)$ over X and

$$\inf \left(\frac{1}{2} *_e (F_1 + F_2)\right) = \frac{1}{2} (\inf F_1 + \inf F_2).$$

Let us illustrate this property on the following elementary example : first recall the formula

$$\left[\frac{1}{2} *_e (F_1 + F_2)\right](u) := \inf\left\{\frac{1}{2} (F_1(u_1) + F_2(u_2)) : \frac{1}{2} (u_1 + u_2) = u\right\}.$$

Take X an Hilbert space and $F_1(u) = a_1 \|u\|^2$, $F_2(u) = a_2 \|u-u_0\|^2 + b$ where $\|.\|$ is the norm of X and $a_1, a_2 \geqslant 0$, $b \in \mathbb{R}$ are given real numbers. (see fig. 1)

An easy computation yields $\left[\frac{1}{2} *_e (F_1 +_e F_2)\right](u) = \dfrac{2a_1a_2}{a_1 + a_2} \|u - u_0/2\|^2 + b/2$. This function achieves its minimum at $u_0/2$ which is the average of 0 and u_0, the respective minimizers of F_1 and F_2, and its minimal value b/2 is the average of the minimal values of F_1 and F_2, that is 0 and b. Note that the coefficients a_1 and a_2 of the above quadratic forms have been replaced by their harmonic mean value.

In the context of convex analysis, F_1 and F_2 being now assumed to be closed convex proper functions on X, the above operation can be viewed as the dual operation of the classical averaging of functions : introducing

$$F^*(u^*) = \sup\{\langle u^*, u\rangle - F(u) : u \in X\}$$

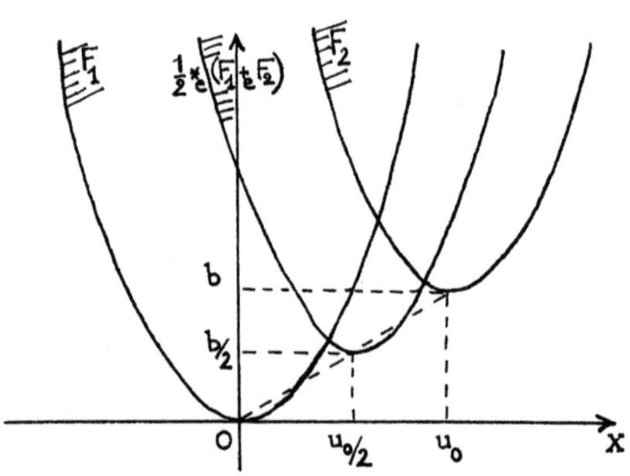

fig. 1

the Legendre-Fenchel transform of F, where $u^* \in X^*$ the topological dual space of X and $\langle u^*,u\rangle$ denotes the duality bracket, we have the following relation

$$\left[\frac{1}{2} *_e (F_1 +_e F_2)\right]^* = \frac{1}{2} (F_1^* + F_2^*)$$

that is

$$\frac{1}{2} *_e (F_1 +_e F_2) = \left[\frac{1}{2} (F_1^* + F_2^*)\right]^* .$$

This formula, combined with the following relation $\partial(F^*) = (\partial F)^{-1}$, where ∂F is the subdifferential operator, allows us to give the interpretation in terms of operators of the above averaging method.

Let $A_1 = \partial F_1$, $A_2 = \partial F_2$, $A = \partial\left[\frac{1}{2} *_e (F_1 *_e F_2)\right]$.

Assuming that $\partial(F_1^* + F_2^*) = \partial F_1^* + \partial F_2^*$ which is satisfied under some qualification assumption, for example, (cf. Attouch & Brezis)

$$\bigcup_{\lambda > 0} \lambda\,(\mathrm{dom}\, F_1^* - \mathrm{dom}\, F_2^*) \quad \text{is a closed subspace of X, then}$$

$$A^{-1}(y) = \frac{1}{2} (A_1^{-1}(y) + A_2^{-1}(y))$$

with $\mathrm{dom}\, A^{-1} = \mathrm{dom}\, A_1^{-1} \cap \mathrm{dom}\, A_2^{-1}$.

Indeed, this formula makes sense in a general setting :

__Proposition 2.2__. Let $A_1, A_2 = X \longrightarrow Y$ two operators (possibly multivalued operators) with X and Y two general normed spaces.

a) The average operator A is defined by

$$A^{-1}y := \frac{1}{2} \left(A_1^{-1}\, y + A_2^{-1}\, y\right)$$

where $A^{-1} = \{(y,x) \in Y \times X : y \in Ax\}$ is the (multivalued) inverse of A

$$A^{-1}y = \{x \in X : (y,x) \in A^{-1}\} = \{x \in X : y \in Ax\}$$

and + stands for the vectorial sum of sets

b) Let us assume that

$$A_1 u_1 \ni 0, \ A_2 u_2 \ni 0$$

Then, denoting $u = \frac{1}{2} (u_1 + u_2)$, we have $Au \ni 0$.

We can notice that in general A is a __multivalued operator__ and that the class of (maximal) monotone operators is preserved by this operation. The theory in such a general setting and in connection with semi-group theory is still in progress.

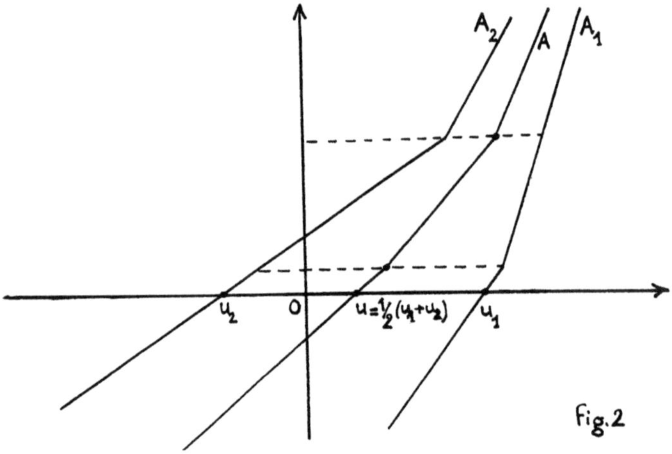

Fig. 2

We have described below an elementary example with $X = Y = \mathbb{R}$. (see fig. 2)

If A_1 et A_2 are linear with slope a_1 et a_2 then A is linear with slope equal to the harmonic mean of a_1 and a_2.
We finally stress the fact that to each type of variational problem corresponds a particular averaging method. Clearly for maximization problems one would have been led to consider hypographs...

3. THE AVERAGE OF STOCHASTIC PROBLEMS

Let us now return to the initial problem. Given a stochastic minimization problem

$$\min\{F(v,\omega) : v \in X\} ,$$

where to each ω belonging to a probability space $(\Sigma,\mathcal{A},P)$ is associated a corresponding functional $F(.,\omega)$, the idea is to sample the problem, let

$$F(.,\omega_1),\ldots,F(.,\omega_n)$$

be independant observations..., then to solve for each ω_i the minimization problem

$$\min\{F(v,\omega_i) : v \in X\}$$

let u_i be such minimizer. The average of these solutions, let

$$\frac{1}{n}(u_1 + u_2 + \ldots + u_n)$$

is, by proposition 2.1, solution of the minimization problem

$$\min\left\{\frac{1}{n} *_e (F_1 +_e \ldots +_e F_n)\right\}$$

where $F_i := F(.,\omega_i)$.

The question is now : what happens when n goes to infinity ? In order to answer this question we have to pass to the limit on a sequence of minimization problems. At this stage, we need to introduce the concept of epigraphical convergence which, in some sense, is the weakest notion of convergence for sequences of functions which allows to pass to the limit on the corresponding minimization problems (see De Giorgi, Mosco, Attouch & Wets... for a complete bibliography refer to [2]).

Definition 3.1. Let (X,τ) be a metrizable topological space and G_n, $G : (X,\tau) \longrightarrow \bar{\mathbb{R}}$ $n = 1,2,\ldots$ a sequence of real extended valued functions. We say that G is the τ-epilimit of G_n if for every u belonging to X the two following properties are satisfied :

$$\begin{cases} \exists\ \hat{u}_n \xrightarrow{\tau} u \quad \text{such that} \quad G(u) = \lim\limits_{n \to \infty} G_n(\hat{u}_n) \\ \\ \forall\ u_n \xrightarrow{\tau} u, \quad G(u) \leqslant \liminf\limits_{n \to \infty} G_n(u_n) \end{cases}$$

We then denote $G = \tau$ - epilim G_n, $G(u) = \min\{\lim G_n(u_n) : u_n \xrightarrow{\tau} u\}$.

Geometrically this is equivalent to say that epi G is the set-limit of epi G_n (convergence of sets is also called Kuratowski-Painlevé convergence). Epi convergence is not, in general, comparable to pointwise convergence. It is a <u>variational convergence</u> :

Proposition 3.2. Let $G_n(u_n) = \min\{G_n(v) : v \in X\}$ and let us assume that $(u_n)_{n\in \mathbb{N}}$ is τ-relatively compact. Then $G = \tau$-epi lim G_n implies that

$$\inf G_n \longrightarrow \min G \text{ as n goes to infinity}$$

and whenever $u_{n_k} \xrightarrow{\tau} u$, then u minimizes G.

In order to go further we need to consider this convergence as attached to a topology. We do not enter into the details of this question, let us just say that it is always possible (this may require some adaptation of the definition 3.1, just like the Mosco-epi topology for convex functions which involves the strong topology in the first sentence of def. 3.1 and the weak topology in the second one).
From now on, we shall consider the stochastic functional F as a random variable

$$F : \omega \in (\Sigma,\mathcal{A},P) \longrightarrow F(.,\omega) \in (SC(X),\text{epitopology})$$

where SC(X) is the hyperspace of lower semicontinuous functions on X equipped with the epitopology. On SC(X) we have now a borelian structure and, given a sequence of random functionals F_1, $F_2,\ldots,F_n,\ldots$ the notions of epi-independance, epi-identically distributed... make sense. We can now state the main theorem (the epigraphical law of large numbers, see A & V & W [5]).

Theorem 3.3. Let $F_1, F_2, \ldots, F_n$ be epi-i.i.d random functionals. Then

$$\frac{1}{n} *_e [F_1 +_e \cdots +_e F_n] \xrightarrow[n \to \infty]{\text{epiconverge}} \oint F \qquad \text{almost surely}$$

where

$$\oint F = \overline{\text{conv}} \inf \left\{ \int_\Sigma F(v(\omega),\omega) P(d\omega) = \int_\Sigma v(\omega)\ P(d\omega) = u \right\}$$

($\overline{\text{conv}}$ is the closed convex regularization). We call $\oint F$ __the epigraphical average of F__.

In view of the previous considerations, we have that, if u minimizes $\oint F$, then

$$\frac{u_1 + \ldots + u_n}{n} \longrightarrow u \text{ and } \frac{1}{n} \sum_{i=1}^{n} \inf F_i \longrightarrow \inf \oint F \qquad \text{as } n \to \infty .$$

Moreover,

$$\text{argmin} \left(\oint_\Sigma F \right) = \int_\Sigma \text{argmin } F(.,\omega)\ P(d\omega)$$

$$\inf \left(\oint_\Sigma F \right) = \int_\Sigma \inf F(.,\omega)\ P(d\omega)$$

i.e. the average of the infima is the infimum of the average functional... we can commute minimization and averaging.

IV. APPLICATION TO RANDOM MEDIA

In stochastic homogenization, typically, there is a small parameter $\epsilon > 0$ which describes the tightness of the structure. For example take $\Omega \subset \mathbb{R}^N$ and

$$F_\epsilon(v,\omega) = \int_\Omega \mathcal{F}(\frac{x}{\epsilon},\omega,Dv(x))\ dx .$$

Under some ergodicity assumptions on the spatial distribution of the materials, one can prove that $F_\epsilon(.,\omega)$ epiconverges to some homogenized functional F_{hom} (see Dal-Maso & Modica [8]). We have the following diagram (convergence = epiconvergence)

$$\frac{1}{n} *_e [F_\epsilon(.,\omega_1) +_e \cdots +_e F_\epsilon(.,\omega_n)]$$

$$\text{a.s.} \Big\downarrow (n \to +\infty) \text{ (epilaw of large numbers)}$$

$$F_\epsilon(.,\omega) \xrightarrow[(\epsilon \to 0)]{\oint} \oint F_\epsilon(.,\omega)$$

$$\text{a.s. (stochastic homogenization)} \Big\downarrow \qquad \swarrow (\epsilon \to 0)$$

$$F_{hom}$$

The idea, numerically, is to compute $\frac{1}{n} *_e [F_\epsilon(.,\omega_1) +_e \cdots +_e F_\epsilon(.,\omega_n)]$ in order to get an approximation of F_{hom} .

In the above example, we sample the problems, then compute for each $i = 1,\dots,n$ via some finite element method ($z \in \mathbb{R}^N$ being fixed)

$$m_{i,\epsilon} = \min\{F_\epsilon(v,\omega_i) : v - \langle z,.\rangle = 0 \text{ on } \partial Y\}$$

where $|Y| = 1$ and take

$$m_{\epsilon,n} = \frac{1}{n} \sum_{i=1}^{n} m_{i,\epsilon}$$

as approximation of $\mathcal{J}_{hom}(z)$ where

$$F_{hom}(v) = \int_\Omega \mathcal{J}_{hom}(Dv)\, dx \quad .$$

A decisive advantage of this method is that is does not require to take ϵ small nor n large, and still provides a good approximation of the effective coefficients of the heterogeneous media.

BIBLIOGRAPHY

[1] H. ATTOUCH, Homogeneisation. Séminaire Nicolas Bourbaki. Exposé n° 686. Nov. 1987. Volume 1987-88.

[2] H. ATTOUCH, Variational convergence for functions and operators. Applicable Math. Series, Pitman, London (1984).

[3] H. ATTOUCH & R. WETS, Epigraphical analysis. Analyse non linéaire. Contributions en l'honneur de J.J. Moreau - Ed. Attouch & Aubin & Clarke & Ekeland, Gauthier-Villars (1989), p. 73-100.

[4] H. ATTOUCH & R. WETS, The average of random minimization problems. Preprint USTL (1989).

[5] H. ATTOUCH & M. VALADIER & R. WETS, The epigraphical strong law of large numbers. Preprint USTL (1989).

[6] M. BALZANO, Random relaxed Dirichlet problems. SISSA Trieste, Working paper 99/86/M (1986).

[7] Ch. CASTAING & M. VALADIER, Convex Analysis and measurable multifunctions. Lecture notes in Math. 580, Springer 1977.

[8] DAL MASO & MODICA, Nonlinear stochastic homogenization. Ann. Mat. Pura Appl. (4) 144 (1986), 347-389.

[9] A.J. KING & R. WETS, Epiconsistency of convex stochastic programs. I.I.A.S.A. Working paper WP 88-57, Laxenburg, Austria May 1988.

[10] G.C. PAPANICOLAOU & S.R. VARADHAN, Boundary value problems with rapidly oscillating random coefficients, in Random fields, Colloquia Mat. S. Janos Bolyai 27, eds E.J. Fritz, J.L. Lebowitz, D. Szasz, North Holland, 1981, 835-873.

Some High Energy Cross Sections from an Inverse Point of View

M. Blažek

Institute of Physics, EPRC, Slovak Academy of Sciences,
84228 Bratislava, Czechoslovakia

Abstract. Recent experimental results on high energy multiplicity distributions serve as motivation for seeking the relation between asymptotic and full probability distributions. Several examples from quantum statistics show successively how the Poisson transform helps to restore the full probability by means of its high energy asymptotics. Also the influence of the leading particle effect on the form of the Poisson transform is mentioned there. It is called for attempts to generalize the approach presented here.

1. Introduction. In physics, to explain a large set of phenomena it is quite natural to assume the validity of some kind of relations arising from the "first principles" like the equations of motion (for instance the Schrödinger equation) which include several free functions (say, potentials). If those functions are specified and the corresponding initial as well as boundary conditions complet the input information, we are able (at least in principle) to obtain the quantities required for the description of the observed phenomena. If this procedure is successful we say that the problem is solved by a direct way.

On the other hand, in an inverse approach the starting point involves the quantities which in the direct way were obtained as the results or consequences and the aim is to determine the "sources" or "potentials". Hence, loosely speaking in the inverse approach the "reason" from its "consequences" should be determined.

An inverse point of view on several problems quite often helps to find new, fresh and attractive ideas. As to the probability and statistics such a situation can be recognized for instance when solving the problem of restoration the probability distributions or structure functions by means of their statistical moments or when a density matrix should be determined in terms of its correlation functions. Other instructive examples can be found among the procedures leading to the formulation of basic equations which solve the inverse problem of quantum scattering theory in terms of the input quantities represented by analytic and asymptotic properties of scattering amplitudes [1], [2], [3] (for more details consult [4]).

In the last case the input scattering amplitude is essentially proportional (up to a phase) to the square root of the corresponding cross section (i.e. to the square root of the corresponding observed probability distribution). However, there exists a special set of probability distributions (very interesting and important in several branches of physics) which can be restored, in an explicit way, directly by means of their (high energy) asymptotics, without reference to any quantity resembling the scattering amplitude (or the "square root" of the asymptotics). Just this type of the inverse problems is discussed in the next part of the present contribution.

2. KNO scaling. At very high energies a collision of two objects (like proton-proton, proton-antiproton, etc.) leads, in general, to the production of many (say "n") secondary particles. Arranging a lot of collisions (at a given energy and with the same species of objects) we can deduce the probability (or cross section) $P(n)$ to observe n-particles and then also different kinds of its moments might be calculated; at present, let us mention at least the average multiplicity $\langle n\rangle = \sum nP(n)$ which depends on the (say, center of mass) energy, W, of the colliding objects.

Experimental data so far clearly express the fact that the average multiplicity $\langle n\rangle$ increases with increasing energy W. Koba, Nielsena nd Olesen (KNO) have shown ([5], however, a similar result can be found already in [6]) that under certain assumptions, at very high energies ($W \to \infty$, i.e. when $n \to \infty$ and also $\langle n\rangle \to \infty$) the validity of the following relation should be expected,

$$\langle n\rangle P(n) = F(z_0) \tag{1}$$

where the ratio

$$n/\langle n\rangle \equiv z_0 \tag{2}$$

is supposed to be finite. This result means that if the multiplicity measurements are plotted for every energy in the form of a curve where we have

$$\langle n\rangle P(n) \quad \text{versus} \quad n/\langle n\rangle , \tag{3}$$

from some energy W on, those curves should approach quite closely an asymptotic form.

(a) From statistical physics point of view there is a well known function which fulfils rel.(1), namely the Bose-Einstein distribution $P_{BE}(n)$. If we assume that the secondary particles are produced in a pure stochastic way from M-centers (cells) and the average occupational number per cell is $\langle n\rangle/M$ then quantum statistics leads to the following form,

$$P_{BE}(n) = \frac{\Gamma(n+M)}{n!\,\Gamma(M)} \cdot \frac{(\langle n\rangle/M)^n}{(1+\langle n\rangle/M)^{n+M}} \equiv P_{BE}(n,\langle n\rangle) \equiv P'(n,W). \qquad (4)$$

In the "KNO limit", i.e. when

$$n \rightarrow \infty, \quad \langle n\rangle \rightarrow \infty \quad \text{but} \quad n/\langle n\rangle \quad \text{is finite,} \qquad (5)$$

rel.(4) leads to the gamma distribution, $P_{ga}(n)$,

$$\left[P_{BE}(n)\right]_{as} = \langle n\rangle^{-1} \cdot \left[M(Mz_0)^{M-1}/\Gamma(M)\right] \cdot e^{-Mz_0} \equiv P_{ga}(n) \qquad (6)$$

where z_0 is given by (2).

Experimental data observed in the proton-proton collisions performed at four energies which have been available at the CERN Intersecting Storage Ring (ISR), $30 \lesssim W \lesssim 60$ GeV, suggested [7] that the KNO scaling is valid (at least approximately) and the application of the dependence (6) with respect to (3) was quite encouraging.

An unexpected result appeared from the CERN Super Proton-Antiproton Synchrotron Collider (Sp$\bar{p}$S): the cross sections [8] obtained at energies W of several hundreds GeV (≈ 500, 900 GeV) failed to follow the curve which appeared at lower, especially at the ISR energies. At that time there was there lack for an adequate reasoning. This fact stimulated a lot of theoretical work (compare e.g. [9]).

In this context, a very interesting direction followed by several authors was concerned with generalizations of the Bose-Einstein distribution (4) in the frame of quantum statistics. There was some hope that by allowing production of secondaries partially also in the coherent way, the resulting distributions will help to remove the aforementioned discrepancy. More concretely, if there is assumed that the secondary particles are produced by a mixture of stochastic and coherent fields (with the corresponding average numbers $\langle n_T\rangle$ and $\langle n_C\rangle$, respectively; the index "T" means "thermal") the following

Glauber-Lachs-Peřina-McGill distribution is obtained, [10],

$$P_{PG}(n) = \left[(n+M-1)!\right]^{-1} . (1+M/\langle n_T\rangle)^{-n}(1+\langle n_T\rangle/M)^{-M}$$
$$. \exp\left[-\langle n_C\rangle M/(\langle n_T\rangle+M)\right]$$
$$.L_n^{M-1}\left\{-\langle n_C\rangle M^2/\left[\langle n_T\rangle(\langle n_T\rangle+M)\right]\right\} \quad (7)$$

where $L_n^m(x)$ means the Laguerre polynomials. If in rel.(7) $\langle n_C\rangle \to 0$ we obtain $P_{BE}(n)$, rel.(4), and with $\langle n_T\rangle \to 0$ the Poisson distribution is recovered. In the KNO limit, rel.(5), it follows from (7),

$$\left[P_{PG}(n)\right]_{as} = \langle n\rangle^{-1}. M(1+R_0{}^2)(\xi_0/R_0)^{M-1}$$
$$. \exp\left[-M(\xi_0{}^2+R_0{}^2)\right] . I_{M-1}(2MR_0\xi_0) \quad (8)$$

with

$$\xi_0 = \left[z_0(1+R_0{}^2)\right]^{1/2} , \quad z_0 = n/\langle n\rangle , \quad R_0{}^2 = \langle n_C\rangle/\langle n_T\rangle . \quad (9)$$

Moreover, in (8), $I_p(x)$ is the modified Bessel function of the first kind and M gives again the number of cells. As it is seen, the distribution (8) does not fulfil the hopes: even if it brings a new parameter, R_0 (with $R_0 \to 0$ rel.(8) gives (6)) it leads in principle to the relation (1). Let us add that the distribution (8) was applied to the high energy data e.g. in [11]; it can be obtained also by a randomization process of the gamma distribution and with M = 1 in the field of quantum optics it is known as the Riccian square or associated to the Laguerre distribution.

(b) Another attempt to find a convenient approach to the aforementioned discrepancy started from the observation that the Bose-Einstein distribution (4) is related to its asymptotic gamma distribution (6) by means of the Poisson transform,

$$P_{BE}(n) = \int_0^\infty P_{ga}(n \to x)P_{Poiss}(x)dx \quad (10)$$

where

$$P_{Poiss}(x) = x^n e^{-x}/n! \quad . \quad (11)$$

(In the field of quantum optics [10], rel.(10) is known as the Mandel' photodetection equation.) For a general case, let us write (10) in two forms, namely

$$P(n,W) = \int_L^\infty f(g,W)\ g^n e^{-g}\ dg/n! \quad (12)$$

or

$$P(n) = \int_L^\infty \left[P(n \to x)\right]_{as}\ x^n e^{-x}\ dx/n! \quad . \quad (13)$$

It is more convenient to apply the first form, rel.(12), for instance in the case when an equation is known which should be satisfied by a distribution $P(n,W)$ under consideration. Especially, the Bose-Einstein distribution (4) satisfies the well-known evolution (difference-differential) equation of the form,

$$dP'(n,W)/dW = V(W).\left[(n+M-1)P'(n-1,W) - (n+M)P'(n,W)\right] \qquad (14)$$

and the G-L-P-G distribution (7) satisfies an analogous equation (see e.g. [12]). In those cases rel.(12) with $L = 0$ allows to derive the corresponding Focker-Planck equtions for the function $f(g,W)$. It can be seen easily that the Bose-Einstein distribution leads to the following Focker-Planck equation (applying the relation $d\langle n\rangle/dW = V(W)(M+n)$ which follows from (14) and $V(W)$ is an arbitrary function),

$$\frac{\partial f(g,W)}{\partial W} = -\frac{\partial}{\partial g}\left[V(W)(M+g)f(g,W)\right] + \frac{\partial^2}{\partial g^2}\left[V(W)gf(g,W)\right]$$

with

$$f(g,W) = \left[(M/\langle n\rangle)^M/\Gamma(M)\right].\, g^{M-1}.\, e^{-gM/\langle n\rangle} \equiv \left[P_{BE}(n \to g)\right]_{as} \,,$$

compare rel.(6) (more details e.g. in [13]; one can learn there how to take into account also the influence of fluctuations). Now, the solution of the aforementioned discrepancy can be looked for by several generalizations of those Focker-Planck equations (some steps can be found in [14]). At this point one can also ask what kind of evolution equations allows a solution exhibiting the KNO scaling: a partial answer can be found e.g. in [15]. Let us add that during the last years the asymptotic gamma distribution (6) as well as the full Bose-Einstein distribution (4) (usually expressed in the form of a negative binomial distribution) were successully applied to several high energy results by a variety of authors (some corresponding references can be found in [9]).

The second form, rel.(13), is more appropriate in the case when the data are described reasonably well by a distribution considered as asymptotic one and then we want to find its "full" form (which is meaningfull also at finite energies). An application of that idea is discussed in the next part of the present contribution.

3. An inverse problem. In early sixtees several physicists observed that even if the multiplicity data are not described very well by a simple distribution like a Poisson one,

$$P_{Poiss}(n) = \langle n\rangle^n \, e^{-\langle n\rangle}/n! \quad , \tag{15}$$

their description is improved if the following change of variables is applied there,

$$n \rightarrow n - \alpha_L \quad \text{and} \quad \langle n\rangle \rightarrow \langle n\rangle - \alpha_L \quad . \tag{16}$$

In rel.(16), α_L is interpreted as a parameter characterizing the leading particle effect. In early days, rels.(16) suggested that from the really observed multiplicity n and the average multiplicity $\langle n\rangle$ the number of colliding objects (equal to 2 in collider and 1 in the fixed target experiments) should be subtracted because they are not new, produced secondary particles. In the next period it was observed that in some other type of collisions α_L might be also not integer and later on the analyses of some experiments led to its negative values. Therefore it will be not surprising if we shall allow that α_L might depend on energy (as well as on the type of the colliding objects).

Now, let us include the transformations (16) into the asymptotic distribution (8); we obtain,

$$(\langle n\rangle - \alpha_L)^{-1} M(1+R_\alpha^2)(\xi_\alpha/R_\alpha)^{M-1} e^{-M(\xi_\alpha^2 + R_\alpha^2)} \; .$$
$$\cdot \; I_{M-1}(2MR_\alpha \xi_\alpha) \equiv [P_{PH}(n)]_{as} \tag{17}$$

where in analogy with (9),

$$\xi_\alpha = [z_\alpha(1+R_\alpha^2)]^{1/2} \quad , \quad z_\alpha = (n-\alpha_L)/(\langle n\rangle - \alpha_L) \quad ,$$
$$R_\alpha^2 = (\langle n_C\rangle - \alpha_L)/\langle n_T\rangle \; . \tag{18}$$

Moreover, let us admit that together with the average multiplicity $\langle n\rangle$ also the parameter α_L changes with the energy W. With those assumptions the distribution (17) was applied to the multiplicity data produced in different kind of collisions [16], [17], [18]: the results (including also the comparison of moments) were sufficiently accurate. Especially, as to the ISR and Collider multiplicity data mentioned above, our procedure helps to restore their scaling property [16] however, now in terms of the variables

$$(\langle n\rangle - \alpha_L)P(n) \quad \text{versus} \quad (n-\alpha_L)/(\langle n\rangle - \alpha_L) \tag{19}$$

instead of (3). The energy dependence of the parameter α_L can be seen in [16], [17], [18] for different collisions; especially Ref. [18]

suggests that some oscillations of α_L with increasing energy cannot be excluded.

The Poisson transform (13) (with non-trivial shift of the lower boundary, $L = \alpha_L$) of the asymptotic distribution (17) leads [22] to the following Peřina-Horák [10] full distribution,

$$P_{PH}(n) = P(0). \sum_{j=0}^{n} \{\alpha_L^{n-j}/[(n-j)!\Gamma(j+M)]\} (1+M/\langle n_T\rangle)^{-j} . \\ .L_j^{M-1}\{-(\langle n_C\rangle - \alpha_L)M^2/[\langle n_T\rangle(\langle n_T\rangle+M)]\} , \tag{20}$$

and

$$P(0) = (1+\langle n_T\rangle/M)^{-M}.\exp[-(\langle n_C\rangle M+\langle n_T\rangle\alpha_L)/(\langle n_T\rangle+M)]$$

(it holds $\langle n_T\rangle + \langle n_C\rangle = \langle n\rangle$). In quantum statistics rel.(20) is obtained in the case when besides the mixture of coherent and stochastic fields also pure coherent fields give rise to the described multiplicity [10]. There are several applications of the distribution (20) in the field of high energy physics [19], [20], [21] as well as quantum optics [10]. Let us note that the procedure leading from (20) to its asymptotic form (17) is not trivial (certain details to that as well as some other procedures involved in the present contribution can be found in [22]).

4. A transverse momentum distribution. In the foregoing part there are presented several multiplicity distributions of electrically charged secondaries. Now, let us consider especially rel.(8) in the case of one cell (M=1); this distribution can be expressed essentially in the form,

$$\langle n\rangle.\{[P_{PG}(n)]_{as}\}_{M=1} = a.e^{-bn}.I_0(cn^{1/2}) \tag{21}$$

where (for a given experiment) a,b,c are constants.

Starting essentially from the same assumptions quantum statistics allows to derive also distributions for the transverse momentum p_T of the secondary particles (let p_T be perpendicular to the direction of one of the incoming objects); some of those distributions were published in [23]. As it can be shown quite easily, a special case of rel.(16) taken from Ref. [23] has the form,

$$P(t=p_T^2) = a'.e^{-b't}.I_0(c'.t^{1/2}) , \tag{22}$$

with a', b', c' being constants. The functional equality of (21) and

(22) suggests that in some types of interactions the charged particle multiplicity and their transverse momentum distributions will be at least very similar. In high energy physics, this similarity was really observed and it might be described in terms of the "jet universality" [24]: existence of this phenomenon is included in quantum statistics quite naturally. Considering eq.(22) as an asymptotic distribution, the Poisson transform allows to associate with it the corresponding full distribution. Let us add that in more general cases than are those which are represented by eq.(22), the calculation of statistical moments leads to the appearance of the hypergeometric series in three variables; unfortunately nothing is known about their (even fundamental) properties and therefore the corresponding research in this field is highly wanted.

5. Conclusions. The scaled asymptotic distributions $\langle n\rangle P(n)$ (with the corresponding free parameters $\{...\}$),

$$\underset{\{M\}}{(6)} \longrightarrow \underset{\{M,\ R_0\}}{(8)} \longrightarrow \underset{\{M, R_\alpha, \alpha_L\}}{(17)} \qquad (23)$$

represent the asymptotics of the full distributions

$$\underset{\{M,\langle n_T\rangle\}}{(4)} \longrightarrow \underset{\{M,\langle n_T\rangle,\langle n_C\rangle\}}{(7)} \longrightarrow \underset{\{M,\langle n_T\rangle,\langle n_C\rangle,\alpha_L\}}{(20)} . \qquad (24)$$

The distributions in line (24) can be obtained from the corresponding distributions in line (23) by application of the Poisson transform. In the present contribution there was outlined a motivation allowing to generalize the distributions in both lines (23) and (24) in the directions shown there by the arrows. It would be very worthwile to find the next step generalization which could give rise to more involved scaling property of the asymptotic distributions.

References

1. Gel'fand I.M. and Levitan B.M., Izvestiya Akad. Nauk CCCP (ser. matem.) 15 (1951) 309.
2. Agranowich Z.S. and Marchenko V.A.: The inverse problem of scattering theory. Kharkov 1960 (english translation: Gordon and Breach, N.Y., 1963).
3. Blažek M., Commun. Math. Phys. 3 (1966) 282.
4. Chadan K. and Sabatier P.C.: Inverse problems in quantum scattering theory. Springer, N.Y., 1977 (russian translation: Mir, Moscow, 1980).
5. Koba Z., Nielsen H.B. and Olesen P., Nucl. Phys. B 40 (1972) 317.
6. Polyakov A.M., Zhur. Eksp. Theor. fiz. 59 (1970) 542 (Soviet Physics JETP 32 (1971) 296).

7. Breakstone A. et al., Phys. Rev. D 30 (1984) 528.
8. Alner G.J. et al., Phys. Lett. 138 B (1984) 304.
9. Carruthers P. and Shih C.C., Int'l J. Mod. Phys. A 2 (1987) 1447.
10. Peřina J.: Quantum statistics of linear and nonlinear optical phenomena. Reidel, Dordrecht, 1984.
11. Biyajima M., Progr. Theor. Phys. 69 (1983) 966.
12. Biyajima M. and Suzuki N., Phys. Lett. 143 B (1984) 463.
13. Gardiner C.V.: Handbook of stochastic methods. Springer, Berlin, 1983.
14. Shih C.C., Phys. Rev. D 33 (1986) 3391.
15. Lam C.S. and Walton A.M., Phys. Lett. 140 B (1984) 246.
16. Blažek M., Z. Phys. C 32 (1986) 309.
17. Blažek M., Phys. Rev. D. 35 (1987) 102.
18. Blažek M., Czech. J. Phys. B 36 (1986) 681.
19. Blažek M., Z. Phys.C 26 (1984) 455 and Czech. J. Phys. B 34 (1984) 838.
20. Blažek M. and Blažek T., Phys. Lett. 159 B (1985) 403.
21. Biyajima M., Fowler G.N., Friedländer E.M. et al., preprint, to be published.
22. Blažek M., Czech. J. Phys. B 38 (1988) 705.
23. Blažek M. and Blažek T., Czech. J. Phys. B (1989) 361.
24. Barth M. et al., Nucl. Phys. B 192 (1981) 189;
Ajinenko I.V. et al., Z. Phys. C 16 (1983) 291.

Exact Shock Wave Solutions for an $8\vec{v}_i$ Discrete Boltzmann Model with Temperature

H. Cornille

Service de Physique Théorique, CEN-Saclay, F-91191 Gif-sur-Yvette, France

We study a $8\vec{v}_i$ discrete Boltzmann model with speeds $1,\sqrt{2}$ and velocities along the two medians and the two diagonals of a square. All the conservation laws being satisfied, we can define a nontrivial temperature. We study the physical properties of the shock wave solutions: positivity of the densities, shock front velocity, sound wave velocity, shock velocity, upstream and downstream domains, subsonic and supersonic inequalities. We observe overshoots of the temperature accross the shock.

1. Introduction

For the discrete Boltzmann models [1] the velocity can take only discrete values $\vec{v}_i$ $i=1,\ldots p$. The known exact solutions [2] correspond to models with only one speed $|\vec{v}_i|=1$ for which the temperature cannot be distinguished from the square of the velocity. On the contrary models with different velocities are investigated for the lattice gas models (time and space being discretized) [3]. Recently [4], exact solutions for a $4\vec{v}_i$ model with two speeds 1,2 were constructed. For nontrivial temperature we have observed overshoots of the temperature for shock wave solutions.

Here we study a $8\vec{v}_i$ planar model with two speeds 1 and $\sqrt{2}$. Four velocities $\vec{v}_1+\vec{v}_5=\vec{v}_3+\vec{v}_7=0$ with speeds $|\vec{v}_1|=|\vec{v}_3|=1$ are along the medians of a square and four others $\vec{v}_2+\vec{v}_6=\vec{v}_4+\vec{v}_8$ with speeds $|\vec{v}_2|=|\vec{v}_4|=\sqrt{2}$ are along the diagonals. In the (x,y) plane the coordinates of $\vec{v}_i$ $i=1,\ldots 5$ are: (1,0),(1,1),(0,1),(-1,1),(-1,0). To each velocity $\vec{v}_i$ is associated a density N_i and we look at the solutions which only depend on x. Then $N_8=N_2, N_7=N_3, N_6=N_4$ and we write down the nonlinear equations for the 5 independent N_i

$$l_1N_1=2C+D,\ l_5N_5=-2C+D,\ l_2N_2=-l_4N_4=-C,\ l_3N_3=-D,\ l_1=l_2=\partial_t+\partial_x,\ l_3=\partial_t$$

$$l_4=l_5=\partial_t-\partial_x,\ C=\sqrt{5}(N_5N_2-N_1N_4),\ D=2d(N_3^2-N_1N_5) \tag{1.1}$$

C and D are the collision terms for particles with speeds $1,\sqrt{2}$ and 1,1. $\sqrt{5}$ in C and 2 in D are the relative velocities and d>0 the cross-section. The conservation laws of mass $M=N_1+N_5+2(N_2+N_3+N_4)$, momentum $J=N_1-N_5+2(N_2-N_4)$ and energy $E=(N_1+N_2)/2+N_3+2(N_2+N_4)$ being linear relations

$$M_t+J_x=0,\ J_t+(N_1+N_5+2(N_2+N_4))_x=0,\ E_t+((N_1-N_5)/2+2(N_2-N_4))_x=0, \tag{1.2}$$

in (1,1) we only have two nonlinear relations and we replace (1,1) by the system

$$1_2N_2+1_4N_4=0,\ 1_1N_1+41_2N_2=1_5N_5,\ 1_1N_1+21_2N_2+1_3N_3=0,\ 1_1N_1+21_2N_2=D,\ 1_2N_2=-C \qquad (1.3)$$

In section 2 we assume both $N_i(\eta=x-\xi t)$ and the existence of two Maxwellian states Ma_o,Ma_s (vanishing of C,D) with parameters n_{0i},s_{0i}. These parameters are constructed from four arbitrary parameters ξ,n_{02},n_{04} and $z_1=(s_{01}-n_{01})/(s_{02}-n_{02})$. We study the positivity of the Maxwellians and an invariance $\xi \leftrightarrow -\xi,\ldots$ allows to study only $\xi>0$. For shock front-speed $|\xi|>1$ the positivity is violated. Physically this means that the shock front cannot propagate along the x-axis faster than the maximum speed 1 of the particles. For Ma_o,Ma_s we study the characteristic velocities ξ_o,ξ_s associated to weak shocks $n_{0i}=s_{0i}$. These ξ_o,ξ_s belong to three intervals limited by ± 1,0 and either $(n_{02}-n_{04})/(n_{02}+n_{04})$ for Ma_o or $(s_{02}-s_{04})/(s_{02}+s_{04})$ for Ma_s. Further ξ lies in the interval ξ_o,ξ_s or ξ_s,ξ_o. Applying the Lax-Whitham stability theory [5], we show that the cubic characteristic velocity polynomial can also be found from a linearization around the Maxwellians, taking into account the lowest third-order differential operator.

In section 3 we construct the similarity shock wave solutions

$$N_i=n_{0i}+n_i/D,\ D=1+e^{\gamma\eta},\quad \eta=x-\xi t,\ s_{0i}=n_{0i}+n_i \qquad (1.4)$$

The previous z_1 arbitrary parameter is replaced by the cross-section $d>0$ with the restriction $\sqrt{5}\xi^2<d$. An additional relation gives γ as a function of the four arbitrary parameters and the above results are applied with d instead of z_1.

In section 4, for the similarity solutions, we introduce the macroscopic quantities: $M=M_0+M/D$, $J=J_0+J/D$, $E=E_0+E/D$, velocity $U=J/M$ and temperature $Te=2E/M-U^2$. To Ma_0 we associate the velocity $U_0=J_0/M_0$, shock velocity $V_o=U_o-\xi$, sound velocity $W_o=U_o-\xi_o$ and similarly for Ma_s. We show that the shocks are compressive (mass increasing accross the shock) and verify that the subsonic and supersonic inequalities are satisfied. Finally we observe that temperature overshoots accross the shock can exist. This effect becomes more pronounced when the shock front-speed $|\xi|$ increases.

2. Densities $N_i(\eta=x-\xi t)$ and two Maxwellian States Ma_o,Ma_s when $|\eta|\to\infty$.

2.1 Construction of Ma_o,Ma_s with parameters n_{oi} and s_{oi}

The three linear relations (1.3) become: $\partial_\eta((1-\xi)N_2-(1+\xi)N_4)=\partial_\eta((1-\xi)(N_1+2N_2)-\xi N_3)=\partial_\eta((1-\xi)(N_1+4N_2)+(1+\xi)N_5)=0$. We integrate from $\eta=-\infty$ to $+\infty$ and define $n_i=s_{0i}-n_{0i}$

$$n_2(1-\xi)=n_4(1+\xi),\ (n_1+2n_2)(1-\xi)=\xi n_3,\ (n_1+4n_2)(\xi-1)=n_5(\xi+1) \qquad (2.1)$$

The vanishing of the collision terms C,D give for Ma_o,Ma_s:

$$s_{02}s_{05}=s_{01}s_{04},n_{02}n_{05}=n_{01}n_{04},\ s_{03}^2=s_{01}s_{05},\ n_{03}^2=n_{01}n_{05} \qquad (2.2)$$

We have 11 parameters ξ,n_{0i},s_{0i} and 7 relations leaving 4 arbitrary parameters. We introduce intermediate parameters $z_i=n_i/n_2$, $i\neq 2$ which from (2.1) are ξ,z_1-dependent

$$z_3=(z_1+2)(1-\xi)/\xi, \quad z_4=(1-\xi)/(1+\xi), \quad z_5=-z_4(z_1+4) \tag{2.3}$$

From (2.2) we find four expressions for n_2 and relations for n_{03}, n_{05}

$$n_2(z_5-z_1z_4)=n_{01}z_4+n_{04}z_1-n_{02}z_5-n_{05}=-s_{01}z_4-s_{04}z_1+s_{02}z_5+s_{05}, \quad z_5-z_1z_4=-2z_4(z_1+2) \tag{2.4a}$$

$$n_2(z_3^2-z_1z_5)=n_{05}z_1+n_{01}z_5-2n_{03}z_3=-s_{05}z_1-s_{01}z_5+2s_{03}z_3, \tag{2.4b}$$

$$n_{05}=n_{01}\bar{n}_{04}, \quad n_{03}=n_{01}\sqrt{\bar{n}_{04}}, \quad \bar{n}_{04}=n_{04}/n_{02} \tag{2.5}$$

Substituting the (2.4a-b) relations into $s_{0i}=n_{0i}+n_2z_i$ we finally obtain:

$$s_{05}/s_{01}=((z_3\sqrt{\bar{n}_{04}}-z_5)/(z_1\sqrt{\bar{n}_{04}}-z_3))^2, \quad s_{01}=n_{01}(z_3-z_1\sqrt{\bar{n}_{04}})^2/(z_3^2-z_1z_5)$$
$$s_{02}=n_{02}(\bar{n}_{04}-z_4)(\bar{n}_{01}-z_1)/(2z_4(z_1+2)), \quad s_{04}/s_{02}=z_4(\bar{n}_{01}+z_1+4)/(\bar{n}_{01}-z_1),$$
$$s_{03}=n_{01}(z_1\sqrt{\bar{n}_{04}}-z_3)(z_3\sqrt{\bar{n}_{04}}-z_5)/(z_3^2-z_1z_5), \quad z_3^2-z_1z_5=z_4(z_1(z_1+4)+4(1-\xi^2))/\xi^2$$
$$\bar{n}_{01}(1-z_4\nu)=z_1+z_4\nu(z_1+4), \quad \nu=s_{01}/s_{05}, \quad \bar{n}_{01}-z_1=\nu z_4(\bar{n}_{01}+z_1+4), \quad \bar{n}_{01}=n_{01}/n_{02} \tag{2.6}$$

We choose for the arbitrary parameters n_{02}, n_{04}, ξ, z_1 and deduce the nonarbitrary ones (i) z_i from (2.3), n_{01} from (2.6), n_{03} and n_{05} from (2.5), (ii) s_{0i} from (2.6). We notice that n_2 and n_{0i} are known either from (2.4a-b) or from $s_{0i}-n_{0i}$.

2.2 Positivity of the two Maxwellians

Theorem 1: <u>If $|\xi|>1$ then Ma_0, Ma_s cannot be positive.</u>

(i) then $z_4<0$ and $z_3^2-z_1z_5>0$ for $s_{01}>0$, (ii) We prove that $-4<z_1<0$ violates positivity In (2.6) $\bar{n}_{01}(1-z_4\nu)>0$ while the r.h.s. is negative. (iii) $z_1<-4$ violates positivity. In that case $z_3>0, z_5<0, z_1\sqrt{\bar{n}_{04}}-z_3<0, z_3\sqrt{\bar{n}_{04}}-z_5>0$ and $s_{03}<0$. (iv) $z_1>0$ violates positivity due to $z_3<0, z_5>0, z_1\sqrt{\bar{n}_{04}}-z_3>0, z_3\sqrt{\bar{n}_{04}}-z_5<0$ and $s_{03}<0$.

We restrict our study to $|\xi|<1$ and assume for the four arbitrary parameters:

$$n_{02}>0, \quad n_{04}>0, \quad z_1>0, \quad 0<\xi<1, \quad \bar{n}_{04}>\sup(z_4, (z_3/z_1)^2) \tag{2.7}$$

Lemma 1: If in addition to (2.7) we assume $\bar{n}_{01}>0$ then Ma_0, Ma_s are positive and $\bar{n}_{01}>z_1$

From (2.3) we find $z_4>0, z_3>0, z_5<0$ and $z_3^2-z_1z_5>0$. Further the r.h.s. of $\bar{n}_{01}-z_1$ is positive in (2.6) and then all n_{0i} and s_{0i} are positive in (2.5-6).

Lemma 2: The condition $n_{01}>0$ or $z_4\nu<1$ with $z_1>0$ and $\sqrt{\bar{n}_{04}}>z_3/z_1$ can be satisfied either if (i) $2\xi^2<1$ or if (ii) $2\xi^2>1$ and $z_1<2(\sqrt{1-\xi^2})(\xi-\sqrt{1-\xi^2})^{-1}$ or (iii) $2\xi^2>1$, $z_1>2(\sqrt{1-\xi^2})/(\xi-\sqrt{1-\xi^2})$, $\sqrt{\bar{n}_{04}}<(z_3\sqrt{z_4}-z_5)/(z_1\sqrt{z_4}-z_3)$, $\bar{n}_{04}>z_4$.

For the (i) proof we see that $\xi<\sqrt{1-\xi^2} \rightarrow 2>z_1(\xi/(\sqrt{1-\xi^2})-1) \rightarrow z_3>z_1\sqrt{z_4} \rightarrow z_3\sqrt{\bar{n}_{04}}-z_5> \sqrt{z_4}(z_1\sqrt{\bar{n}_{04}}-z_3) \rightarrow z_4\nu<1$. For the (ii) proof we see that $\xi>\sqrt{1-\xi^2} \rightarrow z_3>z_1\sqrt{z_4} \rightarrow z_4\nu<1$. For the (iii) proof we see that $z_3<z_1\sqrt{z_4}$ still leading to $z_4\nu<1$ and we note that due to $\sqrt{z_4}(z_1\sqrt{z_4}-z_3)+z_5-z_3\sqrt{z_4}<0$, the lower $\sqrt{z_4}$ bound (2.7) for $\sqrt{\bar{n}_{04}}$ and the present upper bound are compatible.

Theorem 2: <u>If (2.7) and either (i) or (ii) or (iii) of lemma 2 are satisfied then Ma_0 and Ma_s are positive.</u>

2.3 Invariance Transform

We notice that the starting (2.1-2) relations are invariant under the transform T

$$\xi \leftrightarrow -\xi,\ n_{01} \leftrightarrow n_{05},\ n_{02} \leftrightarrow n_{04},\ s_{01} \leftrightarrow s_{05},\ s_{02} \leftrightarrow s_{04},\ n_{03}, s_{03} \leftrightarrow n_{03}, s_{03} \quad (2.8)$$

This means that from any $\xi>0$ solution we can deduce a $\xi<0$ one.

2.4 Characteristic velocities associated to weak shocks

We want to obtain the velocities ξ_0 of Ma_0 and ξ_s of Ma_s when $n_{0i}=s_{0i}$ or $n_2=0$. Two methods exist and here we present the first one.
We assume $n_2=0$, leading to $n_i=0$ or $s_{0i}=n_{0i}$, and fom (2.4a-b) obtain two terms for z_1

$$z_1(n_{05}-n_{01}z_4-2n_{03}(1-\xi)/\xi)=4n_{01}z_4+4n_{03}(1-\xi)/\xi,\quad z_4=(1-\xi)/(1+\xi) \quad (2.9a)$$

$$z_1(n_{04}+z_4n_{02})=n_{05}-z_4(n_{01}+4n_{02}) \quad (2.9b)$$

We eliminate z_1, obtain a cubic ξpolynomial and use (2.5) for the coefficients n_{0i}

$$\hat{N}(\xi,n_{0i})=-n_{01}(\bar{n}_{04}^{-2}\xi(1+\xi)^2+\xi(1-\xi)^2)+2\xi(1-\xi^2)\bar{n}_{04}(n_{01}+4n_{02})$$
$$+2\sqrt{\bar{n}_{04}}(n_{01}+2n_{02})(\xi^2-1)(1-\xi-\bar{n}_{04}(1+\xi)) \quad (2.10)$$

$\hat{N}(\xi_0,n_{0i})=0$ gives three characteristic values ξ_0 for the Ma_0. Similarly starting with (2.4a-b) and s_{0i} instead of n_{0i}, eliminating z_1 we obtain a cubic $\hat{S}(\xi,s_{0i})$ polynomial (2.10) with s_{0i} instead of n_{0i}. $\hat{S}(\xi_s,s_{0i})=0$ gives the characteristic values for Ma_s.

2.5 Intervals for the characteristic values ξ_0 and the shock front ξ

$$\hat{N}(\xi=1)=-4n_{01}\bar{n}_{04}^{-2}<0,\ \hat{N}(\xi=-1)=4n_{01}>0,\ \hat{N}(\xi=0)=2\sqrt{\bar{n}_{04}}(\bar{n}_{04}-1)(n_{01}+2n_{02})$$
$$\hat{N}(\xi=(1-\bar{n}_{04})/(1+\bar{n}_{04})\text{ or }z_4=\bar{n}_{04})=32\bar{n}_{04}^{-2}n_{02}(1-\bar{n}_{04})/(1+\bar{n}_{04})^3 \gtrless 0 \text{ if } 1\gtrless\bar{n}_{04} \quad (2.11)$$

We obtain the three possible intervals for ξ_0:

$$\bar{n}_{04}<1:\ -1<\xi_0<0,\ 0<\xi_0<(1-\bar{n}_{04})/(1+\bar{n}_{04}),\ (1-\bar{n}_{04})/(1+\bar{n}_{04})<\xi_0<1$$
$$\bar{n}_{04}>1:\ -1<\xi_0<(1-\bar{n}_{04})/(1+\bar{n}_{04}),\ (1-\bar{n}_{04})/(1+\bar{n}_{04})<\xi_0<0,\ 0<\xi_0<1 \quad (2.12)$$

Substituting $\bar{s}_{04},\xi_s$ instead of $\bar{n}_{04},\xi_0$ we obtain similar intervals for Ma_s instead of Ma_0. However only one of the three roots correspond to weak shocks. We must locate the shock front ξ. We recall that positivity requires $\bar{n}_{04}>z_4$ or $\xi>(1-\bar{n}_{04})/(1+\bar{n}_{04})$ Still assuming $\xi>0$ we find from (2.11-12) that ξ,ξ_0 belong to the interval:

$$n_{04}<1:\ \xi,\xi_0 \in [(1-n_{04})/(1+n_{04}),1],\ n_{04}>1:\ \xi_0,\xi \in [0,1] \tag{2.13}$$

Similar results hold for ξ_s,ξ.

2.6 Linearization around the Maxwellian Ma_0

In the second method, for the determination of the characteristic ξ_0,ξ_s (Lax-Whitham stability theory [5]), we linearize around Ma_0 or Ma_s. For instance $N_i=n_{0i}(1+X_i)$ with $X_i(x,t)$ for Ma_0 and keeping only terms linear in X_i we find with $D^{\pm}=\partial_t\pm\partial_x$

$$n_{02}D^+X_2+n_{04}D^-X_4=D^+(n_{01}X_1+2n_{02}X_2)+n_{03}\partial_tX_3=D^+(n_{01}X_1+4n_{02}X_2)-D^-X_5=0$$

$$D^+X_2+\sqrt{5}n_{05}(X_2+X_5-X_1-X_4)=0,\quad \partial_tX_3+2dn_{03}(2X_3-X_1-X_5)=0 \tag{2.14}$$

We can rewrite these relations like $\Delta\tilde{X}=0$ with Δ a matrix fifth-order differential operator and $\tilde{X}$ a column vector with $X_1,X_2,\ldots x_5$ components. We find for $\tilde{\Delta}$, the determinant of Δ, a sum of three operators of the fifth, fourth and third order:

$$\begin{aligned}\tilde{\Delta}=&-D^{+2}D^{-2}\partial_t+D^+D^-(4n_{03}dD^-D^++D^+\partial_t(n_{01}(\sqrt{5}+2d)+2\sqrt{5}n_{02})+D^-\partial_t(n_{05}(\sqrt{5}+2d)+2\sqrt{5}n_{04}))\\&+2d\sqrt{5}(D^{+2}\partial_tn_{01}^2+D^-D^+\partial_t(2n_{01}n_{05}+4(n_{01}n_{04}+n_{05}n_{02}))+2D^+D^-n_{03}(D^+(n_{01}+2n_{02})+D^-(n_{05}\\&+2n_{04}))+D^{-2}\partial_tn_{05}^2)\end{aligned} \tag{2.15}$$

We notice that under the transform T, which means here $t\rightarrow -t$, then $D^{\pm}\rightarrow -D^{\mp}$, the odd differential operators change their sign and not the even one. We apply the three operators to a function $h(x-\xi t)$ and find three ξ-dependent polynomials. The first one has roots $\xi=0,\pm1$, the second ±1 and

$$\xi(n_{01}(\sqrt{5}+2d)+2\sqrt{5}n_{02})(1-\xi-(1+\xi)\bar{n}_{04})+4dn_{01}\sqrt{\bar{n}_{04}}(1-\xi^2)=0,\quad \xi^{\pm}\text{roots},\xi^+>0,\xi^-<0,|\xi^{\pm}|<1$$

whereas the third one is just the cubic $2d\sqrt{5}\hat{N}(\xi)/n_{01}$ with $\hat{N}$ defined in (2.11). In conclusion the characteristic velocities ξ_0,ξ_s are associated with the lowest differential operators appearing in the linearizations around Ma_0,Ma_s.

2.7 $\hat{N}(\xi,n_{0i})$ sign for the shock front velocity ξ

We first assume $\bar{s}_{04}-\bar{n}_{04}=\varepsilon$ finite and recall both the rational $\bar{n}_{01}$ relation $\bar{s}_{04}=z_4(\bar{n}_{01}+z_1+4)/(\bar{n}_{01}-z_1)$ and the linear n_2,n_{0i} relation (2.4a). We deduce z_1 as a rational ε relation and n_2 proportional to ε, $n_2\varepsilon<0$ for the solutions of the Theorem 2

$$-n_2=\varepsilon(\bar{n}_{01}-z_1)n_{02}/2z_4(z_1+2),\quad z_1=(\bar{n}_{05}-z_4(\bar{n}_{01}+4)+\varepsilon\bar{n}_{01})/(\bar{n}_{04}+z_4+\varepsilon)=z_1(\varepsilon) \tag{2.16}$$

However (2.4b) is also a n_2 (or ε),n_{0i} relation:

$$\bar{n}_{05}z_1+\bar{n}_{01}z_5-2\bar{n}_{03}z_3=-\varepsilon(\bar{n}_{01}-z_1)(z_3^2-z_1z_5)/2z_4(z_1+2) \tag{2.17}$$

In the l.h.s. we eliminate z_1 given by (2.16) and find a term independent of ε and another ε-dependent so that we rewrite (2.17):

$$N(\xi,n_{0i})\ (\bar n_{01}/\xi(1+\xi)^{2}n_{02})=\varepsilon(B(\varepsilon,\xi,n_{0i})+C(\varepsilon,\xi,n_{0i}), \tag{2.18}$$
$$B=\bar n_{01}(\bar n_{05}-z_4\bar n_{01}-2(1-\xi)\bar n_{03}/\xi)-4(z_4\bar n_{01}+(1-\xi)\bar n_{03}/\xi)$$
$$C=(\bar n_{01}-z_1)(\bar n_{04}+z_4+\varepsilon)(z_3^2-z_1z_5)/2z_4(z_1+2)$$

We recall that our class of positive solutions is for $\xi>0, z_1>0, z_5<0, \bar n_{01}>0, \bar n_{04}>z_4>0$, so that for $\varepsilon>0$ or $\varepsilon>-z_4-\bar n_{04}$ then $C>0$ but the sign of B is unclear. Second we assume ε small and the r.h.s. of (2.18) is $\varepsilon(B(\varepsilon=0)+C(\varepsilon=0))+0(\varepsilon^2)$. We find $z_1(\varepsilon=0)B(0)=4(z_4\bar n_{01}+(1-\xi)\bar n_{03}/\xi)(\bar n_{01}-z_1(0))>0$ so that ε and $\hat N$ have the same sign and $n_2\hat N<0$ in (2.16-18). Third, starting in (2.4a-b) with the s_{0i} instead of the n_{0i} we can perform a similar study. We find in(2.6) $\bar n_{04}=z_4(\bar s_{01}+z_1+4)/(\bar s_{01}-z_1), \bar s_{0i}=s_{0i}/s_{02}$ and $\bar s_{01}>z_1$, $\bar s_{04}>z_4$ for positivity. Then z_1 is given in (2.16) with $\bar n_{0i}\to\bar s_{0i}, \varepsilon\to-\varepsilon$, and we find (2.17) with the same change. In (2.18) we get B,C with $n_{0i}\to s_{0i}$ but with $-\hat S$ instead of $\hat N$, the l.h.s. being: $-\hat S(\xi,s_{0i})\ (\bar s_{01}/\xi(1+\xi)s_{02})$. For ε small we deduce: $\hat S n_2>0$, $\hat S\varepsilon<0$ and $\hat S\hat N<0$ for the shock front velocity ξ. By continuity these results hold for all ξ,ξ_0 (or ξ,ξ_s) belonging to one of the three possible intervals. For $\xi>0$ and (2.13) for ξ,ξ_0 and the corresponding intervals for ξ,ξ_s we get the final result:

$$\xi,\xi_0\ \in[\sup(0,(1-\bar n_{04})/(1+\bar n_{04})),1],\quad \xi,\xi_s\in[\sup(0,(1-\bar s_{04})/(1+\bar s_{04})),1]$$
$$\bar s_{04}>\bar n_{04}\to n_2<0,\ \hat N(\xi)>0, \hat S(\xi)<0,\ \xi_s<\xi<\xi_0;\ \bar s_{04}<\bar n_{04}\to n_2>0,\ \hat N<0, \hat S>0,\ \xi_0<\xi<\xi_s \tag{2.20}$$

3. Similarity Solutions: $N_i=n_{0i}+n_i/D$, $D=1+e^{\gamma\eta}$, $\eta=x-\xi t$

The difference with section 2 is that the $z_1=n_1/n_2$ arbitrary parameter will be replaced by the cross-section d. The three linear (1.3) relations give (2.1)while the nonlinear ones give (2.2) and two new relations:

$$\gamma(1-\xi)(2n_2+n_1)=2d(n_3^2-n_1n_5),\qquad \gamma(1-\xi)n_2=\sqrt5(n_1n_4-n_2n_5) \tag{3.1}$$

Still introducing the $z_i=n_i/n_2$ parameters satisfying (2.3) we obtain from (3.3)

$$\gamma=2\sqrt5 n_2(z_1+2)/(1+\xi)\ \to\ n_2\gamma>0 \text{ for } \xi>0,z_1>0,\ 2d=\sqrt5(2+z_1)(z_1z_4-z_5)/(z_3^2-z_1z_5) \tag{3.2}$$

or $(z_1+2)^2(1-\sqrt5\xi^2/d)=4\xi^2$ or

$$z_1=2(-1\pm\xi/\sqrt\delta),\ \delta=1-\xi^2\sqrt5/d>0\ \to\ \xi^2<d/\sqrt5,\ z_1>0\ \to\ \xi^2>1/(1+\sqrt5/d) \tag{3.3}$$

with the choice + in (3.3) in order that $z_1>0$. Here we choose for the free parameters

$$n_{02}>0,\ n_{04}>0,\ d>0,\ 0<\xi<1 \text{ with } \xi^2<d/\sqrt5 \tag{3.4}$$

and reconstruct the nonarbitrary ones: (i) z_1 by (3.3) and all z_i are functions of ξ,d, (ii) $\bar n_{01}$ is given by (2.6) and we deduce n_{03},n_{05} with (2.5), (iii) n_2 is given by (2.4) and we deduce $n_i=n_2z_i$, (iv) finally (3.2) gives γ. For the positivity of the densities N_i it is sufficient that Ma_0 and Ma_s be positive so that we apply the section 2 results with constraints on d instead of z_1.

The proof of Theorem 1 is trivial: if $|\xi|>1$ then $z_4<0$, $z_3^2-z_1z_5=4z_4\xi^2\sqrt{5}/d\delta<0$ and $s_{01}<0$. For the Theorem 2 we find in the three possible cases:

$$(i): \sqrt{2}\xi<1,\ \sqrt{\bar{n}_{04}}(\xi-\sqrt{\delta})>1-\xi \quad (ii): \sqrt{2}\xi>1,\ \sqrt{\bar{n}_{04}}(\xi-\sqrt{\delta})>1-\xi,\ 1+\xi^4>\xi^2(2+5/4d^2)$$
$$(iii): \sqrt{2}\xi>1,\ 1+\xi^4<\xi^2(2+5/4d^2),\ 1<\sqrt{z_4}\sqrt{\bar{n}_{04}}<(\sqrt{\delta}+\xi+\sqrt{1-\xi^2})/(-\sqrt{\delta}+\xi-\sqrt{1-\xi^2}) \tag{3.5}$$

4. Macroscopic Quantities

We associate the similarity solutions N_i to the mass $M=M_0+M/D$, momentum J, energy E, velocity $U=J/M$, shock velocity $U-\xi$ and temperature $Te=2E/M-U^2$ defined in section 1. To Ma_0 we associate $M_0, J_0, E_0, U_0=J_0/M_0, V_0=U_0-\xi, Te_0=2E_0/M_0-U_0^2$ defined in section 1 and to Ma_s the corresponding quantities: $M_s=M_0+M, J_s=J_0+J, E_s=E_0+E, U_s=J_s/M_s, V_s=U_s-\xi, Te_s=2E_s/M_s-U_s^2$. We define the sound wave velocities $W_0=U_0-\xi_0, W_s=U_s-\xi_s$ and find for the solutions satisfying Theorem 2: $\xi>0$, $\bar{n}_{04}>z_4\ldots$

$$M_0=(n_{01}+2n_{02})(1+\bar{n}_{04})+2n_{03}>0,\ M_s>0,\ J_0=(1-\bar{n}_{04})(n_{01}+2n_{02})$$
$$U_0=J_0/M_0 \gtrless 0 \text{ if } \bar{n}_{04} \lessgtr 1,\ U_s=J_s/M_s \gtrless 0 \text{ if } \bar{s}_{04} \lessgtr 1 \tag{4.1}$$

$$\partial_\eta(-\xi M+J)=0,\ \eta=x-\xi t, \rightarrow J=\xi M,\ M_0V_0=M_sV_s=(1+\xi)(n_{01}+2n_{02})(z_4-\bar{n}_{04})-2\xi n_{03} \rightarrow V_0<0, V_s<0$$
$$W_0M_0=(n_{01}+2n_{02})((1-\xi_0)-(1+\xi_0)\bar{n}_{04})-2\xi_0n_{03}<0 \text{ if } \bar{n}_{04}>(1-\xi_0)/(1+\xi_0)$$
$$\text{or } \xi_0>(1-\bar{n}_{04})/(1+\bar{n}_{04}) \tag{4.2}$$

4.1 Compressive shocks, supersonic and subsonic inequalities for $\xi>0$...(Theorem 2)

(4.1-2) represent the constraints on the two Maxwellians Ma_0, Ma_s, but other relations exist between $M=n_2(z_1+z_5+2+2z_3+2z_4)$, n_2 and γ which appears in $D=1+e^{\gamma\eta}$

$$M=M_s-M_0=2n_2(2+z_1)/\xi(1+\xi)=\gamma/\sqrt{5}d \rightarrow \gamma M>0, \gamma n_2>0 \text{ if } \xi>0, z_1>0 \tag{4.3}$$

Lemma 3: For Ma_0 then either $\xi_0<\xi$ with the supersonic inequality $|V_0|>|W_0|$ or $\xi<\xi_0$ with the subsonic inequality $|V_0|<|W_0|$. We have three cases:
(i) $\xi_0<\xi$ and $\bar{n}_{04}>1$, then $U_0<0$ from (4.1) and $U_0-\xi<U_0-\xi_0<0$ or $-U_0+\xi=|V_0|>-U_0+\xi_0=|W_0|$
(ii) $\xi_0<\xi$ and $\bar{n}_{04}<1$, then $\xi_0>(1-\bar{n}_{04})/(1+\bar{n}_{04})$ from (2.13) and $W_0<0$ from (4.2) still leading to $U_0-\xi<U_0-\xi_0<0$. (iii) $\xi<\xi_0$, then $V_0<0$ and $W_0=U_0-\xi_0<U_0-\xi=V_0<0$ or $|W_0|>|V_0|$.

Lemma 4: Either Ma_0 verifies the supersonic inequality , Ma_s the subsonic one or the converse. First we note that Lemma 3 for Ma_s instead of Ma_0 gives either $\xi_s<\xi$ and the supersonic inequality or $\xi<\xi_s$ and the subsonic one. Second from (2.20) we note that ξ is always in the interval (ξ_0,ξ_s) or (ξ_s,ξ_0).

Theorem 3: For the positive solutions satisfying Theorem 2 then the shocks are compressive and the supersonic inequality for the upstream as well as the subsonic inequality for the downstream are satisfied. We distinguish between M>0 and M<0.

(i) $M>0$: then $n_2>0,\gamma>0$ from (4.3), $\eta\to\infty$ gives Ma_0 and $\eta\to-\infty$ Ma_s, $V_0<0$ and $V_s<0$ → Ma_0 upstream and Ma_s downstream, $M_s>M_0$ → compressive shock (mass increasing accross the shock), $|V_0|>|W_0|$, $|V_s|<|W_s|$ from Lemmas 3-4 → supersonic and subsonic inequalities. (ii) $M<0$: then $n_2<0,\gamma<0$ from (4.3), $\eta\to\infty$ gives Ma_s and $\eta\to-\infty$ Ma_0, $V_0<0,V_s<0$ → Ma_s upstream, Ma_0 downstream, $M_s<M_0$ → compressive shock, $|V_s|>|W_s|$, $|V_0|<|W_0|$ → supersonic and subsonic inequalities.

4.2 Shock Entropy and Local Entropy

We define both the shock entropy $H_1=\Sigma(-\xi+a_i)N_i\log N_i$ and the local entropy $H=\Sigma N_i\log N_i$ for the similarity solutions $N_i(\eta=x-\xi t)=n_{0i}+n_i/D$ where in (1.1) we put $1_i=\partial_t+a_i\partial_x$, $a_1=a_2=a_8=1$, $a_3=a_7=0$, $a_4=a_5=a_6=-1$. From (1.1) we find:

$$\partial_\eta H_1=\bar{C}\log(N_1N_4/N_2N_5)+D\log(N_1N_5/N_3^2)=2\sqrt{5}(N_2N_5-N_1N_4)\log(N_1N_4/N_2N_5)$$
$$+2d(N_3^2-N_1N_5)\log(N_1N_5/N_3^2)<0,\quad \bar{C}=2C$$

$$\partial_t H+\partial_x(N_1\log N_1-N_5\log N_5+2N_2\log N_2-2N_4\log N_4)=2C\log(N_1N_4/N_2N_5)+D\log(N_1N_5/N_3^2) \quad (4.4)$$

From $H_{1\eta}<0$ we see that H_1 is a monotonic η function with $H_1(\eta=-\infty)>H_1(\eta=+\infty)$. On the contrary $H_\eta=-(1/\xi)H_t$ has not a well-defined sign. Consequently overshoots of the local entropy can exist and have been previously observed for other models [6,4].

4.3 Overshoots of the temperature

For the temperature $Te(\eta=x-\xi t)$, corresponding to the similarity solutions, we look both at the $|\eta|\to\infty$ values Te_0,Te_s for Ma_0,Ma_s and at $Te(\eta=0)$:

$$Te(\eta)=2(E_0+E/D)/(M_0+M/D)-U^2,\quad U=(J_0+J/D)^2/(M_0+M/D)^2,\quad D=1+e^{\gamma\eta}$$
$$Ma_0:\ Te_0=2E_0/M_0-U_0^2,\ Ma_s:\ Te_s=2E_s/M_s-U_s^2,\ Te(0)=2(E_0+E/2)/(M_0+M/2)$$
$$-(J_0+J/2)^2/(M_0+M/2)^2 \quad (4.5)$$

For the term $-U^2$ absent, then Te becomes a monotonic η function. When this term is present then overshoots of Te, accross the shock, can occur. For compressive shocks we require that both the mass and the temperature increase accross the shock or $Te_0 \gtrless Te_s$ if $M_0 \gtrless M_s$. A good criterion for the existence of an overshoot is that $Te(0)$ be larger than the two $|\eta|\to\infty$ values. Consequently for compressive shocks the effect will occur if:

$$M>0:\ Te_0<Te_s<Te(0),\quad M<0:\ Te_s<Te_0<Te(0) \quad (4.6)$$

4.4 Numerical Calculations

In FIG.1A-B we choose for ξ a value 0.97 closed to the highest possible value 1, for the cross-section d=8.04 and for Ma_0 $n_{02}=1$ and two values $n_{04}=0.1<1$, $20.1>1$. We present M,H,Te normalised to their highest downstream values. In FIG.1A with $\bar{n}_{04}<\bar{s}_{04}$ we verify that $\xi_s<\xi<\xi_0$ and the supersonic and subsonic inequalities are satisfied.

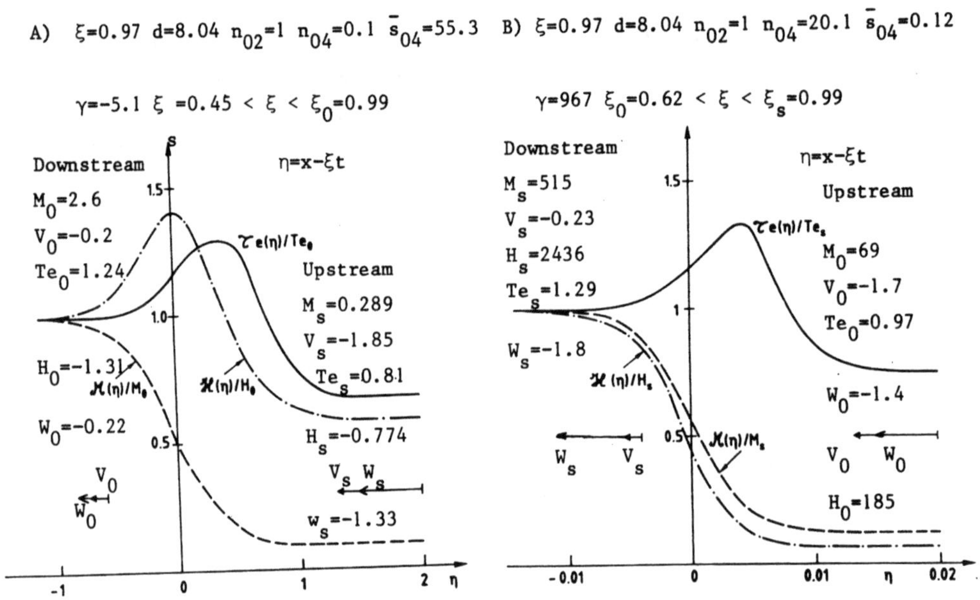

FIG.1A-B: Shock Wave for the $8\vec{v}_i$ Model: Overshoots for $Te(\eta)$ with high shock speed

M<0 leads to Ma_s,Ma_0 for the upstream and downstream. We see overshoots both for Te and H. In FIG.1B with $\bar{n}_{04}>\bar{s}_{04}$ we verify both $\xi_0<\xi<\xi_s$ and the supersonic upstream Ma_0 and subsonic downstream Ma_s inequalities. We see an overshoot for Te but not for H. The great difference concerns the two limits of the local entropy $-H$. In FIG.1A $-H_{up}=-H_s<-H_{down}=-H_0$ the local entropy increases while it decreases in FIG.1B with $-H_{up}=-H_0>-H_{down}=-H_s$. In FIG.2 we choose d=1.2, $\xi=0.64$, $n_{04}=5.6<\bar{s}_{04}$ and we verify both $\xi_s<\xi<\xi_0$ and the supersonic Ma_s and subsonic Ma_0 inequalities. There are no overshoot both for Te, H, M. However (like in FIG.1A) $-H_{up}=-H_s<-H_{down}=-H_0$ with an increase of $-H$ at the limits, corresponds to the physically interesting case.

Calling $R=(\text{Max } Te(0))/\text{Sup}(Te_0,Te_s)$ we have found (for instance for the class of solutions $-H_{down}>-H_{up}$) that R increases with ξ. For $\xi=0.8,0.85,0.9$ we get R=1.08, 1.144, 1.166 with d=1.8, 2.06, 2.4. Finally I notice that exact solutions for a $9\vec{v}_i$ 3 speeds model [3,7] have been found.

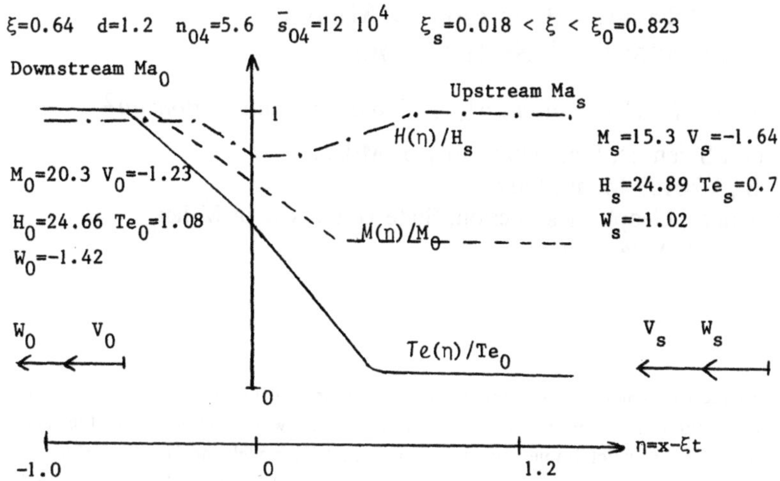

FIG.2: Shock Wave for the $8\vec{v}_i$ Model: No overshoot for $Te(\eta)$

References

1 Gatignol R."Lecture Notes in Physics 36 Springer 1975, Trans.The.Stat.Phys. 16 837 1987; Cabannes H. Trans.The.Stat.Phys.16 809 1987; Platkoski T. and Illner R. Siam Rev. 30 1988; D'Humières D."Proceedings in Phys." 46 Springer Ed. Manneville p. 186

2 Cornille H."Exact Sol. to the Boltzmann Eq." Ed. Conte R. Kluwer Acad.Publish.1990

3 Frisch U. et al Complex Sys.1 649 1987;"Discrete Kinetic Theory,Lattice Gas Dynamics" Ed. Monaco R. World Scient. Singapore 1989: D'Humières et al 102, Rothman D. 286, Higuera J. 162, Hénon M. 160...; Chen S. Lee M. Zhao K. Doolen D. Physica D37 42 1989; Qian Y.H. D'Humières D. Lallemand P. to appear in Jour.Statist.Phys. 1990

4 Cornille H. and Qian Y.H. Comp.Rend.Acad.Scien.309 1883 1989, Saclay SphT/89-113

5 Lax P.D. Com.Pure Ap.Mat. 10 537 1957; Whitham G.B. Com.Pure Ap.Mat. 12 113 1959

6 Platkowski T. Trans.The.Statis.Phys. 18 221 1989

7 Nadiga B. et al"Progr. Astronaut. Aeronaut." Ed. Muntz E.P. vol.118 AIAA 1988,p155

Teaching Direct and Inverse Problems by Means of User Friendly Interfaces: A Status Report

G. Crosta[1], *D. Bertoncelli*[1], *F. Coladonato*[1], *M. Gandolfi*[1], *and G. Ponzini*[2]

[1]Dept. of Information Sciences, State University of Milan,
9, v. Moretto da B., I-20133 Milan, Italy
[2]Dept. of Earth Sciences, Geophysics Section, State University of Milan,
7, v. Cicognara, I-20129 Milan, Italy

Introduction

It is by now widely accepted that a computer is just another means, by which human beings communicate. Consider e.g. the task of communicating to an increasingly wider audience both the scope and limitations of a computer code, which implements some mathematical model: the functions of a computer must then consist of both crunching numbers and handling graphics information. The user-- friendly interfaces we are developing apply to codes, which model some processes in groundwater flow: in this area simulation (the *direct* problem) and parameter identification (the *inverse* problem) are closely related. We shall describe a direct problem solver, which simulates flow in a multi- layered porous medium and a package of inverse problem solvers, where different methods are applied. The users which we expect to "benefit" from our products are graduate students in the College of Science, State University of Milan. The description of each product will consist of an outline of the model, of the corresponding algorithm and its implementation.

1 - The Direct Problem Solver

We consider time dependent groundwater flow in a heterogeneous porous medium, which consists of three stacked subdomains:

i) a *confined aquifer* (subscript C in the quantities which follow), bounded below by an impervious layer and above by

ii) an *aquitard* (subscript A);

iii) a *free surface aquifer* (subscript F) on the top.

In the *domain* $Q := D \times [0, Z] \times [0, T]$, where D is a rectangle, we let $x \in D \subset \mathbf{R}^2$, $z \in [0, Z]$, $t \in [0, T]$. We consider the *initial - boundary value* (IBV) *problem*: find the potentials p_A, p_C, p_F s.t.

$$\text{at } z = 0 \qquad \nabla_T \bullet (a_C \nabla_T p_C) + (\kappa_A \partial_z p_A)\big|_{z=0^+} = f_C + s_C \partial_t p_C \tag{1.1}$$

$$p_C\big|_{\partial D} = p_{CB} ; \qquad p_C(t = 0) = p_{C0}$$

$$\text{at } z = Z \qquad \nabla_T \bullet (h_F \kappa_F \nabla_T p_F) - (\kappa_A \partial_z p_A)\big|_{z=Z^+} = f_F + s_F \partial_t p_F$$

$$p_F\big|_{\partial D} = p_{FB} ; \qquad p_F(t = 0) = p_{F0}; \qquad h_F := p_F - b_F \tag{1.2}$$

$$\text{in } 0<z<Z \qquad \partial_z (\kappa_A \partial_z p_A) = s_A \partial_t p_A \tag{1.3}$$

$$p_A(z = 0) = p_C ; \qquad p_A(z = Z) = p_F ; \qquad p_A(t = 0) = p_{A0}.$$

Here ∇_T stands for differentiation w.r. to x. The system parameters are the position dependent quantities a_C, κ_A, κ_F, s_A, s_C, s_F; b_F is the (known) height of the free-surface aquifer bottom; the source terms are f_C, f_F.

A solution to the above problem is known to exist if the parameters, the source terms, the initial (p_{A0}, p_{C0}, p_{F0}) and boundary (p_{CB}, p_{FB}) conditions are sufficiently regular.

The PDEs are discretised according to a Crank- Nicholson scheme, suitably adapted to the quasi- linear case (Eqn. 1.2) (Gambolati *et al.*, 1984). Details are provided by Ponzini *et al.* (1989b).

The interfaced code is installed on an *ibm6150/125 RT* microcomputer and consists of: an input/output graphics menu module, an algebraic system solver and a graphics transect module.

The displayed interface (Fig. 1) shows:

i) a *function menu* (top row),
ii) a *domain window* (right rectangle),
iii) two *transect windows* (left rectangles),
iv) *online help* and *message area* (bottom row).

All items appearing in the graphics interface are written in modern font.

In the current implementation the number of domain nodes is fixed, as well as the mesh size (square) and side. The data structure relies on domain nodes i.e., all quantities considered by the code are appended to a node.

Data are entered by the update function. First the subdomain (see above) has to be selected, then each node may be addressed by the mouse. A pop-up menu shows which parameters, source terms

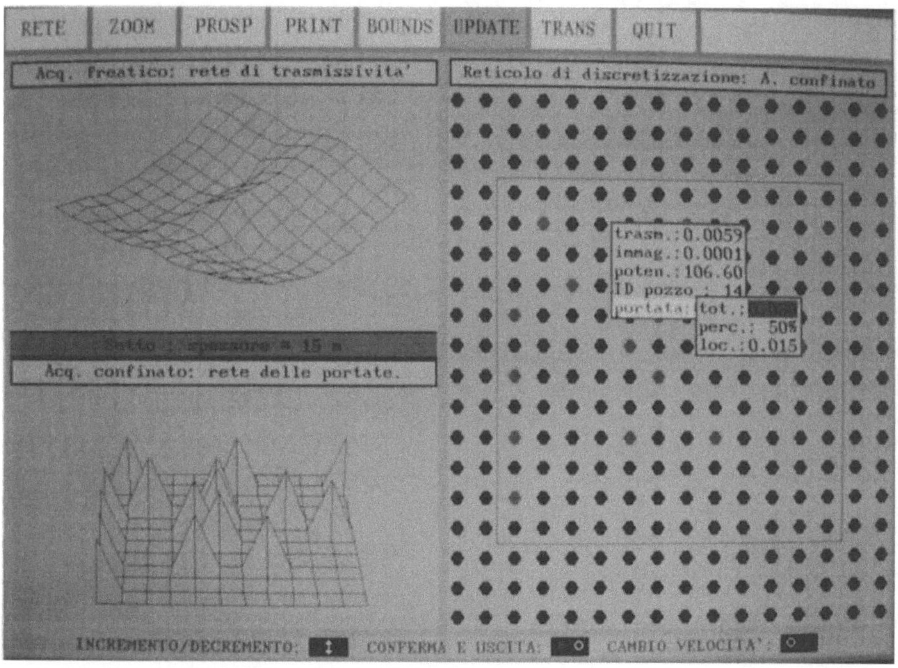

Fig. 1 User – friendly interface to the direct problem solver: the right window shows a data entry pop-up menu; the left windows show transects of some data and parameters.

and initial conditions pertain to that node. A file stores the sets of default values, which correspond to the experimental case study described by Ponzini *et. al.* (1989b). Once a **field** in the menu e.g. the **source term** has been selected, sliding the mouse increases or decreases the total discharge rate at that node. The affected field is highlighted (see Fig. 1). Since fully penetrating wells are being simulated, the total discharge rate can be shared between the free–surface and the confined aquifers according to an arbitrary percentage coefficient. Wherever the discharge rate has been given a nonzero value, the **node** color is changed (red instead of black on the screen).

By selecting the **transect** function (**rete** in Fig. 1) and by going through a sequence of pop–up menus (not shown) the perspective view of the discharge rate in the considered domain can be displayed (**bottom left window**): values are linearly interpolated. The same holds for other quantities: the **top left window** shows a transect of the conductivity coefficient in the free–surface layer, K_F. Some of the transect parameters (viewing azimuth and elevation angles, observer' s distance) can be set by means of the **perspective** (**prosp**) function.

Once data entry is complete i.e., upon exiting the **update** function, steady state potentials are computed, by assuming that the wells have been discharging all the time at the (constant) entered rates. Said potentials are then used as *reference values* during the solution of the time dependent IBV problem. The latter is started by activating the **transient** function. When computed potential values at each node at a given time differ from the respective reference values by less than a prescribed amount, the transient is deemed to have "ended". The above referred machine typically takes 3m30s to compute potentials at 29 unequally spaced time instants in a grid of 16 x 18 nodes. The code is written in *C*. At this stage arrays of computed potentials are available for readout as numerical tables by means of the **bounds** function in the **right window** (partly hidden in Fig. 2) and for off– line animation as transects in the (partly hidden) **left**

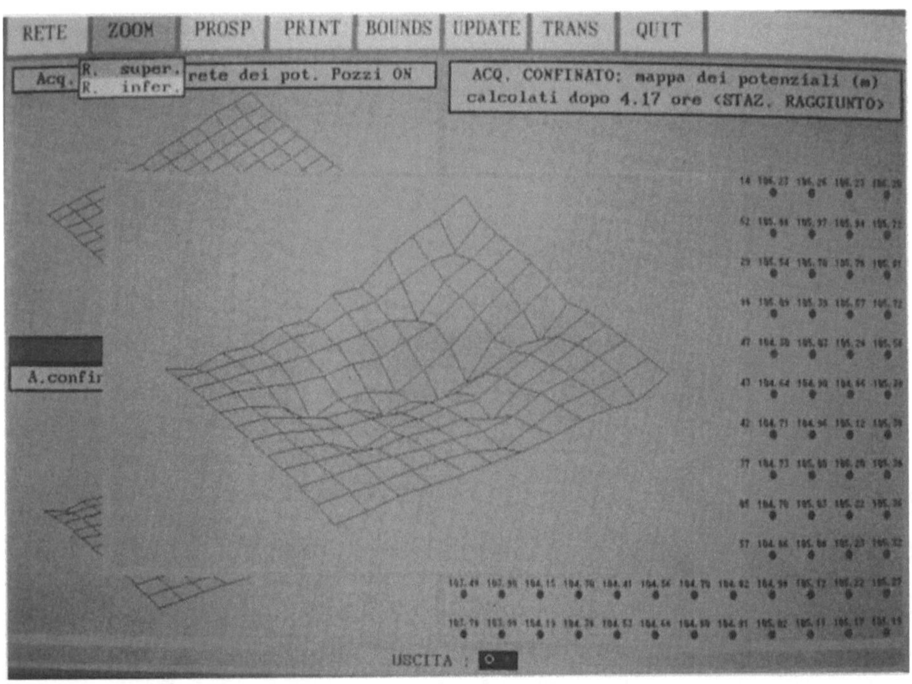

Fig. 2
User – friendly interface to the direct problem solver: **zoom** function activated on the transect of the computed potential. For the contents of the underlying windows see text.

windows. The *pascal* graphics module typically takes 2m20s to display 10 of the just computed potential maps on an *ibm6154* color unit. Fig. 2 also shows the result of activating the **zoom** function on the bottom left transect, which represents p_A at the "end" of the transient.

Other features of this interface, in particular its role in designing the optimal well location, which minimises leakage, are not described herewith for sake of conciseness.

2 - Some inverse problem solvers

2.1 - Identifying the conductivity of a layered medium

Consider the line graph of J (>2) nodes s.t., at each node the following quantities are given: $\{z_j\}$, $\{f_j\}$. The *problem* consists of finding the set of *edge conductivity* values $\{t_{i,j}$, $i:=j-1;\ 2 \leq j \leq J-1\}$ s.t.

$$t_{i,j} > 0,\ \forall j \tag{2.1}$$

$$(z_i - z_j)\,t_{i,j} + (z_{j+1} - z_j)\,t_{j+1,j} = f_j; \qquad 2 \leq j \leq J-1 \tag{2.2}$$

If the data satisfy

$$z_i - z_j \neq 0\ ,\ 2 \leq j \leq J \tag{2.3}$$

then a tentative solution $\{\tilde{t}_{i,j}(b)\}$ is given by (Ponzini and Crosta, 1988)

$$\tilde{t}_{i,j}(b) = \frac{F_i - \langle F \rangle + b(z_J - z_1)/(J-1)}{z_j - z_i} \tag{2.4}$$

where

$$F_i := \sum_{k=2}^{i} f_k\ ;\ F_1 := 0\ ;\ \langle F \rangle := \frac{1}{J-1} \sum_{k=2}^{J} (J-k)\, f_k \tag{2.5}$$

and b is a real parameter, on which $\{\tilde{t}_{i,j}\}$ depends. If a *reference* solution $\{\hat{t}_{i,j}\}$ exists, s.t. (2.1) and (2.2) are met, then there exists at least one b s.t. $\{\tilde{t}_{i,j}(b)\}$ also complies with (2.1).

The code which implements the above algebraic method i.e., solves a discrete inverse problem in one spatial dimension, is being currently interfaced. This implies conversion from *pascal* to *C* in order to improve performance, although speed has never been a problem. Emphasis is being put on exactly solvable problems, where e.g., data are supplied as polynomials w.r. to the index j. As a consequence the architecture of the interface substantially differs from the one described in the previous Section.

On line documentation (papers, theses) is made accessible by *Interleaf*©'s desktop publishing software. Data entry is straightforward: once the value of J has been supplied (Fig. 3) each of the sets $\{z_j\}$, $\{f_j\}$ and $\{\hat{t}_{i,j}\}$ (if available) can be supplied by means of an algebraic **formula**, or read in from a **data** file. A typical formula recognised by the input module is

$$(1/2401)j\hat{\ }2 - (2/4020)j + (1/2) + \ldots \tag{2.6}$$

i.e., the sum of at most 10 monomials s.t. coefficients are ratios of integers (p/q, $q \neq 0$) and powers of j shall be nonnegative integers. Positivity of $\{\hat{t}_{i,j}\}$ is checked on input. If a **procedure** is chosen instead, then some data and parameters must be entered. The **sub-menu** in Fig. 3 pertains to determining

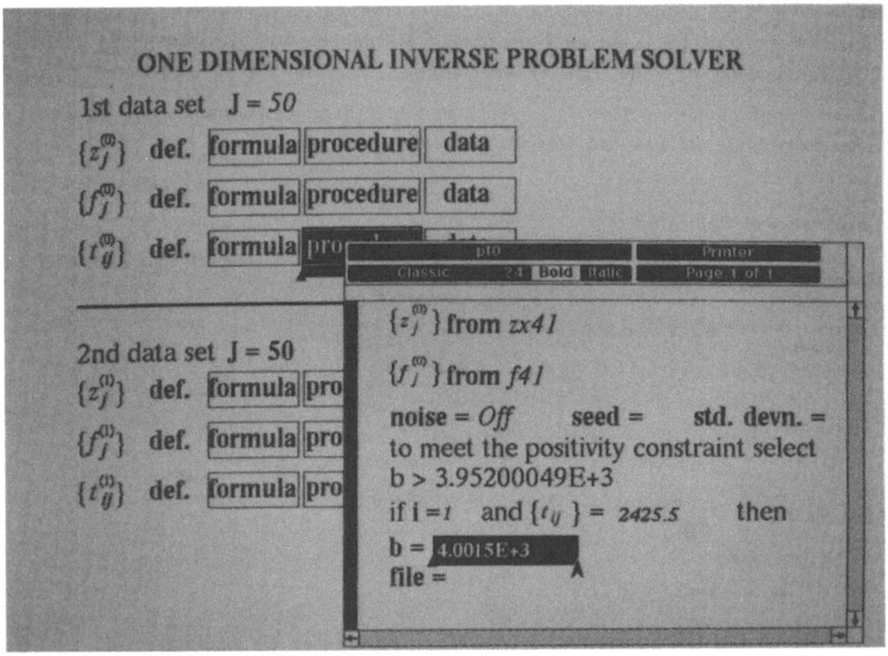

Fig. 3 User – friendly interface to the one dimensional inverse problem solver. Boxes in the main menu show available functions. The submenu pertains to conductivity identification as of Eqn. (2.4).

$\{t^{(0)}_{i,j}(b)\}$ from potential values $\{z^{(0)}_j\}$ and source terms $\{f^{(0)}_j\}$, which shall be read from the corresponding files (first two lines in the sub–menu). On its turn $\{z^{(0)}_j\}$ may have been obtained as the solution to a direct, two– point boundary value problem, where the reference conductivity $\{\hat{t}_{i,j}\}$ had to be supplied.

The code determines the allowed range for the parameter b (message in Fig. 3). The user can either assign a value to b as suggested or have the code determine b from a supplied conductivity value at a given node. This case, where the node index is $i = 1$, is shown in Fig. 3. The corresponding b appears in reverse video and offered for choice (7th line in the submenu).

The resulting values of $\{\tilde{t}_{i,j}(b)\}$ can be made noisy by calling a pseudo– random generator (3rd line in the submenu) and eventually filed. Consistency controls are carried out on file names and contents, in order to avoid e.g., accessing non existing files, overwriting existing ones, mixing data defined on different domains.

The graphics contained in this interface is illustrated by Fig. 4, where the reference solution, a positive constant in the example, is compared with a solution obtained from a set of noisy potential values. Figures along the vertical axis are *differences* between the two conductivities. Dots are joined by vertical bars to enhance readability.

The method described above can in principle make use of such properties as *self–identifiability*, *uniqueness* from two independent sets of data, uniqueness by *parameterisation*, some *stability* estimates and *existence* conditions, which are described in Crosta (1987), Ponzini and Crosta (1988), Ponzini *et. al.*

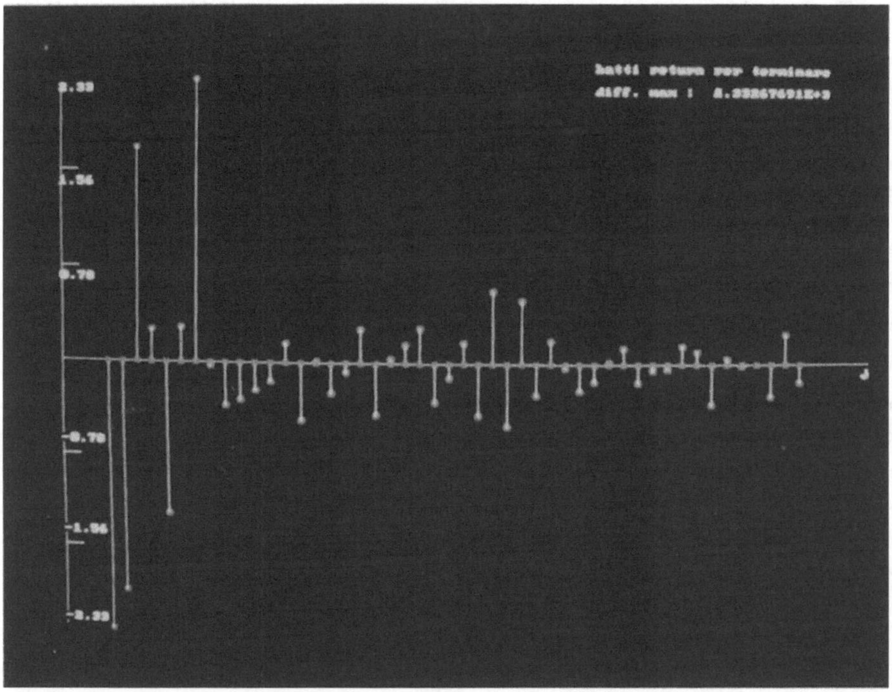

Fig. 4 User – friendly interface to the one dimensional inverse problem solver.
Difference between the reference solution (lying on the J axis) and one obtained from noisy potentials.

(1989a) and have already been applied to some algebraic examples and case studies. The code in its current version does not yet encompass all of these features: they will be gradually inserted as feedback from our students builds up.

2.2 – Identifying the conductivity of a two– dimensional medium

Consider finding

$$a \in \mathcal{A}_{ad} := \{ a \mid a \in L^{\infty}(D),\ \operatorname*{ess\,inf}_{D} a \geq c_I > 0 \}$$

s.t. (2.7)

$$\nabla.(a \nabla z) = f,$$

where D is a rectangle and z, f are sufficiently smooth data.

One method of obtaining a (non unique) output – least– squares solution $\tilde{a}_O$ is the minimisation in $\mathcal{A}_{ad}$ of the error function

$$J_O(a) := \int_D | p(a) - z |^2 \, dD \qquad (2.8)$$

where $p(a)$ is the solution to the Dirichlet BV problem

$$\{ \nabla.(a \nabla p) = f\ ;\ p|_{\partial D} = z|_{\partial D} \} . \qquad (2.9)$$

The method was originally proposed and implemented by Chavent (1971). A family of computer codes, which deal both with the finite difference and finite element discretised counterpart of (2.7), (2.8), (2.9) has been described in Crosta (1989).

The architecture of the interface being currently developed is the same as the one described in § 2.1. Data and solutions of algebraic and transcendental type have priority over experimental ones. The present sequential data entry will be replaced by mouse addressed menus, as shown by Fig. 3. The graphics module is already available, although not yet in its final form. Executable examples are at present restricted to polynomials of 3rd degree. Some domain parameters (width, height, number of nodes) and the transect parameters (see § 1) can be chosen and off- line animation of the solutions computed by the iterative algorithms can be obtained. Fig. 5 shows the reference solution (the transect with straight edges) $\hat{a} = x + y$ in the domain $\{1 \leq x \leq 2,\ 1 \leq y \leq 1.71\}$, divided into 22 by 22 nodes. Superposed to $\hat{a}$ is the transect of a solution $\tilde{a}\ (a^0)$, obtained by minimising the discretised counterpart of (2.8). The finite difference primal (2.9) system and its adjoint have been solved by successive overrelaxation; a Hestenes – Stiefel conjugate direction descent rule has been applied, which is a *discretised* gradient method. In other words the approximate functional derivative is *not* the gradient for the finite – dimensional problem. In the example the algorithm is initialised with $a^0 = 2.75$; the displayed solution has been computed after 30 iterations. The *fortran77* code typically takes 20s per iteration in this case. Graphics functions (see § 1) will benefit most from conversion to *C*.

Other features (Crosta, 1989) which can already be demonstrated by the available codes are:

i) the loss of sensitivity due to "large" values of a^0 ;

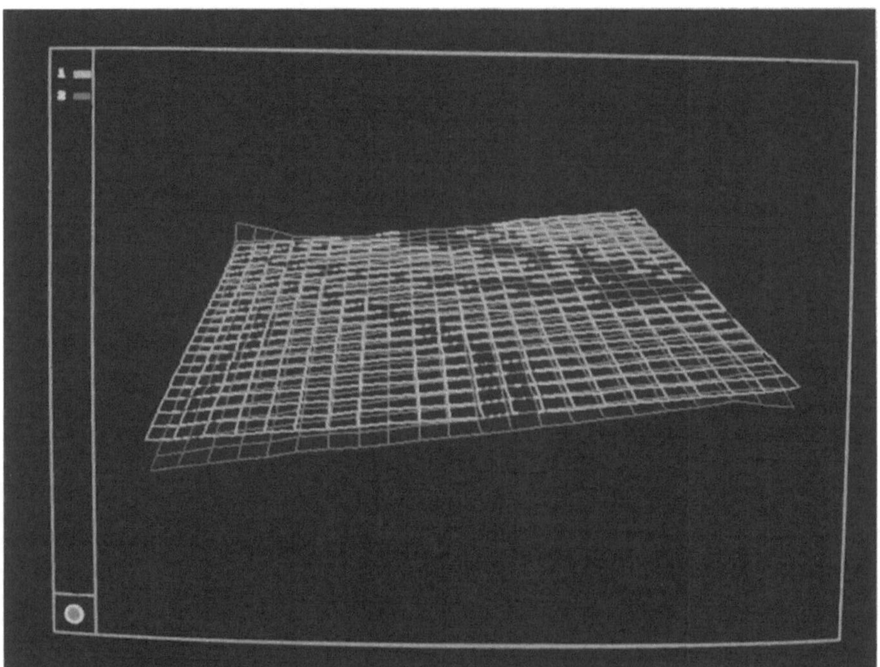

Fig. 5 User – friendly interface to the two dimensional inverse problem solver. Comparison between the reference solution (straight edge transect)) and one yielded by the iterative output least squares algorithm.

ii) the numerical instability predicted when a^0 is "small";

iii) the regularising (smoothing) role of the mesh size;

iv) the role of two – sided constraints applied to nodal values of $a(.)$;

v) the superior noise immunity of discrete gradient over discretised gradient methods.

Acknowledgements

This project has been financially supported by the Italian Ministry of Education grants *MPI 60%* 1987, 1988, 1989. Some of the results were obtained during the IBM Italy – University of Milan joint Contract Study *Teaching Inverse Problems*.

References

Chavent G., 1971: "Analyse fonctionnelle et identification des coefficients répartis dans les équations aux dérivées partielles", *Thèse d' Etat*, Fac. des Sciences, Paris.

Crosta G., 1987: "A Substitution Method Applied to the Identification of the Leading Coefficient Appearing in Linear Equations of Elliptic Type", in P.C. Sabatier, Ed. *Advances in Electronics & Electron Physics, Suppl. 19 – Inverse Problems*, pp 523–32, Academic: London.

Crosta G., 1988: "The Energy Functions of Coefficient Identification Algorithms", in P.C. Sabatier Ed., *Some Topics on Inverse Problems*, pp. 79–92, World Scientific: Singapore.

Crosta G., 1989: "On the Approximate Idenrtification of the Leading Coefficient of a Linear Elliptic Equation by Output Error and Equation Error Minimisation", *Optimal Control*, **10**, 229 253.

Gambolati G., Toffolo F., Uliana F.,1984: "Groundwater response under an electronuclear plant to a river flood wave analized by a nonlinear finite element model", *Water Resources Research*, **20**, (7), 903–13.

Ponzini G. S., Crosta G., 1988: "The Comparison Model Method: a New Arithmetic Approach to the Discrete Inverse Problem of Groundwater Hydrology ", *Transport in Porous Media*, **3**, 415–436.

Ponzini G.S. Crosta G., Giudici M., 1989a: "Identification of Thermal Conductivities by Temperature Gradient Profiles: One–dimensional Steady Flow", *Geophysics*, **54**, 643–53.

Ponzini G.S., Crosta G., Giudici M., 1989b: "The Hydrogeological Role of an Aquitard in preventing Drinkable Water Well Contamination: A Case Study", *Environmental Health Perspectives*, **83** (to appear).

Quantum Inverse Scattering Theory and Nucleon-Nucleon Potentials

Th. Kirst[1], *H. Kohlhoff*[2], *and H.V. von Geramb*

Theoretical Nuclear Physics, University of Hamburg,
Luruper Chaussee 149, D-2000 Hamburg 50, Fed. Rep. of Germany
[1]Supported by BMFT 06-HH-775 and DFG Ge-589/2-1
[2]Supported by the Herbert Quandt-Stiftung of VARTA AG (1989)

The Gel'fand–Levitan and Marchenko quantum inverse scattering theory has been applied to construct nucleon nucleon (NN) potentials. Its aim was to use inversion theory as a mean to construct local and energy independent channel potentials from meson–theoretical predictions of phase shifts, or phase shift analyses results to NN data. Various derivations of the fundamental equations have been studied to facilitate calculations with existing data. A new integral equation has also been found which relates the Gel'fand–Levitan and Marchenko output kernels in a transparent way. For application, a most practical form of the theory is deduced by starting with a rational function representation of the phase shift data and by using Padé approximants for e^z, therewith rational representations of the S-matrix, Jost functions and other useful quantities are determined. The excellent reproductive power was studied using both fundamental equations and with the Reid soft core potential and their phase shifts respectively. Serious results use the meson theoretical phase shifts from the Paris and the Bonn groups, and experimental phases from Arndt et al. Finally, half off–shell T–matrices were generated to compare the exact meson theoretical results with those of inversion and it is found that phase equivalent interactions have essentially the same off-shell behaviour for any physically significant range of momenta.

1. Introduction

Already in its early days started the development of quantum inverse scattering around the fundamental interaction between two nucleons [1]. Meson exchange theory, thought originally as a dynamical theory to explain all phenomena microscopically, turned with the development of QCD from its fundamental aspect into a phenomenological approach with many positive results but also deficiencies. Today, the NN interaction is considered a screened effective residual interaction, similiar in structure as the Van der Waals forces in atomic physics. The true dynamic origin is sought in QCD but this leads to heitherto unsolvable equations of motion. From the meson theoretical approaches, the most comprehensive and sophisticated formulations are the Paris and the Bonn potentials. These groups incorperate all known bosons as mediators of the different components of the NN interaction and which are expressed in terms of the nonrelativistic invariants. We distinguish between central, spin–orbit, quadratic spin–orbit and tensor forces and recall that the potential form factors contain the relative distance and momentum as parameters. The form factors are complicated superpositions of elementary contributions from the individual mesons with adjusted coupling constants and high energy cut off features. The Paris [2] and the Bonn OBEP–R [3] potentials write their parametrization for $T=0$ and $T=1$ in the form

$$V(\vec{r},p^2) = V_0(r,p^2)P_0 + V_1(r,p^2)P_1 + V_{LS}(r)(L\cdot S) + V_{LSQ}(r)L_{12} + V_T(r)S_{12} \qquad (1)$$

with P_0 and P_1 the projection operators onto spin singlet and triplet states, and

$$L_{12} = \frac{1}{2}\left[(\sigma_1 L)(\sigma_2 L) + (\sigma_2 L)(\sigma_1 L)\right], \qquad S_{12} = 3\frac{(\sigma_1\vec{r})(\sigma_2\vec{r})}{r^2} - (\sigma_1\sigma_2) \qquad (2)$$

the standard quadratic spin–orbit and tensor operators; the velocity dependent central components use

$$V(r,p^2) = V^a(r) + \frac{p^2}{m}V^b(r) + V^b(r)\frac{p^2}{m}. \tag{3}$$

These potentials result in strong channel dependences and an also explicit energy dependence.

Inverse scattering theory relates the phase shifts with the potential and we distinguish between inversion at fixed energy and inversion for fixed angular momentum. The explicit energy dependence of the meson theoretical potentials suggests to use fixed energy inversion whereas the operator structure suggests to use fixed angular momentum inversion. However, the preponderance of the non locality in the angular momentum dependence dictates the use of fixed angular momentum inversion with the result that NN potentials are local, energy independent (no explicit momentum dependence) but state dependent. The justification for this restriction is twofold. In the parametrization of the Paris interaction the momentum dependence is a short range feature, essentially only active for $r < 1\ fm$, where the theory is actually replaced by pure phenomenology which sounds plausible but is by no means compulsary. The Bonn group follows the Paris arguments. Within the context of inversion we shall prove that the explicit momentum dependence is, from the physics point of view, not required and the argument of significantly different off–shell behaviour in local and nonlocal potentials is ruled out with calculations where we compare them. It is the explicit momentum dependence of the Paris and Bonn interactions which prohibits an easy use in few and many body nuclear physics and for us, this point was essential in the search for alternatives. Inversion at fixed angular momentum in application to the NN problem allows to achieve several desireable goals at the same time. As the first advantage to mention, it is a technique to generate local, energy independent potentials from meson theory; secondly it reproduces experimental data.

2. Theoretical Backround

The potential of the radial Schrödinger equation is uniquely determined by the Gel'fand–Levitan and Marchenko output kernels

$$V_\ell(r) = 2\frac{d}{dr}K_\ell(r,r) = -2\frac{d}{dr}A_\ell(r,r). \tag{4}$$

Each of these output kernels is a solution of the corresponding fundamental equation

$$K_\ell(r,t) + G_\ell(r,t) + \int_0^r K_\ell(r,s)G_\ell(s,t)\,ds = 0, \tag{5}$$

the Gel'fand–Levitan equation, and

$$A_\ell(r,t) + F_\ell(r,t) + \int_r^\infty A_\ell(r,s)F_\ell(s,t)\,ds = 0, \tag{6}$$

the Marchenko equation. The experimental information is contained in their input kernels

$$G_\ell(r,t) := \frac{2}{\pi}\int_0^\infty j_\ell(kr)\big[|F_\ell(k)|^{-2} - 1\big]j_\ell(kt)\,dk \tag{7}$$

and

$$F_\ell(r,t) := -\frac{1}{2\pi}\int_{-\infty}^\infty h_\ell^+(kr)\big[S_\ell(k) - 1\big]h_\ell^+(kt)\,dk. \tag{8}$$

The Gel'fand–Levitan input kernel (7) uses the Jost function and the Marchenko input kernel

(8) the S–matrix. Experimental data are parametrized in terms of phase shifts to which Jost functions are related

$$S_\ell(k) := e^{2i\delta_\ell(k)} = \frac{F_\ell(-k)}{F_\ell(k)}. \tag{9}$$

For inversion we need the phase shift function at all energies, $k \in [-\infty, \infty]$, and possible bound state data.

The determination of the phase shift function is an ill–posed problem. Phase shift analyses data [4] and phases from meson theory are available for any k–value up to an energy, say 1.4 GeV, but beyond 300 MeV particle creation is important and the Schrödinger equation with (1) becomes inadequate. Some theoretical approaches include isobar excitations to extend the potential concept towards higher energies. As yet, we neglected the isobar excitations completely and use phase shifts in the energy interval without inelasticity defect or unitarize the S–matrix. This approach is consistent with the low energy concept of NN potentials.

With the regularization techniques it is possible to circumvent the explicit specification of the phase shift function. Instead of following the line of regularization, we have choosen a rational function representation for $\delta_\ell(k)$ with a physically motivated asymptotic behaviour $\delta_\ell(k) \sim 1/k$, the symmetry $\delta_\ell(k) = -\delta_\ell(-k)$ and the correct number of bound states measured with Levinson's theorem. With this choice we believe to have a better control on the physics involved. Details about the rational function representations are found in [5,7]. The principle is to start with a fit of the coefficients a_n and b_n such that phase shift data, at typically 300 discret values, agree to better than 0.05 degrees with the rational function

$$\delta_\ell(k) = \frac{\sum_{n=1}^{N/2} a_{2n-1} k^{2n-1}}{1 + \sum_{n=1}^{N/2} b_{2n} k^{2n}}. \tag{10}$$

This rational phase shift function we connect with a symmetric Padé approximation for the exponential function, $L \geq 6$,

$$e^z \simeq \mathcal{P}^{[L|L]}(z) = \frac{\sum_{l=0}^{L} c_l z^l}{\sum_{l=0}^{L} c_l (-z)^l}, \qquad \text{with} \qquad c_l = \frac{(2L-l)!}{l!(L-l)!}. \tag{11}$$

Therewith we obtain a high order and high accuracy representation of the S–matrix, $M = 100 - 200$,

$$S_\ell(k) = 1 + \sum_{m=1}^{M} \frac{S_m}{k - \sigma_m} = \prod_{m=1}^{M} \frac{k + \sigma_m}{k - \sigma_m} = \prod_{m=1}^{M/2} \frac{k + \sigma_m^\downarrow}{k - \sigma_m^\downarrow} \prod_{m=\frac{M}{2}+1}^{M} \frac{k + \sigma_m^\uparrow}{k - \sigma_m^\uparrow}, \tag{12}$$

in which the arrows indicate the poles in the lower half and the upper half respectively; at the same instant we maintain the necessary symmetries – unitarity, reflexion symmetrie, asymptotic behaviour, low energy behaviour and Levinson theorem. It is technically impossible to obtain the same result with a fit of parameters in an S–matrix ansatz.

A further advantage of our technique of a rational S–matrix is the unique factorization into the Jost functions $F_\ell(k)$ and $F_\ell(-k)$

$$F_\ell(k) = \prod_{m=1}^{M/2} \frac{k - \sigma_m^\downarrow}{k + \sigma_m^\uparrow} = 1 + \sum_{m=1}^{M/2} \frac{L_m}{k + \sigma_m^\uparrow}. \tag{13}$$

With the rational representations (12) and (13) the input kernels can be integrated analytically and they become separable. As an immediate consequence, also the output kernels are separable and we write the ansatz

$$K_\ell(r,t) := \sum_{m=1}^{M/2} \Phi_\ell(m,r) j_\ell(\sigma_n^\uparrow t) \tag{14}$$

for Gel'fand–Levitan and

$$A_\ell(r,t) := \sum_{m=1}^{M/2} \Psi_e ll(m,r) h_\ell^+(\sigma_m^\uparrow t) \tag{15}$$

for Marchenko. The fundamental equations are reduced to linear systems of equations

$$h_\ell^-(\sigma_m^\downarrow r) + \sum_{n=1}^{N} W_{m,n}(r)\Phi_\ell(n,r) = 0, \tag{16}$$

and

$$\sum_{n=1}^{M/2} \left[\frac{\delta_{mn}}{iS_m} - M_{m,n}(r)\right] \Psi_e ll(n,r) = h_\ell^+(\sigma_m^\uparrow r). \tag{17}$$

The matrix elements are

$$W_{m,n}(r) := \frac{1}{\sigma_n^{\uparrow 2} - \sigma_m^{\downarrow 2}} \left[\sigma_m^\downarrow j_\ell(\sigma_n^\uparrow r) h_{\ell-1}^-(\sigma_m^\downarrow r) - \sigma_n^\uparrow j_{\ell-1}(\sigma_n^\uparrow r) h_\ell^-(\sigma_m^\downarrow r)\right] \tag{18}$$

and

$$M_{mn}(r) = \frac{1}{\sigma_m^{\uparrow 2} - \sigma_n^{\uparrow 2}} \left[\sigma_n^\uparrow h_\ell^+(\sigma_m^\uparrow r) h_{\ell-1}^+(\sigma_n^\uparrow r) - \sigma_m^\uparrow h_{\ell-1}^+(\sigma_m^\uparrow r) h_\ell^+(\sigma_n^\uparrow r)\right], \tag{19}$$

$$M_{mm}(r) = \frac{r}{2}\left[[h_\ell^+(\sigma_n^\uparrow r)]^2 - h_{\ell-1}^+(\sigma_n^\uparrow r) h_{\ell+1}^+(\sigma_n^\uparrow r)\right]. \tag{20}$$

The whole procedure was tested with the Reid soft core potential and more details are found in [5,7].

An interesting note comes from a theoretical consideration of the inversion technique [6]. Namely, the output kernels are generalized Fourier transforms of the difference between solutions with and without potentials. This fact and the relation between the regular and the Jost solutions yield an integral equation relating the Gel'fand–Levitan and Marchenko output kernels

$$K_\ell(r,t) = Z_\ell(r,t) + \int_r^\infty A_\ell(r,s) Z_\ell(s,t)\, ds, \qquad 0 \le t < r \tag{21}$$

with the *connection kernel*

$$Z_\ell(r,t) := -\frac{i}{\pi} \int_{-\infty}^{\infty} h_\ell^+(kr)[F_\ell(-k) - 1] j_\ell(kt)\, dk. \tag{22}$$

Equation (21) can equally well be used as equation (6) to determine the inversion potential [6]. In case of rational S-matrices and Jost functions the relation connects the auxiliary functions in (14) and (15)

$$\Phi_\ell(m,r) = -2L_m \left[h_\ell^+(\sigma_m^\uparrow r) + \sum_{n=1}^{M/2} M_{mn}(r)\Psi_\ell(n,r)\right]. \tag{23}$$

The last formula is useful for numerical cross checkings.

3. Applications

As input we used the n–p phase shifts of the Paris (dashed) and the Bonn OBEP–R (dotted) potentials which are compared with the Arndt PSA data (full), Fig. 1. In each case the agreement between actual data points and the rational fit is so close that they cannot be distinguished. The potential results are shown in Fig. 2. It was a surprise to see these big differences since

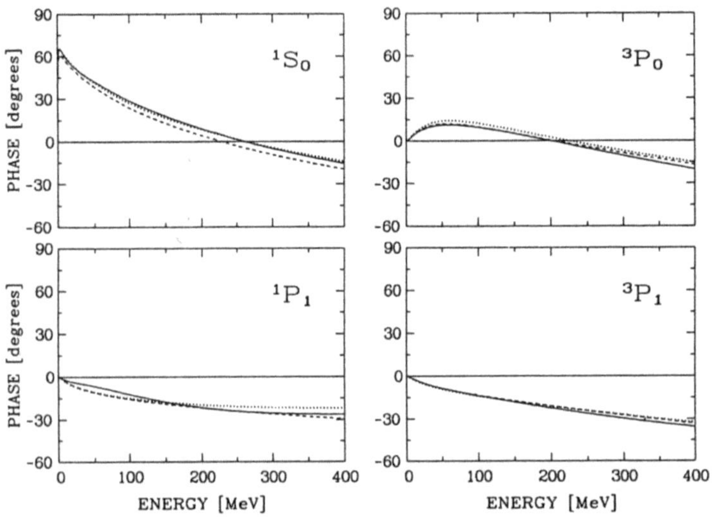

Figure 1. Phase–shifts; Paris (dashed), Bonn (dotted), Arndt (full)

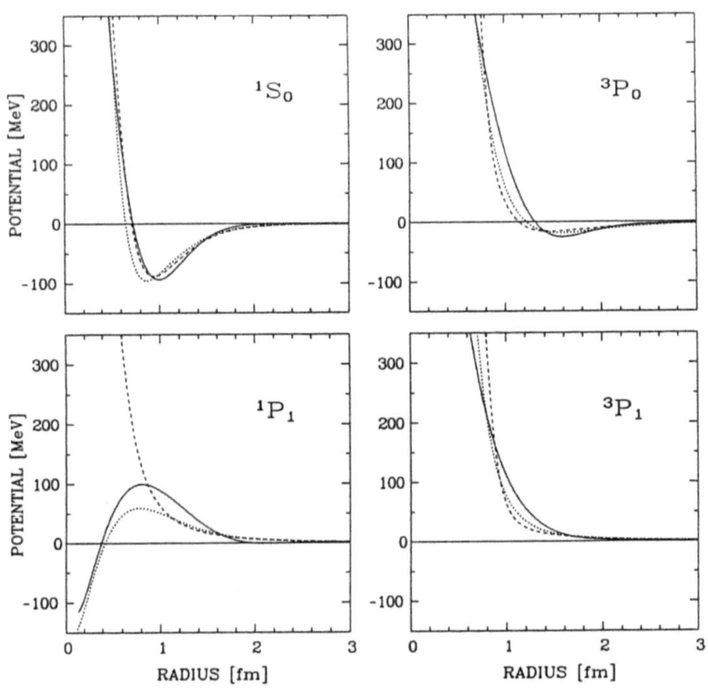

Figure 2. Inversion potentials; Paris (dashed), Bonn (dotted), Arndt (full)

the meson theories claim to be in very close agreement and remaining discrepancies are thought to be minor. The P–channel results of inverse scattering are not only in the $^1P_1$–channel at variance, but all channels vary strongly around $1 - 2\ fm$. This result is thought to lend full justification for the application of inversion to the NN problem. More details and a discussion of the physical implications can be found [5, 7].

4. Off–Shell T Matrices

Clearly by dint of their construction, the inversion potentials give the exact on–shell NN T–matrices, but it may be believed, that their off–shell T–matrices will necessarily differ from those of meson exchange theories. Indeed this has been raised as a criticism of the inverse scattering approach. Specifically, it is thought that the strong energy and momentum dependencies of the meson theory results would be the cause of quite unique off–shell behavior which an inversion potential could not attain. As was shown, these concerns are inapt [8]. As a rule, all interactions, be they local or momentum dependent, give essentially the same off–shell T–matrices whenever they have the same on–shell values. Differences in off–shell T–matrices occur most noticeably when any two interactions give different on–shell values; be they the different meson exchange results or inversions.

To compare off-shell characteristics of T-matrices, we have solved the Schrödinger equation with the Paris, Bonn OBEP–R, and inversion potentials. The scattering wave functions so obtained, were then used to evaluate off–shell T–matrices, per the definition,

$$T_\ell(p,k;E) = -\frac{1}{p}\left(\frac{m}{\hbar^2}\right)\int_0^\infty j_\ell(pr)V_\ell(r)\psi_\ell(k,r)\,dr. \tag{24}$$

As the correlated wave function $\Psi_\ell(k,r)$ vanishes at the origin and asymptotically matches to Riccati functions by

$$\psi_\ell(k,r) = j_\ell(kr) + T_\ell(k,k;E)h_\ell^+(kr), \tag{25}$$

use of the Schrödinger equation enables $V_\ell(r)\psi_\ell(k,r)$ to be replaced in (24) giving

$$T_\ell(p,k;E) = \left(\frac{p}{k}\right)^\ell T_\ell(k,k;E) + \frac{p^2-k^2}{p}\int_0^\infty j_\ell(pr)\zeta_\ell(k,r)\,dr \tag{26}$$

with the healing function

$$\zeta_\ell(k,r) = \psi_\ell(k,r) - j_\ell(kr) - T_\ell(k,k;E)h_\ell^+(kr) = z_\ell(k,r)e^{i\delta_\ell(k)}. \tag{27}$$

The amplitudes of the healing function, $z_\ell(k,r)$, carry the information of the scattering in coordinate space, and thus the first measure of differences in the off–shell character of any two interactions is given by the differences in their healing functions. Nevertheless to display precise variations in off–shell character of T–matrices in physically relevant regions, the Riccati–Bessel transforms must be taken. A convenient measure is then the Kowalski–Noyes f–ratio

$$f_\ell(p,k;E) = \frac{T_\ell(p,k;E)}{T_\ell(k,k;E)} \tag{28}$$

since this is not only a real function of momentum, for uncoupled channels, but also scales out the precise value of the on–shell T–matrix.

The numerical studies start with a comparison of the healing functions obtained from the Paris interaction (dotted) and from its phase equivalent inversion potential, Fig. 3; followed by the resulting f–ratios, Fig. 4. Because in this case the on–shell behaviour is equal, one gets nearly the same off–shell behaviour in the physically important region. In Fig. 5 we compare f–ratios to

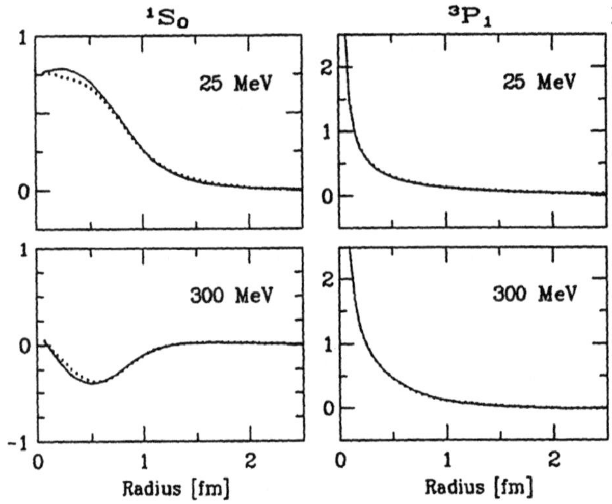

Figure 3. Healing functions; Paris inverted (full), Paris (dotted)

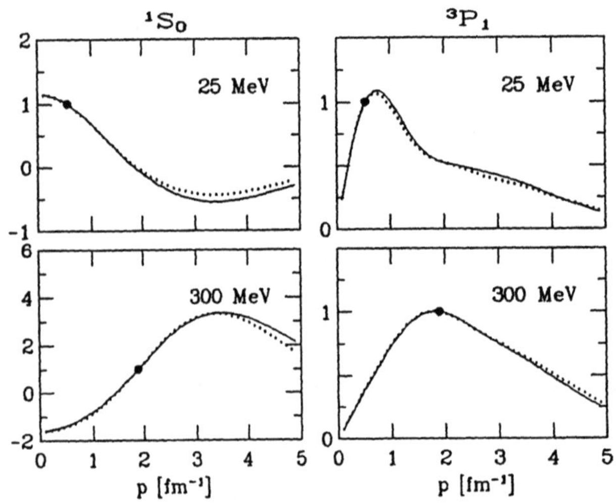

Figure 4. f–ratios; Paris inverted (full), Paris (dotted)

cases where different on–shell behaviour is present. They are the Paris (dashed), Bonn (dotted) and Arndt (full) inversion potentials and, as expected, the differences in their off–shell behaviour in the physically region are significant too.

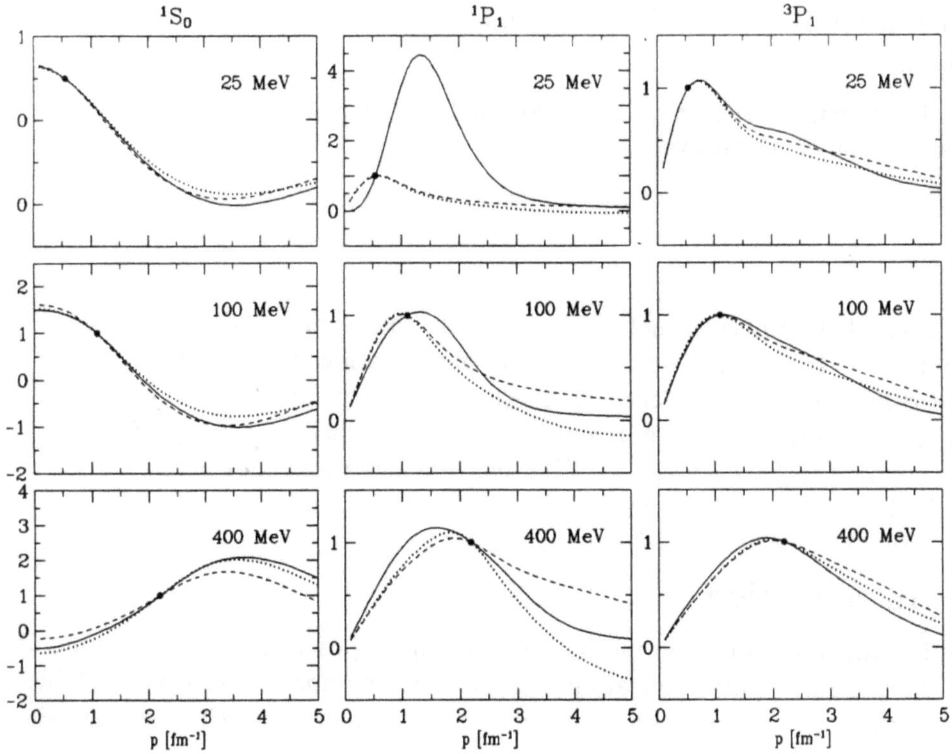

Figure 5. f–ratios; Paris (dashed), Bonn (dotted), Arndt (full)

References

[1] K. Chadan and P.C. Sabatier, *Inverse problems in quantum scattering theory*, (Springer New York 1977, Second Edition 1989).

[2] M. Lacombe, B. Loiseau, J.M. Richard, R. Vinh Mau, J. Côté, P. Pirès and R. de Tourreil, *Parametrization of the Paris Potential*, Phys. Rev. **C21**, 861 (1980).

[3] R. Machleidt, K. Holinde and Ch. Elster, *The Bonn meson–exchange model for the nucleon–nucleon interaction*, Phys. Rep. **149**(1), (1987).

[4] R.A. Arndt, J.S. Hyslop III and L.D. Roper, Nucleon–nucleon partial–wave analysis to 1100 MeV, Phys. Rev. **D35**(1), 128 (1986).
R.A. Arndt and L.D. Roper, *Scattering Analysis Interactive Dial–in SAID*, (Virginia Polytechnic Institute, Blacksbury, 1988).

[5] Th. Kirst, K. Amos, L. Berge, M. Coz and H.V. v. Geramb, *Nucleon Nucleon Potentials from Gel'fand–Levitan and Marchenko Inversions*, Phys. Rev. **C40**, 912 (1989).

[6] Th. Kirst, *Nukleon Nukleon Potentiale aus der Inversion*, PhD thesis, University of Hamburg (1989).

[7] H. Kohlhoff, *Rationale Darstellungen der Nukleon Nukleon S–Matrix*, Diplomarbeit, University of Hamburg (1989).

[8] H.V. v. Geramb and K. Amos *Off–shell T–matrices from inverse scattering*, submitted to Phys. Rev. C (9/1989).

Generalized Backus-Gilbert Methods

P. Maaß

MA 6-1, Technische Universität Berlin, Straße des 17. Juni 135,
D-1000 Berlin 12, Germany

We present a numerical method for the solution of ill-posed linear problems . It is based on a pointwise optimization of the point-spread function, which makes the method specially suitable for local reconstructions. Alternatively it can be described as a projection method in Sobolev spaces. The convergence properties of the method are investigated. Numerical tests verify its usefulness.

1 Introduction

We introduce a numerical method for solving ill-posed integral equations of the first kind. In all applications only a finite number of data is available, thus we are lead to study the so called moment problem, which arise from the operator

$$A: \quad X \to \mathbb{R}^N$$
$$f \to g = \left(\int_\Omega k_i(y) f(y) dy\right)_{i=1,\ldots,N} \quad , \tag{1.1}$$

$\Omega \subset \mathbb{R}^d$, $X \subset L_2(\Omega)$, $k_i \in L_2(\Omega)$. The aim is to approximate the solution f of the inverse problem

$$Af = g \quad .$$

This class of problems includes integral equations of the first kind with discrete data as well as the classical moment problem $k_i(y) = y^i$.

Any linear reconstruction method arises from a linear operator

$$T_N: \quad \mathbb{R}^N \to X$$
$$(g_1, .., g_N)^t \to T_N g = \sum_{i=1}^N g_i \phi_i = < g, \phi >_{\mathbb{R}^N} \quad . \tag{1.2}$$

Different reconstruction methods only differ in their choice of the set of basis functions $\phi_1, .., \phi_N$. Following the ideas of Backus and Gilbert we choose the basis functions such that the resulting point-spread function, that is the outcome of the reconstruction of a point source, is optimized in a specified way. The reconstruction at the point x is given by

$$
\begin{aligned}
f_N(x) \quad &:= T_N g(x) = \Sigma_{i=1}^N g_i \phi_i(x) \\
&= \int -\Omega \underbrace{(\sum_{i=1}^N k_i(y)\phi_i(x))}_{\Phi_N(x,y)} f(y)dy \quad . \qquad (1.3)
\end{aligned}
$$

We see that the point-spread function is simply

$$\Phi_N(x,y) = (A^*\phi(x))(y) =< \phi(x), k(y) >_{\mathbb{R}^N} \quad ,$$

where k denotes the vector of the kernel functions. Obviously the point- spread function should approximate the δ-distribution, $\delta(x-y)$, in a controlled way. This motivation was used in 1968 by Backus and Gilbert, see [1], they proposed to choose the $\{\phi_i\}$ minimizing

$$\int \Phi_N^2(x,y)|x-y|^2 dy,$$

under the normalizing condition $\int \Phi_N(x,y)dy = 1$. Recently this method enjoyed growing interest, for a list of publications on the Backus-Gilbert method see the references given in [2,3]. Backus and Gilbert were interested in applications in geophysics where $|x-y|^2$ appears as a natural weight function. From a mathematical point of view it is more natural to measure the difference $\Phi_N(x,y) - \delta(x-y)$ directly in Sobolev norms. This leads to choose those $\{\phi_i(x)\}_{i=1,N}$ which solve, for fixed x, the minimization problem

$$\min_{\phi \in \mathbb{R}^N} \| \Phi_N(x,\cdot) - \delta(x-\cdot) \|_{H^{-s}} = \min_{\phi \in \mathbb{R}^N} \| \sum_{i=1}^N k_i(\cdot)\phi_i(x) - \delta(x-\cdot) \|_{H^{-s}} \quad ,$$

$s > d/2$. In the next chapter we give a detailed description of the resulting algorithm and state the main convergence estimates. The last chapter shows the results of some numerical tests. Further investigations on this method, in particular relations to standard regularization procedures and all proofs, can be found in [3].

2 Description of the method and convergence results

The generalized Backus-Gilbert method, which we call s-method, approximates the solution f of the inverse problem $Af = g$ at the point x by

$$f_N(x) = \sum_{i=1}^N \phi_i(x)k_i(y) \quad ,$$

where for fixed x the coefficients $\phi(x) = (\phi_1(x), .., \phi_N(x))^t$ are given by the solution of

$$\min_{\phi \in \mathbb{R}^N} \| \sum_{i=1}^{N} k_i(\cdot)\phi_i(x) - \delta(x-\cdot) \|_{H^{-s}} \quad , \quad s > d/2 \ . \tag{2.4}$$

We observe that $\sum_{i=1}^N k_i(\cdot)\phi_i(x) = (A^*\phi(x))(\cdot)$, where $A^* : \mathbb{R}^N \to X$ is the adjoint operator of A. In the minimization problem we use the weaker norm in H^{-s}, hence we denote by X_s the space X equiped with the H^{-s}-norm and consider

$$A^* : \mathbb{R}^N \to X_s \quad .$$

We denote by $A_s : X_s \to \mathbb{R}^N$ the adjoint of A^*, hence the minimization problem (2.4) is solved by the solution of the normal equation

$$A_s A^* \phi(x) = A_s(\delta(x-\cdot) \quad .$$

The matrix $A_s A^*$ and the right hand side $b(x) = A_s(\delta(x-\cdot)$ can be computed explicitly :

$$(A_s A^*)_{ij} = < k_i, k_j >_{H^{-s}} \quad \text{and} \quad b_i(x) = \mathcal{F}^{-1}[(1+|\eta|^2)^{-s}\mathcal{F}k_i(\eta)](x) \ . \tag{2.5}$$

Here $\mathcal{F}, \mathcal{F}^{-1}$ denote the Fouriertransform and its inverse.

The algorithm for reconstructing f at a specified point x from the given data vector g proceeds in two steps :

- solve $A_s A^* \phi(x) = b(x)$
- calculate $f_N(x) = \sum_{i=1}^N g_i \phi_i(x)$.

First we want to note some properties of this algorithm :

- The ill-posed part of the algorithm, step 1, is independent of the data. Thus the linear system can be solved in advance with any desired accuracy. Moreover the matrix is independent of x, in contrast to the classical Backus-Gilbert method, and therefore a decomposition of the matrix has to be computed and stored only once.
- The actual reconstruction, step 2, is fast and stable.
- The reconstructions at different points x are independent, the s-method is purely local.
- No additional discretization in X is introduced.
- Originally the method was restricted to values of $s > d/2$. Looking at the defining linear system, we see, that the method can be used for a wider range of values of s depending on the smoothness of the kernels k_i.

The method was motivated by the aim to optimize the point-spread function in some function space. For a different motivation recall that the reconstruction $f_N(x)$ is a linear combination of the $\phi_i(x)$, which in turn are linear combinations of the $b_i(x)$, see (2.5). Therefore

$$f_N(x) \in X_{N,s} := span\{\mathcal{F}^{-1}[(1+|\eta|^2)^{-s}\mathcal{F}k_i(\eta)](x) \quad , i = 1,..,N\} \quad .$$

We can describe the s-method as a smoothed projection onto the k_i's using a Fourierfilter of the form $(1+|\eta|^2)^{-s}$. To be more precise we can prove

Theorem 2.1 *Let* $k_i \in H^\beta \quad , i = 1,..,N, \quad s > max(-\beta, d/4 - \beta/2)$ *and* $f_N, X_{N,s}$ *defined as above. Then*

f_N *is the orthogonal projection of* f *onto* $X_{N,s}$ *with respect to the* H^s*-Norm.*

Now we can view the s-method in two different ways. On the one hand the s-method was introduced by a pointwise optimization procedure, on the other hand we can view it as a projection method. This is mirrored by different convergence results. The second description leads to convergence estimates in Sobolev spaces, the first gives rise to results on pointwise convergence. We will only state the main result on pointwise convergence, for a complete description see [3]. Standard approximation properties for $span\{k_1(x),..,k_N(x)\}$ as $N \to \infty$ are assumed.

Theorem 2.2 *Let* $f \in H^\nu, \| f \|_{H^\nu} \leq \rho \; , \; \nu \geq s > d/2$. *Then for fixed* $x \in \Omega$

$$|(f - f_N)(x)| = o(N^\alpha) \quad , \quad \forall \alpha > d/2 - s \; .$$

Remark : Given the same situation as in Theorem 2.2 but with noisy data, $\| g - g^\epsilon \| < \epsilon$, one can prove

$$|(f - f_N^\epsilon)| \leq o(\epsilon^\alpha) \quad \text{for any} \quad \alpha > (\nu - d/2)/(\nu - d/2 + 1) \quad .$$

3 Numerical Examples

The first example shows the smoothing effect of the parameter s. We consider the problem of reconstructing an unknown function f from a finite set of its Fourier-moments. The reconstruction splits for the even and the odd part of f. We tested the algorithm with even functions, i.e. $\Omega = [-1,1]$, $k_j(y) = cos(j\pi y)$ and $N = 11$. The critical ill-posed part of the procedure consists in solving $(A_s A^*)\phi(x) = b(x)$. We therefore regularized the linear system according to the machine precision of the computer, which is a bound for the accuracy of $(A_s A^*)_{ij}, b_j(x)$. 5% relative noise were added to the data and we choose $s = 1, 3, 4$.
The reconstructions are shown for the test function

$$f(x) = \chi_{[-1/2,1/2]}(x).$$

Figure 1a) shows the even point-spread function at $x = 0.5$ for $s = 1$. For $s = 0$ the method simply gives the truncated Fourier series and the reconstruction

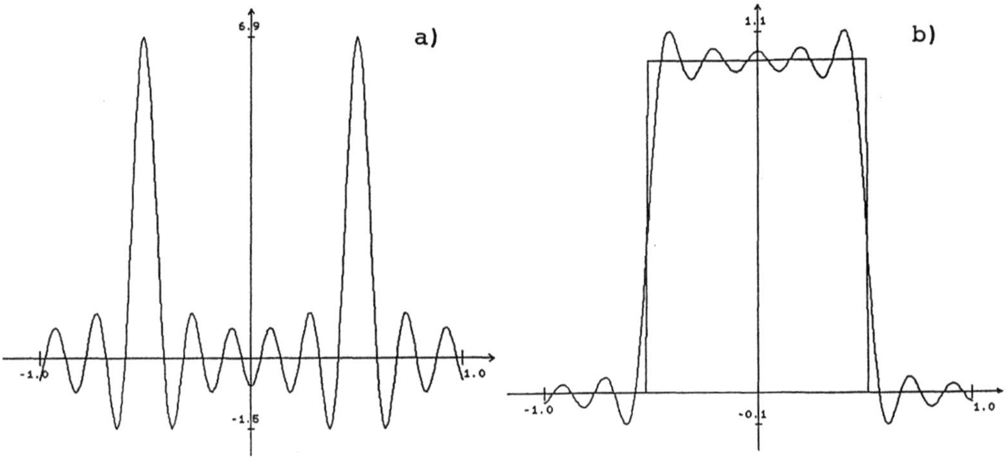

Figure 1a) shows the point-spread function at x=0.5, s=1.
Figure 1b) shows the reconstruction of the test function for s=1.

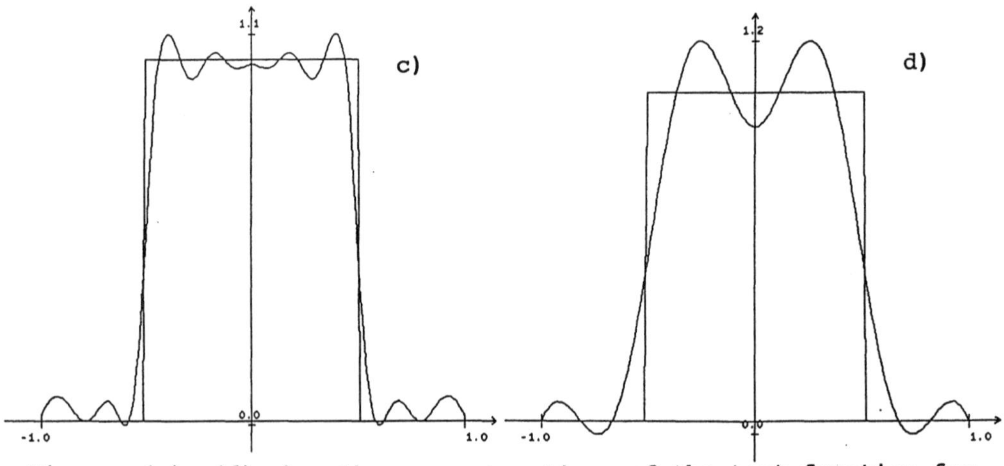

Figures 1c), 1d) show the reconstructions of the test function for s=3,4.

exhibits the wellknown Gibb's effect at the discontinuities. Choosing $s = 1$ already shows the smoothing effect of s, the overshoot decreases slightly, see figure 1b). Reconstructions for $s = 3, 4$ are shown in figure 1c) and 1d).

As a second example we use the integral equation of the first kind

$$g(x) = \int_0^1 k(x,y) f(y) dy \quad ,$$

$$k(x,y) = \begin{cases} y(1-x) & \text{for } x \geq y \\ x(1-y) & \text{for } y \geq x \end{cases}$$

which was used in [4] to compare different regularization procedures. This problem is solved by $f = -g''$, hence we face the ill-posed problem of numerical differentiation. We choose the test function

$$f(x) = x(1-x) \quad .$$

The moment problem arises from an equidistant discretization of the data $g(x)$ with $N = 49$ data points, again 5% noise was added. Figure 2a) and 2b) show the point-spread functions at $x = 0.3$ for $s = 0$ and $s = 1$, again we see the smoothing effect of s. Figure 2c) shows the best reconstruction

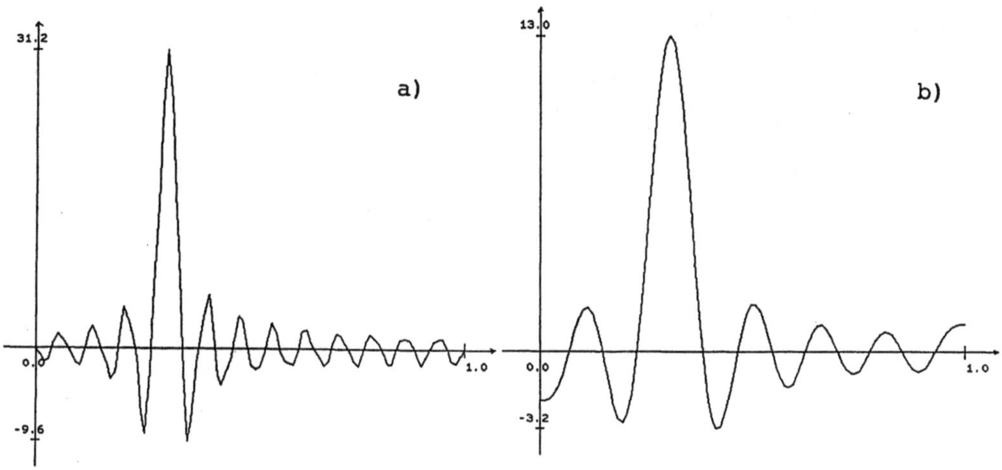

Figure 2a) shows the point-spread function at x=0.3 for the second example with s=0, s=1 is shown in figure 2b).

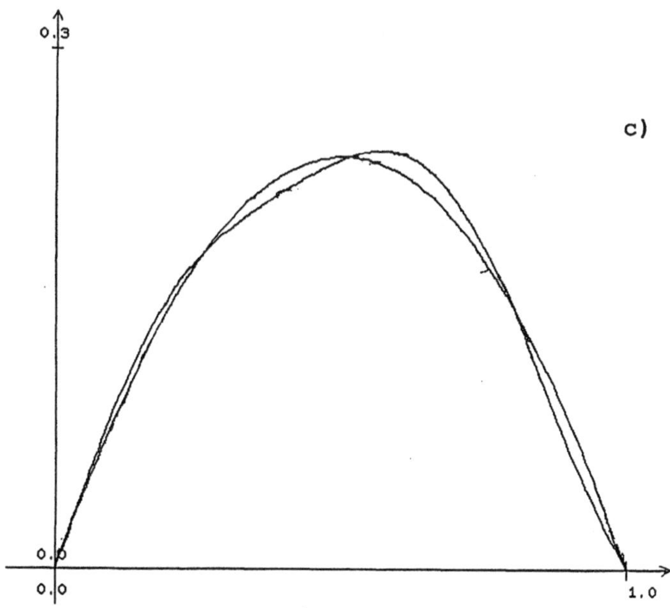

Figure 2c) shows the reconstruction of a test function with s=1.

with $s = 1$, here the linear system $(A_s A^*)\phi(x) = b(x)$ was regularized to a condition number of 10^3.

References

1. G.Backus and F.Gilbert, The resolving power of gross earth data, Geophys. J.R. astr. Soc. 16 (1968), pp.169-205

2. A.Kirsch, B.Schomburg, G.Berendt, The Backus-Gilbert method , Inverse Problems 4(1988)

3. A.K. Louis and P. Maaß , Smoothed projection methods for the moment problem, submitted for publication

4. A.K.Louis, Inverse und schlecht gestellte Probleme, Teubner Studienbücher, Stuttgart (1989)

Inversion of Ionospheric Sounding Data from an HF Backscatter Radar

N. Ruelle and F. Gauthier

Centre National d'Etudes des Télécommunications,
CNET/LAB/MER/GER, Route de Trégastel, F-22301 Lannion, France

1. INTRODUCTION

For radio waves in the high-frequency band (6-30 MHz), the ionospheric plasma is a refractive medium, with an index which depends on local electron density. HF backscatter radars can use this property to take soundings of the ionosphere by transmitting radio pulses at low elevation angles (see fig. 1). These pulses are refracted obliquely in the ionosphere toward a distant area of the earth's surface where a small part of the pulse energy is backscattered in the direction of incidence. This is detected by the radar after a second ionospheric refraction.

This remote sensing operation is interesting for the external geophysicist and for the HF radio engineer because data can be gathered on an extended area of the ionosphere (potentially a circle of more than 2000 km radius) around the site of the sounder. This is particularly useful for areas where no local sensing is possible (oceans, deserts...).

However the analysis of the sounding data in order to derive the ionospheric profile (electron density vs. altitude) is a refractive inverse problem which is nonlinear and for which no general solution has been established. Indeed HF backscatter radar is an area where technical advances have proceeded faster than theoretical work on modelling and data interpretation.

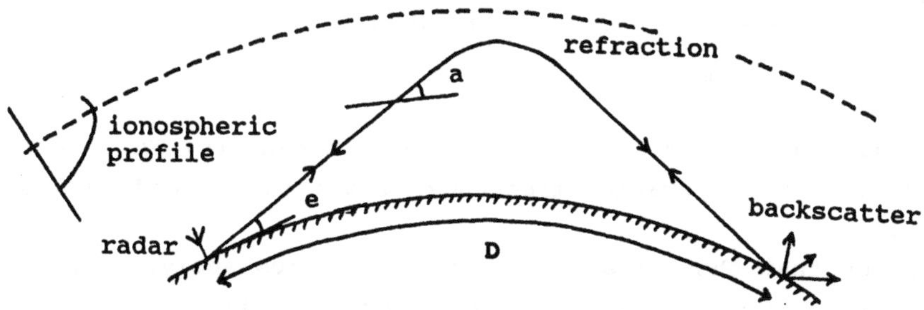

Figure 1 : sounding geometry

The basic modelization of this problem is described in this paper. Three sounding techniques are also described and then compared from the points of view of information content and practical ease of realization.

2. HF PROPAGATION IN THE IONOSPHERE

The ionosphere is a layer of partially ionized gas within the atmosphere; the stratified ionospheric plasma is controlled by factors such as incident sunlight, wind, and the geomagnetic field. For HF waves the ionosphere is a refractive, dispersive, anisotropic, inhomogeneous and time varying medium. From the point of view of HF propagation it can be studied equivalently in terms of local electron density, plasma frequency or refractive index, using magnetoionic theory [1]; the Appleton-Hartree formula gives the necessary relationships.

The stratification of ionospheric electron density varies with time and place, but generally electron density is negligible up to an altitude of 80 km, reaches a peak around 300-400 km and then decreases with altitude. The precise determination of this electron density profile under the peak, vs. place and time is of great importance both for radio communications and for geophysical studies. Further details regarding ionospheric profiles can be found in [1] for example.

3. MODELLING HF BACKSCATTER SOUNDING

In order to model the sounding operation described in the introduction, it is necessary to make simplifying assumptions.

In particular we will consider the ionospheric plasma to be locally homogeneous and isotropic, thus simplifying the Appleton-Hartree formula, giving :

$$Fn^2 = N \cdot \frac{e^2}{4 \cdot \pi^2 \cdot \epsilon_o \cdot m} \quad (Fn^2 \approx 80.6 \cdot N) \quad ; \quad n^2 = 1 - \frac{Fn^2}{F^2}$$

n : local refractive index; N : local electron density; Fn : local plasma frequency; F : operating frequency; e,m : charge and mass of the electron.

We will suppose that the rules of geometrical optics apply, allowing us to treat the forward problem by ray-tracing (fig. 1).

We will assume that the ionosphere is spherically stratified with respect to the center of the earth, so that Bouguer's rule (a form of Snell's law) can be used :

R.n.cos(a) constant along the ray path

R : distance from the center of the Earth, a as defined in fig. 1.
A reasonable approximation of the real electron density profile N(R) is a parabolic function of height. In fact we will model N(R) as a quasi-parabola :

$$N(R) = \frac{Fc^2}{80.6} \cdot \left[1 - \left[\frac{R - Rm}{Ym} \right]^2 \cdot \left[\frac{Rb}{R} \right]^2 \right] \quad ; \quad Rb \leq R \leq Rm \cdot \frac{Rb}{Rb - Ym}$$

Fc : layer critical frequency; Hm : altitude of maximum ionization (Rm = Ro + Hm); Ym = 100 km : layer semi-thickness (Rb = Rm - Ym); Ro : radius of the Earth (6370 km).
Fc and Hm are generally considered to be the most important ionospheric parameters because refraction mostly takes place near the peak (maximum ionization) of the layer.
It has been shown [2] that if the model is quasi-parabolic, then for a ray with known operating frequency F and elevation angle e (see fig. 1) calculations of ground distance D and group delay Tg are analytical; group delay Tg is the time taken by a radio pulse on the circuit of fig. 1 (usually multiplied by c and expressed in km).
With these assumptions we define a 2-parameter ionospheric model which allows us to compute without difficulty the values of D and Tg which we need to complete our simulation of HF backscatter sounding.
An HF backscatter radar transmits at an operating frequency F within the HF band and measures the amplitude of the received echoes as a function of group delay Tg [3]. Most radars project a beam which is narrow in azimuth (the problem can thus be considered 2-dimensional in the vertical plane containing that azimuth, justifying fig. 1) and wide in elevation [4]. In this context it is helpful to imagine the amplitude of the received signal in the operating frequency-ray elevation angle-group delay (F-e-Tg) space. Elevation is in fact more difficult to determine experimentally than operating frequency or group delay because for fine angle determination it is necessary to have receiving arrays which are large compared to the operating wavelength which is of the order of 10 m. This technique is nevertheless being investigated at CNET Lannion [5].
The three experimental sounding methods which we will now describe can be understood as cuts or projections within the F-e-Tg space.

4. DESCRIPTION OF SOUNDING METHODS

We describe briefly the experimental procedures and show simulations of results in fig. 2 (for Fc = 7 MHz and Hm = 300 km).

* First method : the point-to-point or bistatic method involves setting up a receiver or a beacon or a repeater at a fixed ground distance **D** from the sounder site. The operating frequency F is swept and the group delay Tg of the echo is measured (fig. 2, a). In the F-e-Tg space the result can be visualised as a plane cut (for a given ground distance **D**, Tg is roughly a linear function of e).

This well known technique [6] is obviously not a proper backscatter method and we lose one of the main advantages of backscatter radars, which is that they are monostatic; thus it is included mostly for comparison. Fig. 2, a shows that the echo trace is flat, with a "nose" at maximum frequency Fmax. Thus we characterize it sufficiently for our purposes by recording Fmax and Tg(Fmax). These values constitute the data pair used for inversion. Experimental errors are estimated to have a Gaussian distribution with standard deviation : 0.2 MHz for Fmax and 50 km for Tg(Fmax).

* Second method : the swept-frequency backscatter method is monostatic. It has been demonstrated since the 1920s [3]. The operating frequency F is swept and echo amplitude is measured as a function of Tg (projection of the signal on the F-Tg plane, see fig. 2, b). At the leading-edge (minimum Tg curve, see fig. 2, b) a focusing of energy occurs. This curve is used for inversion because it is the most distinctive feature of the data. Since it is roughly linear it can be characterized by two points, at two fixed frequencies (here 10 and 14 MHz). The data is then the pair of values at the leading edge, Tg(10) and Tg(14). This inverse problem has been studied (for 3 frequencies) by Rao [4]. Experimental errors on Tg are estimated to have a Gaussian distribution with standard deviation : 50 km.

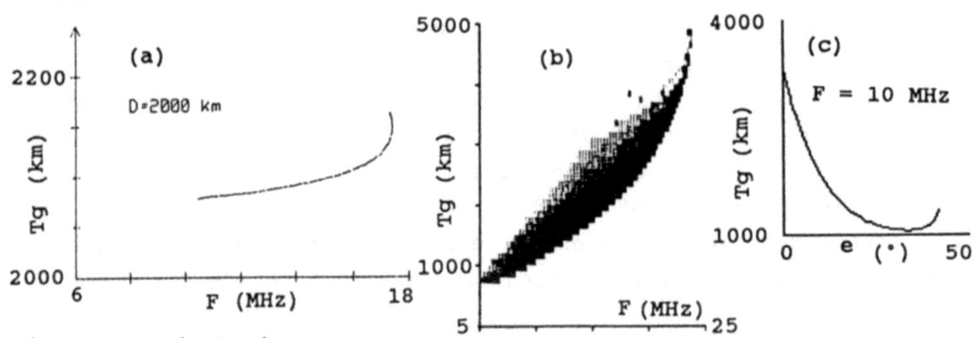

Figure 2 : simulations

* Third method : for the swept-elevation backscatter method, the operating frequency F is fixed and the elevation angle e of the received signal must be controlled.The group delay Tg of the echo is measured as a function of elevation (fig. 2, c). The result corresponds to a plane cut in the F-e-Tg space for a given F. Elevation control is technically difficult but possible through beamforming [5] or interferometry [7]. Fig. 2, c shows that the trace is U-shaped, with a minimum which is used to characterize it. The recorded data is thus Tgmin, e(Tgmin). Experimental errors are estimated to be Gaussian with a standard deviation of 50 km for Tgmin and 5° for e(Tgmin).

It can be noticed that for each method we restrict experimental data to two numerical values, and specify error statistics. This is to provide a more meaningful comparison between the three methods.

5. INFORMATION CONTENT OF THE SOUNDING METHODS

We have described a model which uses two parameters Fc and Hm, and we have explained the nature of the data provided by each method. Now we can compare the three sounding methods from the point of view of ionospheric parameter determination by calculating, in accordance with the formalism described by Tarantola [8], the a posteriori probability density d over the parameter space (Fc,Hm). d(Fc,Hm) represents the probability of Fc and Hm having certain values, after combining prior information on the parameters, experimental data, and the physical model.

Here no a priori information was considered to be known about the ionospheric parameters; the experimental data were restricted to two numerical values D and D'; the experimental errors were assumed to have a Gaussian statistic with standard deviations dD. The model (described previously) was assumed perfect. Thus d(Fc,Hm) is simply a representation of the information content of the data with respect to the model parameters.

The calculation involves choosing a "solution", here the ionosphere characterized by the "true" parameters : Fc = 7 MHz, Hm = 300 km.

Fc is allowed to vary between 5 and 9 MHz in steps of 1 MHz, and Hm between 200 and 400 km in steps of 50 km. For each method, and for each of the 25 pairs (Fc,Hm) we compute the data values (D,D') (direct problem). For the "solution" we compute the "observed" values (Dobs,D'obs), then we calculate the a posteriori probability density:

$$d(Fc,Hm) = \exp\left[\frac{-1}{2}\cdot\left[\frac{D - Dobs}{dD}\right]^2\right]\cdot\exp\left[\frac{-1}{2}\cdot\left[\frac{D' - D'obs}{dD'}\right]^2\right]$$

6. RESULTS AND DISCUSSION

Fig. 3 shows the a posteriori probability density represented for each method over a grid of 25 points.

* For the first method the result (fig. 3, a) shows a well defined peak corresponding to the "solution", indicating a well posed problem. Indeed this method is used experimentally to a high degree of precision for ionospheric parameter determination. Not being truly a backscatter method, it was included for comparison.

* For the second method fig. 3, b shows that there is no well defined peak in the (Fc,Hm) plane. The problem is not well conditioned, which has hindered widespread use of this method, as mentionned in [9]. Possibly with the inclusion of a priori information this could be improved.

* The third method (fig. 3, c) also shows a well defined peak, indicating a well posed problem. This is what we wished to confirm since one of our objectives was to find a backscatter method which did not have the inversion problems of the second method. If technical problems can be solved the swept-elevation method seems promising.

We will now discuss the use of a priori information, and modelization errors : in [8] Tarantola points out the usefulness of a priori information in the inversion process. Although we have not included it in our analysis, there are several possible sources. One is vertical (zenithal) sounding, which gives an accurate representation of the profile over the sounder site at a given time; however the variability of ionospheric parameters is felt over scales of 10 mn in time and 100 km in horizontal distance [10]. Thus vertical sounding parameters can have a real (but limited) use as starting points for an inversion process. Other possible sources of a priori information are global (physical or empirical) models of the ionosphere which are available. This should be investigated, in particular in the case of swept-frequency backscatter.

On the subject of modelization errors, it was supposed that our simple model was a perfect representation of the ionosphere. The real

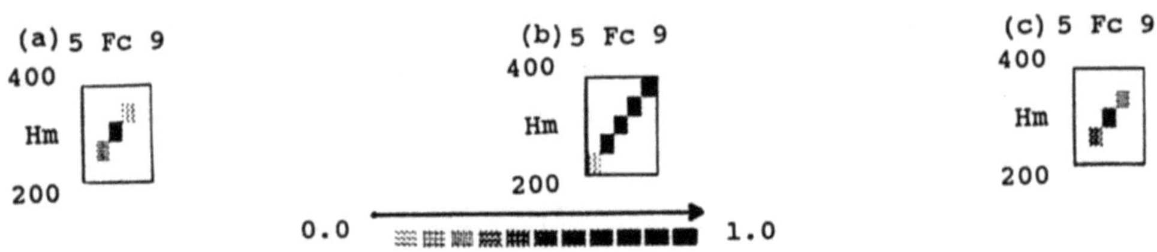

Figure 3 : a posteriori probability density

ionosphere is in fact much more complicated. This should be investigated because the effect of some "secondary" parameters on propagation can be very important in certain circumstances. In particular in daytime the parabola is not a good approximation for the profile due to multiple sun-controlled layers (E, F1). Also horizontal variability of the ionosphere induces tilts and gradients of ionization, so that the layer may not be spherically symetric. To take these points into account, a more complicated model, with more parameters, must be introduced. Data of correspondingly better quality must be obtained from the soundings. It is not at all sure, however, that this is possible.

7. CONCLUSION

The general problem of backscatter data inversion has been introduced, and a simple model of backscatter sounding described. In comparing three basic sounding techniques, it has been shown that swept-frequency backscatter leading edge analysis leads to a badly conditioned inversion. Swept-elevation is shown to be promising, in analogy with the well known bistatic technique, if technical problems can be mastered. Also the use of double sweeping (frequency-elevation) must be investigated.

Most of all it must be emphasized that further theoretical work is necessary in order to devise efficient inversion methods and more realistic models.

In this context, methods generated by work on other inverse problems (internal geophysics, sonar...) may provide useful insights, and should be investigated.

8. REFERENCES

1. Budden, K.G. (1961). "Radio Waves in the Ionosphere". Cambridge University Press.

2. Croft, T.A. and H. Hoogasian (1968). Exact ray calculations in a quasi-parabolic ionosphere with no magnetic field. Radio Science 3, 69-74.

3. Croft, T.A. (1972). Sky-wave backscatter : a means for observing our environment at great distances. Rev. of Geophysics and Space Physics 10, 73-115.

4. Rao, N.N. (1974). Inversion of sweep-frequency sky-wave backscatter leading edge for quasiparabolic ionospheric layer parameters. Radio Science 9, 845-847.

5. Le Saout, J.Y. and L. Bertel (1988). Les systèmes d'aériens utilisés par le radar HF à rétrodiffusion du CNET. JINA 88 Conf. Proceedings, 509-513.
6. Chuang, S.L. and K.C. Yeh (1977). A method for inverting oblique sounding data in the ionosphere. Radio Science 12, 135-140.
7. Baker, K.B. and R.A. Greenwald (1988). The vertical angle of arrival of high-frequency signals propagating from Thule to Goose Bay. Johns Hopkins APL Tech. Digest 9, 121-130.
8. Tarantola, A. (1987). "Inverse Problem Theory". Elsevier, Amsterdam.
9. Ruelle, N., Le Saout, J.Y. and L. Bertel (1989). New perspectives for the inversion of backscatter ionograms. AGARD Conf. Proc. n° 453.
10. Paul, A.K. (1989). Medium scale structure of the F region. Radio Science 24, 301-309.

Generalized Bargmann-Type Potentials

W.A. Schnizer and H. Leeb

Institut für Kernphysik, Technische Universität Wien, A-1040 Wien, Austria

A class of exactly solvable Schrödinger operators, linearly dependent on a parameter γ^2, is presented. The associated potentials are of a generalized Bargmann-type and can contain a contribution with a linear dependence on γ^2. The physical meaning of γ^2 in terms of the energy and the square of the orbital angular momentum is controlled by a function of arbitrary choice. In particular we consider the case of linear mutual dependence of energy and squared angular momentum which allows for a continuous transition between the conventional formulations of the inverse scattering problem at fixed energy and fixed angular momentum. Numerical calculations for this special case have been performed. In the limit of fixed energy the rational scheme of Lipperheide and Fiedeldey is obtained.

1. Introduction

The quantal inverse scattering problem is concerned with the determination of a potential from scattering data. This is the central question of many analyses interpreting experimental data in atomic, nuclear and particle physics. The practical point of view as well as the importance of inversion procedures as basic tools for the solution of nonlinear partial differential equations [1] have been the motivation for a great variety of detailed investigations in the past [2]. Usually the starting point of inversion methods are not the scattering observables but the S-matrix which is more directly related to solutions of the Schrödinger equation.

Because of their simplicty much effort has been devoted to problems with local, spherically symmetric potentials. They can be expanded in partial waves thus reducing the original three-dimensional problem to radial Schrödinger equations for each partial wave. The corresponding expansion of the S-matrix yields matrix elements $S_\ell(k)$ which depend on the orbital angular momentum quantum number ℓ and the energy $E=k^2$. In contrast to the energy which is a continuous variable, ℓ is restricted to integer values. From the mathematical point of view it is most appealing to extend S into the complex λ-plane ($\lambda = \ell + 1/2$) . There is a natural but not unique way to realize this so-called Watson transform $S_\ell(k) \rightarrow S(\lambda,k)$ by extending the radial Schrödinger equation to complex λ [3].

In the inverse problem we look for the unknown potential V(r), which is a function of one continuous variable. Hence it is very plausible that an inversion procedure starting from the two-variable function $S(\lambda,k)$ is in principle overdetermined. This overdetermination is avoided in conventional inversion procedures by fixing one variable. The well known Gel´fand-Levitan and the Marchenko equations [2] are examples for the case of fixed angular momentum and express a unique relationship between $S(\lambda,k)$ and V(r). Similar equations [4,5] as well as other methods [1] have been developed for fixed energy. However, it should be remarked that the situation is less satisfactory at fixed energy. In Fig. 1 we show in a schematic (λ^2,k^2)-plane at which arguments $S(\lambda,k)$ is required in the traditional inversion procedures.

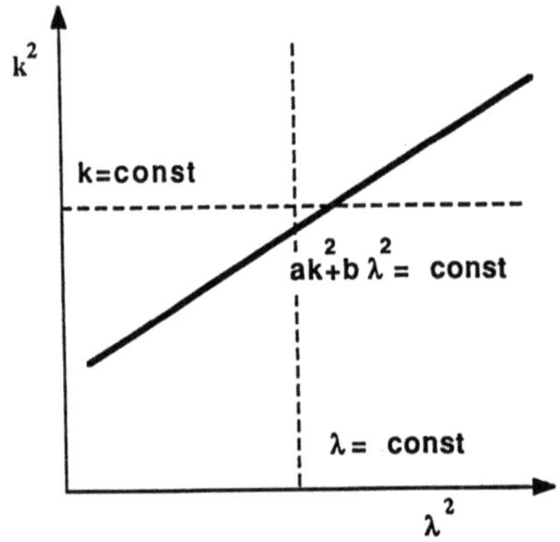

Fig. 1 The (λ,k)-values at which the S-matrix is required in different inversion procedures. The axis for the complex quantities λ^2 and k^2 must be understood symbolically. The dashed lines correspond to the inversion at fixed energy and fixed angular momentum, respectively. The solid line indicates the case of general inversion schemes with linear mutual dependence of λ^2 and k^2 (1).

More general inversion procedures can be obtained by the theory of Coudray and Coz [6] and Cornille [7] using generalized transformation operators. In particular one can construct inversion schemes where S(λ,k) is required at argument values which fulfil the relationship,

$$ak^2 + b\lambda^2 = \text{constant}, \quad a,b \in \mathbf{C}\,. \tag{1}$$

These schemes are represented by straight lines in Fig. 1 with a slope given by the coefficients a,b of (1). It is obvious that the conventional inversion procedures discussed above emerge as special cases of this general class.

The theoretical solutions of the inverse scattering problem are rather complicated and require great numerical effort. Therefore in most applications one relies on exactly solvable models where for a given class of S-matrix elements the associated potential is known analytically. The Bargmann potentials [8] are the best known exactly solvable models. They are characterized by a rational scattering function at fixed λ. For fixed energy the rational scheme of Lipperheide and Fiedeldey [9] is the counterpart to the Bargmann potentials. Recently, using the general inversion scheme as discussed above, Rudyak and Zakhariev [10] found generalized Bargmann potentials.

In this contribution we present a generalization of the rational scheme of Lipperheide and Fiedeldey [9] following the guideline of [6,7]. The resulting potentials can also be considered as generalized Bargmann potentials, but then belong to a wider class than those of Rudyak and Zakhariev [10] and they have been derived in a different way. After a brief sketch of the derivation of the rational scheme in section 2 we generalize this technique in section 3 obtaining a wide class of exactly solvable Schrödinger operators. Restricting ourselves to a mutual linear dependence of λ^2 and E (Fig. 1) we study in section 4 an example numerically. A summary and some concluding remarks are given in section 5.

2. The rational scheme

The rational scheme of Lipperheide and Fiedeldey [9] is based on the Jost solutions $f_0^{\pm}(\lambda,r)$ of the radial Schrödinger equation at fixed energy for given potential $V_0(r)$. Separating the term which depends on λ we obtain

$$\left\{ \frac{d^2}{dr^2} + k_0^2 + \frac{1/4}{r^2} - V_0(r) \right\} f_0^{\pm}(\lambda,r) = \frac{\lambda^2}{r^2} f_0^{\pm}(\lambda,r) , \tag{2}$$

where k_0^2 is the energy considered. The Wronskian of the Jost solutions

$$[f_0^-(\lambda,r) ; f_0^+(\mu,r)] : = f_0^-(\lambda,r) \frac{d}{dr}\left(f_0^+(\mu,r)\right) - \frac{d}{dr}\left(f_0^-(\lambda,r)\right) f_0^+(\mu,r) \tag{3}$$

are used to define new functions

$$g^-(\lambda,r) : = f^-(\lambda,r) - \frac{\alpha^2-\beta^2}{\lambda^2-\beta^2} \frac{[f_0^-(\lambda,r) ; f_0^+(\beta,r)]}{[f_0^-(\alpha,r) ; f_0^+(\beta,r)]} f_0^-(\alpha,r) \quad , \tag{4}$$

$$g^+(\lambda,r) : = f^+(\lambda,r) - \frac{\alpha^2-\beta^2}{\alpha^2-\lambda^2} \frac{[f_0^-(\alpha,r) ; f_0^+(\lambda,r)]}{[f_0^-(\alpha,r) ; f_0^+(\beta,r)]} f_0^+(\beta,r) \quad . \tag{5}$$

It is straightforward to show that these functions are solutions of the Schrödinger equation (2) with the potential

$$V(r) = V_0(r) - \frac{2}{r}\frac{d}{dr} r \frac{d}{dr} \ln[f_0^-(\alpha,r) ; f_0^+(\beta,r)] \quad . \tag{6}$$

From the asymptotic behaviour of the Jost solutions $f_0^{\pm}(\lambda,r)$ it follows directly that the new functions $g^{\pm}(\lambda,r)$ are proportional to the Jost solutions of the Schrödinger equation (2) with the potential (6). Renormalizing the wave functions $g^{\pm}(\lambda,r)$ we can determine the Jost functions $f^{\pm}(\lambda)$ from the behaviour of $g^{\pm}(\lambda,r)$ at the origin, and thus the S-matrix associated with the potential V(r) is found to be

$$S(\lambda) = S_0(\lambda) \frac{\lambda^2 - \beta^2}{\lambda^2 - \alpha^2} \quad . \tag{7}$$

Here, $S_0(\lambda)$ is the S-matrix due to the background potential $V_0(r)$. The modification of the S-matrix caused by the potential $V(r)-V_0(r)$ is purely rational in λ. This is in complete analogy with the Bargmann potentials at fixed λ, which are rational in the wave number k. The quantities α and β are complex numbers, which correspond to fixed values of orbital angular momentum and can be considered as parameters of the potential (6). In principle no condition on α and β is required in the derivation, however, their meaning as poles and zeros of the S-matrix restricts the domain to those values which describe physical Regge poles and zeros [9].

3. A generalized exactly solvable model for the Schrödinger equation

We want to work out exactly solvable models which can be applied in inversion problems with mutual dependence of λ^2 and E as discussed in the introduction (Fig. 1). For this purpose we consider the Jost solutions $f_0^{\pm}(\gamma,r)$ of the equation

$$\left\{ \frac{d^2}{dr^2} + k_0^2 - \frac{\lambda_0^2 - 1/4}{r^2} - V_0(r) \right\} f_0^{\pm}(\gamma,r) = \gamma^2 \, h(r) \, f_0^{\pm}(\gamma,r) , \tag{8}$$

where k_0^2 and λ_0 are fixed values of the energy and orbital angular momentum, respectively. The meaning of the parameter γ^2 is governed by the function h(r). Apart of the fact that h(r) should fulfil the general requirements on potentials in scattering theory, the function h(r) is arbitrary and can be appropriately chosen for a given inversion problem. The problems at fixed energy (2) are obtained by chosing $h(r)=r^{-2}$. In this case the parameter γ is directly related to the orbital angular momentum, $\lambda^2=\lambda_0^2+\gamma^2 a$. With the function $h(r)=-1$ we obtain problems at fixed angular momentum and using e.g. $h(r)=r^{-1}$ we treat the question of variable electric charge at fixed angular momentum and energy, which has been investigated by Poplavsky and Popushoy [11]. Interpreting (2) as a Schrödinger equation it is obvious that for general h(r) the quantities $k(\gamma)$ and $\lambda(\gamma)$ will depend on the parameter γ; furthermore the potential can contain a linear term in γ^2.

Coudray and Coz [6] and Cornille [7] have derived generalized translation operators for the equation (2). The resulting generalized Gel´fand-Levitan equations have been used by Rudyak and Zakhariev [10] to construct exactly solvable models, which yield the S-matrix of the Bargmann potentials at fixed angular momentum. In this paper we want to construct a generalization of the rational scheme [9] at fixed energy. Contrary to Rudyak and Zakhariev we do not use the integral equation but apply the formalism as discussed in section 2.

Again we use the Wronskian of the Jost functions (3) to define new functions $g^{\pm}(\gamma,r)$. The functional relations are exactly the same as displayed in (4,5) besides the fact that the orbital angular momentum quantity λ has to be substituted by the parameter γ. Furthermore, in the context of the general problem, the fixed complex numbers α and β must be interpreted as special values of the parameter γ. A straightforward calculation shows that $g^{\pm}(\gamma,r)$ are solutions of the equation (8) with the potential

$$V(r) = V_0(r) - 2\sqrt{h(r)}\frac{d}{dr}\frac{1}{\sqrt{h(r)}}\frac{d}{dr}\ln[f_0^-(\alpha,r)\,;\,f_0^+(\beta,r)] \quad . \tag{9}$$

In analogy to the rational scheme the wave functions $g^{\pm}(\gamma,r)$ are proportional to the Jost solutions of (8) with the potential (9). The S-matrix associated with the potential (9) can be determined from the Jost functions which are obtained from the behaviour of the Jost solutions at the origin. We obtain the expression

$$S(\gamma) = S_0(\gamma)\,\frac{\gamma^2-\beta^2}{\gamma^2-\alpha^2}\,\frac{\left[1-\dfrac{\alpha^2-\beta^2}{\gamma^2-\beta^2}\,\dfrac{k(\gamma)+k(\beta)}{k(\alpha)+k(\beta)}\right]}{\left[1-\dfrac{\alpha^2-\beta^2}{\alpha^2-\gamma^2}\,\dfrac{k(\alpha)+k(\gamma)}{k(\alpha)+k(\beta)}\right]} \quad , \tag{10}$$

where $S_0(\gamma)$ describes the scattering caused by the background potential $V_0(r)$. The complex numbers α and β are parameters of the potential (9) and are restricted by the requirement of the existence of the background Jost solutions $f_0^{\pm}(\gamma,r)$.

The expression (10) for the S-matrix is valid for general h(r), except for the fixed λ-case. In particular if we use the function $h(r)=r^{-2}$ we obtain the scattering problem at fixed energy, which implies $k(\alpha)=k(\beta)=k(\gamma)=k_0$. In this case the S-matrix (10) reduces to the S-matrix of the rational scheme (7). More general scattering problems are obtained by functions h(r) of the form

$$h(r) = \frac{a}{r^2} + b + h_V(r), \tag{11}$$

where $h_V(r)$ is a nonsingular function of short range and a and b are nonvanishing complex numbers. Assuming (11) the expression (10) for the S-matrix can be reformulated and we obtain

$$S(\gamma) = S_0(\gamma) \frac{k(\gamma)-k(\alpha)}{k(\gamma)+k(\alpha)} \frac{k(\gamma)+k(\beta)}{k(\gamma)-k(\beta)} \quad , \tag{12}$$

which is the S-matrix of the Bargmann potential [8] at fixed angular momentum. However, it must be remarked that the corresponding potentials will be strongly dependent on the scattering problem determined by h(r).

So far we have considered transformations with a single pair of Jost solutions characterized by α, β. Since the new potential (9) fulfils the requirements of a background potential we can formulate an iterative scheme taking V(r) as a new background potential for the next step. The resulting potentials for N pairs (α_i,β_i) represent a wide and flexible class.

4. Numerical Example

In the general scheme we have the possibility to fix the scattering problem by a suitable choice of the function h(r). In particular we can study the transition from the conventional scattering problems at fixed energy to those at fixed angular momentum by

$$h(r) = \frac{a+br^2}{r^2} \quad , \tag{12}$$

where a and b are constants. With the function (12) the functional dependence of k and λ on the parameter γ is given by

$$k^2(\gamma) = k_0^2 - b\,\gamma^2 \quad ,$$
$$\lambda^2(\gamma) = \lambda_0^2 + a\,\gamma^2 \quad , \tag{13}$$

and fulfil the mutual relationship (1).

We have studied numerically the behaviour of the general Bargmann potentials intermediate between the scattering problems at fixed energy and fixed angular momentum. In particular we consider iterated potentials with two complex conjugate parameter pairs,

$$\alpha_1 = \beta_1{}^* = 8.0 + i\,0.7 \quad ,$$
$$\alpha_2 = \beta_2{}^* = 3.0 + i\,0.1 \quad , \tag{14}$$

which lead to real potentials and consequently to a unitary S-matrix. No background potential has been taken into account. In Fig. 2 we show the radial dependence of the potential assuming different values for a and b. The different scattering problems are characterized by the angle arctan($^b/a$) since only the ratio b/a is the relevant quantity entering the potential. At fixed energy $^b/a$=0, and we have the potentials of the rational scheme. Changing $^b/a$ towards the fixed angular momentum case, the potential gets deeper and moves shorterange. In Fig.2 we also show the corresponding phase shifts. In order to test the numerical procedure the potentials have been used to recalculate the phase shifts and complete agreement has been achieved in all cases. The transition to fixed angular momentum, $^b/a=\pm\infty$, is not possible because the Jost functions which are needed in the construction of V(r) have the same irregular behaviour near the origin.

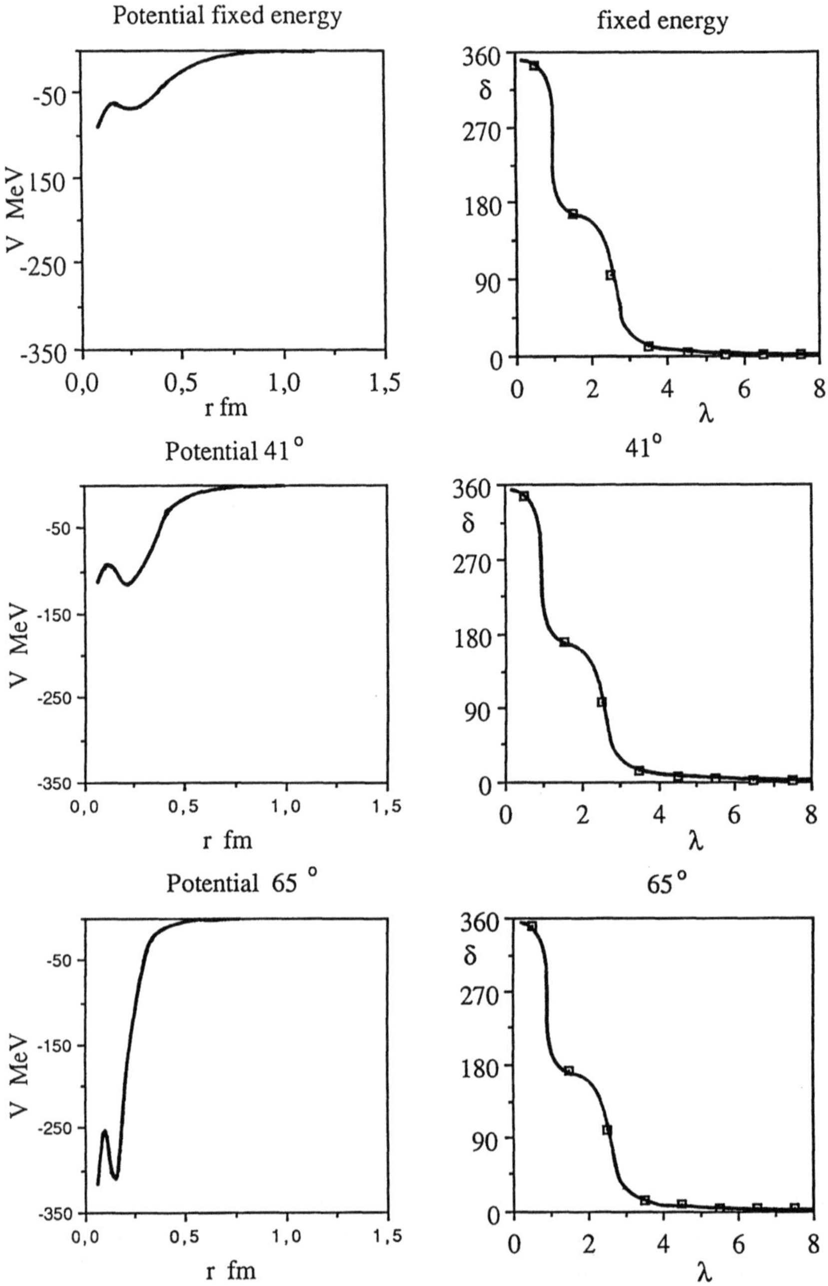

Fig. 2 Generalized Bargmann potentials and corresponding phase shifts for different values of arctan(b/a) describing the transition from fixed energy towards fixed angular momentum scattering problems. Bargmann potentials with two fixed and complex conjugate pairs $\alpha_1 = \beta_1^* = 8.0 + i\,0.7$, $\alpha_2 = \beta_2^* = 3.0 + i\,0.1$ have been used. The points (■) in the phase shift curve denote the half integer values which correspond to physical values.

5. Summary and Conclusions

After a short summary of the derivation of the rational scheme at fixed energy we have extended this technique to more general scattering problems. Requiring only the existence of Jost solutions, we obtain compact analytical expressions for the potential and the S-matrix which in the fixed energy scattering problem reduce to those of the rational scheme [9]. On the other hand, assuming a very general class of scattering problems where both angular momentum and energy are changed, we obtain the S-matrix of the original Bargmann potential, which is purely rational in k. Rudyak and Zakhariev [9] have obtained the same expression for the S-matrix for a smaller class of problems using a degenerate kernel in a Marchenko type equation. From our derivation the transition to the rational scheme at fixed energy is obvious and shows for the first time its relationship with the original Bargmann potential at fixed λ. Despite the formal equivalence of the S-matrix with that of the original Bargmann potential, there are open questions in the transition to the scattering problem at fixed λ because of vanishing Jost functions. Therefore we consider this general class of potentials rather as an extension of the rational scheme and not of the Bargmann potential. A further extension of the class of exactly solvable has been obtained by the formulation of an iteration procedure.

Finally it should be remarked that our procedure is not limited to the use of Jost functions in the Wronskian relations (4,5). Although there are some traps the whole procedure can be formulated with arbitrary solutions in analogy to the fixed energy case [13]. This generalization, which also includes the original Bargmann potential, will be described in detail in a forthcoming paper [14].

Acknowledgement

We thank Prof. Dr. Lipperheide for fruitful discussions. The work has been supported by Fonds zur Förderung der wissenschaftlichen Forschung in Österreich (Project Nr. P6608)

References

[1] R. K. Dodd, J. C. Eilbeck, J. D. Gibbon, H. C. Morris, Solitons and Nonlinear Wave-Equations, Academic Press, Lodon, 1982.
[2] K. Chadan, P. C. Sabatier, Inverse Problems in Quantum Scattering Theory, Springer, 1989.
[3] R. G. Newton, Scattering Theory of Waves and Particles, Springer, 1982.
[4] G. Burdet, M. Giffon, E. Predazzi, Nuovo Cimento **36**, 133 (1965).
[5] V. V. Malyarov, I. V. Poplavsky, N. N. Popushoi, Soviet J. Nucl. Phys. **22**, 445 (1976); **25**, 38 (1977).
[6] C. Coudray and M. Coz, Ann. Phys. (NY) **61**, 488 (1970).
[7] H. Cornille, J. Math. Phys. **17**, 2143 (1976); **18**, 1855 (1977).
[8] W . R. Theis, Z. Naturforsch. **11a**, 889 (1956).
[9] R. Lipperheide, H. Fiedeldey, Z. Phys. **A286**, 45 (1978).
[10] B. V. Rudyak and B. N. Zakhariev, Inverse Problems **3**, 125 (1987).
[11] I. V. Poplavsky, Sov. Ukr. Phys. J. **28**, 1631 (1983); **29**, 977 (1984).
[12] M. N. Popushoy, Sov. J. Theor. Math. Phys. **63**, 340 (1985).
[13] H. Leeb et al., Inverse Problems **5**, 817 (1989).
[14] H. Leeb and W. A. Schnizer, in preparation.

Rigorous Computer Solutions of Infinite-Dimensional Inverse Problems

P.B. Stark

Department of Statistics, University of California, Berkeley, CA 94720, USA

It is safe to say that almost all practical inverse problems are plagued by nonuniqueness. Regularization and other techniques for picking a particular solution fitting the data are useful for construction, but not usually for inference: the "true" model may bear little or no resemblance to the particular model the technique constructs. "Confidence set inference" or "strict bounds" works by looking at properties shared by all *a priori* acceptable models that fit the data adequately; *i.e.* it constructs confidence intervals for functionals of the unknown model. Constructing the confidence intervals involves solving certain infinite-dimensional constrained optimization problems in the model space. The usual way of solving these problems approximates the infinite-dimensional problems in a finite-dimensional subspace of the model space, so that the resulting problem can be solved by computer. Clearly, this can produce confidence intervals that are too short. This report uses Lagrangian duality to show how some of these optimization problems may be solved on a computer to get confidence intervals of the correct length.

1 Introduction

Please see Backus [2], Cuer [3], Lang [5], and Stark [10] for other approaches to the same or similar problems. Here we treat only linear inverse problems, where we know *a priori* that the unknown model x_0 lies in a cone or a ball, and we wish to make inferences about a linear functional of x_0. The model for the experiment is that we observe n data δ_j, $j = 1, \cdots, n$ that are linearly independent linear functionals f_j^* of x_0, plus additive noise ϵ_j. We assume that the joint probability distribution of the noise terms is known, and that all components of the noise have mean zero. In vector form:

$$\delta = \mathbf{f}^*[x_0] + \epsilon. \tag{1}$$

The model x_0 belongs to the real linear vector space X; the functionals f_j^* belong to its algebraic dual space X^*. We know *a priori* that $x_0 \in \mathsf{C}$, where C is a special convex subset of X: either $\mathsf{C} = \mathsf{X}$ (no special prior information); $\mathsf{C} = \mathsf{P}$, a convex cone with vertex at the origin; or $\mathsf{C} = \mathsf{B}$, a ball in some norm in the space X, which will then be assumed to be a Banach space. In the first two cases, no topology on X is needed. Examples of the constraint $x_0 \in \mathsf{P}$ occur in gravimetry, resistivity and magnetotellurics, and seismology; as mass density, electrical conductivity, and seismic velocities must all be nonnegative. The constraint $x_0 \in \mathsf{B}$ may also arise in these fields if an upper bound on the unknown is available (density is less than that of neutronium, velocity less than the speed of light); it also arises naturally in geomagnetism where one may use *a priori* bounds on the energy stored in the main field, or the rate of dissipation of energy in the core (see Backus [2]). The first examples involve infinity-norm bounds, the last a quadratic norm. We wish to find a confidence interval for $h^*[x_0]$, a linear functional of x_0, using the prior information $x_0 \in \mathsf{C}$ and our knowledge of the probability distribution of the errors. That is, for fixed $\alpha \in (0,1)$, we wish to find μ_-, μ_+ such that

$$\wp\{[\mu_-, \mu_+] \ni h^*[x_0]\} \geq 1 - \alpha,$$

where $\wp\{\cdot\}$ is the probability that the event in braces occurs.

Let Ξ be a $1-\alpha$ confidence region for ϵ. Then $\wp\{\delta - \mathbf{f}^*[x_0] \in \Xi\} \geq 1 - \alpha$. Let

$$\mathsf{D} \equiv \{x \in \mathsf{X} : \delta - \mathbf{f}^*[x_0] \in \Xi\}.$$

Then we also have

$$\wp\{\mathsf{D} \ni x_0\} \geq 1 - \alpha,$$

and since $x_0 \in \mathsf{C}$,

$$\wp\{\mathsf{C} \cap \mathsf{D} \ni x_0\} \geq 1 - \alpha. \tag{2}$$

Let

$$\mu_- \equiv \inf_{x \in \mathsf{C} \cap \mathsf{D}} h^*[x]$$

and

$$\mu_+ \equiv \sup_{x \in \mathsf{C} \cap \mathsf{D}} h^*[x]$$

then μ_-, μ_+ satisfy 1. It remains to choose Ξ and solve the resulting optimization problems.

It has been traditional to choose Ξ to be a ball in some norm in $\mathbf{R}^n$; often an infinity norm. Backus [2] advocates a "slab," the region between parallel hyperplanes in $\mathbf{R}^n$. In the case $\mathsf{C} = \mathsf{X}$, it is well known that the only linear functionals that can be bounded by the data are linear combinations of the data functionals f_j^* (for example, [1]). As we have assumed that the f_j^* are linearly independent, there exists at most one representation of h^* as a linear combination of $\{f_j^*\}_{j=1}^n$. Say

$$h^* = \boldsymbol{\lambda} \cdot \mathbf{f}^* \equiv \sum_{j=1}^{n} \lambda_j f_j^* .$$

It is intuitively clear that in this case, if the distribution of errors is symmetric, the optimal choice of confidence regions Ξ to get the narrowest possible confidence interval for $h^*[x_0]$ is

$$\Xi = \{\boldsymbol{\beta} \in \mathbf{R}^n : |\boldsymbol{\lambda} \cdot \boldsymbol{\beta}| \le \chi\}, \tag{3}$$

for an appropriate choice of χ which may be computed from the known distribution of ϵ. This is the "slab" (not his terminology) of Backus. This choice allows D to be unbounded in every direction other than h^*; it concentrates all the statistical power of the data in the relevant direction.

For other sorts of prior information, the optimal choice of sets Ξ to get the shortest confidence intervals in expectation seem to be unsolved problems. In this report, we will use slabs as the confidence regions in data space. The orientation of the slab will be given by the vector γ, which need bear no relation to h^*. In [10] I develop results for general balls in l_p norms, for other functionals, and for slightly more general prior information.

2 Primal Approaches

We specialize to the minimization problem only since μ_+ can be found by minimizing $-h^*$ (also a linear functional). For Ξ as in 3,

$$\mathsf{D} = \{x \in \mathsf{X} : |\gamma \cdot (\mathbf{f}^*[x] - \boldsymbol{\delta})| \le \chi\}.$$

Define $g^* \equiv \gamma \cdot \mathbf{f}^*$ and $\zeta \equiv \gamma \cdot \boldsymbol{\delta}$. Then

$$\mathsf{D} = \{x \in \mathsf{X} : |g^*[x] - \zeta| \le \chi\} \tag{4}$$

In case $\mathsf{C} = \mathsf{X}$, it is well known that the only linear functionals with nontrivial confidence intervals are of the form $h^* = \lambda g^*$, $\lambda \in \mathbf{R}$. In case $\mathsf{C} = \mathsf{P}$ or $\mathsf{C} = \mathsf{B}$, the traditional means of solving the *primal* problem

$$\mu_- = \inf_{x \in \mathsf{C} \cap \mathsf{D}} H[x]$$

is to discretize the problem by approximating x in a finite-dimensional subspace of X. Given $\{x_j\}_{j=1}^m \subset \mathsf{X}$, let $\tilde{\mathsf{X}}$ be the subspace spanned by $\{x_j\}_{j=1}^m$. Then $x \in \tilde{\mathsf{X}}$ can be written $\sum_{k=1^m} \beta_j x_j$, and $x \in \mathsf{D} \cap \tilde{\mathsf{X}}$ if $|\boldsymbol{\phi} \cdot \boldsymbol{\beta} - \zeta| \le \chi$, where the m-vector $\boldsymbol{\phi}$ has elements $\phi_k = g^*[x_k]$ and $\boldsymbol{\beta} = (\beta_1, \cdots, \beta_m)$. Similarly, $h^*[x] = \boldsymbol{\eta} \cdot \boldsymbol{\beta}$, where $\boldsymbol{\eta}$ is the m-vector with components $\eta_j \equiv h^*[x_j]$. One completes the picture by expressing $x \in \mathsf{C}$ in terms of the coefficients β_j. In my experience, it appears to be possible to pick the $\{x_k\}$ so that this is not too difficult. The result is a finite-dimensional optimization problem:

$$\tilde{\mu}_- = \min_{\{\boldsymbol{\beta} : \sum_k \beta_k x_k \in \mathsf{C} \cap \mathsf{D} \cap \tilde{\mathsf{X}}\}} \boldsymbol{\eta} \cdot \boldsymbol{\beta}.$$

It is clear that since the minimization is over the more restrictive set $\mathsf{C} \cap \mathsf{D} \cap \tilde{\mathsf{X}}$, $\tilde{\mu}_- \ge \mu_-$ (and correspondingly $\tilde{\mu}_+ \le \mu_+$), and thus the resulting confidence interval may be too short to achieve the nominal coverage probability. We turn now to another method of solution.

3 Lagrangian Duality

See the excellent book by Luenberger [6]. If (A) there exists $x_1 \in \mathsf{C}$ such that $|g^*[x_1] - \zeta| < \chi$, and (B) $\mu_- > -\infty$, Lagrangian duality says

$$\mu_- = \max_{\lambda \geq 0} \inf_{x \in \mathsf{C}} \{h^*[x] + \lambda(|g^*[x] - \zeta| - \chi)\}. \tag{5}$$

Note that condition (A) can be verified empirically by finding a feasable point for the discretized primal problem; condition (B) can be verified algebraically, as we see below. We now trace the implications of Lagrangian duality for the various choices of C.

3.1 $\mathsf{C} = \mathsf{X}$

We have

$$\begin{aligned} \mu_- &= \sup_{\lambda \geq 0}\{ \inf_{x \in \mathsf{X}} (h^*[x] + \lambda(|g^*[x] - \zeta| - \chi))\} \\ &= \sup_{\lambda \geq 0}\{-\lambda\chi + \inf_{x \in \mathsf{X}} \max[(h^* + \lambda g^*)[x] - \lambda\zeta, (h^* - \lambda g^*)[x] + \lambda\zeta]\}. \end{aligned}$$

Name the second term τ:

$$\tau \equiv \inf_{x \in \mathsf{C}} \max[(h^* + \lambda g^*)[x] - \lambda\zeta, (h^* - \lambda g^*)[x] + \lambda\zeta]\,. \tag{6}$$

For $h^* = \pm\lambda g^*$,

$$\tau = \inf_{x \in \mathsf{X}} \max[\pm\lambda\zeta, \pm 2\lambda g^*[x] \mp \lambda\zeta] = \pm\lambda\zeta.$$

If $h^* \neq \lambda g^*$ and $h^* \neq -\lambda g^*$, write

$$h^* = e^* \pm \lambda g^*,$$

where $e^* \neq 0$ and the sign will be fixed presently. Then

$$\tau \leq \inf_{x \in \mathsf{X}} (e^*[x] \pm \lambda g^*[x] + \lambda|g^*[x]| + \lambda\zeta).$$

Pick the sign so that there exists $y \in \mathsf{X}$ with both $e^*[y] < 0$ and $\pm g^*[y] < 0$ (this is possible since both functionals are nonzero). Then

$$\tau \leq \inf_{x \in \mathsf{X}} (e^*[x] + \lambda\zeta) = -\infty.$$

Thus

$$\begin{aligned} \mu_- &= \sup_{\substack{\lambda \geq 0 \\ h^* = \pm\lambda g^*}} -\lambda\chi \pm \lambda\zeta \\ &= \lambda\zeta - |\lambda|\chi, \quad h^* = \lambda g^*, \ \lambda \in \mathbf{R}. \end{aligned}$$

The only functionals one may make inferences about in this case are multiples of h^*. Note that since g^* is a nonzero linear functional, there always exists x_1 such that $g^*[x_1] = \zeta$, so condition (A) is met. We now have no optimization to perform at all, just the (easy) infinite-dimensional constraint that h^* is proportional to g^*.

3.2 $\mathsf{C} = \mathsf{P}$

The algebra leads us to consider τ (equation 6) again, but for $\mathsf{C} = \mathsf{P}$. Let $\mathsf{P}^{\oplus}$ be the conjugate cone of P in X^*:

$$\mathsf{P}^{\oplus} \equiv \{x^* \in \mathsf{X}^* : x^*[x] \geq 0, \ \forall x \in \mathsf{P}\}.$$

Then if $h^* \pm \lambda g^* \in \mathsf{P}^{\oplus}$,

$$\tau = \inf_{x \in \mathsf{P}} \max[(h^* + \lambda g^*)[x] - \lambda\zeta, (h^* - \lambda g^*)[x] + \lambda\zeta] = \mp\lambda\zeta,$$

while if $h^* + \lambda g^* \notin \mathsf{P}^{\oplus}$ and $h^* - \lambda g^* \notin \mathsf{P}^{\oplus}$, then using the convexity of P it is easy to show that there exists $y \in \mathsf{P}$ such that $(h^* \pm \lambda g^*)[y] < 0$ for *both* choices of sign. Since P is a cone with vertex at the origin, $y \in \mathsf{P}$ implies $\beta y \in \mathsf{P}$, $\beta \geq 0$, and so

$$\tau = \inf_{x \in \mathsf{P}} \max[(h^* + \lambda g^*)[x] - \lambda\zeta, (h^* - \lambda g^*)[x] + \lambda\zeta] = -\infty.$$

We conclude that

$$\begin{aligned}\mu_- &= \max_{\substack{\lambda \geq 0 \\ h^* \pm \lambda g^* \in \mathsf{P}^{\oplus}}} \{-\lambda\chi \mp \lambda\zeta\} \\ &= \max_{\lambda : h^* + \lambda g^* \in \mathsf{P}^{\oplus}} \{-|\lambda|\chi - \lambda\zeta\} \end{aligned} \tag{7}$$

Here we have a one-dimensional optimization problem, with the infinite-dimensional constraint on λ that $h^* + \lambda g^* \in \mathsf{P}^{\oplus}$. In applications, this usually amounts to something like requiring that the sum of a function and a multiple of another is nonnegative everywhere. This sort of constraint can readily be imposed since the two functions are *known*, and so their continuity properties can be exploited to impose the constraint conservatively using only values on a finite grid.

3.3 C=B

Here X is normed; we will assume it is a Banach space and that the functionals h^* and f_j^* (and thus g^*) are continuous. Let B be the ball with radius ν. For this problem, we find

$$\begin{aligned}\tau &= \inf_{x \in \mathsf{B}} \max\left[(h^* + \lambda g^*)[x] - \lambda\zeta, (h^* - \lambda g^*)[x] + \lambda\zeta\right] \\ &\geq \max\left[\inf_{x \in \mathsf{B}}\{(h^* + \lambda g^*)[x] - \lambda\zeta\}, \inf_{x \in \mathsf{B}}\{(h^* - \lambda g^*)[x] + \lambda\zeta\}\right] \\ &= \max\left[-\nu\|h^* + \lambda g^*\| - \lambda\zeta, -\nu\|h^* - \lambda g^*\| + \lambda\zeta\right] \\ &\equiv \sigma,\end{aligned}$$

so

$$\begin{aligned}\mu_- &\geq \max_{\lambda \geq 0}\{-\lambda\chi + \max\left[-\nu\|h^* + \lambda g^*\| - \lambda\zeta, -\nu\|h^* - \lambda g^*\| + \lambda\zeta\right]\} \\ &= \max_{\lambda \in \mathbf{R}}\{-|\lambda|\chi - \nu\|h^* + \lambda g^*\| - \lambda\zeta\}.\end{aligned} \tag{8}$$

We shall see that there exists $y \in \mathsf{B}$ such that

$$h^*[y] + \lambda|g^*[y] - \zeta| = \sigma,$$

and thus equality holds in 8.

Define y_+ such that $-y_+$ is aligned with $h^* + \lambda g^*$, and $\|y_+\| = \nu$. Define y_- such that $-y_-$ is aligned with $h^* - \lambda g^*$, and $\|y_-\| = \nu$. Then

$$(h^* + \lambda g^*)[y_+] - \lambda\zeta = -\nu\|h^* + \lambda g^*\| - \lambda\zeta,$$

and

$$(h^* - \lambda g^*)[y_-] + \lambda\zeta = -\nu\|h^* - \lambda g^*\| + \lambda\zeta.$$

A bit of algebra shows that if $\sigma = -\nu\|h^* + \lambda g^*\| - \lambda\zeta$, then $(h^* + \lambda g^*)[y_+] - \lambda\zeta \geq (h^* - \lambda g^*)[y_+] + \lambda\zeta$, and y_+ is the required vector. On the other hand, if $\sigma = -\nu\|h^* - \lambda g^*\| + \lambda\zeta$, then $(h^* - \lambda g^*)[y_-] + \lambda\zeta \geq (h^* + \lambda g^*)[y_-] - \lambda\zeta$, and y_- is the required vector. □

Note that in this problem, the dual is the unconstrained maximization of a concave function of λ—the infinite-dimensional constrained optimization problem is certainly simplified. However, the evaluation of $\|h^* - \lambda g^*\|$ can involve infinite-dimensional operations such as summation of series or integration. Fortunately, the summation or integration involves the *known* vectors h^* and g^*, and so their properties can be used to bound the accuracy of a truncated sum, or of numerical quadrature. There are many error bounds one may use, depending on how much information about the functionals is incorporated. The error bounds are convex functions of λ. If one of these error bounds is subtracted from the original objective function, and the difference is maximized numerically, the result is a rigorous lower bound on μ_-.

4 Discretizing the Dual Problem

We will look at an example of discretizing the dual problem for the case C = B related to the "ideal bodies" problem in gravimitry [7,8,9]. We make n noisy observations δ_j of the vertical component of gravity at the points $\{\vec{r_j}\}_{j=1}^n$ at a height β above Earth's surface. Let E denote the region of $\mathbf{R}^3$ occupied by Earth, and $x_0(\vec{r})$ be the mass density of Earth as a function of position $\vec{r}$. Then

$$\delta_j = f_j^*[x_0] + \epsilon_j, \quad j = 1, \cdots, n$$

where

$$f_j^*[x] \equiv \int_E \mathcal{G} \frac{\hat{z} \cdot (\vec{r_j} - \vec{r})}{|\vec{r_j} - \vec{r}|^3} x(\vec{r}) d^3\vec{r}.$$

Here $\hat{z}$ is the vertical unit vector in $\mathbf{R}^3$. We shall assume that we know *a priori* that the density is bounded below by zero and above by $v(\vec{r})$. We assume the joint distribution of the noise components is known. We wish to find a 99% confidence interval for the mass in the region $D \subset E$, which is given by the linear functional h^*:

$$h^*[x] = \int_E 1_D(\vec{r}) x(\vec{r}) d^3\vec{r}.$$

Here $1_D(\vec{r}) = \{1,\ \vec{r} \in D;\ 0,\ \text{otherwise}\}$. This problem may be reduced to the form in section 3.3 as follows. We pose the problem in the Banach space $L^\infty(E)$ of bounded functions on the domain E, with the "sup" norm

$$||x|| \equiv \sup_{\vec{r} \in E} |x(\vec{r})|.$$

Let $x'(\vec{r}) \equiv \frac{1}{2} v(\vec{r})$. Pick $\gamma \in \mathbf{R}^n$ (for example, let $\gamma \equiv \arg\min_{\rho \in \mathbf{R}^n} ||h^* - \sum_{j=1}^n \rho_j f_j^*||$, or an approximation to it). Let χ be such that $\wp\{\gamma \cdot \epsilon \leq \chi\} = 0.99$. Then the optimization problem we must solve has the form:

$$\mu_- = \inf_{\substack{x \in \mathrm{L}_\infty : ||\frac{x - x'}{x'}|| \leq 1 \\ |\gamma \cdot \mathbf{f}^*[x] - \gamma \cdot \delta| \leq \chi}} h^*[x].$$

By defining $g^* = \gamma \cdot \mathbf{f}^*$ and $\zeta = \gamma \cdot (\delta - \mathbf{f}^*[x'])$; and renorming the space to take into account the weight $\frac{1}{x'}$, we reduce the optimization problem to the canonical form

$$\mu_- = \inf_{\substack{||x|| \leq 1 \\ |g^*[x] - \zeta| \leq \chi}} h^*[x].$$

A common approach to solving this problem is to divide E into m regions and approximate X by the m-dimensional subspace of functions that are constant in each of the regions; the resulting discretized problem can be solved by linear programming (see [9,4]). The linear programming solution will almost certainly be larger than the true solution to the original infinite-dimensional problem. Similarly, the value for μ_+ determined by the finite-dimensional approximation may be too small. In either case the confidence interval may not be large enough to have 99% coverage probability.

Instead, we use duality to get a different finite (one) dimensional problem. The solution is given in equation 8 to be

$$\max_\lambda \{-|\lambda|\chi - \lambda\zeta - ||h^* + \lambda g^*||\}$$

which is the maximization of a one-dimensional convex function. In practice, it is necessary to approximate numerically the integral

$$||h^* - \lambda g^*|| = \int_E \left| 1_D(\vec{r}) - \lambda \sum_{j=1}^n \gamma_j \mathcal{G} \frac{\hat{z} \cdot (\vec{r_j} - \vec{r})}{|\vec{r_j} - \vec{r}|^3} \right| d^3\vec{r},$$

However, since $\{f_j^*\}_{j=1}^n$ and 1_D are *known* functions, their continuity properties can be used to determine the maximum possible error in the quadrature. As noted in section 3.3, the maximum quadrature error is a convex function of λ (and depends on the quadrature scheme). If the error bound is subtracted from $-|\lambda|\chi - \lambda\zeta - ||h^* - \lambda g^*||$, where the norm is to be approximated by quadrature, and the difference, which is a concave function of λ, is maximized numerically, the resulting number yields a confidence interval of at least the correct length. Note that the dual space of X is the (non-separable) space of finitely additive finite signed measures, elements of which would, in general, be rather difficult to approximate even with a countable number of approximating functions. Fortunately, in the dual approach, we need only consider particular known elements of the dual space. This is true for C = X and C = P as well.

5 Conclusion

A common approach to computing confidence intervals for functionals of a model in linear inverse problems involves approximating the infinite-dimensional model space with a finite-dimensional subspace. The result can be confidence intervals that are too short. Duality provides a tool to formulate finite-dimensional problems that exactly solve the original infinite-dimensional problems. These finite-dimensional problems may have infinite-dimensional constraints, or involve integration or some such infinite-dimensional operation. However, the functions that are involved in the dual problem are known, and so the accuracy of methods to approximate the infinite-dimensional part of the dual problem can be calculated or bounded explicitly. The result is that one may construct confidence intervals for functionals that, under the assumptions of the observation model, are certain have their nominal coverage probability.

Acknowledgements. I am grateful to George E. Backus, Peter J. Bickel and David L. Donoho for helpful and encouraging discussions. This work was supported by National Science Foundation Fellowship DMS-8705843, NSF PYI award DMS-8957573, NSF Grant DMS-8810192, and a grant from Schlumberger-Doll Research.

References

[1] G. Backus. Inference from inadequate and inaccurate data, I. *Proc. Natl. Acad. Sci.*, 65:1–7, 1970.

[2] G.E. Backus. Confidence set inference with a prior quadratic bound. *Geophys. J.*, 97:119–150, 1989.

[3] M. Cuer. Discretisation en $o(h^p)$, $p > 2$ de problèmes de programmation lineare dans $L^\infty(0,1)$. *Manuscript*, 1989.

[4] S.P. Huestis and M.E. Ander. IDB2—a Fortran program for computing extremal bounds in gravity data interpretation. *Geophysics*, 48:999–1010, 1983.

[5] S.W. Lang. Bounds from noisy linear measurements. *IEEE Trans. Info. Th.*, IT-31:498–508, 1985.

[6] D.G. Luenberger. *Optimization by Vector Space Methods.* John Wiley and Sons, Inc., New York, 1969.

[7] R.L. Parker. The theory of ideal bodies for gravity interpretation. *Geophys. J. Roy. astr. Soc.*, 42:315–334, 1975.

[8] P.C. Sabatier. Positivity constraints in linear inverse problems—I. General theory, II. Applications. *Geophys. J. Roy. astr. Soc.*, 48:415–459, 1977.

[9] C. Safon, G. Vasseur and M. Cuer. Some applications of linear programming to the inverse gravity problem. *Geophysics*, 42:1215–1229, 1977.

[10] P.B. Stark. Duality and discretization in linear inverse problems. *Technical Report, Dept. of Statistics, University of California, Berkeley*, 1989.

Seismic Waveform Inversion Using a Layer-Stripping Method

S.C. Singh

Laboratoire de Seismologie, Institut de Physique du Globe de Paris,
4 Place Jussieu, F-75252 Paris Cédex 05, France
Present address: Elf Aquitaine, Direction Recherche Geologie et Geophysique,
F-64018 Pau Cédex, France

In the recent years, idea of layer stripping method has proved to be useful in obtaining elastic parameters for one dimensional media. The idea was first pointed out to seismologist by *Bube and Burridge* (1983) to obtain impedance contrasts from normal incidence data, and was further extended to multiparameters by *Singh et al.* (1989) in the $\tau-p$ domain. The theory should be extended to multidimensional case, but coupling of different ray parameters in the $\tau-p$ domain for laterally inhomogeneous media pose problems. One of the way to decouple these ray parameters is to assume that the medium parameters are smoothly varying in the horizontal directions and are small deviations from the vertical, and then use the WKBJ theory to obtain the causal time (a key point to the layer stripping methods), which allows propagation of surface wavefield downwards and construction of the model parameters. The basic physical assumption in formulating the inverse solution is that the change in amplitude of a plane wave at the causal time is related to the change of the elastic moduli at the spatial position of the plane wave.

In this paper, we derive a solution for inverting model parameters of multi-dimensional media using the layer stripping method. It requires two sources at the two ends of a profile, and an array of receivers along the profile. However, if the medium is laterally invariant, we only need one source and an array of receivers to retrieve both velocity and density uniquely. The possibility of a stable implementation of our methods has been investigated, firstly by using synthetic data for layered acoustic media containing a full range of low frequencies. In this case, the inverted velocity and density profiles are in good agreement with the test models. Secondly, the indeterminacy associated with the lack, in practice, of low frequencies has been studied, and a method has been devised to estimate the missing low-frequency data from travel time information. Finally, the inversion algorithm has then been applied to high-quality marine seismic reflection data obtained by the North Atlantic Transect (NAT) group in the deep ocean using an expanding spread profile (ESP). In this test, only the P wave velocity profile was estimated independently, because the move-out data cannot provide low-frequency information about density. A stable and plausible estimate of the velocity variation was obtained. Unfortunately, no independent measurement of this velocity structure is available with which to verify the method's accuracy.

Highly Dimensional, Highly Non-linear Inverse Problems in Geophysics

A. Tarantola

Institut de Physique du Globe, F-75205 Paris Cédex 05, France

ABSTRACT. Interpretation of seismic data leads to an interesting inverse problem where the parameters to be estimated are the acoustic properties of the Earth. Two sorts of elastic parameters can be resolved, the long wave lengths associated to seismic velocities and the short wavelengths associated to seismic impedances. Due to their oscillatory character, seismograms depend nonlinearly on seismic velocities. Then the Inverse Problem presents many locally optimal solutions. A complete resolution of the Inverse Problem must then involve a full exploration of the model Space.

Using a simulated annealing technique, a pseudorandom collection of solutions can be obtained, each one appearing with a frequency proportional to its probability. Watching a movies whose images are the pseudorandom realizations, allows the human interpreter to grasp the essentials of the probability distribution in the model Space.

Numerical Construction of Nonlinear Wave Train Solutions of the Periodic Korteweg-de Vries Equation*

A.R. Osborne and E. Segré

Istituto di Fisica Generale dell'Università and
Istituto di Cosmo-Geofisica del CNR, Corso Fiume 4, I-10133 Torino, Italy

We discuss a numerical procedure for the construction of exact, nonlinear wave train solutions of the periodic Korteweg-deVries (KdV) equation. We use an approach which is effectively a nonlinear Fourier decomposition of the wave motion formulated in terms of the periodic inverse scattering transform μ-representation. We construct solutions which have both narrow and broad-banded Fourier spectra and discuss some physical consequences of the propagation of complex, nonlinear wave motions.

1. Introduction

The Korteweg-deVries equation describes (among many other physical applications) the motion of long surface waves in shallow water:

$$\eta_t + c_o\eta_x + \alpha\eta\eta_x + \beta\eta_{xxx} = 0 \tag{1.1}$$

Here the constant coefficients are given by: $c_o = (gh)^{1/2}$, $\alpha = 3c_o/2h$ and $\beta = c_oh^2/6$, while subscripts refer to partial derivatives with respect to space x and time t; g is the acceleration of gravity, h is the depth and c_o is the phase speed of the linearized KdV equation (set $\alpha = 0$ in (1.1)). The classical cnoidal wave solution to (1.1) $cn^2(x\text{-}ct|m)$ is well known; when the modulus $m \to 0$ the cnoidal wave reduces to a sinusoid, while for $m \to 1$ one finds the solitary wave solution of KdV.

The discovery of the soliton by Zabusky and Kruskal lead to the subsequent discovery of the exact solution to the KdV equation on the infinite line ($-\infty<x<\infty$) by Gardner, Greene, Kruskal and Muira. The approach, now known as the inverse scattering transform (IST), has since been extended and applied to about one hundred nonlinear wave equations (Ablowitz and Segur (1981), Dodd, et al. (1982), and Degasperis (1989)). The solution $\eta(x,t)$ to the KdV equation for periodic boundary conditions (assume $\eta(x,t) = \eta(x+L,t)$, $0\leq x\leq L$, L the period) is also known (see for example Dubrovin, Matveev and Novikov (1976), Flaschka and McLaughlin (1976) and references cited therein).

A major goal of this paper is to discuss how one can construct simple and complex nonlinear wave train solutions of the KdV equation using *only a few nonlinear Fourier degrees of freedom*. We show that wave trains of this type have instead *many linear Fourier degrees of freedom*. We stress the relative simplicity of the low degree-of-freedom scattering transform representation.

In spite of all the rich mathematical structure known about the KdV equation, almost nothing is known about its description of complex wave trains or wave fields. Knowledge of the dynamical behavior of complex wave motions is necessary for inferring possible motions in the shallow water ocean environment, where "random" waves with many Fourier components are ubiquitous. The purpose and

* Prof. A. R. Osborne was unfortunately unable to attend the meeting to give this invited lecture. The text of the lecture is included herein in its entirety.

scope of this paper is to exploit the periodic, nonlinear Fourier structure of the KdV equation in order to the describe complex wave solutions. We study examples of both simple, low degree-of-freedom motions and complex, many degree-of-freedom motions and find, paraphrasing the last paragraph, that the number of *nonlinear* Fourier modes necessary to describe some wave field is normally quite small compared to the number of modes required by *linear* Fourier analysis. We exploit the nonlinear Fourier structure of KdV in order to understand certain properties of the motion not normally accessible by direct numerical integration of the equation. Several aspects of the periodic KdV spectral representation have been emphasized by Osborne and Bergamasco (1985; 1986):

(1) The general solution to the KdV equation with periodic boundary conditions consists of a linear superposition of *hyperelliptic functions*, $\mu_j(x,t)$, and may be interpreted as a kind of nonlinear Fourier series of these μ_j-function "normal modes."

(2) The nonlinear Fourier spectral amplitudes (i.e. the widths of the *open spectral bands* in the Floquet discriminant or the amplitudes of the hyperelliptic normal modes) are constants in time.

(3) The nonlinear Fourier series, a linear sum of the $\mu_j(x,t)$, reduces to a linear Fourier series in the small amplitude limit.

The original numerical experiments of Zabusky and Kruskal have been reinterpreted in terms of the nonlinear Fourier approach (Osborne and Bergamasco (1986)); in particular the amplitudes of the solitons and the Fermi-Pasta-Ulam (FPU) recurrence time were verified by periodic theory. There are a number of open problems which require study. For example, although many theoretical properties of the μ_j–representation have been found and studied in the past, one still lacks the intuitive grasp of how these functions behave as nonlinear Fourier modes; in fact a major purpose of this paper is to examine several cases for narrow and broad banded modulated wave fields in order to understand μ_j-function behavior when nonlinear effects are moderately large. While it is known that the μ_j reduce to ordinary trigonometric functions in the small amplitude limit (Osborne and Bergamasco, 1985), one can ask a number of other questions regarding their behavior. Do they still resemble the classical cnoidal wave? Can the μ_j be easily and intuitively grasped as physical nonlinear normal modes of the KdV equation? In particular can one exploit the ideas of the nonlinear Fourier analysis and nonlinear filtering of data? These are some of questions which motivate the present work. The algorithm used herein is described in detail in Osborne and Segre (1990), where a complete list of references may also be found.

2. The Construction of Wave Trains with Nonlinear Fourier Series

The solution to the periodic KdV equation may be written as a *linear superposition of nonlinearly interacting, nonlinear waves,* e.g. hyperelliptic functions, $\mu_j(x;x_o,0)$:

$$u(x,t) = \lambda\eta(x,t) = -E_1 + \sum_{j=1}^{N}[2\mu_j(x;x_o,t) - E_{2j} - E_{2j+1}] \tag{2.1}$$

This is the first of the trace formulae for the KdV equation (Flaschka and McLaughlin (1976)). The μ_j evolve in space according to the following system of coupled, nonlinear, ordinary differential equations:

$$\frac{d\mu_j}{dx} = \frac{2i\sigma_j R^{1/2}(\mu_j)}{\prod_{\substack{k=1 \\ j\neq k}}^{N}(\mu_j - \mu_k)} \tag{2.2}$$

and

$$R(\mu_j) = \prod_{k=1}^{2N+1}(\mu_j - E_k) \tag{2.3}$$

The $\sigma_j = \pm 1$ are the signs of the square root of $R(\mu_j)$. The μ_j evolve on two-sheeted Riemann surfaces such that the branch points connecting the surfaces correspond to the band edges E_{2j} and E_{2j+1}. The E_j are eigenvalues of the *main spectrum* and are solutions of $\Delta(E) = \pm 1$, where $\Delta(E)$ is the Floquet discriminant (trace of the monodromy matrix) of the *direct scattering problem* (Osborne and Bergamasco, 1986). The μ_j lie in the intervals $E_{2j} \leq \mu_j \leq E_{2j+1}$ and oscillate between these limits as x is varied. When a μ_j reaches a band edge the associated σ_j changes signs and the motion leaps to the other Riemann sheet. The temporal evolution of the μ_j is given by the following ODEs:

$$\frac{d\mu_j}{dt} = 2[-u(x,t) + 2\mu_j]\, \mu_j' \tag{2.4}$$

where $\mu_j' = d\mu_j/dx$ is given by (2.2) and (2.3). The hyperelliptic function, nonlinear Fourier structure for the KdV equation is now clear. The space (2.2) and time (2.4) ODEs evolve the $\mu_j(x,t)$ and the nonlinear Fourier series (2.1) provides the solutions to the KdV equation $\eta(x,t)$. In what follows we advertise an approach (Osborne and Segre, 1990) for determining numerical solutions of the motion $\mu(x,0)$, independent of time. Once the problems related to the spatial integration are solved the temporal integration follows in a straightforward manner using (2.4).

Generally speaking we refer to the construction of the *main spectrum* (E_k, $1 \leq k \leq 2N+1$) and the *auxiliary spectrum* ($\mu_j(x_o)$, $\sigma_j = \pm 1$, $1 \leq j \leq N$) as the *direct scattering transform*. The construction of the hyperelliptic functions $\mu_j(x;x_o,t)$ by the solution of the nonlinear ODEs (2.2)-(2.4) and the construction of solutions of the KdV equation by the trace formula (2.1) constitutes the *inverse scattering transform.*

5. Examples of Nonlinearly Synthesized Wave Trains

We now give a number of examples of KdV wave trains. We include results from one to eight degrees of freedom. Physically we generate example solutions which (1) resemble simple modulated wave trains, (2) complex (quasi random in appearance) wave trains and (3) the interaction of solitons with a periodic cnoidal wave train.

5.a. One Degree of Freedom

In Figure 1 we give the case of a simple cnoidal wave solution of the KdV equation. This is the classic traveling wave train solution and is well-known. The wave bounded by dashed lines is $\mu_1(x,t)$, while the upper wave is the solution (2.1).

5.b. Two Degrees of Freedom

Figures 2-8 give results for several cases for two degree-of-freedom solutions to the KdV equation. In Figures 2-5 we present results for solutions represented by adjacent main spectrum eigenvalues which are well-separated from each other. The nonlinear effects are therefore not very large. In Figures 6-8 we give examples where adjacent eigenvalues of the main spectrum are quite close to each other. As a result nonlinear effects are quite large. A particular interesting case occurs in Figure 8 which consists of a cnoidal wave train colliding with a single soliton (the largest peak in the figure).

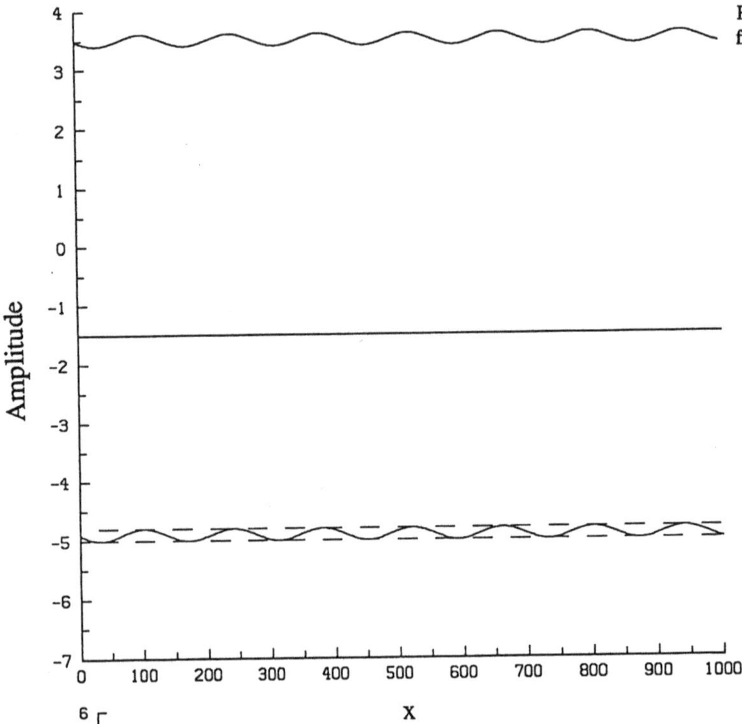

Figure 1. Single degree of freedom cnoidal wave.

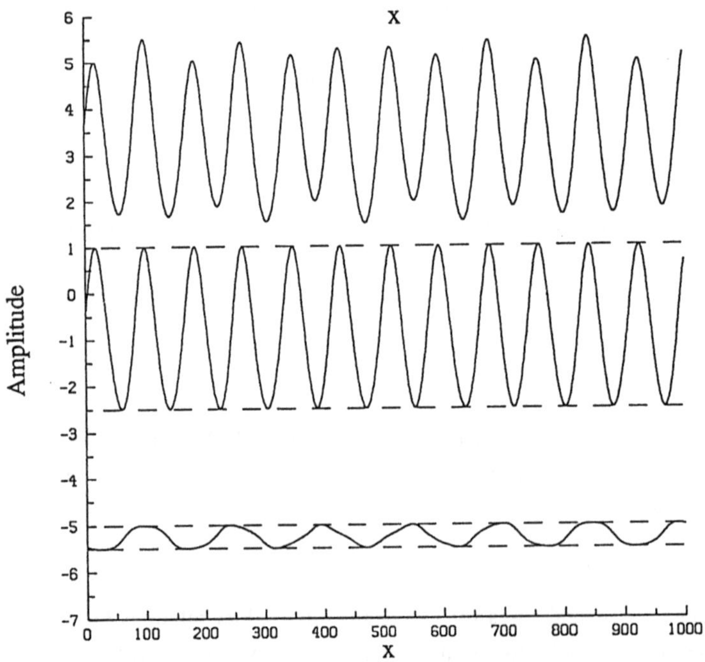

Figure 2. KdV solution consisting of two degrees of freedom. The two lower curves (nonlinear Fourier components, graphed between dotted lines corresponding to the main spectrum E_j) are summed to give the solution to KdV shown in the upper curve. The lower component is quite distorted from the shape of a cnoidal wave due to nonlinear effects. Note that the spacing between adjacent degrees of freedom ("allowed bands" in the spectrum) is quite large.

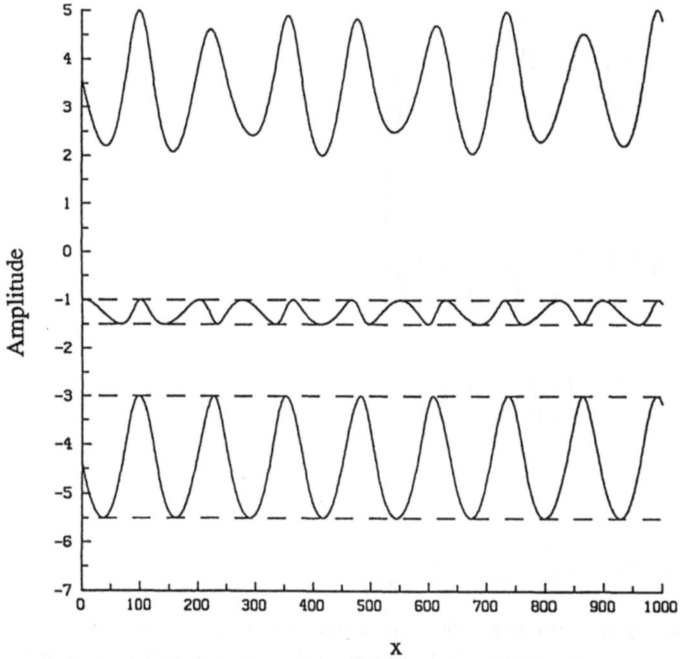

Figure 3. KdV solution consisting of two degrees of freedom. The two lower curves are summed to give the solution to KdV shown in the upper curve. The upper component is quite distorted from the shape of a cnoidal wave due to nonlinear effects.

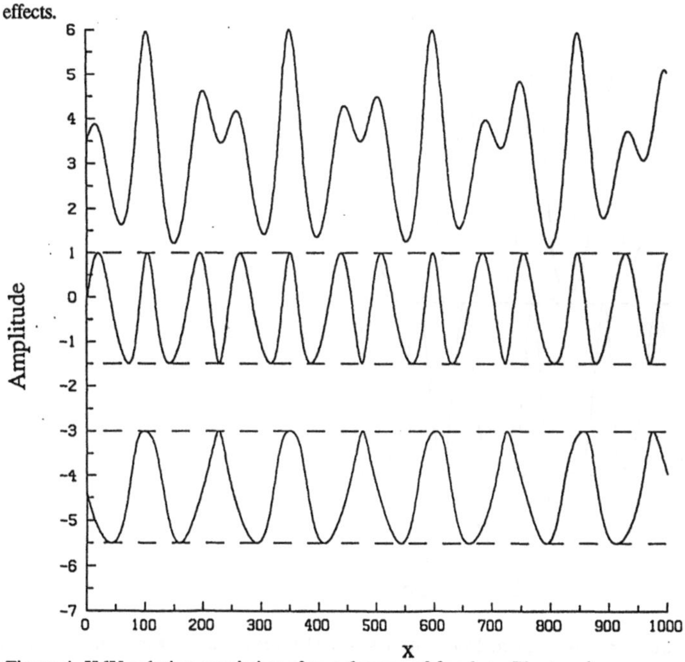

Figure 4. KdV solution consisting of two degrees of freedom. The two lower curves are summed to give the solution to KdV shown in the upper curve. Both the upper and lower degrees of freedom are only subtly different from the shape of a cnoidal wave due to nonlinear effects. Note that the spacing between adjacent degrees of freedom (allowed bands in the spectrum) is quite large.

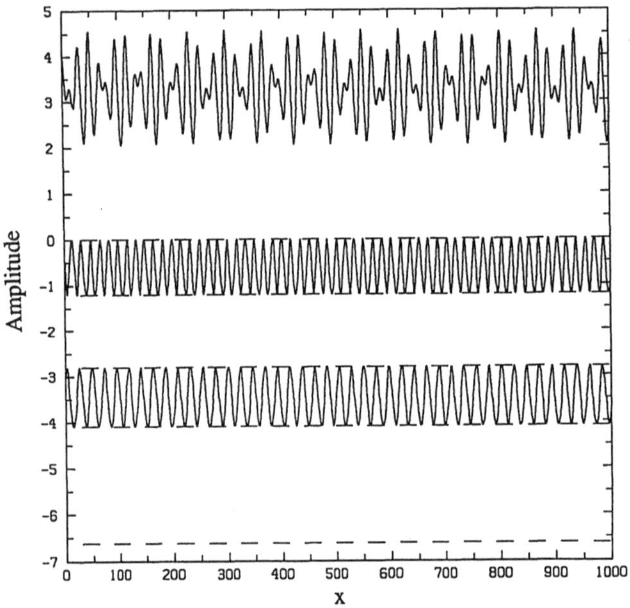

Figure 5. KdV solution consisting of two degrees of freedom at higher frequency than in the examples of Figures 2-4. The two lower curves are summed to give the solution to KdV shown in the upper curve. Both the upper and lower degrees of freedom are only subtly different from the shape of a cnoidal wave due to nonlinear effects. The spacing between adjacent degrees of freedom is rather large.

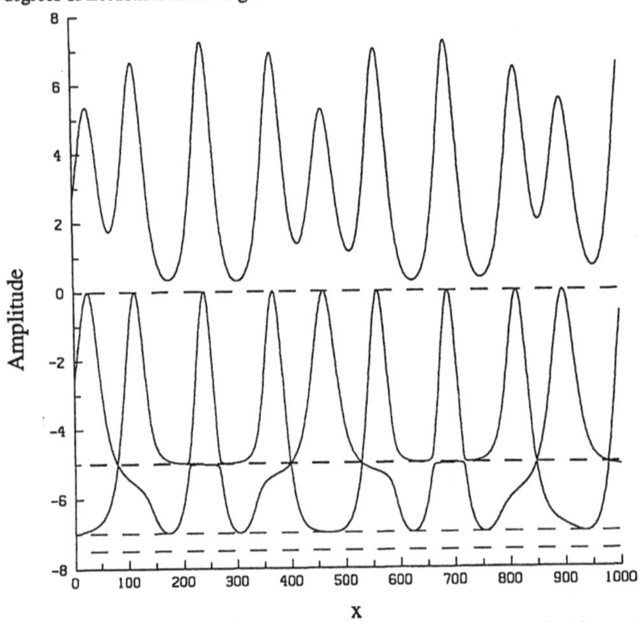

Figure 6. KdV solution consisting of two degrees of freedom. Note that the spacing between adjacent degrees of freedom is quite small, in contrast to the cases in Figures 2-5. The two lower curves (graphed between dotted lines corresponding to the main spectrum E_j) are summed to give the solution to KdV shown in the upper curve. Both the upper and lower degrees of freedom are are greatly different from the usual shape of a cnoidal wave; these distortions are due to nonlinear effects. Note that the two adjacent values of E_j in the main spectrum are quite close together, almost indistinguishable by eye.

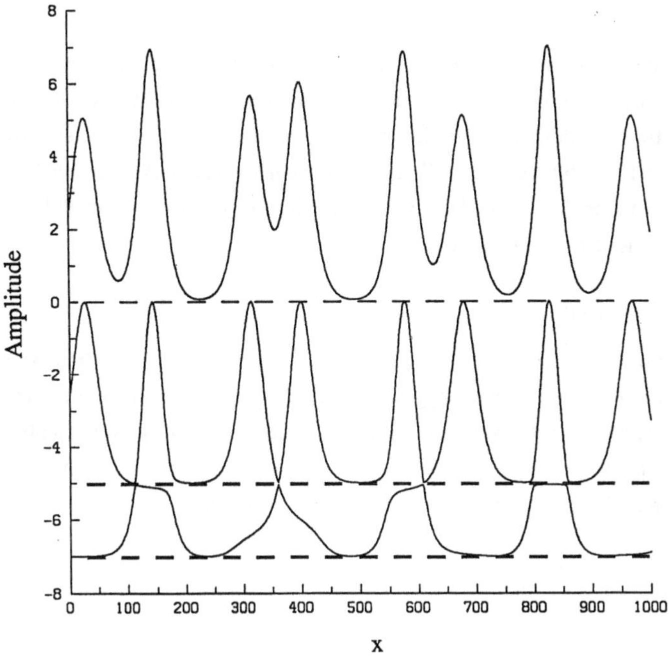

Figure 7. A case similar to Figure 6 with a slightly different spectrum.

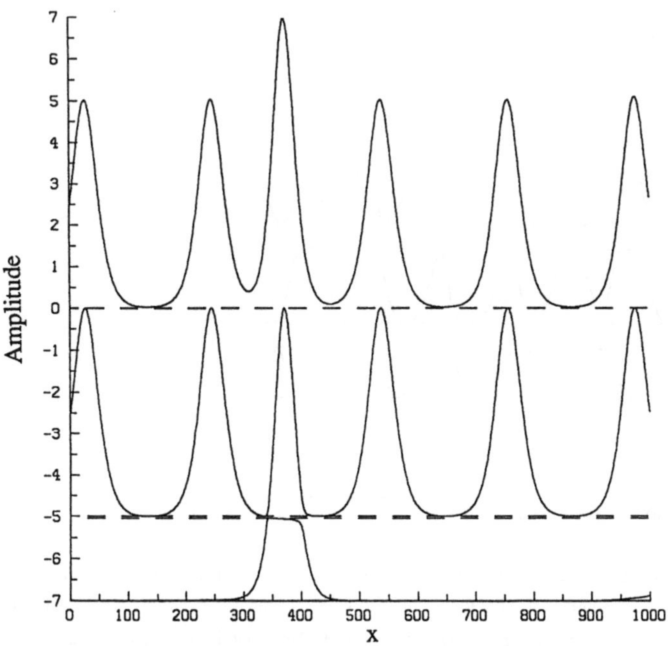

Figure 8. A soliton (largest peak) propagates in an infinite cnoidal wave train.

5.c. Three Degrees of Freedom

Figures 9 and 10 give examples of three degree of freedom solutions of the KdV equation. An interesting result in these simulations is that increasing the number of degrees of freedom increases the level of nonlinearity. The results for three degrees of freedom demonstrate this. In spite of the fact that the adjacent eigenvalues in the main spectrum are quite well separated, the individual degrees of freedom are quite distorted from the shape of a cnoidal wave. Note that Figure 9 results in a slowly modulated wave train, while Figure 10 has a rather erratic appearance.

5.d. Eight Degrees of Freedom

Figure 11 displays a solution to the KdV equation which seems rather like a typical (random) shallow water modulated wave train. These results thus suggest the possibility that "random" wave trains may in fact be representable by only a few nonlinear degrees of freedom, rather that some large number of linear Fourier degrees of freedom. The latter is of course the standard procedure for analyzing data both in the laboratory and the field. Future efforts will be devoted to the analysis of data using the techniques given herein.

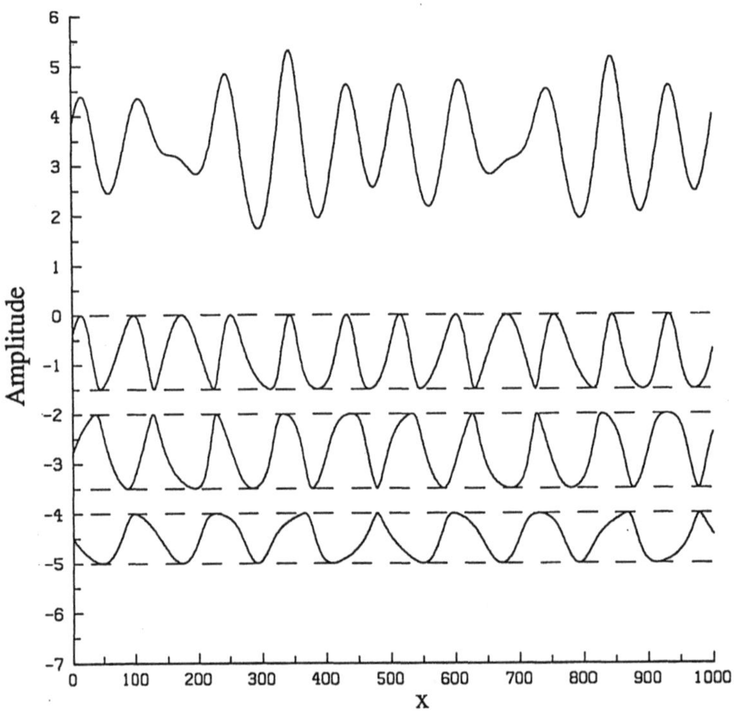

Figure 9. KdV solution consisting of three degrees of freedom. The three lower curves (graphed between dotted lines corresponding to the main spectrum E_j) are summed to give the solution to KdV as shown in the upper curve. All three degrees of freedom differ from the shape of a cnoidal wave due to nonlinear effects. Note that the spacing between adjacent degrees of freedom is rather large. The solution to KdV has the appearance of a slightly modulated wave train.

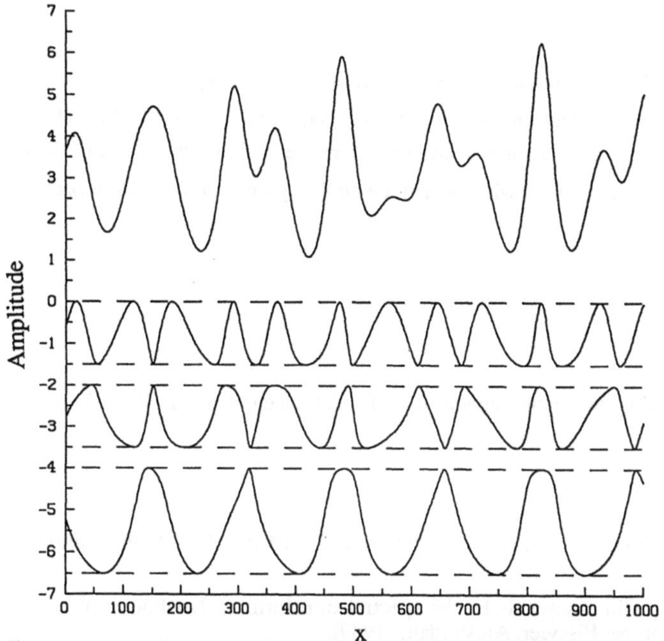

Figure 10. KdV solution consisting of three degrees of freedom. The three lower curves (graphed between dotted lines corresponding to the main spectrum E_j) are summed to give the solution to KdV as shown in the upper curve. All three degrees of freedom differ from the shape of a cnoidal wave due to nonlinear effects. Note that the spacing between adjacent degrees of freedom is rather large. The solution to KdV has a quite erratic appearance.

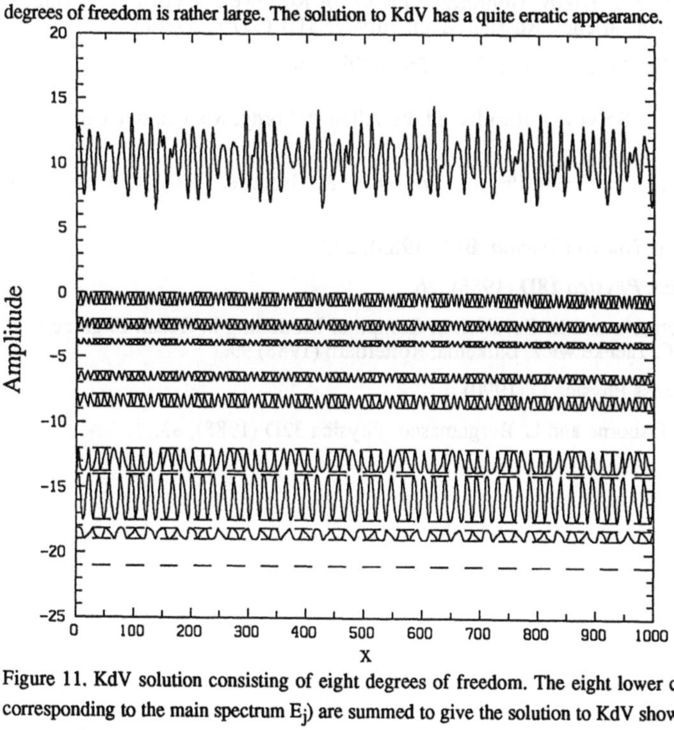

Figure 11. KdV solution consisting of eight degrees of freedom. The eight lower curves (graphed between dotted lines corresponding to the main spectrum E_j) are summed to give the solution to KdV shown in the upper curve. The three lower degrees of freedom differ somewhat from the shape of a cnoidal wave due to nonlinear effects. Note that the spacing between most adjacent degrees of freedom is relatively large.

6. Summary and Conclusions

We have studied numerically the behavior of the μ_j variables in the hyperelliptic function formulation of the periodic inverse scattering transform for the KdV equation, in which solutions are given by a linear superposition of the μ_j. In particular we have considered several examples of KdV wave trains up to eight degrees of freedom. Future work will concentrate on the application of this approach to the analysis of data.

Acknowledgements

We acknowledge the continuing invaluable support of Profs. C. Castagnoli and L. Bergamasco.

References

M. J. Ablowitz and H. Segur, Solitons and the Inverse Scattering Transform, SIAM, Philadelphia, 1981.

A. Degasperis, Nonlinear wave equations solvable by the spectral transform, in Nonlinear Topics in Ocean Physics, ed. by A. R. Osborne, Elsevier, Amsterdam (1989).

R. K. Dodd, J. C. Eilbeck, J. D. Gibbon and H. C. Morris, Solitons and Nonlinear Wave Equations, Academic Press, London, 1982.

B. A. Dubrovin, V. B. Matveev, S. P. Novikov, Nonlinear equation of Korteweg-deVries Type, finite-zone linear operators, and Abelian varieties, Russian Math. Surveys 31:1 (1976), 59-146.

H. Flaschka and D. W. McLaughlin, *Prog. Theoret. Phys.* **55** (1976), 438.

A. R. Osborne, in Soliton Theory: a Survey of Results, ed. by Allan P. Fordy, Manchester University Press (1989).

A. R. Osborne, in Nonlinear Topics in Ocean Physics, ed. by A. R. Osborne, Elsevier, Amsterdam (1989).

A. R. Osborne and L. Bergamasco, *Nuovo Cimento* **B85** (1985), 229.

A. R. Osborne and L. Bergamasco, *Physica* **18D** (1986), 26.

A. R. Osborne, M. Petti, M. Liberatore and L. Cavaleri, in Computer Modelling in Ocean Engineering, ed. by B. A. Schrefler and O. C. Zienkeiwicz, Balkema, Rotterdam (1988) 99.

A. R. Osborne and E. Segre, in press Physica D (1990).

E. R. Tracy, J. W. Larson, A. R. Osborne and L. Bergamasco, Physica 32D (1988), 83.

Part VIII

Inverse Methods and Applications to Nonlinear Problems

The Infeld Factorization in Relation to Various Problems in Mathematical Physics

K. Bleuler

University of Bonn, Theoretical Nuclear Physics Institute,
Nußallee 14–16, D-5300 Bonn 1, Fed. Rep. of Germany

I. INTRODUCTION

Our first aim consists of a renewal and an enlargement of the 'classical' "Infeld factorization" method: With the help of suitable (and simplifying) concepts of present-day theory, we hope to reach a natural embedding into that marvellous network[1] of modern mathematical physics. Our work[2] represents, so to speak, the continuation of a very long and interesting historical development[3] started, in fact, by Frobenius and Darboux and followed up to our days by a large number[3] of well-known scientists from different domains and with different intensions. The present work[4] which should just be considered as an intermediate stage on the way to far deeper insights and more extended connections to basic mathematical structures, is mainly due to my former student and, later, collaborator Alfons Stahlhofen (by now at Stuttgart). The mathematical method to be used is, first of all, the concept of 'deformation' of algebras, an idea, which like many others, stems from that far reaching mathematical developments due to efforts in 'string theory', i.e. an attempt for a universal understanding of physical laws which, after all, turned out to be just an interesting new view-point which, eventually, led to the idea of a much more basic renewal, i.e. A. Connes' non--commutative differential geometry[4].

The second aim of this contribution now consists in establishing connections with various neighbouring domains: We thus emphasize the tight relations to group theory (including supergroups) to more general invariance properties as well as to the basic question of solvability. From a more practical side, inverse scattering and non-linearity play, of course, an important role. At the same time, it should, in fact, be realized that our (more general) algebraic formulation leads on one side to enormous (practical) simplifications in solving classical problems, on the other, however, to far reaching algebraic questions which indicate that factorization might play a central role in modern developments of mathematical physics.

II. THE DEFORMATION OF THE OSCILLATOR ALGEBRA

As stated above, our method is to be based on the idea of deformation, i.e. a principle which applied to the classical Poisson algebra leads (within the framework of geometric quantization) to quantum theory but, at the same time, if applied to Lie algebras, leads to the basic concept of quantum groups.

The algebra to be considered here, i.e. to be 'deformed' (in two steps) corresponds, however, to the most characteristic and most elementary example of factorization, namely the harmonic oscillator! In adapted and simplified notation:

(1) $$H = a^+a^- \qquad [a^-a^+]_- = I \qquad a^-\phi^{(o)} = 0$$

The first step towards our goal consists in replacing (1) by the well-known (first order) system for (normed!) eigenstates $\phi^{(n)}$, $n = 0,1,\ldots$

(2) $$a^+\phi^{(n)} = g_{n+1}\phi^{(n+1)}$$

$$a^-\phi^{(n+1)} = g_{n+1}\phi^{(n)}$$

(2') whereby $g_n = \sqrt{n}$ (yields the representation for $a^{\pm}$!).

By iteration this yields immediately the (normed!1) solutions

(3) $$\phi^{(n)} = \left|\prod^{r=n} \frac{a^+}{r}\right| \phi^{(o)}, \quad (\phi^{(o)} \text{ normed}).$$

Our first deformation just consists in introducing the (eigenvalue) parameter E within the coefficients g:

(4) $$g_n \rightarrow g_n(E) \equiv \sqrt{n-E}$$

in connection with a system of vacuum condition ($n_o = 0,1,\ldots$)

(5) $$g_{n_o}(E) = 0 \quad \text{i.e.} \quad a^-\phi^{(n_o n_o)} = 0$$

$(n_o n_o)$

for the 'new' vacua ϕ with the eigenvalues $E \to E_{n_0} \equiv n_0$. The corresponding system built in the various vacua $\phi^{(n_o n_o)}$ reads:

$$\phi^{n,n_o} = \left| \prod_{r=n_o+1}^{r=n} \frac{a^+}{g_r} \right| \phi^{(n_o n_o)} \quad *) \tag{6}$$

We thus have so-called 'towers' of degenerate (i.e. same E) eigenstates (cp. (4)) $\phi^{(n,n_o)}$, whereby the original eigenstates $\phi^{(n)}$ are represented (trivial in this case!) by the various vacua (according (5)), or, in different words: The relevant eigenstates stem from the (enlarged) vacuum-conditions.

This most elementary reformulation of the oscillator problem suggests, however, the second (decisive!) deformation (whereby this term is used in an enlarged, i.e. 'discrete' sense) which leads to our general formulation of factorization

$$a \to A_n \quad g_n(E) \to G_n(E) \tag{7}$$

$$A^+_\mu = (A^-_n)^* \quad G^+_n = (G^-_n)^* \text{ (for complex values)}$$

whereby the generalized commutation relations read:

$$A^-_{n+1}A^+_{n+1} - A^+_nA^-_n = G^-_{n+1}G^+_{n+1} - G^+_nG^-_n \tag{8}$$

which guarantee that the 'enlarged' solution (6)

$$\phi^{n,n_o} = \left| \prod_{r=n_o+1}^{r=n} \frac{A^+_r}{G_r} \right| \phi^{(n_o n_o)} \tag{9}$$

in connection with the enlarged vacuum condition (5)

$$G_{n_o}(E) = 0 \ , \quad \text{i.e.} \quad A^-_{n_o}\phi^{(n_o n_o)} = 0 \tag{10}$$

leads to the solutions of the naturally enlarged system (2), i.e. to the solutions of the eigenvalue problems for a whole system (i.e. 'towers') of Hamiltonians H^n defined by

$$H^n(E) = A^+_n(E)A^-_n(E) + G^+_n(E)G^-_n(E) \tag{11}$$

Following (i.e. enlarging in a most natural way) the well-known

---

*) we assume here $n>n_0$. For $n<n_0$, i.e. a decreasing index a^+ has to be replaced by a^-.

arguments of proving formula (3), one may, in fact, check that $\phi^{(n,n_o)}$ satisfies

$$H^n(E^{n_o})\,\phi^{(n,n_o)} = 0 \tag{12}$$

whereby the eigenvalue E^{n_o} is determined through the vacuum condition (10). The (degenerate) eigenvalues E^{n_o} are thus determined by an algebraic equation (i.e. the main purpose of the factorization) whereby the eigenstates are explicitely given in most suitable factorized form by (9) and the condition (10) which corresponds to so-called 'extreme' states occurring also within the conventional method as well as in group representation. This formulation generalizes the usual form (used in earlier papers) where the r.h.s. of (8) and, therefore, the A's are independent of E and where the second term in (11) is linear in E, i.e. where (12) appears to be a 'normal' eigenvalue problem.

III. THE HYDROGENE PROBLEM ACCORDING SCHRÖDINGER AND DIRAC

These two well-known cases are only 'sketched' just in order to demonstrate the extreme simplifications with respect to usual methods whereby the - a point which is generally lacking in textbooks - relativistic (Dirac) case appears even more natural and simpler than the unrelativistic one. In fact, Dirac's equation appears, in a way, 'factorized' from the outset!

A. The unrelativistic case

The well-known 'tower' of radial Hamiltonians reads ($n \to l$ and $n_o \to l_o$)

$$-H^l(E) = \partial_r^2 + r^{-1}\partial_r + 2r^{-1} - l(l+1)r^{-2} + E \tag{13}$$

which takes the form given by (11) by putting

$$A_l^{\pm} \equiv \mp\partial_r + lr^{-1} - l^{-1} \tag{14}$$

$$G_l^{\pm} = l^{-1} \mp i\sqrt{E}$$

(to be checked explicitely).

The solution using here the case of decreasing n, i.e. the extreme state has maximal n (now replaced by l_o) wherby the vacuum condition reads:

(15) $$G^{+}_{l_o+1}(E) = 0$$

leading accordig (14) to

$$E = -\frac{1}{(l_o+1)^2}$$

i.e. l_o now represents the N in usual notation. The corresponding condition for the vacuum-state (12) together with the explicite solution of type (9) leads immediately to the well-known system of hydrogene-eigenfunction in 'factorized' form. This last result - all analytical functions are, in fact, factorizable - is the main reason for the enormous formal simplification. It might be remarked that the corresponding angular dependence of the eigenfunctions, i.e. the spherical harmonics are also to be written in this form using for A the angular momentum operators! In both cases this factorized form which, in fact, corresponds to the general representation theory of Lie groups (in this case O(4), cp. next section) has - at the same time - also practical advantages. The conventional polynomials are, of course, immediately obtained by working out the various differenciations contained in (14).

B. The relativistic (Dirac) case

Here, as mentioned above, we have from the outset a linear system for the two radial wave functions F(r) and G(r) for each fixed value of j:

(16) $$\partial_r G - \frac{k}{r} G + \frac{\alpha}{r} F = (1-E)F$$

$$\partial_r F + \frac{k}{r} F - \frac{\alpha}{r} G = (1+E)G$$

whereby (as usual)

(16') $$k = \pm j + \frac{1}{2} \quad (\text{sign according } l = j \pm \frac{1}{2})$$

With the help of a suitable linear combination of F and G as well as the substitution for ϕ

(17) $$\rho = E \cdot r$$

and defining

(17') $$\gamma^o = \sqrt{k^2-\alpha^2}$$

we obtain a system of our general form (2) (however, with non-integer index) which reads in terms of our notation:

(18) $$A^{\pm}_{\gamma^o+1} = \mp\partial_\rho + \frac{\gamma^o}{\rho} - \frac{\alpha}{\gamma^o}$$

(18') $$G^{\pm}_{\gamma^o+1} = \frac{k}{\gamma^o} \mp \frac{1}{E}$$

(this satisfies the basic C.R. (8)). The crucial argument is, however, the enlargement to a full system (with the running index n as contained in (2)) without violating the C.R. (8). It can easily be checked that such a complete (unphysical, i.e. formal) system is obtained (simply!) by the substitution

(19) $$\gamma^o \rightarrow \gamma^o+n$$

in (18), (18') and (17'). The general vacuum-condition (in analogy to (15)) now reads (for an arbitrary fixed value n_o of n):

(20) $$G^{+}_{\gamma^o+n_o}(E) = 0$$

i.e. an algebraic equation for the eigenvalue E leading with (17') and (18'), i.e. the cumbersome elimination of γ^o, the well-known Sommerfeld formula as a consequence of this extremely simple (intrinsic) condition (20). In order to obtain the corresponding (physical) eigenfunction one has (by decreasing n) to go down from the corresponding (unphysical) vaccum state $\phi^{(n_on_o)}$, defined by

(20') $$A^{+}_{\gamma^o+n_o}\,\phi^{n_on_o} = 0$$

to n=0 (according 'inverse' (9)), i.e. to $\phi^{(on_o)}$ representing the physical eigenfunctions, which, again, will be given in product form. In addition to this decisive formal simplification, the quantum-number $n_o \equiv N$ in Sommerfeld's formula obtains a natural interpretation as the 'tower' index in perfect analogy to the unrelativistic case.

IV. INVARIANCE PROPERTIES

It is well-known that integrable systems - our two examples are special cases - are always deeply connected to general invariance properties, i.e. they are to a large extent to be solved on the basis of their inherent invariance group. It has, therefore, to be conjectured that the characteristic operators of our factorization should be directly related to some elements of the corresponding group.

A. For the unrelativistic case Pauli showed (even before Schrödinger's well-known work) that this problem is to be solved directly by introducing the Pauli-Lenz vector $\vec{B}$, or, in other words, by considering the (enlarged) invariance group O(4). It appears, therefore, appealing to check that a suitable (non-trivial) vector $\vec{S}$ constructed out of $\vec{B}$ and the angular momentum L[5] acts (through its z-component) on the full wave function $P_{lm}R_l$ (written in terms of spherical harmonics P and radial functions R) exactly in the expected way, i.e. it reproduces our operator A_l^- of the corresponding factorization:

$$S_3^{op}(P_{lm}R_l) = P_{l-1,m}A_l^- R_l \tag{21}$$

B. The corresponding relation is, in fact, by far more interesting (and not generally known) in the relativistic (Dirac) case: There exists the well-known 2-fold degeneracy, i.e. two states ϕ_k and ϕ_{-k} (on each level) with opposite k-values (as long as $k \neq 0$). Here, k represents the eigenvalue of the conventional operator $K = \rho_3((\vec{\sigma}\vec{L})+1)$. Thus, the degeneracy generating operator $\mathbb{R}$ yields $\phi_{-k} = \mathbb{R}\phi_k$ and, therefore, satisfies $[RK]_+ = 0$ and $[RH]_- = 0$. It reads:

$$\mathbb{R} \equiv \frac{1}{2mc}\rho_3(\vec{\sigma}\vec{B}) + \frac{1}{c^2 r}\rho_2 K \tag{22}$$

(using Dirac's notation). This expression is (apart from the 'relativistic correction' represented by the second term) essentially determined by the scalar product of spin $\vec{\sigma}$ and $\vec{B}$, which (relativistically) is 'slowly' rotating, i.e. an effect compensated by the second term[5]. The first factor ρ_3 also has a natural interpretation: Realizing that $\vec{B}$ is polar, whereas $\vec{\sigma}$ is axial, the factor ρ_3 with its special signs leads to parity conservation, i.e. R is scalar (needed because of the inherent parity conservation of the Dirac equation). If, however, R is rewritten in polar coordinates, i.e.

$$\mathbb{R} = \rho_3 \frac{\vec{\sigma}\vec{r}}{r} \gamma^o \left(\partial_o - \frac{\gamma_o - 1}{\rho} + \frac{\alpha}{\gamma_o}\right) \tag{23}$$

it turns out that the last factor (in parenthesis) is identical with $A^+_{\gamma_o}$ of the factorization. According (18) and the generating property mentioned above this expression had to be expected. The first factor is needed in order to describe the expected space dependence: Enlarging the former notation, i.e. introducing $P^{\pm}_{jl}$ describing, in addition, parity and spin of the 3-dimensional Dirac eigenfunctions, it may, in fact, be checked (in analogy to (21)) that

$$\mathbb{R}(P^{+}_{jl}R) = \text{const.}\ P^{-}_{j,l-1}A^{-}_{j}R\ . \tag{24}$$

It might be worthwhile to include these intuitive relations in conventional treaties. There remains, however, a basic and often discussed problem (in analogy to the unrelativistic case): Where is the still lacking operator, needed in order to generate an (enlarged) invariance (super?) group which would lead to a group-theoretical determination of the eigenvalues?

C. From a more general view-point, it should be (according E. Witten) remarked that the special properties (leading to degeneracy) discussed here, correspond to a (restricted) super-symmetry inherent in all factorizable problems: We conclude from (2), i.e. its generalizations, that

$$H_n \equiv A^{+}_{n}A^{-}_{n} \quad \text{and} \quad \tilde{H}_n \equiv A^{-}_{n}A^{+}_{n} \tag{25}$$

are iso-spectral whereas the corresponding eigenfunctions ϕ and $\tilde{\phi}$ are related through

$$\phi_n = \text{const.}\ A^{+}_{n}\tilde{\phi}_n \tag{26}$$

The corresponding elementary supergroup reads:

$$H^{\text{super}}_{n} = \begin{vmatrix} H_n & 0 \\ 0 & \tilde{H}_n \end{vmatrix} \qquad Q^{+}_{n} = \begin{vmatrix} 0 & A^{+}_{n} \\ 0 & 0 \end{vmatrix} \qquad Q^{-}_{n} = \begin{vmatrix} 0 & 0 \\ A^{-}_{n} & 0 \end{vmatrix} \tag{27}$$

The 2-fold degeneracy of the Dirac-equation thus appears as a special case whereby the 2 wave functions (built out of F and G) form one single state in super space. H^{super} represents a (diagonal) Hamiltonian, whereas Q^{+}_{n} yields the action given by (26), i.e. the radial Dirac equation exhibits super-symmetry in a most natural way!

D. From a much more general structural view-point, our fundamental system of C.R. (8) has some formal similarity to a truncated (i.e. no linear terms, but a 'central extention') Virasoro-type algebra. It appears interesting that the parameter E within this central extention, also exhibits (in analogy to Virasoro) a discrete spectrum. By far more detailed relations to Virasoro within the Schrödinger case will be discussed in a forthcoming paper by A. Stahlhofen.

V. RELATIONS TO INVERSE SCATTERING AND NONLINEAR PROBLEMS

Because of lack of space this far reachig domain is just 'sketched'.

A. An inverse problem is, in fact, inherent within our basic formulas of type (3) which yield

$$\phi^{n} \simeq \frac{A_{n}^{+}}{G_{n}} \phi^{n-1} \tag{28}$$

thus relating (measured) wave function ϕ to the essential operators, or building stones of the generating Hamiltonian H to be expressed explicitely by

$$H \simeq A_{n}^{+}A_{n}^{-} + \lambda I, \qquad \lambda = G_{n}^{+}G_{n}^{-} \tag{29}$$

This relation represents a guide line for the far reaching literature about this problem[7].

B. A general construction for solving special non-linear problems is, again, inherent within the factorization method. Let us consider the fundamental vacuum-condition according (10):

$$A_{n_0}^{-} |\phi^{(n_0 n_0)}\rangle = 0 \; . \tag{30}$$

Herein, ϕ appears to be in general the solution of a linear eigenvalue problem:

$$H\phi = E\phi \; .$$

For given, i.e. calculated ϕ, equation (30) represents in a large class of problems a (linear) differential equation for the characteristic function within the operator A (e.g. $A \simeq \partial_x + k(x)$). H, however, may contain these functions (e.g. $k(x)$) in non-linear form (e.g. $H \simeq \partial_x^2 + V(x)$ where $V(x) \simeq \partial_x k + k^2$)).
It thus follows that (for fixed H) the non-linear problem to find $k(x)$, or, more general functions inherent in A, out of the given structure (i.e. special functions) of H appears reduced to the solutions of 2 linear problems. Whereas the Ricatti equation represents the simplest case, this construction appears to be a

basis of the extended realm of non-linear problems related to the method of Lax-pairs[8].

Summarizing this extremely short survey, it was shown that the (enlarged) factorization method appears related, or, better, yields the central idea for a large number of at first sight perfectly unrelated domains of mathematical physics, e.g. groups, supergroups and quantum groups; non-linear problems and inverse scattering; spectral theory; generalized algebraic structures, a.s.o.

REFERENCES

1) R.K. Bullough: 18th Conf. on Differential Geometical Methods in Theoretical Physics (Lake Tahoe, 1989), Plenum, to be published (An impressive graphical survey exhibits the far-reaching interrelations between a large number of different domains of recent research)
2) A. Stahlhofen, K. Bleuler: An Algebraic Form of the Factorization Method, to be published in Nuovo Cimento B (1990)
   A. Stahlhofen: Dissertation, Univ. of Bonn, Inst. f. Theor. Nucl. Physics, Nussallee 14-16, D-5300 Bonn 1
3) For a historical survey, cp. e.g. L. Schlesinger, Handbuch der Theorie linearer Differentialgleichungen, Teubner (1895) (the very first ideas may be even traced back to Cauchy (1827))
   E. Schoedinger: Proc. Roy. Irish Acad. 46A (1940) p. 9 and p. 183 (i.e. the first paper dealing with eigenvalue problems)
   L. Infeld, T.E. Hull: RMP23 (1951)
   H.S. Green: Matrix Mechanics, Science Publ. Groningen (1966) (a systematic treatment)
   Among an enormous number of subsequent papers, I just indicate two relevant and recent publications:
   H. Grosse: Phys. Lett. 197B (1987), p. 413 (and various earlier publications)
   E. Witten: Com. Math. Phys. 114 (1988) (and earlier papers dealing a.o. with supersymmetry)
4) cp. e.g. D. Kastler: An Introduction to A Connes' non-commut. different. Geometry, Paris, Hermann (1988)
5) cp. A. Stahlhofen: Diploma thesis, Bonn 1983, unpublished
6) the so-called Johnson-Lippmann-Biedenharn operator, cp. A. Stahlhofen, Dissertation cited in[2]
7) cp. the basic monography on 'Inverse Problems in Quantum Mechanics' by K. Chadan and P.C. Sabaties; cp. e.g. p. 65 (the special case for Schrödinger)
   A general survey also in A. Suzko, Physica Scripta, 31 (1985)
8) cp. e.g. H. Grosse, UWThPh-1989-54, i.e. a survey which also contains a large part of our work

Some Ideas on Nonlinear Evolution Equations

F. Calogero

Secretary-General, Pugwash Conferences on Science and World Affairs,
Palazzina dell'Uditorio, Accademia Nazionale dei Lincei,
via della Lungara 229, I-00165 Roma, Italy
Professor of Theoretical Physics (on leave), Dipartimento di Fisica,
Università di Roma "La Sapienza", I-00185 Roma
Research Associate, Istituto Nazionale di Fisica Nucleare, Sezione di Roma, Italy

0. This is a terse review of some ideas and recent results helpful to understand nonlinear evolution equations. The discussion is mainly limited to problems in 1+1 dimensions.

1. S-integrability and C-integrability [1]. S-integrable nonlinear evolution equations are those solvable (linearizable) via the Spectral transform (or the inverse Scattering technique); typical examples are the KdV equation, the Nonlinear Schroedinger equation, the Sine-Gordon equation. C-integrable equations are those solvable (linearizable) via an appropriate Change of variable (generally a change of the dependent variable); typical examples are the Burgers equation, the equation $u_t=u_{xxx}+3u_{xx}u^2+9u_x^2u+3u_xu^4$ [2], the Eckhaus equation [3]. The requirement to be C-integrable is generally more stringent than that to be S-integrable: generally C-integrable equations are also S-integrable (but not viceversa). But there are C-integrable equations that are not S-integrable, if S-integrability is meant to imply the existence of an infinity of local conservation laws; one such instance is the Eckhaus equation [3].

A definition of C- and S-integrability that is a bit more specific than that given above, and which moreover entails that S-integrability necessarily implies C-integrability, reads as follows: [4] "A nonlinear PDE is C-integrable if its solution can be obtained by solving: (i) a finite system of nondifferential (possibly nonlinear, i. e. algebraic or trascendental) equations; (ii) a finite system of linear PDEs (including ODEs and quadratures)". The corresponding definition of S-integrability is closely analogous, except for the replacement (extension) of (ii) as follows: "(ii) a finite system of linear PDEs (including ODEs and quadratures), as well as linear integral equations (Volterra, Fredholm)".

A tricky aspect of this definition has to do with the precise meaning of the term "solution". A heuristic definition of this term would refer to the capability to manufacture a large class of solutions (perhaps all solutions); which however is not quite the same as being able to solve the initial value ("Cauchy") problem (if applicable) [4].

2. C-integrable equations displaying a solitonic behaviour. Since C-integrable equations can be solved simply by a change of variable, they are generally much easier to study than S-integrable equations. It is therefore interesting to know that some such equations display typical solitonic features, such as the presence of solitons that collide purely elastically. One such example is provided by the Eckhaus equation, that in fact displays a very rich solitonic phenomenology [3].

The identification of nontrivial C-integrable equations is a worthwhile task, because of their theoretical and pedagogical interest, as well as their potential importance for applications. For the presentation of techniques suitable to manufacture such equations, as well as the display of several examples, see [4].

3. Boundary/initial value problems for certain nonlinear evolution equations (on the semiline, on a finite interval). Certain C-integrable equations can be solved non only on the whole line, but also on the semiline or finite interval. One such example is the Eckhaus equation [5]; another interesting example (treated only recently) is the Burgers equation [6] (which is however solvable only if the unknown function itself is assigned on the boundary; but cannot be fully linearized if the space derivative of the unknown function is instead assigned on the boundary [7]).

4. Why are certain nonlinear evolution equations both widely applicable and integrable? One might well take the view that this question is metascientific; or, equivalently, one might quote Galileo: "this great book that stands always open before our eyes (I mean the universe)...is written in mathematical language" [8]. We submit that there exist a less methaphysical (if somewhat heuristic) answer: "certain "universal" nonlinear evolution PDEs can be obtained, by a limiting procedure involving rescalings and an asymptotic expansion, from very large classes of nonlinear evolution equations; for instance, from the class of autonomous nonlinear evolution equations whose linear part is dispersive but otherwise arbitrary, and whose nonlinear part depends in an analytic but otherwise arbitrary manner on the dependent variable and its derivatives. Because this limiting procedure is the correct one to evince nonlinear effects, the universal model equations obtained in this manner (of which the nonlinear Schroedinger equation in 1+1 dimensions is the prototype) show up in many, disparate, applicative contexts; they are widely applicable. Because this limiting procedure generally preserves integrability, these universal model equations are likely to be integrable, since in order for this to happen it is sufficient that the very large class from which they are obtainable contain just one integrable equation; indeed, while the fact that an arbitrarily given equation turns out to be integrable may be seen as an exceptional event, the fact that a very large class of equations contain at least one integrable equation may be considered normal, i. e. not exceptional; hence a universal model equation that is obtainable by a limiting procedure from (all!) the equations of a large class is likely to be integrable, provided the limiting procedure preserves integrability" [4].

5. Necessary conditions for integrability of nonlinear PDEs. The argument we have just given "may also be run backwards: if a universal model equation, obtainable via a limiting procedure that preserves integrability from all the equations of a large class, turns out not to be integrable, then none of the

equations contained in the large class is integrable; hence this approach also yields <u>necessary conditions for integrability</u> of wide applicability" [4,9].

6. <u>A novel phenomenon: solitons that change their shape upon collision</u>. Besides explaining "what had hitherto appeared to us a puzzling miracle, namely the fact that certain nonlinear PDEs appear in many applications <u>and</u> are applicable" [1], the approach outlined above "provides moreover a powerful, if heuristic, methodology, to understand the integrability of known equations, to test the integrability of new equations and to obtain novel integrable equations likely to be applicable" [4]. This methodology has already been widely applied (also to problems in more than 1+1 dimensions [10]), and is currently under further investigation. An interesting result in this context has been the study of the nonlinear PDEs describing the interaction of N dispersive waves [11]. A particularly amusing phenomen, discovered in the context of certain (C-integrable) nonlinear PDEs describing the nonresonant interaction of N dispersive waves, is the occurrence of solitons that, whenever they collide, change their shape (in a remarkable way, described by a universal formula); the justification for calling these objects "solitons" is because the effect of the multiple collisions, that occur over time whenever more than two solitons are present, turns out to be independent of the order in which they occur [12].

7. <u>Outlook</u>. A goal is to identify and study interesting multidimensional C-integrable equations (if any).

<u>References</u>

[1] F. Calogero and W. Eckhaus: "Nonlinear evolution equations, rescalings, model PDEs and their integrability. I & II". Inverse Problems <u>3</u>, 229-262 (1987) & <u>4</u>, 11-33 (1988).

[2] F. Calogero: "The evolution PDE $u_t=u_{xxx}+3(u_{xx}u^2+3u_x^2u)+3u_xu^4$". J. Math. Phys. <u>28</u>, 538-555 (1987).

[3] F. Calogero and S. De Lillo: "The Eckhaus PDE $i\,psi_t+psi_{xx}+2[(|psi|^2)_x+|psi|^4]psi=0$". Inverse Problems <u>3</u>, 633-681 (1987).

[4] F. Calogero: "Why are certain nonlinear PDEs both widely applicable and integrable?". Rome preprint n. 582, January 1988. To be published in: V. E. Zakharov (editor): <u>What is integrability (for nonlinear PDEs)?</u>, Springer, 1989 (in press).

[5] F. Calogero and S. De Lillo: "Cauchy problem on the semiline and on a finite interval for the Eckhaus equation". Inverse Problems 4, L33–L37 (1988).

[6] F. Calogero and S. De Lillo: "The Burgers equation on the semi-infinite and finite intervals". Nonlinearity 2, 37–43 (1989).

[7] F. Calogero and S. De Lillo: "Burgers equation on the semiline". Inverse Problems 5, L37L40 (1989).

[8] Galileo Galilei: Il Saggiatore, 1623. [See: Opere di Galileo Galilei, Edizione nazionale, Barbera, Firenze, 1890–1909, vol. VI, p. 232].

[9] F. Calogero and W. Eckhaus: "Necessary conditions for integrability of nonlinear PDEs". Inverse Problems 3, L27–L32 (1987).

[10] F. Calogero and A. Maccari: "Equations of nonlinear Schroedinger type in 1+1 and 2+1 dimensions, obtained from integrable PDEs". In: P. C. Sabatier (editor): Inverse Problems: an Interdisciplinary Study (Proceedings of the meeting on Inverse Problems held in Montpellier, November 1986), Advances in Electronics and Electron Physics, 19, Academic Press, London & New York, 1987, pp. 463-480.

[11] F. Calogero: "Universality and integrability of the nonlinear PDEs describing N-wave interactions". J. Math. Phys. 30, 28–40 (1989).

[12] F. Calogero: "Solutions of certain integrable nonlinear evolution PDEs describing nonresonant N-wave interactions". J. Math Phys. (in press).

Painlevé Singular Manifold Equation and Integrability

R. Conte

Service de Physique du Solide et de Résonance Magnétique,
Centre d'Etudes Nucléaires de Saclay, F-91191 Gif-sur-Yvette Cédex, France

We recall the importance of the truncation procedure in PDE Painlevé analysis. Using the recently introduced invariant analysis, we define the singular manifold equations set, in both integrable and nonintegrable cases, as the condition for a truncated expansion to exist. As an application, we derive the correct form of this set for Sawada-Kotera equation.

1 Introduction

The Painlevé property (PP) of partial differential equations (PDE), at least in its present definition [1], provides only a necessary condition of "integrability". In order to make it sufficient, other hypotheses must be added to the definition of PP, and the best way to get some hints is to concentrate our attention on the PDEs which, up to now, violate this supposed equivalence between PP and integrability. Among them are:
the Mikhailov alias Dodd-Bullough equation

$$u_{xt} - e^{u} + e^{-2u} = 0 \tag{1}$$

the Harry Dym equation

$$u_t - u^3 u_{xxx} = 0 \tag{2}$$

and the Kaup-Kuperschmidt equation

$$u_t + (u_{xxxx} + 30uu_{xx} + \frac{45}{2}u_x^2 + 60u^3)_x = 0 \tag{3}$$

These three equations admit a known Lax pair with an arbitrary spectral parameter, and Painlevé analysis has not yet been able to retrieve it using the singular manifold method developed by Weiss [2], contrarily to many other PDEs.

In this paper, we will emphasize the role of the singular manifold equation (SME). In section 2, we recall the practical tools needed to perform an invariant Painlevé analysis. In section 3, we detail the steps leading to the SME by the truncation procedure. Then, the invariant way leading from the SME to the coefficients of a scattering problem is explained in section 4. In section 5, the correct SME of Sawada-Kotera equation is obtained.

2 Practical invariant Painlevé analysis

For simplicity only, we consider one PDE $E(u, x, t) = 0$ in one dependent variable u and two independent variables (x, t). For (x, t) in a neighborhood of a singular manifold $\varphi(x, t) = 0$, we expand u in powers of a variable

$$\chi = \frac{\alpha\varphi}{\gamma\varphi + \delta} \tag{4}$$

Then χ_x and χ_t are second degree polynomials in χ, and their six coefficients can be made invariant [3]under the group $\mathcal{H}$ of homographic transformations

$$\varphi \rightarrow \frac{a\varphi + b}{c\varphi + d}, \ (a, b, c, d) \text{ complex constants, } ad - bc = 1 \tag{5}$$

provided α, γ and δ are carefully chosen functions of the derivatives $D\varphi$ of φ. The result for expansions

$$u = \sum_{j=0}^{+\infty} u_j \chi^{j+p}, E = \sum_{j=0}^{+\infty} E_j \chi^{j+q} \tag{6}$$

is that u_j and E_j, which are functions of $D\varphi$, depend only on two elementary homographic differential invariants

$$S = \{\varphi; x\} = \frac{\varphi_{xxx}}{\varphi_x} - \frac{3}{2}\left(\frac{\varphi_{xx}}{\varphi_x}\right)^2, \ C = -\frac{\varphi_t}{\varphi_x} \tag{7}$$

linked by the cross-derivative condition

$$\varphi_x^{-1}\left((\varphi_{xxx})_t - (\varphi_t)_{xxx}\right) = S_t + C_{xxx} + 2C_x S + C S_x = 0 \tag{8}$$

In what follows, we adopt the choice:

$$\chi(\varphi_0) = \left(\frac{\varphi_x}{\varphi - \varphi_0} - \frac{\varphi_{xx}}{2\varphi_x}\right)^{-1} \tag{9}$$

$$\chi_x = 1 + \frac{S}{2}\chi^2 \tag{10a}$$

$$\chi_t = -C + C_x\chi - \frac{1}{2}(C_{xx} + CS)\chi^2 \tag{10b}$$

For another choice, see Ref [4].

3 The singular manifold equation

Assume the PDE admits a solution represented by a truncated series, e.g. in the case when p is a negative integer:

$$u = \sum_{j=0}^{-p} u_j \chi^{j+p} \tag{11}$$

without any incompatible resonance in $\{0, ..., -p\}$. Then E is also a truncated series:

$$E = \sum_{j=0}^{-q} E_j \chi^{j+q} \tag{12}$$

Equations $E_j = 0$, $j = 0, ..., -p$, determine the first $-p + 1$ coefficients u_j as equal to those of the infinite expansion. After replacement of u_j by these values, the last equations are

$$E_j(D\varphi, u_r) = 0, \qquad j \in \{-p+1, ..., -q\} \tag{13}$$

$$tj \neq \text{compatible resonance}, \ r = \text{resonances} \in \{0, ..., -p\}$$

Definition. We call equations (13) Painlevé-Darboux equations (PD). The reasons for such a name are detailed elsewhere [4]. No assumption is made about the compatibility of resonances outside $\{0, ..., -p\}$.

Let us solve PD equations as follows. By elimination of all u_r variables between (13), we first get a generically equivalent set of equations, made of a subset yielding u_r, which we are not primarily interested in, and a subset independent of u_r:

$$\begin{cases} u_r = U_r(D\varphi), r = \text{ zero or more values} & (14a) \\ F_j(D\varphi) = 0, j = \text{ one or more values} & (14b) \end{cases}$$

Definition. We call singular manifold equations set $E_{sm}(D\varphi) = 0$ the smallest subset of equations generically equivalent to (14*b*). In this definition, the difficulty lies in "equivalent". Practically, the finding of the SME set amounts to "solving" and "sorting" (14*b*) to get another set generically equivalent to (14*b*)

$$\begin{cases} E_{sm}(D\varphi) = 0 & \text{one or more equations} & (15a) \\ G_j(D\varphi) = 0 & \text{zero or more equations} & (15b) \end{cases}$$

where the second subset (15*b*) is dependent, i.e. is a consequence of the first subset.

Properties of the SME set are:

1) invariance under homography, by construction,

2) unicity, whatever be the integrability of the PDE,

3) the SME set is made of one and only one equation if and only if the PDE is integrable.

The last two properties are only conjectures for the moment.

Let us now give three simple examples, followed by a fourth one which will illustrate most of the difficulties.

Example a: Korteweg-de Vries equation, conservative form [1].

$$bu_t + \left(u_{xx} + \frac{u^2}{2a}\right)_x = 0 \tag{16}$$

$$p = -2, q = -5, \ \text{resonances } (4, 6) \text{ compatible} \tag{17}$$

$$u_0 = -12a, u_1 = 0, u_2 = a(bC - 4S) \tag{18}$$

$$\text{set } (E_j) = \begin{cases} E_3 \equiv 12a(bC - S)_x = 0 \\ E_5 \equiv ab(bC - S)_t + 4a(bC - S)_{xxx} + a(bC - S)_x(bC + 2S) = 0 \end{cases} \tag{19}$$

After E_5 has been put in the above written form, the set (F_j) is immediate to obtain:

$$\text{set } (F_j) = \{F_3 \equiv (bC - S)_x = 0, F_5 \equiv (bC - S)_t = 0\} \tag{20}$$

Then one integration is needed to get the SME set:

$$\text{set } E_{sm} = \{bC - S + 2\lambda = 0, \lambda \text{ arbitrary constant}\}, \text{set } (G_j) = \emptyset \tag{21}$$

We notice that the SME explicitly depends on the spectral parameter, a situation which always occurs at least for second order Lax pairs.

Example b: Boussinesq equation, conservative form [2].

This is the by far simplest example.

$$u_{tt} + e^2 \left((u+a)^2 + \frac{b^2}{3} u_{xx} \right)_x x = 0 \tag{22}$$

$$p = -2, q = -6, \text{ resonances } (4,5,6) \text{ compatible} \tag{23}$$

$$u_0 = -2b^2, u_1 = 0, u_2 = -a - \frac{2}{3} b^2 S - \frac{1}{2e^2} C^2 \tag{24}$$

$$\text{set } (E_j) = \text{set } (F_j) = \text{set } E_{sm} = \left\{ \frac{b^2 e^2}{3} S_x - C_t + C C_x = 0 \right\}, \text{set } (G_j) = \emptyset \tag{25}$$

The SME is independent of the spectral parameter of the third order Lax pair.

Example c: Kuramoto-Sivashinsky equation, conservative form [5].

$$u_t + u u_x + \mu u_{xx} + \nu u_{xxxx} = 0 \tag{26}$$

This equation is nonintegrable and exhibits a chaotic behaviour.

$$p = -3, q = -7, \text{ resonances}: 6 \text{ compatible}, \frac{13 \pm i\sqrt{71}}{2} \tag{27}$$

$$u_0 = 120\nu, u_1 = 0, u_2 = \frac{60}{19}\mu + 60\nu S, u_3 = C - 15\nu S_x$$

$$\text{set } (E_j) = \text{set } (F_j) = \{E_4, E_5, E_7 \text{ eq. } (4.11) - (4.13) \text{ Ref. } [5]\} \tag{28}$$

The SME set is not so easy to obtain, and the result is:

$$\text{set } E_{sm} = \left\{ (S - \frac{\mu}{38\nu})(S + \frac{11\mu}{38\nu}) = 0, C = \text{ arb. constant} \right\}, \text{set } (G_j) = \emptyset \tag{29}$$

The arbitrary constant value for C simply reflects the Galilean invariance.

In the above three examples, the set (G_j) is empty and PD equations have no nongeneric solution. Let us now proceed to an example for which this is no more true.

Example d: Sawada-Kotera equation, potential form, Painlevé branch [6-7].

$$SK(u) \equiv \beta u_t + u_{xxxxx} + \frac{30}{\alpha} u_x u_{xxx} + \frac{60}{\alpha^2} u_x^3 + F(t) = 0 \tag{30}$$

This PDE has two branches, and we analyze the one which has four positive resonances.

$$p = -1, q = -6, \text{ branch } u_0 = \alpha, \text{ resonances } (1,2,3,10) \text{ compatible} \tag{31}$$

Let us denote $v = u_1$. Set (E_j) contains three equations:

$$\begin{cases} E_4 & \equiv & \beta C - 4S^2 + 9S_{xx} + \dfrac{60}{\alpha} S v_x - \dfrac{180}{\alpha^2} v_x^2 - \dfrac{30}{\alpha} v_{xxx} = 0 & (32a) \\ E_5 & \equiv & -\beta C_x - 2SS_x + S_{xxx} + \dfrac{30}{\alpha} S_x v_x = 0 & (32b) \\ E_6 & \equiv & SK(v) + 15S_x \left(\dfrac{v_{xx}}{\alpha} - \dfrac{S_x}{3} \right) + \dfrac{S}{2} E_4 - \dfrac{1}{2} E_{5,x} = 0 & (32c) \end{cases}$$

During the process of elimination of u_1 appears the nongeneric case $S_x = 0$, which we discard here. Set $(14a)$ is:

$$\begin{cases} v_x & = & \alpha S_x^{-1} \left(\dfrac{\beta}{30} C_x - \dfrac{1}{30} S_{xxx} + \dfrac{1}{15} SS_x \right) & (33a) \\ v_t & = & \text{computed from } E_6 & (33b) \end{cases}$$

and, with the notation

$$G \equiv S_{xx} + 4S^2 - bC \qquad (34)$$

set $(14b)$ is:

$$\begin{cases} F_4 & \equiv & \left(S_x^{-1} G_{xx} - G_x S_x^{-2} S_{xx} \right)_x - G - \dfrac{1}{5} S_x^2 G_x^2 + 2SS_x^{-1} G_x = 0 & (35a) \\ F_6 & \equiv & \text{29 terms in } (S, C) (\text{substitution of } v_x, v_t \text{ in } E_{6,x}) & (35b) \end{cases}$$

Despite many attempts, we have not been able to split this set (F_j) into two subsets $(15a)$, $(15b)$ with the second one depending on the first one. Our conjecture is that this step can only be performed when one takes account of the underlying scattering problem, as detailed in next section.

4 Invariant method to find a scattering problem of a given type

Suppose we are given:

- a PDE

$$E(u, x, t) = 0 \qquad (36)$$

- its SME polynomial in $D\varphi$:

$$E_{sm}(D\varphi) = 0 \qquad (37)$$

- a type of scattering problem, e.g. the third order one:

$$\begin{cases} \psi_{xxx} & = & a\psi_x + b\psi & (38a) \\ \psi_t & = & c\psi_{xx} + d\psi_x + e\psi & (38b) \end{cases}$$

The problem we address is to find the coefficients (a, b, c, d, e) of the given type scattering problem underlying the PDE. Let us denote ψ_1 and ψ_2 two independent solutions of $(38ab)$. The direct, noninvariant method [2] is the following:

i) replace φ by $\frac{\psi_1}{\psi_2}$ in $E_{sm}(D\varphi)$ to obtain $E_{sm}(\psi_i, D\psi_i)$

ii) whenever $(m, n) \geq (3, 0)$ or $(0, 1)$, replace $(\psi_i)_{mx,nt}$ using $(38ab)$, to get

$$E_{sm}(\psi_i, \psi_{i,x}, \psi_{i,xx}, (a, b, c, d, e)) = 0 \qquad (39)$$

iii) require that this polynomial of six variables $\psi_i, \psi_{i,x}, \psi_{i,xx}$ identically vanish. This yields a large set of nonlinear coupled PDEs in (a,b,c,d,e) to be solved, which we call determining equations.

This method generates many duplicate or dependent determining equations. The invariant method [8] consists in finding the minimum number of functions $Y_l(\psi_1, \psi_2)$ invariant under the group

$$(\psi_1, \psi_2) \to (\alpha'\psi_1 + \beta'\psi_2, \gamma'\psi_1 + \delta'\psi_2), \alpha'\delta' - \beta'\gamma' = 1 \tag{40}$$

such that, when one takes account of the scattering problem type, the functions S, C, $Y_{l,x}$, $Y_{l,t}$ are polynomials of Y_m with coefficients polynomial in (a, b, c, d, e) and their derivatives. In the case of the third order scattering problem $(38ab)$, such a parametric representation is:

$$Y_1 = \frac{\psi_{1,xx}\psi_2 - \psi_{2,xx}\psi_1}{\psi_{1,x}\psi_2 - \psi_{2,x}\psi_1}, Y_2 = \frac{\psi_{1,xx}\psi_{2,x} - \psi_{2,xx}\psi_{1,x}}{\psi_{1,x}\psi_2 - \psi_{2,x}\psi_1} \tag{41}$$

$$S = -\frac{3}{2}Y_1^2 + 3Y_2 + a, C = -cY_1 - d \tag{42}$$

$$\left\{\begin{array}{lll} Y_{1,x} & = & -Y_1^2 + Y_2 + a \qquad (43a) \\ Y_{2,x} & = & -Y_1Y_2 - b \qquad (43b) \\ Y_{1,t} & = & ((c_x + d)Y_1 - cY_2 + d_x + ac + 2e)_x \qquad (43c) \\ Y_{2,t} & = & -(c_x + d)Y_1Y_2 + cY_2^2 + (bc + e_x)Y_1 + (c_{xx} + 2d_x + ac)Y_2 \\ & & -e_{xx} - 2bc_x - b_xc - bd \qquad (43d) \end{array}\right.$$

with the cross-derivative conditions

$$\left\{\begin{array}{lll} (\psi_t)_{xxx} - (\psi_{xxx})_t & \equiv & X_0\psi + X_1\psi_x + X_2\psi_{xx} = 0 \qquad (44a) \\ (Y_{1,x})_t - (Y_{1,t})_x & \equiv & -X_1 - X_2Y_1 \equiv S_t + C_{xxx} + 2C_xS + CS_x = 0 \qquad (44b) \\ (Y_{2,x})_t - (Y_{2,t})_x & \equiv & X_0 - X_2Y_2 = 0 \qquad (44c) \end{array}\right.$$

$$\left\{\begin{array}{lll} X_0 & \equiv & -b_t - ae_x + e_{xxx} + b_{xx}c + 3bc_{xx} + 3b_xc_x + 3bd_x + b_xd = 0 \qquad (45a) \\ X_1 & \equiv & -a_t + 3e_{xx} + 2b_xc + a_{xx}c + d_{xxx} + 3ac_{xx} + 2ad_x \\ & & +3a_xc_x + 3bc_x + a_xd = 0 \qquad (45b) \\ X_2 & \equiv & (2ac + c_{xx} + 3d_x + 3e)_x = 0 \qquad (45c) \end{array}\right.$$

Hence the invariant method:

i) replace (S, C) given by (42) in $E_{sm}(S, C)$ to get $E_{sm}(Y_l, DY_l, (a, b, c, d, e))$

ii) replace derivatives of Y_l using Riccati equations $(43abcd)$, to get

$$E_{sm} \equiv \sum_{l,m} A_{l,m}(a, b, c, d, e)Y_1^lY_2^m \tag{46}$$

iii) require the vanishing of the determining equations:

$$A_{l,m}(a, b, c, d, e) = 0 \text{ and } (45abc) \tag{47}$$

The replacement of six noninvariant basis variables by two invariant ones results in a considerably shorter set of determining equations.

In the Sawada-Kotera case, when applied to either $S_x^3F_4$ or $S_x^2F_6$, this algorithm yields respectively 35+3 and 35+3 determining equations, which have the same solution, as detailed now for $S_x^3F_4$:

$$A_{8,1} \equiv 18(-9b + 2\beta c - 3a_x) = 0 \tag{48}$$

is solved for coefficient c. Among the remaining equations,

$$A_{9,0} \equiv -27(a^2 + \beta d + a_{xx} + 17b_x) = 0 \tag{49}$$

is solved for coefficient d. Then, in a similar way, we find successively

$$A_{0,4} \equiv -1440 b_{xx} = 0 \tag{50}$$

then

$$\begin{cases} A_{7,0} \equiv 405 a b_x = 0 & (51a) \\ A_{6,0} \equiv 45(15b + a_x) b_x = 0 & (51b) \end{cases}$$

which implies $b_x = 0$, then

$$X_0 \equiv -b_t = 0 \tag{52}$$

which implies that b is an arbitrary constant λ. There remain only two equations,

$$X_2 \equiv (18\beta^{-1}\lambda a + 3e)_x = 0 \tag{53}$$

which yields coefficient e, defined up to an arbitrary additive gauge, and

$$X_1 \equiv -6SK(-\frac{\alpha a}{6}) = 0 \tag{54}$$

which proves that the Lax pair has been obtained.

Let us now return to the question of the unicity of the SME. In the above example, both candidates F_4 and F_6 contain the same information, although F_4 looks "simpler". The dependence we are looking for exists probably not between F_4 and F_6 directly, but between their coefficients $A^{(4)}_{l,m}$ and $A^{(6)}_{l,m}$, and we are currently checking the existence of polynomial functions $P_{l,m}$ such that:

$$\forall (l,m) : A^{(6)}_{l,m} = P_{l,m}(\partial_x, \partial_t, A^{(4)}_{l',m'}) \tag{57}$$

This would establish a functional dependence of F_6 on F_4 *via* the scattering problem, and prove that F_4 is the (unique) SME.

Let us remark that Sawada-Kotera equation admits the sub-SME $G = 0$, with G defined by (34). This sub-SME yields the Miura transformation ($\frac{\alpha}{3}S$ solution of Kaup-Kuperschmidt equation (3) by elimination of C between (34) and (8)), but it cannot yield the Lax pair and therefore cannot be the SME.

Conclusion

We have tried to emphasize the role of the singular manifold equation and to give a working definition for it. For an integrable PDE, we conjecture that the SME is equivalent to the PDE, since the SME should contain the information on the Lax pair. Work currently in progress with M. Musette [8] uses properties of homographic invariance in order to try and uncover the underlying Lax pair without knowing its type in advance.

References

[1] J. Weiss, M. Tabor and G. Carnevale, The Painlevé property for partial differential equations, J. Math. Phys. **24**, 522-526 (1983).

[2] J. Weiss, The Painlevé property for partial differential equations. II: Bäcklund transformation, Lax pairs, and the Schwarzian derivative, J. Math. Phys. **24**, 1405-1413 (1983).

[3] R. Conte, Invariant Painlevé analysis of partial differential equations, Phys. Lett. **A140**, 383-390 (1989).

[4] M. Musette, Painlevé-Darboux transformations in nonlinear partial differential equations, these proceedings.

[5] R. Conte and M. Musette, Painlevé analysis and Bäcklund transformation in the Kuramoto-Sivashinsky equation, J. Phys. **A22**, 169-177 (1989).

[6] K. Sawada and T. Kotera, A method for finding N-soliton solutions of the K.d.V. equation and K.d.V.-like equations, Prog. Theor. Phys. **51**, 1355-1367 (1974).
P. J. Caudrey, R. K. Dodd and J. D. Gibbon, A new hierarchy of Korteweg-de Vries equations, Proc. Roy. Soc. London **A351**, 407-422 (1976).

[7] J. Weiss, Bäcklund transformation and the Painlevé property, J. Math. Phys. **27**, 1293-1305 (1986).

[8] M. Musette and R. Conte, Third order Lax pair of the Hirota-Satsuma equation by a new invariant Painlevé analysis, submitted for publication (1990).

The Davey-Stewartson 1 Equation: A Class of Explicit Solutions Including the Special Case of the Dromions

A. Degasperis

Dipartimento di Fisica Università "La Sapienza", P.le Aldo Moro 2,
I-00185 Roma, Italy and
Istituto Nazionale di Fisica Nucleare, Sezione di Roma, Italy

The Davey-Stewartson 1 equation is a key-equation under many respects. It meets three human desires: universality, applicability and integrability. Moreover, being nonlinear and involving 2+1 (space+time) dimensions, this equation, for the first time, has unveiled to us the way the properties of solitons change when going from the one-dimensional world to the two-dimensional one. By universality we understand that the DS1 equation can be derived from a large class of wave equations, describing nonlinear propagation together with dispersion, by a standard multi-scaling technique; in this general context, it models the slow change of amplitude of a monocromatic wave [1], [2], [3], [4]. Because of this remarkable property, the DS1 equation pops out in various applications such as fluid dynamics [5], plasma physics and nonlinear optics. Finally, the DS1 equation is also integrable since its associated initial-boundary value problem can be solved by a spectral transform method [6], [7].

This lecture is aimed to provide a simple construction of explicit solutions of the DS1 equation. Although we try to keep our presentation as much selfcontained as possible, and despite the fact that our results heavily rest on the inverse spectral approach of A.S.Fokas and P.M.Santini, we will not go into the spectral theory for which we refer the reader to [8]. Our generic solution describes the interaction of a number of bumps which are spatially localized with gaussian, rather than exponential, profile; each bump, however, disperses away with a characteristic dispersion time τ. In the limit in which all dispersion times go to infinity this solution coincides with the multi-dromion solution that turns out to be, therefore, just a special subcase.

The DS1 equation looks like the Schroedinger equation

$$i Z_t + Z_{xx} + Z_{yy} + V(x,y,t) Z = 0, \qquad Z = Z(x,y,t) \quad , \tag{1}$$

where the potential V is given in terms of the unknown function Z itself by the explicit expression

$$V(x,y,t) = v_1(x,t) + v_2(y,t) - \frac{1}{2} \varepsilon \left[\int_{-\infty}^{x} d\xi \, (|Z(\xi,y,t)|^2)_y + \int_{-\infty}^{y} d\eta \, (|Z(x,\eta,t)|^2)_x \right] ,$$
$$\varepsilon = \pm 1 . \tag{2}$$

Hereafter the function Z(x,y,t) is assumed to vanish (fast enough) as $x^2 + y^2 \to \infty$, while the potential (2) has instead a nonvanishing boundary value which is characterized by two functions $v_1(x,t)$ and $v_2(y,t)$. The so called initial-boundary value problem associated with equations (1) and (2) is that of finding Z(x,y,t) for a given initial datum $Z(x,y,t_o)$ and given boundary values $v_1(x,t)$ and $v_2(y,t)$; here we are not concerned with it, but with the construction of particular boundary functions v_1 and v_2 and corresponding solutions Z of equation (1).

Let us start by observing that the linearized form of the DS1 equation (1), with (2), is the Schroedinger equation

$$i\, z_t + z_{xx} + z_{yy} + [v_1(x,t) + v_2(y,t)]\, z = 0, \qquad z=z(x,y,t)\ , \tag{3}$$

where z is the linear approximation of Z. The point here is that the DS1 equation has (spatially) localized solutions already in its linear approximation, quite in contrast with the one-dimensional case. Indeed, since the potential in (3) is the sum of two functions of one single space variable only, there certainly exist solutions of (3) wich are the product, $z(x,y,t) = z_1(x,t)\, z_2(y,t)$, of two localized solutions of the 1+1 Schroedinger equation with the potential given by the boundary functions, namely

$$iz_{1t}+z_{1xx}+v_1(x,t)\, z_1 = 0, \qquad z_1=z_1(x,t), \tag{4.1}$$

$$iz_{2t}+z_{2yy}+v_2(y,t)\, z_2 = 0, \qquad z_2=z_2(y,t)\ . \tag{4.2}$$

In fact, we will consider only solutions of the DS1 equation whose linear approximation is a linear superposition of products of this type, i.e.

$$z(x,y,t) = \sum_{n=1}^{N_1} \sum_{m=1}^{N_2} a_{nm}\, z_1^{(n)}(x,t)\, z_2^{(m)}(y,t) = (z_1, a z_2) \tag{5}$$

where the vector notation is conveniently used: (u,w) is the symmetric scalar product,

$$(u,w) = \sum_{n=1}^{N} u^{(n)}\, w^{(n)}\ , \tag{6}$$

z_j is a N_j-dim vector, whose components $z_j^{(n)}$ are solutions of the Schroedinger equation (4.j), j=1,2, and a is a constan $N_1 x N_2$ rectangular matrix with elements a_{nm}. Of course, the bumps in the solution (5) may or may not disperse away at large t, but, in any case, they will not interact with each other because of the linearity of the evolution in this approximation. Instead, they do interact with each other if they evolve according to the fully nonlinear DS1 equation. Their dynamics can be investigated by the important "dressing theorem": to any solution of the form (5) of the linear equation (3) there corresponds a solution of the DS1 equation whose explicit expression is

$$Z(x,y,t) = (z_1(x,t),\, A(x,y,t)\, z_2(y,t))\ , \tag{7}$$

where the $N_1 x N_2$ rectangular matrix A is constructed via the following formulae

$$A(x,y,t) = [\, 1\!\!1 - \varepsilon\, a B_2(y,t)\, a^+ B_1{}^*(x,t)]^{-1} a = \\ = a\, [\, 1\!\!1 - \varepsilon\, B_2(y,t)\, a^+ B_1{}^*(x,t) a]^{-1}\ , \tag{8a}$$

$$B_1(x,t) = \frac{1}{2} \int_{-\infty}^{x} d\xi\, z_1(\xi,t)\, z_1^+(\xi,t)\ , \tag{8b}$$

$$B_2(y,t) = \frac{1}{2} \int_{-\infty}^{y} d\eta\, z_2(\eta,t)\, z_2^+(\eta,t)\ ; \tag{8c}$$

hereafter * and + mean complex and, respectively, hermitian conjugation, and $B_1(B_2)$ is a N_1xN_1 (N_2xN_2) square matrix. Note that, since a is the constant matrix which appears in (5), the linear approximation of the solution (7) is precisely the expression (5). This theorem can be either proved by the spectral method [8] or verified by direct computation. A remarkable corollary of this theorem is that the constant of the motion

$$C = \int\int_{-\infty}^{+\infty} dxdy\, |Z(x,y,t)|^2 , \tag{9}$$

for the DS1 equation has the following expression

$$C = - 4\varepsilon \ln \det [\, \mathbb{1} - \frac{\varepsilon}{4} a\, C_2 a^+ C_1^*] \tag{10}$$

in terms of the time-independent hermitian matrices

$$C_1 = \int_{-\infty}^{+\infty} dx\, z_1(x,t)\, z_1^+(x,t) \quad , C_2 = \int_{-\infty}^{+\infty} dy\, z_2(y,t)\, z_2^+(y,t) \, . \tag{11}$$

This result obtains via direct computation of the integral (9) by using the explicit expression (7), with (8). The relevant consequence of the "dressing theorem", as implied by (7), (8) and (4), is that the construction of solutions of the DS1 equation reduces to finding explicit solutions of the linear one-dimensional Schroedinger equation

$$if_t + f_{xx} + v(x,t)\, f = 0 \, , \quad f = f(x,t) \quad , \tag{12}$$

or, more precisely, to the construction of solvable potentials v(x,t).

Classes of solvable potentials are already available; the best known case is that of time-independent potentials, i.e. v(x,t) = v(x). Several examples of such potentials are given in [9]; their solvability is due to the possibility of obtaining either the general solution of the corresponding stationary Schroedinger equation, $k^2\varphi+\varphi_{xx}+v(x)\varphi=0$, in terms of special functions, or the explicit solution of the inverse spectral problem associated with the stationary Schroedinger equation (these are the reflectionless potentials or, more generally, potentials with a rational reflection coefficient [9]). Simple instances of solvable time-dependent potentials are given by functions of the form v(x,t) = u(x+ct), where c is a real constant and u(x) is a solvable potential; or by a piece-wise time-independent potential, namely $v(x,t) = v_j(x)$ for $t_j<t<t_{j+1}$, j=0,1,...,N, each $v_j(x)$ being a solvable potential and $t_0=-\infty<t_1<t_2<..<t_N<t_{N+1}=+\infty$. In the following we provide an algebraic way of generating solvable potentials v(x,t); because of the simplicity of the method, we limit our presentation to the display of the relevant formulae (further details and proofs are given in [10]).

If v(x,t) is a solvable potential, and f(x,t) is a N-dimensional vector whose components $f^{(n)}(x,t)$, n=1,2,..,N, are solutions of the Schroedinger equation (12), then

$$v'(x,t) = v(x,t) + 2\, \{\ln \det [\mathbb{1} - M(x,t)\, L]\}_{xx} \tag{13}$$

is also a solvable potential, where L is a constant NxN matrix and M(x,t) is the matrix

$$M(x,t) = \int_{-\infty}^{x} dy\, f(y,t) f^+(y,t) \quad . \tag{14}$$

Moreover, if L is hermitian, $L = L^+$, and v is real, $v=v^*$, then also v' is real. In fact, if g(x,t) is a solution of (12), and (see (6))

$$K(x,y,t) = (f^*(y,t), L\,[\mathbb{1} - M(x,t)L]^{-1} f(x,t)), \tag{15}$$

then

$$g'(x,t) = g(x,t) + \int_{-\infty}^{x} dy\; K(x,y,t)\, g(y,t) \tag{16}$$

solves the Schroedinger equation (12) with v(x,t) replaced by v'(x,t). In particular, if g in (16) is replaced by the component $f^{(n)}$ of f, then the l.h.s. of (16) g', is the component $f^{(n)\prime}$ of the N-dimensional vector f ' whose explicit expression is

$$f'(x,t) = [\mathbb{1} - M(x,t)\, L]^{-1} f(x,t) \quad . \tag{17}$$

We also note that if v=v(x) is time-indipendent, $f^{(n)}(x,t) = \varphi^{(n)}(x) \exp(-iE_n t)$, with $E_n=E_n^*$, and the matrix L is diagonal, then also v'=v'(x) is time-independent, and the formulae given above coincide with those obtained via the Marchenko formulation of the inverse problem associated with the stationary Schroedinger equation [9].

These results easily apply to the construction of solutions of the DS1 equation via the formulae (7), (4) and (8). Indeed, suppose $f_1(x,t)$ ($f_2(y,t)$) is a N_1-dim. (N_2-dim.) vector whose components are solutions of the Scroedinger equation (12) with v(x,t) = 0, and L_1(L_2) is a $N_1 x N_1$($N_2 x N_2$) constant matrix, then

$$z_1(x,t) = [\mathbb{1} - M_1(x,t)\, L_1]^{-1} f_1(x,t) \tag{18a}$$

$$z_2(y,t) = [\mathbb{1} - M_2(y,t)\, L_2]^{-1} f_2(y,t) \tag{18b}$$

are solutions of (4.1) and (4.2), where (see (13) and (14))

$$v_1(x,t) = 2\,\{\ln \det [\mathbb{1} - M_1(x,t)L_1]\}_{xx}, \tag{19a}$$

$$v_2(y,t) = 2\,\{\ln \det [\mathbb{1} - M_2(y,t)L_2]\}_{yy}, \tag{19b}$$

$$M_1(x,t) = \int_{-\infty}^{x} d\xi\, f_1(\xi,t) f_1^+(\xi,t), \quad M_2(y,t) = \int_{-\infty}^{y} d\eta\, f_2(\eta,t) f_2^+(\eta,t) \quad . \tag{20}$$

Furthermore, the matrices $B_1(x,t)$ and $B_2(y,t)$, defined by (8.b,c), can be easily computed and read

$$B_1(x,t) = \frac{1}{2}\,[\mathbb{1} - M_1(x,t)\, L_1]^{-1} M_1(x,t) \quad , \tag{21}$$

$$B_2(y,t) = \frac{1}{2}\,[\mathbb{1} - M_2(y,t)\, L_2]^{-1} M_2(y,t) \quad ; \tag{22}$$

therefore, see (8.a) and (7), the solution of the DS1 equation finally takes the expression

$$Z(x,y,t) = (f_1(x,t), \tilde{A}(x,y,t) [\mathbb{1} - \frac{\varepsilon}{4} M_2(y,t)\, a^+ M_1^*(x,t)\, \tilde{A}(x,y,t)]^{-1} f_2(y,t), \tag{23}$$

where $\tilde{A}$ is the $N_1 x N_2$ matrix

$$\tilde{A}(x,y,t) = [\mathbb{1} - L_1^* M_1^*(x,t)]^{-1}\, a\, [\mathbb{1} - M_2(y,t)\, L_2]^{-1}, \tag{24}$$

and a is a constant $N_1 x N_2$ matrix. The virtue of the expression (23) is that of producing solutions of the DS1 equation directly from N_1+N_2 solutions of the Schroedinger equation with vanishing potential (i.e. $if_{jt}+f_{jxx} = 0$, $j=1,2$); one should also notice that the solution $Z(x,y,t)$, once the N_1+N_2 functions $f_j^{(n)}$, $n=1,2,..,N_j$, $j=1,2$, are fixed, depends on the matrices L_1,L_2 and a, namely on $N_1^2 + N_2^2 + 2N_1N_2 = (N_1+N_2)^2$ real parameters. Of course, the condition that the integrals (20) exist is sufficient for the expressions (23) and (24) to make sense. If, in addition, we assume that also the integrals

$$\Gamma_1 = \int_{-\infty}^{+\infty} dx\, f_1(x,t)\, f_1^+(x,t)\,, \quad \Gamma_2 = \int_{-\infty}^{+\infty} dy\, f_2(y,t)\, f_2^+(y,t) \tag{25}$$

are finite, then the time-independent matrices (11) turn out to have the expressions (use (18))

$$C_j = [\mathbb{1} - \Gamma_j L_j]^{-1} \Gamma_j = \Gamma_j [\mathbb{1} - L_j \Gamma_j]^{-1} \quad , \ j=1,2..., \tag{26}$$

which, together with (10), imply that the constant of the motion (9) explicitly reads

$$C = -4\varepsilon \ln \det [\mathbb{1} - \frac{\varepsilon}{4} a (\mathbb{1} - \Gamma_2 L_2)^{-1} \Gamma_2\, a^+ \Gamma_1^* (\mathbb{1} - L_1^* \Gamma_1^*)^{-1}]\,. \tag{27}$$

A particularly interesting choice of the building-block functions $f_j^{(n)}(x,t)$ is the gaussian profile

$$G(x,t;S,T,w,\tau) = \{(\tau - iT)/[\tau + i(t-T)]\}^{1/2}\, F \exp[i\varphi(x,t)] \exp\{-\frac{\tau}{4}[x-S-w(t-T)]^2/[\tau^2+(t-T)^2], \tag{28a}$$

$$F = \exp\{\frac{1}{4}[\tau(S-wT)^2 + i(2\tau^2 wS + S^2 T - w^2\tau^2 T)]/(\tau^2+T^2)\}, \tag{28b}$$

$$\varphi(x,t) = \frac{1}{4}\{(t-T)[(x-S)^2 - w^2\tau^2] + 2w(x-S)\}/[\tau^2+(t-T)^2], \tag{28c}$$

which is characterized by four real parameters, i.e. S,T, w and $\tau > 0$, whose physical meaning is quite transparent: $S+w(t-T)$ is the position of the maximum of $|G|$ at time t, w is its translation velocity, T is the time at which the width of the profile is smallest and τ is the dispersion time. This solution of the Schroedinger equation (with v=0) is very well localized in space, but it disperses away with its characteristic dispersion time τ; in the limit as $\tau \to \infty$, to say

$$\lim_{\tau\to\infty} G(x,t;S,-\alpha\tau,w,\tau) = \exp(i\lambda x - i\lambda^2 t), \qquad \lambda = \frac{w}{2}/(1+i\alpha), \tag{29}$$

the localization is lost (here α is a real parameter with the condition $w\alpha>0$), and the function G goes into the usual exponential solution that depends on two real parameters only.

If $S_j^{(n)}$, $T_j^{(n)}$, $w_j^{(n)}$ and $\tau_j^{(n)}$, $n=1,2,..,N_j$, $j=1,2$, is a collection of $4(N_1+N_2)$ real numbers, the solution (23) of the DS1 equation with

$$f_j^{(n)}(x,t) = G(x,t;S_j^{(n)},T_j^{(n)},w_j^{(n)},\tau_j^{(n)}), \qquad (30)$$

describes the interaction of N_1N_2 gaussian bumps, each one fading with its own characteristic dispersion time [10]; quite naturally these solutions may be termed gaussons (i.e. the solution (23) with (30) is a (N_1,N_2) gausson). The simplest solution of this class, which obtains with $N_1=N_2=1$ and $L_1=L_2=0$ (corresponding, therefore, to vanishing boundary values, $v_1=v_2=0$) has been already given in [11]. On the other hand, if all dispersion times $\tau_j^{(n)}$ go to infinity, according to the limit (29), and L_1 and L_2 are non vanishing, the solution (23) remains localized and becomes dispersionless: it coincides with the multi-dromion solution discussed in [8], [6].

Collisions of gaussons with gaussons and dromions can be easily investigated in this formalism [10].

References

[1] V.E. Zakharov and E.A.Kuznetsov, Physica 18D, (1986) 455.

[2] F.Calogero and A.Maccari in Inverse Problems, P.C.Sabatier (editor) Academic Press, New York, 1987.

[3] S.V.Manakov, NEEDS conference, Kolimbari, Grece, 1989.

[4] F.Calogero, this conference and references quoted there.

[5] M.J.Ablowitz and H.Segur, J.Fluid Mech. 92, (1979) 691.

[6] P.M.Santini, this conference and references quoted there.

[7] J.JP.Leon, this conference and references quoted there.

[8] A.S.Fokas and P.M.Santini, Physica D, to be published.

[9] F.Calogero and A.Degasperis, Spectral Transform and Solitons I, North-Holland, Amsterdam, 1982, appendix A19.

[10] A.Degasperis (in preparation).

[11] H.Cornille, J.Math. Phys. 20, (1979) 199.

The "IST" as a Diagnostic Tool for Perturbed Nonlinear Schrödinger Systems

Th. Eickermann and K.H. Spatschek

Institut für Theoretische Physik I, Heinrich-Heine-Universität Düsseldorf, D-4000 Düsseldorf, Fed. Rep. of Germany

Abstract: A method for analyzing the dynamics of near-integrable nonlinear Schrödinger (NLS)-equations is presented. By solving the Zakharov-Shabat scattering problem numerically, for both infinite and periodic settings, the parameters of solitons and radiation (or the corresponding periodic structures) can be determined. As an example a driven and damped NLS-equation is studied. The results justify a model with finite degrees of freedom.

1 Introduction

The investigation of the transition from ordered to turbulent motion is a topic of still growing interest. Whereas most of the rigorous mathematical investigations apply only to finite degrees of freedom [ordinary differential equations (ODE's) and maps] the physical situation often requires more complicated models which lead to partial differential equations (PDE's). Nevertheless, even in such situations low-dimensional attractors are often observed. Then a truncated description which takes into account only a few modes might be successful.

In this contribution the problem of finding appropriate modes is treated for the special case of a perturbed soliton system of the nonlinear Schrödinger type. In the near-integrable case the best choice of the set of basis functions consists of the modes of the unperturbed system, i.e. solitons and radiation. Starting from the numerical solution of the PDE we estimate the parameters of both, solitons and radiation, using inverse scattering transform (IST) techniques. We then decide which of these modes seem to be relevant for the dynamics of the PDE solution. It should be mentioned that a similar approach proved to work fine for sine-Gordon-systems [1]. The equation under investigation,

$$iq_t + q_{xx} + 2|q|^2 q - \omega q = -i\epsilon - i\gamma q, \tag{1}$$

models the envelope of collisionally damped nonlinear Langmuir-waves in a plasma driven by an external rf capacitor field [2]. While the unperturbed NLS equation is a generic model for weakly nonlinear and weakly dispersive wave phenomena, the physical results of the perturbed equation will depend on the specific model [3],[4]. The diagnostics developed here can be applied to all such equations in order to find out (i) whether a simpler model of the dynamics can be developed, and (ii) which modes have to be included.

2 Dynamical behavior of the PDE solutions

The study of (1) shows that for $\gamma > \frac{\pi}{2\sqrt{\omega}}\epsilon$ the system is dissipation dominated and any spatial structure of an initial distribution will be damped out in time. In this case all the solutions

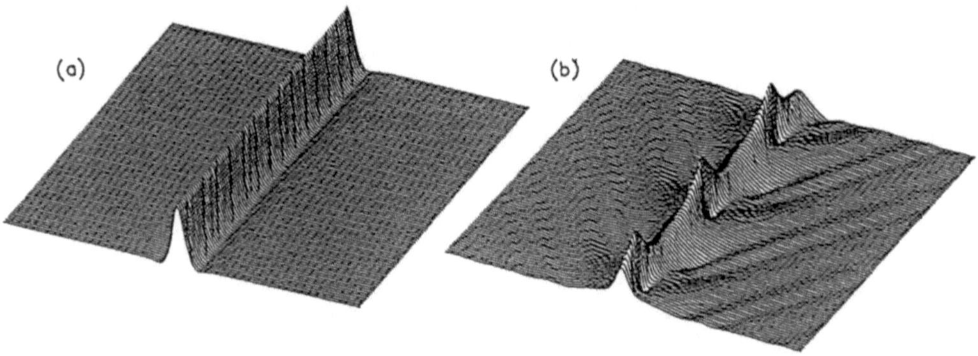

Fig. 1: Time evolutions of the PDE solutions. In each line $|q|$ is plotted against x from -30 to 30. The time between two consecutive lines is 0.2. Figure a is for $\epsilon = 0.1, \gamma = 0.11$, i.e. a fixed point; figure b is for $\epsilon = 0.14, \gamma = 0.11$, i.e. a limit cycle.

tend to the stationary space-independent solutions of (1). It can be shown [5] that for $\gamma \leq \frac{\pi}{2\sqrt{\omega}}\epsilon$ a localized phase-locked soliton-like structure appears as an additional attractor. Numerical investigations verify this result (Fig. 1a) [2],[6]. They also show that when the driving parameter ϵ is further increased the fixed point becomes unstable via a Hopf-bifurcation. The new attractor is a periodically oscillating state (limit cycle) which is still spatially localized (Fig. 1b). The periodicity of this solution can be visualized in a reduced phase portrait when the amplitude of the main bump is plotted against its phase. The limit cycle corresponds to a simple closed curve in this picture (Fig. 2a). Further increase of the driver amplitude ϵ leads to a transition to chaos via the Feigenbaum route: after a sequence of period doubling bifurcations (Fig. 2b shows as an example the period-8 cycle) temporal chaos can be observed. Within the caotic regime subharmonic windows appear. Performing a Poincaré-map by plotting only the points of minimal phase leads to a more compact picture (Fig. 3). From these data we were able to obtain first approximations of the Feigenbaum constant $\delta_\infty = 4.6692..$ to which the relative spacings between the bifurcation values of ϵ should converge. The numerically calculated values $\delta_2 = 4.16$ and $\delta_3 = 4.52$ are already quite good.

As far as the temporal evolution of the solutions is concerned, the numerical results fit quite well the facts known from simpler systems such as maps and ODE's. We now focus on the new PDE-aspect of the problem: the spatial structure of the solutions. There is a striking similarity between the limit-cycle solution of (1) and the bounded two-soliton state of the unperturbed NLS equation (Fig. 4). This might lead to the suggestion, that together with the temporal bifurcation from a fixed point to a limit cycle a spatial bifurcation from a soliton into a two-soliton state takes place. To decide wether this suggestion is correct or not we used IST-techniques which are briefly reviewed – as far as we need them – in the next section.

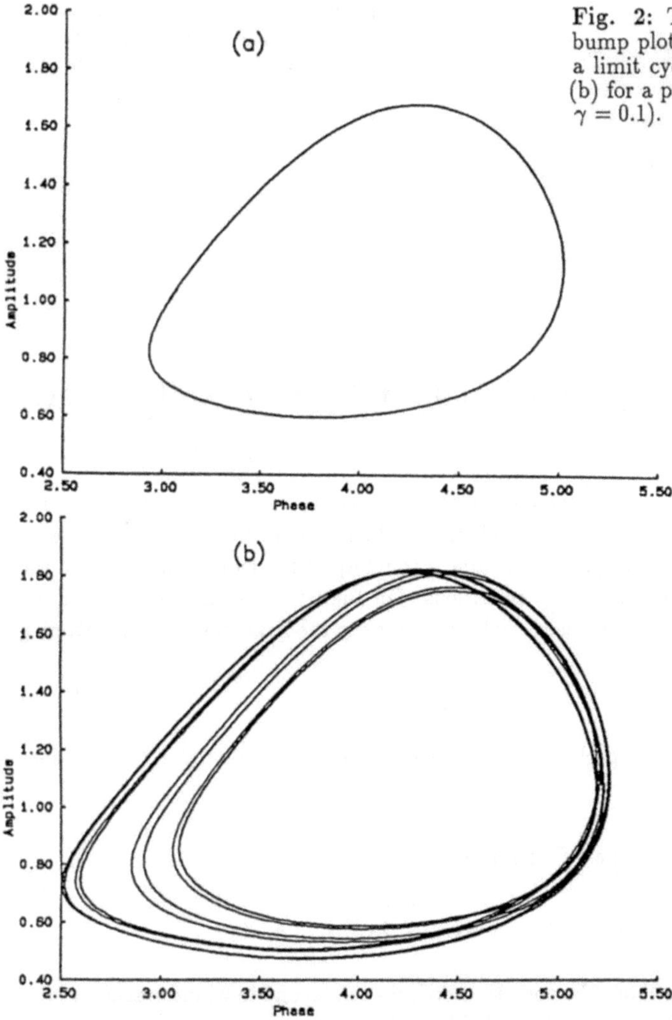

Fig. 2: The amplitude of the main bump plotted against its phase; (a) for a limit cycle ($\epsilon = 0.14, \gamma = 0.11$) and (b) for a period-8 cycle ($\epsilon = 0.147$, $\gamma = 0.1$).

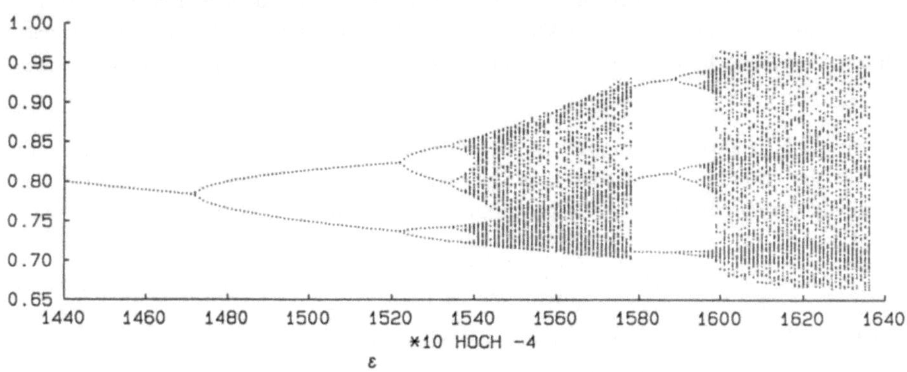

Fig. 3: Bifurcation diagram for the PDE with ϵ as bifurcation parameter.

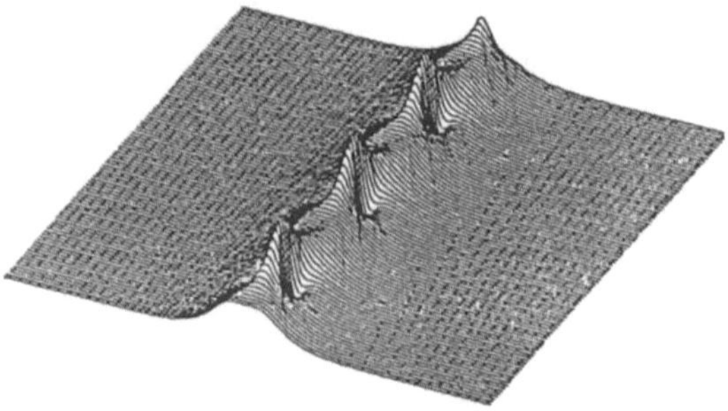

Fig. 4: Time evolution of a bounded two-soliton solution with soliton-amplitudes $\eta_1 = 0.5$ and $\eta_2 = 0.2$ (x and t are as in Fig. 1).

3 The scattering problem in the infinite as well as periodic setting

Since the calculations are much simpler for the infinite line setting, our *analytical* estimates are presented for this case only. But our numerical code [6] based on a Crank-Nicholson-scheme forces us to deal with a finite length system. Since our aim is to interpret the numerical data with the help of an IST we choose periodic boundary conditions. For the latter - as for the infinte line - an IST is available for the unperturbed case [7],[8]. In both cases the Zakharov-Shabat scattering problem

$$\phi_x = \begin{pmatrix} -i\zeta & q \\ -q^* & i\zeta^* \end{pmatrix} \phi \tag{2}$$

has to be solved. In the infinite-line problem, with a spatially localized potential q, the functions $a(\zeta), b(\zeta)$ are defined via

$$\begin{aligned} \phi(x,\zeta) &\sim \begin{pmatrix} 1 \\ 0 \end{pmatrix} e^{-i\zeta x}, \text{ for } x \to -\infty, \\ \phi(x,\zeta) &\sim a(\zeta)\begin{pmatrix} 1 \\ 0 \end{pmatrix} e^{-i\zeta x} + b(\zeta)\begin{pmatrix} 0 \\ 1 \end{pmatrix} e^{i\zeta x}, \text{ for } x \to \infty, \end{aligned}$$

They contain all the information about the potential. Solitons correspond to bound states $a(\zeta) = 0$ and the amplitude of wavenumber-k-radiation is $b^*/a(\zeta = k)$.

In the periodic setting, the main spectrum consists of the edges of the so-called stable bands, i.e. those places in the complex ζ-plane where the transfer matrix M has eigenvalues of modulus one. To establish a connection between the two boundary conditions we consider the following cases [9] :

Let the potential q have a compact support $[x_1, x_2]$ which is a subinterval of some interval of length L. If we treat the latter, in the first case, as the periodicity interval, and in a second case, as part of an infinite line problem, with zero potential outside the interval, then the following exact relation holds

$$M(\zeta) = \begin{pmatrix} a(\zeta)e^{-i\zeta L} & -b^*(\zeta^*) \\ b(\zeta) & a^*(\zeta^*)e^{i\zeta L} \end{pmatrix}. \tag{3}$$

In this equation, M is the transfer matrix for the periodic setting (first case) and a as well as b are evaluated for the infinite line problem (second case). For a given discrete eigenvalue ζ the length of the system can be chosen large enough [$\text{Im}(\zeta)L \gg 1$] so that close to this eigenvalue a stable

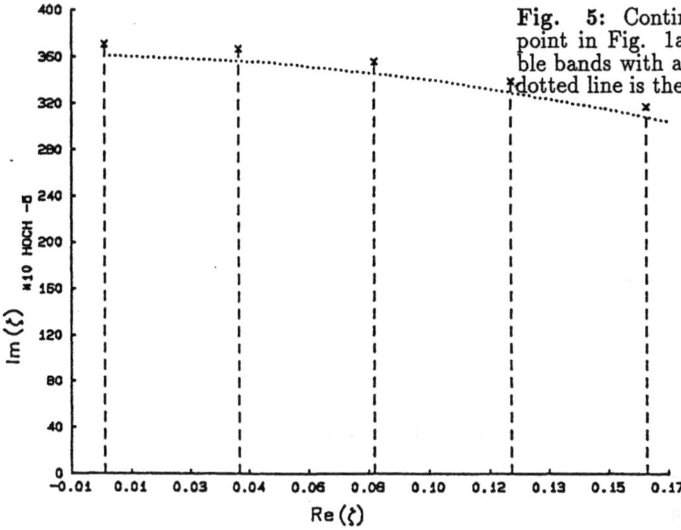

Fig. 5: Continuous spectrum of the fixed point in Fig. 1a. The dashed lines are stable bands with a cross at the band edges; the dotted line is the radiation amplitude $\left|\frac{b^*}{a}\right|/L$.

band of length

$$\Delta\zeta \approx \left|\frac{4e^{i\zeta L}}{\frac{\partial a}{\partial \zeta}}\right| \tag{4}$$

appears in the periodic case. It corresponds to a soliton. Similarly associated with a stable band of length $\Delta\zeta$, being connected to the real ζ-axis at $\zeta = \xi$, is a radiation amplitude

$$b^*/a(k = \xi) \approx \Delta\zeta L \text{ , for } \xi L \gg 1 \text{ .} \tag{5}$$

The reason for this close connection is that structures which are small compared to the length L of the system do not feel the boundaries. Of course, close to the origin of the complex ζ-plane this analogy breaks down and a description in terms of solitons and radiation is no longer correct.

In our simulations we took $r(x) = q(x) - c$ as the potential for the scattering problem; c is the space-independent fixed point of (1) and r evolves according to a perturbed NLS-equation

$$ir_t + r_{xx} + 2|r|^2 r = -4|r|^2 c - 2r^2 c - 4rc^2 - 2c^2 r^* - i\gamma r \text{ .} \tag{6}$$

Its solution has (in contrast to q) the advantage of being localized. In the following, a soliton-like fixed point ($\epsilon = 0.1, \gamma = 0.1$), computed in an interval of length $L = 80$, is analyzed for demonstration. As expected, we find a discrete eigenvalue at $\zeta = i0.603$ correspondig to a narrow band of length $\Delta\zeta \approx 4e^{i\zeta L} = 4.5 \cdot 10^{-21}$ at the same position in the ζ-plane. In Fig.5, the radiation bands are plotted together with the associated amplitudes of the radiation $|b^*/a|$. The estimate (5) is fulfilled quite well, as long as ξ is not too small.

4 Results and consequences

We now come back to the interpretation of the spatial structures of the solutions of (1). Fig.6a shows the stable bands of a limit cycle at a fixed time. As one can see the spectrum is not qualitatively different from that for the fixed point. It consists of a narrow soliton band and some small radiation bands. Fig 6b shows the time-evolution of a soliton and a $k = 0$-radiation

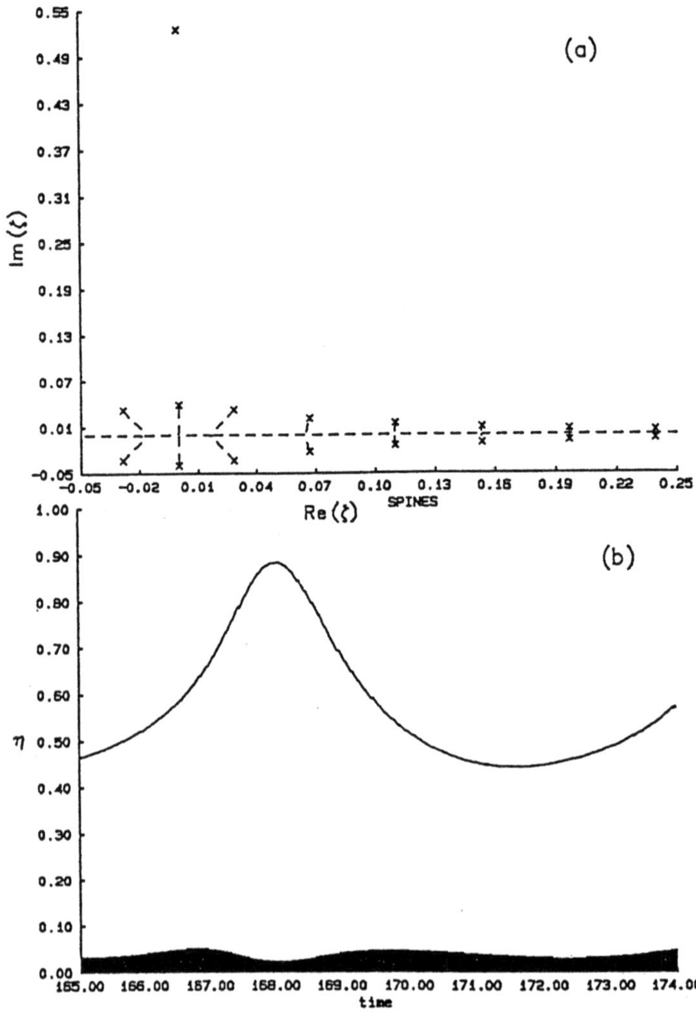

Fig. 6: Spectrum of the limit cycle in Fig. 1b: (a) stable bands (see Fig. 5) at a fixed time; (b) time-evolution of the stable bands (filled area) with Re(ζ) = 0.

band for the limit cycle corresponding to Fig.1b. The dominating structure is a soliton with time-dependent parameters. The time-evolution of the spectral data is completely different from that of the bounded two-soliton (dotted line in Fig.6b) and shows clearly that the interpretation in terms of a bifurcation from a one-soliton into a two-soliton state is not correct. From these results it is clear that a finite-degree-of-freedom model for the dynamics of the PDE-solutions should contain one soliton and at least one of the radiation modes. In fact a model of 4 ODE's [2] based on one soliton and $k = 0$-radiation is very successful in predicting the Hopf-bifurcation and the Feigenbaum-route to chaos. In Fig.7 a limit cycle constructed from the scattering data is compared to a limit cycle based on the ODE-model. However in the chaotic regime the ODE-model is less precise. For example, we calculated the correlation dimension of the chaotic attractor for both, the PDE and ODE's. While for the PDE the dimension is typically between 1.5 and 2.7 (depending on the parameters) it never exeeds 1.9 for the ODE's. The spectral data give a

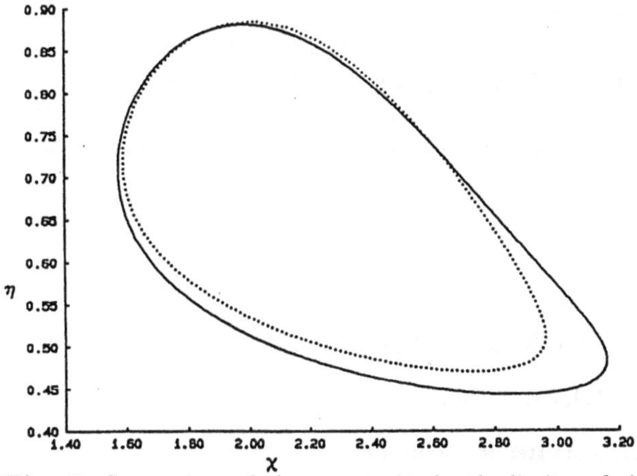

Fig. 7: Comparison of phase portraits for the limit cycle in Fig. 1b: the soliton amplitude η is plotted versus its phase χ. The full line is estimated from the scattering data of the PDE solution whereas the dotted line is the prediction from the ODE-model.

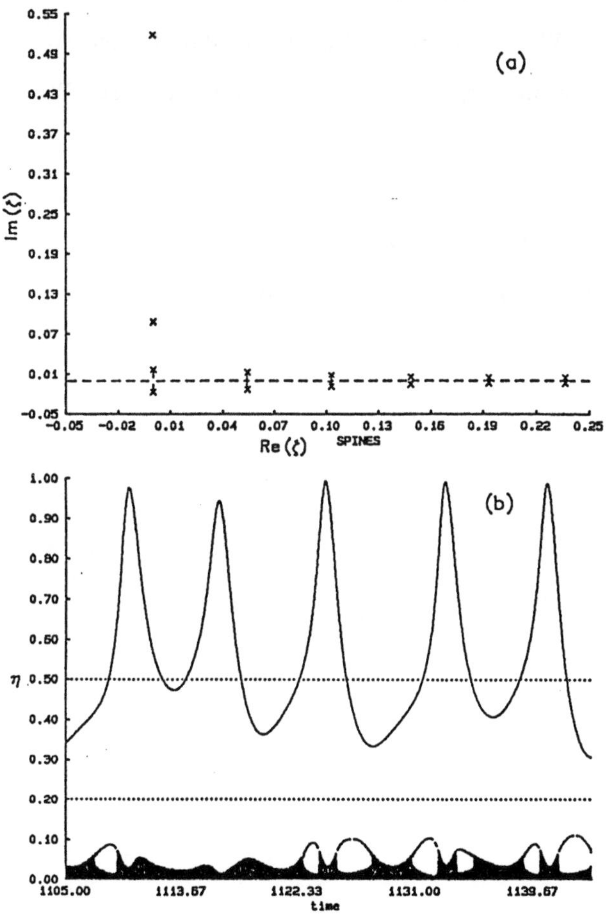

Fig. 8: Spectrum of the temporal chaotic solution with $\epsilon = 0.156, \gamma = 0.11$ (see Fig. 6). The dotted lines are the spectrum of the unperturbed two-soliton.

hint why the value obtained from the ODE's is too small. The appearance of a second discrete band in the chaotic regime (Fig.8) shows that in this case the two modes of the ODE-model are not sufficient to give a detailed description of the PDE-dynamics. In conclusion, we can say that the IST diagnostics, which we applied to the NLS-equation for the first time, is a useful tool for analyzing the solutions of near-integrable soliton systems and constructing models with finite degrees of freedom.

References

[1] Overman II, E.A., McLaughlin, D.W., and Bishop, A.R., Physica **19D**, 1 (1986).

[2] Nozaki, K., and Bekki, N., Physica **21D**, 381 (1986).

[3] Laedke, E.W., and Spatschek, K.H., submitted for publication.

[4] Pietsch, H., Spatschek, K.H., and Laedke, E.W., submitted for publication.

[5] Kaup, D.J., and Newell, A.C., Proc.R.Soc.Lond. **361A**,413 (1978)

[6] Spatschek, K.H., Pietsch, H., Laedke, W. and Eickermann, Th., in **Singular Behavior and Nonlinear Dynamics**, eds. Pnevmatikos,St., Bountis,T., and Pnevmatikos,Sp. (World Scientific Singapore 1989), p.555.

[7] Ma, Y., and Ablowitz, M.J., Stud.Appl.Math. **65**, 113 (1981).

[8] Tracy, E.R., and Chen, H.H., Phys.Rev. **37A**, 815 (1988).

[9] Forest, M.G., and McLaughlin, D.W., J.Math.Phys. **23**, 1248 (1982).

Gaps of One-Dimensional Periodic AKNS Systems

B. Grebert[1] *and J.C. Guillot*[2]

[1]Centre de Mathématiques Appliquées de l'Ecole Polytechnique,
F-91128 Palaiseau Cedex, France
[2]Dept. Mathématiques, Université Paris-Nord, F-93430 Villetaneuse, France

Consider the periodic AKNS operator (see [AKNS] and [Zak-Sha])

$$\begin{pmatrix} 0 & -1 \\ 1 & 0 \end{pmatrix} \frac{d}{dx} + \begin{pmatrix} -q(x) & p(x) \\ p(x) & q(x) \end{pmatrix}, \quad x \in \mathbb{R} \quad (1)$$

where p(x) and q(x) are real valued potentials in $H^1_{loc}(\mathbb{R})$ or in $H^2_{loc}(\mathbb{R})$ such that

$$\begin{aligned} p(x+1) &= p(x) \\ & \qquad x \in \mathbb{R} \qquad (2) \\ q(x+1) &= q(x) \end{aligned}$$

The operator H with domain $H^1(\mathbb{R},\mathbb{C}^2)$ is self adjoint in $L^2(\mathbb{R},\mathbb{C}^2)$.

A complex number λ is called a periodic eigenvalue of H if there exists a non trivial solution $F(x,\lambda ; p,q)$ in $H^1_{loc}(\mathbb{R},\mathbb{C}^2)$ to

$$HF = \lambda F \qquad (3)$$

satisfying

$$F(x + 2, \lambda ; p,q) = F(x,\lambda ; p,q) \qquad x \in \mathbb{R}, \qquad (4)$$

Repetitions of arguments concerning the periodic Schrödinger operator show that those λ's form a non decreasing sequence of real numbers $\lambda_k(p,q))_{k\in Z}$ written with multiplicities. Each λ_k has multiplicity one or two. $(\lambda_k(p,q))_{k\in Z}$ is called the periodic spectrum of H and one has

$$\leq \lambda_{-2}(p,q) < \lambda_{-1}(p,q) \leq \lambda_0(p,q) < \lambda_1(p,q) \leq \lambda_2(p,q) < \ldots < \lambda_{2k-1}(p,q) \leq \lambda_{2p} < \qquad (5)$$

The λ_{2k-1}'s and λ_{2k}'s for even k correspond to nontrivial solutions $F(x, \lambda ; p,q)$ in $H^1_{loc}(\mathbb{R},\mathbb{C}^2)$ to (3) satisfying

$$F(x+1, \lambda ; p,q) = F(x,\lambda ; p,q), \; x \in \mathbb{R} \qquad (6)$$

(6) is the so-called periodic boundary value problem.

The λ_{2k}'s and λ_{2k-1}'s for odd k correspond to non trivial solutions $F(x,\lambda ; p,q)$ in $H^1_{loc}(\mathbb{R},\mathbb{C}^2)$ to (3) satisfaying

$$F(x + 1, \lambda ; p,q) = -F(x, \lambda ; p,q), \; x \in \mathbb{R}, \quad (7)$$

(7) is the so called antiperiodic boundary value problem.

The operator H is self adjoint in $L^2(\mathbb{R},\mathbb{C}^2)$ and has an absolutely continuous band spectrum : $\bigcup_{k\in\mathbf{Z}} [\lambda_{2k},\lambda_{2k+1}]$

The gaps are the open intervals $(\lambda_{2k-1}, \lambda_{2k})$ when $\lambda_{2k-1} \neq \lambda_{2k}$ and the points $\{\lambda_{2k}\}$ when $\lambda_{2k-1} = \lambda_{2k}$. When $\lambda_{2k-1} \neq \lambda_{2k}$, λ_{2k-1} and λ_{2k} have multiplicity one. When $\lambda_{2k-1} = \lambda_{2k}$ the eigenvalue has multiplicity two. The simplicity of all periodic eigenvalues is a generic property.

It is natural to pose the periodic problem in the context of the whole real line. But one usually does not have to look outside the interval [0,1].

One shows that λ is in the periodic spectrum of H if and only if there is a nontrivial solution F to (3) satisfying, with respect to the interval [0,1], either the periodic boundary condition

$$F(0,\lambda\,;\,p,q) = F(1,\lambda\,;\,p,q) \qquad (8)$$

or the antiperiodic boundary condition

$$F(0,\lambda\,;\,p,q) = -F(1,\lambda\,;\,p,q) \qquad (9)$$

The λ_{2k}'s and λ_{2k-1}'s for even k correspond to the periodic boundary condition and the λ_{2k}'s and λ_{2k-1}'s for odd k to the antiperiodic boundary condition. Thus, instead of considering the space of periodic potentials p and q in $H^1_{loc}(\mathbb{R})$ or $H^2_{loc}(\mathbb{R})$, one introduces the following spaces

$$\mathcal{H}^1 = \{(p,q) \in H^1_{\mathbb{R}}([0,1]) \times H^1_{\mathbb{R}}([0,1])\,;\, p(0) = p(1) \text{ and } q(0) = q(1)\}$$

$$\mathcal{H}^2 = \{(p,q) \in H^2_{\mathbb{R}}([0,1]) \times H^2_{\mathbb{R}}([0,1])\,;\, p(0) = p(1),\ p'(0) = p'(1),\ q(0) = q(1), ,\ q'(0) = q'(1)\}$$

The potentials p and q in $\mathcal{H}'^1$ or $\mathcal{H}^2$ are then identified with their periodic extension on the whole real line. The inverse periodic problem consists of two parts. The first part is the description of the periodic isospectral sets associated with potentials in $\mathcal{H}^1$ or in $\mathcal{H}^2$. The second part is the characterization of all possible periodic spectra arising from potentials in $\mathcal{H}^1$ or in $\mathcal{H}^2$. The point of view that one takes towards characterizing a periodic spectrum is to consider its sequence of gaps. (See [Mar.] and [Mar. Ost. 1,2], [Mis. 1,2] for an another point of view).

Let (p_0,q_0) be two periodic real-valued potentials in $\mathcal{H}'^1$(resp. $\mathcal{H}^2$). Let $L(p_0,q_0)$ be the set of potentials (p,q) in $\mathcal{H}^1$(resp $\mathcal{H}^2$) such that

$$\lambda_k(p,q) = \lambda_k(p_0,q_0) \quad \text{for every k in } \mathbf{Z} \qquad (10)$$

The isospectral set $L(p_0,q_0)$ is realized as a product of circles as follows. To every open gap $(\lambda_{2k-1}, \lambda_{2k})$ it corresponds a circle of radius $\frac{\lambda_{2k} - \lambda_{2k-1}}{2}$ centered at $\frac{\lambda_{2k} + \lambda_{2k-1}}{2}$ on the real line.

The isospectral set $L(p_0,q_0)$ is homeomorphic to the product of such circles, one for each non trivial gap $(\lambda_{2k-1}, \lambda_{2k})$.

Therefore one has the following theorem (see also [Pre] and references in [Fad. Takh]).

Theorem 1 For $(p_0,q_0) \in \mathcal{H}^1$(resp.$\mathcal{H}^2$), $L(p_0, q_0)$ is a compact torus, generically of infinite dimension.

In order to get parameters which uniquely determine $(p,q) \in L(p_0, q_0)$ and to characterize all possible periodic spectra one must introduce auxiliary spectra which one now describes.

Let

$$F_1(x,\lambda\,;\,p,q) = \begin{pmatrix} Y_1(x,\lambda\,;\,p,q) \\ Z_1(x,\lambda\,;\,p,q) \end{pmatrix} \qquad (11)$$

be the solution to (3) on [0,1] such that

$$F_1(0,\lambda\,;\,p,q) = \begin{pmatrix} 1 \\ 0 \end{pmatrix} \qquad (12)$$

For every x in [0,1] and p and q in $H^1_{\mathbf{R}}([0,1])$, $F(x,\lambda\,;\,p,q)$ is an entire function of λ. As in [Pö -Tru] one shows that the zeroes of $Z_1(1,.\,;\,p,q)$ in $\mathbb{C}$ is a strictly increasing sequence of real numbers $(\mu_k(p,q))_{k\in \mathbf{Z}}$. Each $\mu_k(p,q)$ is a simple zero. $(\mu_k(p,q))_{k\in\mathbf{Z}}$ is the spectrum of a self adjoint operator in $L^2([0,1], \mathbb{C}^2)$ associated with a restriction of H to $H^1([0,1], \mathbb{C}^2)$ satisfying a Dirichlet boundary condition.
That is the first auxiliary spectrum.

The second auxiliary spectrum , denoted by $(\widetilde{\mu}_k(p,q))_{k\in\mathbf{Z}}$, is defined by

$$\widetilde{\mu}_k(p,q) = \mu_k(q,-p) \quad k \in \mathbf{Z} \qquad (13)$$

The map $(p,q) \rightarrow (q,-p)$ is isospectral.

One has $\lambda_{2k-1}(p,q) \leq \mu_k(p,q), \widetilde{\mu}_k(p,q) \leq \lambda_{2k}(p,q)$ (see [Ma - Abl]).

Now let

$$F_2(x,\lambda\,;\,p,q) = \begin{pmatrix} Y_2(x,\lambda\,;\,p,q) \\ Z_2(x,\lambda\,;\,p,q) \end{pmatrix} \qquad (14)$$

be the solution to (3) on [0,1] such that

$$F_2(0,\lambda\,;\,p,q) = \begin{pmatrix} 0 \\ 1 \end{pmatrix} \qquad (15)$$

The zeroes of $Y_2(1,.\,;\,p,q)$ in $\mathbb{C}$ is a strictly increasing sequence of real numbers $(\upsilon_k(p,q))_{k\in\mathbf{Z}}$ and each $\upsilon_k(p,q)$ is a simple zero. Again $(\upsilon_k(p,q))_{k\in\mathbf{Z}}$ is the spectrum of a self adjoint operator in $L^2([0,1], \mathbb{C}^2)$ associated with the restriction of H to $H^1([0,1], \mathbb{C}^2)$ satisfying a Dirichlet boundary condition $(\upsilon_k(p,q))_{k\in\mathbf{Z}}$ is the third auxiliary spectrum and one has $\lambda_{2k-1}(p,q) \leq \upsilon_k(p,q) \leq \lambda_{2k}(p,q)$, $k \in \mathbf{Z}$.

For real-valued periodic potentials p and q, let $\gamma_k(p,q)$ be the k-th gap length , i.e.,

$$\gamma_k(p,q) = \lambda_{2k}(p,q) - \lambda_{2k-1}(p,q) \quad , \quad k \in \mathbf{Z} \qquad (16)$$

Let E^1 (resp E^2) be the set of potentials (p,q) in $\mathcal{H}^1$ (resp $\mathcal{H}^2$) such that

$$p(x) = -p(1 - x) \text{ and } q(x) = q(1 - x) \text{ for every x in } [0,1] \qquad (17)$$

The restriction of $\gamma_k(.,.)$ to E^1 is given by

$$\gamma_k(p,q) = |\mu_k(p,q) - \upsilon_k(p,q)|\,,\ (p,q) \in E^1 \qquad (18)$$

For $(p,q) \in E^1$. Let $\sigma_k(p,q)$ be the k-th signed gap length, i.e.,

$$\sigma_k(p,q) = \mu_k(p,q) - \upsilon_k(p,q) \ , \ k \in \mathbf{Z} \ , \qquad (19)$$

Now set

$$\sigma(p,q) = (\sigma_k(p,q))_{k \in \mathbf{Z}} \qquad (20)$$

σ is a map from E^1 (resp. E^2) into $l_1^2(\mathbf{Z})$ (resp $l_2^2(\mathbf{Z})$) σ is the so-called signed gap length map. σ is a real analytic map and one then has

Theorem 2

$d_{(p,q)}\sigma$ is an isomorphism from E^1 (resp E^2) onto $l_1^2(\mathbf{Z})$ (resp $l_2^2(\mathbf{Z})$).

Theorem 3

(i) If $(p,q) \in \mathcal{H}^1$, σ is one-to-one and is a local real analytic isomorphism at every point in $\mathcal{H}^1$.
(ii) If, furthermore, $(p,q) \in \mathcal{H}^2$, σ is onto · σ and σ^{-1} are real analytic maps.
To prove theorem 2 one uses functional analysis in Hilbert spaces in the spirit of [Pö - Tru] and [Gar - Tru 2]. As a corollary of theorem 2 ii) one gets the following result which is the first step characterizing all possible periodic spectra. Let $(l_2^2)^+$ be the set of sequences $(a_k)_{k \in \mathbf{Z}}$ in l_2^2 such that $a_k \geq 0$ for every k in .Z.

Theorem 4

Given any sequence $(\gamma_k)_{k\in \mathbf{Z}}$ in $(l_2^2)^+$ there exist (p,q) in E^2 such that

$$\gamma_k(p,q) = \gamma_k$$

for every k in $\mathbf{Z}$.

In other words the map $\gamma(p,q) = (\gamma_k(p,q))_{k \in \mathbf{Z}}$ from the set of potentials p and q in $\mathcal{H}^2$ to $(l_2^2)^+$ is onto. The following result is the second step towards characterizing all possible periodic spectra. It is shown that the periodic spectrum is completely determined by the sequence of its gap lengths.

Theorem 5

Let $(\gamma_k)_{k\in \mathbf{Z}}$ be anyt sequence in $(l_2^2)^+$. From theorem 4 one knows that there exist (p_0, q_0) in E^2 such that $\gamma_k(p_0,q_0) = \gamma_k$ for every k in $\mathbf{Z}$. Then the set of potentials p and q in $\mathcal{H}^2$ such that $\gamma_k(p,q) = \gamma_k$ for every k in $\mathbf{Z}$ is $L(p_0,q_0)$, i.e., the isospectral set of (p_0,q_0).

Following [Gar - Tru 1] and [Tru 2], one defines a tile of length γ as an open interval of length γ when γ is positive and a point when γ is zero. Theorem 5 shows that, given any sequence of tiles such that the sequence of their lengths is in $l_2^2(\mathbf{Z})$, there is an unique way of placing them on the real axis such that their endpoints are the periodic spectrum of potentials in $\mathcal{H}^2$.

The proof of theorems 1 - 5 is very long. A very important tool for proving these theorems are the so-called trace formulas :

Let $(T_x u)(t) = u(x+t)$

Recall that p and q are identified with their periodic extension to the whole real line.

One has for $(p,q) \in \mathcal{H}^1$ (resp $\mathcal{H}^2$) and $x \in [0,1]$.

$$q(x) = -\frac{1}{2}\sum_{k\in \mathbf{Z}} (\lambda_{2k} + \lambda_{2k-1} - 2\tilde{\mu}_k (T_x p, T_x q))$$

$$q(x) = \frac{1}{2}\sum_{k\in \mathbf{Z}} (\lambda_{2k} + \lambda_{2k-1} - 2\mu_k (T_x p, T_x q))$$

The series converge in $C([0,1])$ (resp $C^1([0,1])$) if $(p,q) \in \mathcal{H}^1$ (resp $\mathcal{H}^2$). One easily deduce from the trace formulas the generalisation of a theorem of Borg characterizing potentials.

Theorem 6

Suppose that $\lambda_{2k} = \lambda_{2k-1}$ for every $k \in \mathbb{Z}$. Then $p(x) = q(x) = 0$ for every x in $[0,1]$.

Finally we are able to describe the set of potentials p and q having only one open gap $\gamma_0 = \gamma > 0$:

Theorem 7

The set of potentials p(x) and q(x) such that $\gamma_k(p,q) = 0$ for every $k \neq 0$ and $\gamma_0(p,q) = \gamma > 0$ is the set of isospectral potentials such that $p(x) = p, q(x) = q$, for every $x \in [0,1]$ and $p^2 + q^2 = \frac{\gamma^2}{4}$.

Furthermore

$$\lambda_0 = -\lambda_{-1} = (p^2 + q^2)^{1/2}$$

$$\lambda_{2k} = \lambda_{2k-1} = (k^2\pi^2 + p^2 + q^2)^{1/2} \qquad k > 0$$

$$\lambda_{-2k} = \lambda_{-2k-1} = -(k^2\pi^2 + p^2 + q^2)^{1/2}$$

REFERENCES

[A.K.N.S.] :

M.J. Ablowitz, D.J. Kaup, A.C. Newell and H. Segur. The inverse scattering transform - Fourier analysis for non linear problems. Stud. Appl. Math. 53,1974, p.249-315.

[Fa - Takh] :

L.D. Faddeev and L.A. Takhtajan. "Hamiltonian methods in the theory of solitons". Springer . 1987.

[Ga - Tru 1] :

J. Garnett and E.T. Trubowitz. Gaps and brands of one dimensional Schrödinger operators. Comment. Math. Helvetici, 59 (1984), p.258-312.

[Ga - Tru 2] :

J. Garnett and E. Trubowitz.Gaps and bands of one dimensional periodic Schrödinger operators. Comment. Math. Helvetici. 62, (1987). p. 18 - 37.

[Ka] :

T. Kappeler. On the periodic spectrum of the one-dimensional Schrödinger equations (to be published in Comment. Math. Helvetici.

[Ma - Abl] :

Y.C. Ma and M.J. Ablowitz. The periodic cubic Schrödinger equation. Stud. Appl. Math. 65 , (1981), p.113 - 158.

[Mar] :

V.A. Marchenko "Sturm-Liouville operators and applications" Operator theory : Advances and Applications, vol 22, Birkhaüser, 1986.

[Mar - Ost 1] :

V.A. Marchenko and I.V. Ostrovsky. A characterization of the spectrum of Hill's operator. Math U.S.S.R. - Sbornik, 97 (1975), p.493 - 554.

[Mar - Ost 2]

V.A.. Marchenko and I.V. Ostrovsky. Approximation of periodic by finite zone potentials . Selecta Math. Sovietici 6, 1987, p.101 - 136.

[Mis.1] :

T.V. Misyura . Properties of the spectra of periodic and antiperiodic boundary value problems generated by Dirac operators I, II (in Russian). Teor. Funktsü Funktsional Anal. i Prilozhen.30 , (1978), p.90 -101 ; 33, (1980), p. 107 - 111 (in Russian).

[New] :

R.G. Newton "Scattering Theory of Waves and Particles" Mc Graw-Hill, 1966.

[PRE] :

E. Previato : Hyperelliptic quasi-periodic and solitons solutions of the non linear Schrödinger equation. Duke Math. J., 52, 1985 , p.329 -377.

[Pö - Tru] :

J. Pöschel and E. Trubowitz "Inverse Spectral Theory" Academic Press, 1987.

Combinatorial Aspects of Some Classes of Linearizable and Bilinearizable NLPDEs

F. Lambert and R. Willox

Dienst Theoretische Natuurkunde, Vrije Universiteit Brussel,
Pleinlaan 2, B-1050 Brussels, Belgium

1. Introduction.

An important intermediate step towards KdV-type integrability [1] of a nonlinear partial differential evolution equation (NLPDE) is the existence of two-soliton solutions. This property is known to hold for those equations which are expressible in the bilinear form [2]:

$$F(D_t, D_x) = 0\,, \tag{1}$$

where F is a polynomial satisfying the conditions F(-a,-b) = F(a,b) and F(0,0)=0, and where the D-operators are defined on a pair of functions by the rule:

$$D_t^n D_x^m f.g = (\partial_t - \partial_{t'})^n (\partial_x - \partial_{x'})^m \; f(x, t) g(x', t')|_{x'=x,\; t'=t}. \tag{2}$$

Despite the impressive success of the Hirota formalism in obtaining multisoliton solutions to a large number of equations - the method is also connected with other remarkable properties of soliton equations [3] - there are still some open questions about the precise conditions which underlie the *bilinearizability* of an equation and about the role of the seemingly artificial D-operators...

In this note we investigate the properties that an equation must have in order to be expressible in the Hirota-form (1). We show that the *bilinear two-soliton balance* between dispersion and nonlinearity which is associated with sech squared solitary waves can be characterized by simple combinatorics in close analogy with the conditions which underlie the Cole-Hopf linearizability of the Hopf-Burgers (HB) hierarchy [4]. This analogy is most striking when the equations are written in terms of the q-variable which corresponds to the *potential* from which the potential two-soliton solutions can be derived. The analysis gives a new insight into that part of the bilinear machinery which corresponds precisely to the dependent variable transformation which seemed originally motivated by the linearization of the Burgers equation [3]. The result is that both the *linearizable* and the *bilinearizable* equations that we consider are intimately related to the combinatorial problem of placing labeled objects into indistinguishable non-empty boxes. Other realizations of the two-soliton balance suggestive of more elaborate (but yet undisclosed) combinatorics are briefly mentioned.

2. Linearizable equations of the generalized BH-hierarchy.

It is well-known that the Burgers nonlinear diffusion equation :

$$V_t - V_{xx} + VV_x = 0 \tag{3}$$

belongs to the BH-hierarchy of NLPDE's which are linearized by the Hopf-Cole transformation:

$$V = -2\partial_x \ln f. \tag{4}$$

Setting $V = -2q_x$ it is worth noticing that the corresponding "potential equations" for the q-variable:

$$q_t - (q_{xx} + q_x^2) = 0, \tag{5}$$

$$q_t + (q_{xxx} + 3q_x q_{xx} + q_x^3) = 0 \tag{6}$$

...

which are linearized by the transformation $q = \ln f$, are all of the form:

$$q_t + \alpha[q_{mx} + Q_m(q_x, ...q_{(m-1)x})] = 0 \ , \qquad \alpha = \text{constant} \tag{7}$$

where $q_{mx} = \partial_x^m q$ and where the polynomial nonlinearity $Q_m(q_x, ... q_{(m-1)x})$ is simply related to the combinatorial problem of placing m labeled objects into non-empty indistinguishable boxes. Thus,

$$Q_m(q_x, ...q_{(m-1)x}) = \sum w_m(r_1, r_2, ..., r_{p-1})\, q_{r_1 x}\, q_{r_2 x} \cdots q_{r_p x} \ , \tag{8}$$

where $w_m(r_1, r_2, ... r_{p-1})$ denotes the combinatorial weight of the partition:

$$m = r_1 + r_2 + ... + r_p \quad , r_1 \le r_2 \le ... \le r_p \quad , r_i = \text{integer} < r \tag{9}$$

(i.e. the number of ways of placing m labeled objects into p indistinguishable boxes containing $r_1, r_2, ...$ and r_p objects), and where the sum is taken over all partitions of m involving two or more parts.

This combinatorial property follows from the identities [5]:

$$q_{mx} + Q_m(q_x, ...) \equiv \exp(-q)\partial_x^m \exp q \equiv \sum \left(c_1, c_2, ...c_m\right) \prod_{p=1}^{m} \left(\frac{q_{px}}{p!}\right)^{c_p} , \tag{10}$$

where the sum goes over all partitions of m that are written as follows:

$$m = c_1 + 2c_2 + ... + mc_m \quad , c_p = \text{number of parts equal to p.} \tag{11}$$

They imply that the various fractions which arise from $Q_m(q_x, ...)$ by setting $q = \ln f$ cancel all the fractional terms of $q_{mx} = \partial_x^m \ln f$, except the highest order term f_{mx}/f:

$$q_{mx} + Q_m(q_x, ...)|_{q=\ln f} = f^{-1}\, \partial_x^m f. \tag{12}$$

The polynomials $p_m = q_{mx} + Q_m(q_x, ... q_{(m-1)x})$, written in terms of the m variables $x_r = q_{rx}$, are known as the complete (exponential) Bell polynomials [5].

To any mixed derivative $q_{mx,nt} = \partial_t^n \partial_x^m q$, $m+n \ge 2$, there corresponds a polynomial $p_{m,n}(q_x, q_t, ...)$ which is similarly related to the problem of placing m + n labeled objects (m of type x and n of type t) into non-empty indistinguishable boxes:

$$p_{m,n} = \exp(-q)\partial_t^n\partial_x^m \exp q \equiv \sum (\prod_j p_j!)^{-1}\left(r_1,\overset{m}{\ldots} r_p\right)\left(s_1,\overset{n}{\ldots} s_p\right)\prod_{i=1}^{p} q_{r_i x, s_i t}\,, \quad (13)$$

the sum being now taken over all generalized partitions of m+n:

$$m+n = \sum_{i=1}^{p}(r_i + s_i) = \sum_j p_j\,(r_j + s_j) \text{ with } \sum_{i=1}^{p} r_i = \sum_j p_j\, r_j = m \text{ and } \sum_{i=1}^{p} s_i = \sum_j p_j\, s_j = n \quad (14)$$

where each part $r_i + s_i > 0$ corresponds to an ordered couple (r_i, s_i) ,r_i, s_i = integer or zero, and where the integer p_j denotes the multiplicity of each of the different couples (r_j, s_j) which appear in the set $\{(r_i, s_i), 1 \le i \le p)\}$.
The compensation between the linear term $q_{mx,nt}$ of $p_{m,n}$ and its nonlinear part $Q_{m,n}$ can be expressed by a relation which generalizes the rel.(12):

$$q_{mx,nt} + Q_{m,n}(q_x, q_t, \ldots)|_{q=\ln f} = f^{-1}\partial_t^n\partial_x^m f. \quad (15)$$

This compensation yields the following:

<u>Proposition 1.</u>
Any NLPDE of the form:

$$\alpha q_x + \bar{\alpha} q_t + \sum_i \alpha_i [q_{m_i x, n_i t} + Q_{m_i,n_i}(q_x, q_t, \ldots)] = 0, \quad m_i, n_i = \text{integer or zero} \quad (16)$$

α, $\bar{\alpha}$, α_i = constants, can be linearized by the transformation $q = \ln f$ into:

$$\alpha f_x + \bar{\alpha} f_t + \sum_i \alpha_i\, f_{m_i x, n_i t} = 0. \quad (17)$$

<u>Corollary.</u>
The generalized BH equations for the variable $V = -2q_x$ which are derived from the equations (16) have "degenerate" N-soliton potential solutions:

$$V_d^{(N)} = -2\partial_x \ln f_d^{(N)}\,, \qquad f_d^{(N)} = 1 + \sum_{j=1}^{N} \exp\theta_j\,, \qquad \theta_j = -k_j x + \omega_j t\,, \quad (18)$$

with

$$\bar{\alpha}\omega_j - \alpha k_j + \sum_i (-)^{m_i} \alpha_i\, k_j^{m_i}\, \omega_j^{n_i} = 0 \quad (19)$$

which generalize the "N-step solutions" of the Burgers equation (3).

<u>3. Bilinearizable equations of KdV-type.</u>

Let us now examine whether a combinatorial compensation between dispersion and nonlinearity does also characterize classes of NLPDE's which cannot be linearized, but which possess two-soliton solutions of the potential KdV-type:

$$V^{(2)} = -2\partial_x \ln f^{(2)}, \quad f^{(2)} = 1 + \exp\theta_1 + \exp\theta_2 + A_{12}\exp(\theta_1+\theta_2), \quad \theta_j = -k_j x + \omega_j t. \tag{20}$$

Setting $V = -q_x$ we look for corresponding NLPDE's for the potential variable $q(x,t)$ with solutions: $q^{(2)} = 2 \ln f^{(2)}$.

A hint at a possible rule for getting families of equations with such solutions can be obtained by rewriting some of the well-known (potential) soliton equations with solutions of type $V^{(2)}$, such as the potential KdV equation, the potential Boussinesq equation and the potential Caudrey-Dodd-Gibbon-Sawada-Kotera equation, in terms of the q-variable:

$$q_{x,t} + (q_{4x} + 3q_{2x}^2) = 0 \qquad \text{(PKdV)} \tag{21}$$

$$[q_{tt} - q_{xx} + (q_{4x} + 3q_{2x}^2)]_x = 0 \qquad \text{(PBq)} \tag{22}$$

$$q_{x,t} + (q_{6x} + 15q_{2x}\, q_{4x} + 15q_{2x}^3) = 0 \qquad \text{(PCDGSK)} \tag{23}$$

We observe that each of the linear dispersive terms q_{2mx}, $m \geq 2$, is "compensated " by a specific polynomial nonlinearity $\hat{Q}_{2m}(q_{2x}, \ldots)$ which involves only even-order derivatives of q, and which is simply related to the problem of placing 2m labeled objects into non-empty indistinguishable boxes under the condition that each box should contain an even number of objects:

$$q_{4x}: \quad 4 = 2+2, \quad w_4(2) = \frac{1}{2!}\binom{4}{2,2} = 3 \quad : \; 3q_{2x}^2 \tag{24}$$

$$q_{6x}: \quad 6 = 2+4, \quad w_6(2) = \binom{6}{2,4} = 15 \quad : \; 15q_{2x}q_{4x} \tag{25}$$

$$6 = 2+2+2, \quad w_6(2,2) = \frac{1}{3!}\binom{6}{2,2,2} = 15 \quad : \; 15q_{2x}^3 \tag{26}$$

Thus, with each even-order derivative q_{2mx}, we may associate a polynomial nonlinearity $Q_{2m}(q_{2x}, \ldots q_{(2m-2)x})$ which is determined by the partitions of 2m into even parts:

$$\hat{Q}_{2m}(q_{2x}, \ldots q_{(2m-2)x}) = \sum_{r_1+r_2+\ldots+r_p=m,\, p\geq 2} w_{2m}(2r_1, \ldots 2r_{p-1})\, q_{2r_1x}\, q_{2r_2x} \cdots q_{2r_px}. \tag{27}$$

We shall now see that the corresponding polynomial:

$$P_{2m}(q_{2x}, \ldots q_{2mx}) = q_{2mx} + \hat{Q}_{2m} \equiv \sum_{2c_2+\ldots+2mc_{2m}=2m} \binom{2m}{c_2, c_4, \ldots c_{2m}} \prod_{p=1}^{m} \left[\frac{q_{2px}}{(2p)!}\right]^{c_{2p}} \tag{28}$$

can be expressed in a compact form similar to that of the Bell polynomials (10). This form can easily be inferred from the analogy between the equations (21-23), with solutions $q^{(2)} = 2 \ln f^{(2)}$, and the linearizable equations (15) with solutions $q_d^{(2)} = \ln f_d^{(2)}$. We verify by direct substitution that:

$$(q_{4x} + 3q_{2x}^2)\big|_{q=2\ln f} = f^{-2}(2ff_{4x} - 8f_x f_{3x} + 6f_{2x}^2) \tag{29}$$

$$(q_{6x} + 15q_{2x}q_{4x} + 15q_{2x}^3)\big|_{q=2\ln f} = f^{-2}(2ff_{6x} - 12f_x f_{5x} + 30f_{2x}f_{4x} - 20f_{3x}^2). \tag{30}$$

The binomial structure of the quadratic forms at the r.h. side of equ.(29) and (30) suggests the following counterpart to the compensation property (12):

$$q_{2mx} + \hat{Q}_{2m}(q_{2x},...q_{(2m-2)x})|_{q=2\ln f} = f^{-2}B_{2m}(f.f) \tag{31}$$

with $$B_{2m}(f.f) = 2\sum_{p=0}^{m-1}(-)^p\binom{2m}{p}f_{px}f_{(2m-p)x} + (-)^m\binom{2m}{m}f_{mx}^2. \tag{32}$$

The r.h. side of equ.(32) is the binomial expansion of the expression:

$$B_{2m}(f.f) = f(\overleftarrow{\partial}_x - \overrightarrow{\partial}_x)^{2m}f, \tag{33}$$

where $\overrightarrow{\partial}_x$ and $\overleftarrow{\partial}_x$ denote x-differentiations acting to the right and to the left. The presumed compensation (31) corresponds, therefore, to the property:

$$P_{2m}(q_{2x},...q_{2mx}) = \exp(-q)[\exp(q/2)(\overleftarrow{\partial}_x - \overrightarrow{\partial}_x)^{2m}\exp(q/2)]. \tag{34}$$

A straightforward (but tedious) proof of this formula, based on standard combinatorics, can be obtained by relating P_{2m} to the Bell polynomials:

$$P_{2m}(2q_{2x},...2q_{2mx}) = 2[p_{2m} + \sum_{r=1}^{m}(-)^r w_{2m}(r)p_r p_{2m-r}]. \tag{35}$$

A shorter proof was recently given [6] by generalizing the Faà di Bruno formula [5].

At this stage it is obvious that the D-operators (2) are precisely the tool needed for expressing the full analogy between the formulas (31) and (34) and the corresponding formulas (12) and of sect.2:

$$P_{2m}(q_{2x},...q_{2mx}) = \exp(-q)D_x^{2m}\exp(q/2).\exp(q/2). \tag{36}$$

$$q_{2mx} + \hat{Q}_{2m}(q_{2x}, ...q_{(2m-2)x})|_{q=\ln(f^2)} = f^{-2}\, D_x^{2m}\, f.f \tag{37}$$

We may also generalize the polynomials P_{2m}, in the same way as we generalized the Bell polynomials (10), by considering the "bilinear polynomials":

$$P_{m,n}(q_x, q_t,...) = \exp(-q)D_t^n D_x^n \exp(q/2).\exp(q/2), \quad m+n = \text{even}, \tag{38}$$

which are similarly related to the problem of placing m+n = 2r objects - m of type x and n of type t - into non-empty indistinguishable boxes under the condition that each box should contain an even number of objects:

$$P_{m,n} \equiv q_{mx,nt} + \hat{Q}_{m,n} \equiv \sum (\prod_j p_j!)^{-1}\binom{m}{r_1,...r_p}\binom{n}{s_1,...s_p}\prod_{i=1}^{p} q_{r_ix,s_it}, \tag{39}$$

where the sum goes over the generalized partitions (14) of m+n = 2r into even parts $r_i + s_i$.

This yields the following counterpart to the proposition 1:

Proposition 2.

Any NLPDE of the form: $\sum_i \alpha_i P_{m_i,n_i}(q_x, q_t, ...) = 0$, with $m_i + n_i$ = even, and α_i= constant, can be transformed into the bilinear equation:

$$\sum_i \alpha_i D_t^{n_i} D_x^{m_i} f.f = 0 \tag{40}$$

by the transformation: $$q = \ln(f^2) \tag{41}$$

Corollary.

The bilinearizable equations for the variable $V = -q_x$ which are derived from the equations (40) have two-soliton potential solutions of the PKdV-type (20). The proof of this corollary follows from the Hirota theorem [2].

The class (40) comprises, among many other bilinearizable equations, the following m-families:

i) $$q_{xt} + \alpha q_{2x} + [q_{2mx} + \hat{Q}_{2m}(q_{2x}, ...q_{(2m-2)x})] = 0 \tag{42}$$

which may be rewritten in the variable $V = -q_x$ as the "bilinearizable equations of higher PKdV-type":

$$V_t + \alpha V_x + V_{(2m-1)x} = \sum_{e_1+e_2+...+e_p=2m} w_{2m}(e_1,...e_{p-1}) \prod_{i=1}^{p} (-V_{(e_i-1)x}), \tag{43}$$

with e_i = even integer.

ii) $$q_{2t} - q_{2x} + [q_{2mx} + \hat{Q}_{2m}(q_{2x},...q_{(2m-2)x})] = 0 \tag{44}$$

which are the potential equations from which one may derive the "bilinearizable equations of higher PBq-type":

$$V_{2t} - V_{2x} + V_{2mx} = \Big[\sum_{e_1+e_2+...+e_p=2m} w_{2m}(e_1,...e_{p-1}) \prod_{i=1}^{p} (-V_{(e_i-1)x})\Big]_x \tag{45}$$

Each of these equations for V(x,t) has two-soliton potential solutions (20) with a two-soliton coupling factor A_{12} which is expressible in terms of the "dispersion polynomial" $\phi(k,\omega) = \exp(kx - \omega t)\, L\, \exp(-kx + \omega t)$, where $L(V)$ stands for the linear part of the equation:

$$A_{12} = -\frac{\phi(k_1 - k_2, \omega_1 - \omega_2)}{\phi(k_1 + k_2, \omega_1 + \omega_2)}\Big|_{\phi(k_j, \omega_j) = 0,\ j=1,2}. \tag{46}$$

4. Other two-soliton generating nonlinearities.

The "bilinear compensation" between dispersion and nonlinearity:

$$q_{mx,nt} + \hat{Q}_{m,n}(q_x, q_t, ...)|_{q = \ln(f^2)} = f^{-2} D_t^n D_x^m f.f, \quad m+n = \text{even} \tag{47}$$

is but one particular recipe for realizing a "two-soliton balance" which satisfies a linear superposition principle. It is the simplest possible way of producing two-soliton solutions which, through the phaseshift parameter A_{12}, depend specifically on the dispersive properties of the linear terms in the equation.

To each equation of the family (43) there corresponds also a potential soliton equation, with linear part $L_m(V) = V_t + \alpha V_x + V_{(2m-1)x}$, which belongs to the (potential) KdV-hierarchy [3] and which has potential two-soliton solutions (20) with the KdV-coupling $A_{12}^{(KdV)} = (k_1-k_2)^2(k_1+k_2)^{-2}$.
The simplest members of this hierarchy for q(x,t) are the PKdV equ.(21) and the following alternative to the PCDGSK equation (23):

$$q_{xt} + (q_{6x} + 10q_{2x}q_{4x} + 5q_{3x}^2 + 10q_{2x}^3) = 0 \qquad \text{(PKdV5)} \qquad (48)$$

A further (lacunar) hierarchy of integrable NLPDE's, associated with $L_m(V)$, are the higher PCDGSK equations [7], with $m \geq 3$ and $2m-1 \neq$ multiple of 3, which have two-soliton solutions with the CDGSK coupling:

$$A_{12}^{(CDGSK)} = A_{12}^{(KdV)} (k_1^2 - k_1k_2 + k_2^2)\,(k_1^2 + k_1k_2 + k_2^2)^{-1} \qquad (49)$$

The two first members of this hierarchy for q(x,t) are the equ.(23) and the $PCDGSK_7$ equation:

$$q_{xt}+(q_{8x}+21q_{2x}q_{6x}+21q_{4x}^2+21q_{3x}q_{5x}+126q_{2x}^2q_{4x}+63q_{2x}q_{3x}^2+63q_{2x}^4) = 0 \qquad (50)$$

It is worth noticing that the nonlinearities which appear in these hierarchies include polynomial terms in $q_{2x}, q_{3x}, \ldots q_{(2m-2)x}$ which correspond to all possible partitions of 2m involving odd parts as well as even parts, with the exclusion of parts equal to 1. The coefficients of these terms (which are no longer equal to the combinatorial weight of the partition) suggest that possible extensions of the bilinear formalism (such as the one used for writing the fifth-order KdV equation in bilinear form) should also correspond to some combinatorics. This question is under investigation, together with the problem of connecting the various possible two-soliton generating nonlinearities with necessary two-soliton conditions [8].
Finally, it is striking to compare the following two equations which correspond to two different realizations of the two-soliton balance with the same type of terms:

$$V_{tt} - V_{6x} + 15V_xV_{4x} + 15V_{2x}V_{3x} - 45V_x^2V_{2x} = 0 \qquad (51)$$

$$V_{tt} - V_{6x} + 12V_xV_{4x} + 18V_{2x}V_{3x} - 36V_x^2V_{2x} = 0 \qquad (52)$$

The former one belongs to the bilinearizable equations of higher PBq type (45) and has two-soliton potential solutions (20) with Boussinesq two-soliton coupling: $A_{12}^{(Bq)}$

$$A_{12}^{(Bq)} = A_{12}^{(KdV)}(2k_1^2 - k_1k_2 + 2k_2^2)\,(2k_1^2 + k_1k_2 + 2k_2^2)^{-1} \qquad (53)$$

The latter equation (52) can be derived from the PKdV equation $V_t + V_{3x} = 3V_x^2$ and has therefore all its solutions, including two-soliton potential solutions (20) with KdV coupling: A_{12}^{KdV}.

References.

1. Hietarinta J. (1987) "A search for bilinear equations passing Hirota's three-soliton condition. I. KdV-type bilinear equations", J. Math. Phys. 28 , 1732.
2. Matsuno Y. (1984) "Bilinear transformation method", Acad. Press, Orlando, 14.
3. Newell A.C. (1985) "Solitons in mathematics and physics", SIAM, Philadelphia, ch.4.
4. Choodnovsky D.V., Choodnovsky G.V (1977) "Pole expansions of nonlinear partial differential equations", Nuovo Cimento 40 B , 339.
5. Comtet L. (1974) "Advanced combinatorics", D. Reidel, Dordrecht, 133.
6. Tu G.Z. (1989) "A combinatorial formula relating to Hirota's bilinear equations", preprint, to appear in J. Discr. Math.
7. Weiss J. (1984) "On classes of integrable systems and the Painlevé property", J. Math. Phys. 25 , 13.
8. Lambert F. and Willox R. (1989) "Standard nonlinearities associated with KdV-like two-soliton solutions, to appear in Proc. of the Int. Conf. on Nonlinear Phys., Shanghai, Springer (Heidelberg).

An Integrable Model for Small Amplitude Nonlinear Coupled Waves Produced During Laser-Plasma Interaction

A. Latifi and J. Leon

Laboratoire de Physique Mathématique, USTL,
F-34095 Montpellier Cédex 5, France
Equipe de Recherche Associée au CNRS, No. URA 040768

ABSTRACT : To describe the dynamics of the interaction of a (small amplitude) laser beam with a plasma in the fluid approximation, taking into account the stimulated Brillouin back-scattering, the broadening of the pump wave and the ponderomotive effects (nonlinear coupling), we build a one-dimensional model by means of multiscale expansions. The electronic density (ion acoustic wave) is nonlinearily coupled to the electrostatic incoming and outgoing waves through the ponderomotive force. The model derived is shown to be integrable and to lead to a continuous transfer of energy from the laser beam to the ion-acoustic wave.

I - INTRODUCTION

Evolution problems resulting from the coupling of high frequency waves (plasma waves) to low frequency waves (ion-acoustic waves) in a plasma have been widely studied in recent years [1]. On a physical level the main consequence of this coupling is to make possible an energy transfer from long wave length domain to the dissipative short wave length range [2].

A first progress in finding an integrable model describing the nonlinear coupling has been achieved by KARPMAN [3] who obtained an equation later generalized by KAUP [4].

We derive and study here a one dimensional model for the propagation of a **wave packet** of electrostatic (Langmuir) polarized waves in a uniform warm-electrons/cold-ions plasma, including altogether nonlinear coupling, nonlinear dynamical effects and dispersion. After rescaling , the system reads in dimensionless variables T and X :

$$\left\{\begin{array}{ll} -iq_T+\frac{1}{2}q_{XX}-q|q|^2 = i\,\frac{\omega_0 K_e}{6K_s}\displaystyle\int_{-\infty}^{+\infty} d\zeta\; a_1(\zeta,X,T)\; a_2^*(\zeta,X,T) & \text{(I.a)} \\ a_{1,X}+i\zeta a_1 = q\,a_2 & \text{(I.b)} \\ a_{2,X}-i\zeta a_2 = q^* a_1 & \text{(I.c)} \end{array}\right.$$

We shall see in section II that q is related to the fractional change in the plasma density and, a_1 and a_2 are respectively proportional (after scaling) to the amplitude of the pump wave-packet and the wave-packet scattered through stimulated Brillouin back-scattering (SBS). The coupling term in the right hand side of (I.a) takes its origin in the ponderomotive force and the integral comes from the fact that we consider a wave **packet** instead of a monochromatic wave. Finally the parameter ζ is proportional to the wave number.

It is worth remarking that, contrary to what happens for the decoupled models, all constants cannot be scaled away and it is very surprising that all the coefficients that appear in (I) have precisely the value for which the system is integrable ! Moreover we have shown [5] that, neglecting SBS phenomena would lead to a system "parametrically non integrable" : the Korteveg-de-Vries equation coupled to the Shroedinger eigenvalue problem through the ponderomotive term.

In section III we establish the spectral transform for the system (I), making use of the general formalism introduced in [12] and further developped in [13]÷[16], based on the $\bar{\partial}$-approach of integrable nonlinear evolution. The nonlinear coupled system (I) corresponds, in the spectral transform scheme, to a **singular dispersion relation** [11]÷[16].

One consequence of the integrability is the existence of an infinite sequence of **conservation laws**. This is also true for singular dispersion relations but these conservation laws do not lead to **conserved quantities**, We establish the conservation laws related to (I) in section IV and obtain the explicit time-dependence of the quantities which would be conserved for $a_1 = a_2 = 0$ in (I). We may then describe the transfer of energy from the laser pump wave to the acoustic wave.

We finally mention that the system (I) is of general interest in physics as soon as it is necessary to deal with **nonlinear interaction between different types of waves**. For instance, a similar system of equation has been recently used to explain the reduction of reflectivity in SBS [17].

II - THE MODEL

We consider a linearily polarized electrostatic ($\vec{B}=0$) field vector $\vec{E}=(0,0,E(z,t))$, propagating along the z direction with a frequency close to plasma frequency ω_0. Moreover we take into account the presence of some broadening of the frequency of the pump wave (with linewidth $\Delta(\omega)$ centered on ω_0) by writing E as the following wave packet (we work throughout with complex quantities) :

$$E(z,t) = \int_{-\infty}^{-\infty} d\omega\ \Delta(\omega)\ \tilde{E}(\omega,z,\tau)\ e^{-i\omega t} \tag{II.1}$$

where τ denotes a slow time (see (II.5) below).

In order to write down the equations of the plasma fluid approximation, we first evaluate the low-frequency effect of the high frequency field E in (II.1) on the electrons (ponderomotive force). The first order variation of the field E around the average position z is $\delta z\frac{\partial E}{\partial z}$ where δz is the (complex) displacement due to the high-frequency electrostatic field : $m_e(\delta z)_{tt} = eE$. This last equation, gives after integration (the effects of the slow variation of $\tilde{E}$ are considered separately)

$$\delta z = -\frac{e}{m_e}\int_{-\infty}^{+\infty} d\omega\ \frac{\Delta(\omega)}{\omega^2}\ \tilde{E}(\omega,z,\tau)\ e^{-i\omega t} \tag{II.2}$$

The resulting pondromotive force is obtained by considering the low frequency terms of $\mathrm{Re}\left[z\frac{\partial E}{\partial z} \right]$ around the average position z

$$f_p = -\frac{e^2}{2m_e}\frac{\partial}{\partial z}\int_{-\infty}^{+\infty} d\omega\ \omega^{-2}\ \Delta^2(\omega)\ |\tilde{E}(\omega,z,\tau)|^2 \tag{II.3}$$

The small parameter is the ratio of the electron and the ion masses, more precisely we set

$$\varepsilon_1 = \left(\frac{m_e}{m_1} \right)^{\frac{1}{2}} \tag{II.4}$$

and we make the following change of variables [6]÷[8]

$$\left.\begin{aligned} t &\longrightarrow t' = \varepsilon_1 \omega_0 t \longrightarrow \tau = \varepsilon_1^{\frac{3}{2}}\ t' \\ z &\longrightarrow z' = \frac{1}{\lambda_D} z \longrightarrow \xi = \varepsilon_1^{\frac{1}{2}}\ (z'-t') \end{aligned}\right\} \tag{II.5}$$

where λ_D is the electron Debeye wave lengh

$$\lambda_D^2 = \frac{K_B T_e}{4\pi n_0 e^2} \tag{II.6}$$

$n_0 e$ is the average charge density and ω_0 is the characteristic frequency [18]

$$\omega_0^2 = 4\pi e^2 \frac{n_0}{m_e} \quad . \tag{II.7}$$

We redefine the fields as

$$\tilde{E}' = \Delta(\omega)\tilde{E}\left(\frac{e^2}{2m_e\omega^2 K_B T_e}\right)^{\frac{1}{2}}, \quad \varphi' = \frac{e}{K_B T_e}\varphi, \quad v_i' = v_i\left(\frac{m_i}{k_B T_e}\right)^{\frac{1}{2}} \tag{II.8}$$

where φ is the electrostatic potential and v_i the ions velocity. Writing $n_e = n_0(1+q_e)$ and $n_i = n_0(1+q_i)$, the momentum transfer equation for the electrons reads [18]

$$\frac{\partial\varphi'}{\partial z'} - \frac{1}{1+q_e}\frac{\partial q_e}{\partial z'} - \frac{\partial}{\partial z'}\int_{-\infty}^{-\infty} d\omega \; |\tilde{E}'(\omega,z',\tau)|^2 = 0 \tag{II.9}$$

and for the ions

$$\frac{\partial v_i'}{\partial t'} + v_i'\frac{\partial v_i'}{\partial z'} = -\frac{\partial\varphi'}{\partial z'} \tag{II.10}$$

The above system is completed with the continuity equation

$$\frac{\partial q_i}{\partial t'} + \frac{\partial}{\partial z'}\left((1+q_i)v_i'\right) = 0 \tag{II.11}$$

and the Poisson equation

$$\frac{\partial^2\varphi'}{\partial z'^2} = q_e - q_i \tag{II.12}$$

Then after integration of (II.9) we go to the variables (ξ,τ) and expand all functions in power of ε_1 as

$$\varphi' = \varepsilon_1\varphi^{(1)} + \varepsilon_1^2\,\varphi^{(2)} + O(\varepsilon_1^3) \tag{II.13.a}$$

for $\{\varphi', q_e, q_i, v_i\}$, and

$$\tilde{E}' = \varepsilon_1^{\frac{1}{2}}\left(\mathcal{E} + O(\varepsilon_1)\right) . \tag{II.13.b}$$

The system (II.9)÷(II.12) then gives at order ε_1

$$r \doteq q_e^{(1)} = q_i^{(1)} = \varphi^{(1)} = v_i^{(1)} \tag{II.14}$$

and at order ε_1^2, we obtain by eliminating $q_i^{(2)}$, $q_e^{(2)}$, $v_i^{(2)}$ and $\varphi^{(2)}$ the following evolution for r :

$$2r_\tau + 2rr_\xi + r_{\xi\xi\xi} = -\frac{\partial}{\partial\xi}\int_{-\infty}^{+\infty} d\omega' \; |\mathcal{E}|^2 \tag{II.15}$$

On the other hand, the plasma wave has the following dispersion relation [18]

$$\omega^2 = \omega_p^{\ 2} + 3V_{Te}^{\ 2}\, K^2 \qquad \text{(II.16)}$$

where V_{Te} is the termal electron velocity and where ω_p stands for the plasma frequency (of the electrons)

$$\omega_p^{\ 2} = \omega_0^{\ 2}\,(1+q_e) \qquad \text{(II.17)}$$

with ω_0 defined in (II.7).

The fractional change q_e in the electron density of average value n_0 is function of space and slowly varies in time. The Maxwell's equation resulting from (II.16) reads

$$\left(\frac{\partial^2}{\partial t^2} - 3V_{Te}^{\ 2}\,\frac{\partial^2}{\partial z^2}\right) E(z,t) = -\,\omega_0^{\ 2}\left(1+q_e(z,\tau)\right) E(z,t) \qquad \text{(II.18)}$$

in which τ represents the slow (scaled) time defined in (II.5). Using the scalings (II.5) and (II.13) the equation (II.18) gives at first order (use $V_{Te}=\lambda_D\omega_0$)

$$\frac{\partial^2}{\partial \xi^2}\,\mathcal{E} + (\nu^2 - \frac{r}{3})\,\mathcal{E} = 0 \qquad \text{(II.19)}$$

where the parameter ν defines the relative difference between the frequency ω of the applied laser beam and the plasma characteristic frequency ω_0 :

$$\frac{\omega^2-\omega_0^{\ 2}}{3\omega_0^{\ 2}} = \varepsilon_1\,\nu^2 \qquad \text{(II.20)}$$

We have shown in [5] that the system (II.19), (II.15) representing the nonlinear prapagation of an electrostatic wave coupled to ion acoustic wave when ignoring SBS is not integrable.

To take SBS into account we allow the electric field envelope $\mathcal{E}$ to contain at first order a left-going wave $\exp\,[i(\omega_1\tau + K_1\xi)]$ and a right going one $\exp\,[i(\omega_2\tau + K_2\xi)]$ with small slowly varying amplitudes $\varepsilon_2 a_1(\xi,\tau)$ and $\varepsilon_2 a_2(\xi,\tau)$ respectively. The wave numbers moduli are close to the same average value K_e namely

$$K_1 = K_e + O(\varepsilon_2) \quad , \quad K_2 = -K_e + O(\varepsilon_2) \qquad \text{(II.21)}$$

Then we have

$$\mathcal{E}(z,\tau)=\varepsilon_2\left(a_1\, e^{i(\omega_1\tau+ K_1 z)} + a_2\, e^{i(\omega_2\tau+ K_2 z)}\right) + O(\varepsilon_2^{\ 2}) \qquad \text{(II.22)}$$

and the frequencies are related through

$$\omega_1 = \omega_2 + \omega_s \quad , \quad K_1 = K_2 + K_s \tag{II.23}$$

ω_s and K_s stand for the ion sound wave parameters and the dispersion relation is in the frame (ξ,τ), $\omega_s = \frac{1}{2} K_s^3$. We consider also r to be a superposition of its harmonic modes expanded in power of ε_2 as the formal series

$$r(\xi,\tau) = \varepsilon_2^2\, q_0(\xi,\tau) + \sum_{n=1}^{N} \varepsilon_2^n . e^{in(\omega_s\tau + K_s\xi)} \sum_{j=0}^{M} \left[\varepsilon_2^j\, q_{nj}(X,T) + O(\varepsilon_2^{M+1})\right] +$$

$$+ \text{c.c.} + O(\varepsilon_2^{N+1}) \tag{II.24}$$

where X, T are the new variables defined by

$$X = \varepsilon_2\, (\xi + \frac{3}{2} K_s \tau) \qquad T = 3\varepsilon_2^2\, K_s \tau \tag{II.25}$$

The NLS equation is obtained by examining of the leading orders with N = 2 and M ≤ 3 .

It is worth remarking that the KdV and NLS scalings have to be consistent which is guaranteed by the constraint on ε_2 :

$$\varepsilon_1 \ll \varepsilon_2^2 \ll \varepsilon_2 \ll 1 \tag{II.26}$$

Substituting (II.22), (II.23) and (II.24) in (II.15), we obtain the defocusing NLS with a coupling term (I.a), and from (II.19) we obtain the Zakharov-Shabat equation (I.b,c) where ξ is defined by

$$\frac{K_e^2 - \nu^2}{2K_e} = \varepsilon_2\, \xi + O(\varepsilon_2^2) \tag{II.27}$$

The left-going wave of amplitude $a_1(X,T)$ is the incident (pump) laser beam of given amplitude, say $A(\zeta)$ proportional to the linewidth Δ in (II.1). The right-going wave of amplitude $a_2(X,T)$ is the wave reflected by SBS. To the above system of equation we therefore add the following requirements on the boundary values

$$\left.\begin{array}{llll} a_1(\zeta,X,T) \longrightarrow A(\zeta)\, e^{-i\zeta X} & , & X \longrightarrow +\infty \\ a_2(\zeta,X,T) \longrightarrow 0 & , & X \longrightarrow -\infty \end{array}\right\} \tag{II.28}$$

We will show in the following section that, for a given initial datum $q(X,0)$ in $L^1(\mathbb{R})$ and the boundary values (II.28) we are able to reduce the solution of the nonlinear system (I) to that of a linear integral equation. In particular we show that the normalized reflection coefficient α and transmission coefficient β, defined by

$$\left.\begin{array}{lll} a_1(\zeta,X,T) \longrightarrow \beta(\zeta,T)A(\zeta)e^{-i\zeta X} & , & X \longrightarrow -\infty \\ a_2(\zeta,X,T) \longrightarrow \alpha(\zeta,T)A(\zeta)e^{i\zeta X} & , & X \longrightarrow +\infty \end{array}\right\} \qquad \text{(II.29)}$$

can be computed for all time from $q(X,0)$ and $A(\zeta)$ only by solving a system of Volterra equations. In short, the system (I) is integrable.

III - SINGULAR GENERAL EVOLUTION RELATED TO THE ZAKHAROV-SHABAT SPECTRAL PROBLEM

We show in this section that the system (I) is integrable in the scheme of the spectral transform with **nonanalytic dispersion relation.**

Because we shall see later that the usual scattering theory is useless, we use here as a basic tool the $\bar{\partial}$-formulation of the spectral transform.

We start with the $\bar{\partial}$-problem

$$\frac{\partial}{\partial \bar{k}} \Phi(k) = \Phi(k)\, R(k) \qquad , \quad k \in \mathbb{C} \qquad \text{(III.1)}$$

$$\Phi(k) = 1 + O(\tfrac{1}{k}) \qquad , \quad |k| \longrightarrow \infty. \qquad \text{(III.2)}$$

A parametric (x, t) dependence of Φ is given by [20]

$$R_x = [\, R \,,\, \Lambda \,] \qquad \text{(III.3)}$$

$$R_t = [\, R \,,\, \Omega \,] \qquad \text{(III.4)}$$

Where the matrices Λ and Ω are given distributions of $k \in \mathbb{C}$, functions of x, $t \in \mathbb{R}$.

In order to obtain the general nonlinear evolution equation, we deduce from the above equations the following relations

$$\frac{\partial}{\partial \bar{k}} (\Phi_x \Phi^{-1} - \Phi\, \Lambda\, \Phi^{-1}) = -\, \Phi \frac{\partial \Lambda}{\partial \bar{k}} \Phi^{-1} \qquad \text{(III.5)}$$

$$\frac{\partial}{\partial \bar{k}} (\Phi_t \Phi^{-1} - \Phi\, \Omega\, \Phi^{-1}) = -\, \Phi \frac{\partial \Omega}{\partial \bar{k}} \Phi^{-1} \quad . \qquad \text{(III.6)}$$

After integration, (III.5) and (III.6) give

$$\Phi_x = U\, \Phi + \Phi\, \Lambda - \frac{1}{2i\pi} \left(\iint_{\mathbb{C}} \frac{d\lambda \wedge d\bar{\lambda}}{\lambda - k}\, \Phi \frac{\partial \Lambda}{\partial \bar{\lambda}} \Phi^{-1} \right) \Phi \qquad \text{(III.7)}$$

$$\Phi_t = V\, \Phi + \Phi\, \Omega - \frac{1}{2i\pi} \left(\iint_{\mathbb{C}} \frac{d\lambda \wedge d\bar{\lambda}}{\lambda - k}\, \Phi \frac{\partial \Omega}{\partial \bar{\lambda}} \Phi^{-1} \right) \Phi \qquad \text{(III.8)}$$

where $U(k,x,t)$ and $V(k,x,t)$ are polynomials in k of same order as the polynomial part of $\Omega(k,x,t)$ (by identification of the successive positive orders of k in (III.7) and (III.8)) [20].

When

$$\Lambda = i\,k\,\sigma_3 \tag{III.9}$$

The relation (III.7) becomes the Zakharov-Shabat spectral equation :

$$\Phi_x = -ik\,[\sigma_3\,,\,\Phi] + Q\,\Phi \tag{III.10}$$

Where Q is a off-diagonal matrix.

A polynomial form (in k) of Ω leads to the usual AKNS hierarchy. The addition of a nonalytical part to this polynomial form introduces a coupling between the field and its corresponding eigenfunction. We choose here

$$\Omega(k,t) = -ik^2\sigma_3 + \frac{1}{2i\pi}\iint_{\mathbb{C}} \frac{d\lambda\wedge d\bar{\lambda}}{\lambda-k}\,\frac{\partial\Omega(\lambda)}{\partial\bar{\lambda}} \quad , \tag{III.11}$$

(the monome $ik^2\sigma_3$ gives NLS).

The general evolution equation is given by the compatibility condition between the pair of ODE (III.7) and (III.8). The particular choices (III.9) for Λ and (III.11) for Ω lead to

$$Q_t - V_x + [-ik\sigma_3 + Q\,,\,V] = i\left[\sigma_3\,,\,\frac{1}{2i\pi}\iint_{\mathbb{C}} d\lambda\wedge d\bar{\lambda}\;\Phi\,\frac{\partial\Omega(\lambda)}{\partial\bar{\lambda}}\,\Phi^{-1}\right] \tag{III.12}$$

The link between the pair of coupled equations (III.10), (III.12) and the physical coupled system (I) is realised through the following particular choices :

1- the reduction

$$Q = \begin{pmatrix} 0 & q \\ q^* & 0 \end{pmatrix} \tag{III.13}$$

2- the choice

$$\frac{\partial\Omega}{\partial\bar{k}} = g(k)\,\sigma_3 \quad , \qquad g(k) = \frac{i\pi}{24}\,\frac{\omega_0 k_e}{k_s}\,\delta(k_I)\,|A(k_R)|^2 \tag{III.14}$$

3- the relation between Φ and the electrostatic wave (see (II.28))

$$\begin{pmatrix} a_1 \\ a_2 \end{pmatrix} = A\begin{pmatrix} \Phi_{11} \\ \Phi_{21} \end{pmatrix} e^{-ikx} \tag{III.15}$$

Here Φ_{11} and Φ_{21} represent the first column elements of the matrix Φ, with the following behaviours (see (II.28) and (II.29))

$$e^{-ikx}\begin{pmatrix}\Phi_{11}\\ \Phi_{21}\end{pmatrix} \xrightarrow[x\to-\infty]{} \begin{pmatrix}\beta\ e^{\cdots\cdots}\\ 0\end{pmatrix} \begin{matrix}\text{transmitted wave going to } -\infty\\ \text{no pump wave coming from } -\infty\end{matrix}$$

$$e^{-ikx}\begin{pmatrix}\Phi_{11}\\ \Phi_{21}\end{pmatrix} \xrightarrow[x\to+\infty]{} \begin{pmatrix}e^{-ikx}\\ \alpha\ e^{ikx}\end{pmatrix} \begin{matrix}\text{pump wave coming from } +\infty\\ \text{reflected wave going to } +\infty\end{matrix} \qquad \text{(III.16)}$$

The reduction (III.13) implies

$$\sigma_1\ \overline{\Phi(\bar{k})}\ \sigma_1 = \Phi(k) \qquad \text{(III.17)}$$

which relates the first column of Φ to its second column. Consequently [20]

$$R(k,x,t)= \frac{i}{2} \lim_{\varepsilon\to 0^+} \begin{pmatrix} 0 & -\alpha^*(k^*,t)\delta(k_I+\varepsilon)\\ \alpha(k,t)\delta(k_I-\varepsilon) & 0\end{pmatrix} \exp[2ik\sigma_3 x] \qquad \text{(III.18)}$$

where α is the reflection coefficient (see (II.29)).

Then using (III.14) and (III.11) we obtain from (III.4) :

$$\alpha_t(k_R+i0,t)= -2i\left[k_R^{\,2}+ \frac{1}{4}\,\frac{\omega_0}{12}\int_{-\infty}^{+\infty}\frac{d\lambda}{\lambda-(k_R+i0)}\,|A(\lambda)|^2\right]\ \alpha(k_R+i0,t) \qquad \text{(III.19)}$$

$$\bar{\alpha}_t(k_R+i0,t)= 2i\left[k_R^{\,2}+ \frac{1}{4}\,\frac{\omega_0}{12}\int_{-\infty}^{+\infty}\frac{d\lambda}{\lambda-(k_R+i0)}\,|A(\lambda)|^2\right]\ \bar{\alpha}(k_R+i0,t) \qquad \text{(III.20)}$$

From now on we set $\alpha(k_R) = \alpha(k_R+i0)$; $\bar{\alpha}(k_R) = \alpha^*(k_R-i0)$.

Using the Plemelj-Sokhotski formula we obtain

$$\alpha(k_R,t) = \alpha(k_R,0)\ e^{-i\Theta t}\ e^{\rho t} \quad , \qquad \text{(III.21)}$$

$$\bar{\alpha}(k_R,t) = \bar{\alpha}(k_R,0)\ e^{i\Theta t}\ e^{-\rho t} \quad , \qquad \text{(III.22)}$$

with (the slashed integral denotes the Cauchy principal value)

$$\Theta = 2k_R^{\,2} + \frac{\omega_0}{24} \fint_{-\infty}^{+\infty}\frac{d\lambda}{\lambda-k_R}\ |A(\lambda)|^2 \quad , \qquad \text{(III.23)}$$

$$\rho = \pi\ \frac{\omega_0}{24}\ |A(k_R)|^2 \quad . \qquad \text{(III.24)}$$

We end this section by remarking that the use of the $\bar{\partial}$-formulation (III.1) for the analytical properties of the solution Φ of the Zakharov-Shabat system (III.10) is essential here. Indeed it is due to the particular structure

(III.18) of R(k,x,t) (see in particular the two δ functions) that we obtain the evolutions (III.21) and (III.22) which do not violate the unitarity relation $\alpha\bar{\alpha} + \beta\bar{\beta} = 1$.

IV - CONSERVATION LAWS

It is well known that an integrable evolution has an infinite sequence of conservation laws when the dispersion relation is polynomial in the spectral parameter k [22]. Moreover when the field has vanishing asymptotic behaviours, the conservation laws give rise to infinitely many conserved quantities. This is not necessarily true when the dispersion relation is not everywhere analytic.

In order to obtain an infinite sequence of conservation laws related to (I) we apply the method of [21]. In our case the recurrent form of the conservation laws is given by equating to zero the coefficients of k^{-n} successively for n = 1, 2, ... (for more details see [19])

$$\partial_t\left\{\sum_{n=1}^{\infty}\frac{1}{(2ik)^n}(qZ_n)\right\} = \partial_x\left\{\frac{i}{2}\sum_{n=1}^{\infty}\frac{1}{(2ik)^{n-1}}(qZ_n) - \frac{i}{2}\sum_{n=1}^{\infty}\frac{1}{(2ik)^n}(q_xZ_n) - \right.$$
$$-\frac{1}{i\pi}\sum_{n=1}^{\infty}\sum_{m=1}^{\infty}\frac{(2i)^n.Z_m}{(2ik)^{n+m}}\iint_{\mathbb{C}} d\lambda\wedge d\bar{\lambda}\ \lambda^{n-1}g(\lambda)\Phi_{12}\Phi_{11} \quad + \frac{i}{2}|q|^2 \ -$$
$$\left. -\frac{1}{2i\pi}\sum_{n=1}^{\infty}\frac{(2i)^n}{(2ik)^n}\iint_{\mathbb{C}} d\lambda\wedge d\bar{\lambda}\ \lambda^{n-1}g(\lambda)(\Phi_{11}\Phi_{22}+\Phi_{12}\Phi_{21})\right\} \tag{IV.1}$$

with

$$Z_1 = -q^* \tag{IV.2}$$

We consider now the first conservation law (n = 1). If we assume

$$q \longrightarrow 0 \quad \text{for} \quad x \longrightarrow \pm\infty \tag{IV.3}$$

we obtain after integration (using the asymptotic behaviour of Φ (III.16) and the unitarity relation)

$$\frac{\partial}{\partial t}\int_{-\infty}^{+\infty}|q|^2\,dx = \frac{\omega_0}{12}\int_{-\infty}^{+\infty}d\lambda\ |A(\lambda)|^2\ \alpha(\lambda,t)\ \bar{\alpha}(\lambda,t) \tag{IV.4}$$

wich gives with (III.21), (III.22)

$$\int_{-\infty}^{+\infty}\left(|q(x,t)|^2-|q(x,0)|^2\right)dx = \frac{\omega_0}{12}\,t\int_{-\infty}^{+\infty}d\lambda\,|A(\lambda)|^2\ \alpha(\lambda,0)\ \bar{\alpha}(\lambda,0) \tag{IV.5}$$

The above linear growth is the signature of the continuous transfer of energy from the input laser beam (HF wave) to the electronic density (LF wave).

The detailed mechanism of this transfer requires the study of the asymptotic solution (in time) when the reflection coefficient $\alpha(k,t)$ grows exponentially (see (III.21)). This is now under study and will be the subject of a forthcoming publication [19].

REFERENCES

[1] See f.i. M.V. GOLDMAN, Rev. Mod. Phys. 56 709 (1984)

[2] V.E. ZAKHAROV, Sov. Phys. JETP 35 908 (1972) [Zh. Eksp. Teor. Fiz. 62 1745 (1972)]

[3] V.I. KARPMAN, Phys. Scripta 11 263 (1975)

[4] D.J. KAUP, Phys. Rev. Lett. 59 2062 (1987)

[5] J. LEON, A. LATIFI, "solution of an initial/boundary value problem for coupled nonlinear waves". Preprint Montpellier PM 89/23 (1989), to appear in J. Phys. A

[6] P.K. KAW, K. NISHIKAWA, J. Phys. Soc. Japan 38 1753 (1975)

[7] K. NISHIKAWA, H. HOJO, K. MIMA, H. IKEZI Phys. Rev. Lett. 33 148 (1974)

[8] V.G. MAKHANKOV, Phys. Lett. A 50 42 (1974)

[9] F. CALOGERO, J. Math. Phys. 30 (3) 639-55 (1989)

[10] V.E. ZAKAROV, E.A. KUZNETSOV Physica D 18 455 (1986)

[11] D.J. KAUP, A.C. NEWELL, Adv. Math. 31 67 (1979)

[12] J. LEON, Phys. Lett. A 50 42 (1984)

[13] J. LEON, Jour. Math. Phys. 29 2012 (1988)

[14] M. BOITI, J. LEON, L. MARTINA, F. PEMPINELLI, J.Phys. A 21 (1988)

[15] J. LEON, F. PEMPINELLI, in "Nonlinear Evolution Equations : integrability and spectral methods" Eds: A.DEGASPERIS, A.P. FORDY, M. LAKSHMANAN (Manchester Univ. Press (1989))

[16] J. LEON, Phys. Lett. A 131 79 (1988)

[17] M. CASANOVA, G. LAVAL, R.PELLAT, D. PESME Phys. Rev. Lett. 54 2230 (1985)

[18] N.A. KRALL, A.W. TRIVELPIECE, "Principles of Plasma Physics" MC Graw Hill (1973)

[19] A. LATIFI, J. LEON, in preparation

[20] J. LEON, "Nonlinear Evolution with Singular Dispersion Laws and Forced systems". Preprint Montpellier PM 89/24

[21] K. KONNO, H. SANUKI, Y.H. ICHIKAWA, Prog. Theor. Phys. 52 886 (1974)

[22] F. CALOGERO, A. DEGASPERIS, "Spetral Trasform and Solitons", North Holland (1982)

On the Spectral Theory of the Davey-Stewartson Equation*

M. Boiti[1], *J. Leon, and F. Pempinelli*[1]

Laboratoire de Physique Mathématique+, USTL, Place Eugène Bataillon, F-34095 Montpellier Cédex 5, France

[1]Permanent address: Dipartimento di Fisica, Sezione INFN, I-73100 Lecce, Italy

+Unité de recherche associée au CNRS No. URA 040768

ABSTRACT. A spectral theory for the 2+1 nonlinear Davey-Stewartson equation is proposed. In this scheme the solitons (wave, semi-wave and localized solitons) have a spectrum consisting in points in the complex plane. This result allows the establishment of the $\bar{\partial}$ method (sometimes called the dressing method) to construct the N×M soliton solution.

1 - INTRODUCTION

The discovery of the existence of localized coherent structures, solutions of integrable nonlinear evolutions in two spacial dimensions [1], has helped in understanding precisely the working mechanism of the spectral theory.

The first remark was that the spectral transform is not unique [2], more precisely that the integral equations defining the Jost-like solutions in 2D can be written with two arbitrary functions of each space variable. Then the nature of the soliton spectrum depends deeply on the choice of these functions as shown in [2].

Then followed the discovery that some non-vanishing boundary values are essential for the existence of localized coherent structures in 2D [3]. There the spectral transform for the Davey-Stewartson eq. [4] has been reconsidered, more precisely the time-evolution of the eigenfunctions has been reconstructed by taking into account the non-vanishing nature of the boundary-values. The approach of [3] has been proved very useful for constructing and studying the soliton solutions with in particular the discovery of energy exchanges between the different localized structures underlying a general N×M soliton solution [5].

With the help of the discovery in [3] and the results of [2], we were able then to propose a new spectral transform of the solutions of the D.S. eq. with the following properties [6].

i) the spectrum is made of a continuous and a discrete part. The continuous part represents the "radiation" and it is related to the discontinuity of the eigenfunction as the complex spectral parameter k crosses the real axis. The discrete part represents the "solitons" and it is related to the isolated singularities in the complex k-plane of the eigenfunctions.

* This work has been done partly under contract RCP 264 and with support of INFN Sezione di LECCE.

ii) the isolated singularities come from the arbitrary functions previously mentioned which play in 2D the role played in 1D by the transmission coefficient.

iii) the time evolution of the spectral transform is explicitly integrable.

The results of [6] give the general structure under which the $\bar{\partial}$-problem is explicitly solvable and allows us to construct all known explicit solutions of the Davey-Stewartson eq. This construction is sometimes referred to as the dressing method and we report here for the first time the general structure under which we are able to solve the $\bar{\partial}$-problem explicitly.

In the following sect. 2 we write down the Lax pair from which the D.S-eq. is obtained and define the space of eigenfunctions we are interested in.

The sect. 3 is devoted to the problem of defining the spectral transform ("direct problem"). This problem consists in deriving, from the integral equation written in sect.2, a nonlocal Riemann-Hilbert problem with singularities. We shall only sketch the results and report to previous works for proofs and details.

In sect. 4 we solve the inverse problem and show that the solitons are particular solutions for which the inversion equation (Cauchy-Green integral equation) is explicitly solvable as an algebraic system. We give the example of the 1-soliton which can be a wave-soliton, a semi-wave soliton or a localized soliton, depending on the choice of the parameters.

2 - FROM THE LAX PAIR TO THE INTEGRAL EQUATIONS

The commutativity condition

$$[T_1 , T_2] = 0 \tag{2-1}$$

between the following pair of operators

$$T_1 = \begin{pmatrix} \partial_x & \frac{1}{2} q \\ \frac{1}{2} r & \partial_y \end{pmatrix} \tag{2-2}$$

$$T_2 = i\,\partial_t + \begin{pmatrix} \partial_x - \partial_y & q \\ r & \partial_y - \partial_x \end{pmatrix} (\partial_x - \partial_y) + \begin{pmatrix} \nu_1 & -q_y \\ r_x & \nu_2 \end{pmatrix} \tag{2-3}$$

is equivalent to the following set of eqs. for q(x,y,t) and r(x,y,t) (which are supposed to vanish as $x^2 + y^2 \longrightarrow \infty$)

$$i\,q_t + q_{xx} + q_{yy} + (\nu_1 - \nu_2)q = 0 \quad ,$$

$$-\,i\,r_t + r_{xx} + r_{yy} + (\nu_1 - \nu_2)r = 0 \quad , \tag{2-4}$$

$$\nu_{1x} = -\frac{1}{2}(rq)_y \quad , \quad \nu_{2y} = \frac{1}{2}(rq)_x \quad . \tag{2-5}$$

As a consequence, the Lax pair does not determine uniquely the integrable system. Indeed $\nu_1(x,y,t)$ and $\nu_2(x,y,t)$ are defined through (2-5) only up to arbitrary functions of (y,t) for ν_1 and of (x,t) for ν_2.

We fix here the following determination

$$\nu_1 = -\frac{1}{2}\int_{-\infty}^{x} dx'(rq)_y + v_1(y,t) \quad , \quad \nu_2 = -\frac{1}{2}\int_{-\infty}^{y} dy'(rq)_x + v_2(x,t) \qquad (2\text{-}6)$$

where v_1 and v_2 are arbitrary functions.

As noticed in [3] the spectral transform machinery provides the eqs (2-4) (2-6) with the solution of the initial-boundary value problem : data of $q(x,y,0)$, $r(x,y,0)$ $v_1(y,t)$ and $v_2(x,t)$.

We define the eigenspace of (T_1,T_2) as the set of solutions of the following spectral problems (k is the spectral parameter) :

$$T_1\,\psi = 0 \quad , \qquad (2\text{-}7)$$

$$T_2\psi = k^2\,\psi\,\sigma_3 \quad , \qquad (2\text{-}8)$$

with the behaviour for large $|k|$:

$$\psi(k;x,y,t) = \left[\mathbb{1} + O\left(\frac{1}{k}\right)\right]\begin{pmatrix} e^{iky} & 0 \\ 0 & e^{-ikx}\end{pmatrix} \quad . \qquad (2\text{-}9)$$

Under the above requirements and the eq.(2-6), we can fix the asymptotic behaviours of ψ in the (x,y) plane and write finally ψ as the solution of the following system of Volterra equations for the two column vectors ψ_1 and ψ_2 (the time enters these eq. as a parameter) in terms of some arbitrary scalar functions $\psi_{01}^{\pm}$ and $\psi_{02}^{\pm}$:

$$\psi_1^+(k;x,y) = \begin{pmatrix}1\\0\end{pmatrix}\psi_{01}^+(k;y) + G_1^+\,\psi_1^+(k;.,.) \qquad \text{Im } k > 0 \qquad (2\text{-}10,a)$$

$$\psi_1^-(k;x,y) = \begin{pmatrix}1\\0\end{pmatrix}\psi_{01}^-(k;y) + G\,\psi_1^-(k;.,.) \qquad \text{Im } k < 0 \qquad (2\text{-}10,b)$$

$$\psi_2^+(k;x,y) = \begin{pmatrix}0\\1\end{pmatrix}\psi_{02}^+(k;x) + G\,\psi_2^+(k;.,.) \qquad \text{Im } k > 0 \qquad (2\text{-}10,c)$$

$$\psi_2^-(k;x,y) = \begin{pmatrix}0\\1\end{pmatrix}\psi_{02}^-(k;x) + G_2^-\,\psi_2^-(k;.,.) \qquad \text{Im } k < 0 \;. \qquad (2\text{-}10,d)$$

In the above equation the integral operators are defined by

$$G = \frac{1}{2}\begin{pmatrix} 0 & -\int_{-\infty}^{x} dx'\; q(x',y) \\ -\int_{-\infty}^{y} dy'\; r(x,y') & 0 \end{pmatrix} \quad , \qquad (2\text{-}11,a)$$

$$G_1^+ = G + \begin{pmatrix} 0 & 0 \\ 1 & 0 \end{pmatrix} \int_{-\infty}^{+\infty} dy' \ r(x,y') \quad , \tag{2-11,b}$$

$$G_2^- = G + \begin{pmatrix} 0 & 1 \\ 0 & 0 \end{pmatrix} \int_{-\infty}^{+\infty} dx' \ q(x',y) \quad . \tag{2-11,c}$$

Scaling off the exponentials in (2-10), it is easy to prove that if $\psi_{01}^+(k;y)$ is meromorphic in Im k > 0 with N first order poles λ_n and $\psi_{01}^-(k;y)$ holomorphic in Im k < 0, then ψ_1^+ and ψ_1^- have respectively the same property. The same holds for $\psi_2^\pm$ which will have M simple poles μ_m in Im k < 0.

We are then left with defining the $\psi_{0j}^\pm$ s. This is done by using (2-8) at the limits $x \to -\infty$ for $\psi_{11}^\pm$ and $y \to -\infty$ for $\psi_{22}^\pm$. We obtain with help of (2-6) and (2-10)

$$(i\, \partial_t + \partial_{yy} + v_1)\, \psi_{01}^\pm = -k^2\, \psi_{01}^\pm \quad , \tag{2-12,a}$$

$$(-i\, \partial_t + \partial_{xx} - v_2)\, \psi_{02}^\pm = -k^2\, \psi_{02}^\pm \quad . \tag{2-12,b}$$

The $\psi_{0j}^\pm$ s are then the solutions of two time-dependent Schrödinger spectral problems with the following behaviours ar large $|k|$:

$$\psi_{01}^\pm(k;y,t) = \left[1 + O\left(\frac{1}{k}\right)\right] e^{iky} \quad , \tag{2-13,a}$$

$$\psi_{02}^\pm(k;x,t) = \left[1 + O\left(\frac{1}{k}\right)\right] e^{-ikx} \quad . \tag{2-13,b}$$

We do not report here the spectral theory for (2-12) which has been worked out in [7] for v_j vanishing as $|x|$ (or $|y|$) *and* t go to infinity and in [8] for v_j vanishing as $|x|$ (or $|y|$) go to infinity for any t except in some N directions $x - 2\,\mathrm{Re}(\lambda_n)t$ for v_1, and some M directions $y - 2\,\mathrm{Re}(\mu_n)t$ for v_2, where they go to some constants.

We shall see later that to the choice (2-12) for ψ_{oj} (or equivalentely to the choice (2-8)) it corresponds the simplest explicitly integrable time-evolution of the spectral transform.

In the work [3] ψ_{01} and ψ_{02} have been chosen equal to 1 and consequently (2-8) had to be modified. The result is that ψ defined out of (2-10) has no singular points and that the spectral transform evolves in a way still linear but not explicitly integrable.

3 - FROM THE INTEGRAL EQUATION TO THE $\bar\partial$-PROBLEM

The general procedure consists in expressing the $\bar\partial$-derivative ($\bar\partial$ stands for $\dfrac{\partial}{\partial \bar k} = \dfrac{1}{2}\left(\dfrac{\partial}{\partial k_R} + i\,\dfrac{\partial}{\partial k_I}\right)$ in the sense of the distributions) of ψ in terms of ψ itself, out of the integral equation (2-10). More precisely we have proved [6] that (2-10)-(2-12) imply

$$\frac{\partial}{\partial\bar{k}}\Phi(k;x,y,t) = \int\int_{\mathbb{C}} d\ell \wedge d\bar{\ell}\; \Phi(\ell;x,y,t)\; \mathcal{R}(k,\ell;x,y,t) \tag{3-1}$$

with

$$\Phi(k;x,y,t) = \psi(k;x,y,t)\begin{pmatrix} e^{-iky} & 0 \\ 0 & e^{ikx}\end{pmatrix} , \tag{3-2}$$

such that the behaviour (2-9) leads to the following solution of (3-1) :

$$\Phi(k;x,y,t) = \mathbb{1} + \frac{1}{2i\pi}\int\int_{\mathbb{C}} \frac{d\zeta \wedge d\bar{\zeta}}{\zeta - k}\frac{\partial}{\partial\bar{\zeta}}\Phi(\zeta;x,y,t) . \tag{3-3}$$

We may summarize the result of [6] by giving hereafter the structure of $\mathcal{R}$

$$\mathcal{R}(\zeta,\ell;x,y,t) = \begin{pmatrix} e^{i\ell(y-\ell t)} & 0 \\ 0 & e^{-i\ell(x-\ell t)}\end{pmatrix} R(\zeta,\ell) \begin{pmatrix} e^{-i\zeta(y-\zeta t)} & 0 \\ 0 & e^{i\zeta(x-\zeta t)}\end{pmatrix} \tag{3-4}$$

$$R(\zeta,\ell) = R_c(\zeta,\ell) + R_s(\zeta,\ell) . \tag{3-5}$$

The radiative part R_c (continuum) represents the discontinuity of Φ on the real axis and reads

$$R_c(\zeta,\ell) = \frac{i}{2}\begin{pmatrix} \delta(\ell_I+0)\;\delta(\zeta_I+0) & 0 \\ 0 & \delta(\ell_I-0)\;\delta(\zeta_I-0)\end{pmatrix}\begin{pmatrix} r_1(\zeta,\ell) & \rho_2(\zeta,\ell) \\ \rho_1(\zeta,\ell) & r_2(\zeta,\ell)\end{pmatrix} \tag{3-6}$$

The functions r_1 and r_2 are the spectral transform (with no discrete part) of ψ_{01} and ψ_{02} defined in [7]. The matrix $\begin{pmatrix} 0 & \rho_2 \\ \rho_1 & 0\end{pmatrix}$ is the spectral transform of ψ itself (with no discrete part) as defined in [3].

The discrete part R_s (solitons) represents the singularities of Φ in the complex plane (eigenvalues of (2-12)) and reads

$$R_s(\zeta,\ell) = 2\pi\sum_{n=1}^{N}\sum_{m=1}^{M}\begin{pmatrix} \delta(\ell-\bar{\lambda}_n) & 0 \\ 0 & \delta(\ell-\bar{\mu}_m)\end{pmatrix}\begin{pmatrix} \alpha_{nm} & \eta_{nm} \\ \rho_{nm} & \beta_{nm}\end{pmatrix}\begin{pmatrix} \delta(\zeta-\lambda_n) & 0 \\ 0 & \delta(\zeta-\mu_m)\end{pmatrix} \tag{3-7}$$

where α_{nm}, η_{nm}, ρ_{nm} and β_{nm} are some constants related to the amplitudes and initial position of the solitons.

The above described structure of $\mathcal{R}(\zeta,\ell;x,y,t)$ has been derived in [6] where the essential tool was the discovery of the orthogonality relations for the eigenfunctions of the time-dependent Schrödinger equation [7]. (In [6] R_s has a different but equivalent structure).

It is worth remarking that the time evolution of $\mathcal{R}$, as explicitely written in (3-4), is obtained by applying (2-8) to (3-1) and it is only for the choice (2-8) that this evolution is simple and explicit.

We may also point out that the $\bar{\partial}$ equation (3-3) *with* the datum of (3-4) in (3-1) *implies* that ψ given by (3-2) is solution of the spectral problems (2-7) and (2-8). This result can be obtained by computing $T_1\psi$ and $T_2\psi$ out of (3-3) and by inspecting the coefficients of the different powers of $\frac{1}{k}$. This result is essential for the construction of explicit solutions of the Davey-Stewartson eq. from the solutions of the integral eq. (3-3) by means of some simple convenient choices of R. This method is sometimes referred to as the "dressing method" and we use it in the following section.

4 - FROM THE $\bar{\partial}$-PROBLEM TO THE SOLITONS

The problem here is to compute the set $\{r(x,y,t),\ q(x,y,t),\ \nu_1(x,y,t),\ \nu_2(x,y,t)\}$ from the datum of $\mathfrak{R}(\zeta,\ell;x,y,t)$ having the structure given in the previous section.

The solution of this problem is given by solving first the integral equation (3-3) to get $\Phi(k,;x,y,t)$. Then, inserting (3-3) in the spectral problem (2-7) and inspecting the different coefficients of the resulting polynomial in $\frac{1}{k}$ we obtain

$$\begin{pmatrix} 0 & q \\ r & 0 \end{pmatrix} = i\left[\sigma_3,\ \Phi^{(1)}\right] , \tag{4-1}$$

$$\begin{pmatrix} \nu_1 & 0 \\ 0 & \nu_2 \end{pmatrix} = -2i \begin{pmatrix} \partial_y & 0 \\ 0 & -\partial_x \end{pmatrix} \operatorname{diag} \Phi^{(1)} . \tag{4-2}$$

In the above eq. $\Phi^{(1)}$ is the first order term of the series

$$\Phi(k;x,y,t) = \mathbb{1} + \sum_{1}^{n} \frac{1}{k^n} \Phi^{(n)}(x,y,t) + 0\left(\frac{1}{k^{n+1}}\right) \tag{4-3}$$

obtained out of the integral equation (3-3). Namely

$$\Phi^{(1)}(x,y,t) = -\frac{1}{2i\pi}\int_{\mathbb{C}} d\zeta \wedge d\bar{\zeta} \int_{\mathbb{C}} d\ell \wedge d\bar{\ell}\ \Phi(\ell;x,y,t)\ \mathfrak{R}(\zeta,\ell;x,y,t) . \tag{4-4}$$

The integral eq. (3-3) is not in general explicitely solvable except for some particular choices of $\mathfrak{R}$. This is the case for intance when $\mathfrak{R}$ has only a discrete part as given in (3-7). In that case the solution of (3-3) reduces to the solution of an algebraic system for the unknowns N+M 2-vectors $\Phi_1\left(\bar{\lambda}_n\right)$ and $\Phi_2\left(\bar{\mu}_m\right)$. In the simplest case N=M=1 we obtain the one soliton solution

$$q(x,y,t) = \frac{2\eta}{\Delta(x,y,t)}\, e^{i\theta(x,y,t)} \tag{4-5}$$

$$r(x,y,t) = \frac{-\ 2\rho}{\Delta(x,y,t)}\, e^{-i\theta(x,y,t)} \tag{4-6}$$

$$\nu_1(x,y,t) = -2\frac{\partial}{\partial y}\left\{\frac{1}{\Delta}\left[\alpha e^u\left(1+\frac{\beta}{2\mu_I}e^v\right) - \frac{\eta\rho}{2\mu_I}e^{u+v}\right]\right\} \tag{4-7}$$

$$\nu_2(x,y,t) = -2\frac{\partial}{\partial x}\left\{\frac{1}{\Delta}\left[\beta e^v\left(1+\frac{\alpha}{2\lambda_I}e^u\right) - \frac{\eta\rho}{2\lambda_I}e^{u+v}\right]\right\} \tag{4-8}$$

in which we have defined $\lambda = \lambda_R + i\,\lambda_I$, $\mu = \mu_R + i\,\mu_I$,

$$u = -2\,\mu_I\,(x - 2\mu_R t) \qquad v = 2\,\lambda_I(y - 2\lambda_R t) \tag{4-9}$$

$$\theta = \mu_R x + \lambda_R y - \left(\lambda_R^2 - \lambda_I^2 + \mu_R^2 - \mu_I^2\right) t \tag{4-10}$$

$$\Delta = e^{-\frac{1}{2}(u+v)} + \frac{\alpha}{2\lambda_I}e^{-\frac{1}{2}(u-v)} + \frac{\beta}{2\mu_I}e^{\frac{1}{2}(u-v)} + \frac{\alpha\beta - \eta\rho}{4\lambda_I\mu_I}e^{\frac{1}{2}(u+v)} \quad . \tag{4-11}$$

To get a regular solution we must choose $\alpha > 0$ (because $\lambda_I > 0$), $\beta < 0$ (because $\mu_I < 0$) and $\alpha\beta - \eta\rho < 0$. This solution depends on 8 real parameters which define the velocity in the plane (2) the initial position (2) the amplitudes of the fields r and q (2) and 2 more parameters give the shape (slope of Δ^{-1}).

With $\alpha=\beta=0$ the solution (q,r) reduces to the wave soliton with argument $u(x,t) + v(y,t)$.

With only α (or β) vanishing, the solution describes a semi-wave.

With the full matrix ($\alpha\beta\eta\rho \neq 0$) we obtain the localized soliton originally discovered in [1].

For the case N=M=2 we recover the 2×2 soliton which has been described in [1] for the particular choice

$$\alpha_{12} = \alpha_{21} = \beta_{12} = \beta_{21} = \eta_{12} = \eta_{21} = \rho_{12} = \rho_{21} = 0 \tag{4-12}$$

and was constituted with *two* localized bumps. Releasing (4-12) we recover the solution described in [3] and [5] constituted with *four* localized bumps. We may also play with many other different choices of the parameters and for instance recover the solutions described recently in [9] made of four bumps at (say) $t = -\infty$ and only three at $t = +\infty$.

We finally mention that, for a convenient choice of the parameters, in the 2×1 soliton case (eigenvalues λ_1,λ_2, and μ_1) we obtain a solution consisting in *two* localized structure having DIFFERENT amplitudes at $t = +\infty$ and $-\infty$. This solution describes the interaction of two solitons with both a shift in the positions and a transfer of energy.

REFERENCES

[1] M. BOITI, J. LEON, L. MARTINA, F. PEMPINELLI, Phys. Lett. A 132 432 (1988).

[2] M. BOITI, J. LEON, F. PEMPINELLI "Multidimensional solitons and their spectral transforms" Preprint 88-44, Montpellier (Octobre 1988).

[3] A.S. FOKAS, P.M. SANTINI, Phys. Rev. Lett. 63 1329 (1988).

[4] A.S. FOKAS, M.J. ABLOWITZ, J. Math. Phys. 25 8 (1984).

[5] P.M. SANTINI, Preprint 662, March 1989 ROMA, to appear in Physica D. see also P.M. SANTINI in the present volume and the references quoted therein.

[6] M. BOITI, J. LEON, F. PEMPINELLI, Phys. Lett. A 141 (1989).

[7] M. BOITI, J. LEON, F. PEMPINELLI, Phys. Lett. A 141 96 (1989).

[8] M. BOITI, F. PEMPINELLI, A. POGREBKOV (in preparation).

[9] J. HIETARINTA, R. HIROTA, Preprint TURKU (Finland) Nov. 14, 1989.

Multidimensional Solitons in Terms of Fermions*

M. Jaulent[1] *and L. Martinez-Alonso*[2]

[1]Laboratoire de Physique Mathématique, USTL,
F-34095 Montpellier Cédex 5, France
[2]Departamento de Métodos Matemáticos de la Física, Facultad de Ciencias Físicas,
Universidad Complutense, E-28040 Madrid, Spain

ABSTRACT. Multidimensional solitons of the Davey-Stewartson equation are analyzed by means of the bilinear formalism of the Kyoto School.

1. Introduction

This paper is a part of a research program on integrable systems which aims to obtain realizations of asymptotic modules [1-2] by means of the elegant bilinear formalism provided by M. Sato, Y. Sato, E. Date, M. Jimbo, M. Kashiwara and T. Miwa (SDJKM) [3]. We are here concerned with the multidimensional solitons of the Davey-Stewartson (DS) equation discovered by Boiti et al [4]. Despite the inverse scattering transform method for studying these solutions is already available [5], it may be important to investigate alternative methods in order not only to characterize these dynamical objects but also to determine the class of nonlinear models in which they arise. Thus, we show how the τ-functions for soliton solutions of the DS equation in Hirota's bilinear form can be written as expectation values of certain elements of a Clifford algebra. As a consequence the characterization of these solutions in terms of the Lie group $G\ell(2\ \infty)$ is obtained. Furthermore, the formalism provides a simple method for deriving the phase-shift formulas for soliton-soliton interactions.

2. The bilinear formulation of the DS equation

We consider the following expression of the DS equation

$$iq_t + q_{uu} + q_{vv} + 2(U_{uu} + U_{vv})q = 0 \quad , \quad 4\, U_{uv} = |q|^2 \quad . \tag{1}$$

If we use the new dependent variables τ and σ defined by

$$q = \sigma/\tau \quad , \quad U = \log \tau \quad , \tag{2}$$

then (1) can be written in bilinear form as ([6])

$$\left(- i\, D_t + D_u^2 + D_v^2\right)\tau\sigma = 0 \quad , \quad 2\, D_u D_v \tau\tau = |\sigma|^2 \quad . \tag{3}$$

This bilinear equation is a reduction of a member of the 2-component KP hierarchy introduced by SDJKM (see [3]) in terms of the following system of bilinear

---

(*) *Action Intégrée Franco-Espagnole "Modèles intégrables non linéaires"* identities on a circle γ ($|k| = R$)

$$\oint_{\gamma} f(k,x)\, g(k,x')\, dk = 0 \quad , \quad \text{for any } x,x' \quad . \tag{4}$$

Here f and g denote 2×2 matrix-valued functions depending on a complex variable k and on an infinite set of complex variables $x = (x^{(1)},x^{(2)})$ with $x^{(j)} = \left(x_1^{(j)},\, x_2^{(j)},\dots\right)$ (j=1,2), which are of the form

$$f = \frac{1}{\tau_1(x)} \begin{pmatrix} \tau_1(x - \epsilon_1(k)) & k^{-1}\tau_2(x - \epsilon_2(k)) \\ -k^{-1}\tau_3(x - \epsilon_1(k)) & \tau_1(x - \epsilon_2(k)) \end{pmatrix} \begin{pmatrix} \exp \xi(k,x^{(1)}) & 0 \\ 0 & \exp \xi(k,x^{(2)}) \end{pmatrix},$$

$$g = \frac{1}{\tau_1(x)} \begin{pmatrix} \exp\left(-\xi(k,x^{(1)})\right) & 0 \\ 0 & \exp\left(-\xi(k,x^{(2)})\right) \end{pmatrix} \begin{pmatrix} \tau_1(x+\epsilon_1(k)) & -k^{-1}\tau_2(x+\epsilon_1(k)) \\ k^{-1}\tau_3(x+\epsilon_2(k)) & \tau_1(x+\epsilon_2(k)) \end{pmatrix},$$

where

$$\epsilon_1(k) = \left(\epsilon_1^{(1)}(k),\, \epsilon_1^{(2)}(k)\right) \quad , \quad \epsilon_1^{(j)} = \delta_{1j}\left(\frac{1}{k},\, \frac{1}{2k^2},\dots,\, \frac{1}{nk^n},\dots\right) \quad ,$$

$$\xi(k,x^{(1)}) = \sum_{n=1}^{\infty} k^n\, x_n^{(1)} \quad .$$

By inserting these expressions in (4) it follows that the functions $\tau_i(x)$ (i=1,2,3) satisfy an infinite system of bilinear equations. In particular, if we denote

$$w = 2x_1^{(1)} \quad , \quad v = 2\, x_1^{(2)} \quad , \quad t = 2i\left(x_2^{(1)} - x_2^{(2)}\right)$$

$$\tau = \tau_1 \quad , \quad \sigma = \tau_3 \quad ,$$

then the bilinear equations (3) are obtained provided that

$$\tau_1^* = \tau_1 \quad , \quad \tau_3^* = \tau_2 \quad . \tag{5}$$

In what follows, this condition will be referred to as the DS reduction condition.

3. Clifford algebra

The main tool used by SDJKM in order to construct solutions to (4) is Clifford algebra. Let us see in brief some of the more important aspects of their analysis. The starting point is the complex Clifford algebra A with generators $\psi_n^{(i)}$, $\psi_n^{(i)*}$ ($i = 1,2,\ -\infty < n < \infty$) satisfying the fermion anticommutation relations

$$\left\{\psi_n^{(i)},\ \psi_m^{(j)}\right\} = \left\{\psi_n^{(i)*},\ \psi_m^{(j)*}\right\} = 0 \quad ,$$

$$\left\{\psi_n^{(i)},\ \psi_m^{(j)*}\right\} = \delta^{ij}\ \delta_{nm} \quad . \tag{6}$$

The standard representation of A can be defined in terms of a set of operators on a Fock space. For our purposes two Fock spaces $\mathcal{F}$ and $\mathcal{F}^*$ are required, which are associated with two vacuum vectors $|vac\rangle$ and $\langle vac|$, respectively, defined by

$$\psi_n^{(i)}\ |vac\rangle = 0\ (n < 0)\ ,\ \psi_n^{(i)*}\ |vac\rangle = 0 \quad (n \geqslant 0)\quad ,$$

$$\langle vac|\ \psi_n^{(i)} = 0\ (n \geqslant 0)\ ,\ \langle vac|\ \psi_n^{(i)*} = 0 \quad (n < 0) \quad . \tag{7}$$

Thus $\mathcal{F} = \{\alpha\ |vac\rangle\ :\ \alpha \in A\}$ and $\tilde{\mathcal{F}} = \{\langle vac|\alpha\ :\ \alpha \in A\}$. It can be proved that these is a unique bilinear form (,) on $\tilde{\mathcal{F}} \times \mathcal{F}$ such that

$$(\langle vac|,\ |vac\rangle) = 1 \quad ,$$

$$(\langle vac|\alpha,\ \beta|vac\rangle) = (\langle vac|\alpha\beta,\ |vac\rangle) = (\langle vac|,\ \alpha\beta|vac\rangle) \quad .$$

Henceforth, we will use the notation conventions

$$\langle\alpha\rangle = \langle vac|\alpha|vac\rangle = (\langle vac|\alpha,\ |vac\rangle) = (\langle vac|,\ \alpha|vac\rangle) \quad .$$

Let us denote by $V^{(i)}$ and $V^{(i)*}$ the linear subspaces of A generated by $\left\{\psi_n^{(i)}\ :\ -\infty < n > \infty\right\}$ and $\left\{\psi_n^{(i)*} = -\infty < n < \infty\right\}$ ($i = 1,2$), respectively, and let us introduce "charge" operators

$$Q^{(i)} = \sum_{n=-\infty}^{\infty}\ :\ \psi_n^{(i)}\ \psi_n^{(i)*}\ : \quad , \qquad i = 1,2, \tag{8}$$

where the normal order operation :: is defined as

$$:\psi_n^{(i)}\ \psi_m^{(j)*}: = \psi_n^{(i)}\ \psi_m^{(j)*} - \left\langle vac\ \middle|\psi_n^{(i)}\ \psi_m^{(j)*}\middle|\ vac\right\rangle$$

It is easy to see that given $w_n \in \bigcup_{j=1}^{2} (V^{(j)} \cup V^{(j)*})$ ($n=1,\dots,N$), then

$$\left[Q^{(i)},\ w_1 w_2 \dots w_N\right] = (N^{(i)} - N^{(i)*})\ w_1 w_2 \dots w_N \quad , \tag{9}$$

with $N^{(i)}$ and $N^{(i)*}$ being the number of factors w_n in $V^{(i)}$ and $V^{(i)*}$, respectively.

Therefore, taking into account that from the definition (8)

$$Q^{(i)}\ |vac\rangle = 0 \qquad i = 1,2,$$

it is a simple task to construct vectors in $\tilde{\mathfrak{F}}$ and $\mathfrak{F}$ with definite charges (see [3])

$$Q^{(i)}|\ell_2\ell_1\rangle = \ell_i\,|\ell_2\ell_1\rangle \quad , \quad \langle\ell_1\ell_2|\ Q^{(i)} = \ell_i\,\langle\ell_1\ell_2| \quad .$$

These is an important Lie algebra in A provided by the set of quadratic elements

$$\mathcal{G} = \left\{ \sum a^{ij}_{nm}\ \psi^{(i)}_n\ \psi^{(j)*}_m \ :\ a^{ij}_{nm} \in \mathbb{C} \right\} \quad .$$

Observe that the total charge $Q = Q^{(1)} + Q^{(2)}$ commutes with all the elements of $\mathcal{G}$. In order to deal with infinite linear combinations of quadratic elements it is required to define appropriate completions of $\mathcal{G}$. By assuming that one of such completions has been performed, we denote by $\mathcal{G}\ell\,(2\infty)$ the direct sum of $\mathcal{G}$ with the set of constants $\mathbb{C}.1$ of A, and by $G\ell(2\infty)$ the corresponding Lie group. There is a relevant infinite-dimensional Abelian subalgebra of $\mathcal{G}\ell(2\infty)$ given by the elements of the form

$$H(x) = \sum_{i=1,2}\ \sum_{n=1}^{\infty}\ \sum_{m=-\infty}^{\infty} x^{(i)}_n\ \psi^{(i)}_m\ \psi^{(i)*}_{m+n} \quad , \quad x = (x^{(1)},\ x^{(2)}) \in \mathbb{C}^{2\infty}. \tag{10}$$

It must be noticed that the following two properties hold

$$H(x)\ |vac\rangle = 0 \quad , \tag{11}$$

$$[H(x),\ Q^{(i)}] = 0 \quad , \quad i = 1,2 \quad . \tag{12}$$

We are now ready to describe the main result of the SDJKM method, it may be stated as follows. Every element g of the group $G\ell(2\infty)$ determines a set of solutions of the 2-component KP hierarchy of bilinear equations given by

$$\begin{aligned}
\tau_1(x) &= \left\langle \ell_1\ell_2-1\ |e^{H(x)}g|\ \ell_2-\ell-1\ \ell_1+\ell \right\rangle \quad , \\
\tau_2(x) &= \left\langle \ell_1+1\ \ell_2-2\ |e^{H(x)}g|\ \ell_2-\ell-1\ \ell_1+\ell \right\rangle \quad , \\
\tau_3(x) &= \left\langle \ell_1-1\ \ell_2\ \ |e^{H(x)}g|\ \ell_2-\ell-1\ \ell_1+\ell \right\rangle \quad ,
\end{aligned} \tag{13}$$

where ℓ_1, ℓ_2, ℓ are arbitrary integers.

4. Solitons solutions

Henceforth we adopt the simplest choice $\ell_1 = \ell = \ell_2 - 1 = 0$ in (13), so that we have

$$
\begin{aligned}
\tau_1(x) &= \left\langle \text{vac} \left| e^{H(x)} g \right| \text{vac} \right\rangle , \\
\tau_2(x) &= \left\langle 1 -1 \left| e^{H(x)} g \right| \text{vac} \right\rangle , \\
\tau_3(x) &= \left\langle -1\ 1 \left| e^{H(x)} g \right| \text{vac} \right\rangle .
\end{aligned}
\tag{14}
$$

The analysis of one-dimensional solitons for KP hierarchies requires the introduction of the following k-dependent fermion fields

$$
\psi^{(i)}(k) = \sum_{n=-\infty}^{\infty} \psi_n^{(i)} k^n \quad , \quad \psi^{(i)*}(k) = \sum_{n=-\infty}^{\infty} \psi_n^{(i)*} k^{-n} \quad . \tag{15}
$$

It follows easily that

$$
[H(x), \psi^{(i)}(k)] = \xi(k, x^{(i)})\, \psi^{(i)}(k) \quad , \quad [H(x), \psi^{(i)*}(k)] = -\xi(k, x^{(i)}) \psi^{(i)*}(k) \quad ,
$$

and consequently

$$
\begin{aligned}
e^{H(x)}\, \psi^{(i)}(k)\, e^{-H(x)} &= e^{\xi(k, x^{(i)})}\, \psi^{(i)}(k) \quad , \\
e^{H(x)}\, \psi^{(i)*}(k)\, e^{-H(x)} &= e^{-\xi(k, x^{(i)})}\, \psi^{(i)*}(k) \quad .
\end{aligned}
\tag{16}
$$

We look for soliton solutions to the DS equation through the consideration of elements $g \in G\ell(2\ \infty)$ of the form

$$
g = \exp\left[\sum_{\substack{i,j=1,2 \\ \ell=1,\dots,N}} a_\ell^{ij}\, \psi^{(i)}\left(k_\ell^{(i)}\right) \psi^{(j)*}\left(p_\ell^{(j)}\right) \right] \quad . \tag{17}
$$

Because of (11) we have

$$
e^{H(x)} g |\text{vac}\rangle = e^{H(x)} g\, e^{-H(x)}\, |\text{vac}\rangle \quad , \tag{18}
$$

and by using (16) we find

$$
e^{H(x)} g\, e^{-H(x)} = \exp\left[\sum_{\substack{i,j=1,2 \\ \ell=1,\dots,N}} A_\ell^{ij}(x) \right] \quad , \tag{19}
$$

where

$$A_\ell^{ij}(x) = a_\ell^{ij}\, \psi^{(i)}\left(k_\ell^{(i)}\right)\, \psi^{(j)*}\left(p_\ell^{(j)}\right)\, \exp\left[\xi\left(k_\ell^{(i)},x^{(i)}\right) - \xi\left(p_\ell^{(j)},x^{(j)}\right)\right] \quad . \tag{20}$$

The method for evaluating (14) is greatly simplified by using Wick's theorem. In this way, the crucial rule to be taken into account is that for arbitrary elements $w_n \in V^{(1)} \oplus V^{(2)}$, $w_n^* \in V^{(1)*} \oplus V^{(2)*}$ $(n=1,\dots,N)$

$$\left\langle w_1 w_1^*\, w_2 w_2^* \,\dots\, w_N w_N^*\right\rangle = \det \alpha$$

$$\alpha_{ij} = \left\langle w_i\, w_j^*\right\rangle \quad . \tag{21}$$

As a consequence, from (18)-(20) it follows that the functions (14) remain invariant if we substitute $e^{H(x)}g$ by

$$\sum_{\alpha,\beta,\sigma,\delta=0,1}\ \prod_{\ell=1}^{N} A_\ell^{11}(x)^\alpha\, A_\ell^{12}(x)^\beta\, A_\ell^{21}(x)^\sigma\, A_\ell^{22}(x)^\delta \quad , \tag{22}$$

for any order prescription of the product symbol for ℓ,m. The resulting vacuum expectation values are readily computed from (21) by using the formulas

$$\left\langle \psi_n^{(i)}\, \psi_m^{(j)}\right\rangle = \left\langle \psi_n^{(i)*}\, \psi_m^{(j)*}\right\rangle = 0 \quad ,$$

$$\left\langle \psi_n^{(i)}\, \psi_m^{(j)*}\right\rangle = \begin{cases} \delta^{ij}\, \delta_{nm} & (n < 0) \\ -\, \delta^{ij}\, \delta_{nm} & (n \geq 0) \end{cases} \quad ,$$

$$\langle \psi^{(i)}(k)\, \psi^{(j)*}(p)\, \rangle = \delta^{ij}\, \frac{p}{k-p} \quad , \quad |p| < |k| \quad .$$

Further simplifications of (22) arise from the fact that the vacuum expectation value (21) vanishes provided $w_i = w_j$ or $w_i^* = w_j^*$ for some pair $i \neq j$. For example, if we consider

$$g = \exp\left[\sum_{i,j=1,2} a^{ij}\, \psi^{(i)}(k^{(i)})\, \psi^{(j)*}(p^{(j)})\right] \quad , \tag{23}$$

then, it is enough to take

$$1 + A^{11} + A^{22} + A^{12} + A^{21} + A^{12}A^{21} + A^{11}A^{22} \quad . \tag{24}$$

Moreover, charge conservation implies that A^{12} and A^{21} are the only relevant terms for calculating τ_2 and τ_3, respectively, while they are both irrelevant for calculating τ_1. If we require, in particular, that

$$p^{(j)} = -k^{(j)*} \quad , \quad q^{21}\, k^{(1)*} = -\, a^{12*}\, k^{(2)} \quad , \tag{25}$$

and we make

$$x_1^{(1)}= u/2 \quad , \quad x_1^{(2)}= v/2 \quad , \quad x_2^{(1)}= - it/4 \quad , \quad x_2^{(2)}= it/4 \quad , \quad x_n^{(1)}= 0 \ (n\geqslant 3),$$

then the corresponding τ-functions turn out to be

$$\tau_1 = 1 + d_1 \ e^{2\lambda_R u(t)} + d_2 \ e^{2\mu_R v(t)} + d_3 \ e^{2\lambda_R u(t) + 2\mu_R v(t)} ,$$
$$\tau_3 = \tau_2^* = d_4 \ \exp\left[\lambda u(t) + \mu v(t) + i(|\lambda|^2 + |\mu|^2) t\right] , \tag{26}$$

where

$$\lambda = \lambda_R + i\lambda_I = \frac{1}{2} k^{(1)*} \quad , \quad \mu = \mu_R + i\mu_I = \frac{1}{2} k^{(2)} ,$$

$$d_i = - \frac{k^{(i)*}a^{ii}}{2k_R^{(i)}} \ (i = 1,2) \quad , \quad d_3 = \frac{k^{(1)*} \ k^{(2)*}}{4 \ k_R^{(1)} \ k_R^{(2)}} (a^{11}a^{22} - a^{12}a^{21}) ,$$

$$d_4 = -k^{(1)*} \ a^{21} \quad , \quad u(t) = u - 2\lambda_I t \quad , \quad v(t) = v - 2\mu_I t \quad .$$

Provided $d_i > 0$ (i=1,2,3) the solution (26) describes the single bidimensional soliton solution of the DS equation [4]-[5].

Multisoliton solutions are associated with elements $g \in G\ell(2\ \infty)$ of the form (17) satisfying an appropriate set of constraints in order to ensure that the DS reduction condition (5) is verified. For instance, we must require

$$p_\ell^{(j)} = - k_\ell^{(j)*} \quad , \quad a_\ell^{21} \ k_\ell^{(1)*} = - a_\ell^{12*} \ k_\ell^{(2)} ,$$

for all $\ell = 1,\dots,N$. By assuming

$$\left(k_1^{(1)}\right)_I < \dots < \left(k_N^{(1)}\right)_I \quad , \quad \left(k_1^{(2)}\right)_I > \dots > \left(k_N^{(2)}\right)_I$$
$$\left(k_\ell^{(i)}\right)_R > 0 \quad , \quad \ell = 1,\dots,N \quad , \quad i=1,2 ,$$

it is rather easy to prove that the corresponding solution of the DS equation decomposes asymptotically into a superposition of N solitons with velocities $\vec{c}_\ell = \left(- \left(k_\ell^{(1)}\right)_I , \left(k_\ell^{(2)}\right)_I\right)$. Thus, if we calculate the expression for $\tau_1\left(\vec{x}+\vec{c}_\ell t,t\right)$ as $t \to +\infty$, the dominant contribution comes from the term

$$B_\ell \left(1 + A_\ell^{11} + A_\ell^{22} + A_\ell^{12} \ A_\ell^{21} + A_\ell^{11} \ A_\ell^{22}\right) , \tag{27}$$

where

$$B_\ell = \sum \left\{ \prod_{\ell' \in I'} A^{11}_{\ell'} A^{22}_{\ell'} \right\} \left\{ \prod_{\ell'' \in I''} A^{12}_{\ell''} A^{21}_{\ell''} \right\} ,$$

with the sum symbol being extended to all partitions $I' \cup I''$ of the set $\{\ell+1,\ldots,N\}$. Now, since $|q|^2 = 4\, \partial_{uv} \log \tau_1$ and in view of (24) and (26), we deduce that (27) determines a simple soliton of the form (26) with

$$d_i = \left\langle B_\ell A^{ii}_\ell \right\rangle (B_\ell)^{-1} |_{u=v=t=o} \qquad i = 1,2,$$

$$d_3 = \left\langle B_\ell \left(A^{11}_\ell A^{22}_\ell + A^{12}_\ell A^{21}_\ell \right) \right\rangle (B_\ell)^{-1} |_{u=v=t=0} \quad .$$

The calculation of the phase-sift formulas for soliton-soliton interactions is now a simple task in view of Wick's theorem. Details will be given elsewhere [7].

REFERENCES

[1] M. Jaulent, M. Manna and L. Martinez-Alonso. Inverse Problems 5, 573 (1989).

[2] M. Jaulent, M. Manna, L. Martinez-Alonso and E. Medina, On the $A_1^{(1)}$ reduction of the two-component KP hierarchy, preprint Montpellier 1989.

[3] M. Jimbo and T. Miwa, Publ. RIMS, Kyoto Univ. 19, 943 (1983).

[4] M. Boiti, J. JP Léon, L. Martina and F. Pempinelli, Physics Lett. A 132, 432 (1988).

[5] A.S. Fokas and P.M. Santini, Phys. Rev. Lett. 63, 1329 (1989).

[6] J. Satsuma and M.J. Ablowitz, J. Math. Phys. 20, 1496 (1979).

[7] M. Jaulent, M. Manna, L. Martinez-Alonso and E. Medina, in preparation.

On the AMCP Functional Equation for the Solvable Chiral Potts Model

V.B. Matveev and A.O. Smirnov*

Department of Mathematics, Leningrad Institute of Aviation Instrumentation, Gertzena 67, SU-190000 Leningrad, USSR
*Present address: Department of Physics, Division of Mathematical Physics of Leningrad University, Pervomaiskaya 100, SU-198904 Leningrad, USSR

Quite recently Albertiny,McCoy and Perk /AMCP/ derived an exact functional equation satisfied by the eigenvalues of the transfer-matrix of the 3-states solvable chiral Potts model. Here we precise some analytic properties of the objects related with this functional equation.Particularly we find that the function f_{pq} introduced in the work of Baxter ,Perk and Au-Yang on the new solutions of the star-triangle equations is multivalued on the generalized Fermat curve exept the case when p or q are the branch points,but the coefficients of AMCP equation are singlevalued.

1.Introduction

Quite recently new integrable 2D lattice model of statistical physics - the solvable N-states chiral Potts model was discovered [1,2,3] .The crucial property of the model different from all previosly known integrable models is that associated Boltzmann weights are parametrized by the points of the algebraic curve Γ of the genus g defined by the number of states N per site: $g=N^3-2N-1$.The curve Γ itself in the homogenous projective coordinates a,b,c,d is represented [3] by the intersection of the two Fermat surfaces fixed by the choice of the "temperature" parameter k :

$$a^N + kb^N = k_1 d^N ,$$
$$ka^N + b^N = k_1 c^N , \quad k^2 + k_1^2 = 1 . \tag{1}$$

Let p and q are two points lying on the curve (1). The Boltzmann weights $w_{pq}(n)$ and $\bar{w}_{pq}(n)$ related to two kinds of neighbouring spin pairs are of the form

$$w_{pq}(n) = \prod_{j=1}^{n} \left(\frac{d_p b_q - a_p c_q \omega^j}{b_p d_q - c_p a_q \omega^j} \right) , \tag{2}$$

$$\bar{w}_{pq}(n) = \prod_{j=1}^{n} \left(\frac{\omega a_p d_q - d_p a_q}{c_p b_q - b_p c_q \omega^j} \right) , \quad \omega = \exp(2\pi i/N). \qquad (3)$$

The Boltzmann weights (2),(3) satisfy the periodicity requirement

$$w_{pq}(n+N) = w_{pq}(n), \ \bar{w}_{pq}(n+N)=\bar{w}_{pq}(n)$$

and the normalization conditions $w_{pq}(0)=1, \bar{w}_{pq}(0)=1$.

Now the transfer matrix T_{pq} is given by :

$$(T_{pq})_{l_1 \ldots l_M}^{l'_1 \ldots l'_M} = \prod_{j=1}^{M} \bar{w}_{pq}(l_j - l'_j) \, w_{pq}(l_j - l'_{j+1}), \qquad (4)$$

$$l'_{M+1} = l'_1,$$

M defines a number of columns of the lattice. Albertiny, McCoy and Perk proved that for the case N=3 the eigenvalues of T_{pq} satisfy to the following functional equation [16]

$$T_{pq}T_{pRq}T_{pR^2q} = (f^M_{pRq} f^M_{Rqp} T_{pq} + f^M_{pq} f^M_{qp} T_{pR^2q} + f^M_{pR^5q} f^M_{R^2qp} T_{pR^4q}), \qquad (5)$$

$$f^N_{pq} = \prod_{j=1}^{N} (\hat{w}_{pq}(j) / w_{pq}(j)), \qquad (6)$$

$$\hat{w}_{pq}(j) = \sum_{k=1}^{N} \omega^{kj} \bar{w}_{pq}(k), \qquad (7)$$

$$\hat{w}_{pq}(n) = \hat{w}_{pq}(0) \prod_{j=1}^{n} \frac{c_p b_q - a_p d_q \omega^j}{b_p c_p - d_p a_q \omega^j}. \qquad (8)$$

R in (5) is the conformal automorphism of the curve Γ :

$$R(a,b,c,d) = (b, \omega a, d, c), \ R^{2N} = id,$$

$$\lambda = \lim_{q \to p} T_{pRq}, \ \lambda^M = 1.$$

The main aim of this paper is to study the properties of f_{pq} and T_{pq} as the functions defined on the Riemann surface Γ .

2. Function f_{pq} and its analytical properties

Boltzmann weights $w_{pq}(n), \bar{w}_{pq}(n)$ are the solutions of the Yang-Baxter /star-triangle/ equation which can be written in the following matrix form ,introduced by the authors [18] :

$$\bar{W}_{pq} W_{pr} \bar{W}_{qr} = R_{pqr} \; W_{qr} \, W_{pr} \, W_{pq} \; . \tag{10}$$

In (10) W and $\bar{W}$ are $N \times N$ matrices with the following matrix elements:

$$(\bar{W}_{pq})_{ij} = \bar{w}_{pq}(i-j) \; , \quad (W_{pq})_{ij} = \delta_{ij} w_{pq}(i) \; . \tag{11}$$

W and $\bar{W}$ satisfy the following conditions

$$\bar{W}_{pq} \bar{W}_{qp} = \hat{w}_{pq}(0) \hat{w}_{qp}(0) I, \quad W_{pq} W_{qp} = I \; . \tag{12}$$

Taking traces from the RHS and LHS of (10) we conclude [ ] that the function $R_{pqr} = f_{pq} f_{qr} f_{pr}^{-1}$ may be represented in the form

$$R_{pqr} = \frac{Sp(\bar{W}_{pq} \, W_{pr} \, W_{qr})}{Sp(W_{qr} \bar{W}_{pr} W_{pq})} \; , \tag{13}$$

and hence is singlevalued on the curve Γ .

Below we shall show that the relations (6),(13) completely determine the analytical properties of the functions f_{pq} and of the coefficients $f_{pq} f_{qp}$, $f_{pq} f_R N_{qp}$ of the AMCP equation.

Further it would be convinient to perform some renormalization introducing the functions μ_{pq}, U_{pq} differing from f_{pq}, $\bar{W}_{pq}$ on some meromorphic factors :

$$\mu_{pq} = f_{pq} \, / \, \hat{w}_{pq}(0) \; , \tag{14}$$

$$U_{pq} = \bar{W}_{pq} / \hat{w}_{pq}(0), \tag{15}$$

$$\frac{\mu_{pq}\,\mu_{qr}}{\mu_{pr}} = \frac{Sp(U_{pq}\,W_{pr}\,U_{qr})}{Sp(W_{qr}U_{pr}W_{pr})} \tag{16}$$

Below we prefere to use nonhomogenous coordinates on /c=1/. Introduce the following notations :

$$p=(a_p, b_p, 1, d_p), \qquad P_q(l,s,t)=(\omega^{-l}b_q/d_q,\ \omega^{s}a_q/d_q\ ,1,\ \omega^{t}/d_q), \tag{17}$$

$$Q_q(l,s,t)=(\omega^{l}a_q,\ \omega^{-s}b_q, 1,\ \omega^{t}d_q)\ . \tag{18}$$

Compute the divisor D of poles and zeroes of the meromorphic function μ_{pq}^{N} considered as a function depending on p . If q is not a branch point /i.e. we are in the case of general position/ ,the divisor D is an N-th multiple of some other divisor $\widetilde{D}$. The zeroes of D are the points $P_q(l,s,t)$ specified by the conditions: $1 \leq t \leq N-1$,$1 \leq s \leq t, 1 \leq l \leq N-t$. The poles are the points $P_q(l,s,t), 1 \leq t \leq N-1$,$1+t \leq s \leq N, N-t+1 \leq s \leq N$.Despite the fact that D is N-th multiple the following proposition holds :

<u>Theorem 1</u> Function μ_{pq} /and hence f_{pq} / is not singlevalued on Γ under the condition that q is not a branch point.

<u>Proof</u> Assume that μ_{pq} is meromorphic on Γ . From the first Riemann bilinear identity it follows that

$$\int_{\partial\tilde{\Gamma}} \mu_{pq}\, d\eta_p = 0$$

where $\partial\tilde{\Gamma}$ is the boundary of the fundamental polygone of Γ and $d\eta_p$ is an arbitrary holomorphic differential.The basis $d\xi_{jkm}$ of holomorphic differentials on Γ may be constructed as follows

$$d\xi_{jkm} = a_p^{j} b_p^{-k} d_p^{-m} da_p\ ,$$

$$0 \leq k \leq N-1,\quad 0 \leq m \leq N-1,\ 0 \leq j \leq k+m-2\ . \tag{19}$$

Now choose $d\eta_p$ in the form :

$$d\eta_p = \frac{(d_p^N\, d_q^N-1)(b_p^N b_q^N - a_q^N)d_p^{1-N}\ b_p^{1-N}}{(d_p d_q-1)(d_p d_q-\omega)(b_p b_q-a_q\omega)(b_p b_q-\omega\, a_q)}\ da_p\ , \tag{20}$$

It is well known that deg(Ω) =2g-2 for an arbitrary meromorphic differential Ω defined on the algebraic curve of the genus g. Hence $d\eta_p$ has exactly $2N^2(N-2)$ zeroes: $N^2(N-2)$ zeroes located at the points $P_q(l,s,t)$, $1 \leq l \leq N$, $1 \leq t \leq N$, $2 \leq t \leq N-1$ and $N^2(N-2)$ zeroes at the points $P_q(l,s,t)$, $1 \leq l \leq N$, $2 \leq s \leq N-1$, $1 \leq t \leq N$. If the point $P_q(l,s,t)$ belongs to the intersection of these two sets it represents a double zero of $d\eta_p$. It is easy to check that $\mu_{pq} d\eta_p$ admits a unique pole of the first order situated at the point $P_q(N,N,1)$ and hence

$$\int_{\partial \tilde{\Gamma}} \mu_{pq} d\eta_p = 2\pi i \operatorname{res}(\mu_{pq} d\eta_p)\Big|_{P_q(N,N,1)} \neq 0$$

This contradicts to singlevaluedness of μ_{pq} and hence proves the Theorem 1.

Corollary 1 The function μ_{pq} as a function depending on q for a fixed p is not singlevalued on Γ under the condition that p is not one of the branch points.

Proof From (2),(3),(11) and (15) it follows that

$$W_{pp} = I, \quad U_{pp} = I. \tag{21}$$

From (16) it is now evident that

$$\mu_{pp} = 1, \quad \mu_{pq} \mu_{pq} = 1 \tag{22}$$

By the theorem I μ_{qp} is multivalued as a function depending on q and hence by virtue of (22) μ_{pq} is also nonsinglevalued as a function depending on q.

Remark 1 In the case N=2 theorem I reduces to the well known fact from the theory of elliptic functions ,saying that there is no double periodic meromorpfic functions with a single simple pole in the parallelogramm of periods.

Corollary 2. f_{pq} as a function depending on p is not singlevalued on Γ .

Theorem 2 The products $f_{pq} f_{qp}$ and $f_{pq} f_{Rqp}$ are the singlevalued meromorphic functions of p and q on Γ .

Proof Formulae (2),(3) and (7) include the relations

$$W_{qRq}(n) = \delta_{no}, \tag{23}$$

$$\bar{W}_{qRq}(n) = 1, \tag{24}$$

$$W_{qRq}(n) = N_{no}. \tag{25}$$

Using the relations (14),(22),(8) and taking the limit $p \to Rq$ in (6) we obtain $\mu_{Rqq}=k^{(N-1)/N}$, $\mu_{qRq}=k^{(1-N)/N}$. (26)

Now we can apply the formula (16) in which we consider another limit $r \to Rq$. Taking into account (11),(12),(15),(22)-(26) and the relations [15]

$$w_{pRq}(j) = \bar{w}_{qp}(j) \tag{27}$$

we obtain

$$\mu_{pq} \ \mu_{Rqp} = \frac{1}{N} k^{(N-1)/N} \hat{w}_{pRq}(o) \ \hat{w}_{qp}(o) \tag{28}$$

So far we have proved that the products $\mu_{pq} \ \mu_{Rqp}$ are meromorphic functions. Hence $f_{pq} f_{Rqp}$ and $f_{pq} f_{qp}$ are also meromorphic and can be represented in the following form:

$$f_{pq} f_{qp} = \hat{w}_{pq}(o) \ \hat{w}_{qp}(o) \tag{29}$$

$$f_{pq} f_{Rqp} = k^{(N-1)/N} \hat{w}_{pq}(o) \hat{w}_{qp}(o) \hat{w}_{pRq}(o) \hat{w}_{Rqp}(o)/N \tag{30}$$

<u>Corollary 3</u> The meromorphic functions $f_{pq} f_{R^N qp}$ may be expressed in terms of the Boltzmann weights :

$$f_{pq} f_{R^N qp} = k^{N-1} \prod_{j=o}^{N} \hat{w}_{pR^j q}(o) \hat{w}_{R^j qp}(o)/N^N \tag{31}$$

<u>Corollary 4</u>. Functions $\hat{w}_{pq}(o)$ satify the relations:

$$\prod_{j=1}^{2N} \hat{w}_{p_R j_q}(o) \ \hat{w}_{R^j qp}(o) = N^{2N} k^{2(1-N)} \tag{32}$$

Surely (31) and (32) follows directly from (22) and (28).

3. Analytical properties of the solutions of the functional equations

The main property of the integrable models of statistical mechanics on the lattice is the commutativity of transfer matrices for different values of the spectral parameter. In the solvable Potts model case this property reads :

$$[T_{pq}, T_{pr}] = 0 , \tag{33}$$

$$[T_{pq}, T_{rq}] = 0. \tag{34}$$

Using these relations it is easy to check the following proposition.

<u>Lemma 1</u> The eigenvalues of the transfer matrix T_{pq} /which are denoted by the same letter T_{pq}/ are meromorpfic functions of p on the curve Γ .

<u>Proof</u> From (34) it follows That the eigenvalues are meromorphic functions of p by the reason that associated eigenvector are p-independent. Quite similarly we can deduce that these eigenvalues are meromorphic functions depending on q.

It is necessary to mention that due to the relations (33), (34) the eigenvalues of the transfer matrix satify to the same functional equation (5) as the transfer matrices themselves.

<u>Theorem 3</u> Eigenvalues of the 3-states transfer matrix are the meromorphic solutions of the following equation with meromorphic coefficients

$$T_{pq}\, T_{pRq}\, T_{pR^2q} = \lambda \,\big(S^M(p,Rq)\, T_{pq} + S^M(p,q) T_{pR^2q} + 27^M k^{-2M} T_{pR^4q} \cdot$$
$$\cdot S^{-M}(p,R^3q) S^{-M}(p,R^4q)\big), \tag{35}$$

$$S(p,q) = \hat{w}_{pq}(o)\, \hat{W}_{qp}(o) . \tag{36}$$

Theorem 3 follows directly from (5), theorem 2, Lemma 1 and equality (32)

<u>Remark 2</u> Taking into account that the eigenvalues T_{pq} of the transfer matrix and the coefficients $S(p,q)$ of the functional equation (35) are meromorpfic functions on Γ we can obtain from (35) five similar equations replacing q by R^jq, j= 0,...,5. However only three of the obtained equations are independent For example we get three independent equations adding to (35) the following relations:

$$T_{pR^2q}\ T_{pR^3q}T_{pR^4q} = \lambda\Big((27)^{M_k-2M}\, S^{-M}(p,R^5q)S^{-M}(p,q)\, T_{pq} + S^M(p,R^3q)T_{pR^2q} + S^M(p,R^2q)T_{pR^4q}\Big), \quad (37)$$

$$T_{pR^4q}\ T_{pR^5q}T_{pq} = \lambda\Big(S^M(p,R^4q)T_{pq} + (27)^{M_k-2M}\, S^{-M}(p,Rq)S^{-M}(p,R^2q)\cdot T_{pR^2q} + S^M(p,R^5q)T_{pR^4q}\Big). \quad (38)$$

Three other equations follows from (35),(37),(38) and the relation (32).

Appendix 1 Zeroes and poles of $\hat{w}_{pq}(o)$.

From the relations (2) ,(7),(8) we get the following representation for $\hat{w}_{pq}(o)$ and $\hat{w}^{-1}_{pq}(o)$:

$$\hat{w}_{pq}(o) = \prod_{j=1}^{n} \frac{b_p c_q - d_p a_q \omega^j}{c_p b_q - a_p d_q \omega^j} \left[\sum_{n=1}^{N} \omega^{mn} \prod_{j=1}^{n} \frac{\omega\, a_p d_q - d_p a_q \omega^j}{c_p b_q - b_p c_q \omega^j}\right],$$

$$\hat{w}^{-1}_{pq}(o) = \prod_{j=1}^{n} \frac{c_p b_q - b_p c_q \omega^j}{a_p d_q - d_p a_q \omega^j} \cdot \left[\frac{1}{N}\sum_{n=1}^{N} \omega^{-mn} \prod_{j=1}^{m} \frac{c_p b_q - a_p d_q \omega^j}{b_p c_q - d_p a_q \omega^j}\right]$$

Using these representations it is easy to prove that the zeroes of $\hat{w}_{pq}(o)$ are simple and located at the points $P_1(l,s,t), 1 \leq t \leq N$, $2 \leq l \leq N, t+1 \leq s \leq l+t-1$. The poles of $w_{pq}(o)$ are situated at the points $Q_q(l,s,t)$, $1 \leq t \leq N, t+1 \leq l \leq t+N-1$, $1 \leq s \leq l-t$. Hence the zeroes of the function $\hat{w}_{pq}(o)\hat{w}_{qp}(o)$ are also simple and located at the points $P_q(l,s,t), s \neq l+t \pmod N$, and the poles are described like so: double poles are situated at the points $Q_q(l,s,t)$ $s=l-t \pmod N, s \neq o \pmod N$, $l-t \neq o \pmod N$, the simple poles are located at the points $Q_q(l,s,t), s \neq o \pmod N, l-t \neq o \pmod N$, $s \neq l-t \pmod N$.

Aknowledgments We wish to thank Professors B.McCoy and J.H.H.Perk for sending their very interesting work [16].The main content of this a article was borned trying to clarify some points of their study.We are also grateful to V.Korepin for discussion of results and information about the work of Bazanov and Stroganov [19] ,containing among other results some derivation of the functional equation (5) different from the work [16], and starting from the relation with the six-vertex model.When this work was finished I.Korepanov kindly informed us about his works deponized by VINITY and containing the description of the solutions of Yang-Baxter equation related with the algebraic curves of genus greater then one[20-22] arising from the study of hidden symmetries of the six-vertex model .His study was mainly based on the ideas of the work of Krichever[23] in which however the solutions of the genus greater than one have not been discovered.

REFERENCES

[1] Au-Yang, H., McCoy, B., Perk, J.H.H., and Yan M.L., Commuting transfer matrices in the chiral Potts models : solutions of star-triangle equations with genus, Phys. Lett. A 123, 219-223 (1987).

[2] McCoy, B., Perk, J.H.H., and Sah, C-H., Commuting transfer matrices for the four-state self-dual chiral Potts model with a genus-three uniformizing Fermat curve, Phys. Lett. A 125, 9-14 (1987).

[3] Baxter, R.J., Perk, J.H.H., and Au-Yang, H., New solutions of the star-triangle relations for the chiral Potts model. Phys. Lett. A 128, 138-142 (1988).

[4] Baxter, R.J., Free energy of the solvable chiral Potts model, J. Stat. Phys. 52, 639-667 (1988).

[5] Baxter, R.J., The superintegrable chiral Potts model, Phys. Lett A 133, 185-189 (1988).

[6] Perk, J.H.H., Onsager star-triangle equation : master key to integrability, Proc. Tuniguchi Conference, Kyoto, October, 1988 (to appear).

[7] Baxter, R.J., Solving models in statistical mechanics, Preprint University of Canberra 1988, to appear in Proc. of Tuniguchi Conf. Kyoto, October 1988.

[8] Albertini, G., McCoy, B., and Perk, J.H.H., First-order transition in the ground state of the superintegrable chiral Potts model, Preprint ITP-SB-88-58 of Stony Brook University, 1988.

[9] Bazhanov, V.V. and Reshetikhin, N. Yu., Critical RSOS models and conformal field theory, Preprint Institute for High Energy Physics, IFVE 87-102, 1987.

[10] Matveev, V.B. and Smirnov. A.O., On the Riemann theta function of as trigonal curve and solutions of the Boussinesq and KP equations, Lett. Math. Phys. 14, 25-31 (1987).

[11] Belokolos, E.D., Bobenko, A.I., Matveev, V.B., and Enol'sky, V.Z. Algebrogeometrical superposition principles for the finite-gap solutions of the nonlinear integrable evolution equations, Russian Math. Surveys 41, 1-49 (1986).

[12] Smirnov, A.O., Matrix version of the Appel's theorem and the reductions of the multidimensional theta-functions. Math. Sb. (in Russian) 133, 382-391 (1987).

[13] Smirnov, A.O., Finite-gap solutions of the Abelian Toda lattice of the genus 4 and 5 in elliptical functions, Theor. Math. Phys. 78, 11-21 (1989).

[14] Babitch,M.V., Bobenko, A.I., and Matveev, V.B., Reductions of the Riemann theta functions of the genus g to the theta-functions of the lower generee and the symmetries of the algebraic curves, DAN SSSR (Soviet Doklady) 272, 13-17 (1983) (in Russian).

[15] Baxter, R., Exactly Solved Models in Statistical Mechanics, Academic Press, London, 1982, pp. 1-500.

[16] Albertini, G., Mc Coy, B., and Perk, J.H.H., Eigenvalue spectrum of the superintegrable chiral Potts model, Preprint ITP-SB-89-1.

[17] Albertini, G., McCoy, B., and Perk, J.H.H., Commensurate-incommensurate transitions in the ground state of the superintegrable chiral Potts model, Phys. Lett. A. 135, 159-167 (1989).

[18] Matveev V.B., Smirnov A.O. Some comments on the solvable chiral Potts model LMP 18-1989 to be published.

[19] Bazhanov V.V. Stroganov Yu G. Chiral Potts model as a descendant of the six-vertex model June 1989 Serpuhov Preprint.

[20] Korepanov I.G. Vacunme curves of L-operators related with the six-vertex model Dep. VINITY N2271-B86, 1986 (In Russian).

[21] Korepanov I.G. Hidden symmetries of the six-vertex model ibid. N1472-B87

[22] Korepanov I.G. On the spectrum of the transfer-matrix of the six-vertex model ibid N 3268-B87.

[23] Krichever I.M. On the solution of the Baxter equation.

Painlevé-Darboux Transformation in Nonlinear PDEs

M. Musette

Dienst Theoretische Natuurkunde, Vrije Universiteit Brussel,
Pleinlaan 2, B-1050 Brussels, Belgium

1. Invariant Painlevé analysis.

Let us first compare some results concerning invariance properties in the Painlevé analysis of a partial differential equation

$$E(x, t, u, Du) = 0 \tag{1}$$

which depends polynomially on its solution u and its partial derivatives Du with respect to x and t. Considering a series expansion for u in the neighbourhood of the singular manifold $\varphi(x,t) = 0$, we are looking for, as a solution of (1), the formal expression

$$u = \sum_{j=0}^{\infty} u_j \chi^{j+p} \quad , \; (p \text{ constant}, \quad u_0 \neq 0) \tag{2}$$

where the expansion variable χ goes to zero as φ in the following way:

$$\chi = \frac{\alpha\varphi}{\beta\varphi + \gamma} \quad , \quad \frac{\alpha}{\gamma} \neq 0 \tag{3}$$

and the coefficients α, β, γ and u_j are only functions of the derivatives $D\varphi$ of φ. It is assumed that, at least, the first derivative φ_x or φ_t is different from zero. Then, the substitution of the series (2) in equ.(1) leads to the corresponding expansion for E

$$E = \sum_{j=0}^{\infty} E_j \chi^{j+q} \qquad (q \text{ constant}) \tag{4}$$

where q plays in (4) the same role as p in (2).
The equation $E_0 = 0$ defines u_0 while, for $j > 0$, the equation $E_j = 0$ represents a recurrence relation between the coefficient u_j and all the preceding coefficients $\{u_k\}$ for $k \leq j - 1$.
The invariance properties of the (u_j, E_j) for a transformation on φ depends on the suitable choice of the expansion variable χ. Indeed, if $\beta = 0$ in (3), the choice considered by Weiss, Tabor and Carnevale [1],

$$\chi = \frac{\varphi}{\varphi_x} \tag{5}$$

implies that the functions (u_j, E_j) are invariant for the transformation:

$\varphi \to a\varphi + b$ (a, b constants).

If $\beta \neq 0$ in (3), the choice considered by Conte [2]

$$\chi = \left(\frac{\varphi_x}{\varphi} - \frac{1}{2} \frac{\varphi_{xx}}{\varphi_x} \right)^{-1} \tag{6}$$

has the great advantage to extend the invariance of the (u_j , E_j) to the group of the homographic transformation acting on φ:

$$H : \varphi \to \frac{a\varphi + b}{c\varphi + d} \quad , \quad ad - bc \neq 0 \qquad (a , b , c , d \text{ constants}) \tag{7}$$

This result implies that the coefficients (u_j , E_j) are only functions of two differential elementary invariants:

$$S = \left(\frac{\varphi_{xx}}{\varphi_x} \right)_x - \frac{1}{2} \left(\frac{\varphi_{xx}}{\varphi_x} \right)^2 \tag{8}$$

$$C = - \frac{\varphi_t}{\varphi_x} \tag{9}$$

linked by the cross derivative condition $(\varphi_t)_{xxx} = (\varphi_{xxx})_t$ which is expressed as:

$$S_t + C_{xxx} + 2C_xS + CS_x = 0 \tag{10}$$

Consequently, important technical simplifications occur in the different steps of the Painlevé analysis, making easier the obtaining of the general or particular solutions of the p.d.e. under consideration [3].

The invariance property for the transformation (7) on φ of every result of a p.d.e. Painlevé analysis has been rigorously proven by Conte. It is a consequence to the fact that, as the expression (6) is a particular solution of a Riccati system which, for the variable $v = \chi^{-1}$, has the form

$$v_x = -v^2 - \frac{S}{2} \tag{11.a}$$

$$v_t = Cv^2 - C_xv + \frac{1}{2}(CS + C_{xx}) \tag{11.b}$$

every derivative of χ can be replaced by a polynomial in χ with invariant coefficients. Let us remark that the general solution of (11.a,b) is

$$v(\varphi_0) = \frac{\varphi_x}{\varphi - \varphi_0} - \frac{1}{2} \frac{\varphi_{xx}}{\varphi_x} \quad , \quad (\varphi_0 \text{ complex constant}) \tag{12}$$

and, by the transformation $v = \frac{\psi_x}{\psi}$, the system (11.a,b) is equivalent to the linear system

$$\psi_{xx} + \frac{S}{2}\psi = 0 \tag{13.a}$$

$$\psi_t + C\psi_x = \left(\frac{C_x}{2} + f(t) \right)\psi \tag{13.b}$$

where f(t) is an arbitrary function and $\psi = \psi_1 - \varphi_0 \psi_2$ is given in terms of two particular solutions ψ_1, ψ_2 with wronskian equal to 1: $\psi_1 = (\varphi_x)^{-1/2}\varphi$, $\psi_2 = (\varphi_x)^{-1/2}$.

2. Darboux transformation in Painlevé analysis.

If one sets

$$\varphi = \frac{\Psi_1}{\Psi_2} \tag{14}$$

the transformation (7) on φ is equivalent to the following linear transformation acting on the couple (Ψ_1, Ψ_2)

$$(\Psi_1, \Psi_2) \rightarrow (a\Psi_1 + b\Psi_2, c\Psi_1 + d\Psi_2) \tag{15}$$

Therefore, one may assume that Ψ_1 and Ψ_2 are two solutions of a couple of linear equations corresponding, if it exists, to the Lax pair associated to the p.d.e. under consideration. Let us remark that even if the p.d.e. is partially integrable, the linear system (13.a,b) can it be associated, provided one knows an explicit form of the invariants S and C.
Replacing in (12) the expression (14) for φ, we obtain that

$$v(\varphi_0) = \frac{\Psi_x}{\Psi} - \frac{1}{2}\frac{W_x}{W} \tag{16}$$

where $\quad \Psi = \Psi_1 - \varphi_0 \Psi_2 \quad$ and $\quad W = \Psi_{1,x}\Psi_2 - \Psi_{2,x}\Psi_1 \neq 0$

From (15), it follows that

$$Y = \frac{W_x}{W} \tag{17}$$

is a homographic invariant function.
In order to determine the Lax pair, we think that the appropriate expansion variable for u is the function Ψ_x/Ψ, related to $v(\varphi_0)$ by

$$\frac{\Psi_x}{\Psi} = v(\varphi_0) + \frac{1}{2}Y \tag{18}$$

where v satisfies the Riccati system (11.a,b) and Y is an adjustable function. The Painlevé analysis is performed in assuming the existence of a Darboux transformation [4] between 2 different solutions of the p.d.e.
Let us assume that the Painlevé property of a p.d.e. should be equivalent to the 3 following requirements:

1. The p.d.e. under consideration $E(x, t, u, Du) = 0$ is or can be transformed, after a change of variable, into a *polynomial expression in u and Du*, where the maximum *order* of the partial derivatives Du is assigned to $k > 1$.

2. After the substitution in E of the formal expansion for u, the recurrence relation $\{E_j\} = 0$ imply the existence of *k-1 arbitrary functions* u_r (compatible resonances).
3. It exists a *truncated expansion for u,* referred to as the expression u_T, which is a *solution of the equ. E = 0.* This requirement can be realized if the power p in (2) corresponds to a negative integer value. Therefore, one can define

$$u_T = \sum_{j=0}^{-p} u_j \left(\frac{\Psi_x}{\Psi}\right)^{-j-p}$$

and the corresponding truncated expansion for E

$$E_T = \sum_{j=0}^{-q} E_j \left(\frac{\Psi_x}{\Psi}\right)^{-j-q}$$

implying that p - q equations $\{E_j\} = 0$ for $j \geq -p +1$ must be satisfied. Those equations shall be named *Painlevé-Darboux equations* (P-D) because they produce a transformation between u and the "constant" coefficient "$\tilde{u}_{-p}$" of the truncated series under the form:

$$u - \tilde{u}_{-p} = D(\ln \Psi) \tag{19}$$

where D represents a linear differential operator acting on the logarithm of Ψ which corresponds to the general solution of the linear system associated to the nonlinear p.d.e.

Then the proposition linking the Painlevé property to the integrability will consist in obtaining *the Lax pair without assuming its order.* For this purpose, the equation satisfied by the invariant function Y should be determined in function of $\tilde{u}_{-p}$ under the condition that u and $\tilde{u}_{-p}$ are both solutions of the p.d.e.

3. Comparison between the three Painlevé analyses.

In the Painlevé analysis introduced by Weiss, Tabor and Carnevale, if one associates the following truncated expansion to the solution u:

$$u(\varphi_0) = \sum_{j=0}^{-p} u'_j \left(\frac{\varphi_x}{\varphi - \varphi_0}\right)^{-j-p} \qquad \text{(p negative integer)} \tag{20}$$

the "constant" coefficient u'_{-p} is automatically an other solution of the equ. E = 0 if the solutions of the P-D equations can be determined. However, as the coefficients E'_j of the expansion for E are only invariant for a subgroup of H, in many cases it is a difficult task to solve the P-D equations.

In the "invariant" analysis considered by Conte, the P-D equations only depend on two invariant functions S and C by means of differential equations simpler to solve than the p.d.e. under consideration. The truncated series for u

$$u(\varphi_0) = \sum_{j=0}^{-p} \bar{u}_j \, v^{-j-p} \tag{21}$$

is written in terms of the expansion variable (12) solution of the system (11.a,b). The coefficients $\bar{u}_j$ are invariant expressions of S and C but, in general, the "constant" coefficient $\bar{u}_{-p}$ is not a solution of the p.d.e. Two particular solutions corresponding to the values $\varphi_0 = 0$ and $\varphi_0 = \infty$ are related by means of a linear differential operator acting on log φ:

$$u(0) - u(\infty) = \sum_{\substack{i,j\geq 0 \\ i+j=1}}^{-p} a_{ij}^{(p)}(S,C)\,(\log \varphi)_{ix,jt} \tag{22}$$

where the coefficients a_{ij} are functions of the invariant coefficients $\bar{u}_j$ of the expansion (21).
For p = -1, one has that

$$u(0) - u(\infty) = \bar{u}_0\,(\log \varphi)_x \tag{23}$$

Several conservative p.d.e.'s written in their potential form, like KdV, KdV5, Sawada-Kotera and Boussinesq equations have in their solutions the dominant exponent p equal to -1. In all cases, $\bar{u}_0 = c$ (c: constant) and the relation between 2 solutions is simply given by

$$u(0) - u(\infty) = c\,(\log \varphi)_x \tag{24}$$

For p = -2, two equivalent representations, deduced from the Riccati system (11.a,b), are

$$u(0) - u(\infty) = \begin{cases} -\bar{u}_0\,(\log \varphi)_{xx} + \bar{u}_1\,(\log \varphi)_x & (25.a) \\[2ex] \dfrac{\bar{u}_0}{C}\,(\log \varphi)_{xt} + \left(\bar{u}_1 + \bar{u}_0\,\dfrac{C_x}{C}\right)(\log \varphi)_x & (25.b) \end{cases}$$

The value p = -2 is, for example, associated to the solution of the Sine-Gordon equation ($U_{xt} - \sin U = 0$) written, after the change of variable exp(iU) = u, in the polynomial form:

$$2uu_{xt} - 2u_x u_t - u^3 + u = 0 \tag{26}$$

The Painlevé analysis determines the coefficients $\bar{u}_0 = -4C$, $\bar{u}_1 = 4C_x$ and from (25.b), we obtain the following relation

$$u(0) - u(\infty) = -4(\log \varphi)_{xt} \tag{27}$$

For p = -3, one of the representations including only the dependence of S is

$$u(0)-u(\infty)=\frac{\bar{u}_0}{2}(\log\varphi)_{xxx}-\bar{u}_1(\log\varphi)_{xx}+\left(\bar{u}_2-\frac{S}{2}\bar{u}_0\right)(\log\varphi)_x \tag{28}$$

The value $p = -3$ is, for example, associated to the solution of the Kuramoto-Sivashinsky equation:

$$u_t + u_{xx} + u_{xxxx} + uu_x = 0 \tag{29}$$

for which the Painlevé analysis [5] gives as values for the coefficients of the truncated series (21): $\bar{u}_0 = 2c_1$, $\bar{u}_1 = 0$, $\bar{u}_2 = c_2 + c_1 S$ ($c_1 = 60$, $c_2 = 60/19$). From the relation (28), they also give:

$$u(0) - u(\infty) = c_1 (\log\varphi)_{3x} + c_2 (\log\varphi)_x \tag{30}$$

In the third analysis we here consider, one introduces in the truncated expansions for u and E an explicit dependence on the function Ψ which represents the general solution of the yet unknown linear system associated to the p.d.e. The expressions are:

$$u(\varphi_0) = \sum_{j=0}^{-p} u_j \left(\frac{\Psi_x}{\Psi}\right)^{-j-p} \tag{31}$$

$$E(\varphi_0) = \sum_{j=0}^{-q} E_j \left(\frac{\Psi_x}{\Psi}\right)^{-j-q} \tag{32}$$

with the expansion variable related to $v(\varphi_0)$ and Y by (18) and the coefficients (u_j, E_j) depending on the 3 homographic invariants S, C and Y.
The function $u(\varphi_0)$ also possesses the following development

$$u(\varphi_0) = \sum_{\substack{i,j\neq 0\\ i+j=1}}^{-p} \tilde{a}_{ij}^{(p)}(S, C, Y)\,(\log\Psi)_{ix,jt} + \tilde{u}_{-p}(S, C, Y) \tag{33}$$

and we impose the constraint that

$$E(\tilde{u}_{-p}) = 0 \tag{34}$$

Comparing the expressions obtained for the difference $u(0) - u(\infty)$ by exploiting relation (33) or (22), we deduce the following identity between the coefficients:

$$\tilde{a}_{ij}^{(p)}(S, C, Y) = a_{ij}^{(p)}(S, C) \tag{35}$$

which, therefore, must be constant.
From (18), the relation between the coefficients of the two expansions (21) and (31) is

$$\bar{u}_j = \sum_{k=0}^{j\leq -p} u_k \frac{(-k-p)!}{(j-k)!\,(-j-p)!}\left(\frac{1}{2}Y\right)^{j-k} \tag{36}$$

Therefore, for $p = -1$, we obtain between u and u_1 the following transformation:

$$u(\varphi_0) = u_0\,(\ln \Psi)_x + u_1 \tag{37}$$

with the constraint

$$E(u) = E(u_1) = 0$$

and in this special case, the relation (36) gives:

$$\bar{u}_0 = u_0 \quad , \quad \bar{u}_1 = u_1 + u_0 Y/2.$$

For $p = -2$, two representations of the Darboux transformation are:

i) $$u(\varphi_0) = -u_0(\log \Psi)_{xx} + (u_0 Y + u_1)(\log \Psi)_x + \tilde{u}_2 \tag{38}$$

$$\tilde{u}_2 = \frac{u_0}{2}(Y_x - \frac{Y^2}{2} - S) + u_2 \tag{39}$$

ii) $$u(\varphi_0) = \frac{u_0}{C}(\log \Psi)_{xt} + [u_0(Y + \frac{C_x}{C}) + u_1](\log \Psi)_x + \tilde{u}_2 \tag{40}$$

$$\tilde{u}_2 = \frac{u_0}{2}(-S - \frac{C_{xx}}{2} - \frac{1}{C}Y_t - \frac{Y^2}{2} - \frac{C_x}{C}Y) + u_2 \tag{41}$$

with the constraint

$$E(u) = E(\tilde{u}_2) = 0$$

while the relation (36) gives:

$$\bar{u}_0 = u_0 \quad , \quad \bar{u}_1 = u_0 Y + u_1 \quad , \quad \bar{u}_2 = u_0(\frac{Y}{2})^2 + u_1\frac{Y}{2} + u_2$$

As an example of application, we consider the potential KdV equation:

$$u_t + u_{xxx} - 3u_x^{\,2} = 0 \tag{42}$$

for the function u, whose truncated expansion (31) might be a solution if $p = -1$ and $u_0 = -2$. The conditions of existence are:

$$Y_t + 3Y_x^2 + Y_{xxx} - 6Y_x\, u_{1,x} = 0 \tag{43}$$

$$u_{1,t} + u_{1,3x} - 3u_{1,x}^2 = 0 \tag{44}$$

the last equation (44) being introduced as a constraint in the analysis to determine the function Y.

The equation (43) is transformed in a fourth degree polynomial in Y by assuming that the first derivatives Y_x and Y_t satisfy a Riccati system. By identifying to zero each coefficient of this polynomial, one obtains:

$$Y_x = -\frac{1}{2}Y^2 + 2u_{1,x} + \lambda \tag{45}$$

$$Y_t = (-u_{1,x} + \lambda)Y^2 + 2u_{1,xx}Y + 4u_{1,x}^2 - 2\lambda u_{1,x} - 2u_{1,xxx} - 2\lambda^2 \qquad (46)$$

where λ is an arbitrary constant parameter.
We remark that the cross-derivative condition $Y_{xt} = Y_{tx}$ is equivalent to:

$$(u_{1,t} + u_{1,xxx} - 3u_{1,x}^2)_x = 0 \qquad (47)$$

such that the function Y plays in this analysis the role of a pseudopotential, as introduced by Wahlquist and Estabrook [6].
On the other side, the couple of equations (45),(46) is equivalent to:

$$Y_t + Y_{3x} - \frac{3}{2}Y^2 Y_x + \frac{3}{2}\lambda Y_x = 0$$

which is the modified KdV equation for Y.
Less trivial applications to other p.d.e.'s are presently undertaken in collaboration with R. Conte. Definite results shall be published elsewhere.

References

[1] Weiss J., Tabor M. and Carnevale G., J. Math. Phys. 24 , 522 (1983).
[2] Conte R., P.L.A. 140 , 383 (1989).
[3] Conte R. and Musette M., J. Phys. A: Math. Gen. 22 , 169 (1989).
[4] Levi D., Rep. on Math. Phys. 23 , 41 (1986)
Levi D. and Ragnisco A., Inverse Problems 4 , 815 (1989).
[5] Musette M., "Painlevé analysis for integrable and nonintegrable p.d.e.'s" presented at the VI International Workshop "Solitons and applications", 25-27 august 1989, Dubna, USSR - VUB/TENA/89/10.
[6] Wahlquist H.D. and Estabrook F.B., J. Math. Phys. 16 , 1 (1975)
Estabrook F.B. and Wahlquist H.D., J. Math. Phys. 17 , 1293 (1976).

Solitons and Spectral Transform for the Kadomtsev-Petviashvili Equation

F. Pempinelli

Laboratoire de Physique Mathématique, USTL, F-34095 Montpellier Cédex 2, France
Permanent address: Dipartimento di Fisica dell'Università and Sezione INFN di Lecce, I-73100 Lecce, Italy

ABSTRACT. A new spectral transform for the Kadomtsev-Petviashvili equation allows us to include wave solitons in the theory of the Inverse Spectral Transform. The Cauchy problem is completely solved. The orthogonality relations for the eigenfunctions of the associated time dependent Schrödinger equation are obtained.

1. The Direct Problem

The Kadomtsev-Petviashvili I (KPI) equation

$$(u_t - 6uu_x + u_{xxx})_x + 3\, u_{yy} = 0 \quad ; \quad u = u(x,y\, ;\, t) \tag{1.1}$$

generalizes the Korteweg-de Vries equation in 2+1 dimensions. It admits wave solitons as solutions.

The principal spectral problem, which is used to linearize the initial value problem for the KPI, is given by the "time" dependent Schrödinger equation

$$\Phi_{xx} - i\, \Phi_y - u\, \Phi = 0 \ . \tag{1.2}$$

The spectral parameter k is introduced by requiring

$$\Phi(x,y,k)\, e^{ikx-ik^2y} \longrightarrow \mathbb{1} + O\left(\frac{1}{k}\right) , \quad \text{as} \quad k \to \infty \ . \tag{1.3}$$

Manakov [1] discovered that the eigenfunction Φ is sectionally holomorphic in the spectral parameter k and that, in contrast with the one dimensional case, one has to introduce a *non local* Riemann-Hilbert problem. Therefore the spectral transform (ST) $f(k,\ell)$ of the potential $u(x,y)$ depends on *two* spectral parameter k and ℓ. Subsequently, Fokas and Ablowitz [2], by using the so called $\bar{\partial}$ approach, were able to express the spectral data $f(k,\ell)$ in terms of the potential $u(x,y)$ and of an additional *non-analytic* eigenfunction. Recently, Boiti, Léon and Pempinelli [3] proposed a new ST $r(k,\ell)$, which does not need this non analytic eigenfunction and can be expressed explicitly in terms of the potential $u(x,y)$ and the eigenfunction Φ of Manakov. By means of this new ST we can also solve the direct problem in the case when the potential describes n interacting wave solitons, i.e. it goes to a constant in some directions $x-2\lambda_i\, y = \text{const}$ ($\lambda_i \in \mathbb{R}$, $i=1,2,\dots,n$).

By using the Manakov's Green function [1] for eq. (1.2), one gets the following integral equation

$$\Phi^{\sigma}(x,y,k) = e^{-ikx+ik^2y} - \int dp\, \mathfrak{R}^{\sigma}(k,p,y)\, e^{-ipx+ip^2y} \tag{1.4}$$

($\sigma = \pm$)

where

$$\mathfrak{R}^{\sigma}(k,p,y) = \frac{\sigma}{2\pi i}\, \epsilon(p-k) \int d\xi \int d\eta\, \theta(\sigma(y-\eta)\ (p-k)) \ .$$

$$.\ e^{ip\xi-ip^2\eta}\, u(\xi,\eta)\, \Phi^{\sigma}(\xi,\eta,k) \ . \tag{1.5}$$

These equations furnish the boundary values $\Phi^{\pm}$ of the eigenfunction Φ on both sides Im $k \lessgtr 0$.

If $u(x,y)$ is not going to zero as $x^2+y^2 \to \infty$, the function $\mathfrak{R}^{\sigma}$ is not well defined in $k = p$ and it must be replaced by the following formula obtained by using the Green function proposed by Pogrebkov [4].

$$\mathfrak{R}^{\sigma}(k,p,y) = \frac{\sigma}{2\pi i}\frac{1}{p-k-i0}\left\{|p-k| \int d\eta \int d\xi\, \theta(\sigma(y-\eta)\ (p-k))\ .\right.$$

$$\left. .\ e^{ip\xi-ip^2\eta}\, u(\xi,\eta)\, \Phi^{\sigma}(\xi,\eta,k)\right\} \ . \tag{1.6}$$

Products of distributions and integrations have to be done in the order indicated from the right to the left. An additional bracket has been introduced to stress this fact. When it is not differently indicated the integration is performed all along the real axis from $-\infty$ to $+\infty$. For a potential u going sufficiently fast to zero for large x^2+y^2 the order is unessential and the distribution sign $(p-k)$ can be substituted for $\frac{1}{p-k-i0}|p-k|$ recovering the Green function used by Manakov [1] and Fokas, Ablowitz [2].

The main quantity to study in order to get complete information on the ST of the KPI equation is the scalar product (u is chosen to be real)

$$\langle \Phi^{\sigma}(k),\ \Phi^{\sigma'}(q) \rangle \equiv \frac{1}{2\pi}\int dx\, \Phi^{\sigma}(x,y,k)\ \bar{\Phi}^{\sigma'}(x,y,q) \ . \tag{1.7}$$

The bar means complex conjugate.

By inserting above the integral equation (1.4) for Φ^{σ} and $\bar{\Phi}^{\sigma'}$ we get

$$\langle \Phi^{\sigma}(k),\ \Phi^{\sigma'}(q) \rangle - \delta(k-q) =$$
$$= -\mathfrak{R}^{\sigma}(k,q,y) - \bar{\mathfrak{R}}^{\sigma'}(q,k,y) + \int dp\, \mathfrak{R}^{\sigma}(k,p,y)\, \bar{\mathfrak{R}}^{\sigma'}(q,p,y) \ . \tag{1.8}$$

By deriving with respect to y the r.h.s. and by using once more the integral equations for Φ^{σ} and $\bar{\Phi}^{\sigma'}$ we prove that the scalar product is y independent. Then, by equating the values of the r.h.s. at $y = +\infty$ and at $y = -\infty$ we derive the orthogonality relations and the characterization equations for the ST of the KPI equation.

Let us first consider the case $u(x,y) \to 0$ as $x^2+y^2 \to \infty$.

We introduce the ST [3] of $u(x,y)$

$$r^\sigma(k,\ell) = \frac{1}{2\pi i} \int_{-\infty}^{+\infty} d\xi \int_{-\infty}^{+\infty} d\eta\, e^{i\ell\xi - i\ell^2\eta}\, u(\xi,\eta)\, \Phi^\sigma(\xi,\eta,k) \quad . \tag{1.9}$$

By equating the values of the scalar product (1.7) at $y = \pm\infty$, we get the characterization equations for the ST

$$r^+(k,q) + \bar{r}^-(q,k) = \int_k^q d\ell\, r^+(k,\ell)\, \bar{r}^-(q,\ell) \quad , \tag{1.10}$$

$$r^\sigma(k,q) + \bar{r}^\sigma(q,k) = \sigma \left\{ \int_k^\infty - \int_{-\infty}^q \right\} d\ell\, r^\sigma(k,\ell)\, \bar{r}^\sigma(q,\ell) \tag{1.11}$$

and the orthogonality relations

$$\langle \Phi^+(k), \Phi^-(q) \rangle = \delta(k-q) \quad , \tag{1.12}$$

$$\langle \Phi^\sigma(k), \Phi^\sigma(q) \rangle = \delta(k-q) + \sigma f^{-\sigma}(k,q) \quad . \tag{1.13}$$

$f^\sigma(k,q)$ measures the lack of analyticity of Φ

$$\Phi^+(k) - \Phi^-(k) = \int_{-\infty}^{+\infty} d\ell\, f^\sigma(k,\ell)\, \Phi^\sigma(\ell) \quad . \tag{1.14}$$

It is the ST of Manakov [1] and Fokas-Ablowitz [2]. We can also express $f^\sigma(k,q)$ in terms of the new ST $r^\sigma(k,q)$

$$f^{-\sigma}(k,q) = r^\sigma(k,q)\,\theta(k-q) + \bar{r}^\sigma(q,k)\,\theta(q-k) + $$
$$+ \sigma \int_{-\infty}^k d\ell\, \theta(q-\ell)\, r^\sigma(k,\ell)\, \bar{r}^\sigma(q,\ell) \quad . \tag{1.15}$$

Characterization equations for $f^\sigma(k,q)$ can also be obtained

$$f^+(k,q) - f^-(k,q) = - \int_{-\infty}^{+\infty} d\ell\, f^\sigma(k,\ell)\, f^{-\sigma}(\ell,q) \tag{1.16}$$

together with a symmetry relation

$$\bar{f}^\sigma(k,q) = f^\sigma(q,k) \quad . \tag{1.17}$$

We consider now the one-dimensional case. In fact the integral equation (1.4), with the $\mathfrak{R}$ function defined in (1.6), contains as a special case the unidimensional case when $u(x,y) \equiv \tilde{u}(x-2\lambda y)$ $(\lambda \in \mathbb{R})$. By introducing

$$\tilde{\Phi}^{\sigma}(x-2\lambda y,\ k-\lambda) = e^{i\lambda x-i[(k-\lambda)^2+\lambda^2]y}\ \Phi^{\sigma}(x,y,k) \tag{1.18}$$

we recover the stationary Schrödinger equation $\tilde{\Phi}_{xx}(x,k)+[k^2-\tilde{u}(x)]\tilde{\Phi}(x,k)=0$. Moreover, after the change of variable $\xi' = \xi-2\lambda\eta$, the integration over η in (1.4) can be explicitly performed and we obtain the familiar integral equations

$$\tilde{\Phi}^{+}(x,k) = e^{-ikx} + \int d\xi\ \frac{e^{ik|x-\xi|}}{2i(k+i0)}\ \tilde{u}(\xi)\ \tilde{\Phi}^{+}(\xi,k)$$

$$\tilde{\Phi}^{-}(x,k) = e^{-ikx} - \int d\xi\ \theta(\xi-x)\ \frac{\sin[k(x-\xi)]}{k-i0}\ \tilde{u}(\xi)\ \tilde{\Phi}^{-}(\xi,k) \tag{1.19}$$

defining the sectionally meromorphic eigenfunction $\tilde{\Phi}$ of the stationary Schrödinger equation.

The limits $\mathfrak{R}^{\pm}$ of $\mathfrak{R}$ as $y \to \pm\infty$ can be explicitly computed ($R(k)$ and $T(k)$ are the reflection and transmission coefficients)

$$\mathfrak{R}^{+}(k,p,y) \xrightarrow[y\to\pm\infty]{} -\delta(p-k)\ \theta(\mp(k-\lambda))\ [T(k-\lambda)-1] -$$

$$-\ \delta(p+k-2\lambda)\ \theta(\pm(p-k))\ R(k-\lambda)$$

$$\mathfrak{R}^{-}(k,p,y) \xrightarrow[y\to\pm\infty]{} -\delta(p-k)\ \theta(\mp(k-\lambda))\ \left[\frac{1}{\bar{T}(k-\lambda)}-1\right] +$$

$$+\ \delta(p+k-2\lambda)\ \theta(\mp(p-k))\ \frac{R(k-\lambda)}{T(k-\lambda)}\quad . \tag{1.20}$$

We recover the well known orthogonality relations

$$\langle\Phi^{+}(k),\ \Phi^{-}(q)\rangle = \delta(k-q)$$

$$\langle\Phi^{+}(k),\ \Phi^{+}(q)\rangle = \delta(k-q) + R(k-\lambda)\ \delta(q+k-2\lambda)$$

$$\langle\Phi^{-}(k),\ \Phi^{-}(q)\rangle = \left|\frac{1}{T(k-\lambda)}\right|^2\ [\delta(k-q) - R(k-\lambda)\ \delta(q+k-2\lambda)] \tag{1.21}$$

and characterization equations

$$R(k) = \bar{R}(-k)\quad,\quad T(k) = \bar{T}(-k)$$

$$|R(k)|^2 + |T(k)|^2 = 1\quad . \tag{1.22}$$

Finally, we consider the 2-dimensional case where wave solitons are present. We have

$$\lim_{y\to\pm\infty}\ u(x+2\lambda_i y,\ y) = u_{\pm i}(x) \neq 0\quad , \tag{1.23}$$

$(\lambda_i \in \mathbb{R},\quad i = 1,2,\dots,n)$.

In this case we must separate $\mathcal{R}^{\sigma}(k,p,y)$ in two parts, one corresponding to the directions for which $u(x,y)$ goes to zero and another one corresponding to the direction $x-2\lambda_1 y = \text{const}$ for which $u(x,y)$ does not go to zero.

For this very complicate case, we report here only the principal results (for details, see [5]).

The limits of $\mathcal{R}^{\sigma}$ at large y can be explicitly computed. We have

$$\mathcal{R}^{\sigma}(k,p,y) \xrightarrow[y \to \pm\infty]{} \delta(k-p) \left(1 + Z^{\sigma}_{\pm}(k)\right) \pm \theta(\pm\sigma\,(p-k))\; r^{\sigma}(k,p) \quad , \tag{1.24}$$

where the $Z^{\sigma}_{\pm}(k)$ are some complicated functions, which satisfy some characterization equations [5]. The most relevant is

$$Z^{+}_{+}(k)\; \bar{Z}^{-}_{+}(k) = Z^{+}_{-}(k)\; \bar{Z}^{-}_{-}(k) = 1 \quad . \tag{1.25}$$

However is worth noting that for the potential $u(x,y)$ going to zero sufficiently fast at infinity $Z^{\sigma}_{\pm} = -1$ and the first term in (1.24) desappears.

The orthogonality relations take the form

$$\langle \Phi^{+}(k),\; \Phi^{-}(q) \rangle = \delta(k-q) \quad , \tag{1.26}$$

$$\langle \Phi^{\sigma}(k),\; \Phi^{\sigma}(q) \rangle = \delta(k-q) - \sigma\, g^{\sigma}(k,q) \tag{1.27}$$

where

$$g^{\sigma}(k,q) = Z^{\sigma}_{-}(k)\; \bar{r}^{\sigma}(q,k)\; \theta(q-k) + \bar{Z}^{\sigma}_{+}(q)\; r^{\sigma}(k,q)\; \theta(k-q)$$
$$+ \sigma \int_{-\infty}^{k} dp\; \theta(q-p)\; r^{\sigma}(k,p)\; \bar{r}^{\sigma}(p,q) \quad .$$

2. The inverse problem for the n-soliton solution

We suppose that Φ has n simple poles in the k-plane ($k = \mu_j$), where it satisfies the $\bar{\partial}$-equation

$$\frac{\partial}{\partial \bar{k}}\, \Phi(k) = 2\pi \sum_{j=1}^{n} \rho_j c_j\; \Phi(\bar{k})\; \delta(k-\mu_j) \quad , \tag{2.1}$$

with

$$\mu_j = \lambda_j + i\,\rho_j \quad , \quad \rho_j > 0 \quad , \quad \lambda_1 \leqslant \lambda_2 \leqslant \ldots \leqslant \lambda_n ,$$

$$c_j = \exp\left[-i\left(\mu_j - \bar{\mu}_j\right) x_{oj} + i\left(\mu_j^2 - \bar{\mu}_j^2\right) y_{oj}\right] \tag{2.2}$$

x_{oj} and y_{oj} are arbitrary real constants.

The eigenfunction $\Phi(x,y,k)$ admits the representation

$$\Phi(x,y,k) = e^{-ikx+ik^2y} \left[1 + \sum_{j=1}^{n} \frac{\varphi_j(x,y)}{k-\mu_j}\, e^{\alpha_j}\right] \quad , \tag{2.3}$$

where

$$\alpha_j = i\,\mu_j(x-x_{oj}) - i\,\mu_j^2(y-y_{oj}) \quad . \tag{2.4}$$

By inserting this representation in the $\bar{\partial}$-equation (2.1), we obtain the matrix formula for the φ_j

$$\varphi_j = 2i\rho_j\, e^{\bar{\alpha}_j} \left[1 + \sum_{\ell=1}^{n} \frac{\varphi_\ell}{\bar{\mu}_j - \mu_\ell}\, e^{\alpha_\ell} \right] \tag{2.5}$$

and the n-wave soliton solution

$$u(x,y) = -\,2i\partial_x \sum_{j=1}^{n} \varphi_j\, e^{\alpha_j} \quad . \tag{2.6}$$

We recover in this way the result of [6]. The lump solutions can be obtained by the limiting procedure indicated in [1].

3. The time evolution

Let us consider the auxiliary spectral problem for the KPI equation

$$\Phi_t + 4\,\Phi_x^3 - 6\,u\,\Phi_x - 3u_x\,\Phi + a\Phi = 4i\,k^3\Phi, \; a_x = u_y \quad . \tag{3.1}$$

If we derive this linear equation with respect to $\bar{k}$, we easily get the time evolution of the ST $r^\sigma(k,q)$

$$r_t^\sigma(k,q) = 4i(k^3-q^3)\, r^\sigma(k,q) \quad . \tag{3.2}$$

References

[1] S.V. Manakov, Physica D3, 420 (1981).

[2] A.S. Fokas and M.J. Ablowitz, Stud. Appl. Math. 69, 211 (1983).

[3] M. Boiti, J. JP Léon and F. Pempinelli, "Spectral Transform and Orthogonality Relations for the KPI equation", to appear in Phys. Lett. A.

[4] M. Boiti, J. JP. Léon, F. Pempinelli and A. Pogrebkov, "Solitons and Spectral Transform for the DSI and KPI equations", to appear in the Proceedings of the Dubna Workshop on "Solitons and their applications" (August 1989).

[5] M. Boiti, F. Pempinelli and A. Pogrebkov, preprint in preparation.

[6] S.V. Manakov, V.E. Zakharov, L.A. Bordag, A.R. Its and V.B. Matveev, Phys. Lett. A63, 205 (1977).

Dromions, a New Manifestation in Nonlinear Phenomena

P.M. Santini[1] *and A.S. Fokas*[2]

[1]Dipartimento di Fisica, Università "La Sapienza", I-00185 Roma, Italy
[2]Department of Mathematics and Computer Science, Clarkson University, Potsdam, NY 13676, USA

1. The Davey-Stewartson 1 Equation and the Notion of Dromion

In these notes we concentrate our attention on the Davey - Stewartson 1 (DS1) equation [1](in light- cone coordinates $\xi = x + y$, $\eta = x - y$)

$$iq_t + q_{\xi\xi} + q_{\eta\eta} + (U_1 + U_2)q = 0, \qquad q = q(\xi, \eta, t), \tag{1a}$$

$$U_{1\xi} = -\frac{\epsilon}{2}(|q|^2)_\eta, \quad U_{2\eta} = -\frac{\epsilon}{2}(|q|^2)_\xi, \quad \epsilon = \pm 1, \tag{1b}$$

an integrable 2+1 dimensional generalization of the celebrated nonlinear Schroedinger (NLS) equation

$$iq_t + q_{xx} - \epsilon|q|^2 q = 0, \quad q = q(x,t). \tag{2}$$

DS1 is the compatibility condition between the following linear equations [2] (Lax pair)

$$\begin{pmatrix} \partial_\xi & q/2 \\ \epsilon q^*/2 & \partial_\eta \end{pmatrix} V = 0, \qquad V_t = \hat{P}V, \tag{3a}$$

$$\hat{P} := i\begin{pmatrix} 1 & 0 \\ 0 & -1 \end{pmatrix}(\partial_\xi - \partial_\eta)^2 - i\begin{pmatrix} 0 & q \\ \epsilon q^* & 0 \end{pmatrix}(\partial_\xi - \partial_\eta) + \begin{pmatrix} iU_1 & -iq_\eta \\ i\epsilon q^*_\xi & -iU_2 \end{pmatrix}, \tag{3b}$$

for the 2x2 matrix eigenfunction V.

In a fluid dynamical context DS1 describes the evolution of waves of slowly varying amplitude q on a two - dimensional water surface under gravity and surface tension (with surface tension dominating gravity) and its interaction with the associated velocity field described by (U_1, U_2) [3],[4],[1]. DS1 is a universal model equation, since it can be obtained via an asymptotically exact technique of broad applicability from large classes of nonlinear evolution equations in 2+1 dimensions [5],[6]; this also suggests its potential applicability in other areas of physics, like plasma physics and nonlinear optics.

Spatially confined solutions of dispersive type of equation (1) were constructed in [7] using a Gel'fand-Levitan-Marchenko equation. An initial value problem for the DS1 equation (1) was investigated in [8] using the inverse scattering (or spectral) transform (IST) method (see also [9] and [10]); it has been recently shown in [11] that the case studied in [8] corresponds to $U(\xi, -\infty, t) = U(-\infty, \eta, t) = 0$. In this particular situation arbitrary initial data $q(\xi, \eta, 0)$ disperse away. An existence theorem which does not make use of IST has recently been proved [12].

We were motivated to reexamine DS1 and its associated IST formalism because of the important discovery of Boiti et al. [13] that DS1 and its associated class of nonlinear equations admit nondispersive exponentially localized solutions. The above authors used Backlund transformations (BT) to obtain these solutions and investigated the dynamics of a certain two-soliton solution obtained using a nonlinear superposition formula, showing that the only effect of the interaction of the two solitons is a two-dimensional phase-shift.

In References [14]-[16] we have extended the class of localized solutions of DS1, we have established their spectral interpretation, their genericity and their remarkably novel properties. We have called the localized coherent structures of DS1 **dromions** (the reason will be clear in what follows); they represent a new manifestation in nonlinear phenomena and, in the rest of this section, we list their properties emphasizing the intrinsic differences with respect to solitons.

i) Taking NLS and DS1 as prototype examples of integrable systems in 1+1 and 2+1 dimensions respectively, solitons arise generically from the solution of the following initial - boundary value problem [17]

$$iq_t + q_{xx} - \epsilon|q|^2 q = 0, \qquad q = q(x,t), \;\; \epsilon = -1, \tag{4a}$$

$$q(x,0) = q_0(x), \qquad\qquad q(x,t) \to 0, \; |x| \to \infty. \tag{4b}$$

An arbitrary initial condition q_0 decomposes into a number of solitons created by the balance of nonlinearity and dispersion; then **solitons are nonlinear modes** of NLS.

As we shall see in the following sections, dromions arise generically from the solution of an initial- boundary value problem for DS1 in which the complex initial condition $q_0(\xi,\eta) := q(\xi,\eta,0)$ and the real boundary conditions $u_2(\xi,t) := U_2(\xi,-\infty,t), u_1(\eta,t) := U_1(-\infty,\eta,t)$ are given, are suitably decaying for large values of ξ,η and are bounded in t [14],[15], i.e. they arise generically from

$$iq_t + q_{\xi\xi} + q_{\eta\eta} + [u_1(\eta,t) + u_2(\xi,t) - \frac{\epsilon}{2}\int_{-\infty}^{\xi} d\xi'(|q|^2)_\eta - \frac{\epsilon}{2}\int_{-\infty}^{\eta} d\eta'(|q|^2)_\xi]q = 0, \tag{5a}$$

$$q(\xi,\eta,0) = q_0(\xi,\eta), \qquad q(\xi,\eta,t) \to 0, \quad |\xi|,|\eta| \to \infty. \tag{5b}$$

Arbitrary nonzero boundary flows u_1, u_2 cause a transfering of energy from the mean flow to the surface waves where focusing effects balancing dispersion can be created. The mathematical manifestation of this phenomenon is the fact that DS1 admits localized solutions which decay exponentially in both ξ and η [14],[15]. The basic mechanism for generating dromions is due to the boundary conditions u_2, u_1, while the initial condition q_0 plays a secondary role, affecting the amplitudes only. Such a mechanism holds even in the linear limit

$$iq_t + q_{\xi\xi} + q_{\eta\eta} + [u_1(\eta,t) + u_2(\xi,t)]q = 0, \tag{6a}$$

$$q(\xi,\eta,0) = q_0(\xi,\eta), \qquad q(\xi,\eta,t) \to 0, \; |\xi|,|\eta| \to \infty. \tag{6b}$$

of (5), then **dromions are nonlinear distorsions of the linear modes** of the linearized DS1 equation.

ii) In two space dimensions the presence of nontrivial boundary conditions makes it possible for (5a,b) to exhibit regular localized solutions even in the defocusing case $\epsilon = 1$ [15], [16]; this is not allowed to NLS.

iii) In the process of interaction dromions not only exhibit a **two-dimensional phase shift**, but also a **change of form**; such a change of form is associated with an **exchange of energy** [16]. Although exponentially localized in the plane, **dromions are not solitary** since they are connected through the mean flow (U_1, U_2) which describes a traveling rectangular network in the (ξ,η) plane; the nodes of this network coincide with the positions of the dromions [16].

iv) **Dromions can be driven** everywhere in the (ξ,η) plane choosing a suitable motion of the boundaries u_2, u_1. The price payed for this important freedom (which is absent in one dimension) is that **dromions radiate energy** if their motion is not uniform [15].

The well-known notion of **soliton** is not adeguate to capture the above novel features associated with coherent structures in multidimensions and summarized in i)-iv). We have called them **dromions** to emphasize that they travel on the tracks (in ancient greek **dromos**) described by the mean flow (U_1, U_2) and to stress their ability to be **driven** by the boundaries.

The mathematical aspects of the theory are given in the following sections, these notes provide a summary of the results contained in [14]-[16]. For a more exhaustive review of these results, see [18].

2. The IST Formalism Associated with the Initial - Boundary Value Problem for the DS1 Equation

To discover the properties listed in §1 we have used the inverse scattering formalism (IST) introduced in [8] and based on the use of the eigenfunctions $\Psi = exp(ik\begin{pmatrix}\eta & 0\\ 0 & -\xi\end{pmatrix})M$ of the spectral problem (3.a) defined by the integral equations [14],[15]

$$M_{11}^{\pm} = 1 - \frac{1}{2}\int_{-\infty}^{\xi} d\xi' q M_{21}^{\pm}; \quad M_{12}^{\pm} = -\frac{1}{2}\int_{\mp\infty}^{\xi} d\xi' q M_{22}^{\pm} e^{-ik(\xi'-\xi)}; \tag{7a}$$

$$M_{21}^{\pm} = -\frac{\epsilon}{2}\int_{\pm\infty}^{\eta} d\eta' q^* M_{11}^{\pm} e^{ik(\eta'-\eta)}; \quad M_{22}^{\pm} = 1 - \frac{\epsilon}{2}\int_{-\infty}^{\eta} d\eta' q^* M_{12}^{\pm}. \tag{7b}$$

Equations (7) (defined for $k \in \Re$) can be analytically extended to the complex k- plane and imply that M^+ and M^- are analytic in the upper and lower half k-plane respectively. In addition these eigenfunctions satisfy the following scattering equations

$$\mu^+(k) - \mu^-(k) = \int_R dl T(k,l) e^{-i(l\xi + k\eta)} \hat{\mu}^+(l), \tag{8a}$$

$$\hat{\mu}^-(k) - \hat{\mu}^+(k) = \int_R dl S(k,l) e^{i(l\eta + k\xi)} \mu^-(l), \tag{8b}$$

where $\mu^{\pm}, \hat{\mu}^{\pm}$ denote the vectors $\mu^{\pm} := (M_{11}^{\pm}, M_{21}^{\pm})^T, \hat{\mu}^{\pm} := (M_{12}^{\pm}, M_{22}^{\pm})^T$, and the scattering data S, T are given by

$$S(k,l) := \int_{R^2} \frac{d\xi d\eta}{4\pi} q M_{22}^{-}(k) e^{-i(k\xi + l\eta)}, \quad T(k,l) = \epsilon S^*(l,k). \tag{9}$$

Given $S(k,l)$, equations (8) and the analyticity properties of $M^{\pm}$ define a linear Riemann-Hilbert problem in the complex k- plane, from the solution of which we obtain the reconstruction formulae of the solution q, U_1, U_2 of the DS1 equation:

$$q = \frac{1}{\pi}\int_{R^2} dkdl S(k,l) e^{i(l\eta+k\xi)} M_{11}^{-}(l) = \frac{1}{\pi}\int_{R^2} dkdl S(k,l) e^{i(k\xi+l\eta)} M^{+*}_{22}(k), \tag{10}$$

$$U_2 = u_2 + \partial_\xi \int_{R^2} \frac{dkdl}{\pi} S(k,l) e^{i(k\xi+l\eta)} M_{21}^{-}(l), \tag{11a}$$

$$U_1 = u_1 + \epsilon\partial_\eta \int_{R^2} \frac{dkdl}{\pi} S(k,l) e^{i(k\xi+l\eta)} M_{12}^{+*}(k). \tag{11b}$$

The IST scheme is completed noticing that if q evolves according to equation (5), then the Fourier transform $\hat{S}(\xi,\eta)$ of $S(k,l)$ evolves according to [14],[15]

$$i\hat{S}_t + \hat{S}_{\xi\xi} + \hat{S}_{\eta\eta} + (u_1+u_2)\hat{S} = 0; \quad \hat{S}(\xi,\eta) := \int_{R^2} \frac{dkdl}{2\pi} S(k,l) e^{i(k\xi+l\eta)}. \tag{12}$$

Notice that if $|q| << 1$ then $q \sim 2\hat{S}$ and equation (12) is the weak field limit (6a) of (5).

An IST alternative to this one has been recently proposed in [19].

3. Degenerate Scattering Data and the (N,M) Dromion Solution

Degenerate scattering data play an important role in the construction of closed form solutions and, as we shall see in §4, in the determination of the asymptotic behavior in t of solutions of DS1.

If the scattering data $S(k,l,t)$ are degenerate, then $\hat{S}(\xi,\eta,t)$ is also degenerate, namely

$$\hat{S} = \sum_{n=1}^{N}\sum_{m=1}^{M} \rho_{nm} X_n(\xi,t) Y_m(\eta,t), \quad \rho_{nm} \in C, \tag{13}$$

where

$$iX_{jt} + X_{j\xi\xi} + u_2 X_j = 0, \quad j = 1,..,N; \quad iY_{jt} + Y_{j\eta\eta} + u_1 Y_j = 0, \quad j = 1,..,M; \tag{14}$$

then the inverse problem of §2 reduces to the algebraic equation [14],[15]

$$(I - \epsilon A) Z = \rho \tag{15}$$

which defines matrix Z in terms of matrices ρ and A, where

$$A(\xi,\eta,t) = \rho\beta(\eta,t)\rho^{+}\alpha^{*}(\xi,t), \quad (\rho)_{ij} = \rho_{ij}, \; (\rho^{+})_{ij} = \rho^{*}_{ji}, \tag{16a}$$

$$\beta_{ij}(\eta,t) := \int_{-\infty}^{\eta} d\eta' Y_i(\eta',t) Y_j^{*}(\eta',t), \quad \alpha_{ij}(\xi,t) := \int_{-\infty}^{\xi} d\xi' X_i(\xi',t) X_j^{*}(\xi',t). \tag{16b}$$

Equations (10), (15) finally yield

$$q(\xi,\eta,t) = 2\sum_{n=1}^{N}\sum_{m=1}^{M} Z_{nm}(\xi,\eta,t) X_n(\xi,t) Y_m(\eta,t), \tag{17}$$

$$|q|^2 = -4\epsilon\partial_\xi\partial_\eta ln \; det\,(I - \epsilon A), \tag{18a}$$

$$U_2 = u_2 + 2\partial_\xi{}^2 ln \; det\,(I - \epsilon A), \quad U_1 = u_1 + 2\partial_\eta{}^2 ln \; det\,(I - \epsilon A). \tag{18b}$$

We remark that the solution (17) of the DS1 equation (5) is obtained from the solution (13) of the linearized DS1 equation (6) replacing the constant amplitudes ρ_{nm} by the bounded functions $Z_{nm}(\xi,\eta,t)$. The localization properties, described by the eigenfunctions $X_n(\xi,t)$, $Y_m(\eta,t)$ of equations (14), are the same for both solutions. Then localized solutions of DS1 are nonlinear distorsions of the bound states of the linearized equation.

The solution (17),(18) of DS1 is in closed form if the eigenfunctions X_j, Y_j are in closed form; if the boundary conditions u_2, u_1 are reflectionless potentials of the Schroedinger equations (14), then X_j and Y_j can be found in closed form through the system [14],[15]

$$\phi_n + \sum_{j=1}^{N} \frac{c_n c_j^*}{p_n + p_j^*} e^{-(p_n+p_j^*)[x-i(p_n-p_j^*)t]}\phi_j = c_n e^{-p_n(x-ip_nt)}, \quad Re(p_n) := p_{nR} > 0, \quad (19a)$$

$$u(x,t) = -2\partial_x \sum_{j=1}^{N} c_j^* e^{-p_j^*(x+ip_j^*t)}\phi_j, \quad (19b)$$

(where $\phi_i = X_i$, $u = u_2$ if $x \to \xi$, $p_i \to \mu_i$ and $\phi_i = Y_i$, $u = u_1$ if $x \to \eta$, $p_i \to \lambda_i$, $c_i \to \tilde{c}_i$). In this case equations (19),(15)-(18) yield what we call the **(N,M) dromion solution** of DS1 [14],[15]. In the subcase $Im(p_j) := p_{jI} = 0$ and $c_j \in R$, then $\phi_j(x,t) = f_j(x)e^{ip_j{}^2t}$, and

$$f_n + \sum_{j=1}^{N} \frac{c_n c_j}{p_n + p_j} e^{-(p_n+p_j)x} f_j = c_n e^{-p_n x}; \quad u(x) = -2\partial_x \sum_{j=1}^{N} c_j e^{-p_j x} f_j; \quad (20)$$

then the solution q of DS1 oscillates in time and we obtain what we call an **(N,M) breather solution** [14],[15].

It is well-known that the total energy $E := \int_{\Re^2} d\xi d\eta |q|^2$ of the wave packet of amplitude q is a constant of the motion for the DS equation. It turns out that the energy associated with the (N,M) dromion solution of DS1 is given by [15],[16]

$$E = -4\epsilon \; ln \; det(I - \epsilon\rho\rho^+) \quad (21)$$

and depends on the spectral matrix ρ only.

3.1 The (1,1) dromion solution of DS1

If $N = M = 1$ equations (19) yield $X(\xi,t) = \Phi(\xi,t;\mu,\nu(\mu,c))$, $Y(\eta,t) = \Phi(\eta,t;\lambda,\nu(\lambda,\tilde{c}))$, $u_2(\xi,t) = \bar{u}(\xi,t;\mu,\nu(\mu,c))$, $u_1(\eta,t) = \bar{u}(\eta,t;\lambda,\nu(\lambda,\tilde{c}))$, where

$$\Phi(x,t;p,\nu(p,c)) := \sqrt{\frac{p_R}{2}} \frac{e^{i[-p_I(x+2p_It)+|p|^2t+arg\nu]}}{cosh[p_R(x+2p_It - \frac{1}{p_R}ln|\nu|)]}, \quad (22a)$$

$$\bar{u}(x,t;p,\nu(p,c)) := \frac{2p_R{}^2}{cosh^2[p_R(x+2p_It - \frac{1}{p_R}ln|\nu(p,c)|)]}, \quad \nu(p,c) := \frac{|c|}{\sqrt{2p_R}}. \quad (22b)$$

The degenerate datum $\hat{S} = \rho XY, \rho \in C$ gives rise, via equations (15)-(17) to the (1,1) dromion solution

$$q = 2Z(\xi,\eta,t)X(\xi,t)Y(\eta,t), \quad (23)$$

where

$$Z := \rho[1 - \epsilon|\rho|^2 \tau(\xi, t; \mu, \frac{1}{\mu_R} ln|\nu(\mu, c)|)\tau(\eta, t; \lambda, \frac{1}{\lambda_R} ln|\nu(\lambda, \tilde{c})|]^{-1}, \tag{24}$$

$$\tau(x, t; p, \bar{x}) := \frac{1}{2}(1 + tanh(p_R(x + 2p_I t - \bar{x})) = (1 + e^{-2p_R(x+2p_I t - \bar{x})})^{-1}. \tag{25}$$

In equation (31) the factor XY indicates that the motion of the (1,1) dromion solution of DS1 in the (ξ, η) plane is completely described by the motion of the boundaries u_2 and u_1 on the ξ and η axes respectively. The ξ and η components of the velocity of the dromion, given by $-2\mu_I$ and $-2\lambda_I$ respectively, are independent, can be positive or negative, and are not related to the amplitude and width, which depend on μ_R, λ_R. If $\epsilon = -1$, Z is bounded for any choice of the spectral parameter ρ; if $\epsilon = 1$, Z is bounded if $|\rho| < 1$; the bounded function Z determines an asymmetry of q with respect to its center.

3.2 The (N,M) dromion solution

The (N,M) dromion solution of DS1 describes the interaction of localized coherent structures in multidimensions. We refer to [16] for an exhaustive study of the asymptotic behavior (as $t \to \pm\infty$) of such a solution, based on the asymptotic investigation of equations (19) and (15); here we only report the final results.

If we order the imaginary part of the eigenvalues of the Schroedinger equations (14): $\mu_{1I} < \mu_{2I} < .. < \mu_{NI}$, $\lambda_{1I} < \lambda_{2I} < .. < \lambda_{MI}$, and we transform to a moving reference frame with spatial coordinates $\xi + 2\mu_{kI}t$, $\eta + 2\lambda_{hI}t$, $1 \le k \le N$, $1 \le h \le M$, then

$$q \xrightarrow[t \to \pm\infty]{} q_{kh}^{\pm} := 2Z_{kh}^{\pm}(\xi, \eta, t) X_k^{\pm}(\xi, t) Y_h^{\pm}(\eta, t), \tag{26}$$

where $X_k{}^{\pm} = \Phi(\xi, t; \mu_k, \gamma_k{}^{\pm}(\mu, c, N))$, $Y_h{}^{\pm} = \Phi(\eta, t; \lambda_h, \gamma_h{}^{\pm}(\lambda, \tilde{c}, M))$ are the asymptotic solutions of equations (19) with

$$\gamma_k^{\pm}(p, c, L) := \frac{c_k}{\sqrt{2p_{kR}}} \prod_{i\pm}^{L,k} \frac{p_k - p_i}{p_k + p_i^*}; \quad \prod_{i+}^{N,k} := \prod_{i=1}^{k-1}, \prod_{i-}^{N,k} := \prod_{i=k+1}^{N}, \prod_{i=1}^{0} = \prod_{i=N+1}^{N} = 1,$$

and the bounded amplitudes $Z_{nm}^{\pm}$ are solutions of the asymptotic linear systems (for $1 \le r \le N$)

$$Z_{rs}^{\pm} - \epsilon\tau_k^{\pm}(\tilde{\tau}_h^{\pm}\rho_{rh}\rho_{kh}^* - \epsilon\sum_{j\pm}^{M,h}\rho_{rj}\rho_{kj}^*)Z_{ks}^{\pm} + \sum_{i\pm}^{N,k}(\tilde{\tau}_h^{\pm}\rho_{rh}\rho_{ih}^* + \sum_{j\pm}^{M,h}\rho_{rj}\rho_{ij}^*)Z_{is}^{\pm} = \rho_{rh}, \tag{27}$$

$$\tau_k^{\pm} := \tau(\xi, t; \mu_k, \frac{1}{\mu_{kR}} ln|\gamma_k^{\pm}(\mu, c, N)|), \quad \tilde{\tau}_h^{\pm} := \tau(\eta, t; \lambda_h, \frac{1}{\lambda_{hR}} ln|\gamma_h{}^{\pm}(\lambda, \tilde{c}, M)|), \tag{28a}$$

$$\sum_{j+}^{N,k} := \sum_{j=k+1}^{N}, \quad \sum_{j-}^{N,k} := \sum_{j=1}^{k-1}, \quad \sum_{j=1}^{0} = \sum_{j=N+1}^{N} = 0. \tag{28b}$$

The factor $X_k^{\pm}Y_h^{\pm}$ in (26) indicates that the dynamics of the (N,M) dromion solution of DS1 in the plane (ξ, η) is completely characterized (and generated) by the one-dimensional dynamics of the boundaries $u_2(\xi, t)$ and $u_1(\eta, t)$ which describe the motion of N, respectively M one-dimensional interacting solitons [16]. Then in the limit $t \to \pm\infty$ the (N,M) dromion solution decomposes into N times M widely separated bumps locally represented by the functions $q^{\pm}_{ij}$ defined in (26), with $1 \le i \le N, 1 \le j \le M$, and the $(i,j)^{th}$-dromion travels with velocity $(-2\mu_{i_I}, -2\lambda_{j_I})$ and exhibits the phase shift $(\delta_i, \varepsilon_j)$ defined by $\delta_i = \varsigma_i(\mu, N)$, $\varepsilon_j = \varsigma_j(\lambda, M)$, where

$$\varsigma_i(p, L) := \frac{1}{p_{i_R}} (\sum_{l=1}^{i-1} - \sum_{l=i+1}^{L}) ln \left| \frac{p_i - p_l}{p_i + p_l^*} \right|, \tag{29}$$

which indicates the typical pairwise character of the interaction.

The nonlocal nature of the interaction implies that $Z^{+}_{ij} \neq Z^{-}_{ij}$ (the asymptotic systems $(27)^+$ and $(27)^-$ are essentially different); then the $(i,j)^{th}$-dromion **changes form** during the interaction with all the others! It turns out that such a change of form is associated with an **exchange of energy** [16].

If $\rho_{ij} = \delta_{ij}\rho_i$, then

$$Z^{\pm}_{kh} = \delta_{kh}\rho_k[1 - \epsilon|\rho_k|^2 \tau^{\pm}_k \tilde{\tau}^{\pm}_h]^{-1} \tag{30}$$

and the number of dromions emerging at $t = \pm\infty$ is the minimum between N and M, the form of each dromion is the same at $t = \pm\infty$ and coincides with the one of the (1,1) dromion solution (23); then dromions do not exchange energy and the only effect of the interaction is a two-dimensional phase shift. This special subcase corresponds to the solutions obtained in [13] via BT and superposition formulae.

Computer graphs describing the evolution of the (2,2) dromion solution for generic spectral data and for the special case (30) in which $\rho_{ij} = \delta_{ij}\rho_i$ are presented in [16]. In [16] we also study the asymptotics of the velocity field (U_1, U_2) which describes a traveling rectangular network on the (ξ, η) plane whose nodes coincide with the positions of the dromions. The velocity field (U_1, U_2) is constant on each branch between two nodes and it has a discontinuity at the nodes (notice the analogy with an electric network). Then dromions, although exponentially localized in the (ξ, η) plane, **are not solitary**, being connected through the network described by the mean field (U_1, U_2).

3.3 Other explicit solutions of DS1

The (N,M) dromion and breather solutions are not the only explicit, exponentially decaying coherent structures of DS1. Using the formalism of this §3 we have obtained in [16] several examples of closed form solutions of DS1 describing finite - energy coherent structures associated with physically relevant boundary flows u_1, u_2 chosen among the following generalized functions: i) $u(x) = a\delta(x - x_0)$, $x_0 \in R$, $a > 0$ (δ- boundary); ii) $u(x) = a(\theta(x - x_1) - \theta(x - x_2)), x_1 < x_2, a > 0$ (rectangular boundary); iii) $u(x) = ae^{-2|x|}$ (exponential boundary); iv) $u(x) = a\theta(x - x_0)$ (step boundary). The corresponding solutions of DS1 are explicit generalized functions (distributions).

4. The Evolution of the Scattering Data

The time-dependent Schroedinger equation (12) in two space dimensions describes the evolution of the scattering data and then plays a central role in the theory associated with DS1. In the following we investigate the initial-value problem for equation (12) for suitably decaying initial conditions $\hat{S}(\xi, \eta, 0)$.

4.1 Time independent boundaries

We first consider the case when u_1, u_2 are time independent and suitably decaying. Looking for separable solutions $\hat{S} = T(t)X(\xi)Y(\eta)$ of (12), it follows that

$$T' + i(k^2 + k'^2)T = 0, \quad X'' + (u_2 + k^2)X = 0, \quad Y'' + (u_1 + k'^2)Y = 0; \tag{31}$$

then the analysis of (12) is intimately related to the stationary Schrodinger equation

$$\psi_{xx} + (u(x) + k^2)\psi = 0 \tag{32}$$

for the potentials u_2, u_1.

The solution of the initial-value problem associated with equation (12) can be written in the form

$$\hat{S}(\xi,\eta,t) = \sum_{n=1}^{N}\sum_{m=1}^{M} \rho_{nm}X_n(\xi)Y_m(\eta)e^{i(\mu_n^2+\lambda_m^2)t} +$$

$$\int_{R^2} dkdl\rho(k,l)X(\xi,k)Y(\eta,l)e^{-i(k^2+l^2)t} +$$

$$\int_R dk[\sum_{n=1}^{N}\rho_n(k)e^{i(\mu_n^2-k^2)t}X_n(\xi)Y(\eta,k) + \sum_{m=1}^{M}\tilde{\rho}_m(k)e^{i(\lambda_m^2-k^2)t}X(\xi,k)Y_m(\eta)], \tag{33}$$

where $\{X(\xi,k), X_j(\xi), \mu_j, 1 \le j \le N\}$ and $\{Y(\eta,k), Y_j(\eta), \lambda_j, 1 \le j \le M\}$ are the orthonormal bases of eigenfunctions and eigenvalues of (32), corresponding to $u_2(\xi)$ and $u_1(\eta)$ respectively. Furthermore ρ_{nm}, $\rho(k,l)$, $\rho_n(k)$ and $\tilde{\rho}_m(k)$ are given in terms of the initial condition via $\int_{R^2} d\xi d\eta \hat{S}(\xi,\eta,0)F(\xi,\eta,k,l)$, where $F = X_n(\xi)Y_m(\eta)$, $X(\xi,k)Y(\eta,l)$, $X_n(\xi)Y(\eta,k)$ and $X(\xi,k)Y_m(\eta)$ respectively. The stationary phase method implies the following asymptotics in time

$$\hat{S} \to \sum_{n=1}^{N}\sum_{m=1}^{M} \rho_{nm}X_n(\xi)Y_m(\eta)e^{i(\mu_n{}^2+\lambda_m{}^2)t}, \quad t \to \infty, \tag{34}$$

thus the scattering data are asymptotically degenerate and equations (15)-(18) yield a localized solution q of DS1 which oscillates in time (breather)[14],[15].

From the above it follows that if at least one of the two boundaries does not give rise to bound states of the Schroedinger operator, then every initial condition q_0 will disperse away. If bound states do exist, the asymptotic behaviour is essentially determined by these bound states; the initial condition only fixes the constants ρ.

4.2 Time dependent boundaries and driven coherent structures

If u_1, u_2 are time-dependent, then separation of variables implies that the solution of equation (12) is intimately related to the time-dependent Schroedinger equation

$$i\psi_t + \psi_{xx} + u(x,t)\psi = 0. \tag{35}$$

We recall that this equation plays an important role in the integration of the Kadomtsev-Petviashvili (KP) equation where t is replaced by y [20]-[22]. However we now demand u to be decreasing in x only as opposed to the case of KP where u is decreasing in both x and y (see [23] for recent results on this problem). Equation (35) was also used to construct explicit solutions of several physical models [24]. We expect that a completeness result is also valid for the above equation, so that equation (33) is appropriately generalized, however interesting results can be obtained considering special mathematical experiments.

A standard asymptotic analysis of equation (35) shows that, if $u_2(\xi,t)$, $u_1(\eta,t)$ are arbitrary jet - like boundary flows satisfying the asymptotics

$$u_2(\xi,t)\to\bar{u}_2(\xi-vt-\xi_0),\quad u_1(\eta,t)\to\bar{u}_1(\eta-wt-\eta_0),\quad v,w,\xi_0,\eta_0\in\Re,\quad t\to\infty, \tag{36}$$

then

$$\hat{S}\to e^{i\frac{v}{2}(\xi-\frac{v}{2}t-\xi_0)+i\frac{w}{2}(\eta-\frac{w}{2}t-\eta_0)}$$

$$\sum_{n=1}^{N}\sum_{m=1}^{M}\rho_{nm}X_n(\xi-vt-\xi_0)Y_m(\eta-wt-\eta_0)e^{i(\mu_n^2+\lambda_m^2)t},\quad t\to\pm\infty \tag{37}$$

where $\{X_j(\xi),\mu_j,1\le j\le N\}$ and $\{Y_j(\eta),\lambda_j,1\le j\le M\}$ are the eigenfunctions and eigenvalues of (32), corresponding to $\bar{u}_2(\xi)$ and $\bar{u}_1(\eta)$ respectively and the coefficients ρ_{nm} depend on the history of the process, namely on $\hat{S}(\xi,\eta,0)$ and on $u_2(\xi,t)$, $u_1(\eta,t)$ [15].

Equations (37) show that, asymptotically, the spectral data are degenerate and then equations (15)-(18) give the asymptotics of the corresponding DS1 solution in closed form. Such a solution is asymptotically localized where u_1 and u_2 are localized, namely dromions follows the motion of the boundaries u_2, u_1 or, equivalently, they are driven by the boundaries [15]. This mathematical experiment contains of course the particular case in which the boundaries are asymptotically stationary ($v=w=0$); furthermore it can be easily generalized to the case in which

$$u_2(\xi,t)\to\sum_j\bar{u}_2^{(j)}(\xi-v_jt),\quad u_1(\eta,t)\to\sum_j\bar{u}_1^{(j)}(\eta-w_jt),\quad v_j,w_j\in R,\quad t\to\infty.$$

4.3 Dromions and radiation of energy

In order to show that driven coherent structures radiate energy if their motion is not uniform, we consider the following simplified version of the mathematical experiment of §4.2, in which the traveling boundaries u_2, u_1 are 1- soliton solutions (22b) which change their velocity at a finite time t_0, i.e.

$$u_2=\frac{2\mu_R{}^2\theta(t_0-t)}{cosh^2(\mu_R(\xi-v_{in}t-\xi_{in}))}+\frac{2\mu_R{}^2\theta(t-t_0)}{cosh^2(\mu_R(\xi-v_{fin}t-\xi_{fin}))}, \tag{38a}$$

$$u_1=\frac{2\lambda_R{}^2\theta(t_0-t)}{cosh^2(\lambda_R(\eta-w_{in}t-\eta_{in}))}+\frac{2\lambda_R{}^2\theta(t-t_0)}{cosh^2(\lambda_R(\eta-w_{fin}t-\eta_{fin}))}, \tag{38b}$$

where θ is the Heaviside function. We also assume the continuity of u_2, u_1 at $t=t_0$ which implies $(v_{in}-v_{fin})t_0=\xi_{fin}-\xi_{in}$, $(w_{in}-w_{fin})t_0=\eta_{fin}-\eta_{in}$, and we choose the corresponding solution of (12) for $t<t_0$ to be given by the (1,1) dromion $\hat{S}(\xi,\eta,t)=\rho_0XY$, where

$$X=\sqrt{\frac{\mu_R}{2}}\frac{e^{i[\frac{v_{in}}{2}(\xi-\frac{v_{in}}{2}t)+\mu_R{}^2t]}}{cosh(\mu_R(\xi-v_{in}t-\xi_{in}))},\quad Y=\sqrt{\frac{\lambda_R}{2}}\frac{e^{i[\frac{w_{in}}{2}(\eta-\frac{w_{in}}{2}t)+\lambda_R{}^2t]}}{cosh(\lambda_R(\eta-w_{in}t-\eta_{in}))}. \tag{39}$$

Then it is possible to show that, driven by the boundaries u_2, u_1, the dromion of DS1 has a transition from an initial state of energy $E_0=4\ ln\ det(1+|\rho_0|^2)$ to a final state of energy $E_1 = 4\, ln\, det(1 + |\rho_1|^2)$, where

$$|\rho_1| = |\rho_0| \frac{\tilde{v}}{sinh\tilde{v}} \frac{\tilde{w}}{sinh\tilde{w}}, \quad \tilde{v} := \frac{\pi}{4} \frac{v_{in} - v_{fin}}{\mu_R}, \quad \tilde{w} := \frac{\pi}{4} \frac{w_{in} - w_{fin}}{\lambda_R}. \tag{40}$$

Then $|\rho_1| < |\rho_0|$ and the dromion radiates energy. We observe that radiation does not occur if and only if $v_{in} = v_{fin}$ and $w_{in} = w_{fin}$, namely when the motion is uniform, while positive or negative accelerations give rise to radiation. We also observe that complicated trajectories in the (ξ, η) plane can be obtained composing the elementary straight motions of this experiment [15].

In conclusion, **dromions can be driven everywhere in the (ξ, η) plane choosing a suitable motion of the boundaries; the price payed for this important freedom is that dromions radiate energy if their motion is not uniform.**

References

1. A. Davey and K. Stewartson, Proc. R. Soc. london Ser. A **338**, (1974) 101.
2. M.J. Ablowitz and R. Haberman, Phys. Rev. Lett., **35**, (1975) 691.
   V. E. Zakharov and A. B. Shabat, Funct. Anal. Appl. **8**, (1974) 43.
3. M.J. Ablowitz and H. Segur, J. Fluid Mech., **92**, (1979) 691.
4. D.J. Benney and G.J. Roskes, Stud. Appl. Math. **48**, (1969) 377.
5. V.E. Zakharov and E.A. Kuznetsov, Physica **18D**, (1986) 455. V. E. Zakharov and A. M. Rubenchik, Prikl. Mat. Techn. Phys. **5**, (1972) 84. V. E. Manakov, talk presented at the Vth Workshop NEED's 1989, Kolymbari, Crete.
6. F. Calogero and V. Eckhaus, Inverse Problems **3**, (1987) 27; F. Calogero and A. Maccari, on Inverse Problems; an Interdisciplinary Study, ed. P. Sabatier, Academic Press, New York, 1987;
7. H. Cornille, J. Math. Phys. **20**, (1979) 199.
8. A.S.Fokas, Phys. Rev. Lett. **51**, (1983) 3.
9. A.S. Fokas and M.J. Ablowitz, Phys. Rev. Lett., **51**, (1983) 7.
   A.S. Fokas and M.J. Ablowitz, J. Math. Phys., **25**, (1984) 8.
10. D.J. Kaup, Physica **1D**, (1980) 45.
11. C. Shultz and M.J. Ablowitz, Phys. Lett., **135** (1989) 433.
12. J.-M. Ghidaglia and J.-C. Saut, Sur le Problem de Cauchy pour les Equations de Davey- Stewartson, C. R. Acad. Sci. Paris, t. 308, Serie I, (1989) 115.
13. M.Boiti, J.J-P. Leon, L. Martina and F. Pempinelli, Phys. Lett. A, **132**,(1988) 432.
14. A.S Fokas and P.M.Santini, Phys. Rev. Lett., **63** (1989) 1329.
15. A.S. Fokas and P.M. Santini, Coherent Structures and a Boundary-Value Problem for the Davey-Stewartson 1 Equation, Preprint INS 121, Clarkson University, 1989. Physica D (submitted to).
16. P.M. Santini, Energy Exchange of Interacting Coherent Structures in Multidimensions, Preprint 662, Dipartimento di Fisica, Università di Roma I, 1989; Physica D (in press).
17. V.E. Zakharov and P.B. Shabat, Sov. Phys. JETP, **34** (1972) 62.
18. P. M. Santini and A. S. Fokas, The Initial - Boundary Value Problem for the Davey - Stewartson Equation; how to Generate and Drive Localized Coherent Structures in Multidimensions, Preprint 684, Dipartimento di Fisica, Università di Roma I, 1989. Proceedings of the Ecole de Physique Theorique on "Partially Integrable Nonlinear Equations and their Applications" , Les Houches 1989. To appear in the NATO A.S.I., series C, edited by Kluwer.

19. M. Boiti, J.J.-P. Leon and F. Pempinelli, A New Spectral Transform for the Davey-Stewartson Equation, PM /89 -10, Montpellier.
20. S.V. Manakov, Physica **3D**, (1981) 420.
21. A.S. Fokas and M.J. Ablowitz, Stud. Appl. Math., **69**, (1983) 211.
22. M. Boiti, J. J.-P. Leon and F. Pempinelli, A new Spectral Transform for the Kadomtsev - Petviashvili I Equation, Preprint PM/ 89-09, Montpellier.
23. M. Boiti, F. Pempinelli and A. Pogrebkov, talk presented at this Conference.
24. B. A. Dubrovin, T. M. Malanyuk, I. M. Krichever and V. G. Makhan'kov, Sov. J. Part. Nucl. **19**, (1988) 252.

The Phoenix Equation

A.P. Csizmazia

Mathematical Sciences Research Institute, 1000 Centennial Dr.,
Berkeley, CA94720, USA

The CS equations (C : congruent ; S : Schroedinger operators) are strongly 2+1-D generalizations of the Korteweg-de Vries (KdV) eqs. of soliton theory. The CS eqs. are integrable nonlinear evolution equations which share an infinite number of independent time invariants. These invariants and other entities of central importance derive from an operator K satisfying a commutivity condition. An explicit expression for K is given. A proof of invariance consists of identifying the gradients of the candidate "invariants" and then proving that each gradient is both weakly and strongly orthogonal to each evolution equation's density. These orthogonality conditions are established herein using a new fundamental approach. In particular, strong orthogonality is obtained as a consequence of the "Phoenix equation".

Let the space manifold M be the 2-D plane with standard coordinates x,y. Set $d=\partial_x + i\,\partial_y = 2\,\partial_{\bar{z}}$ with $z = x + i\,y$. Then $d^* = -\,\partial_x + i\,\partial_y = -2\,\partial_z$. For each real t (= time) let $u<t>$ be a real valued C^∞ function on M. Require that for each t : ($u<t>$ and) each spatial derivative of $u<t>$ vanishes rapidly at infinity. Say b is a complex valued function in C^∞ (M). Set $b_{\underline{k}} = (d^*)^k\,(b)$. The Veselov-Novikov eq., or CS eq., is $\dot{u} = 2\ \mathrm{Re}\ d^*\ (u_{\underline{2}} + 3\ s\ u)$. Here $s = d^{-1} d^*(u)$ with s vanishing at infinity [1,2]. Say $u = \Delta v$ with v vanishing at infinity. Set $f = (-\partial_x^2 + \partial_y^2)\,(v)$ and $g = 2\,v_{xy}$. The CS eq. becomes $-\frac{1}{2}\,\dot{u} = (\partial_x^2 - 3\,\partial_y^2)\,(u_x) + \partial_x\,(3\,u\,f) + \partial_y\,(3\,u\,g)$.

Consider linear operators on a Hilbert space. The C product (C : congruence)(L,H) is $LH+H^*\circ L$. The C eq. is $-\dot{L}_t = \langle L_t, H_t \rangle$. On M take $L_t = -\,\Delta + u<t>$ with $\Delta = \partial_x^2 + \partial_y^2 = -\,d^*\,d = -\,d\,d^*$. Say each H_t is a differential operator in the canonical form described next. Let $\partial = d^*$. Allow $\ell = 0,1,\ldots$ Set $r = 2\ell + 1$. Take $H = 2\,\mathrm{Re}\ K_r$ with $K_r = T_\ell \partial$ and T_ℓ in the central form $T_\ell = \Sigma_n\,\partial^{\ell-n} \circ a_n\,\partial^{\ell-n}$, where $0 \le n \le \ell$. Each $a_n = a_n <\ell> (x,y)$ is (for fixed t) a complex valued function. Normalization : $a_0 = 1$. For $n \ge 1$ require that a_n vanish at infinity. The $CS_{\ell+1}$ eq. $\dot{u} = 2\,\mathrm{Re}\,[T_\ell\,(u)\,]$ arises from the C eq. with $H = H<r>$ in canonical form. The CS eq. is CS_2 [3,4].

Set $L' = \partial^{-1} \circ(-\Delta + u) = d + \partial^{-1} \circ u$ and $P = -\,\partial \circ[\,L'\,, K_r\,]$. Now $\langle L, H<r> \rangle$ is a function iff P is a function. The latter condition is the reduced C eq. The proof uses $\langle L,H \rangle = \langle L,K+\bar{K} \rangle = P + P^*$ with K,P d^*- differential operators. Therefore $\bar{K}$, P^* are d- differential operators.

The (reduced) C eq. holds iff the $a_n<\ell>$ for n = 0,1,... satisfy the central recursion eq. $d(a_n) = T[p,p](a_{n-1}) + \ldots + T[p,\alpha](a_{\ell-\alpha}) + \ldots + T[p\ell](a_0)$ with $p = \ell+1-n$. $T[p,\alpha] = T[p,\alpha]<u>$ is a d*-differential operator depending linearly on u. Allow $\alpha = 0, \pm 1, \pm 2 \ldots$ The T-generating eq. is

$$(\partial^{\alpha+1} \cdot f \partial^{\alpha} \cdot u)_+ = \sum_{p \leq \alpha} \partial^p \cdot \{T[p,\alpha]<u>(f)\} \partial^p \tag{1}$$

Here $A_+ = A + A^{\tau}$ with τ = transpose without conjugation. $T[p,\alpha]$ involves d* a total (free + bound occurrences) of $2(\alpha-p)+1$ times.

$$\{ T[p,\alpha] = \sum_{w=0}^{q} \left[\binom{q}{w} + \binom{q-1}{w} \right] \binom{\alpha}{k} \partial^w \cdot u_{\underline{k}}, \text{ with } q = \alpha - p + 1 \text{ and } k = 2q-1-w \}$$

Set $I_{\ell+1} = r^{-1} \int_M T\ (u)$.

Examples: $I_1 = \int u$; $I_2 = \frac{1}{3} \int su$; $I_3 = \frac{1}{5} \int u(2s^2 + s_{\underline{2}})$.

Each CS_j eq. has the infinite sequence of independent time invariants $I_1, I_2, \ldots, I_k \ldots$
See the Main Theorem below.
Say $\dot{I}_{\ell+1} = \int_M \beta_\ell \dot{u}$. The gradient function β_ℓ is determined as follows. Let h = - 1,0,1... Fix h.. $\beta_{h+1} = g_{h+1}<h>$ is a member of the sequence $g_0<h>, g_1<h>, \ldots g_{h+1}<h>, \ldots$ satisfying the cô-recursion eq. $d(g_n) = \sum_{\alpha=p}^{h+1} T'[p,\alpha](g_{h+1-\alpha})$ with $p = h+2-n$. Take $g_0 = 1$. The T' - generating eq. is

$$(\partial^{\alpha} \circ \langle g \rangle \partial^{\alpha-1} \circ u)_- = \sum_{p \leq \alpha} \partial^{p-1} \circ \langle T'[p,\alpha]<u>(g) \rangle \partial^{p-1} \tag{2}$$

Definitions: $\langle g \rangle = g\,\partial + 2^{-1} g_{\underline{1}}$; $A_- = A - A^{\tau}$. $T'[p,\alpha]$ involves d* a total of $2(\alpha-p)+1$ times.

$$\left\{ T'[p,\alpha] = \sum_{w=0}^{q} \left[\binom{q}{w}\binom{\alpha-1}{k} + \binom{q-1}{w}\binom{\alpha}{k} \right] \partial^w \circ u_{\underline{k}} \text{ with } q = \alpha-p+1 \text{ and } k = 2q-1-w. \right\}$$

The stationary C eq. is $\langle L,H \rangle = 0$. We consider real solutions $H = K + \bar{K}$ and imaginary solutions $H = K - \bar{K}$. $K = T\partial$ with T a formal d* - ψ.d.0. (ψ.d.0. : pseudodifferential operator) of the following form.

For H real, require $T = T<\ell> = \sum_{-\infty<p\leq\ell} \partial^p \circ a_{\ell-p} \partial^p$ for some natural number ℓ and complex valued functions $a_0(x,y), \ldots a_n(x,y) \ldots$ T is symmetric : $T^{\tau} = T$. For H imaginary require $T = iT'$ with $T' = T'<h> = \sum_{-\infty<p\leq h} \partial^p \circ \langle g_{h-p} \rangle \partial^p$ for some integer $h \geq -1$ and complex valued function $g_n(x,y)$. T' is skew symmetric : $T'^{\tau} = -T'$.

Let $\sigma = \pm 1$. Say $T^{\iota} = \sigma\ T$. $K + \sigma\ K$ and $i\ K + \sigma(i\ K)$ are both solutions of the stationary C eq. precisely when K commutes with $L' = \partial^{-1} L$. When T is symmetric, the commutivity condition translates into the central recursion equation on $a_0, a_1 \ldots a_n \ldots$ When T is skew symmetric, the commutivity condition amounts to the corecursion eq. on $g_0, g_1 \ldots g_n \ldots$.

We normalize a_n . Take $a_0 = 1$. For each $n \geq 1$ require that a_n vanish at infinity. Similarly normalize g_0, g_n. Then $a_n = a_n <\ell>$ and $g_n = g_n <h>$.

The case $\ell = 0$ is fundamental. Assume there exists a $K = (1 + \partial^{-1} \circ a_1 \partial^{-1} + \partial^{-2} \circ a_2 \overline{\partial^{-2}} + \ldots)\partial$ which commutes with L'. The "canonical" solutions of the stationary C eq. are : $H = K^r + \overline{K^r}$, for r odd and $r \geq 1$; and $H = i(K^r + \overline{K^r})$, for r even and $r \geq 0$.

For r odd, the canonical differential operator solution $H<r>$ of the C eq. with H of degree r is $2\,\mathrm{Re}\,(K^r)^+$. Here A^+ is the differential operator part of A.

Set $\alpha_{\ell+1} = a_{\ell+1} <\ell>$. This $\alpha_{\ell+1}$ is the "$\ell+1^{th}$ equation density". The $CS_{\ell+1}$ eq. is $\dot{u} = 2\mathrm{Re}\ d(\alpha_{\ell+1})$.

An "explicit" expression is given for K^r in terms of u, next. Let E be a formal ψ.d.0. on M. Say $\sigma = +$ or $\sigma = -$. Let $V = (E\,\partial^{-1} \circ u)_\sigma$. Then $V^\tau = \sigma V$. So V can be put in central form. Assume there exists a formal ψ.d.0. S (in the same central form) with $d(S) = V$ and each coefficient of S vanishing at infinity on M. Set $\theta(E) = S$. Define $N<\sigma> = (1-\theta)^{-1}$. Then $K^r = \partial^{-1} \circ [N<\sigma> (\partial^{r+1})]$ with $\sigma = +$ for $r = 2\ell + 1$ and $\sigma = -$ for $r = 2(h+1)$. This follows directly from the commutivity condition.

Let λ be a real scaler. Take $T(\lambda) = \Sigma_\ell T <\ell> \lambda^\ell$ with $\ell \geq 0$. Then $T(\lambda = (1 - \lambda K^2)^{-1} K \partial^{-1} = \partial^{-1} \cdot [N <+> (J\partial)] \partial^{-1}$ with $J = (1 - \lambda \partial^2)^{-1} \partial$. Now take $T'(\lambda) = \Sigma_h T' <h> \lambda^{h+1}$. Then $T'(\lambda) = (1 - K^2 \lambda)^{-1} \partial^{-1} = \partial^{-1} \cdot [N <-> (J)] \partial^{-1}$.

Each of the $CS_{\ell+1}$ eqs. has the infinite sequence of (independent) time invariants $I_1, I_2 \ldots$ provided : (1) the gradient of I_{h+2} is β_{h+1} ; and (2) $(\beta_{h+1}, d(\alpha_{\ell+1})) + <\beta_{h+1}, d(\alpha_{\ell+1})> = 0$. Here $(a,b) = \int_M a\,b$ and $<a,b> = \int_M \bar{a}\,b$. The I_k for which (1) above holds are said to be "normal".

Main Theorem (Weak orthogonality) $\big(\beta_{h+1}, d(\alpha_{\ell+1})\big) = 0$. (Strong orthogonality) $<\beta_{h+1}, d\alpha_{\ell+1}> = 0$.

The Main Theorem is proven below using a novel fundamental approach. This does not involve the powerful but elaborate theory of recursion operators (set in a formalism of generalized functions with bilocal arguments) used to generate integrable nonlinear evolution equations [5,6].

Say μ is a real variable. Let A_p, G_p be the formal series

$A_p = \Sigma_\ell a_n <\ell> \lambda^{\ell+1}$ with $\ell = p - 1 + n$: $\ell, n \geq 0$.

$G_p = \Sigma_h g_n <h> \mu^{h+2}$ with $h = p - 2 + n$: $h \geq -1$; $n \geq 0$.

$d(A_p) = \Sigma_\alpha T[p,\alpha](A_{\alpha+1})$ with $\alpha \geq p$, since the $a_n <\ell>$ satisfy the central recursion eq. The co-recursion eq. has the global version $d(G_p) = \Sigma T'[p,\alpha](G_{\alpha+1})$ with $\alpha \geq p$.

$\Sigma_p \partial^{p-1} \cdot A_p \partial^{p-1} = \lambda T(\lambda)$. Also $\Sigma_p \partial^{p-2} \cdot \langle G_p \rangle \partial^{p-2} = \mu T(\mu)$.

Weak orthogonality amounts to $(G_1, d A_0) = 0$. Strong orthogonality amounts to $< G_1, d A_0> = 0$. The proofs of orthogonality given later use the following lemma.

(Transpose) Lemma. $T^\tau[p,\alpha] = - T'[-\alpha,-p]$.

Proof - The touchstone of the Lemma's validity will be revealed to be an algebraic identity. A generating equation for the $T^\tau[p,\alpha]$ with $p = \alpha, \alpha-1, \alpha-2$ ... is derived by performing the following steps on each side of the T-generating Eq. (1). Apply the operator to g. The result is the value at f of a formal ψ.d.0. Transpose this ψ.d.0. Multiply by -1. Let E_α be the new generator obtained. In the new expansion replace each $-T^\tau[p,\alpha]$ with $T'[-\alpha,-p]$. Say J_α is the resulting expansion. $J_\alpha(h) = \Sigma_p T'[-\alpha,-p][(-1)^p g_p h_p]$.

The lemma holds if $J_\alpha(h) = E_\alpha(h)$ for all u,g,h. Assemble the $J_\alpha(h)$, with $\alpha \geq 0$, into a "known" expression by using the T'-generating Eq. (2) as follows. (The case with $\alpha < 0$ is similar). Set $J = \Sigma \partial^{-\alpha-1} \cdot \langle J_\alpha(h) \rangle \partial^{-\alpha-1}$. Obtain E from J by replacing $J_\alpha(h)$ with $E_\alpha(h)$. In J interchange the order of summation over α,p. We get $J = E$ with $E' = \Sigma_p E'_{-p} < (-1)^p g_p h_p >$. Here $E'_\alpha < f >$ is the T'-generator of Eq. (2). Thus $J_\alpha(h) = E_\alpha(h)$ for $\alpha \geq 0$, if $E' = E$. Now $E' = E$ holds for all u, g, h, if it holds for $u = e^{k_1 z}$, $g = e^{k_2 z}$, $h = e^{k_3 z}$ with k_1, k_2, k_3 arbitrary constants. Set $A = e^{-kz} E'$ and $B = e^{-kz} E$ with $k = k_1 + k_2 + k_3$. A,B are independent of z and are rational functions of ∂. A = B.

(T*-) Corollary. $(T[d^* ; p,\alpha] < u >)^* = T'[d ; -\alpha, -p] < u >$.

Proof. $T^* = \overline{T^\tau}$. Apply the Transpose Lemma. A generating eq. for $\overline{T}'$ is obtained by taking the conjugate of each side of that for T'.

Proof of weak orthogonality. Set $W = (G_1, d(A_0))$. $W = 0$ is established by successively, relative to $\ell = 0,1,2\ldots$, eliminating the local occurrences from W of $a_n <\ell>$ for all n. The recursion equations for A_p, $G_{p'}$ are used for $p \geq 0$, $p' \leq 0$ respectively. Set $S_{p'} < q > = \Sigma_\alpha T'[p',\alpha](G_{\alpha+1})$, with $0 \leq \alpha \leq q - 1$. Then $d(G_{p'}) = S_{p'} < 1 - p' >$, $< G, T[p,\alpha](A) > = < -T'[-a,-p](G), A >$ follows by the Transpose Lemma. $W = (G_{1-p}, d A_p) + \sum_{p'>p} (-S_{1-p'} < p >, A_{p'})$ follows by induction on p.

The "Phoenix equation" is the sufficient condition for strong orthogonality derived below. The strong orthogonality condition involves the nonlocal operator $d^{-1}d^*$ which enters in building the gradient packet G_1 and equation density packet A_0 from u. The Phoenix eq. has the virtue of avoiding nonlocal dependence on u.

Each differential operator in d^*,d has a unique "thin" L (respectively : "thin" R) representation as $d^* \cdot R + Ld + r$ with R, L differential operators, r a function and L a d (respectively : R a d^*) differential operator. The "remainder" r is the same for the two representations.The example $A < z,d^*> B < \bar{z},d >$, with the variables $z, \bar{z}$ separated, has the remainder $A^{\tau}(1)\, B(1)$.

The Phoenix eq. concerns the remainders defined next.

Allow $n,q \geq 1$ and $k \geq 2$. The kth (representation) coefficients $R_{n,q}< k >, L_{n,q}< k >, r_{n,q}<k>$ are defined recursively in k as follows. Require $L_{n,q} < 2j >$, $R_{n,q} < 2j+1 >$ to be thin.

Consider : q' with $1\text{-}q \leq \text{-}q' \leq \text{-}1$; and n' with $1\text{-}n \leq \text{-}n' \leq \text{-}1$.
Set $T'(q,q') = T' [d\,;\, 1\text{-}q, \text{-}q'] < u >$. Let $T(n, n') = T [d^*\,;\, \text{-}n, \text{-}1\text{-}n'] < u >$.
$T'(q,0)$ has remainder zero. {Proof : The $T'[p,0]$ (f) with $p \leq 0$ have a generator which vanishes at f = 1}.Set $B_{n,q} < 2 > = -T'(q\text{-}1,0)\, d^{-1} \cdot T^{\tau}(n,0)$. The representation of $B_{n,q} < 2 >$ yields the 2nd coefficients.

Suppose the kth coefficients have been obtained. Define $B_{n,q}< k+1 > = \hat{R}_q - \check{L}_n$ with : $R_q = R_{n,q}< k >$; $L_n = L_{n,q} < k >$; $\hat{R}_q = \Sigma_{q'}\, T'(q,q')\, R_{q'}$; and $\check{L}_n = \Sigma_{n'}\, L_{n'}\, T^{\tau}(n,n')$.The representation of $B_{n,q} < k+1 >$ produces the $k+1^{th}$ coefficients. $B_{n,q} < k > = 0$, for $k > 2q$.
Proof : The order of homogeneity in d of $B_{n,q} < k >$ is 2 q-k.

<u>Theorem</u>. The "Phoenix equation", $r_{n,q} < k > = 0$, implies strong orthogonality.

<u>Proof</u>. It suffices to show the following.

<u>Claim</u>. $< G_1, d\, a_j > = 0$, if $r_{n,q} <k> = 0$ for $k \leq 2j$.

<u>Corollary of the Claim</u>. The CS_j eq. has every normal I_v as an invariant if the kth remainders vanish for $k \leq 2j$.

<u>Proof of the Claim</u>. Set $S = < G_1, d(A_0) >$. Take $S_k = \Sigma_{n,q} < B_{n,q} <k> G_{1+n}, A_q >$ with $q \geq \frac{k}{2}$. Then $< G_1, d\, \alpha_j > = 0$, if $S = S_{2j+1}$. (Proof : Equate the coefficients of λ^j). S is transformed into S_2 using the steps (1) , (2) stated below. $S_k = S_{k+1}$, if (') each kth remainder vanishes. <u>Proof</u>. In S_k replace each $B_{n,q} < k >$ with its representation and assume (') above. Then applying to S_k the following algorithm yields S_{k+1}.

(0) $< (d^* \cdot R + Ld)\, G, A > = < R, dA > + < L\, d\, G, A >$.

(1) Say $p \geq 0$. Replace each $d(A_p)$ using the A-recursion eq. This introduces T $(= T[p,\alpha])$.

(2) Use $< G, T(A) > = < T^* (G), A >$. Replace T^* with T' using the T^* - Corollary.

(3) Say $p \geq 1$. Replace each $d(G_p)$ by employing the G-recursion eq. This introduces T'. Replace T' with $-T^{\tau}$ using the Transpose Lemma.

Each of the kth coefficients is homogeneous in u of order k.

We seek to complete the proof of invariance under the aegis of the Phoenix eq. The possibility of establishing the Phoenix eq. seems consumed in the equation's firestorm of nonlinearities of arbitrary high order. Remarkably, the Laplace transform enables us to reduce the Phoenix equation to an algebraic identity.

Theorem. Each remainder $r_{n,q} < k >$ vanishes.

Proof. u can be expressed as a Laplace transform for $x, y > 0$. We can restrict x, y to be > 0 in the recursion eq. for the kth coefficients since the equation is invariant under a rotation of coordinates. Temporarily assume : $u = e(z)\, e'(\bar{z})$; and for a given k each $R_{n,q} < k >$, $L_{n,q} < k >$ is in the variables separated form $A_n [z,d^*]\, A'_q [\bar{z},d]$. Then $B_{n,q}^{\cdot} < k+1 >$ is a sum of such variables separated terms. Thus, so are the $k+1^{th}$ coefficients.

Each kth representation coefficient has a multilinear version obtained by replacing u with $\sum_{v=1}^{k} \lambda_v u_v$ (λ_v being a real scaler) and reading the coefficient of $\lambda_1 \cdot \lambda_2 \cdots \lambda_k$. The multilinear versions are connected by a recursion formula similarly obtained from the original representation recursion. The Phoenix eq. is true, if for each k the multilinear remainders $\tilde{r}_{n,q} < k > = r_{n,q} [u_1,\ldots u_k]$ vanish when (#) : Each u_v has the form $u_v = \exp(-\bar{c}_v z - c_v \bar{z})$ with c_v a complex constant having real and imaginary parts > 0.

Claim. $\tilde{r}_{n,q} < 2 > = 0$.

Proof of Claim. Take $u_v = e_v(z)\, e'_v(\bar{z})$ for $v = 1, 2$. Then $r_{n,q} [u_1,u_2] = \Sigma\, r_n\, r'_q$. Here $r_n = -T[-n, -1] <e_2> (e_1)$ and $r'_q = T' [1-q,0] < e'_1 > [d^{-1} (e'_2)]$. The summation is over the permutations of 1,2.

The crucial step is to efface the occurrences of T, T' in r_n, r'_q using the generating equations for T, T' with $\alpha = -1$, $\alpha = 0$ respectively. Let $n \rightarrow f_n$ be the sequence $f = (f_1, f_2, \ldots f_n, \ldots)$ of complex valued functions $f_n(\bar{z})$. Define $D_* (n \rightarrow f_n) = D_* (f) = \Sigma_n\, \partial^{-n} \cdot f_n\, \partial^{-n}$. Similarly, for functions $f'_q(z)$, define $D(q \rightarrow f'_q) = D(f') = \Sigma_q\, d^{-q} \cdot [f'_q\, d + \frac{1}{2} d\, (f'_q)]\, d^{-q}$. Say $f_{n,q}$ depends on $\bar{z}$, z. $f_{n,q} = 0$ for all $n,q \geq 1$, if $D_* D\, [(f_{n,q})]$ {which means $D_* [n \rightarrow D(q \rightarrow f_{n,q})]$} vanishes. Apply this with $f_{n,q} = \tilde{r}_{n,q} < 2 >$. Now $D_* \cdot D\, \{[r_n(z)\, r'_q (\bar{z})]\} = D_* [(r_n)]\, D[(r'_q)] = F_{1,2}$ with $F_{1,2} = -(c_1\, \partial^{-1} \cdot e_2)_+ \cdot (c'_1\, d^{-1} \cdot e'_2)_-$. Set $F = F_{1,2} + F_{2,1}$. Then $D_* D\, [(\tilde{r}_{n,q} < 2 >)] = F$.

Say $e_v(z) = \exp(-c_v z)$ and $e'_v = \overline{e_v(z)}$. Then $F = gF'$ with F' independent of z, $\bar{z}$ and $g(\bar{z}, z) \neq 0$. This F' is a rational expression in the commuting elements d^*, d (and c_1, $\bar{c}_1$, c_2, $\bar{c}_2$). $F' = 0$ is an algebraic identity.

Claim. $\tilde{r}_{n,q} < 3 > = 0$.

Proof of Claim. We may assume each u_v is as in (#) above. Part of $\tilde{r}_{n,q} < 3 >$ vanishes because it involves $\tilde{r}_{n,q} < 2 >$ as a factor. Apply $D_* D$ to the rest of $\tilde{r}_{n,q} < 3 >$. The resulting series is summable (with sum F say) by iterated use of the T,T' generating eqs for $\alpha = -1$, $\alpha = 0$ respectively. The relevant identities are stated next. Say $b = b(z)$. Define $\hat{f_q} = \Sigma_{q'} T'(q,q') < b > (f_{q'})$. Then $D[(\hat{f_q})] = [D(f)\, d^{-1} \cdot b]_-$. Say $a = a(z)$.

Define $\breve{f}_n = \Sigma_{n'} T(n,n') < a > (f'_{n'})$. Then $D_* [(\breve{f'}_n)] = [\, D_*[(f'_{n'})]\, \partial^{-1} \cdot a\,]_+$. Let F' be the residual part of F after the (exponential) dependence on z, z has been extracted. The claim follows from the algebraic (in d^*,d) identity $F' = 0$.

The proof that $\tilde{r}_{n,q} < k > = 0$ for general k is similar. We omit the details lest the continuance of the proof of the theorem itself seem time invariant.

References

[1] Veselov, A.P. and Novikov, S.P., "Finite-Zone, Two-Dimensional, Potential Schrödinger Operators. Explicit Formulas and Evolution Equations", Sov. Math. Dokl. 30, 588-591 (1984).

[2] Csizmazia, A.P., "Invariants under Congruence for Infinite Dimensional Operators", Thesis, University of California, Berkeley (1984).

[3] Csizmazia, A.P., "The Heisenberg - Lax Equation Generalized", in "Some Topics on Inverse Problems, Proceedings of the XVIth Workshop on the Interdisciplinary Study of Inverse Problems" (Montpellier, France, Nov. 30-Dec. 4, 1987), World Scientific Publishing Co., Singapore (1988).

[4] Csizmazia, A.P., "Sylvester's Law in Infinite Dimensions", in "Proceedings of the International Workshop on Nonlinear Evolution Equations : Integrability and Spectral Methods", (Como, Italy, July 4-15, 1988), Series : Proceedings in Nonlinear Science, Manchester Univ. Press.

[5] Boiti, M., Léon, J. J-P. and Pempinelli, F., "Canonical and Non Canonical Recursion Operators in Multidimensions", Preprint PM : 87/03, Laboratoire de Physique Mathématique, Université des Sciences et Techniques du Languedoc, Montpellier, 1987.

[6] Santini, P.M., "Solvable Nonlinear Equations as Concrete Realizations of the same Abstract Algebra", in "Integrable Systems and Applications : Proceedings of a Workshop Held at Oléron, France, June 20-24, 1988", Lecture Notes in Physics, Springer-Verlag, Berlin.

On the Dressing Method

V.E. Zakharov

L.D. Landau Institute for Theoretical Physics, GSP-1,
Moscow V-334, Kosygina 2, SU-117940 USSR

Introduction

The purpose of this lecture is to present the development of the so-called "dressing method" in the theory of integrable systems. This method, allowing to find both new integrable systems and their exact solutions, is one of the most poweffful tools in the nonlinear mathematical physics. It is clear now that a natural field for its applications is the theory of integrable nonlinear equations in 2+1 dimentions, in other words, the theory of integrable evolutional equations on (x,y) plane. We use here and further the words "integrable system" in a colloquial meaning – most of this systems are not integrable in a strict Louville's sense. Nevertheless they have infinite sets of motion integrals and exact solutions and they could be studied effectively. There are two different versions of the dressing method in 2+1 dimensions. The first one, based on Volterra-type linear integral equations, was introduced in 1974 in the article of A.B.Shabat and the author [1]. The second version, using $\bar{\partial}$-problem on the complex plane was developed by S.V.Manakov and the author in 1985 [2]. One of the purposes of my lecture is to show that both mentioned approaches are equivalent in some sense. This fact plays a key role in the further progress of the theory.

My second aim is to outline some possibilities for generalization of the dressing method. We shall see that this method can be expanded to cover broad new classes of integrable equations. It is a serious work to explore all the new capacities. We shall do just the first steps in this direction. I will display several new integrable equations mostly with space- and time-dependent coefficients and show also some new interesting solutions of previosly well known integrable equations.

1 The "old" dressing method.

Let us describe shortly the principal idea of article [1]. We start from the equation:

$$K(x,z) + F(x,z) + \int_x^\infty K(x,s)F(s,z)ds = 0 \qquad (1.1)$$

Here F is a known and K is unknown $N \times N$ matrix functions, depending also on some additional variables y_j $(j = 1, ..., n)$. The equation (1.1) could be rewritten in the symbolic form

$$K + F + K * F = 0 \tag{1.2}$$

The first step in the dressing method is finding ouf a pair of operators $D, \tilde{D}$ satisfying the equation

$$\tilde{D}K + DF + \tilde{D}K * F + K * DF = 0 \tag{1.3}$$

The operator D is *bare* while the operator $\tilde{D}$ is *dressed* one. An operator D is cold *dressible* if a corresponding $\tilde{D}$ could be found. The most simple dressible operators are differential. Let us put $N = 1$ (K and F are scalar functions) and choose the operator D in the form

$$DF = \alpha \frac{\partial F}{\partial y} + \frac{\partial^2 F}{\partial x^2} - \frac{\partial^2 F}{\partial z^2} \tag{1.4}$$

Here α is an arbitrary constant. Applying then operator D to (1.1), we find after a simple calculation that (1.3) is valid if

$$\tilde{D}K = \alpha \frac{\partial K}{\partial y} + \frac{\partial^2 K}{\partial x^2} + U(x)K - \frac{\partial^2 K}{\partial z^2} \tag{1.5}$$

Here $U(x) = -2\frac{d}{dx}K(x,x)$. So the operator (1.4) is dressible. More general example of dressible operators was found in [1]. Let N be arbitrary and

$$L_0 = l_0(x, y, ...y_n)\frac{\partial^n}{\partial x^n} + ... \tag{1.6}$$

be a linear ordinary differential operator of order n. All its coefficients $l_i (i = 0, 1, ...n)$ are $N \times N$ matrix functions. Let us consider the operator

$$DF = \alpha \frac{\partial F}{\partial y} + L_0 F - F L_0^+ \tag{1.7}$$

Here L_0^+ is a coadjoned operator acting from right side on the variable z. The central result of the article [1] is formulated as a following: *the operator (1.7) is dressible.* The dressed operator $\tilde{D}$ has a form

$$\tilde{D}K = \alpha \frac{\partial K}{\partial y} + LK - KL_0^+ \tag{1.8}$$

Here $L = L_0 + \tilde{L}$ and

$$\tilde{L} = \tilde{l}_0 \frac{\partial^{n-1}}{\partial x^{n-1}} + ... \tag{1.9}$$

is a linear differential operator of the order $n-1$. Its coefficients $\tilde{l}_i$ depend upon the function $K(x,z,y_i)$. Let us consider the set of functions

$$K_i(x,y) = (\frac{\partial}{\partial x} - \frac{\partial}{\partial z})^i K(x,y,z)|_{z=x} \qquad (i = 0,1,...,n-1)$$

Here coefficients $\tilde{\ell}_i$ are polinomial on $K_j (0 \leq j \leq i)$ and on their x-derivatives. In the previous example we had

$$L_0 = \frac{\partial^2}{\partial x^2} \qquad L = \frac{\partial^2}{\partial x^2} + U(x) \qquad U = -2\frac{d}{dx}K_0$$

In this case $\tilde{\ell}_0 = 0$, $\tilde{\ell}_1 = U$. To construct an integrable system we should have two *commuting* dressible operators

$$\begin{aligned} D_1 F &= \alpha_1 \frac{\partial F}{\partial y_1} + L_0^1 F - F L_0^{1+} \\ D_2 F &= \alpha_2 \frac{\partial F}{\partial y_2} + L_0^2 F - F L_0^{2+} \end{aligned} \qquad (1.10)$$

The condition of commutativity $[D_1 D_2] = 0$ gives

$$\alpha_1 \frac{\partial L_0^2}{\partial y_1} - \alpha_2 \frac{\partial L_0^1}{\partial y_2} + [L_0^1, L_0^2] = 0 \qquad (1.11)$$

Assume now that

$$D_1 F = 0 \qquad D_2 F = 0 \qquad (1.12)$$

In accordance with (1.11) these equations are compatible. Supposing that the equation (1.2) could be resolved uniquely, we have

$$\tilde{D}_1 K = 0 \qquad \tilde{D}_2 K = 0 \qquad (1.13)$$

and hence

$$\alpha_1 \frac{\partial L^2}{\partial y_1} - \alpha_2 \frac{\partial L^1}{\partial y_2} + [L^1, L^2] = 0 \qquad (1.14)$$

From (1.14) and (1.11) we find

$$\alpha_1 \frac{\partial \tilde{L}^2}{\partial y_1} - \alpha_2 \frac{\partial \tilde{L}^1}{\partial y_2} + [L_0^1, \tilde{L}^2] + [\tilde{L}^1, L_0^2] + [\tilde{L}^1, \tilde{L}^2] = 0 \qquad (1.15)$$

This is a system of nonlinear equations on the coefficients of operators $\tilde{L}^{1,2}$. This system is imposed on the finite set of functions K_i in fact. After solving the equation (1.1) we can get $K(x,z,y,...y_n)$ and some definite solutions of this system. Assuming that

$$L_0^2 = \frac{\partial^3}{\partial x^3} \qquad (1.16)$$

we have after some computations the expression

$$L^2 = \frac{\partial^3}{\partial x^3} + 6U\frac{\partial}{\partial x} + W \tag{1.17}$$

Here

$$W = 12\frac{\partial^2}{\partial x^2}K_0 + 12\frac{\partial}{\partial x}K_1 - GUK_0 = 3U_x + 6\alpha\frac{\partial}{\partial y}K_0$$

Considering further $\alpha_2 = -1, y_2 = t, \alpha_1 = \alpha, y_1 = y$ one can find from (1.15) that the function $U(x,t,y)$ satisfies the Kadomtzev-Petviashwili (KP) equation

$$\frac{\partial}{\partial x}(\frac{\partial U}{\partial t} + 6U\frac{\partial U}{\partial x} + U_{xxx}) + 3\alpha^2 U_{yy} = 0 \tag{1.18}$$

In the case $\alpha^2 = -1$ it is the KP-1 equation, in the case $\alpha^2 = 1$ it is the KP-2 equation.

2 Two comments.

1. The developed theory is still correct if the operation $K * F$ is understood in more general sense:

$$K * F = (1-\mu)\int_x^\infty K(x,s)F(s,z)ds - \mu\int_{-\infty}^x K(x,s)F(s,z)ds \tag{2.1}$$

In particulary, instead of (1.1) we can start from the equation (2.1) with $\mu = 1$

$$K(x,z) + F(x,z) - \int_{-\infty}^x K(x,s)F(s,z)ds = 0 \tag{2.2}$$

or from the equation

$$K(x,z)+F(x,z)+\frac{1}{2}\int_x^\infty K(x,s)F(s,z)ds - \frac{1}{2}\int_{-\infty}^x K(x,s)F(s,z)ds = 0 \tag{2.3}$$

Varying μ at the same F we will get *different solutions* of the same nonlinear equation (1.15). In the matrix case μ could be replaced by a matrix, commuting with operator $L_0^{1,2}$ ($[L_0^{1,2}, \mu] = 0$).

2. The system (1.14) could be considered as an integrable nonlinear system instead of (1.15), if the unknown functions are the coefficients of operators $L^{1,2}$. The coefficients of "bare operators" $L_0^{1,2}$ satisfies the same system of equations (1.11). It means that we can do dressing procedure starting from any given solution of the nonlinear system (1.14). If the coefficients of operators $L_0^{1,2}$ are constant we speak about dressing over a trivial background. In an opposite case we have dressing over a nontrivial background. It was used first by E.A.Kuznetzov and A.V.Mikhailov [3] to describe a propagation of a soliton on a knoidal wave in the KdV equation. Nontrivial dressing for 2+1 dimension equations will be studied in the article of A.Fokas and the author [4].

3 The generalized N-wave equations.

Suppose now that we have *three or more* commuting dressible operator $D_i([D_i, D_j] = 0, i = 1, ..., n)$, $n \geq 3$. Suppose also that a function F satisfies the set of equations

$$D_i F = 0 \qquad i = 1, ...n. \tag{3.1}$$

After the dressing procedure we get the system of $\frac{1}{2}n(n-1)$ equations

$$\alpha_i \frac{\partial L^j}{\partial y_i} - \alpha_j \frac{\partial L^i}{\partial y_j} + [L^i, L^j] = 0 \tag{3.2}$$

Only $(n-1)$ of these equations are independent, all others are satisfied due to the Jacoby identity. The system (3.2) is overdeterminated but compatible. It could be used for constructing of new integrable systems. It is possible , for instance, to exclude all x-derivatives from the system (3.2). Let

$$\alpha_i = 1 (i = 1, 2, 3) \qquad L_0^i = I^i \frac{\partial}{\partial x} \qquad [I^i, I^j] = 0$$

Here I^i is a diagonal commuting matrix. It is easy now to get for dressed operators

$$L^i = I^i \frac{\partial}{\partial x} + [I^i, Q]$$

$$Q = K(x, y_1, y_2, y_3)$$

And the system (3.2) became the following

$$\frac{\partial}{\partial y_j}[I^i, Q] - \frac{\partial}{\partial y_i}[I^j, Q] + I^j \frac{\partial Q}{\partial x} I^i - I^i \frac{\partial Q}{\partial x} I^j + [[I^j, Q], [I^i, Q]] = 0 \tag{3.3}$$

Multiplying the equation (3.3) by I_k from the left side and doing the cyclic permutation we drop out all the terms containing $\frac{\partial Q}{\partial x}$, and get as a result

$$\varepsilon_{ijk}(I_i \frac{\partial Q}{\partial y_j} I_k - I_i Q I_j Q I_k) = 0 \tag{3.4}$$

Here ε_{ijk} is a totally antisymmetric unit tensor. The equation (3.4) is a natural generalization of so-called "N-wave" system. This equation is invariant under the transformation

$$I_i \rightarrow I_i + \alpha I_j + \beta I_k$$

If the matrix $I_k = 1$, the system (3.4) coincides to (3.3) after the change $\frac{\partial}{\partial y_k} \rightarrow \frac{\partial}{\partial x}$. For $N = 3$ we can put one of the matrixes I_i $(i = 1, 2, 3)$ to be scalar and get the equation (3.3). For $N \geq 4$ the system (3.4) is really more general then (3.3). About the system (3.3) see also [4].

4 Shock-wave on solution.

Let us now concentrate our attention on the application of the "old" dressing method to the equation KP-2

$$\frac{\partial}{\partial x}(U_t + U_{xxx} + 6UU_x) + 3U_{yy} = 0 \tag{4.1}$$

We will use the integral equation (1.1). The function F satisfies the system

$$\begin{aligned} \frac{\partial F}{\partial y} + \frac{\partial^2 F}{\partial x^2} - \frac{\partial^2 F}{\partial z^2} &= 0 \\ \frac{\partial F}{\partial t} + 4(\frac{\partial^3 F}{\partial x^3} + \frac{\partial^3 F}{\partial z^3}) &= 0 \end{aligned} \tag{4.2}$$

One can find the solution (4.2) in the form

$$F = \phi(x,y,t)\theta(z,y,t) \tag{4.3}$$

$$\frac{\partial \phi}{\partial y} + \frac{\partial^2 \phi}{\partial x^2} = 0 \qquad \frac{\partial \theta}{\partial y} - \frac{\partial^2 \theta}{\partial z^2} = 0 \tag{4.4}$$

$$\frac{\partial \phi}{\partial t} + 4\frac{\partial^3 \phi}{\partial x^3} = 0 \qquad \frac{\partial \theta}{\partial y} + 4\frac{\partial^3 \theta}{\partial z^3} = 0 \tag{4.5}$$

To provide a convergence of the integral in (1.1) let us demand F vanishing at $x \to \infty$, $z \to \infty$. We may find ϕ, θ in the form

$$\phi = \int_0^\infty f(\lambda) e^{-\lambda x - \lambda^2 y + 4\lambda^3 t} d\lambda \tag{4.6}$$

$$\theta = \int_0^\infty h(\lambda) e^{-\lambda y + \lambda^2 y + 4\lambda^3 t} d\lambda \tag{4.7}$$

The equation (1.1) could be solved easily

$$K(x,z) = \psi(x,y,t)\theta(z,y,t) \tag{4.8}$$

Here

$$\begin{aligned} \psi &= -\frac{\phi}{1 + \int_x^\infty \phi\theta dx} \\ U &= 2\frac{d}{dx}K(x,x) = -2\frac{d}{dx}(\frac{\phi\theta}{1 + \int_x^\infty \phi\theta dx}) \end{aligned} \tag{4.9}$$

This solution could be rewritten in the following form

$$U = 2\frac{d^2}{dx^2}\log R \tag{4.10}$$

$$R = 1 + \int_x^\infty \phi\theta dx \tag{4.11}$$

We found an exact solution of the KP-2 equation. Let the functions $f(\lambda), h(\lambda)$ now be *real and positively definite* and vanishing if $\lambda \to \infty$ faster than $e^{-A\lambda^2}$ at any A. Under this assumptions the solution (4.9) has no singularities on (x,y) plane. In the particular case $f(\lambda) = h(\lambda)$ this solution is symmetric with respect to changing sign of y : $U(x, y) = U(x, -y)$.

Choosing

$$f(\lambda) = h(\lambda) = \sqrt{2\nu}\delta(\lambda - \nu) \tag{4.12}$$

we have

$$U(x,t) = \frac{2\nu^2}{ch^2\nu(x - 4\nu^2 t)} \tag{4.13}$$

It is a solution of amplitude ν^2 moving in x-direction. It is very interesting to study a more general solution in some sense close to the soliton. Let us assume

$$f(\lambda) = h(\lambda) = \sqrt{2\nu}g(\lambda - \nu) \tag{4.14}$$

The function $g(\xi) > 0$ is concentrated on the support $-\Delta \leq \xi \leq \Delta$. At condition $\frac{\Delta}{\nu} \to 0$ we find

$$R = 1 + e^{-2\nu x'}\Phi(x' - 2\nu y - 4\nu^2 t)\Phi(x' + 2\nu y - 4\nu^2 t)$$

Here $x' = x - 4\nu^2 t$ is a coordinate in the moving frame, while

$$\Phi(z) = \int_{-\Delta}^{\Delta} g(\xi)e^{-\xi z}\partial\xi$$

The solution (4.9) describes the process at "straghtening" of initially curved solution, as it is drawn on the picture 1 (Section 5). This process is corresponded by propagation of two "shock-waves of compression", producing in the negative half-plane $x' < 0$ some extended weak radiation. To get a more general solution, let us put

$$F = \sum_{n=1}^{N} \phi_n(x,y,t)\theta_n(z,y,t) \tag{4.15}$$

$$\phi_n(x,y,t) = \int_0^\infty f_n(\lambda)e^{-\lambda x - \lambda^2 y + 4\lambda^3 t}d\lambda \tag{4.16}$$

$$\theta_n(x,y,t) = \int_0^\infty h_n(\lambda)e^{-\lambda x + \lambda^2 y + 4\lambda^3 t}d\lambda \tag{4.17}$$

The solution is given by the formula (4.10)

$$R = det \parallel \delta_{ij} + \int_x^\infty \phi_i\theta_i dx \parallel \tag{4.18}$$

This solution describes a nonlinear superposition on N curved and straghtening solutions.

One can use, instead of (1.1), the equation (2.2). In this case a simple solution has the form

$$\begin{aligned} U &= 2\frac{d^2}{dx^2}\ln R \\ R &= 1+\int_{-\infty}^{x}\tilde{\phi}\tilde{\theta}dx \end{aligned}$$

Here

$$\begin{aligned} \tilde{\phi} &= \int_0^{\infty} f(\lambda)e^{\lambda x-\lambda^2 y-4\lambda^3 t}d\lambda \\ \tilde{\theta} &= \int_0^{\infty} h(\lambda)e^{\lambda x+\lambda^2 y-4\lambda^3 t}d\lambda \end{aligned}$$

One can see that this equation could be obtained from the previous one by the single transformation $x \to -x, t \to -t$. Performing this transformation in (4.14), we have a more general solution

$$U = 2\frac{d^2}{dx^2}\log\det \| \delta_{i,j} + \int_{-\infty}^{x} \tilde{\phi}_i\tilde{\theta}_j\partial x \| \tag{4.19}$$

5 Towards the "new" dressing method.

The obtained solutions are interesting enough from the view point of physics. But they are very far from some kind of general solutions of KP-2. The general solution could not be found by means of the "old" dressing method. Any attempt to choose the function F, different from described above, gives the singular solution on (x,-y) - plane. It seems attractive to choose functions ϕ and θ in the form

$$\phi = \int_{-\infty}^{\infty} f(k)e^{ikx+k^2y-4ik^3t}dk \tag{5.1}$$

$$\theta = \int_{-\infty}^{\infty} h(k)e^{ikx-k^2y-4ik^3t}dk \tag{5.2}$$

But one can see that when $y \to \infty$ the function ϕ grows exponentially fast, while θ decreases slowly. As a result the determinant R becomes zero on some line $x = x_0(y)$, where the solution has severe singularities $U \sim c/(x-x_0(y))^2$. Using the previousely described version of the dressing method we could not find even such fundamental solutions of KP-2 that are small in some proper norm and close to solutions of the linearized equation.

It is clear that the dressing method should be seriously generalized. To do so we will start from the equation (2,3). (Note that the framework of the "old" dressing method is almost useless, because it produces only singular solutions.) To generalize the dressing method, we should release the restrictions on the asymptotic behavior of $F(x,z)$ at $x \to \pm\infty$, $z \to \pm\infty$. Let us study the operator

$$\partial^{-1} f(x) = \frac{1}{2}\int_{-\infty}^{x} f(x)dx - \frac{1}{2}\int_{x}^{\infty} f(x)\text{x} \tag{5.3}$$

This operator is well-defined not only if $f(x) \to 0$ at $x \to \pm\infty$ fast enough. Let $f(x)$ be presented by the Fourier transformation

$$f(x) = \int_{-\infty}^{\infty} f(k)e^{ikx}dk \tag{5.4}$$

In this case

$$\partial^{-1} f(x) = ⨍_{-\infty}^{\infty} \frac{f(k)}{ik} e^{ikx}dk \tag{5.5}$$

Here $⨍_{-\infty}^{\infty}$ is a principal value of the integral. Assume now that $f(x)$ is represented by the integral

$$f(x) = \int_{\Omega} f(k,\bar{k})e^{ikx}dk \tag{5.6}$$

Here Ω is some domain in k–plane. In the general position $f(x)$ grows exponentially at $x \to \pm\infty$.

The operator (5.3)could be extended to functions (5.5) by the following way

$$\partial^{-1} = ⨍_{\Omega} \frac{f(k,\bar{k})}{ik} e^{ikx}dkd\bar{k} \tag{5.7}$$

The symbbol $⨍$ means that we understand $1/k$ as a limit

$$\frac{1}{k} = \lim_{\epsilon\to 0} \frac{\bar{k}}{|k|^2 + \epsilon^2} \tag{5.8}$$

In the particular case when Ω is a real axis, (5.6) coincides with (5.4). If the function $f(k,\bar{k})$ is decreasing fast enough at $| k | \to \infty$, the whole k-plane could be chosen instead of Ω. The only thing that should be done to extend tremendously the possibilities of the dressing method is to understand the integrals in (2.3) as it was described above. Returning to the equation (2.3),let us represent F and K in the following form

$$F = -\frac{1}{\pi}\int T(\nu,\bar{\nu},\lambda,\bar{\lambda})e^{i(\nu x-\lambda(x-z)}d\nu d\bar{\nu}d\lambda d\bar{\lambda} \tag{5.9}$$

$$K = -\frac{1}{\pi}\int Q(x,\lambda,\bar{\lambda})e^{i\lambda(x-z)}d\lambda d\bar{\lambda} \tag{5.10}$$

and denote that

$$G = Te^{i(\nu-\lambda)x} \tag{5.11}$$

Let us substitute(5.8) and (5.9) into (2.3) and do integration according to the rule(5.6). The equation (2.3) is satisfied if T and Q obey the following equality

$$Q(x,\lambda,\bar{\lambda})-\int G(\nu,\bar{\nu},\lambda,\bar{\lambda})d\nu d\bar{\nu}+\frac{1}{\pi i}\int\frac{Q(x,\mu,\bar{\mu})G(\nu,\bar{\nu},\lambda,\bar{\lambda})}{\nu-\mu}d\nu d\bar{\nu}=0 \quad (5.12)$$

Introducing the function

$$\chi(\lambda,\bar{\lambda})=1+\frac{1}{\pi}\int\frac{Q(x,\mu,\bar{\mu})}{\lambda-\mu}d\mu d\bar{\mu} \qquad (5.13)$$

and using the well-known formula

$$\frac{1}{\pi}\frac{\partial}{\partial\bar{\lambda}}\frac{1}{\lambda-\nu}=\delta(\lambda-\nu) \qquad (5.14)$$

we get

$$Q(x,\lambda,\bar{\lambda})=i\frac{\partial\chi}{\partial\bar{\lambda}}$$

Hence the equation (5.11) could be rewritten in a form

$$i\frac{\partial\chi}{\partial\bar{\lambda}}=\int\chi(x,\nu,\bar{\nu})G(\nu,\bar{\nu},\lambda,\bar{\lambda})d\nu d\bar{\nu} \qquad (5.15)$$

It means that the generalized equation (2.3) is equivalent to the equation which provides the solving of "nonlocal $\bar{\partial}$ – problem"(5.14). It is assumed that the $\bar{\partial}$ – problem is normalized by the condition

$$\chi\to 1 \qquad at \qquad |\lambda|\to\infty \qquad (5.16)$$

The other possibility to solve this problem is using the equation on a function χ

$$\chi=1-\frac{i}{\pi}\int\frac{\chi(x,\nu,\bar{\nu})G(\nu,\bar{\nu},\mu,\bar{\mu})}{\lambda-\mu}d\nu d\bar{\nu}d\mu d\bar{\mu} \qquad (5.17)$$

In these cases we can assume that all functions are $N\times N$ matrix functions. If the function F satisfies the equation (3.1), and functions $l_k^i(i=0,1,...)$ don't depend on x,the function G obeys the equation

$$\frac{\partial G}{\partial y_i}+L_i(i\nu)G-GL_i^+(i\lambda)=0 \qquad (5.18)$$

Here $L_i(i\nu)$ is the symbol of the operator L_i. As a function of parameter x, the function G obeys also the equation

$$\frac{\partial G}{\partial x}=i(\lambda-\nu)G \qquad (5.19)$$

In particular, for the KP-2, we have

$$G=G_0(\nu,\bar{\nu},\lambda,\bar{\lambda})e^{i(\lambda-\nu)x+(\lambda^2-\nu^2)y-4i(\lambda^3-\nu^3)t} \qquad (5.20)$$

Suppose $G_0(\nu,\bar{\nu},\lambda,\bar{\lambda})=G_0(\lambda,\bar{\lambda})\delta(\lambda+\nu)\delta(\bar{\lambda}+\bar{\nu})$. In this case the dependence on the variable y drops out, and we move from the KP-equation to the KdV-equation

$$U_t + 6UU_x + U_{xxx} = 0$$

Hence, we found that in order to solve the KdV equation we should solve the following $\bar{\partial}$- problem

$$i\frac{\partial \chi}{\partial \bar{\lambda}} = G_0(\lambda, \bar{\lambda})e^{2i\lambda x - 8ix^3 t}\chi(-\lambda, -\bar{\lambda}) \tag{5.21}$$

This fact was established in the article [2] and reopened after by many authors.

The function $K_0 = K(x, \lambda)$ is connected with the function Q by the formula

$$K_0 = \frac{1}{\pi i}\int Q(x, \mu, \bar{\mu})d\mu d\bar{\mu} \tag{5.22}$$

There is one obvious difference between the old and the new versions of the dressing method. The function F has two independent variables — as arbitrary as the general solution of an evolution equation . On the contrary , the function G has four independent variables $\lambda, \bar{\lambda}, \nu, \bar{\nu}$. It leads to a conjecture that this version of the dressing method could be used for solving an equation in more than $2+1$ dimensions. This is an illusion . The different choice of T could lead to a different Q but to the same K_0 and to the same solution of the equation. Only the function F is important, it means that the function T is defined up to adding an arbitrary $\tilde{T}$

$$T \to T + \tilde{T} \tag{5.23}$$

where

$$\int \tilde{T}(\nu, \bar{\nu}, \lambda, \bar{\lambda})e^{i(\nu x - \lambda z)}d\nu d\bar{\nu}d\lambda d\bar{\lambda} = 0 \tag{5.24}$$

Note, that if Ω in (5.5) is bounded domain with boundary Γ, an integral on the domain could be replaced by the integral over Γ

$$\int_\Omega f(k, \bar{k})e^{ikx}dx = \frac{1}{2\pi i}\int_\Gamma \tilde{f}(\xi, \bar{\xi})e^{i\xi x}d\xi \tag{5.25}$$

where

$$\tilde{f}(\xi, \bar{\xi}) = \int_\Omega \frac{f(k, \bar{k}}{\xi - k}dkd\bar{k} \tag{5.26}$$

A similar formula is also correct for F.

6 The new dressing method

In the advanced variant of the dressing method we starts directly from the $\bar{\partial}$-problem (5.14) with the condition of normalization (5.15).We will assume this problem to be solvable by the unique way.It means that the function χ obeying equation (5.14) and vanishing at large λ

$$\chi(\lambda) \to 0 \qquad |\lambda| \to \infty \tag{6.1}$$

is identically equal to zero : $\chi(\lambda) = 0$.

Let us introduce three commuting first order operators $D_i (i = 1, 2, 3)$

$$D_i \chi = \frac{\partial \chi}{\partial y_i} + i \chi I_i(\lambda) \tag{6.2}$$

using three commuting $[I_i(\lambda), I_j(\lambda)] = 0$ matrix functions of the spectral parameter λ. In the first step we will assume them to be polynomial. The function G could be consider as the kernel of the integral operator $\hat{G}$, required to commute with D_i. This requirement imposes on G the system of equations

$$\frac{\partial G}{\partial y_i} = i(I_i(\nu) G - G I_i(\lambda)) \tag{6.3}$$

This system could be solved

$$G = e^{i\Phi(\nu)} G_0(\nu, \bar{\nu}, \lambda, \bar{\lambda}) e^{-i\Phi(\lambda)} \tag{6.4}$$

Here $\Phi(\lambda) = \sum_k I_k(\lambda) y_k$.

Let us rewrite (5.14) in a symbolic form

$$i \frac{\partial \chi}{\partial \bar{\lambda}} = \chi * G \tag{6.5}$$

Polyvomial on λ operators D_i, commute with the operators $\partial / \partial \bar{\lambda}$. Due to (6.3) one has

$$i \frac{\partial}{\partial \bar{\lambda}} D_i \chi = D_i \chi * G \tag{6.6}$$

Let $R = R(D_1, D_2, D_3)$ be any differential operator polynomial on D_i. Its coefficients are a matrix function of y_i acting on χ from the left side. Obviously $R \partial / \partial \bar{\lambda} = \partial / \partial \bar{\lambda} R$, we have

$$i \frac{\partial}{\partial \bar{\lambda}} R \chi = R \chi * G \tag{6.7}$$

The function $R\chi$ has at $\lambda \to \infty$ a polynomial-type behavior and in general it is not equal to zero. *But for any given function* χ it is possible to find a set of operators $\tilde{R}$ such that $\tilde{R}\chi \to 0$ at $|\lambda| \to \infty$. It means that

$$\tilde{R} \chi = 0 \tag{6.8}$$

It is obvious that the operator $\tilde{R}$ constitute a left ideal in the noncomutative ring of operators R. It was shown in [2] that it is possible to find two linearly independent operators $\tilde{R}_{1,2}$ serving as a basis in $\tilde{R}$. The function χ nearby the infinite point could be represented by the expansion

$$\chi = 1 + \frac{\chi_0}{\lambda} + \frac{\chi_1}{\lambda^2} + \ldots \tag{6.9}$$

The coefficients of $\tilde{R}_{12}$ are polynomial on χ_i and on their derivatives. Let us consider the very simple example

$$I_k(\lambda) = I_k \lambda \qquad [I_i, I_j] = 0 \tag{6.10}$$

Now

$$D_i \chi = \frac{\partial \chi}{\partial y_i} + \chi I_i \lambda \tag{6.11}$$

Let us introduce the operators

$$L_{ij} = I_i D_j - I_j D_i - Uij \tag{6.12}$$
$$U_{ij} = i(I_i \chi_0 I_j - I_j \chi_0 I_i) \tag{6.13}$$

One can see that $L_{ij}\chi \sim 0(\frac{1}{\lambda})$.Hence

$$L_{ij}\chi \equiv 0 \tag{6.14}$$

With order $\frac{1}{\lambda}$ we have from (6.14)

$$I_j \chi_1 I_i - I_i \chi_1 I_j = I_i \frac{\partial \chi_0}{\partial y_j} - I_j \frac{\partial \chi_0}{\partial y_i} - U_{ij}\chi_o \tag{6.15}$$

Multiplying from the right side by I_k and performing a cyclic permutation we drop out all the members containing χ_1. As a result, we get

$$\varepsilon_{ijk}(I_i \frac{\partial \chi_0}{\partial y_j} I_k - i I_i \chi_0 I_j \chi_0 I_k) = 0 \tag{6.16}$$

One can see that we get the equation (3.4) where $Q = i\chi_0$.

The question arises — how many of the new integrable equations one can find using the new dressing method instead of the old one. Note that if one of the functions I_i is scalar and linear ($I_3 = \lambda$)we get the complete set of equations that is given by the old method. In the general case we could get some new equations . But the comparision of equations (3.4) and (6.16) show that using the procedure of excluding x–derivatives from an initially overdeterminated system, as was described in section 3, one can get by the old method the same equations that may be obtained by the new one.

The great advantage of the new method is the possibility to put functions I_1, I_2, I_3 be rational instead of polynomial. In this case

$$\frac{\partial}{\partial \bar{\lambda}} D_i \neq D_i \frac{\partial}{\partial \bar{\lambda}}$$

If for instance

$$I_i(\lambda) = \frac{I_i}{\lambda - \lambda_i}$$

then

$$\frac{\partial}{\partial \bar{\lambda}} D_i - D_i \frac{\partial}{\partial \bar{\lambda}} = \pi I_i \delta(\lambda - \lambda_i) \tag{6.17}$$

For a general operator $R = R(D_1, D_2, D_3)$ the commutator

$$\frac{\partial}{\partial \bar{\lambda}} R - R \frac{\partial}{\partial \bar{\lambda}} = \Delta(\lambda) \tag{6.18}$$

is not equal to sero. In general $\Delta(\lambda)$ is a sum of δ–like singularities in all poles of R.

In order to find the operator $\tilde{R}$ (6.8) obeying the equation (6.18) one should require $\tilde{R}\chi \to 0$ at $\lambda \to \infty$ and

$$\tilde{R} \frac{\partial}{\partial \bar{\lambda}} = \frac{\partial}{\partial \bar{\lambda}} \tilde{R}$$

It was shown in [2] that both these conditions could be satisfied for at least two linearly independent operators $\tilde{R}_1, \tilde{R}_2$. These operators are direct analogs of "dressed" operators L_1, L_2 in the old version of the method. See also [11,12].

7 Small solutions and dispersion laws

In the old dressing method the variable x is the special one. In the new method all the variables y_i have the same abilities. In the old method we must consider the global behavior of the solution on the x-axis, in particular at $x \to \pm\infty$. That is an inheritage of the Inverge Scattering Transform. The new version of the method cuts any connections with the Inverge Scattering Approach. A solution obtained due to the $\bar{\partial}$-problem is essensially *local* in the space (y_1, y_2, y_3).In general it could grow arbitrarily fast at $|y| \to \infty$. Moreover, it could have singularities (as described in section 5) on some set S in the space (y_1, y_2, y_3). In the general position this set is a surface, it is obviousely that S coincides with the set where the $\bar{\partial}$-problem (5.14) is unresolvable. The question arises —how to choose the "dressing function" $G_0(\nu, \bar{\nu}, \lambda, \bar{\lambda})$ to produce a "good enough" solution of the nonlinear system. In order to formulate the problem more accurately, let us suppose that one of the variables is "time"($y_3 = t$, for instance), while two others (y_1, y_2) are "spatial" variables. In other words , we consider our nonlinear system as an evolutional one. (Note, that a corresponding Cauchy problem may be ill-posed.)

Now let us try to find G_o in a such way that the initial data (at $y_3 = 0$) are regular in the plane (y_1, y_2) and vanish at $|y| =\!\!\longrightarrow \infty$. This problem could easily be solved in the limit $G_0 \to 0$. In this case the solution of $\bar{\partial}$– problem is given by the first iteration of the integral equation (5.17). We have

$$\chi = 1 + \frac{1}{\pi i} \int \frac{G(\nu, \bar{\nu}, \xi, \bar{\xi})}{\lambda - \xi} d\nu d\bar{\nu} d\xi d\bar{\xi} \tag{7.1}$$

Comparing (7.1) and (6.9) one can find

$$\chi_n = \frac{1}{\pi i} \int \xi^n G(\nu, \bar{\nu}, \xi, \bar{\xi}) d\nu d\bar{\nu} d\xi d\bar{\xi} \tag{7.2}$$

Assuming that the matrixes $I_i(\lambda)$ are diagonal

$$I_i^{\alpha\beta}(\lambda) = \delta_{\alpha\beta} I_i^{\alpha}(\lambda)$$

we have for $\chi_n^{\alpha\beta}$

$$\chi_n^{\alpha\beta} = \frac{1}{\pi i} \int \xi^n G_0^{\alpha\beta}(\nu, \bar{\nu}, \xi, \bar{\xi}) e^{i \sum_{k=1}^{3} (I_k^{\alpha}(\nu) - I_k^{\beta}(\xi)) y_k} d\nu d\bar{\nu} d\xi d\bar{\xi} \tag{7.3}$$

To provide the regularity of $\chi_n^{\alpha\beta}$ at $y_3 = 0$ it is enough to require

$$Im(I_1^{\alpha}(\nu) - I_1^{\beta}(\xi)) \equiv 0 \tag{7.4}$$

$$Im(I_2^{\alpha}(\nu) - I_2^{\beta}(\xi)) \equiv 0 \tag{7.5}$$

These conditions define a two-dimensional manifold $\Lambda^{\alpha\beta}$ in four-dimensional space $\nu_R, \nu_I, \xi_R, \xi_I$ ($\xi = \xi_R + i\xi_I, \nu = \nu_R + i\nu_I$). The function $G_0^{\alpha\beta}(\nu, \bar{\nu}, \xi, \bar{\xi})$ should be concentrated on this manifold. If $I_{1,2}^{\alpha}(\nu)$ are real at real ν, the manifold $\Lambda^{\alpha\beta}$ includes the product of two real axis $R^1 \times R^1$. It could also include other components. Let us consider some simple examples .

1. KP-1.

$$\alpha = \beta = 0 \qquad I_1(\nu) = \nu \qquad I_2(\nu) = \nu^2$$

From (7.4), (7.5) one can get

$$\nu_R = \xi_R = 0$$

and

$$G_0(\nu, \bar{\nu}, \xi, \bar{\xi}) = G_0(\nu_R, \xi_R) \delta(\nu_I) \delta(\xi_I)$$

So the manifold Ω is the product of two real axes. The function χ is analytical in both half-planes $Im > 0, Im < 0$ and have a jump on the real axis. The equation (5.14) describes a nonlocal Riemann problem on the real axis.

2. KP-2.

$$\alpha = \beta = 0 \qquad I_1(\nu) = \nu \qquad I_2(\nu) = \alpha\nu^2 \qquad \alpha = \pm i$$

From (7.4), (7.5) one can get

$$\xi_I = \nu_I \qquad \xi_R = -\nu_R \qquad or \qquad \xi = -\bar{\nu} \tag{7.6}$$

So

$$G_0(\nu,\bar{\nu},\xi,\bar{\xi}) = G_0(\nu,\bar{\nu})\delta(\xi+\bar{\nu}) \tag{7.7}$$

Instead of (5.14) we now have

$$i\frac{\partial\chi}{\partial\bar{\lambda}} = G_0(\lambda,\bar{\lambda})\chi(-\bar{\lambda})e^{i(\lambda+\bar{\lambda})x-(\lambda^2-\bar{\lambda}^2)y+4i(\lambda^3+\bar{\lambda}^3)t} \tag{7.8}$$

This result was first obtained in [5].

The formula (7.3) allows to solve one rather important problem. All nonlinear equations under consideration have one trivial solution

$$\chi_n \equiv 0 \qquad (n = 0,\ldots)$$

These equations could be linearized in the vicinity of this solution. After the linearization one can get N^2 independent linear equations with constant coefficients for different elements of the matrix $\chi_i^{\alpha\beta}$.These equations could be solved by the Fourier transformation method. Assuming

$$\chi_i^{\alpha\beta} \cong e^{i(py_1+qy_2+\omega y_3)} \tag{7.9}$$

one should find N^2 relations

$$R_{\alpha\beta}(\omega,p,q) = 0 \tag{7.10}$$

The formula (7.3) shows that this relation could be found directly before calculating the explicit form of a nonlinear system. Supposing that (7.4) and (7.5) are satisfied, one can find (7.9)in a parametric form

$$p = Re(I_1^\alpha(\nu) - I_1^\beta(\xi)) \tag{7.11}$$
$$q = Re(I_2^\alpha(\nu) - I_2^\beta(\xi)) \tag{7.12}$$
$$\omega = I_3^\alpha(\nu) - I_3^\beta(\xi) \tag{7.13}$$

Equations (7.10)—(7.12) show that $R_{\alpha\beta} = R_{\beta\alpha}$ and we have $N(N+1)/2$ independent relations. Expressing ν and ξ from (7.10),(7.11) and substituting them into (7.12) one can find the full set of "dispersion laws".

$$\omega = \omega_{\alpha\beta}(p,q) = \omega_{\beta\alpha}(p,q) \tag{7.14}$$

Finding the whole set of dispersion laws hidden in a given nonlinear system is the first step for this system to find any applications (see also [6]).

8 On extention of the dressing method

We shall start from the $\bar{\partial}$-problem

$$i\frac{\partial\chi}{\partial\bar{\lambda}} = \chi * G = \int \chi(\nu,\bar{\nu})G(\nu,\bar{\nu},\lambda\bar{\lambda})d\nu d\bar{\nu} \tag{8.1}$$

and introduce a set of operators

$$D_i\chi = \frac{\partial \chi}{\partial y_i} + \chi * V_i \qquad i = 1, ..., n \tag{8.2}$$

Operators V_i in (8.2) could be some integral operators in general. In the general case their kernels $V_i(\nu, \bar{\nu}, \lambda, \bar{\lambda}, y_1...y_n)$ depend on the variables y_i. In the previous case we have

$$V_i = I_i(\lambda)\delta(\lambda - \nu)\delta(\bar{\lambda} - \bar{\nu}) \tag{8.3}$$

Now, suppose that G satisfies the system

$$\frac{\partial G}{\partial y_i} + G * V_i - V_i * G = 0 \tag{8.4}$$

This system is assumed to be compatible. A sufficient (perhabs, not necessary) condition is commutativity for the operators

$$[D_i, D_j] = 0 \tag{8.5}$$

We now have

$$iD_k \frac{\partial \chi}{\partial \bar{\lambda}} = D_k \chi * G \tag{8.6}$$

Again, let us consider that $R = R(D_1, ...D_n)$ is polynomial on D_i. Its coefficients are the $N \times N$ matrix function on $y_1, ...y_n$. Obviously

$$iR \frac{\partial \chi}{\partial \bar{\lambda}} = R\chi * G \tag{8.7}$$

suppose we have found an operator R satisfying the conditions

$$\tilde{R} \frac{\partial \chi}{\partial \bar{\lambda}} = \frac{\partial}{\partial \bar{\lambda}} \tilde{R}\chi \qquad R\chi \to 0 \qquad at|\lambda| \to \infty \tag{8.8}$$

Assuming that $\bar{\partial}$–problem (7.1) has a unique solution,we now have

$$\tilde{R}\chi \equiv 0 \tag{8.9}$$

The condition (8.8) could be satisfied for only for concrete χ. So coefficients of R depend on χ and (8.9) is a nonlinear equation for the function χ, having a known solution, depending on λ. To exclude this dependence one should have a set of equations

$$\tilde{R}_i\chi = 0 \qquad i = 1, ...k \qquad 2 \le k \le n \tag{8.10}$$

It is obvious that all the equations (8.10) are compatible.

It is hard to develop this theory for general integral operators. So we shall assume that V_i are *differential* operators on λ with *polynomial* on λ coefficients.

Even this very special case is not so simple. Let us consider an example (N=1)

$$D_1\chi = \frac{\partial\chi}{\partial x} + iV\chi \qquad V = \lambda^2\frac{\partial}{\partial\lambda} \tag{8.11}$$

$$D_2\chi = \frac{\partial\chi}{\partial y} + V^2\chi \tag{8.12}$$

It is possible to construct two compatible equations

$$R_1\chi = \frac{\partial\chi}{\partial x} + i\lambda^2\frac{\partial\chi}{\partial\lambda} + i\chi_0\chi \equiv 0 \tag{8.13}$$

$$R_2\chi = (D_2 + D_1^2 + 2i\frac{\partial\chi_0}{\partial x})\chi \equiv 0 \tag{8.14}$$

The last equation may be rewritten in the form

$$\frac{\partial\chi}{\partial y} + \frac{\partial^2\chi}{\partial x^2} + 2i\frac{\partial^2\chi}{\partial\lambda\partial x} + 2i\frac{\partial\chi_0}{\partial x}\chi = 0 \tag{8.15}$$

Let us consider $\lambda \to \infty$ and use the expansion (6.9). In the order $\frac{1}{\lambda}$ we have

$$2i\chi_1 = \frac{\partial\chi_0}{\partial x} + i\chi_0^2$$

$$\frac{\partial\chi_0}{\partial y} + \frac{\partial^2\chi_0}{\partial x^2} - 2i\frac{\partial\chi_1}{\partial x} + 2i\chi_0\frac{\partial\chi_0}{\partial x} = 0$$

or

$$\frac{\partial\chi_0}{\partial y} = \frac{\partial^2\chi_0}{\partial x^2} + 2i\chi_0\frac{\partial\chi_0}{\partial x} \tag{8.16}$$

It means that function $W = i\chi_0$ satisfies the Burgers equation. Adding to (8.11),(8.12) the third operator

$$D_3\chi = \frac{\partial\chi}{\partial t} + 4iV^3\chi \tag{8.17}$$

one can find that χ_0 satisfies two other Burgers-like equations. It is interesting that at the same time the function $U = 2i\frac{\partial\chi_0}{\partial x}$ satisfies the KP-2 equation.

In the previous example, D_i belongs to the simplest class of commuting differential operators with polynomial coefficients

$$D_i^0\chi = \frac{\partial\chi}{\partial y_i} + P_i(V_0)\chi \tag{8.18}$$

Here P_i are polynomials on some elementary operator V_0 with constant coefficients. In the matrix case, all coefficients of P_i must commute with V_0. To get more general polynomial differential operators one could apply to D_i^0 a gauge transformation

$$D_i = e^{-i\Phi}D_i^0e^{i\Phi} \tag{8.19}$$

Here Φ is polynomial on λ with y_i independent coefficients. This case includes the previously described (in section 6) variant of the dressing method. Assuming

$$D_i^0 = \frac{\partial}{\partial y_i} \qquad \Phi = \sum I_i(\lambda) y_i \tag{8.20}$$

one can get

$$D_i \chi = \frac{\partial \chi}{\partial y_i} + i\chi I_i(\lambda) \qquad i = 1,2,3 \tag{8.21}$$

To obtain a more advanced example, put in the $N \times N$ matrix case

$$D_i^0 \chi = \frac{\partial \chi}{\partial y_i} + \frac{\partial \chi}{\partial \lambda}(1 + \alpha\lambda) A_i \tag{8.22}$$

$$i = 1,2,3 \qquad [A_i A_j] = 0 \qquad [A_i, \alpha] = 0$$

After the transformation (8.19), using

$$\Phi = \lambda \sum I_i y_i \qquad [A_i, I_j] = 0 \qquad [I_i, I_j] = 0$$

we have

$$\tilde{D}_i \chi = D_i \chi - i\Phi A_i \chi = \frac{\partial \chi}{\partial y_i} + \frac{\partial \chi}{\partial \lambda}(1 + \alpha\lambda) A_i + i\lambda\chi \tilde{I}_i + i[\chi_1 \Phi A_i] \tag{8.23}$$

$$\tilde{I}_i = I_i + \alpha \Phi A_i$$

It is easy to show that

$$L_{ij}\chi = 0 \tag{8.24}$$

with

$$\begin{aligned} L_{ij} &= \tilde{I}_i \tilde{D}_j - \tilde{I}_j \tilde{D}_i - U_{ij} \\ U_{ij} &= i(\tilde{I}_i \chi_0 \tilde{I}_j - \tilde{I}_j \chi_0 \tilde{I}_i) \end{aligned} \tag{8.25}$$

Acting like we did in section 6 one can get the equation for χ_0

$$\varepsilon_{ijk}(\tilde{I}_i \frac{\partial \chi_0}{\partial x_j} \tilde{I}_k - i\tilde{I}_i \chi_0 \tilde{I}_j \chi_0 \tilde{I}_k + i\tilde{I}_i [\chi_1 \Phi A_j] \tilde{I}_k - \alpha \tilde{I}_i \chi_0 A_j \tilde{I}_k) = 0 \tag{8.26}$$

This equation is a generalization of the N–wave system (3.4),(6.16). Matrixes $\tilde{I}_i$ are now linear functions of coordinates. The system (8.26) may have some physical applications. Using the operators (8.18), (8.19) we can get a very broad class of "integrable" systems. To show that , let us consider the operators (8.20), (8.21) producing some nonlinear integrable systems, as it was described in section 6. Let us add to the operator D_i^0 another one operator

$$D_1^0 \chi = \frac{\partial \chi}{\partial \tau} + V_0 \chi \tag{8.27}$$

Here V_0 is any differential operator

$$V_0\chi = \frac{d^2\chi_0}{\partial\lambda^n} l_0 + ...$$

commuting with the matrixes I_i.

Let us also perform transformation

$$\Phi \to \tilde{\Phi} = \Phi + P(\lambda)\tau \qquad [P(\lambda), I_i] = 0$$

Here $P(\lambda)$ is an arbitrary polynomial. Now the operator

$$D_4 = e^{-i\tilde{\Phi}} D_4^0 e^{i\tilde{\Phi}} \tag{8.28}$$

commutes with operators (8.21). Using any triplet of operators (D_1, D_2, D_4),(D_1, D_3, D_4),(D_2, D_3, D_4) as a basic system one can get three new nonlinear systems, compatible with the initial equations. All of them are *symmetries* of the initial systems. If $V_0 \neq 0$, these symmetries have coordinate-depending coefficients. It is plausible that on this way one may find all the symmetries of $2+1$ dimension integrable systems.

9 Some perspectives

A further developing of the dressing method depends upon a progress in the classification of the operators D_i. Besides the class of these operators, described in the section 8, one can separate two other natural classes.

1. The class of operators D_i not including λ-derivatives

$$D_i\chi = \frac{\partial\chi}{\partial y_i} + \chi U_i(\lambda, y_i) \qquad i = 1, 2, 3 \tag{9.1}$$

Here $U_i = U_i(\lambda, y)$ are rational functions on λ with fixed poles. In fact, poles of U_i may depend on y_i. The commutativity condition $[D_i D_j] = 0$ gives

$$\frac{\partial U_i}{\partial y_j} - \frac{\partial U_j}{\partial y_i} + [U_j, U_i] = 0 \tag{9.2}$$

The equations (9.2) could be solved by $1+1$ variant of the dressing method, using a local $\bar{\partial}$-problem or a local Riemann problem. This variant of the dressing method was described in article [7].

2. The class of operators D_i including only first λ-derivatives with a scalar coefficients.

$$D_i\chi = \frac{\partial\chi}{\partial y_i} + \frac{\partial\chi}{\partial\lambda} W_i + \chi U_i \tag{9.3}$$

Here $U_i = U_i(\lambda, y)$ are matrixes, while $W_i = W_i(\lambda, y)$ are scalar rational on λ functions. They have the same (for a given i) y-dependent poles. The commutativity conditions impose the well definite system of equations on this poles and on the residues of W_i and U_i ([8], see also [9]). This system also could be solved by some extention of the dressed method, based on the local $\bar{\partial}$-problem. This variant of the method was called in article [8] "The Inverse Scattering Method with a Moving Spectral Parameter".

To go further we should consider that $W_i(\lambda, y)$ are matrix functions, or introduce λ-derivatives of a higher order. The only known class of such D_i-operators was described in section 8. To find more general classes we should use the old dressing method, maybe in its modernized form, introduced in this article. Let us suppose that the variable x is in fact a spectral parameter ($x = \lambda$). To find the D_i-operators we should find a solutions of the system (3.2) *that are rational functions of* x. In general, their poles depend on variables y_i. To classify the solutions of this type is a very intriguing mathematical problem. For the KP-equation this problem was solved by Krichever (see, for instance [10]). It was shown that in the y-dependent case, the system governing moving of poles coincides with the well-known Calogero– Moser system of particle of the axis with pair interaction $U_{ij} \sim \frac{c}{(x_i - x_j)^2}$.

Let us do one concluding remark. A mathematician, using the dressing method to find a new integrable system, could be compared with a fishman, plunging his net into the sea. He does not know what a fish he will pull out. He hopes to catch a goldfish, of course. But too often his catch is something that could not be used for any known to him purpose. He invents more and more sophysticated nets and equipments and plunge all that deeper and deeper. As a result he pulls on the shore after a hard work more and more strange creatures. He should not despair, nevertheless. The strange creatures may be interesting enough if you are not too pragmatic. And who knows how deep in the sea do goldfishes live?

References

1. V. E. Zakharov and A. B. Shabat, "A scheme for integrating the nonlinear equations of mathematical physics by the method of the inverse scattering transform I," Funct. Anal. Appl. **8**, 226–235 (1974).

2. V. E. Zakharov and S. V. Manakov, "The many–dimensional integrable systems and their solutions," Zap. Nauch. Sem. LOMI, **133**, 76 (1984); "Construction of the many–dimensional integrable systems and their solutions," Funct. Anal. Appl. **19**, 11–25.

3. E. A. Kuznetsov and A. V. Mikhailov, "Stability of stationary waves in nonlinear weakly dispersive media," Sov. Phys. YETP **40**, 855–859.

4. A. S. Fokas and V. E. Zakharov, "The dressed method revised," to be published.

5. M. Y. Ablowitz, D. Bar Yaacov, and A. S. Fokas, "On the inverse scattering transform for the Kadomtsev–Petviashvili equation," Stud. Appl. Math. **69**, 135–143.

6. L. V. Bogdamov and V. E. Zakharov, "The small solutions and the dispersion laws of (2+1) dimension integrable equations," Algebra and Analysis, to be published.

7. V. E. Zakharov and A. B. Shabat, "A scheme for integrating the nonlinear equations of mathematical physics by the method of the inverse scattering problem II," Funct. Anal. Appl. **13**, 166–174 (1979).

8. S. P. Burtzev, V. E. Zakharov, and A. V. Mikhailov, "The inverse scattering method with a movable spectral parameter," Theor. Math. Phys. **70** (3), 323–341 (1987).

9. V. E. Zakharov and V. A. Belinskii, "Integration of the Einstein equations by the inverse scattering method and calculation of the exact soliton solutions," Sov. Physics YETP **48**, 985–994 (1978).

10. V. E. Zakharov, S. V. Manakov, S. P. Novikov, and L. P. Pitayevsky, "Theory of solitons. The method of the inverse scattering," (in Russian) Nauka, Moscow (1980); (in English) Plenum Press, New York (1984).

11. L. V. Bogdanov and S. V. Manakov, "The non-local $\bar{\partial}$–problem and (2+1) dimensional soliton equations," Y. Phys. A. and Math. Gen. **21**, L537–L544 (1988).

12. V. E. Zakharov, "Multidimensional integrable systems," *Proceedings of the International Congress of Mathematicians*, Warsaw, Poland, August 16–24 (1983).

Subject Index

Reference is made to the *first* page of relevant articles

Index of Contributors

r.c.p. 264

du 5 au 9 décembre 1988 - MONTPELLIER (FRANCE)

rencontre interdisciplinaire
problèmes inverses

LUNDI 5 DECEMBRE (aprés-midi)

13.45 - 14.30 Accueil

14.30 - 15.30 **P.C. SABATIER**, Lab. Phys. Math., U.S.T.L., Montpellier.
Conférence d'introduction.

15.30 - 16.20* **J. JP. LEON**, Lab. Phys. Math., U.S.T.L., Montpellier.
An integrable non-conservative model in plasma theory.

BREAK 16.30 - 16.50

16.50 - 17.10* **V. RICHARD**, Institut Français du Pétrole,Rueil Malmaison.
Inversion with a priori information: an approach to integrated stratigraphic interpretation (en coll. avec J. Brac, P.Y. Dequirez, F. Hervé, C. Jacques, P. Lailly, D. Tran Van Nhieu).

17.20 - 18.00* **P.M. SANTINI**, Università di Roma, Roma, Italy.
Topics in soliton theory: the initial boundary value problem for the Davey-Stewardson equation and solvable algebraic systems.

18.10 - 18.40* **H. CORNILLE**, CEN-Saclay, Orme des Merisiers, Gif-sur-Yvette.
Solutions to discrete Boltzman models.

DEPART DU BUS VERS LA VILLE - 19.00

MARDI 6 DECEMBRE (matin)

8.50 - 9.40* **M. BERTERO**, Università di Genova, Italy.
Image restoration and super-resolution in confocal light microscopy.

9.50 - 10.20* **M. PIDCOCK**, Dept. of Comp. and Math. Sciences, Oxford Polytech.
Electrical impedance tomography using adaptative currents.

BREAK 10.30 - 10.50

10.50 - 11.40* **A. FARIDANI**, Westfälische Wilhelms- Universität Münster, F.R.G.
Sampling and resolution in diffraction tomography.

11.50 - 12.20* **K. HOPCRAFT**, Kings College London, Strand, London, U.K.
Inversion and phase recovery of electromagnetic amplitude data.

12.30 - **REPAS**

MARDI 6 DECEMBRE (aprés-midi)

14.00 - 14.40* **J. CORONES**, Iowa State University, Ames, U.S.A.
An electromagnetic inverse problem.

14.50 - 15.20* **M. CUER**, Lab. Phys. Math., U.S.T.L., Montpellier.
Problèmes de programmation linéaire portant sur des fonctions (de $L^{\infty}(a, b)$) et non des vecteurs: nécessité de résoudre des problèmes discrétisés non linéaires pour construire des schémas d'ordre plus grands que 2.

15.30 - 16.00* **M. FIDDY**, Univ. of Lowell, U.S.A.
An optimisation technique for the inversion of scattered field data.

16.10 - 16.50* **A.R. OSBORNE**, Ist. di Cosmogeofisica, Torino, Italy.
Nonlinear Fourier analysis of laboratory and oceanic surface waves.

17.00 - **VIN D'HONNEUR** offert par le Président de l'Université.

DEPART DU BUS VERS LA VILLE - 18.30

MERCREDI 7 DECEMBRE (matin)

8.50 - 9.40* **F. CALOGERO**, Università di Roma, Italy.
Universality, integrability and solutions of the non-linear PDEs describing N-wave interactions.

9.50 - 10.20* **W.M. BOERNER**, Univ. of Illinois, Chicago, U.S.A.
Direct inverse methods in radar polarimeters./
The depolarization term in vector diffraction tomography.

BREAK 10.30 - 10.50

10.50 - 11.20* **B.G. KONOPELCHENKO**, USSR Acad. of Sciences, Novosibirsk.
Titre non parvenu.

11.30 - 12.10* **E. BERETTA**, Ist. di Analisi Globale e Appl., Firenze.
Stability results for an inverse problem in potential theory.

12.20 - **REPAS**

MERCREDI 7 DECEMBRE (aprés-midi)

14.00 - 14.40* **B. DeFACIO**, Univ. of Missouri-Columbia, U.S.A.

14.50 - 15.10* **De-Biao GE**, Dept. of Physics, Xi'an, China.
On determination of permittivity and conductivity profites in lossy media.

15.20 - 15.50* **A.K. JORDAN**, Naval Research Lab., Washington D.C., U.S.A.
a) Perturbation analysis of inverse scattering problems (avec D. Ladouceur); b) Design of optical waveguides by inverse scattering theory (avec S. Lakshmanasamy).

16.00 - 16.20* **H. LADOUCEUR**, Naval Research Lab., Washington D.C., U.S.A.
a) The Curtis-Godson approximation and spectral profile reconstruction; b) Perturbation solutions of the Gelfand-Levitan equation (avec A.K.Jordan).

BREAK 16.30 - 16.50

16.50 - 17.20* **K. SEIP**, Universitet I Trondheim, Norge.
Mean value theorems and concentration operators in Bargmann and Bers spaces with applications to sampling and interpolation of wavelet transforms.

17.30 - 18.00* **M. GRASSELLI**, Università di Milano, Italy.
An inverse problem for a nonlinear hyperbolic equation.

18.10 - 18.40* **P.D. STEFANOV**, Inst. of Math. of Bulgarian Acad. of Sciences.
Uniqueness of the inverse scattering problem for the wave equation with time-dependent.

DEPART DU BUS VERS LA VILLE - 19.00

JEUDI 8 DECEMBRE (matin)

8.50 - 9.30* **R. CONTE**, CEN-Saclay, Gif-sur-Yvette.
New analytic solutions of the nonintegrable Kolmogorov-Petrovsky-Piskunov (KPP) equation.

9.40 - 10.20* **D.J. KAUP**,Univ. of Minnesota, Minneapolis, U.S.A.
Solitons in cosmology.

BREAK 10.30 - 10.50

10.50 - 11.30* **M. BOITI**, Università degli Studi di Lecce, Lecce, Italy.
Solitons in the plane and their spectral transform.

11.40 - 12.20* **F. PEMPINELLI**, Università degli Studi di Lecce, Lecce, Italy.
Solitons interacting in the plane.

12.30 - **REPAS**

JEUDI 8 DECEMBRE (aprés-midi)

14.00 - 14.50* **J.-Cl. FERNANDEZ, G. REINISCH**, Observatoire de Nice.
IREM-film: Mechanics of the Sine-Gordon soliton.

15.10 - 15.40* **M. JAULENT**, Lab. Phys. Math., U.S.T.L., Montpellier.
Modules asymptotiques et équations nonlinéaires.

15.50 - 16.20* **L. MARTINEZ-ALONSO**, Universidad Complutense de Madrid.
Asymptotic modules, infinite groups and Riemann surfaces.

BREAK 16.20 - 16.40

16.40 - 17.10* **F. GONZALEZ GASCON**, Universidad Complutense de Madrid.
New results on the inverse problem of Lie groups and differential equations.

17.20 - 17.50* **F. LAMBERT**, Vrije Universiteit Brussel, Belgique.
On the balance between dispersion and nonlinearity: necessary conditions for the existence of sech squared solitons.

18.00 - 18.20* **Z. JIANG**, University of Manchester, U.K.
The inverse scattering methods for a class of multi-dimensional spectral equations.

18.30 - 18.55* **H. LANGE**, Mathematisches Inst., Universität Kôln, F.R.G.
Non integrable nonlinear Schrödinger equations.

DEPART DU BUS VERS LA VILLE - 19.00

VENDREDI 9 DECEMBRE (matin)

8.50 - 9.30* **A. RONVEAUX**, Univ. Notre-Dame de la Paix, Namur, Belgique.
Construction d'equations différentielles satisfaites par des polynômes ortogonaux non classiques.

9.40 - 10.20* **K. BLEULER**, Université de Bonn.
La factorisation selon Infeld dans le cadre des développements rècents.

BREAK 10.30 - 10.50

10.50 - 11.10* **S. LAKSHMANASAMY**, University of Texas at Dallas, U.S.A.
Design of optical waveguides by inverse scattering theory.

11.20 - 11.45* **I. PINTO**, Ist. di Elettronica, Università di Salerno.

VENDREDI 9 DECEMBRE - (aprés-midi)

14.00 - 14.50* **M. FEIX**, PMMS/CNRS, Orleans.
Time-frequency representations: Vignerville and other transformations.

15.00 - 17.00* **A.G. RAMM**, Kansas State University, Manhattan, Kansas,U.S.A.
Uniqueness theorems and numerical solution of multidimensional inverse problems with data at fixed energy.

17.10 - 17.50* **P.M. KOLESNIKOV**, BSSR Academy of Sciences, Minsk, USSR.
Some inverse problems of non-linear heat and mass transfer theory and non-linear electrodynamics.

DEPART DU BUS VERS LA VILLE - 18.00

* = temps pour discussion et questions.

J. M. Combes, A. Grossmann, P. Tchamitchian, CNRS, Marseille (Eds.)

Wavelets

Time-Frequency Methods and Phase Space

Proceedings of the International Conference, Marseille, France, December 14–18, 1987

1989. IX, 315 pp. 88 figs. Hardcover DM 148,–
ISBN 3-540-51159-8

Time-frequency methods and phase space are as well known to most physicists, engineers and mathematicians as traditional Fourier analysis, which has recently found for many applications a competitor in the concept of wavelets. Crudely speaking a wavelet decomposition is an expansion of an arbitrary function into smooth localized contributions labeled by a scale and a position parameter.
The meeting recorded in this volume brought together people exploring and applying these concepts in an interdisciplinary framework. Topics discussed range from purely mathematical aspects to signal and speech analysis, seismic and acoustic applications, and wavelets in computer vision.

K. Chadan, Université de Paris-Sud, Orsay; **P. C. Sabatier,** Université des Sciences et Techniques du Languedoc, Montpellier

Inverse Problems in Quantum Scattering Theory

With a Foreword by R. G. Newton

2nd rev. and expanded ed. 1989. XXXI, 499 pp. 24 figs. (Texts and Monographs in Physics) Hardcover DM 138,–
ISBN 3-540-18731-6

Contents: Some Results from Scattering Theory. – Bound States – Eigenfunction Expansions. – The Gel'fand-Levitan-Jost-Kohn Method. – Applications of the Gel'fand–Levitan Equation. – The Marchenko Method. – Examples. – Special Classes of Potentials. – Nonlocal Separable Interactions. – Miscellaneous Approaches to the Inverse Problems at Fixed 1. – Scattering Amplitudes from Elastic Cross Sections. – Potentials from the Scattering Amplitude at Fixed Energy: General Equation and Mathematical Tools. – Potentials from the Scattering Amplitude at Fixed Energy: Matrix Methods: Potentials from the Scattering Amplitude at Fixed Energy: Operator Methods. – The Three-Dimensional Inverse Problem. – Miscellaneous Approaches to Inverse Problems at Fixed Energy. – Approximate Methods. – Inverse Problems in One Dimension. – Problems Connected with Discrete Spectra. – Numerical Problem. – Reference List. – Subject Index.

Springer-Verlag Berlin Heidelberg New York London Paris Tokyo Hong Kong

Springer

B. N. Zakhariev, Lab. of Theoretical Physics, Dubna;
A. A. Suzko, Luikov Heat and Mass Transfer Institute, Minsk

Direct and Inverse Problems

Potentials in Quantum Scattering

1990. Approx. 200 pp. 42 figs. Softcover, in prep.
ISBN 3-540-52484-3

This textbook can almost be viewed as a "how-to" manual for solving quantum inverse problems, that is, for deriving the potential from spectra and/or scattering data. The formal exposition of inverse methods is paralleled by a discussion of the direct problem. In part differential and finite-difference equations are presented side by side. A variety of solution methods is presented. Their common features and (dis)advantages are analyzed.
To foster a better understanding, the physical meaning of the mathematical quantities are discussed in detail. Wave confinement in continuum bound states, resonance and collective tunneling, and the spectral and phase equivalence of various interactions are some of the physical problems covered.

A. G. Sitenko, Academy of the Ukrainian SSR

Scattering Theory

1990. Approx. 320 pp. 32 figs. Hardcover, in prep.
ISBN 3-540-51953-X

This mathematically rigorous introduction to nonrelativistic scattering theory addresses upper level undergraduates in physics. The relationship between the scattering matrix and physical observables is discussed in detail. Among the emphasized topics are the stationary formulation of the scattering problem, the inverse scattering problem, dispersion relations, three-particle bound states and their scattering, collisions of particles with spin and polarization phenomena. The analytical properties of the scattering matrix are discussed. Problems are included to help the reader to gain some experience and more expertise in scattering theory.

Springer-Verlag Berlin
Heidelberg New York London
Paris Tokyo Hong Kong

Springer

MIX
Papier aus verantwortungsvollen Quellen
Paper from responsible sources
FSC® C105338

If you have any concerns about our products,
you can contact us on
ProductSafety@springernature.com

In case Publisher is established outside the EU,
the EU authorized representative is:
Springer Nature Customer Service Center GmbH
Europaplatz 3, 69115 Heidelberg, Germany

Printed by Libri Plureos GmbH
in Hamburg, Germany